Active Volcanoes of the World

More information about this series at http://www.springer.com/series/10081

Patrick Bachèlery · Jean-François Lénat
Andrea Di Muro · Laurent Michon
Editors

Active Volcanoes of the Southwest Indian Ocean

Piton de la Fournaise and Karthala

Springer

Editors
Patrick Bachèlery
Laboratoire Magmas et Volcans
Université Blaise Pascal—Clermont-Ferrand
Clermont-Ferrand
France

Andrea Di Muro
Piton de la Fournaise Volcano Observatory
Institut de Physique du Globe de Paris
La Plaine des Cafres
France

Jean-François Lénat
Laboratoire Magmas et Volcans
Université Blaise Pascal—Clermont-Ferrand
Clermont-Ferrand
France

Laurent Michon
Laboratoire GeoSciences
University La Réunion
Saint-Denis
France

ISSN 2195-3589 ISSN 2195-7029 (electronic)
Active Volcanoes of the World
ISBN 978-3-642-31394-3 ISBN 978-3-642-31395-0 (eBook)
DOI 10.1007/978-3-642-31395-0

Library of Congress Control Number: 2015947405

Springer Heidelberg New York Dordrecht London

Printed on acid-free paper

Springer-Verlag GmbH Berlin Heidelberg is part of Springer Science+Business Media
(www.springer.com)

Preface

The two active volcanoes in the south-west Indian Ocean are Piton de la Fournaise on La Réunion and Karthala on Grande Comore in the Comorian archipelago. This book provides a summary of the geological evolution of these two major volcanoes of the south-western Indian Ocean. Both are classified as basaltic shield volcanoes. Yet if their similarities are many, there are also many differences. Extensive research has been carried on these two volcanoes, their geodynamic and magmatic settings, as well as on the socio-economic consequences of their volcanic activity. This book presents a synthesis encompassing both the history of researches and our current understanding of these two volcanoes.

Piton de la Fournaise volcano is one of the most active volcanoes on Earth, and the only one in the Indian Ocean whose activity has been explicitly linked with a deep mantle plume. Since 2010, the 'Pitons, cirques and ramparts of La Réunion Island' have been made a World Heritage Natural Site by the United Nations Educational, Scientific, and Cultural Organization (UNESCO), by virtue of the '*rugged and dramatic landscape of striking beauty*' created by the combined action of volcanism, tectonic, erosion, heavy rainfalls and a mosaic of ecosystems with a high degree of endemism. Piton de la Fournaise was a keystone of this inscription.

Karthala is a magnificent volcano whose access is complicated by the lack of road and the dense tropical forest that covers much of its flanks. Its summit caldera and crater are preserved spaces of great beauty and danger. Although its eruptions are less frequent than those of Piton de la Fournaise, it, nonetheless, constitutes a serious threat to the population living close to its shores and on its flanks, as illustrated by recent eruptions.

This book includes 25 chapters: 20 on Piton de la Fournaise, 5 on Karthala. It addresses a wide set of disciplines. For Piton de la Fournaise, the first chapters deal with historic and general geodynamic aspects concerning La Réunion and Piton de la Fournaise, including hydrogeology and deep marine environment. Then, eruptive history, magmatic and geochemical aspects, geophysical monitoring and volcanic hazards are reviewed. For Karthala, the less studied of the two volcanoes, geodynamical, volcanological and petrological questions are addressed together with hydrogeology, volcanic risk and crisis management. Our hope is that this book will represent a clear review and a strong incentive for further researches on both volcanoes.

Karthala

The four main islands of the Comorian archipelago are distributed along ca. 280 km, defining a generalised NW–SE volcanic lineament. The morphology of the islands indicates a progressive younging to the NW that could be attributed to plate motion towards the SE across a localised mantle hot spot or alternatively to a fracture system in the underlying lithosphere that is propagating towards the NW with de-pressurisation of the asthenosphere triggering decompression melting as it advances. Karthala, in Grande Comore (Ngazidja), the youngest of the Comorian volcanoes is situated at 11°75′S and 43°38′E and rises from the floor of the Mozambique Channel.

Grande Comore rises as a volcanic doublet comprising the coalescing shields of La Grille and Karthala. Both are undissected and present youthful characteristics. There are, however, no historic eruptions from La Grille, whereas Karthala has erupted at least twenty times since records began in 1857. Karthala is 2361 m high, rising above its northern neighbour La Grille (1067 m). It is an elongate shield volcano with flanks dipping up to a maximum of 24°. Up to 1900 m, the volcano is covered by tropical rainforest that gives way in the uppermost 400 m to *Philippia comorensis* scrub vegetation.

The volcano grew across a NW-SE-trending fissure zone. The latter is manifest in the overall trend of the Comorian islands, in the morphology of the volcano itself and in its prominent crater chains. The eruptive zone marked by crater chains and eruptive fissures is some 6–8 km broad and can be traced along the island for 45 km. The summit region is truncated by a complex caldera (ca. 4 × 3 km) that consists of several intersecting collapse pits. This is considered to be one of the largest calderas of any active basaltic volcano on Earth. Occasionally, lava-lakes develop as in 2005 and 2006. Choungou Chahale, the main crater, itself is 1300 × 800 m across and deep.

Whilst eruptions have occurred all along the fissure zone, those of the past century have tended to be localised around the summit region. The caldera collapses appear to have resulted as a consequence either of flank eruptions or magma withdrawal along dyke systems in the underlying crust. In 1858, a lava originating in the summit region travelled 13 km north-west, reaching the coast to the north of the capital town, Moroni. In April–June 1948, a relatively violent strombolian eruption hurled large blocks to heights of over 100 m. More recent eruptions include those of 1991 that produced a water lake. Two eruptions in April and November 2005 led to the evacuation of thousands of people. A further eruption took place in May 2006, and the last one was in January 2007.

Karthala appears to have erupted only basaltic lavas of which the great majority are ankaramitic, i.e. they are olivine- and augite-phyric. Flows also carrying plagioclase phenocrysts are distinctly subordinate. As it is clear from their phenocryst characteristics, the Karthala lavas are alkaline olivine basalts (average normative nepheline, 3.7 %) with the most primitive melt compositions rarely above 8 wt.% MgO. These have dominantly been erupted as ‘a’ā type lava flows although there are very prominent pāhoehoe lava flows on its north-western flanks.

A number of prominent tuff-rings, the products of phreato-magmatic eruptions through interaction with the sea, occur close to the western coast. The tuff-rings are largely composed of palagonite with olivine + augite crystals. An eruption in the Chougou Chahale summit crater in July 1991 is inferred to have been triggered by lava/water interaction, and an active hydrothermal system has been postulated.

Compositionally Karthala contrasts markedly with La Grille. The latter is composed of basanitic to nephelinitic lavas, principally products of a less productive fissure zone lying sub-parallel and to the north-east of the Karthala fissure zone. Parasitic ash/scoria cones are more abundant than on Karthala. They are not only more primitive (melts with up to 13.5 wt.% MgO and bearing mantle xenoliths) but representing distinctly smaller mantle melt-fractions than those of its big neighbour. Thus, Karthala magmas indicate either higher thermal input or greater depressurisation controlling the mantle source.

Piton de la Fournaise

The island of La Réunion, 55.5°E and 21.0°S, is slightly larger than Grande Comore. Although it lies only 170 km WSW of Mauritius, it rises from the deep ocean floor (>4000 m) that separates it from Mauritius and thus not directly from the Mascarene Ridge.

The arrival (or 'impact') of a mantle plume at about 66 Ma, approximately at the time of the Cretaceous–Palaeogene boundary, is inferred to have given rise to the great eruptions of the Deccan region in NW India. It is widely supposed that subsequent motions of the Indian plate and later the African plate across the residual 'plume tail' led to a hot spot trail through the Laccadive and Maldive islands, the Chagos Bank and much of the Mascarene Plateau to its current siting beneath La Réunion. The ocean crust beneath La Réunion has an age of ~60 Ma and the absolute plate velocity over the hot spot is estimated at 1.7 cm/year. A seismic section across the island showed three main features of the infrastructure of the island: (1) The core of the island has a significantly higher velocity (density) than the flanks; (2) there is no pronounced lithospheric flexure, contrary to most oceanic islands; and (3) underplating is present beneath the crust in the SW part of the island.

La Réunion is the emergent part of a great, approximately conical, volcanic pile with a basal diameter of *ca.* 250 km, that rises *ca.* 7 km above the oceanic crust. The whole structure is an enormous elliptical shield with an overall volume of approximately 45,000 km^3 so that it ranks among the planet's largest intraplate oceanic volcanoes. Although it may be regarded as a single entity grown over at least 4 Ma, several volcanic shields have been recognised in the construction of La Réunion edifice. Morphologically, the island is composed of a volcanic doublet, Piton des Neiges and Piton de la Fournaise. Piton des Neiges is deeply dissected and seems to have been inactive during the last 10,000 years. It consists of a primitive shield of olivine basalts overlain by more differentiated products (grading to silica-oversaturated trachytes) originating from residual magma trapped within the volcano superstructure.

The younger edifice (Piton de la Fournaise) is crowned by a system of concentric caldera-collapse structures. The overall shield morphology is modified by these collapses, which have strongly affected the drainage pattern. The summit cone (at 2631 m above sea level), containing intersecting craters, is situated within a U-shaped caldera that is closed to the west but open to its eastern (seaward) side. Most historic eruptions have occurred within pit craters (Cratère Dolomieu and Cratère Bory) in the summit area. Eruptions occur, on average, every nine months. Three rift zones, each between 5 and 7 km across trending NW-SE, NNW-SSE and SW-NE have been identified.

Piton de la Fournaise's 'plumbing system' may comprise at least three storage zones: (1) A deep reservoir below the oceanic crust at ~15 km depth, (2) one at intermediate depth (5–7 km) and (3) a shallow reservoir at approximately 2.5 km depth, i.e. close to sea level.

The overall morphology of the volcano reflects large-scale slumping on its unsupported eastern flank. The youngest of these slumps has been dated as of less than 4745 years old and may have coincided with large-scale phreato-magmatic eruptions. The collapse created a horseshoe-shaped structure 8 km broad (the Enclos Fouqué) within which most historic eruptions have occurred. Outside of the Enclos Fouqué, there are earlier caldera walls that appear to represent successive collapses accompanying an eastward or south-eastward migration of the eruptive centre. Debris from large-scale slumps can be identified by side-scan sonar several tens of kilometres out to sea.

The volcano has been deeply incised by river gorges, the siting of which is inferred to have been controlled by faults. The gorge of the Rivière des Remparts may follow an old caldera wall. In the course of the eastward slumps, the focus of volcanic activity has moved 6–10 km east from close to La Plaine des Sables to its present position over the past 150 ka. Such gorges provide excellent stratigraphic sequences dating back 530 ka.

The lavas are mainly olivine basalts. Some are distinguished by very high modal concentrations (up to 60 % modal) of olivine crystals and were called 'oceanites'. A cyclic eruptive pattern was discerned involving emission of oceanite-type lava from major flank eruptions every 20–40 years. The oceanites can carry ~60 wt.% olivine crystals, close to the theoretical maximum for flow. The last eruption of oceanite in 2007 was the greatest since the island was first colonised in the seventeenth century.

Although the bulk of the Piton de la Fournaise shield consists of basalts containing olivine and subordinate spinel phenocrysts, the Ancient Piton de la Fournaise Series also contains relatively differentiated lavas. Some contain augite and/or plagioclase together with olivine and there is gradation to hawaiites and mugearites. Comparable lavas are absent from the recent Piton de la Fournaise Series, a contrast that has been ascribed to an overall higher magmatic supply rate. Compositional variation within the less differentiated basalts of Piton de la Fournaise (and those of the early Piton des Neiges shield) has mainly been dictated by olivine fractionation. Despite their comparatively primitive nature, the olivine basalts of La Réunion (with olivine compositions Fo_{89-82}) cannot be regarded as 'primary' magmas. Geochemical evidence, however, shows that *all* of the erupted Piton de la Fournaise magmas have undergone ol-cpx-sp fractionation at depth. However, no corresponding cumulates have ever been recorded among the enclaves carried to the surface inside the Enclos Fouqué caldera and this high-pressure fractionation may have been confined to 'underplated' magmas held temporarily at, or close to, the crust-mantle boundary defined by the discontinuity where p-wave velocities change from ~7 km/s to mantle values of >8 km/s.

The importance of olivine-rich cumulates within the carapace of Piton de la Fournaise is indicated by deep drilling on the east coast of Piton de la Fournaise that revealed the presence of ultrabasic cumulates at >1 km. Further evidence comes from the occurrence of cognate olivine cumulate enclaves. In conjunction with evidence from Piton des Neiges, these point to derivation from a source or sources some 70 km across. Nested bodies of differing age but similar character are inferred presenting idealised successions from dunites via wehrlites to gabbro cumulates. Dunite, however, is likely to be the dominant lithology, probably produced from successive replenishments of relatively primitive magma in this open system. The ultramafic cumulates may form a body of dense rock at a depth of 4 or 5 km beneath the summit. Such a body may be susceptible to deformation by gravitational creep and slumping. Listric down-faulting towards the east may be facilitated by the incompetent behaviour of olivine-rich rocks; infiltration of a CO_2-rich fluid phase may also be rheologically important. The functioning of the Piton de la Fournaise's 'plumbing system' is far from fully understood.

There is close compositional similarity between the Piton de la Fournaise basalts and those of the Piton des Neiges primitive shield. All have compositions that are neither typical alkali olivine basalts nor typical tholeiites. The SiO_2 activity is too low for crystallisation of any low-Ca pyroxene and the basalts fall into the intermediate category of transitional olivine basalts. They have lower silica and higher alkalis than those of the Hawaiian volcanoes, with their K_2O content being roughly twice as high. The implication is that generation of La Réunion basalts was a near steady-state process involving relatively small per cent melting of a fertile mantle source. The inference that the latter is a garnet lherzolite may require reconsideration and a more pyroxenitic source may be responsible. Whatever the truth, it appears that the (relative) uniformity of magma product throughout the growth of the 'La Réunion volcano' implies remarkable continuity of the genetic process throughout its history. The very significant difference in basalt chemistry between La Réunion and Mauritius is of interest, those of the latter being distinctly more alkalic, with one series having a markedly more primitive character. Clearly, the relative uniformity of La Réunion basalts is not shared by its neighbouring ocean island volcano with implications possibly concerning sources differences and/or changes in the melt processes.

The productivity of Piton de la Fournaise is approximately 0.01 $km^3\ a^{-1}$. This is about ten times greater than that of Karthala but only about a tenth of

Kilauea's 0.1 km^3/a^{-1}. These estimates are in harmony with the concept that the degree of silica saturation relates to depth of derivation and that the deeper the melt-regime, the smaller the volume of magma and consequently the lower the surface volcano productivity.

Brian G.J. Upton
Emeritus Professor of Petrology
and Senior Honorary Professorial Fellow
Grant Institute
Edinburgh, UK

Contents

Part I

Piton de la Fournaise Volcano

1 A Brief History of the Observation of the Central Area of Piton de la Fournaise

Jean-François Lénat

Abstract

The first road to reach Piton de la Fournaise was built in the 1960s, so until that point the central area and summit was seldom visited due to access problems. In addition, not all the visitors recorded their observations. However, since 1755 a number of reports have been written in various journals, or preserved as unpublished documents. Thanks to the recent availability of scans of old journals and documents on public websites, it has become easier to consult the original reports, and therefore to transcribe, as accurately as possible, their geological information. Drawings, maps and photographs at different periods are as precious as the texts themselves. Here we present an analysis of the historical documents to highlight the main volcanic and volcano-tectonic events that occurred in the central area of Piton de la Fournaise from 1751 to the first half of the 20th century. The best descriptions and documentation from this period come from the works of Bory de Saint Vincent and Alfred Lacroix, although numerous other accounts contribute to the reconstruction of the volcanological events. Beyond an obvious historical interest, the accurate analysis of the past behaviour of Piton de la Fournaise is of paramount importance in the understanding its functioning.

1.1 Introduction

The very first reference to the island of La Réunion is found on a Portuguese planisphere from 1502 (known as the Cantino's Planisphere, a world map compiled by Portuguese explorers and cartographers). However, it is highly probable that Arab navigators, trading along the African coast, knew of the existence of this island,

J.-F. Lénat (✉)
Laboratoire Magmas et Volcans, Université Blaise Pascal - CNRS - IRD, OPGC, 5 rue Kessler, 63038 Clermont-Ferrand, France
e-mail: J.F.Lenat@opgc.univ-bpclermont.fr

P. Bachèlery et al. (eds.), *Active Volcanoes of the Southwest Indian Ocean*, Active Volcanoes of the World, DOI 10.1007/978-3-642-31395-0_1

although it not was located on the traditional maritime routes of the Mozambique Channel. Several accounts exist of ships making stopovers on the western coast of La Réunion for fresh water and food (fruits and various animals) at the beginning of the 17th century. After the French took possession of the island in 1642, they began colonizing it in 1663. The historical period of La Réunion's volcanic activity is thus restricted to a few centuries, and, as we will see below, observations of Piton de la Fournaise remained scarce until the 20th century.

1.2 Early Observations at the Beginning of the Island's Colonization

A comprehensive analysis of published and unpublished reports, private letters, administrative documents, and newspapers dealing with various observations of Piton de la Fournaise between the beginning of colonization and 1935 was carried out by Lacroix (1925, 1936, 1938), and more recently by Stieltjes and Moutou (1989), Peltier et al. (2012) and Michon et al. (2013). Although indications of volcanic activity were loosely suggested at the end of the 17th century and at the beginning of the 18th century, the first eruption to be confidently dated was in 1708. Throughout the 18th century activity was mainly recorded in unpublished reports by individuals when large lava flows reached the coast or when tephra fell in inhabited areas (Lacroix 1925, 1936, 1938). In this way thirty or so eruptions were reported, with more or less precision, during the 18th century, with four of them (1708; 1774; 1776; 1800) outside the Enclos Fouqué caldera and at least seven reaching the sea in the Grand-Brûlé. The state of the summit remained poorly known, except through a few visits.

The first reported visit to the summit of Piton de la Fournaise was made on September 20, 1751 by Andoche Dolnet de Palmaroux and a party composed of several Creoles and slaves (in d'Herguerty 1755). It was preceded by a visit to the Grand Brûlé by Fréri (a naval surgeon) on August 16, 1751. Fréri was escorted by a naval quarter-master, four Creoles and four slaves. Having reached the Grand Brûlé area a few days before, they then started the ascent of the volcano from the northern part of the Grand Brûlé. Fréri's report (in Guettard 1757) indicates an ongoing eruption from a cone located to the north of the Central Cone, and both recent and active lava flows, with the latter reaching the sea. The expedition stopped at less than two kilometers from the active cone. The September visit to the summit is not precisely described by Dolnet de Palmaroux (in d'Herguerty 1755). What he calls the "mouth" of the volcano is probably the Bory crater, although his estimate of the diameter ("près d'une lieue", nearly four kilometers) is amazingly unrealistic. He writes that the heat from the crater prevented him from approaching closer than 40 m to the rim, although "no eruption" (he probably means "no explosion/ejection") was observed. This suggests some kind of activity inside the crater, possibly a lava lake, because he later mentions observing a "fire abyss" from a small mound. While walking around the crater, Dolnet de Palmaroux encountered steam emitted from fractures, caves (probably lava tubes) and "horrible chasms" (large fractures?). He was also struck by the field of fresh-looking lava flows in the Enclos (their surface being described as composed of an accumulation of clinker, some strongly reflecting the sun beams). Bory de Saint Vincent (1804, Vol. 3, p. 4) indicates that he read Dolnet de Palmaroux's report in a manuscript where a crude drawing showed a small, apparently active, vent beneath the summit. This latter may be the active vent observed by Fréri in August.

In 1753, Piton de la Fournaise was visited by a group of 15 Creoles led by Jean Dugain (well known as a fugitive slave hunter). This expedition was requested by Jean Baptiste Charles Bouvet de Lozier, the governor of the island, following what was apparently a large eruption of the volcano. A short report is given by de Crémont (1770) who does not provide the exact dates of the events. The eruption was associated with an earthquake, whose tremors were felt, and noticeable ash falls over a large part of the island (St. Benoît, St. Denis). This suggests violent explosive activity at the summit, but we cannot rule out that what is described as cinders could in fact be Pele's hair, because falls of Pele's hair at a large distance from the volcano have been reported several times in the volcano's history. Dugain apparently made his observations from the rim of the Enclos. He described active lava flows in a large part of the western Enclos, eventually reaching the sea to the east.

The second visit to the summit was made on October 29, 1768 by a party comprising the governor and the intendant of the island (Léonard de Bellecombe and Honoré de Crémont), three residents (Germain, Joseph Hubert and his younger brother Hubert de Monfleury), a guide (Jean Dugain) and 20 or so slaves. A report was published by de Crémont (1770). On October 28, they observed a weak intermittent plume at the summit from the rim of the Enclos. The same day, one slave, named Jacob, discovered a pathway by which to climb down the wall of the Enclos Caldera. This pathway, still used nowadays, was named "Bellecombe's Descent" (Pas de Bellecombe in French), although the governor himself did not make the descent to the caldera floor (because of suffering from stomach ache). The next day, part of the group made the ascent to the summit. At the base of the rampart, they first observed the Formica Leo scoria cone, which may have formed during the 1753 eruption, according to Dugain's citation in de Crémont (1770). At the base of the Central Cone, they observed two spatter cones, or hornitos, now called "Chapelle de Rosemond". After this, de Crémont's geographic description becomes confusing. He probably made the ascent via the north-western flank of the Central Cone. At the summit he observed an active volcanic construction surrounded by a flat area. Although he does not mention the presence of two craters, we suspect that the active crater was probably located to the east of Bory crater, if we take into consideration the drawing made four years later by Lislet-Geoffroy (Fig. 1.1). This crater may have formed during the 1753 episode and/or the 1766 one (Sage in Lacroix 1936, p. 64) when Pele's hair was dispersed over a large area. Although de Crémont did not dare get too close to the crater rim ("à petite portée de pistolet", within pistol-shot range), he estimated the diameter to be about 60 m. He could hear loud noises (comparable to those of a "rough sea") coming from the crater and periodically observed incandescent projections in a thick dark plume. He noticed the presence of "four or five small volcanoes" to the north of the active cone, some of them emitting steam. Those could have been spatter cones along the fracture of a recent eruption. An important observation is that the surface of the flat area around the active cone was not made of lava flows, but covered by highly vesiculated tephra ("the pieces were ... light and easily crunched in our hand"). The cone is probably what was later described by Bory de Saint Vincent (1804) as the "Mamelon Central".

Two subsequent visits were made on November 18, 1772 (there is an ambiguity about the year, which could have been 1771, but we follow here the original publication that mentions 1772), and on October 29, 1791. In 1772, the expedition was carried out by Philibert Commerson, Honoré de Crémont, and a 21-year old illustrator (Paul Sauguin de Jossigny), a 16-year old herbalist (Lislet-Geoffroy), and his father, a surveyor (Jean-Baptiste Bank), a notary (Philippe Leclerc de Saint-Lubin), a guide (Payet), four residents and 32 slaves (Lislet-Geoffroy 1890). For this visit to the summit we have descriptions by Lislet-Geoffroy (1890) and Commerson (p. 154 in de Montesus 1889) and a drawing by Lislet-Geoffroy (Fig. 1.1). Although the shape of the Central Cone is not realistically reproduced in Fig. 1.1, two craters were present at that time. On November 17, from Piton de Bert, the visitors clearly observed steam from the western one (Bory crater) and glowing ejecta from the eastern one (probably the one observed during the 1768 visit). The diameter of the latter was roughly estimated to be about 40 m. On November 18, they probably went down into the Enclos from Piton de Bert (on the southern rim of the Enclos Fouqué caldera) and Lislet-Geoffroy noted that the surface of the Enclos was covered by both

Fig. 1.1 Pencil sketch by Lislet-Geoffroy (reproduced in Lacroix 1936) showing the Enclos and the Central Cone as seen from the southwest in 1772

smooth (pahoehoe) and rough (aa) lava flows. From the rim of Bory crater, they could observe steam flowing from fractures at the bottom of this crater and they had a view of the eastern crater, where the top of the lava column was close to the rim.

Hubert Dumorier and Patu de Rosemont attempted a new ascent in 1789, but they were trapped in bad weather and forced to abandon the expedition.

The October 1791 visit was motivated by the desire to explore the changes at the summit following major effusive and explosive activity (on July, 14) observed from the coast from June to July. The group was composed of Alexis Bert (a cultured military officer), Joseph Hubert, Jean-Joseph Patu de Rosemont, two guides and nine slaves. Their expedition is reported by Bert in an unpublished manuscript reproduced in Lacroix (1936). On October 27, from the Piton de Bert (which was subsequently named this by Bory de Saint Vincent), they observed steam emanating from a fracture apparently extending from the summit to the mid-slope of the Central Cone. During his ascent to the summit, on October 27, Bert describes the Formica Leo and the Chapelle de Rosemont (that he appropriately proposes naming "Rosement Cave", because Patu de Rosemont came with him to make a drawing of this volcanic construction). At the summit, the "old vent" (later called the "Bory crater" by Bory de Saint Vincent) was remarkably similar to how it remained until 1953: about 240 m in diameter and 50 m in depth (much less to the south where it was possible to get inside the crater), with steep walls and a flat bottom covered with scoria. At that time, what he calls the "modern crater" (later called the "Mamelon central" by Bory de Saint

Vincent) was located about 400 m to the east, at lower elevation than the Bory crater. This is the crater which was active during the previous visit in 1772. Its shape was that of a truncated cone, nearly 50 m high, with extremely steep slopes (Bert noted "often greater than 80°"). Its crater was circular, about 80 m in diameter and 25 m in depth. The walls were sub-vertical, showing a pile of lava flows and scree near and at the bottom. It was inactive, but Bert notes the presence of weak steam vents with a sulphurous smell in the upper part of the crater. Finally, Bert observes a "new crater" (later called the "Dolomieu" by Bory de Saint Vincent). From his more or less precise description, we can deduce that this latter was probably located at about 400 m from the Bory crater, at the southern border of the summit area, close to the "modern crater". Bert suggests that this new crater formed during the earlier 1791 effusive activity. Interestingly, he notes that going into the new crater, he had to cross a place where the ground was composed of what may be tentatively interpreted as thin shelly pahoehoe lava flows. The new crater was about 200 m in diameter and 240 m deep. The walls exhibited piles of often reddish layers (from what we can nowadays see in the section created by the 2007 collapse, these layers are composed of reddish and yellowish tephra, and lava flows), and the bottom was occupied by scree. Sulphurous-smelling steam was emitted from vents in the walls and the bottom, sometimes associated with yellow deposits. The crater was surrounded by scoriaceaous tephra, up to a distance of about 200 m. This new crater can be tentatively associated with a now filled pit crater named "Petit Plateau", in the southern part of the summit, where Bory de Saint Vincent observed activity in 1801. Bert suggests, very ingeniously, that the collapse of the new crater might be related to the emission of a large volume of lava during the April–July eruption that would have drained magma from beneath the summit. The main phase of the collapse would have happened on July 14, when an unusually dense plume was observed above the summit. The drawing made by Patu de Rosemont (Fig. 1.2, top) from a viewpoint comparable to that of Fig. 1.1 is more correct in the representation of the Central Cone (see Fig. 1.2, bottom). The cone to the east of Bory crater is not visible. The darker lava flows to the southeast of the Central Cone could be the 1791 lava flows.

1.3 Visits and Descriptions by Inhabitants and Naturalists During the 19th Century

The observations and drawings of Bory de Saint-Vincent in 1801 (Bory de Saint-Vincent 1804) mark a milestone in the study of Piton de la Fournaise. Bory de Saint-Vincent was a young (23-year old) naturalist hired for a marine expedition in the "Southern seas" which he left in Mauritius (then called "Isle de France"), because of health problems. After spending several months on Mauritius Island, he stayed on La Réunion Island (then called "Isle Bourbon") from August 1801 to March 1802. During this period, he made two visits to the summit of Piton de la Fournaise, one in late October, and the other less than a month later in November. His descriptions and drawings from Piton de la Fournaise are definitely the first scientific documents of the structure and activity of the summit area of Piton de la Fournaise. The text of Bory de Saint-Vincent is remarkably rich in analysis and thoughts about the volcanic phenomena. In addition, he made a fairly good map of the island (probably with the help of earlier documents by Chisny and Lislet Geoffroy) and of the volcano (Fig. 1.3), and gave names to many places and features of the volcano (Plaine des Osmondes, Piton de Faujas, Cratère Dolomieu, Cratère Commerson, Piton Guichard, Piton Du Petit Thouars, Piton Chisny, Cratère Hubert, Cratères Ramond, Cratère Haüy, etc.).

For his first visit, the ascent was made from the eastern coast, starting in the Grand Brûlé. Bory de Saint-Vincent was accompanied by two inhabitants, named Jouvancourt and Cochinard, and by some slaves. The summit was very similar to the 1791 description of Bert (in Lacroix 1936), with three craters. From Bory de Saint-Vincent's map, the Mamelon Central Cone

Fig. 1.2 *Top* Pencil sketch by Patu de Rosemont (reproduced in Lacroix, 1936) showing the Central Cone as seen from the southwest in 1791. *Bottom* A present-day view from the same place (Piton de Bert) as Figs. 1.1 and 1.2 *top*, for comparison

may have been situated in the median part of the present day Dolomieu crater, and it had not changed since Bert's visit. The drawing made by Bory de Saint-Vincent (Fig. 1.4a) allows us to better identify this construction as a spatter cone similar, for example, to the one formed in the Dolomieu crater in 2006 (Fig. 1.4b). The small lava flows on the Mamelon Central Cone indicate that the lava column periodically overflowed the rim of the crater.

The Bory crater had not changed since Bert's visit ten years earlier. Jouvancourt proposed to name it Bory during this visit, and comparison with the drawing of Bory de Saint-Vincent shows that it probably did not change much between that period and 1953.

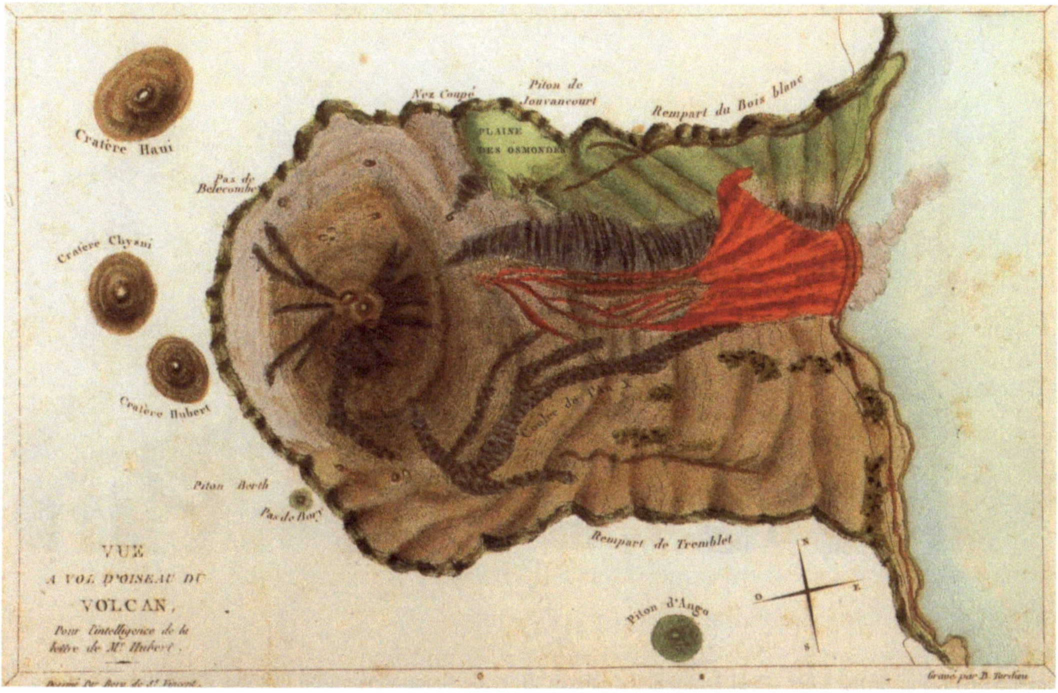

Fig. 1.3 Map of the Enclos by Bory de Saint-Vincent. The "Coulée de l'an X" is the lava flow emitted at the base of the Central Cone during Bory's visit to the summit. The *red one* is 1802 lava flow. North of the latter is the 1787 lava flow

Fig. 1.4 **a** The drawing of the Mamelon Central by Bory de Saint-Vincent; **b** For comparison, a similar construction in 2006

Fig. 1.5 Drawing of the 1791–1801 Dolomieu crater by Bory de Saint-Vincent on October 27, 1801

The main change since Bert's visit was the occurrence of sustained activity in Bert's "new crater", which Bory de Saint-Vincent named "Dolomieu". He provides a drawing (Fig. 1.5) and a description of the activity in the crater. Steam was emitted from several vents while explosions were produced at one vent that emitted a narrow lava flow. The latter flowed for a short distance before disappearing into a fracture. At the same time, a large lava flow was emitted from the base of the Central Cone.

Bory de Saint-Vincent made a second ascent to the summit on November 20, 1801, less than one month after his first visit. This time he approached the volcano from the east, because he was travelling to the Piton des Neiges summit. While waiting a few days for some of his companions in the Plaines des Cafres, he noticed a strong glow above Piton de la Fournaise, and therefore decided to pay a second visit to the summit. Bory de Saint-Vincent was accompanied by Jouvancourt, Déjean, Cochinard, Guichard and some slaves. The night before the ascent they camped at Piton de Bert, from where they could observe the ongoing eruptive activity. Bory de Saint-Vincent describes more or less continuous explosive activity. During the night, he observed two short-lived small lava flows originating from a fracture close to the crater. When he reached the summit, the day after, he was struck by the changes to Dolomieu crater since his previous visit. The diameter had increased to about 400 m by collapse along concentric faults which he had observed during the first visit. The crater was now almost filled by a lava lake. Bory de Saint-Vincent's report is remarkable in describing features that are often seen in present-day basaltic lava lakes: a dark crust broken by zigzags of glowing lava, concentric waves or oscillations on the surface of the lake, and small vents (Fig. 1.6).

An interesting anecdote in Bory de Saint-Vincent's report is the way he measured the distance between Piton de Bert and the summit. At a given signal (a large blanket waved at night in the light of the lava lake and two gun shots) someone at Piton de Bert fired a shot. Bory de Saint-Vincent measured the time between the light of the shot and the arrival of the sound to compute the distance.

Fig. 1.6 Drawing by Bory de Saint-Vincent of the lava lake inside the 1791–1801 Dolomieu crater on November 20, 1801

Fig. 1.7 Section through the 1791–1801 Dolomieu crater preserved in the rim of the present-day Dolomieu crater. The 1801 Dolomieu crater was filled by horizontal thick lava flows. Discontinuous lines show the rims of the filled crater

Note that the location of the present day Dolomieu crater does not correspond to the 1801 Dolomieu crater. The 1791–1801 Dolomieu pit crater can now be observed in section on the rim of the present day Dolomieu crater (Fig. 1.7), and its preserved southern half constitutes what is locally called the "Petit Plateau".

Fig. 1.8 Drawing by Bory de Saint Vincent from Piton de Bert (same viewpoint as for Fig. 1.2b) during the 19–20 November 1801 eruption. The 3 summit craters (Bory, Mamelon Central, and the 1791–1801 Dolomieu) are clearly shown

Figure 1.8, drawn by Bory de Saint-Vincent from Piton de Bert, perfectly illustrates the state of the summit at this time. The two inactive craters, Bory and the Mamelon Central, are clearly shown beside the active 1791–1801 Dolomieu pit crater.

Surprisingly, given that Bory de Saint Vincent's work might have been expected to have sparked a wave of scientific curiosity for the evolution of the summit area, we have relatively few reported visits to the summit during the rest of the 19th century. Only 10 or so ascents to the summit figure in the literature after that of Bory de Saint Vincent: in 1807, by an Englishman named Montagu; between 1822 and 1825, by a surveyor named Schnieder; in 1831, by Mézières-Lepervanche, a naturalist from La Réunion accompanied by F. Bellier-Beaumont; in 1852, by the engineer Maillard, and again in 1860; in 1862, by Jacob de Cordemoy, a botanist; then by two geologists, Velain in 1874 and von Drasche in 1875; in 1889 and 1890, by Hermann, a notary and naturalist; and, finally, a hydrographer, Fichot, in 1890. It is highly probable that more people went to the summit but did not write reports. For example, Hugoulin (1862) reports that an inhabitant from Saint Joseph went to the summit in 1858. However, observations of volcanic events from places located near the volcano were frequently reported. Combining the observations of the summit and those from farther afield allows us to reconstruct the main phases of the summit evolution, albeit with uncertainties.

It is highly probable that volcano-tectonic events occurred at the summit in 1802, because ash falls were reported as far away as Saint Denis (nearly 50 km from the summit) on January 7, and then on April 14 in the vicinity of Sainte Rose (about 15 km from the summit) (Hubert, in Bory de Saint-Vincent 1804). Explosions capable of sending ashes such distances have always been associated with large collapses and/or phreatic or phreatomagmatic events at Piton de la Fournaise, occurring often during emission of large oceanite lava flows in the Enclos (see, for example, Bachèlery 1981; Lacroix 1936). The March 1802 lava flow was indeed a large oceanite lava flow (Fig. 1.3) (Hubert, in Bory de Saint-Vincent 1804). Although Hubert (in Bory de Saint-Vincent 1804) reports smoke and glows above the summit (probably from the 1791–1801 Dolomieu pit crater) in December 1801 and early January 1802, and then smoke during the March lava flow, it is not possible to determine the changes at the summit during these 1802 events. Nothing is said by Montagu about the 1791–1801 Dolomieu pit crater during his visit to the summit in 1807 (Hubert in Lacroix 1919). Instead, he noticed a new active vent near the "Mamelon Central". This vent constructed a small cone (Montagu crater) probably between March or May and June 13. Hubert (in Lacroix 1919) gives indications about the summit activity between 1807 and 1817, but these are somewhat ambiguous since he uses unspecific terms such as the "old crater", the "crater", the "new crater" or sometimes simply "the volcano". The main reason is that the observations were made either from Saint Benoît (from which the summit is not visible) or from the area of the Grand Brûlé. We can tentatively infer the following events at the summit: (1) in 1807: 23 March, eruption at the summit with a lava flow; 22–26 May, glows observed at the summit; 10 June, strombolian activity at the summit (Montagu crater); (2) in 1809: 17 July–8 August, glows at the summit; 20–22, then again on 24–26 November; (3) in 1812: glows at the summit in March, glows and smoke 28–31 August, the Montagu crater had disappeared; large flank oceanite lava flow from 1–16 September; during this eruption, ashfalls and Pele's hair reported, but their origin is not described. However, dull noises are heard on the 10 and 11, concurrently with ashfalls mixed with Pele's hair; in October and November, glows and smoke at the summit. These events (large flank oceanite lava flow, dull noises and ashfalls, disappearance of Montagu crater) strongly suggest that collapses and/or phreatic explosion occurred at the summit in 1812. From 1813 to 1817, Hubert's description of the activity (in Lacroix 1919) is even less detailed. We can note simply that glows at the summit are frequently observed between inferred eruptions and that ashfalls occurred in September 1814. If Hubert's reports of frequent observations of glows at the summit between 1807 and 1817 are reliable, this suggests more or less continuous magmatic activity at the summit (lava lake). During the first reported visit to the summit after this sequence of events, in 1831, Mézières-Lepervanche (Lacroix 1938; Vinson, 1865) found no trace of the "Mamelon Central", but instead a crater much wider than the 1791–1801 Dolomieu crater described by Bory de Saint Vincent. Lacroix (1938) speculates that the "Mamelon Central" must have disappeared between 1817 and 1822–23, a period that brackets the observations of Hubert (in Lacroix 1919), who would probably have noticed its disappearance, and the geodesy survey at the summit by Schnieder (in Lacroix 1938) when only the summit of the Bory and of an eastern crater were encountered. Although this hypothesis is reasonable, it is difficult to exclude the possibility that the summit modifications occurred mainly during the 1802 and 1812 crises, accompanied by ashfalls outside the Enclos.

The next period, until about the 1860s, is mostly described by Maillard and Hugoulin (Hugoulin 1860, 1862; Maillard 1853, 1862) with additions by Lacroix (1936). Maillard, a civil engineer in La Réunion from 1842 to 1861, wrote a book covering many topics on La Réunion (Maillard 1862) and published the first accurate map of the island. Hugoulin was a navy pharmacist who stayed in La Réunion (dates unknown). Before Maillard's summit visit in 1852, lava flows were reported every few years, with some large ones reaching the sea (1821, 1830, 1832, 1844 and 1850). During the 1821 crisis, large explosions, associated with ashes and Pele's hair, occurred at the summit. Similarly, violent explosions were observed at the summit at the beginning of the 1844 eruption. Maillard (1853) found, in 1852 (the date is given by Hugoulin 1862), the eastern part of the summit considerably modified since the last detailed description by Bory de Saint Vincent (1804). The Bory crater to the west was apparently not changed. Maillard notes that its bottom was occupied by a flat lava flow where steam was emitted from fractures. The 1791–1801 Dolomieu crater and the Mamelon Central no longer existed, but a new crater, called "Brûlant", had formed at the eastern part of the summit. It is probable that this new crater had already been observed (whether from the Grand Brûlé or during unreported visits) because Maillard's writing suggests that its presence was known. He describes the Brûlant crater as a pit crater, 150 m in diameter and 200–300 m deep, with vertical walls (Fig. 1.9). At the bottom he observed an active lava lake emitting sulphurous gases. This account of the structure of the summit is of paramount importance, because it supports that: (1) as suspected, the explosive events observed from the surroundings of the volcano in 1802, 1812, 1814 and 1844 were associated with volcano-tectonic events (disappearance of the 1791–1801 Dolomieu crater and Mamelon Central, collapse of the new Brûlant crater); and (2) that the glows frequently reported at the summit from at least 1807 (Hubet in Lacroix 1919) were indeed associated

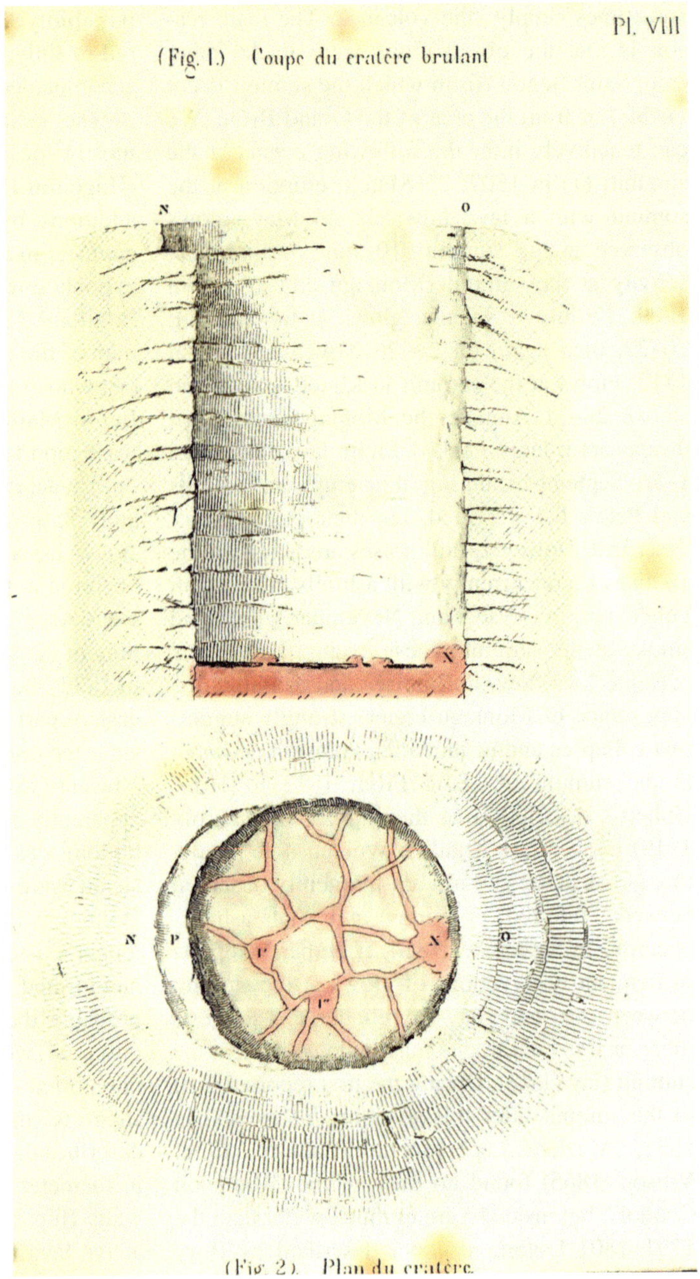

Fig. 1.9 Cross-section of the Brûlant crater and view of its basal lava lake by Maillard (1862) in 1852

with more or less continuous activity in the old and then new craters. The occurrence of large lava flow emissions during this period fully agree with the current models in which the main summit collapses happen when the shallow reservoirs are drained by voluminous flank lava flows (e.g. Michon et al. 2009).

A new major event happened at the summit in 1860. It was preceded by a large flank eruption in 1858 (Hugoulin 1862) and a summit eruption in 1859 (Maillard 1862). The 1860 sequence of events started in January with a summit eruption, with lava flows and explosions, followed two weeks later by a large flank eruption lasting about 40 days (Maillard 1862). Two days later a large explosive event occurred at the summit, accompanied by seismic activity (Hugoulin 1860) and a large plume laden with ashes. This latter divided in two, with one part drifting to the south and the other to the north-east. Ashfalls were observed up to a distance of about 30 km to the north-east (on a sailing boat) and lapilli fell outside of the Enclos to the north-east (Rempart de Bois Blanc). Maillard (1862) described the Enclos as being covered with ashes, lapilli and blocks. Up to more than 1 km from the Brûlant crater, large angular blocks, up to several cubic meters, were observed (including in the Bory crater). All these features indicate that a larger than usual phreatic explosive activity had occurred at the summit. The source was the Brûlant crater, which Maillard described soon after the eruption as a large, circular, funnel-shaped, collapse whose bottom was filled with angular blocks and emitted steam and sulphurous gases. Two years later, a description of the summit was provided by Camille Jacob de Cordemoy (Jacob de Cordemoy 1867), a civil engineer and an expert on the local botany. Most of the ashes and lapilli in the Enclos seem to have been washed away, because Jacob de Cordemoy describes the Enclos as the "domain of lava". But the blocks at the summit were still there and he observed a large one, which he estimated to be about 10 m^3 in volume, resting in an impact crater and surrounded by impact ejecta. Jacob de Cordemoy described a structure that Maillard did not mentioned. The Brûlant crater was a small cone located inside a semi-circular enclosure bounded by vertical walls, which he interpreted as an ancient, filled crater, with fractures emitting steam. This seems to be the first observation of what would later be called Enclos Velain by Lacroix (1936). Although Maillard did not describe this structure, we can infer that it had probably formed before the formation of the new Dolomieu crater, and possibly as early as the 1802 or 1812 summit events (Hubert in Bory de Saint-Vincent 1804; Hubert in Lacroix 1919). For the Brûlant crater, instead of the funnel-shaped collapse seen by Maillard two years before, Jacob de Cordemoy described a circular pit with vertical walls and a bottom filled by blocks and emitting steam.

In the following years, large lava flows were reported in 1863–64, 1865, 1868 and 1871. Velain's visit of the summit at the beginning of September 1874 was preceded by explosive activity at the summit and a flank lava flow in July. He describes the Bory crater as filled with scoria and lava blocks, with no steam emission. Two small cones, "partially destroyed", were also present. A very interesting part of Velain's report (Velain 1878) is his description of the Enclos Velain. From his description, we can assume that it was pretty similar to the structure observed by Lacroix in 1911 (1936, plates IX and X; Lacroix 1938, Fig. 1). At that time, steam was emitted from the base of the walls delimiting the semi-circular Enclos Velain. The lower parts of the cone in the Brûlant crater were buttressed by the walls of the Enclos Velain. The surface of the cone was covered by scoria and small lava flows. The interior is well illustrated in Fig. 1.10. Velain estimates its diameter to have been 400 m, with a depth of 150–160 m.

The following year, von Drasche made two ascents of the volcano (von Drasche 1878). During the first one, approaching from the west, he got to the summit. His description of the summit does not provide critical additional information to Velain's report. The second ascent was made from the east (Grand Brûlé). This time von Drasche did not go to the summit, but visited the Enclos. He made an interesting drawing of the Central Cone from the north (Fig. 1.11).

Major eruptive activity occurred in June to August 1889, just before the next reported visits to the summit by Hermann (1890) and Fichot (1892) in 1890, associated with large flank lava flows and explosive activity at the summit producing Pele's hair. Hermann went to the summit in May 1890, but in his report he also says that he had made a visit in September 1889. During his approach to the Enclos, he witnessed explosions at the summit and an overflowing of lava. At the summit, he was struck by the huge changes since his September 1889 visit. Apparently, the Enclos Velain had changed significantly. From his imprecise description, we can

Fig. 1.10 Drawing of the Brûlant crater in 1874 by Velain (1878)

infer that it had collapsed more and had been further filled with lava flows. The Brûlant crater that he saw as a 150 m-diameter and 200 m-deep crater with steep walls and lava at the bottom in September 1889, was replaced by an active 50 m-high cone to the northeast. This new cone was "covered" with fumaroles. Hermann took what is apparently the first photograph of an explosion at Piton de la Fournaise, but this picture (which he claims was successful) was not published. After an explosion, he climbed to the rim of the cone and saw a magma column 5–6 m beneath the rim. He suggested naming this new cone Mahy crater (from the name of a contemporaneous politician and scholar). Fichot climbed to the summit one month later, in early June. His guide, who had visited the summit two months before (in April), confirmed the changes described by Hermann. The Brûlant crater had been completely filled over that time and the new cone was located further to the northeast (no distance is given by Fichot or Hermann). The Mahy crater had a similar strombolian activity to that during Hermann's visit.

1.4 Observations During the 20th Century

The beginning of the 20th century is deeply marked by the work of Lacroix (1925, 1936, 1938) who visited the volcano in 1911, and followed the activity until 1942, with the help of Maurice Jean, a teacher at the Leconte de l'Isle high school in Saint Denis. Maurice Jean was put in charge of observing the volcano by the governor.

Lacroix visited the Piton de la Fournaise summit in 1911. Since Hermann and Fichot's visits in 1890, about 15 flank eruptions had occurred (Lacroix 1936, 1938; Stieltjes and Moutou 1989). In addition, summit or Central Cone eruptions were reported in 1905 and 1910. Figure 1.12 shows the morphology of the summit in 1911. This map is very important in the understanding of the previous descriptions of the Enclos Velain by Jacob de Cordemoy, Velain, Hermann and Fichot. As noted above, it is probable that the collapse of the Enclos Velain occurred early in the 19th century, possibly

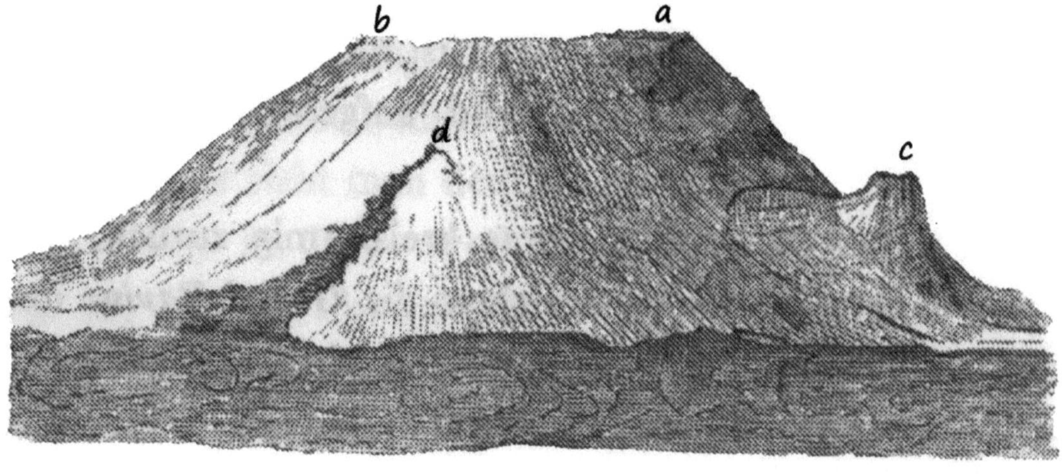

Fig. 1.11 Drawing of the Central Cone, as seen from the north, by von Drasche (1878) in 1875: **a** Bory crater; **b** Brûlant crater; **c** Puy mi-Côte; **d** A recent lava flow

corresponding with the disappearance of the Mamelon Central. Ten or so eruptions were reported between Lacroix's visit and the great July–August 1931 eruption (Jean 1935).

This latter eruption is the second most voluminous historical eruption, after the 2007 eruption (Bachèlery et al. 2010), with an estimated volume of 135×10^6 m^3, which is one order of magnitude greater than the average for the "usual" eruptions. Following this huge emission of lava, a major volcano-tectonic event would have happened at the summit, with a large collapse of the summit (Fig. 1.13). Unlike in 2007 (Michon et al. 2007, 2009; Staudacher et al. 2009), when the summit collapse occurred simultaneously with a huge flank eruption, the collapse following the 1931 large flank eruption seems to have occurred progressively over the following years. The observations of the summit were very discontinuous at that time, therefore it is difficult to establish the dynamics of the collapse. However, reliable reports of Jean (in Lacroix 1936) suggest that the collapse may have even started as early as 1927 and lasted until 1935, with major phases in 1933–35.

The last eruption reported by Lacroix was that of 1942 (Lacroix 1942), whereas Jean reported only the July 1946 one (Jean 1946). After that, Marcel Ducrot, the curator of the Natural History Museum of Saint Denis, was formally in charge of reporting the activity of Piton de la Fournaise. His reports cover the period 1943–1959 (Ducrot 1958, 1959, 1960). Until the start of the Piton de la Fournaise Volcano Observatory, in late 1979, the activity was reported by various authors (Aubert de la Rüe 1965; Kieffer et al. 1977; Kieffer and Vincent 1978; Krafft and Gerente 1977a, b; Montaggioni et al. 1972). In the 1950s and 1960s new works on La Réunion and Piton de la Fournaise volcanism began to be published (Aubert de la Rüe 1965; Bussière 1958; Defos du Rau 1959; Rivals 1950), as well as geological maps (Billard 1974; Bussière 1967). Near the end of the 1960s, publications in international scientific reviews marked the start of the modern approach to Piton de la Fournaise volcanism.

1.5 Modern Observations: Piton de La Fournaise Volcanological Observatory

Visiting the volcano remained difficult for a long time because there was no road. It took at least three days from the nearest village, Bourg Murat. In 1957, the Forest Service started building a road, but it was not until 1968 that it reached the Pas de

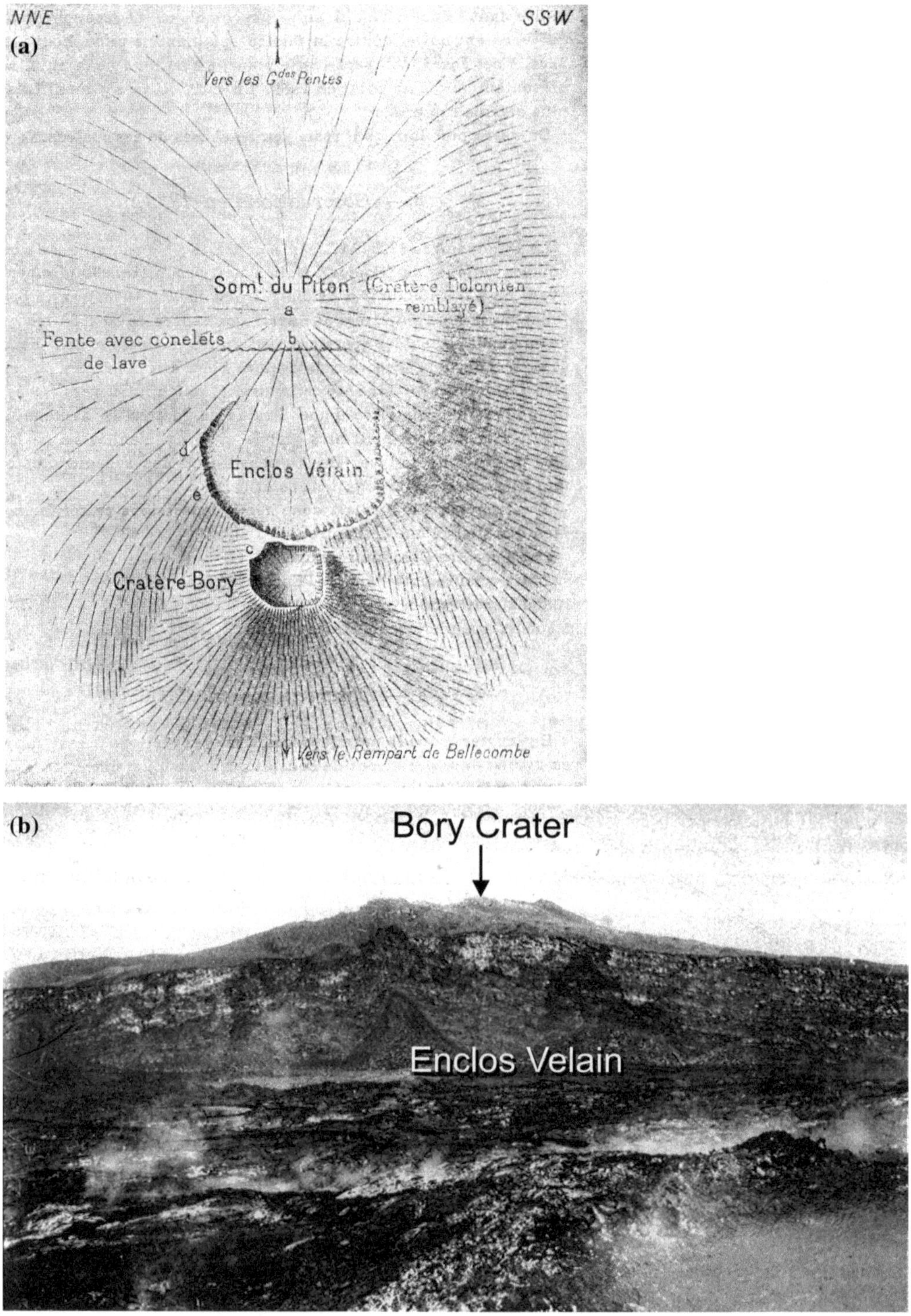

Fig. 1.12 The summit in 1911 by Lacroix (1936, 1938). *Top* sketch map. *Bottom* picture taken from point *a* on the map)

Fig. 1.13 Oblique air photo from 1936 showing the collapsed summit. This is one from the first set of aerial pictures of Piton de la Fournaise. It was taken by Jean Arnoux from an airplane piloted by Squadron Leader Wachenheim in September 1936 (in Lacroix 1936)

Bellecombe. However, before the road was built, Marcel Ducrot organized a periodic surveillance which included a visit to the volcano every 2–3 months. This surveillance was funded by the local council (Conseil Général) (Ducrot 1960). The curator of the museum in Saint Denis was formally in charge of the surveillance, and he was entitled to help from the Forest Service, the police and the school teachers. Reports were distributed to the prefect, to the Geological Survey of Madagascar, and to the Madagascar academy. Petrographical and geochemical analyses of samples from the eruptions were performed by the Geological Survey of Madagascar. Aerial photos, taken by small airplanes from the local flying club, could be requested by the prefect. This was the birth of the surveillance of Piton de la Fournaise, which had been eagerly awaited for years, in particular by Lacroix (1938).

The April 1977 eruption, when lava flows partially destroyed the village of Sainte Rose, prompted the National Center for Scientific Research (CNRS) to establish a permanent and modern observatory. France had long had permanent volcanological observatories in the Lesser Antilles; from 1950 at the Soufrière de Guadeloupe and from the early 1930s at Mount Pelé. The Piton de la Fournaise Volcanological

Observatory (OVPF) is located in Bourg Murat and is part of the Institut de Physique du Globe in Paris. The first monitored eruption by the OVPF was in 1981 (Bachèlery 1981; Chevallier et al. 1981). In the last 35 years or so, the OVPF has been developed dramatically, and it is now one of the best volcanological observatories in the world. In parallel, increasingly numerous scientific projects have been carried out in the domains of volcanology, geodynamics, geothermics and hydrogeology. As a result of all these efforts, Piton de la Fournaise is now one of the best studied volcanoes in the world. This volume is intended to provide a summary of the current knowledge of the volcano. It also gives an idea of how much is still left to fully understand the functioning of such a volcanic system.

Acknowledgments We thank Andrea Di Muro for his accurate suggestions that helped to improve this contribution.

References

Aubert de la Rüe E (1965) Le Piton de la Fournaise, volcan actif de l'île de la Réunion. Bulletin Volcanologique. Bull Volcanol 28(1):311–320

Bachèlery P (1981) Le Piton de la Fournaise (Ile de la Reunion). Etude volcanologique, structurale et petrologique. Piton de la Fournaise, Reunion; volcanologic, structural and petrographic study. Doctoral thesis, University of Clermont-Ferrand, Clermont-Ferrand, p 255

Bachèlery P, Saint-Ange F, Villeneuve N, Savoye B, Normand A, Le Drezen E, Barrère A, Quod J-P, Deplus C (2010) Huge lava flows into the sea and caldera collapse, April 2007, Piton de la Fournaise volcano. In: Michon L (ed) IAVCEI third workshop on collapse calderas. Laboratoire GéoSciences Réunion, La Réunion, pp 73–74

Billard G (1974) Carte géologique de la France—La Réunion. Echelle 1/50 000 encollaboration avec P.M. Vincent. Carte (4 feuilles) et notice explicative. BRGM

Bory de Saint-Vincent, J-B-G-M (1804) Voyage dans les quatre principales îles des mers d'Afrique, pendant les années neuf et dix de la République (1801 et 1802), Marseille, Laffitte. Buisson F., Paris

Bussière P (1958) Étude géologique de l'île de La Réunion. Travaux du Bureau Géologique de Madagascar, 98, p 64

Bussière P (1967) Carte géologique de La Réunion à l'échelle de 1/100000. Notice explicative. Bureau Géologique de Madagascar

Chevallier L, Lalanne F-X, Bachèlery P, Vincent PM (1981) L'éruption du mois de février 1981 au Piton de la Fournaise (Ile de la Réunion, Océan Indien). Phénoménologie et remarques structurales. C R Acad Sci Paris 293(II):187–190

d'Herguerty M (1755) Observations sur le volcan de l'Ile Bourbon. Mémoires de la Société royale des sciences, et belles-lettres de Nancy, 3:218–235

de Crémont H (1770) Excursion au Volcan de la Fournaise en 1768. In: É -C Fréron (ed), L'Année littéraire. Delalain, Paris, vol 7, pp 73–97

de Montesus F -B (1889) Martyrologe et biographie de Commerson, médecin botaniste et naturaliste du roi, médecin de Toulon-sur-Arroux (Saône-et-Loire) au XVIIème siècle. Extrait des Bulletins de la Société des sciences naturelles de Saône-et-Loire. Marceau, L., Châlon-sur-Saône, p 224

Defos du Rau J (1959) Le relief de l'ile de la Réunion: étude de morphologie volcanique. Ph. D. thesis, Bordeaux

Ducrot M (1958) Le volcan de La Réunion (période 1952–1957). Travaux du Bureau Géologique, Service Géologique Tananarive, vol 88, p 101

Ducrot M (1959) Le volcan de La Réunion (période 1943–1952). Travaux du Bureau Géologique, Service Géologique Tananarive, vol 94, p 28

Ducrot M (1960) Le volcan actif de l'île de la Réunion. Surveillance et étude de l'activité. Possibilités de l'utilisation de l'énergie émise. Bull Volcanol 23 (1):109–120

Fichot E (1892) Une excursion au volcan de la Réunion. Revue scientifique, Paris 50:44–48

Guettard M (1757) Mémoire sur plusieurs morceaux d'histoire naturelle tirés du cabinet de S. A. S. M. le duc d'Orléans. Histoire de l'Académie royale des sciences, année 1753 (Mémoires): 376–382

Hermann J (1890) Lettre à Monsieur le Président de la Société des Sciences et Arts de St Denis. Bull Soc Sc et Arts de la Réunion, Saint Denis, p 77

Hugoulin F (1860) Dernière éruption du volcan de l'île de la Réunion (19 mars 1860). Revue Algérienne et Coloniale 2:483–487

Hugoulin F (1862) Le volcan de La Réunion; éruption de novembre 1858. Revue maritime et coloniale 6:284–308

Jacob de Cordemoy C (1867) Itinéraire d'un voyage au volcan en 1862. In: A Roussin (ed), Album de l'Ile de la Réunion, Saint-Denis, vol 4, p 118

Jean M (1935) Les éruptions de 1931–1934 à la Réunion. An Phys du Globe de la France d'outre-mer: 147–154

Jean M (1946) Sur l'éruption intra-cratèrienne du 18 juin-juillet 1946 du Piton de la Fournaise (La Réunion). C R Acad Sci Paris T223:913–915

Kieffer G, Tricot B, Vincent PM (1977) Une éruption inhabituelle (avril 1977) du Piton de la Fournaise (Ile de la Réunion): ses enseignements volcanologiques et structuraux. C R Acad Sci Paris, 285 série D: 957–960

Kieffer G, Vincent PM (1978) L'éruption d'octobre-novembre 1977 du Piton de la Fournaise (île de la Réunion): une éruption terminale sans cratère terminal. C R Acad Sci Paris 286(II):1767–1770

Krafft M, Gerente A (1977a) L'activité du Piton de la Fournaise entre novembre 1975 et avril 1976 (île de la Réunion, océan indien). C R Acad Sci Paris, 284 série D: 2091–2094
Krafft M, Gerente A (1977b) L'activité du Piton de la Fournaise entre octobre 1972 et mai 1973 (île de la Réunion, océan indien). C R Acad Sci Paris, 284 série D: 607–610
Lacroix A (1919) L'activité éruptive du volcan de La Réunion de 1802 à 1817, daprès les observations d'un témoin oculaire. Bull Soc Geol Fr 19:3–10
Lacroix A (1925) Succession des éruptions et bibliographie du volcan actif de la Réunion. Bull Volcanol 1 (3):20–56
Lacroix A (1936) Le volcan actif de l'île de la Réunion et ses produits. Gauthier-Villars, Paris, p 297
Lacroix A (1938) Le volcan actif de l'île de la Réunion (supplément) et celui de la Grande-Comore. Gauthier-Villars, Paris, p 57
Lacroix A (1942) Une éruption du Piton de la Fournaise (Ile de la Réunion) en 1942. C R Acad Sci Paris 215:517–518
Lislet-Geoffroy JB (1890) Voyage au volcan de Bourbon en 1772 (note: or more probably in 1771). Revue historique et littéraire de l'Ile Maurice: archives coloniales 3(2):361–365
Maillard L (1853) Note sur l'île de la Réunion. Bulletin de la Société Géologique de France. Bull Soc Geol Fr 10:499–505
Maillard L (1862) Notes sur l'île de la Réunion (Bourbon). Dentu, Paris, pp 344 + appendices
Michon L, Di Muro A, Villeneuve N, Saint-Marc C, Fadda P, Manta F (2013) Explosive activity of the summit cone of Piton de la Fournaise volcano (La Réunion island): a historical and geological review. J Volcano Geotherm Res 264:117–133
Michon L, Staudacher T, Ferrazzini V, Bachelery P, Marti J (2007) April 2007 collapse of Piton de la Fournaise: a new example of caldera formation. Geophys Res Lett 34:L21301. doi:10.1029/2007GL031248
Michon L, Villeneuve N, Catry T, Merle O (2009) How summit calderas collapse on basaltic volcanoes: new insights from the April 2007 caldera collapse of Piton de la Fournaise volcano. J Volcano Geotherm Res, 184(1–2): 138–151. doi.org/10.1016/j.jvolgeores.2008.11.003
Montaggioni L, Nativel P, Billard G (1972) L'activite actuelle du Piton de la Fournaise (Ile de la Reunion, Ocean Indien). C R Acad Sci Paris 275:2615–2618
Peltier A, Massin F, Bachèlery P, Finizola A (2012) Internal structures and building of basaltic shield volcanoes: the example of the Piton de La Fournaise terminal cone (La Réunion). Bull Volcanol 74:1881–1897. doi:10.1007/s00445-012-0636-7
Rivals P (1950) Histoire géologique de l'île de La Réunion. Ph. D. thesis, Paris, p 143
Staudacher T, Ferrazzini V, Peltier A, Kowalski P, Boissier P, Catherine P, Lauret F, Massin F (2009) The April 2007 eruption and the Dolomieu crater collapse, two major events at Piton de la Fournaise (La Réunion Island, Indian Ocean). J Volcano Geotherm Res 184(1–2):126–137. doi:10.1016/j.jvolgeores.2008.11.005
Stieltjes L, Moutou P (1989) A statistical and probalistic study of the historic activity of Piton de la Fournaise, Réunion Island. J Volcanol Geotherm Res 36:67–86
Velain C (1878) Description géologique de la presqu'ile d'Aden, de l'ile de la Réunion, des iles Saint-Paul et Amsterdam. Hennuyer, Paris, p 356
Vinson A (1865) Un pieux et savant créole de l'Ile Bourbon. In: M Mézières Lépervanche. Almanach religieux de l'Ile Bourbon ou de la Réunion, pp 181–188
von Drasche R (1878) Die Insel Réunion (Bourbon) im Indischen Ocean: Ein geologisch-petrographische Studie mit einem Angange über die Insel Mauritius. Hölder A., Wien, p 85

Geodynamic Setting of La Réunion

2

Jean-François Lénat

Abstract

The island of La Réunion is located in the complex oceanic environment of the Marscarene Basin. The latter was created by oceanic expansion between the late Cretaceous and the Paleocene. The expansion of the Basin stopped at magnetic chron 27 when the spreading jumped north, between the Seychelles and India, to become the Central Indian Ridge. Dating and geodynamic reconstructions support a hot spot origin for La Réunion volcanism. La Réunion lies in the middle of an approximately 350 km-wide zone bounded by transform faults (Mauritius and Mahanoro-Wilshaw Fracture Zones) and a trough identified as the extinct spreading center (chron 27). This compartment exhibits a substantial depth anomaly (several hundreds of meters), possibly related to the interaction with the hot spot plume. Gravity and seismic data show the virtual absence of flexure beneath and around La Réunion. Near La Réunion the organization of the oceanic crust magnetic anomalies is very different from the classic parallel striped oceanic organization. It has been suggested that this pattern might be interpreted in terms of a paleo-triple junction of oceanic rifts. Many seamounts protrude from the sediments in the La Réunion segment, some associated with the transform faults, others associated with a series of parallel ridges and a few isolated others. All these constructions postdate the oceanic crust, but most of them appear to be much older than La Réunion, with the possible exception of some small seamounts close to La Réunion or on its submarine flanks.

2.1 Regional Tectonic Setting

The Indian Ocean started to open up around 175 Ma (Schettino and Scotese 2005). The early stages of Gondwanan break-up separated Africa from a block composed of Madagascar, India, Australia and Antartica (Coffin and Rabinowitz

J.-F. Lénat (✉)
Laboratoire Magmas et Volcans, Université Blaise Pascal-CNRS-IRD, OPGC, 5 rue Kessler, 63038 Clermont-Ferrand, France
e-mail: J.F.Lenat@opgc.univ-bpclermont.fr

P. Bachèlery et al. (eds.), *Active Volcanoes of the Southwest Indian Ocean*,
Active Volcanoes of the World, DOI 10.1007/978-3-642-31395-0_2

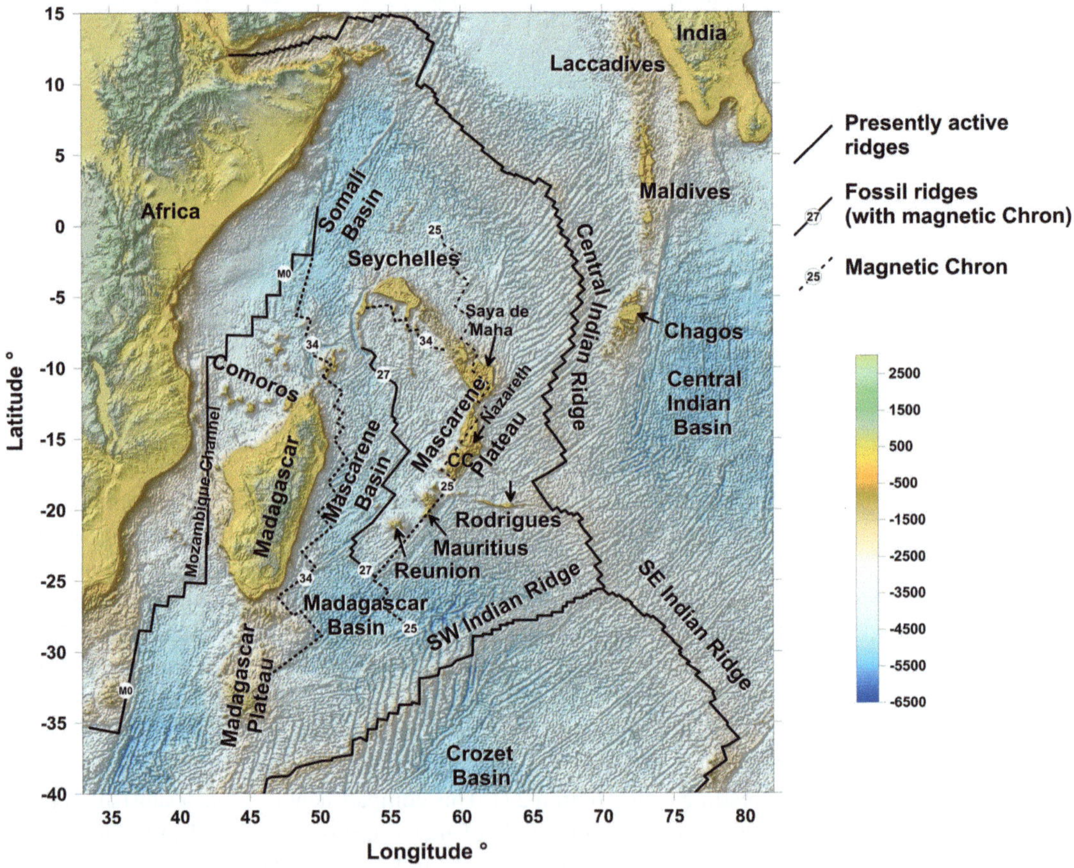

Fig. 2.1 Geodynamic setting of the Comoros and La Réunion. The magnetic chrons are taken from Müller et al. (1997)

1987; Smith and Sandwell 1997). This first rift became inactive at about 120 Ma (magnetic chron M0; Fig. 2.1). It opened the Mozambic Channel in which the magmatism of the Comoros archipelago would develop later (Michon 2016, Chap. 21). Around 83 Ma (magnetic chron 34), the Mascarene Basin (Fig. 2.1) opened, marking the separation of Madagascar and India (Bernard and Munschy 2000; Dyment 1991a). The Mascarene Basin rift became extinct at anomaly 27 ($\sim$61 Ma), when the spreading jumped to a new location, between the Seychelles and India, to become the Central Indian Ridge. Nowadays, La Réunion and the Comoros Archipelago lie on the Somalian plate, which is bounded by the Central and SW Indian ridges to the east and south and by the East African Rift to the west (Stamps et al. 2008).

It is easy to pick out the plateau features on Fig. 2.1, and they can be divided into continental and oceanic areas. The Seychelles and Madagascar are micro-continents isolated by the episodes of sea floor spreading described above. Torsvik et al. (2013) suggest that additional continental fragments are present beneath part of the Saya de Maha and Nazareth Banks, Cargados-Carjos and Mauritius, forming an arc of anomalously thick crust that extends northwards from Mauritius to the Seychelles. The other plateaus, Laccadives, Maldive, Chagos, Mascarene, and Mauritius and La Réunion islands, are widely regarded as volcanic constructions of the La Réunion hotspot which can be traced back to the emission of the Deccan Trapps in the late Cretaceous era. This hypothesis is supported by dating and geodynamic reconstructions (Duncan 1990; Duncan et al. 1989; Georgen et al. 2001; Storey 1995). When the hotspot was near the mid-ocean spreading ridge, it produced complex structural and geochemical interactions (Ito et al. 2003; Morgan 1978; Sleep 2002). For example, the Rodrigues Ridge is interpreted as the result of the interaction between the Central Indian Ridge and the Réunion hotspot (Füri et al. 2011; Morgan 1978; Nauret et al. 2006).

The structure of the Mascarene Basin has been the subject of numerous studies since the 1970s because it is a key area to understand the geodynamics of the early stages of the Indian Ocean. The structure and evolution of the northwestern end of the Basin is still poorly known, although recent works (Bernard and Munschy 2000; Collier et al. 2004; Fretzdorff et al. 1998; Todal and Eldholm 1998) have provided significant new knowledge. Various models have been proposed over the years. Bernard and Munschy (2000) suggest that the Mascarene Basin is composed of 9 zones separated by major transform faults. Their geometry implies two different rotation poles (between 83 and 67.7 Ma, and between 67.7 and 59 Ma). The magnetic anomaly pattern suggests a progressive southward extinction of the spreading center (67 Ma in the north and about 61 Ma in the south).

2.2 La Réunion and the Surrounding Lithosphere

2.2.1 Characteristics of the Oceanic Compartment of La Réunion

La Réunion lies in the middle of an approximately 350 km-wide zone bounded by the Mauritius transform fault (Mauritius Fracture Zone or FZ) to the SE and by the Mahanoro-Wilshaw transform fault (Mahanoro-Wilshaw Fracture Zone or FZ) to the NW (Fig. 2.2a). Another structural boundary, to the SW, is a trough identified as an extinct spreading center (magnetic anomaly 27).

2.2.1.1 Bathymetry Anomaly

A distinctive characteristic of this compartment is its anomalous bathymetry (Fig. 2.2a). The shallow bathymetry of the compartment was first mentioned by Bonneville et al. (1988). Lénat et al. (2009) have analysed this anomaly by comparing the actual bathymetry corrected for sediment load with a theoretical bathymetry based on the thermal subsidence of the oceanic lithosphere as a function of its age. They found a substantial depth anomaly, reaching several hundreds of meters, bounded by steep gradients corresponding to the transform faults and the fossil ridge. These authors have carried out analogue modelling to explore the deformation resulting from a mantle plume rising beneath a faulted lithosphere. The scale models suggest that the most important factor controlling the buoyant flow of the plume material is the presence of mechanical discontinuities (transform faults and an extinct ridge) at the lithospheric scale, which allow a complete decoupling of the deformation on their two sides. Such a model could account for the anomalous bathymetry of the La Réunion compartment where the upward flow of the plume material would be trapped in the zone bounded by the lithospheric discontinuities and accommodated by deformation within this zone. As we will see later, this mechanism could contribute to the dynamic support of La Réunion, where virtually no lithospheric flexure exists. Note that Bonneville et al. (1997), Michon et al. (2007) also suggested a dynamic support by the mantle plume beneath La Réunion.

2.2.1.2 Gravity and Magnetic Patterns

Gravity

Figure 2.2b shows a free air gravity map from Sandwell and Smith (2009). Gravity is better than bathymetry for highlighting the features of the oceanic crust where there is a blanket of lightweight sediments (abyssal plains,

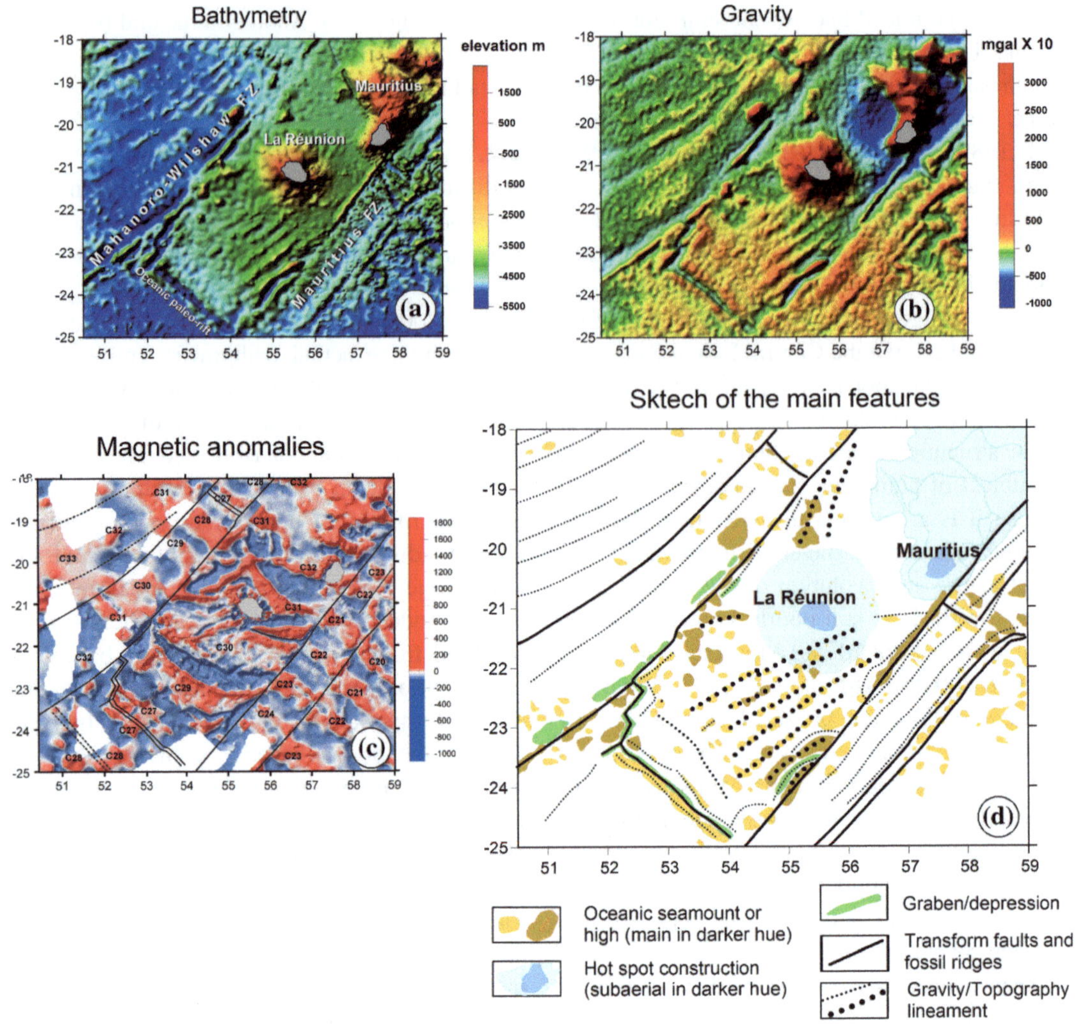

Fig. 2.2 **a.** Shaded topography of the La Réunion zone (Smith and Sandwell 1997). **b** Shaded free air gravity (Smith and Sandwell 1997). **c** Magnetic anomaly map. Magnetic chrons from Bernard and Munschy (2000), Dyment (1991b). **d** Structural plan of the area

topographic depressions filled with sediment, etc.). The short wavelength features observed on the bathymetry are therefore better defined on the gravity, which also reveals some lineaments that were less perceptible on the bathymetry (2.2a). The large range of gravity values in this area of the map is related to the presence of the islands and of large seamounts for the positive values, and to the presence of sedimentary basins along transform faults and around Mauritius for the negative ones. Note that to the SW of La Réunion (and further to the S and SE) there are positive values that suggest poor isostatic compensation. The load of Mauritius has induced a flexure of the lithosphere which has been filled by sediments. This is well documented by the gravity and seismic data acquired during the FOREVER cruise in 2006 (de Voogd et al. 2009). Conversely, it has been shown by gravity and seismic data acquired during various cruises (Charvis et al. 1999; de Voogd et al. 1999, 2009; Gallart et al. 1999) that no flexure and corresponding sediment filling is observed beneath and around La Réunion.

Magnetics

In the La Réunion compartment, the authors (Bernard and Munschy 2000; Dyment 1991a; Fretzdorff et al. 1998) were confronted with difficulties in trying to establish the location and identification of the magnetic chrons, partly because of inadequate data coverage, but above all because of the complexity of the anomaly pattern. Recently, Bissessur (2011) has used newly acquired data from the FOREVER and ERODER 1 and 2 cruises (Deplus et al. 2007; Savoye et al. 2008) to propose a more complete analysis and interpretation of the history of the oceanic crust in this area. Figure 2.2c shows a magnetic anomaly map of the area. The anomalies are now very well defined near La Réunion, and to different degrees in the other areas, depending on the available data. Figure 2.2c clearly shows that near La Réunion the organization of the oceanic crust anomalies departs drastically from the classic oceanic organization into parallel stripes, as is observed beyond the La Réunion compartment. In the bottom left of the figure the authors identify a traditional pattern of anomalies with a ridge jump at chron 28 (~63 Ma) and an extinct rift at chron 27 (~61 Ma). Then towards the northeast of the compartment, the pattern of anomalies 29 to 32 becomes complex, with different and varying orientations. Bissessur (2011) interprets this pattern as the result of a triple junction of oceanic rifts. He identifies this junction with the Indian Ocean Triple Junction (its present location can be seen on Fig. 2.1) that would have been located in the Mascarene Basin before anomaly 29 (~64–65 Ma).

2.2.1.3 Seamounts and Sediments of the Oceanic Plate

As compared to the predicted bathymetry (Smith and Sandwell 1997), from which the structural plan of Fig. 2.2d is drawn, the description of the submarine part of La Réunion and its surrounding oceanic environment has been drastically improved recently thanks to high resolution bathymetric and acoustic mapping and seismic profiling. FOREVER and ERODER 1 and 2 cruises (de Voogd et al. 2009; Deplus et al. 2007; Savoye et al. 2008) in 2006 and 2008 were instrumental in providing a homogeneous

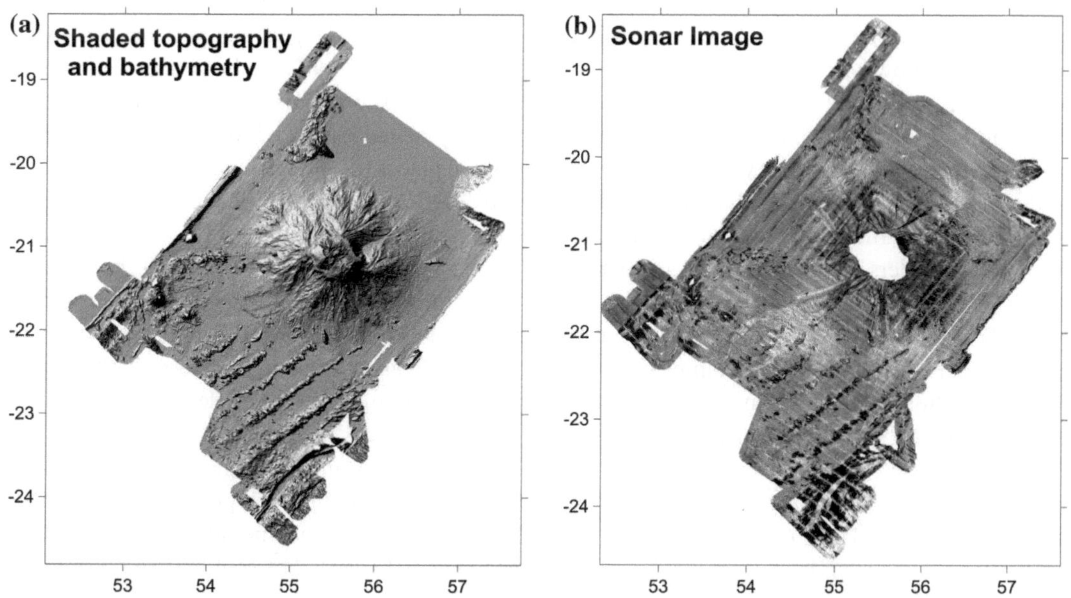

Fig. 2.3 **a** shaded bathymetry, and **b** acoustic images from forever and eroder cruises (Deplus et al. 2009; Savoye et al. 2008)

coverage of a large part of the La Réunion compartment (Fig. 2.3).

Many seamounts protrude from the sediments in the La Réunion segment (Figs. 2.2a, d and 2.3). We can distinguish those associated with the transform faults and those associated with a series of parallel ridges. A few others, however, are more difficult to link to those categories.

The Mahanoro-Wilshaw and Mauritius transform faults are only partially mapped by high resolution bathymetry (Fig. 2.3), but the predicted bathymetry of Smith and Sandwell (1997) provides complete low resolution coverage. Linear ridges and grabens are observed along the transform faults. Studies in various places suggest that magmatic activity can take place at transform faults near spreading centers or by interaction with hotspot plumes (Hekinian and Bideau 1995; Lowrie et al. 1986; Pockalny et al. 1997). Another classic interpretation for uplift and subsidence along oceanic transform faults is differential thermal subsidence and bending (Sandwell and Schubert 1982) or shear heating and lateral heat conduction across the fault (Chen 1988) when two lithospheres of different age are brought into contact. The difference in age between the La Réunion zone and that to the NW of the Mahanoro-Wilshaw FZ ranges from 0 to 10 Ma, whereas contrasts of 15–25 Ma occur across the Mauritius FZ. However, similar age contrasts exist across other transform faults in the vicinity of the La Réunion zone without creating such marked signatures in the topography, except for the branch of the Mahanoro-Wilshaw FZ that goes to Madagascar. The constructions on the transform faults forming the borders of the La Réunion segment are mostly straight ridges, apparently incised longitudinally by the faults. Theses ridges may act as barriers to the sediments. For example, we note that the bathymetry between the foot of La Réunion and the Mauritius transform fault is not flat but decreases gently toward the fault, suggesting that deposits from La Réunion extend down to the foot of the ridge, possibly as fine sediments derived from landslides.

To the S-SW of La Réunion, a series of ~50°N trending ridges (Figs. 2.2a, d and 2.3) constitutes one of the main topographic characteristics of the oceanic crust in the La Réunion compartment. They are quasi-periodic features, separated from each other by ~35 km. They are more or less continuous and apparently formed by continuous coalescence of individual seamounts of variable dimension and height. Some of the seamounts are isolated features, but this may be an artifact due to the burial of the area by sediments, under which they may be coalescent. Near the Mahanoro-Wilshaw fracture zone, the ridge pattern becomes less clear. Two very large seamounts are observed in this area. At the latitude of La Réunion, a less focussed ~N90 ridge extends from near the Mahanoro-Wilshaw fracture zone and disappears beneath the western submarine flank of La Réunion. On the other side of the island, a large elongated seamount is in line with the ~90°N ridge. Along this line, we also observe a seamount protruding from the western flank of La Réunion. It has been studied by Oehler et al. (2008) and named Sonne Seamount. The ridges and their sedimentary environment have been studied by Fretzdorff et al. (1998), de Voogd et al. (2009), Deplus et al. (2009). The ridges and seamounts protrude from a flat seafloor covered by a few hundred meters of sediments. The morphology of their flanks, their sediment cover and the coating of the dredged samples by Mn-oxides (Fretzdorff et al. 1998) indicate that they are old seamounts. The seismic data show that the sedimentary sequence is deformed and disrupted in several places (de Voogd et al. 2009). The ~50°N and ~90°N elongation of the ridges is significantly different from that of the transform faults (~35–40°N). They clearly intersect the magnetic fabric of the oceanic crust and must therefore be regarded as being younger than the oceanic crust. Linear ridges, oblique to transform faults and the oceanic ridge, have been described in several places (e.g. in the South Pacific between the Louisville chain and the ridge (Small 1995), or the Musicians Seamounts (Kopp et al. 2003)). Most authors now agree that the ridges represent plume channels between a hot spot and an oceanic rift. Thus Bissessur (2011) proposes that these ridges would result from interaction between the nascent

La Réunion hot spot, around 65 Ma, and the Mascarene Basin rifting, until the latter vanished when the rifting jumped north to the Carlsberg ridge.

It is clear that some of the ridges intersect La Réunion and should be present beneath the edifice, but they do not reappear at the north. Processing of the seismic data acquired during the FOREVER and ERODER cruises should show whether or not they do reach that far, or if they are buried by the sediments which are thicker there (de Voogd et al. 2009). We also note the presence of a large seamount partialy buried by the edifice of La Réunion to the south. This seamount is in the continuation of a 50°N ridge, but has an elongation at nearly right angles to the ridge. The interferences between these oceanic constructions and the construction of La Réunion are still unexplored. At the least these positive reliefs have probably acted as topographic barriers during the evolution of the submarine slopes of La Réunion, and Oehler et al. (2008) document some cases where La Réunion products are impounded by ocean seamounts. Finally, to the north, near the Mahanoro-Wilshaw fracture zone, a large ridge could be related to the transform fault system. At its southern tip there is a substantial seamount with a clear 120°N elongation.

A different type of individual seamount is observed to the north-east of the island (Oehler et al. 2008) (Fig. 2.3). These seem to have a pristine morphology and are well-shaped flat-topped seamounts. Oehler et al. describe two of them that appear to have grown on the lower submarine flank of La Réunion and must therefore be young constructions.

References

Bernard A, Munschy M (2000) Were the Mascarene and Laxmi Basins (western Indian Ocean) formed at the same spreading centre? C R Acad Sci Paris Sci de la Terre et des Planètes Earth Planet Sci 330:777–783

Bissessur, PD (2011) Structure, age and evolution of the Mascarene Basin, Western Indian Ocean. Ph.D. Thesis, Institut de Physique du Globe de Paris, Paris, 168 pp

Bonneville A, Barriot JP, Bayer R (1988) Evidence from geoid data of a hotspot origin for the southern Mascarene Plateau and Mascarene Islands, (Indian Ocean). J Geophys Res 93:4199–4212

Bonneville A, Von Herzen RP, Lucazeau F (1997) Heat flow over Reunion hot spot track: additional evidence for thermal rejuvenation of oceanic lithosphere. J Geophys Res 102(B10):22731–22748

Charvis P, Laesanpura A, Gallart J, Hirn A, Lepine J, de Voogd B, Minshull TA, Hello Y, Pontoise B (1999) Spatial distribution of hotspot material added to the lithosphere under La Reunion, from wide-angle seismic data. J Geophys Res B 104(2):2875–2893

Chen Y (1988) Thermal model of oceanic transform faults. J Geophys Res 93:8839–8851

Coffin MF, Rabinowitz PD (1987) Reconstruction of Madagascar and Africa: evidence fron the Davie fracture zone and western Somali basin. J Geophys Res 92(B9):9385–9406

Collier JS, Minshull TA, Kendall J-M, Whitmarsh RB, Rümpker G, Joseph P, Samson P, Lane CI, Sansom VP, Vermeesch M, Hammond J, Wookey J, Teanby N, Ryberg TM, Dean SM (2004) Rapid continental breakup and microcontinent formation in the western Indian Ocean. EOS Trans Am Geophys Union 85(46):481–496

de Voogd B, Deplus C, Sisavath E, Depuiset F, Mercier M (2009) Vertical movements of the oceanic lithosphere above La Reunion hotspot. In: EGU General Assembly 2009. Geophysical Research Abstracts, Vienna, pp EGU2009-5754-3

de Voogd B, Palome S, Hirn A, Charvis P, Gallart J, Rousset D, Danobeitia J, Perroud H (1999) Vertical movements and material transport during hotspot activity: seismic reflection profiling offshore La Reunion. J Geophys Res B 104(2):2855–2874

Deplus C, de Voogd B, Dyment J, Bissessur PD, Sisavath E, Depuiset F, Mercier M (2009) New insights on the oceanic lithosphere at La Réunion hotspot volcano, EGU General Assembly 2009. Geophysical Research Abstracts, Vienna, pp. EGU2009-5728

Deplus C, de Voogd B, Dyment J, Depuiset F, Sisavath E, Forever cruise scientific party (2007) Does the Reunion hotspot volcano emplace on a fossil ridge or a fracture zone? In: Eos Transactions Supplement, AGU fall meeting abstracts V31F-05, San Francisco

Duncan RA (1990) The volcanic record of the Reunion hotspot. In: Duncan RA, Backman J, Dunbar RB, Peterson LC (eds) Proceedings of ocean drilling program science results, vol 115, pp 3–10

Duncan RA, Backman J, Peterson L (1989) Reunion hotspot activity through tertiary time: initial results from the ocean drilling program, Leg 115. J Volcanol Geoth Res 36:193–198

Dyment J (1991a) Structure et évolution de la lithosphère océanique dans l'Océan Indien : apport des anomalies magnétiques. Thesis Thesis, IPGS, Université Louis-Pasteur, Strasbourg, 374 pp

Dyment J (1991b) Structure et évolution de la lithosphère océanique dans l'Océan Indien: apport des anomalies magnétiques. Ph.D. Thesis, Strasbourg, 374 pp

Fretzdorff S, Stoffers P, Devey CW, Munschy M (1998) Structure and morphology of submarine volcanism in the hotspot region around Reunion Island, western Indian Ocean. Mar Geol 148(1–2):39–53

Füri E, Hilton DR, Murton BJ, Hémond C, Dyment J, Day JMD (2011) Helium isotope variations between Réunion Island and the Central Indian Ridge (17°–21°S): new evidence for ridge–hot spot interaction. J Geophys Res 116:B02207. doi:10.1029/2010JB007609

Gallart J, Driad L, Charvis P, Sapin M, Hirn A, Diaz J, de Voogd B, Sachpazi M (1999) Perturbation to the lithosphere along the hotspot track of La Reunion from an offshore-onshore seismic transect. J Geophys Res B 104(2):2895–2908

Georgen JE, Lin J, Dick HJB (2001) Evidence from gravity anomalies for interactions of the Marion and Bouvet hotpots with the Southwest Indian Ridge: effects of transform offsets. Earth Planet Sci Lett 187:283–300

Hekinian R, Bideau D (1995) Magmatism in the Garrett transform fault (East Pacific Rise near 13° 27′ S). J Geophys Res 100:10163–10185

Ito G, Lin J, Graham D (2003) Observational and theoretical studies of the dynamics of mantle plume–mid-ocean ridge interaction. Rev Geophys 41(4):3–1, 3–24. doi:10.1029/2002RG000117

Kopp H, Kopp C, Phipps M, Flueh J, Ernst R, Weinrebe W, Morgan WJ (2003) Fossil hot spot-ridge interaction in the Musicians Seamount Province: geophysical investigations of hot spot volcanism at volcanic elongated ridges. J Geophys Res 108(B3). doi:10.1029/2002JB002015

Lénat JF, Merle, Lespagnol L (2009) La Réunion: an example of channeled hot spot plume. J Volcanol Geoth Res 184:1–13. doi:10.1016/j.jvolgeores.2008.12.001

Lowrie A, Smoot C, Batiza R (1986) Are oceanic fracture zones locked or strong and weak? New evidence for volcanic activity and weakness. Geology 14:242–245

Michon L (2016) The volcanism of the Comoros archipelago integrated at a regional scale. In: Bachèlery P, Lénat J-F, Di Muro A, Michon L (eds) Active Volcanoes of the Southwest Indian Ocean: Piton de la Fournaise and Karthala. Active Volcanoes of the World. Springer, Berlin

Michon L, Saint-Ange F, Bachelery P, Villeneuve N, Staudacher T (2007) Role of the structural inheritance of the oceanic lithosphere in the magmato-tectonic evolution of Piton de la Fournaise volcano (La Réunion Island). J Geophys Res 112(B04205). doi:10.1029/2006JB004598

Morgan WJ (1978) Rodriguez, Darwin, Amsterdam, …, a second type of hot spot islands. J Geophys Res 83:5355–5360

Müller RD, Roest WR, Royer JY, Gahagan LM, Sclater JG (1997) Digital isochrons of the world's ocean floor. J Geophys Res 102(B2):3211–3214

Nauret F, Abouchami W, Galer SJG, Hofmann AW, Hémond C, Chauvel C, Dyment J (2006) Correlated trace element-Pb isotope enrichments in Indian MORB along 18–20°S, Central Indian Ridge. Earth Planet Sci Lett 245:137–152. doi:10.1016/j.epsl.2006.03.015

Oehler JF, Lénat JF, Labazuy P (2008) Growth and collapse of the Reunion Island volcanoes. Bull Volcanol 70:717–742. doi:10.1007/s00445-007-0163-0

Pockalny RA, Fox PJ, Fornari DJ, Macdonald KC, Perfit MR (1997) Tectonic reconstruction of the Clipperton and Siqueiros Fracture Zones: evidence and consequences of plate motion change for the last 3 Myr. J Geophys Res 102(B2):3167–3181

Sandwell D, Schubert G (1982) Lithospheric flexure at fracture zones. J Geophys Res 87:4657–4667

Sandwell DT, Smith WHF (2009) Global marine gravity from retracked Geosat and ERS-1 altimetry: ridge segmentation versus spreading rate. J Geophys Res 114:B01411. doi:10.1029/2008JB006008

Savoye B, Saint-Ange F, Sisavath E, Deplus C, Bachèlery P (2008) How an Intraplate Volcanic Island may generate true Volcanoclastic Deep-Sea Fans - La Réunion Island. Indian Ocean, AAPG, Cape town

Schettino A, Scotese CR (2005) Apparent polar wander paths for the major continents (200 Ma to the present day): a palaeomagnetic reference frame for global plate tectonic reconstructions. Geophys J Int 163:727–759. doi:10.1111/j.1365-246X.2005.02638.x

Sleep NH (2002) Ridge-crossing mantle plumes and gaps in tracks. Geochem Geophys Geosyst 3(12). doi:10.1029/2001GC000290

Small C (1995) Observations of ridge/hot spot interactions in the Southern Ocean. J Geophys Res 100:17931–17946

Smith W, Sandwell D (1997) Global sea floor topography from satellite altimetry and ship depth soundings. Science 277:1956–1962

Stamps DS, Calais E, Saria E, Hartnady C, Nocquet J-M, Ebinger CJ, Fernandes R (2008) A kinematic model for the East African Rift. Geophys Res Lett 35:L05304. doi:10.1029/2007GL032781

Storey BC (1995) The role of mantle plumes in continental breakup: case histories from Gondwanaland. Nature 377:301–308

Todal A, Eldholm O (1998) Continental margin off western India and Deccan large igneous province. Marine Geophys Res 20:273–291

Torsvik TH, Amundsen H, Hartz EH, Corfu F, Kusznir N, Gaina C, Doubrovine PV, Steinberger B, Ashwal LD, Jamtveit B (2013) A Precambrian microcontinent in the Indian Ocean. Nat Geosci 6(3):223–227. doi:10.1038/ngeo1736

Construction of La Réunion

3

Jean-François Lénat

Abstract

La Réunion is a huge volcanic edifice, 220–240 km in diameter and 7 km in height, lying on a ~4000 m deep ocean floor. Unlike for most oceanic volcanoes, the lithosphere beneath La Réunion exhibits, no, or very weak, flexure. A large body of underplated material has been identified beneath the southwestern part of the island. Two volcano edifices currently compose the island: Piton des Neiges to the NW and Piton de la Fournaise to the SE. The main internal feature of the Piton des Neiges volcano is a huge intrusive complex with a lateral extent coinciding at the surface with that of three depressions called Cirques. At the eastern side of the island, a similar, though smaller, complex is associated with Les Alizés volcano, an old edifice no longer visible in the landscape. The existence of the latter is inferred by gravity, drill hole and dating data. Piton de la Fournaise is a relatively young (450–400 ka) and thin (2 km on average) edifice lying on the presumed remnants of Les Alizés edifice. Piton des Neiges and Les Alizés are inferred to be the primary volcanoes of La Réunion activity. In the submarine domain, debris avalanche deposits are regarded as the most extensive and voluminous formations. They have built four huge submarine bulges separated by submarine canyons offshore of the main rivers. The submarine bulges are essentially built of mass wasting products and sediments. The island is surrounded by a belt of hyaloclastites and pillow lavas. A few volcanic constructional features (cones, eroded piles of lava flows) have been found on the submarine flanks, with the majority being located close to the coast in continuation of the on land southwest and northeast rift zones of Piton de la Fournaise. Two exceptions are

J.-F. Lénat (✉)
Laboratoire Magmas et Volcans, Université Blaise Pascal-CNRS-IRD, OPGC, 5 rue Kessler, 63038 Clermont-Ferrand, France
e-mail: J.F.Lenat@opgc.univ-bpclermont.fr

P. Bachèlery et al. (eds.), *Active Volcanoes of the Southwest Indian Ocean*,
Active Volcanoes of the World, DOI 10.1007/978-3-642-31395-0_3

constructions, mostly buried beneath mass wasting and sedimentary deposits, to the SW and E of the island, extending tens of kilometres offshore. The first one is interpreted as an old rift zone of Piton des Neiges and the second as the flank of Les Alizés.

3.1 La Réunion Edifice

La Réunion edifice is almost centrally located between the transform faults that bound the oceanic compartment on which it lies (Lénat 2016, this volume Chap. 2). The island is elliptical in shape (50 × 70 km) with a NW–SE orientation, and rises to 3070 m above sea level. It is composed of two volcanoes, Piton des Neiges, occupying the NW two thirds of the island, and Piton de la Fournaise. The whole edifice appears as a flattened cone, 220–240 km in diameter and 7 km in height. Its contact with the quasi-flat sediments of the ocean bottom is clear, at depths of around −4000 m. In the transition zone between the lower flanks of the edifice and the abyssal plain, there are almost continuous sediment waves, concentric to the edifice, except where major canyons reach the abyssal plain (Le Friant et al. 2011; Oehler et al. 2008). These concentric features are classically attributed to turbidity currents (Le Friant et al. 2011; Wynn et al. 2000), although Oehler et al. (2008) suggest that they could correspond to the folding of ductile pelagic sediments in response to compressive strain created at the front of advancing debris avalanches. Although the base of the whole island is almost circular (Le Friant et al. 2011), the volcanic edifice does not correspond to a perfectly shaped cone. Labazuy (1996) and Oehler et al. (2008) describe the submarine cone as being composed of four main fan-shaped bulges which are separated by major canyon systems.

The geology of the submarine flanks of La Réunion was established by Oehler et al. (2008), before the recent data from FOREVER (Deplus et al. 2009) and ERODER (Savoye et al. 2008) cruises were available. Preliminary interpretations using the most recent data have been published by Le Friant et al. (2011), Lebas (2012). It is likely that the submarine geology will be significantly refined in the near future, although the major units defined by Oehler et al. (2008) will probably remain largely unchanged. The geological maps of Oehler et al. (2008), Lebas (2012) (for the eastern slope) show surface rocks resulting from both constructional and destruction processes.

Few constructional features (cones, eroded piles of lava flows) have been recognized on the submarine surface, and, of these, most are located close to the coast in continuation of the on land southwest and northeast rift zones of Piton de la Fournaise. The associated magnetic anomalies indicate that both the Brunhes and Matuyama (and older) formations are present (Gailler and Lénat 2010). This is well supported by ages of between 3.77 and 0.44 Ma determined for dredged samples from the offshore continuation of the north-east rift zone of Piton de la Fournaise (Lénat et al. 2009a; Smietana 2011; Smietana et al. 2010) and between 0.07 and 0.127 Ma for the Cone Elianne, on the southern submarine flank of Piton de la Fournaise (Lénat et al. 2009a; Smietana 2011; Smietana et al. 2010). In addition, an age of 2.48 Ma was found (Smietana 2011) at the surface of a ridge (“Ride d’Etang Salé”) at the southwest of the island. Pristine, small monogenic cones and flat-topped cones on the north and northeast flanks (Gailler and Lénat 2010; Oehler et al. 2008) indicate small eruptions outside of the main volcano centres during the construction of La Réunion. The fact that they are not buried by sediments or debris flows argues for a relatively young, albeit unconstrained, age.

According to Oehler et al. (2008), debris avalanche deposits are by far the most extensive and voluminous formations in the submarine domain. They have built four huge submarine bulges to the east, north, west, and south of the island. They form fans 20–30 km wide at the coastline and 100–150 km wide at their ends, 70–80 km offshore. For Oehler et al. (2008), they were built gradually by the superimposition and/or juxtaposition of products moved during landslide episodes, involving up to several hundred cubic kilometres of material. Around 50 individual event deposits have been recognized by Oehler et al. (2008). About one third are interpreted as secondary landslides, affecting previously emplaced debris avalanches. The location and geometry of some mass wasting units suggest source areas on land, in accordance with the fact that dredged samples at the surface of the deposits often have a clear subaerial origin (Cochonat et al. 1990; Lénat et al. 2009a). Recently, Le Friant et al. (2011) proposed that debris avalanches were more developed offshore of Piton de la Fournaise than offshore of Piton de Neiges, where the flanks are more likely to have resulted from a general spreading of the edifice. The submarine bulges are dissected and bound by canyons up to 200 m deep, filled with coarse sediments, and generally connected to onshore streams (Babonneau et al. 2013; Saint-Ange et al. 2011; Sisavath et al. 2011). In the abyssal plain, these canyons evolve into huge volcanoclastic deep-sea fans, clearly visible on the acoustic image on Fig. 2.3b of Lénat (2016, Chap. 2). These sediment systems have been studied by several authors (Ollier et al. 1998; Saint-Ange et al. 2011; Saint-Ange 2009; Sisavath et al. 2011, 2012) and a synthesis is found in this volume (Babonneau et al. 2016, Chap. 6).

3.2 Basement Geometry

La Réunion is built on Upper Cretaceous to Paleocene oceanic crust. Although the onset of La Réunion volcanism is unknown, it is probably younger than 10 Ma and therefore a thick sequence of oceanic sediments should already have blanketed the basaltic crust when the hotspot volcanism started. As described by Lénat (2016, Chap. 2), seamounts were built up after the formation of the oceanic crust and some are probably buried beneath La Réunion.

The seismic studies from Charvis et al. (1999), de Voogd et al. (1999), Gallart et al. (1999) show that there is virtually no, or very weak, lithospheric flexure beneath La Réunion, whereas a large body of underplating material has been demonstrated beneath the southwest part of the island (Fig. 3.1). The absence of a large flexure contrasts with most similar oceanic volcanoes (e.g. Minshull and Charvis 2001; Watts 1994; Watts and ten Brink 1989) and suggests a dynamic equilibrium where positive buoyancy of the plume would contribute to the dynamic support of the load of La Réunion. This had been suggested by Bonneville et al. (1997) and Lénat et al. (2009b). The gravity data support this absence of flexure (Lénat 2016, Chap. 2). On the base of this geometry of the ocean floor, de Voogd et al. (1999) estimate the subaerial part of La Réunion as being only 3 % by volume of the whole volcanic system.

3.3 Infrastructure of the Submarine Flanks

The interior of the submarine flanks of La Réunion have been studied using different geophysical approaches. Gailler and Lénat (2010) have based their study on the interpretation of a residual Bouguer anomaly map of the whole La Réunion edifice and a composite reduced-to-the-pole (RTP) magnetic anomaly map (Fig. 3.2a, b). Charvis et al. (1999), de Voogd et al. (1999), Gallart et al. (1999), Le Friant et al. (2011), Lebas (2012) used seismic data.

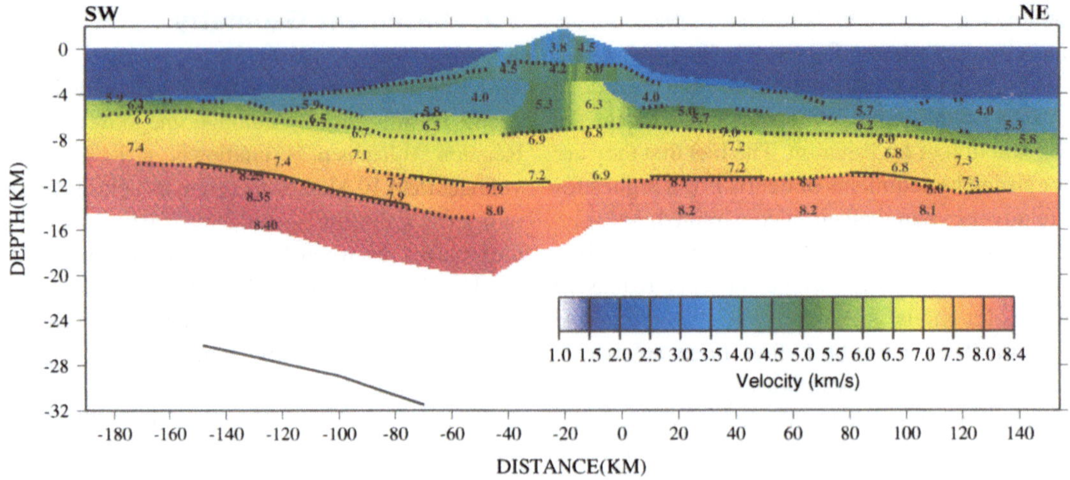

Fig. 3.1 P velocity model along a SW–NE transect through La Réunion (Gallart et al. 1999)

The strong magnetization and low density of the coastal shelf (Fig. 3.2) is interpreted by Gailler and Lénat (2010) as being due to the presence of a belt of hyaloclastites and pillow lavas around the island. In the case of La Montagne Massif (see location on Fig. 3.2a), offshore anomalies are explained by the continuation of a rift zone (age ≥ 2.1 Ma (McDougall 1971)) with additional pillow lavas and hyaloclastites.

In two locations on the submarine flanks, the Etang Salé (ES) and eastern areas, well defined positive and negative, short- to medium wavelength, anomalies extend tens of kilometres offshore (they are easily differentiated from the larger wavelength anomalies created by the oceanic crust, see Fig. 2.2c in Lénat (2016, Chap. 2). In this context, this type of anomaly is created by massive, coherent (i.e. not chaotic or brecciated) volcanic constructions. The ES submarine magnetic anomalies (Fig. 3.2b) are not restricted to the bathymetric ridge observed in this area (Fig. 3.2), but extend farther offshore (about 38 km) and cover a wider area (about 36 km near the seashore). Both positive and negative anomalies are observed, although the latter cover a larger surface. The presence of negative and positive anomalies indicates that the ES construction spans successive paleomagnetic epochs. Gailler and Lénat (2010) propose that this elongated array of magnetic anomalies is caused by long- lasting rift zone activity of the Piton des Neiges volcano, starting at least as far back as the Matuyama paleomagnetic epoch (2.59–0.78 Ma). However, the age of 2.48 Ma obtained by Smietana (2011) on a sample dredged at the top of the submarine ridge suggests that this structure is still older, possibly dating from the Gilbert reverse paleomagnetic epoch ($\sim$5–3.3 Ma). Most of this large structure is buried by subsequent mass wasting and sedimentary products.

Using the same approach, Gailler and Lénat (2010) establish that the magnetic anomalies on the eastern submarine flank are caused by the presence of large, now buried volcanic constructions spanning at least the Brunhes and Matuyama paleomagnetic epochs. However, similarly to the ES zone, the ages of 3.34 and 3.77 Ma (Smietana 2011) for samples dredged a few kilometres offshore of the Piton de la Fournaise NE rift zone, as well as an age of 1.8 Ma for a sample dredged about 20 km from the coast (Lénat et al. 2009a), suggest that the construction of this structure started much earlier, during the Gauss and Gilbert paleomagnetic epochs. These offshore constructions are identified by Gailler and Lénat (2010) as the remnant flank of the now concealed edifice of Les Alizés volcano (see below).

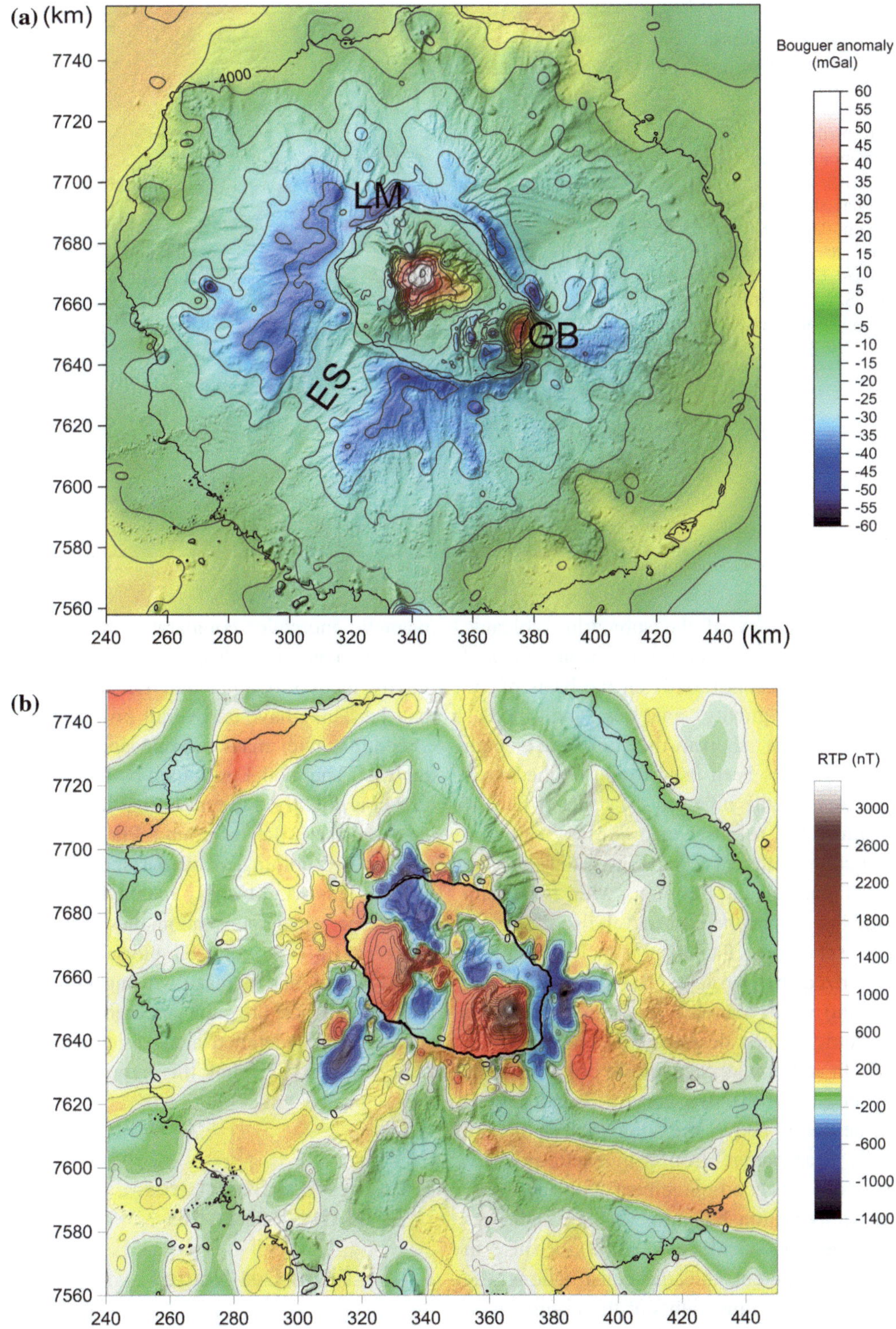

Fig. 3.2 **a** Residual Bouguer anomaly map of the whole La Réunion edifice (contour interval of 10 mGal). The black solid contours refer to the 0 and −4000 m isobaths. Coordinates in kilometers (WGS84, UTM 40S). *ES* Etang Salé ridge; *LM* La Montagne Massif; *GB* Grand Brûlé. **b** Composite reduced to the pole magnetic anomaly map (contour interval of 200 nT): the marine part is computed at sea level and the on-land part is computed at 3500 m. From Gailler and Lénat (2010)

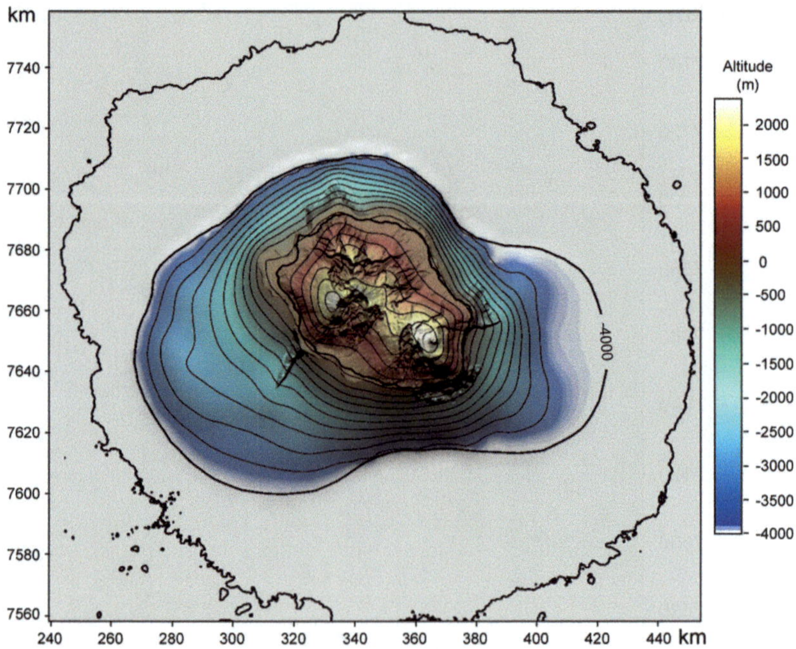

Fig. 3.3 3-D morphology of the constructed volcanic edifice integrating the on-land part and the submarine parts inferred from magnetic interpretation (Gailler and Lénat 2010). The present-day shoreline and the present-day −4000 m contours are shown with black lines. Coordinates in km, WGS84 UTM 40S. From Gailler and Lénat (2010)

The morphology of the submarine volcanic constructions of La Réunion has been recreated by Gailler and Lénat (2010) on the basis of the magnetic and gravity analyses and models (Fig. 3.3). The short-to-medium wavelength anomaly belt has been used to define the maximum extent of the offshore constructed part of the edifice. The reconstructed edifice exhibits a dramatically different shape to that of the emerged part of La Réunion, with two main outgrowths toward the east and the southwest.

The submarine flanks of La Réunion are characterized by negative Bouguer anomalies (Fig. 3.2a) which more or less mimics the four bathymetric bulges described by Oehler et al. (2008). In addition, except for the proximal zone of the coast and the two areas, ES and eastern flank, described above, the submarine flanks are virtually "transparent" from a magnetic point of view. This is in good agreement with their postulated construction by mass wasting and sedimentary deposits (Oehler et al. 2008), because, in contrast to coherent volcanic constructions, the thermoremanent components of the fragments in such deposits are randomly oriented, leaving only a relatively weak induced magnetization component. Therefore, they do not generate magnetic anomalies comparable in amplitude to that of massive volcanic constructions. The gravity models (Gailler and Lénat 2010) show that the flanks have an overall low density, also compatible with mass wasting and sedimentary deposits. The model presented by Le Friant et al. (2011) is slightly different in that it suggests that the submarine flanks around Piton des Neiges volcano are dominated by the spreading of the edifice. This would imply magnetic and gravity signatures (magnetic anomalies created by coherent, though rotated, blocks and higher density than that of mass wasting deposits and sediments) that are not apparently observed, although more detailed analyses would be necessary to reject or refine this hypothesis.

3.4 Internal Structures of the Volcanoes

3.4.1 Piton Des Neiges

Piton des Neiges, in the north-west of the island, is a dormant volcano whose oldest recognized outcropping lava flows are 2.2 Ma old basalts

(McDougall 1971; Quidelleur et al. 2010). Salvany et al. (2012) have made a recent chronological compilation of the subaerial evolution of Piton des Neiges. It started with the edification of a basaltic shield between 2.2 and 0.43 Ma (Oceanic Series). After a period of apparent quiescence, Piton des Neiges then erupted differentiated alkaline lavas (Differentiated Series) between 350 and 29 ka. The evolution of Piton des Neiges volcano over more than 2 Ma, involving volcanic and volcano-tectonic activity and high erosion rates, has resulted in a complex geology exposed in deep valleys and in three main depressions (Fig. 3.4): the Mafate, Salazie and Cilaos Cirques. To the east of Salazie, a fourth cirque, the Marsouins Cirque, has been partially filled by lava and pyroclastite flows (Rocher and Westercamp 1989) giving rise to the present Bébour-Bélouve Plain (Fig. 3.4).

The internal structure of Piton des Neiges volcano has been studied with gravity, magnetic and resistivity methods as well as by a 2108 m-deep drill hole (Demange et al. 1989; Gailler and Lénat 2012; Malengreau et al. 1999). The main feature of the interior of the volcano is the presence of a huge intrusive complex which creates a large positive gravity anomaly (Fig. 3.2a). This complex emerges at various places at the bottom of the Salazie and Cilaos Cirques as gabbro outcrops (some syenites and micro-syenites are also observed). The models of (Gailler and Lénat 2012) suggest that the dense complex may extend down to the base of the edifice. It is overlaid by a complex volcanic construction comprising lava flows, pyroclastites, landslide deposits and a very high density of sills and dykes in the central part (e.g. Chevallier and Vatin-Pérignon 1982; Maillot 1999; Rançon 1982). The gravity models show that the lateral extent of the dense complex more or less mimics those of the present Cirques and the paleo-Cirque, suggesting that its loading effect may have guided

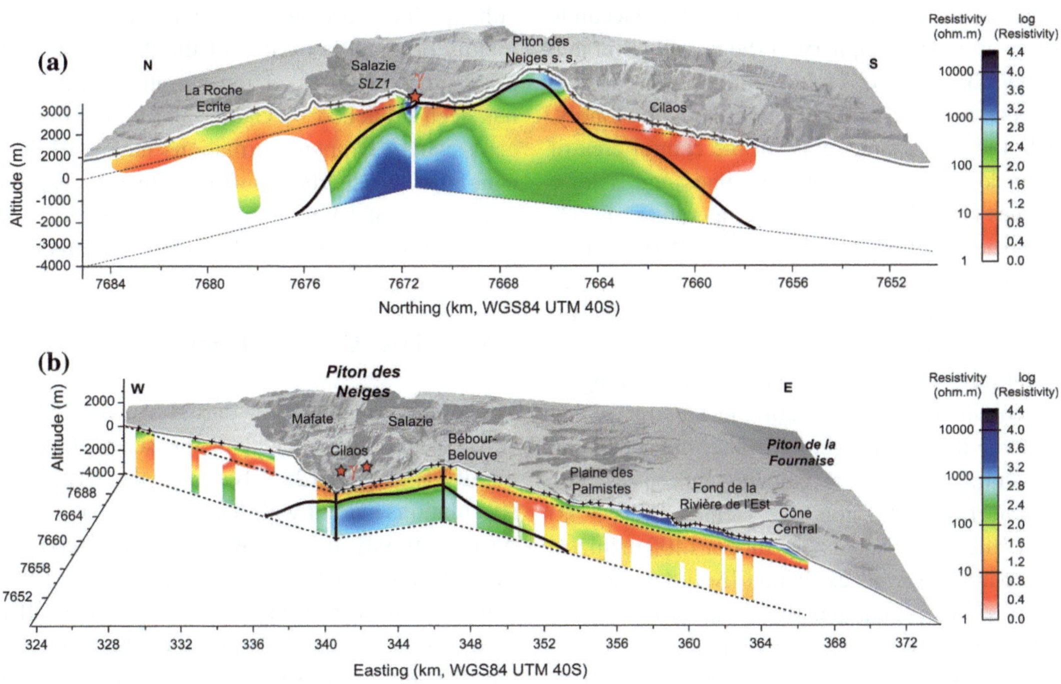

Fig. 3.4 **a** 2D magnetotelluric model crossing Piton des Neiges superimposed onto a 3D view of the topography. The *red star* locates the outcrop of gabbro (γ) in Salazie and Cilaos Cirques. The location of the deep exploration well is shown (SLZ1). The *thick black* contour represents the *top* of the Piton des Neiges hypovolcanic dense complex, as obtained through gravity modelling. The *black crosses* mark the soundings locations. Areas unconstrained by the soundings are whitened; **b** same as **a** along a northwest–southeast profile extending to Piton de la Fournaise. From Gailler and Lénat (2012)

erosion (however, the absence of centripetal dipping of the various formations around the complex precludes significant subsidence of the dense body). The general shape of the dense complex model is that of a flattened dome culminating in the center of Piton des Neiges. Beneath the central area it rises to elevations significantly higher than that of the bottom of the Cirques, and therefore well above the observed outcrops of gabbro. It is thus probable that the gravity models incorporate the effect of the shallow complex of dikes and sills of this area. The dense complex is also characterized by high resistivity values (Fig. 3.4) while being surrounded by more conductive terrains. There are several explanations for the generally weak resistivity of the overlying and surrounding terrains. Firstly, the shallow layers in the Cirques, mostly breccias of various origins overlying lava flows and hyaloclastites, are extensively hydro thermally altered. Besides a zeolites sequence centred on Piton des Neiges (Rançon 1985), a more pervasive hydrothermal alteration (zeolites, clays, chlorite, etc.) extends into the Oceanite Series up to the periphery of the massif (Bret and Join 2003; Folio 2001). In addition, they can also become seawater-saturated at depth. Secondly, hydrothermal springs exist in the central area of Piton des Neiges and in the Cilaos and Salazie Cirques (Join et al. 1997). A temperature of 192° has been measured at the bottom of a 2108 m-deep drill hole in Salazie (Demange et al. 1989). All these observations argue for the presence of a still active hydrothermal system driven by the cooling of the intrusive complex. Note the deepening of the conductive cap beneath the summit which might be due to stronger hydrothermal activity in the central zone of the volcano. In contrast, the summit ridge of Piton des Neiges (Fig. 3.4a) shows high resistivity layers (about 1400 m thick). This pattern is well explained by the geological nature of this area, mostly composed of water unsaturated, young (Salvany et al. 2012) lava flows, scoriaceous formations and a dense network of dykes and sills.

The magnetic anomaly map of Piton des Neiges (Fig. 3.2b) broadly reflects the age distribution of surface rocks (Salvany et al. 2012). Negative magnetic anomalies are observed where the largest outcrops of rocks older than the Brunhes-Matuyama reversal (0.780 Ma) are mapped. However, in detail, the magnetic map appears to provide a smoother or more integrated image of the pre- and post-Brunhes-Matuyama reversal formations. This pattern can be easily understood if we consider that the bulk of the massif is constructed of rocks older than 0.780 Ma blanketed by more recent formations. The thickness of the Brunhes period rocks must be significant in order to create a positive anomaly which overides the generally negative signal created by the bulk of the massif at the altitude of the aeromagnetic survey (3500 m). The major Brunhes period constructions (in terms of thickness; Gailler and Lénat 2012) are in the central and west-southwest areas, and to a lesser extent to the north. The magnetic signature of the poorly magnetized breccias, hydrothermally altered and erosion products, recognized in the cirque bottoms and in some peripheral areas (Bret et al. 2003; Demange et al. 1989), is difficult to find, because the magnetic signal is mainly dominated by that of the thick lava flow piles. The thickness of the Brunhes formations has been modelled by Gailler and Lénat. For the areas where Brunhes formations have a significant thickness, the model suggests average thicknesses of 1 km on the western flank, 0.5 km on the northern flank and a few hundreds of meters on the eastern flanks.

3.4.2 Les Alizés Volcano

Les Alizés volcano is a buried volcano whose existence is still debated in the scientific community (Lebas 2012; Salvany et al. 2012). The speculations started with the discovery of a large positive gravity anomaly near the eastern coast of Piton de la Fournaise volcano (Gérard et al. 1980; Rousset et al. 1989) (Fig. 3.2b), in the area called Grand Brûlé. A drill hole provided the proof that the structure responsible for the gravity anomaly was actually a hypovolcanic intrusive complex, mostly composed of gabbros and cumulates (Rançon et al. 1989) encountered between about 1000 and 3000 m b.s.l. (bottom of the drill hole). The geothermal gradient and the relatively low temperature at the bottom of the drill hole ($\sim 140°$) indicated that the complex was old and inactive (Rançon et al. 1989). The 5–8 km-wide complex is elongated in a north-south direction (15–18 km) in plan view, with its NS axis located 1.5 km inland. Subsequent gravity measurements and modelling showed that the dense structure was independent from the structures of Piton de la Fournaise volcano (Gailler et al. 2009). On the basis of such arguments, it became indisputable to some authors (Lénat et al. 2001; Malengreau et al. 1999; Rançon et al. 1989) that the dense complex could only have grown in the core of a pre-existing volcano that they named Les Alizés volcano (or Proto-Fournaise for Rançon et al.). Their main argument was that such intrusive complexes usually develop during the edification of shield volcanoes, as is the case, for example, for Piton des Neiges volcano (see above) and the Hawaiian volcanoes (Kauahikaua et al. 2000; Ryan 1987).

The presence of negative magnetic anomalies (Fig. 3.2b) in the eastern part of Piton de La Fournaise indicate that rocks older than the Brunhes-Matuyama magnetic reversal (0.78 Ma) are present at fairly shallow depths beneath the younger formations of Piton de la Fournaise and its submarine flank (Gailler and Lénat 2012). On land, the oldest dated rocks of the Piton de la Fournaise volcano (up to 535 ka; Gillot and Nativel 1989; Kluska 1997) are found at the bottom of deep valleys (Rivière des Remparts, Rivière Langevin, Rivière de L'Est) and in La Plaine des Palmistes. Up to 400–450 ka the lavas are plagioclase-rich basalts (locally called "Roches Pintades Series"; Albarède et al. 1997), and they are significantly different from the subsequently-emitted basaltic lavas of Piton de la Fournaise. Similar plagioclase-rich basalts are found in Piton des Neiges, but they are about 200 ka younger than the Piton de la Fournaise ones. Smietana (2011) suggests that, as in the case of Piton des Neiges (Kluska 1997; Upton and Wadsworth 1972), these plagioclase-rich basalts represent the transition from a basaltic shield building stage to a more differentiated stage. Accordingly, the Roches Pintades Series of Piton de la Fournaise would more likely mark the evolution of a pre-existing basaltic shield than represent the early stage of Piton de la Fournaise volcano. Considering this body of information, several authors have tried to construct a model of evolution that includes the construction of Les Alizés volcano (Lénat et al. 2001; Oehler et al. 2008; Rançon et al. 1989; Smietana 2011). Considering the age of about 3.5 Ma for pillows lava near 2000 m b.s.l, we can speculate that Les Alizés volcano could have emerged at about the same period as Piton des Neiges, around 2 Ma. If the "Roches Pintades Series" is associated with Les Alizés, the latter would have reached a stage of transition from a basaltic shield-building stage to a more differentiated stage between at least 535 and 400–450 ka. The evolution of Les Alizés volcano is still more speculative after this period. With the top of its dense intrusive complex at about 1000 m b.s.l., the summit of the edifice could have reached at least 1500–2500 m in elevation if we compare to similar sites such as Piton des Neiges or Hawaii (Ryan 1987). With no remnants of such an edifice above about 1000–1200 m in elevation ("Roches Pintades"), it must be assumed that there were catastrophic destruction processes involved. Merle et al. (2010) and Oehler et al. (2008) have inferred that large flank landslides were responsible for the massive destruction of the superstructures of Les Alizés volcano. Note that Lebas (2012) questions this interpretation, and instead proposes that a volcano (that she names Proto-Fournaise), much smaller than Les Alizés, grew on the flank of a "Proto-Piton des Neiges" which extended up to this area.

Les Alizés volcano is one of the main issues in the understanding of the evolution of La Réunion. The model presented above is coherent with the currently available information, but large uncertainties remain on a number of points.

3.4.3 Piton de La Fournaise

Piton de la Fournaise, on the south-eastern part of the island, is a highly active basaltic shield volcano, with most of its activity restricted to effusions from its Central Cone and along rift zones (e.g. Michon et al. 2016, Chap. 7). Deep valleys have been incised into the areas not covered by young volcanic deposits, providing exposures that enable the reconstruction of certain parts of the volcano's evolution. Piton de la Fournaise has been extensively studied, mostly since the 1980s, using geological, petrological, geochemical and geophysical monitoring approaches. Most of these results are reviewed or synthesized in the next chapters of this volume. In this chapter we focus on the internal structure, and the model shown here (Fig. 3.5) is directly inspired by the compilation made by Lénat et al. (2012a). One of the main difficulties in trying to construct a model which is consistent with all the known data is that the various works provide heterogeneous information, in terms of scale, detail, and nature.

- The lithosphere beneath Piton de la Fournaise is not significantly flexed, and the crust is underlain by an underplating body (Fig. 3.1), which might represent the deep magma reservoir for La Réunion volcanism.
- Piton de la Fournaise is a relatively thin volcano lying on a huge volcanic construction attributed mostly to Les Alizés volcano.
- The endogenous constructions (intrusive complexes) related to Les Alizés and Piton de la Fournaise volcanoes represent a large volume (Gailler et al. 2009).
- The huge intrusive complex of Les Alizés volcano probably rests on top of the oceanic crust and appears to have a buttressing effect for the present eastern volcano-tectonic activity of Piton de la Fournaise.
- The early Piton de la Fournaise edifice was built around a focus located beneath the Plaine des Sables area (Bachèlery and Lénat 1993; Bachèlery and Mairine 1990). The

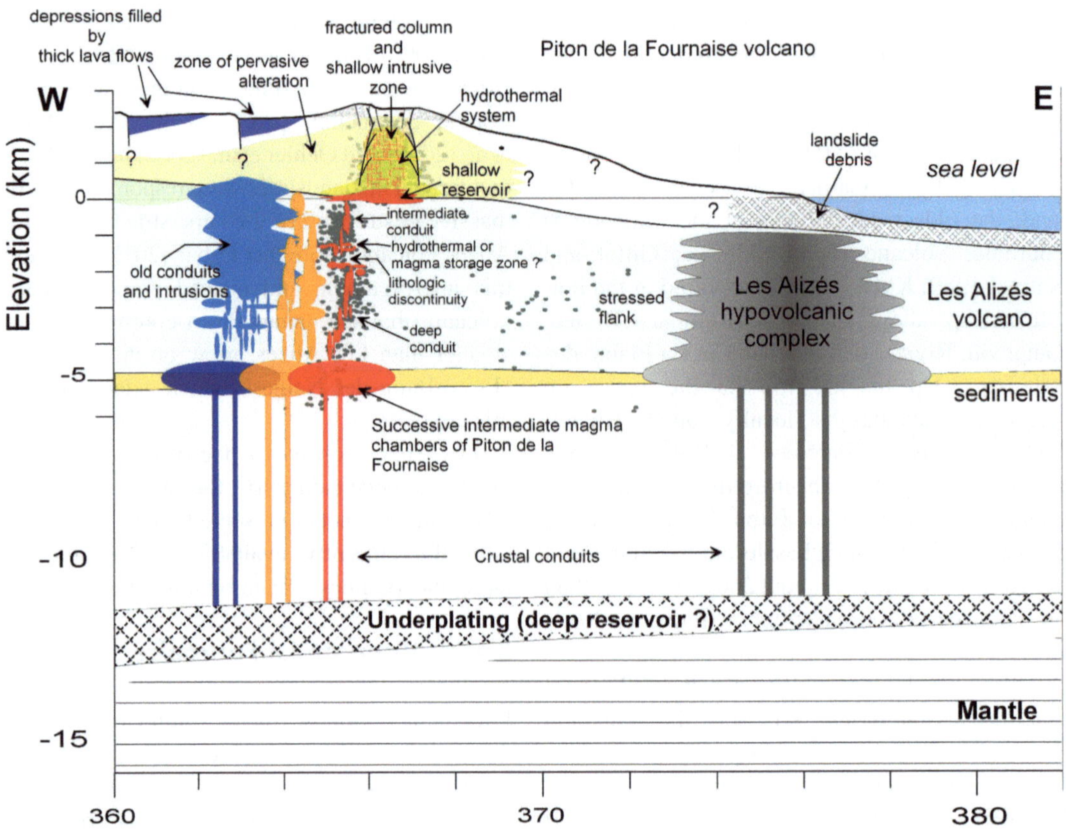

Fig. 3.5 Interpreted W–E geological section of Piton de la Fournaise from Lénat et al. (2012a). Horizontal coordinates: UTM km WGS84

centre subsequently moved 5–6 km eastward to its current location. The dense, high-velocity body beneath the Plaine des Sables and the western part of the Enclos probably corresponds to the hypovolcanic intrusive complex that developed before the volcanic centre shifted to its present-day position.

- Magma reservoirs may have existed, and may still exist, as illustrated by the March 1998 crisis (Battaglia et al. 2005), at the mechanical and density interface between the oceanic crust and the Les Alizés edifice.
- Strong evidence also exists for the presence of a shallower magma reservoir located near sea level beneath the summit (Lénat and Bachèlery 1990; Peltier et al. 2009). The March 1998 pre-eruptive seismic pattern seems to be evidence for magma transfer between the two reservoirs.
- The dominant structural feature of the central zone is a collapse structure beneath the summit craters, above the inferred magma reservoir near sea level. The collapsed column constitutes a major mechanical heterogeneity and concentrates most of the seismic, intrusive, and hydrothermal activity because of its higher permeability and weaker mechanical strength (Lénat et al. 2012b).

3.5 Synthesis of the Construction of La Réunion

The internal structure of the edifice is dominated by two large dense intrusive complexes located beneath the centres of Piton des Neiges volcano and an inferred buried volcano named Les Alizés. Such large intrusive complexes are commonly found beneath large oceanic basaltic shields. According to Ryan (1987), Walker (1993) these complexes grow upwards as a volcano builds up. The fact that both of the La Réunion complexes are deeply rooted suggests that Piton des Neiges and Les Alizés are primary volcanoes that started to form in the early stages of La Réunion volcanism. Piton des Neiges is obviously the larger volcanic centre of the two, if we consider the geometry of its intrusive complex as well as the visible part of the edifice. In addition, the offshore magnetic data show that Piton des Neiges constructions extend 25–30 km offshore from the coast to the southwest (about 50 km from the volcano centre). The case of Les Alizés volcano is different, because proof of its existence rests only on the presence of a dense intrusive complex and on the possible association of some rocks on land and offshore to this volcano. If this association is correct, Piton des Neiges and Les Alizés volcanoes would have had a large period of synchronous activity. The latter edifice must have been largely destroyed in order to explain why more remnants are not visible. It has often been suggested that large volcano landslides have an effect on the subsequent volcanic activity and magmatic evolution, because the rapid unloading of the lithosphere induces pressure changes in the plumbing system, the magma reservoirs and melt sources (Francalanci et al. 1989; Hildenbrand et al. 2004; Longpré et al. 2009; Smith and Wessel 2000). In this context, Smietana (2011) proposes that a large collapse of Les Alizés volcano may have occurred around 450 ka, causing a period of quiescence of Piton des Neiges and the birth of Piton de la Fournaise.

Whereas the subaerial part of La Réunion is almost entirely built up of volcanic constructions,

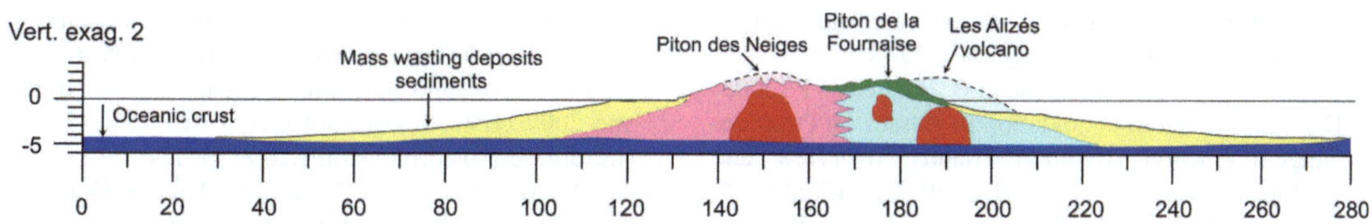

Fig. 3.6 Generalized section of La Réunion showing the volcanic constructions, the hypovolcanic dense complexes and the accumulation of mass wasting deposits and sediments on the submarine flanks. Note the virtual absence of flexure on the underlying lithosphere

most of the submarine flanks appear to be composed of an accumulation of mass wasting products and sediments (with the exception of the buried constructions to the southwest and to the east). A summary view of the internal structure of La Réunion is shown on Fig. 3.6.

Acknowledgements This synthesis benefited from discussions with many colleagues. Among them, Lydie Gailler has played a major role in the interpretation of the geophysical data. An anonymous review helped to clarify some important issues in the text.

References

Albarède F, Luais B, Fitton G, Semet M, Kaminski E, Upton BGJ, Bachelèry P, Cheminée JL (1997) The geochemical regimes of Piton de la Fournaise volcano (Réunion) during the last 530,000 years. J Petrol 38 (2):171–201

Babonneau N, Delacourt C, Cancouët R, Sisavath E, Bachèlery P, Mazuel A, Jorry SJ, Deschamps A, Ammanna J, Villeneuve N (2013) Direct sediment transfer from land to deep-sea: Insights into shallow multibeam bathymetry at La Réunion Island. Mar Geol 346:47–57. doi:10.1016/j.margeo.2013.08.006

Babonneau N, Villeneuve N, Mazuel A, Bachèlery P (2016) Erosion and volcaniclastic sedimentation from source to sink In: Bachèlery P, Lénat JF, Di Muro A, Michon L (eds) Active volcanoes of the Southwest Indian Ocean: Piton de la Fournaise and Karthala. Active Volcanoes of the World. Springer, Berlin

Bachèlery P, Lénat JF (1993) Le Piton de la Fournaise. Mémoires de la Soc Géol France, Nouvelle Série, 163:221–229

Bachèlery P, Mairine P (1990) Evolution volcano–structurale du Piton de la Fournaise depuis 0.53 Ma. In: Lénat JF (ed) Le volcanisme de la Réunion, Monographie. Cent. Rech. Volcanol., Clermont-Ferrand, France, pp 213–242

Battaglia J, Ferrazzini V, Staudacher T, Aki K, Cheminée JL (2005) Pre-eruptive migration of earthquake at Piton de la Fournaise volcano (Réunion Island). Geophys J Int 161:549–558

Bonneville A, Von Herzen RP, Lucazeau F (1997) Heat flow over Reunion hot spot track: Additional evidence for thermal rejuvenation of oceanic lithosphere. J Geophys Res 102(B10): 22731–22748

Bret L, Fèvre Y, Join JL, Robineau B, Bachèlery P (2003) Deposits related to degradation processes on Piton des Neiges Volcano (Reunion Island): overview and geological hazard. J Volcanol Geoth Res 123:25–41

Bret L, Join J-L (2003) Argiles et zéolites dans l'altération d'un volcan bouclier en milieu tropical (Le Piton des Neiges, La Réunion). Clays and zeolites in the weathering of a basaltic tropical shield volcano (Le Piton des Neiges, Reunion Island). Comptes Rendus Géosci 335(14):1031–1038

Charvis P, Laesanpura A, Gallart J, Hirn A, Lepine J, de Voogd B, Minshull TA, Hello Y, Pontoise B (1999) Spatial distribution of hotspot material added to the lithosphere under La Reunion, from wide-angle seismic data. J Geophys Res B 104(2):2875–2893

Chevallier L, Vatin-Pérignon N (1982) Volcano-structural evolution of Piton des Neiges, Reunion Island, Indian Ocean. Bull Volcanol 45:285–298

Cochonat P, Lénat JF, Bachèlery P, Boivin P, Cornaglia B, Deniel C, Labazuy P, Le Drezen E, Lipman PW, Ollier G, Savoye B, Vincent PM, Voisset M (1990) Importance des dépôts gravitaires dans la mise d'un système volcano-sédimentaire sous-marin (volcan de la Fournaise, Ile de la Réunion). C R Acad Sci Paris, 311 Série II: 679–686

de Voogd B, Palome S, Hirn A, Charvis P, Gallart J, Rousset D, Danobeitia J, Perroud H (1999) Vertical movements and material transport during hotspot activity: seismic reflection profiling offshore La Reunion. J Geophys Res B 104(2):2855–2874

Demange J, Chovelon P, Puvilland P (1989) Geothermal model of the Salazie Cirque (Reunion Island): volcanic and structural implications. J Volcanol Geoth Res 36:153–176

Deplus C, de Voogd B, Dyment J, Bissessur PD, Sisavath E, Depuiset F, Mercier M (2009) New insights on the oceanic lithosphere at La Réunion hotspot volcano. In: EGU General Assembly 2009. Geophysical research abstracts, Vienna, pp EGU2009–5728

Folio JL (2001) Distribution de la perméabilité dans le massif du Piton de la Fournaise: Apport à la connaissance du fonctionnement hydrogéologique d'un volcan bouclier., Univ. de La Réunion, St Denis, France, p 148

Francalanci L, Manetti P, Peccerillo A (1989) Volcanological and magmatological evolution of Stromboli volcano (Aeolian Islands): The roles of fractional crystallization, magma mixing, crustal contamination and source heterogeneity. Bull Volcanol 51(5):355–378

Gailler L, Lénat JF (2010) Three-dimensional structure of the submarine flanks of La Réunion inferred from geophysical data. J Geophys Res 115 B12105. doi:10.1029/2009JB007193

Gailler L, Lénat JF (2012) Architecture of La Réunion inferred from geophysical data. Bull Volcanol 221–222:83–98. doi: 10.1016/j.jvolgeores.2012.01.015

Gailler L, Lénat JF, Lambert M, Levieux G, Villeneuve N, Froger JL (2009) Gravity structure of Piton de la Fournaise volcano and inferred mass transfer during the 2007 crisis. J Volcanol Geoth Res 184(1–2):31–48. doi:10.1016/j.jvolgeores.2009.01.024

Gallart J, Driad L, Charvis P, Sapin M, Hirn A, Diaz J, de Voogd B, Sachpazi M (1999) Perturbation to the lithosphere along the hotspot track of La Reunion from an offshore-onshore seismic transect. J Geophys Res B 104(2):2895–2908

Gérard A, Lesquer A, Lachaud JC, Louis P, Mennechet C (1980) Etude gravimétrique de la moitié sud-est de l'île de la Réunion. C R Acad Sci Paris 290:139–142
Gillot PY, Nativel P (1989) Eruptive history of the Piton de la Fournaise volcano, Réunion Island, Indian Ocean. J Volcanol Geoth Res 36:53–65
Hildenbrand A, Gillot PY, Le Roy I (2004) Volcano-tectonic and geochemical evolution of an oceanic intra-plate volcano: Tahiti-Nui (French Polynesia). Earth Planet Sci Lett 217:349–365
Join JL, Coudray J, Longworth K (1997) Using principal components analysis and Na/Cl ratios to trace groundwater circulation in a volcanic island: the example of Réunion. J Hydrol 190:1–18
Kauahikaua J, Hildenbrand T, Webring M (2000) Deep magmatic structures of Hawaiianvolcanoes, imaged by three-dimensional gravity models. Geology 28 (10):883–886
Kluska JM (1997) Evolution magmatique et morpho-structurale du Piton des Neiges (Ile de la Réunion) au cours des derniers 500000 ans. Ph.D. Thesis, Université Paris XI, Orsay, 125 pp
Labazuy P (1996) Recurrent landslides events on the submarine flank of Piton de la Fournaise Volcano (Reunion Island). In: McGuire WJ, Jones AP, Neuberg J (eds) Volcano instability on the Earth and other planets. Geological Society of London, London, pp 295–306
Le Friant A, Lebas E, Clément G, Boudon G, Deplus C, de Voogd B, Bachèlery P (2011) A new model for the evolution of La Réunion volcanic complex from complete marine geophysical surveys. Geophys Res Lett 38:L09312. doi:10.1029/2011GL047489
Lebas E (2012) Processus de démantèlement des édifices volcaniques au cours de leur évolution: application à La Réunion et Montserrat et comparaison avec d'autres édifices. Ph.D. Thesis, Paris Diderot, Paris, 407 pp
Lénat JF (2016) Geodynamic setting of La Réunion. In: Bachèlery P, Lénat JF, Di Muro A, Michon L (eds), Active volcanoes of the Southwest Indian Ocean: Piton de la Fournaise and Karthala. Active Volcanoes of the World. Springer, Berlin
Lénat JF, Bachèlery P (1990) Structure et fonctionnement de la zone centrale du Piton de la Fournaise. In: Lénat JF (ed) Le volcanisme de la Réunion, monographie. Centre de Recherches Volcanologiques, Clermont-Ferrand, pp 257–296
Lénat JF, Bachèlery P, Merle O (2012a) Anatomy of Piton de la Fournaise volcano (La Réunion, Indian Ocean). Bull Volcanol 74:1945–1961. doi:10.1007/s00445-012-0640-y
Lénat JF, Bachèlery P, Peltier A (2012b) The interplay between collapse structures, hydrothermal systems and magma intrusions: the case of the central area of Piton de la Fournaise volcano. Bull Volcanol 74 (2):407–421. doi:10.1007/s00445-011-0535-3
Lénat JF, Boivin P, Deniel C, Gillot PY, Bachèlery P, Fournaise-2 T (2009a) Age and nature of lavas on the submarine flanks of Piton de la Fournaise (Reunion island). J Volcano Geoth. Res 184:199–207. doi.org/10.1016/j.jvolgeores.2009.01.013
Lénat JF, Merle, Lespagnol L (2009b) La Réunion: an example of channeled hot spot plume. J Volcano Geoth Res 184:1–13. doi:10.1016/j.jvolgeores.2008.12.001
Lénat JF, Gibert-Malengreau B, Galdeano A (2001) A new model for the evolution of the volcanic island of Reunion (Indian Ocean). J Geophys Res B 106 (5):8645–8663
Longpré MA, Troll VR, Walter TR, Hansteen TH (2009) Volcanic and geochemical evolution of the Teno massif, Tenerife (Canary Islands): ssome repercussions of giant landslides on ocean island magmatism. Geochem Geophys Geosyst 10:Q12017. doi:10.1029/2009GC002892
Maillot E (1999) Les systèmes intrusifs des volcans boucliers océaniques: Ile de la Réunion (Océan Indien). Approche structurale et expérimentale. Ph.D. Thesis, Université de La Réunion, Saint Denis, 289 pp
Malengreau B, Lénat JF, Froger JL (1999) Structure of Reunion Island (Indian Ocean) inferred from the interpretation of gravity anomalies. J Volcanol Geoth Res 88(3):131–146
McDougall I (1971) The geochronology and evolution of the young volcanic island of Réunion (Indian Ocean). Geochim Cosmochim Acta 35:261–288
Merle O, Mairine P, Michon L, Bachèlery P, Smietana M (2010) Calderas, landslides and paleo-canyons on Piton de la Fournaise volcano (La Réunion Island, Indian Ocean). J Volcanol Geoth Res 189:131–142. doi:10.1016/j.jvolgeores.2009.11.001
Michon L, Ferrazzini V, Di Muro A (2016) Magma paths at Piton de la Fournaise volcano. In: Bachèlery P, Lénat J-F, Di Muro A, Michon L (eds) Active Volcanoes of the Southwest Indian Ocean: Piton de la Fournaise and Karthala. Active Volcanoes of the World. Springer, Berlin
Minshull TA, Charvis P (2001) Ocean Island densities and estimates of the effective elastic thickness of the oceanic lithosphere. Geophys J Int 145:731–739
Oehler JF, Lénat JF, Labazuy P (2008) Growth and collapse of the Reunion Island volcanoes. Bull Volcanol 70:717–742. doi: 10.1007/s00445-007-0163-0
Ollier G, Cochonat P, Lenat JF, Labazuy P (1998) Deep-sea volcaniclastic sedimentary systems; an example from La Fournaise volcano, Reunion Island. Indian Ocean Sedimentol 45(2):293–330
Peltier A, Bachèlery P, Staudacher T (2009) Magma transport and storage at Piton de La Fournaise (La Réunion) between 1972 and 2007: a review of geophysical and geochemical data. J Volcanol Geoth Res 184(1–2):93–108. doi.org/10.1016/j.jvolgeores.2008.12.008
Quidelleur X, Holt JW, Salvany T, Bouquerel H (2010) The double structure of the Réunion geomagnetic event based on new K-Ar ages from the type locality, massif de la Montagne, Réunion Island (Indian Ocean) and assessment of the global context. Geophys J Int 182:699–710

Rançon JP (1985) Hydrothermal history of Piton des Neiges volcano (Réunion Island, Indian Ocean). J Volcanol Geoth Res 26:297–315
Rançon JP, Lerebour P, Augé T (1989) The grand brule exploration drilling; new data on the deep framework of the Piton de la Fournaise Volcano; Part 1, Lithostratigraphic units and volcanostructural implications. J Volcanol Geoth Res 36(1–3):113–127
Rançon JP (1982) Contribution à l'étude des minéralisations hydrothermales liées à un système géothermique récent dans l'île de la Réunion. Ph.D. Thesis, University of Paris-Sud XI, Orsay, 225 pp
Rocher P, Westercamp D (1989) The Salazie Cirque ignimbrite (Piton des Neiges volcano, Réunion Island): chronostratigraphy, description and significance of lithic fragments and eruptive mechanisms. J Volcanol Geoth Res 36(1–3):177–191
Rousset D, Lesquer A, Bonneville A, Lénat J-F (1989) Complete gravity study of Piton de la Fournaise Volcano Reunion Island. J Volcanol Geoth Res 36(1–3):37–52
Ryan MP (1987) Elasticity and contractancy of Hawaiian olivine tholeiite and its role in the stability and structural evolution of sibcaldera magma reservoirs and rifts systems. U.S. Geol Surv Prof Pap 1350:1395–1447
Saint-Ange F, Savoye B, Michon L, Bachelery P, Deplus C, De Voogd B, Dyment J, Le Drezen E, Voisset M, Le Friant A, Boudon G (2011) A volcaniclastic deep-sea fan off La Reunion Island (Indian Ocean): Gradualism versus catastrophism. Geology 39(3):271–274. doi:10.1130/g31478.1
Saint-Ange F (2009) La sédimentation volcanoclastique en contexte de point chaud (île de la Réunion, Océan Indien). Thèse d'Université Thesis, Université de La Réunion
Salvany T, Lahitte P, Nativel P, Gillot P-Y (2012) Geomorphic evolution of the Piton des Neiges volcano (Réunion Island, Indian Ocean): competition between volcanic construction and erosion since 1.4 Ma. Geomorphology 136:132–147. doi:10.1016/j.geomorph.2011.06.009
Savoye B, Saint-Ange F, Sisavath E, Deplus C, Bachèlery P (2008) How an intraplate volcanic Island may generate true volcanoclastic deep-sea fans—La Réunion Island. Indian Ocean, AAPG, Cape town
Sisavath E, Babonneau N, Saint-Ange F, Bachèlery P, Jorry S, Deplus C, De Voogd B, Savoye B (2011) Morphology and sedimentary architecture of a modern volcaniclastic turbidite system: the Cilaos fan, offshore La Réunion Island. Mar Geol. doi:10.1016/j.margeo.2011.06.011
Sisavath E, Mazuel A, Jorry SJ, Babonneau N, Bachèlery P, De Voogd B, Salpin M, Emmanuel L, Beaufort L, Toucanne S (2012) Processes controlling a volcaniclastic turbiditic system during the last climatic cycle: example of the Cilaos deep-sea fan, offshore La Réunion Island. Sed Geol 281:180–193. doi:10.1016/j.sedgeo.2012.09.010
Smietana M (2011) Pétrologie, géochronologie (K-Ar) et géochimie élémentaire et isotopique (Sr, Nd, Hf, Pb) des laves anciennes de La Réunion . Implications sur la construction de l'édifice volcanique. Université Thesis, Université de La Réunion, 362 pp
Smietana M, Bachèlery P, Hémond C (2010) Heterogeneity in the mantle source of La Réunion Island. In: Geochimica et Cosmochimica Acta, Goldschmidt conference 2010 abstract, vol 74(12 (s1)), pp A972
Smith JR, Wessel P (2000) Isostatic consequences of giant landslides on the Hawaiian Ridge. Pure Appl Geophys 157:1097–1114
Upton BGJ, Wadsworth WJ (1972) Aspect of magmatic evolution of Réunion Island. Philos Trans R Soc London Ser A Math Phys Sci 271:105–130
Walker GPL (1993) Basaltic-volcano systems. Geol Soc London Spec Publ 76:3–38. doi:10.1144/GSL.SP.1993.076.01.01
Watts AB (1994) Crustal structure, gravity anomalies and flexure of the lithosphere in the vicinity of the Canary Islands. Geophys J Int 119:648–666
Watts AB, Ten brink US (1989) Crustal structure, flexure and subsidence history of the Hawaiian Islands. J Geophys Res 94:10473–10500
Wynn RB, Masson DG, Stow DAV, Weaver PPE (2000) Turbidity current sediment waves on the submarine slopes of the western Canary Islands. Mar Geol 163:185–198. doi:10.1016/S0025-3227(99)

Geology and Morphostructural Evolution of Piton de la Fournaise

4

Laurent Michon, Jean-François Lénat, Patrick Bachèlery and Andrea Di Muro

Abstract

The morphology of Piton de la Fournaise volcano results from the succession of construction, destruction and deformation processes that occurred since at least 530 ka. The chaotic surface of the gently dipping submarine flanks indicates that volcaniclastic deposits related to massive flank landslides and erosion cover most of the submarine flanks. Only a few seamounts like Cône Elianne and the submarine continuation of the rift zones are built by lava flows. In the subaerial domain, Piton de la Fournaise exhibits deeply incised canyons evidencing intense erosion and eastward verging scarps whose origin is still controversial. The different interpretations invoking flank landslides and/or summit collapse calderas are summarized. Geological data indicate a twofold construction of Piton de la Fournaise. Between 530 and 60 kyrs, the volcanic centre located in the current Plaine des Sables led to the building of the western part of the massif. The volcanic centre migrated eastwards to its current location, possibly at 60–40 kyrs. Then Piton de la Fournaise experienced caldera collapses and recurrent phreatomagmatic eruptions especially between

L. Michon (✉)
Laboratoire Géosciences Réunion, Institut de Physique du Globe de Paris, Université de La Réunion, Sorbonne Paris Cité, CNRS, F-97744 Saint Denis, France
e-mail: laurent.michon@univ-reunion.fr

J.-F. Lénat · P. Bachèlery
Laboratoire Magmas et Volcans, UMR CNRS-IRD 6524, Observatoire de Physique du Globe de Clermont-Ferrand, Université Blaise Pascal, F-63038 Clermont-Ferrand, France

A. Di Muro
Institut de Physique du Globe de Paris, Observatoire Volcanologique du Piton de la Fournaise (OVPF), Sorbonne Paris Cité, UMR 7154 CNRS, Université Paris Diderot, 75005 Paris, France

P. Bachèlery et al. (eds.), *Active Volcanoes of the Southwest Indian Ocean*, Active Volcanoes of the World, DOI 10.1007/978-3-642-31395-0_4

4880 and 2340 yr BP as evidenced by the Bellecombe ash deposit. Most of the recent volcanic activity is now currently focused restricted inside the Enclos Fouqué caldera where lava flow accumulation and rare explosive events built the 400-m-high Central Cone.

4.1 Introduction

The morphology of volcanoes results from construction, destruction and deformation processes that interact during their evolution (e.g., Moore 1964; Moore and Mark 1992; Merle and Borgia 1996; Rowland and Garbeil 2000). Thus the analysis of the morphology allows the identification of structures whose development is related to internal processes and/or to a specific eruption history (e.g., Tort and Finizola 2005). At Piton de la Fournaise, the succession of construction and dismantling phases (i.e., erosion and landslide) led to the development of a complex morphology (e.g., Bachèlery 1981; Rowland and Garbeil 2000). One of the most striking features is the E-W elongated horseshoe-shaped structure in which the currently active Central Cone developed (Fig. 4.1). This structure is composed, from west to east, by the Enclos Fouqué depression, the Grandes Pentes (upper volcano flank), and the Grand Brûlé (lower volcano flank) and is bounded by 100- to 200-m-high subvertical escarpments, the Bois Blanc, Bellecombe and Tremblet cliffs in the north, west and south, respectively. The formation of the U-shaped structure is one the greatest scientific controversies on Piton de la Fournaise. Thus, the present study aims at

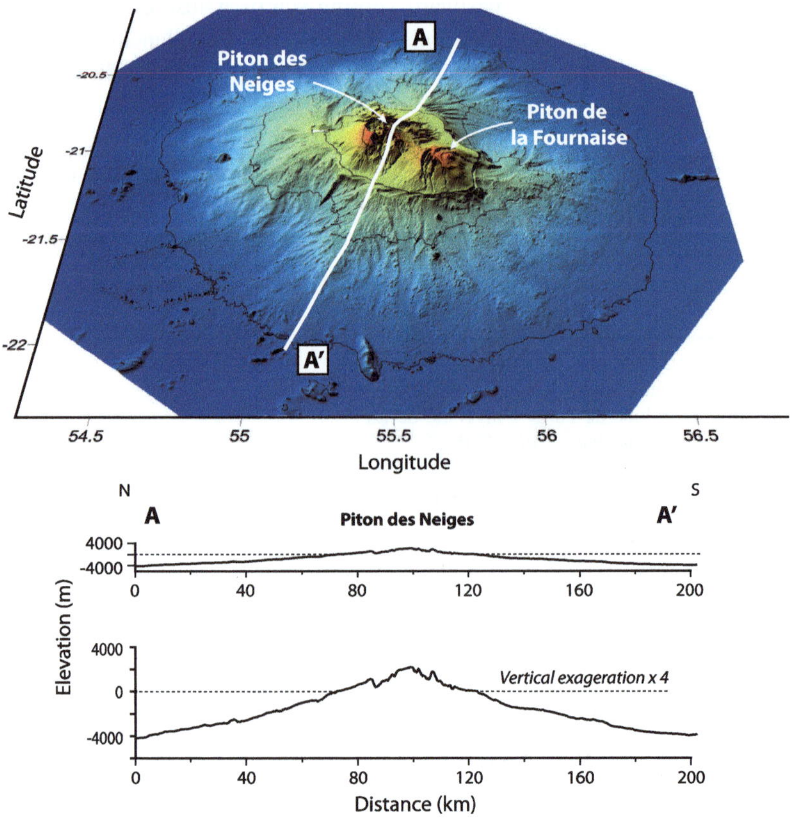

Fig. 4.1 Morphology of the volcanic edifice of La Réunion. Shaded bathymetric map compiled from FOREVER and ERODER oceanic surveys. The submarine flanks are characterized by slopes <5° on average

describing the submarine and subaerial morphologies of Piton de la Fournaise and to present the different interpretations, which have been published to explain the development of the main volcano-tectonic structures.

4.2 Morphology

The whole volcanic complex of La Réunion is a large cone of about 51,500 km^3, from the sea floor (about 4200 m below sea level) to the summit of the dormant Piton des Neiges volcano (3070 m above sea level). The morphology of the very large submarine base (volcano diameter of about 210 km) is well known since the FOREVER and ERODER oceanographic cruises in 2006 (Fig. 4.1; Saint-Ange et al. 2011, 2013; Le Friant et al. 2011; Sisavath et al. 2011, 2012; Babonneau et al. 2013). Submarine flanks have relatively gentle slopes ranging between 2° and 5° on average (Fig. 4.1; Saint-Ange et al. 2011; Lénat, Chap. 3). This morphology of the submarine slopes of La Réunion clearly differs with that observed in Hawaii (Big Island and Lo'ihi for instance), where the submarine flanks are much steeper than those of La Réunion (10 to 15°; Malahoff 1987; Bachèlery and Villeneuve 2013). In Hawaii, the slopes are widely constructed by lava-flows, mainly along the rift zones, and the destabilization products are restricted to well-defined structures. Conversely, in La Réunion, products of flank destabilization and erosion are widely distributed on the submarine slopes (Gailler and Lénat 2010; Saint-Ange et al. 2013). These deposits are related to slow deformation (spreading) and rapid mass movements (debris avalanches, or sediment transfer by shallow landsliding and turbidity currents; Oehler et al. 2008; Le Friant et al. 2011). Only a few portions of the submarine flanks of Piton de la Fournaise are directly built by lava-flows. The importance of remobilization processes on La Réunion island results in a widespread extension of volcaniclastic products (Babonneau et al. 2016, Chap. 6) and a very wide submarine base of the volcanic edifice compared to the size of the island itself; the subaerial part of the edifice representing only 4 % of the total volume.

Among volcaniclastic deposits, large volume of debris avalanche deposits (Fig. 4.2; Oehler et al. 2004; Lénat et al. 2009; Le Friant et al. 2011; Saint-Ange et al. 2013) have been identified on the submarine flanks of Piton de la Fournaise. They have been identified from their chaotic morphology, acoustic facies on backscatter imagery, seismic reflexion profiles, deep-water pictures and dredging. These deposits reveal the occurrence of old large flank collapse events Piton de la Fournaise. Some K/Ar ages of subaerial lava flows dredged on the submarine east flank of Piton de la Fournaise (Lénat et al. 2009) suggest a main flank landslide about 40–60 ka, which could be correlated with the development of the Plaine des Sables scarp (Oehler et al. 2008).

Submarine debris avalanche deposits are cut by several canyons and depressions, like Chenal Vincent, edged with abrupt walls (Fig. 4.2). In the valley floors of these depressions, the volcaniclastic sedimentation testifies to sedimentary processes that are mainly dominated by erosion and sediment transport due to coastal and submarine gravity instabilities (Cochonat et al. 1990; Ollier et al. 1998; Saint-Ange et al. 2013; Babonneau et al. 2016, Chap. 6). Furrows, scours, sediment waves, current ripples and a wide range of gravity flows are identified, suggesting a high sediment supply. South of Piton de la Fournaise, in the continuation of the Rivière des Remparts and Rivière Langevin (Fig. 4.2), recurrent gravity events and efficient (fast?) sediment transfer related to flash floods result into turbidity currents incising progressively deeper submarine canyons across the upper slope and forming deep-sea fans composed by turbidite deposits (see Babonneau et al. 2016, Chap. 6 and references herein).

As previously indicated, the portions of the submarine flanks built by lava-flows accumulations are fairly rare. They essentially correspond to the submarine extensions of the subaerial NE and SE rift zones (Fig. 4.2). The smooth surface that characterizes the shallowest part (from sea

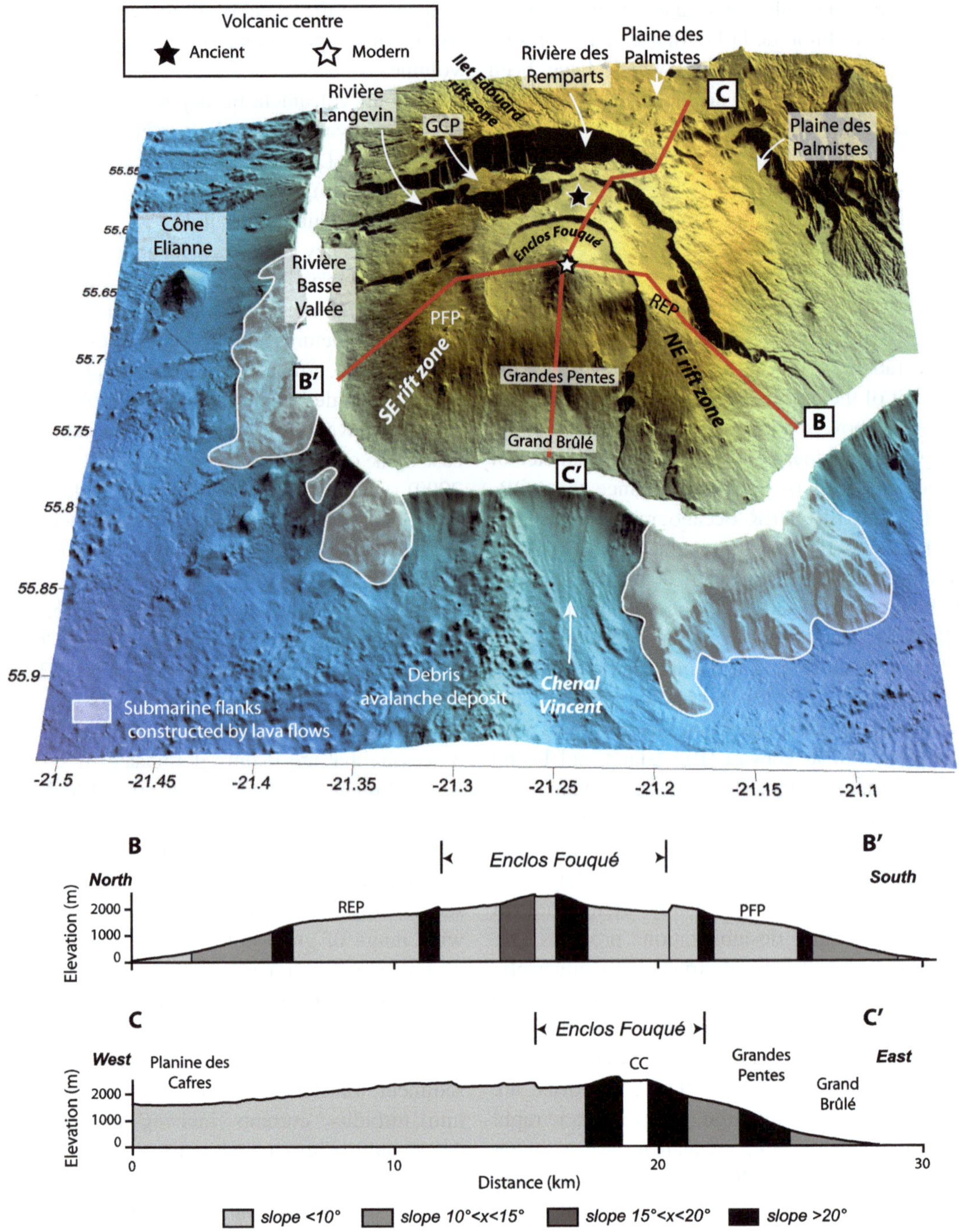

Fig. 4.2 Morphology of Piton de la Fournaise volcano. Subaerial (*black*) and submarine (*white*) contour lines every 200 m. The subaerial part of the edifice shows a "overturned soup plate" geometry. *CC* Central Cone. *PFP* Piton de Fourche plateau. *REP* Rivière de l'Est plateau

level to 1800 m deep) of the NE rift zone submarine extension is in a good morphologic continuity with the subaerial slopes. However, the incision of deepest relief together with two datings at 3.34 ± 0.07 and 3.77 ± 0.08 Ma obtained on pillow lavas dredged 1400 and 2100 m depth on this area (Smietana 2011) clearly indicate that the NE rift zone is built on an older edifice, probably Les Alizés volcano (Lénat et al. 2001a; Lénat, Chap. 3).

In addition, several large volcanic cones have been identified between 1500 and 2500 m below sea level. Cône Elianne, on the south submarine flank, stands out because of its height (about 700 m; Fig. 4.2). In comparison with the today's summit cone of Piton de la Fournaise volcano, Cône Elianne is twice larger. TOBI acoustic imagery allows to clearly identified lava flows on the southern flank of this cone (Fig. 4.3). Thus, Cône Elianne was probably built by the accumulation of several eruptive phases. Samples of alkali basalt, with a Th–U age estimated between 70 and 105 ka, were dredged on the western flank of this cone during the FOURNAISE 2 cruise (Lénat et al. 2009). Smietana (2011) obtained a K–Ar age of 127 ± 8 ka on the same sample.

In the subaerial domain, Piton de la Fournaise reaches a maximum elevation of 2630 m above sea level at the summit of the Central Cone that built in the Enclos Fouqué caldera (Fig. 4.2). The massif is characterized by three main morphological features (Fig. 4.2): (1) The flanks, which are continuous from the summit to the sea in the North, in the East and in the South, whereas the western flank is buttressed by the Piton des Neiges edifice (2), very deeply incised valleys, which dissect the western oldest part of the edifice and (3) two asymmetric caldera cliffs, the Plaine des Sables caldera and the Enclos Fouqué caldera, bounded by up to 100-m-high escarpments (Michon and Saint-Ange 2008; Lénat et al. 2012).

The flanks of Piton de la Fournaise are characterized by distinct slope domains, from the

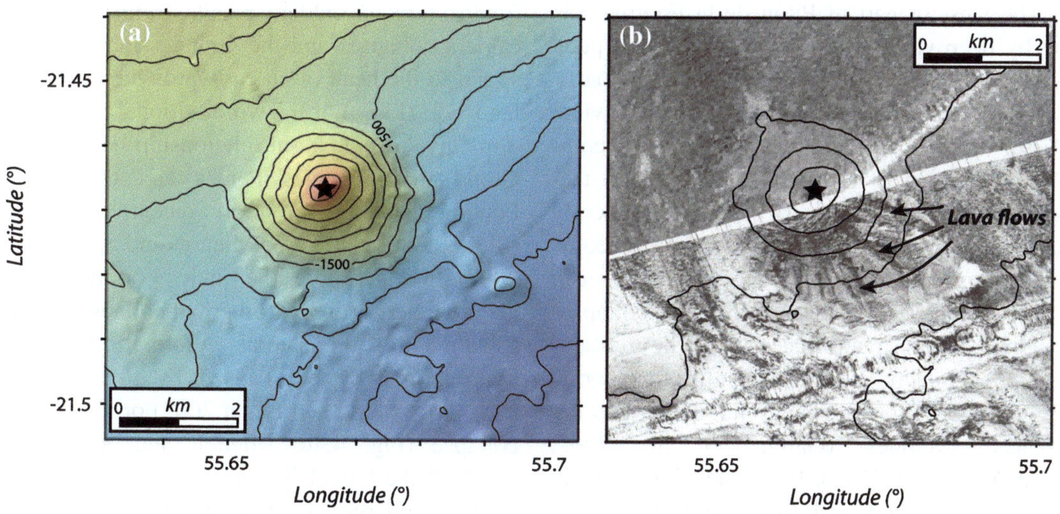

Fig. 4.3 Cône Elianne on the southern submarine flank of Piton de la Fournaise (see Fig. 4.2 for location). **a** Shaded bathymetry from ERODER1 oceanic survey. Contour lines every 100 m. **b** Backscatter acoustic imagery from ERODER1 survey (*upper part*) and TOBI side-scan sonar (*lower part*—ERODER3 data). Representation of TOBI images on the backscatter map requires the inversion of their reflectivity scale. High reflectivity facies (*black*) are assumed to be formed by more heterogeneous coarse products than those shown by low reflectivity (*white*). Cône Elianne's summit reaches the depth of 800 m bsl and its base diameter is approximately 3000 m. Chaotic lava flows are clearly visible on the TOBI acoustic imagery. Pillow-lavas dated at 127 ± 8 ka (see text) were dredged on the northern flank of Cône Elianne during the FOURNAISE 2 cruise (Lénat et al. 2009). Contour lines every 200 m

coast to the summit (Fig. 4.2; Rowland and Garbeil 2000; Michon and Saint-Ange 2008). At low elevations, the slope value range between 8° and 15°, corresponding to typical values for subaerial slopes of basaltic oceanic volcanoes (Mark and Moore 1987; Hürlimann et al. 2004). At higher elevations, the topography of the southern, eastern and northern flanks is characterized by steep slope zones ranging between 20° and 35°, while the summit shows slopes between 2° and 8°, if the escarpments and the active cone are disregarded. Such a slope distribution is strikingly similar to what is observed on Fernandina, Wolf, and Cerro Azul volcanoes in the western Galapagos archipelago (Rowland 1996; Rowland and Garbeil 2000) and remarkably differ from Hawaiian volcanoes where subaerial slopes do not exceed 15° (Bachèlery and Villeneuve 2013). In contrast, this slope distribution is not observed on the western flank of Piton de la Fournaise where the edifice is buttressed by Piton des Neiges (Fig. 4.2). The resulting morphology is characterized by a geometry resembling an "overturned soup plate".

The western part of Piton de la Fournaise is cut by deep valleys that incised the oldest parts of the volcano. The Rivière des Remparts, Rivière Langevin, Rivière de l'Est and Rivière Basse Vallée have been sculptured by a complex interaction of erosion, volcanism and tectonics (Merle et al. 2010). As frequently observed in La Réunion, the valleys become wider upstream. They are bounded by steep walls, which experience repeated collapses yielding an upstream recession. The destabilization products are subsequently transported by torrential river, which floods in during tropical cyclones (Garcin et al. 2005; Saint-Ange et al. 2011). Plaine des Palmistes, in the northwest flank of Piton de la Fournaise, is a complex and wide area, with a depression open to the sea edged by steep scarps and partially filled in by young lava flows. The current morphology of Plaine des Palmistes reveals the prevalence of erosion in its shaping.

4.3 Volcano-Tectonic Structures

4.3.1 Collapse Structures

Several collapse-related structures can be identified in the morphology of Piton de la Fournaise massif. All of them have been interpreted either as the scar of huge landslides or the scarps of collapse calderas (Bachèlery 1981; Chevallier and Bachèlery 1981; Duffield et al. 1982; Lénat et al. 1989; Labazuy 1996; Merle and Lénat 2003; Oehler et al. 2004, 2008; Michon and Saint-Ange 2008). The youngest one corresponds to the poly-lobate Enclos Fouqué caldera, which is opened in the east and connected to the Grand Brûlé depression (Fig. 4.4). The U-shaped structure of the eastern flank is bounded by the Tremblet and Bois Blanc scarps in the South and the North, respectively. The horseshoe-shaped geometry was interpreted as the result of a huge flank landslide (Fig. 4.4a; Duffield et al. 1982; Lénat et al. 1989; Labazuy 1996; Oehler et al. 2004, 2008). This interpretation contrasts with the pioneer one of Bachèlery (1981) in which the Enclos Fouqué caldera results from coalescent caldera collapses and the Grandes Pentes correspond to the head of an eastwards directed flank landslide (Fig. 4.4b). Two additional interpretations invoking a causal relationship between the Grand Brûlé/Grandes Pentes and the Enclos Fouqué caldera were more recently proposed (Merle and Lénat 2003; Michon and Saint-Ange 2008). In both models the Enclos Fouqué caldera is interpreted as an hydrothermal collapse caldera triggered by the lateral deformation of the hydrothermal system caused by the landslide (Fig. 4.4c; Merle and Lénat 2003) or the vertical collapse (Fig. 4.4d; Michon and Saint-Ange 2008) of the Grand Brûlé.

Another larger structure, the Plaine des Sables limited by a north-south trending scarp, is well developed west of the Enclos Fouqué (Fig. 4.5). This structure would result from a vertical collapse of the edifice (Fig. 4.5a; Bachèlery 1981), a large flank landslide toward the east (Fig. 4.5b, c;

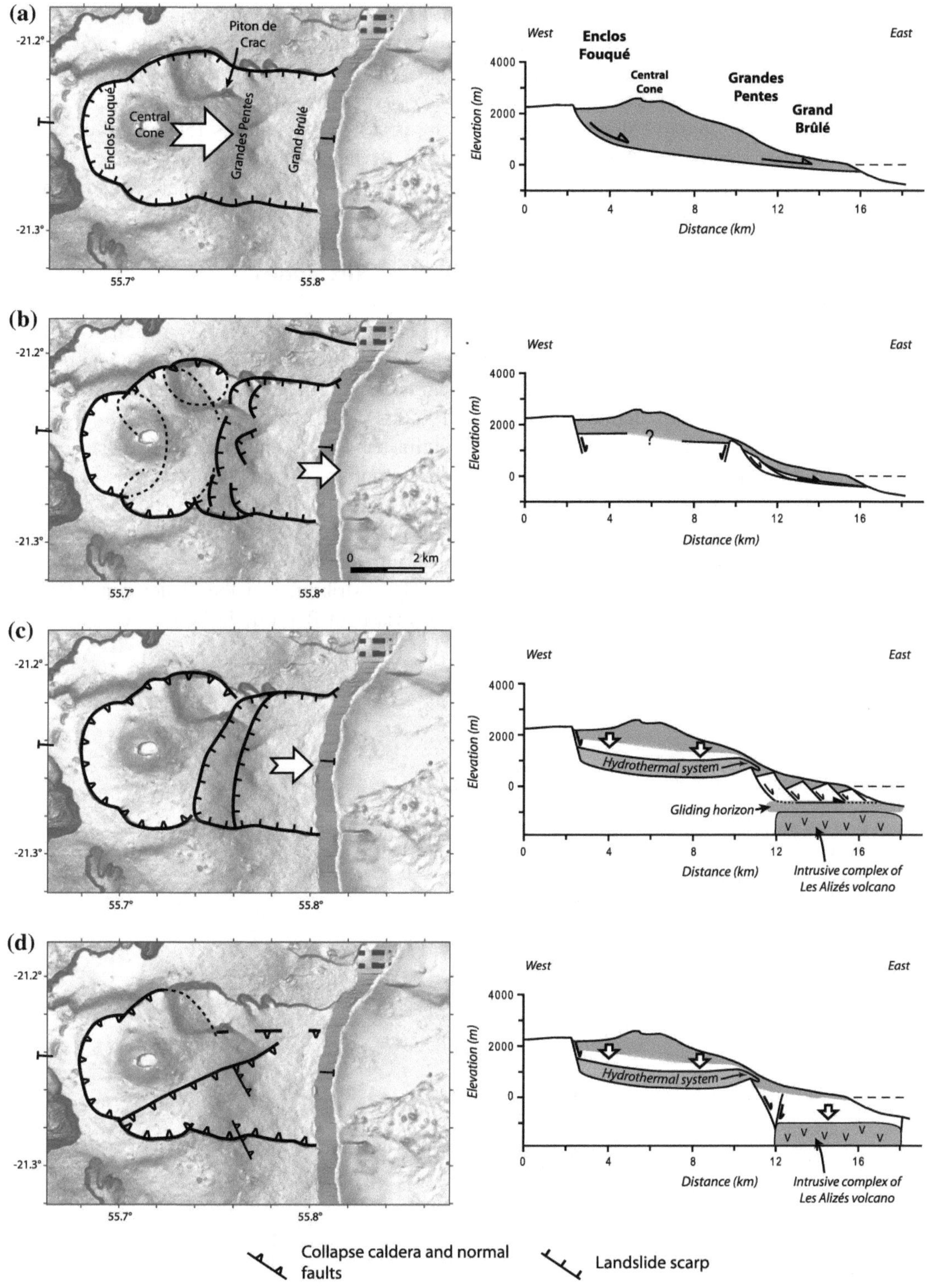

Fig. 4.4 Interpretations of the formation of the U-shaped structure composed of the Enclos Fouqué, the Grandes Pentes and the Grand Brûlé areas. **a** Duffield et al. (1982), Gillot et al. (1994), Labazuy (1996), Oelher et al. (2004, 2008). **b** Bachèlery (1981). **c** Merle and Lénat (2003). **d** Michon and Saint-Ange (2008)

Duffield et al. 1982; Gillot et al. 1994; Oehler et al. 2004, 2008), or the deformation of the hydrothermal system in a way similar to the one proposed for the Enclos Fouqué (Fig. 4.5d; Merle and Lénat 2003). The massive debris avalanche deposits on the submarine flank containing blocks dated between 110 and 45 ka suggests that a large volcano destabilization is compatible with the development of the Plaine des Sables scarp (Labazuy 1996).

Geological data suggest the occurrence of older collapse structures related to the Ancient Piton de la Fournaise, i.e. older than 60 ka. However, except for the Morne Langevin caldera whose western limit is well defined, the existence, the type (vertical or lateral collapses) and the limits of the other structures are still poorly constrained (Oehler et al. 2008; Merle et al. 2010). The atypical orientation of the Rivière des Remparts canyon, almost concentric instead of radial to the volcano's summit, would have been controlled by an initial landslide (Duffield et al. 1982; Gillot et al. 1994; Merle et al. 2010). After this event whose age is estimated at about 290 ka ago, the Ancient Piton de la Fournaise would have experienced two successive caldera collapses at ~250 ka and ~150 ka, corresponding to the Rivière des Remparts and Morne Langevin calderas, respectively (Bachèlery and Mairine 1990; Merle et al. 2010).

4.3.2 Rift Zones

On ocean basaltic volcanoes, magma transfer at shallow levels generally occurs along preferential intrusion paths named rift zones (e.g. MacDonald 1972). At the surface, volcanic rift zones are outlined by a concentration of pyroclastic cones and eruptive fissures that results from their overall history of activity. At depth, their existence is indicated by dense networks of subvertical intrusions (Walker 1986).

The distribution of the cinder cones built on the massif of Piton de la Fournaise suggests the

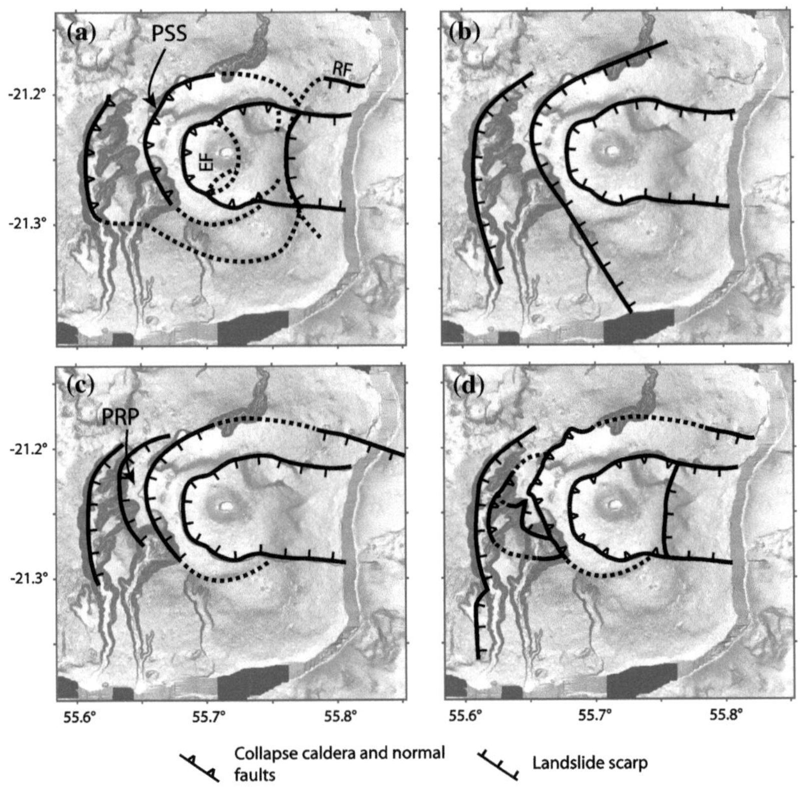

Fig. 4.5 Origin of the main volcano-tectonic structure of Piton de la Fournaise. **a** Summit collapse calderas and landslide of the lower eastern flank (Bachèlery 1981). **b** Successive huge landslides (Duffield et al. 1982; Gillot et al. 1994). **c** Successive huge landslides (Oehler et al. 2004, 2008). **d** Summit collapse calderas and landslides (Merle and Lénat 2003; Merle et al. 2010). *PSS* Plaine des Sables scarp. *EF* Enclos. *RF* Ravine Ferdinand

occurrence of two NE and SE rift zones (Fig. 4.2) and a NW-SE rift zone located between the Enclos Fouqué caldera and the summit of the Piton des Neiges inactive volcano (Bachèlery 1981; Chevallier and Bachèlery 1981; Villeneuve and Bachèlery 2006; Michon et al. 2016, this book for a review). The NE and SE rift zones are characterized by a smooth convex surface. Unlike the rift zones of the volcanoes of Hawaii, the average eruption frequency of Piton de la Fournaise rift zone is low for the historical period (post 1638 CE), with about one eruption every 50 years. This rate is even lower for the NW-SE rift zone for which radiocarbon data suggest eruptive occurrences of one event every 200 years (Morandi et al. 2016, this book, Chap. 8).

For the Ancient Piton de la Fournaise, which mostly crops out in the deeply incised Rivière des Remparts, Rivière Langevin and Rivière de l'Est, dyke distribution suggests the occurrence of a SW rift zone active between 530 and 290 ka (Mairine and Bachèlery 1997). This rift zone also named the Ilet Edouard rift zone is composed of a dense network of dykes in the western scarp of the Rivière des Remparts (Fig. 4.2; Bachèlery and Mairine 1990).

4.4 Geological Evolution

After the pioneering works of Bachèlery (1981) and Chevalier and Bachèlery (1981) at the beginning of the 1980's, the old part of the massif has been studied by Bachèlery and Mairine (1990) and Mairine and Bachèlery (1997) and summarized by Bachèlery and Lénat (1993) in the 1990's. More recently, the geology of the Piton de la Fournaise has been revisited by Michon and Saint-Ange (2008) and Merle et al. (2010).

4.4.1 The Ancient Piton de la Fournaise

The oldest on-land outcrops are dated at about 530 ka (Gillot and Nativel 1989) and have been grouped into a single *Pintades* lava unit (Bachèlery 1981; Bachèlery and Mairine 1990; Mairine and Bachèlery 1997; Merle et al. 2010), observed at the base of all lava sequences in the deep valleys incising Piton de la Fournaise (Rivière des Remparts, Rivière de l'Est and Rivière Langevin). According to recent interpretations (Smietana 2011), the *Pintades* lava unit, composed of feldspar-rich basalt to mugearite, does not belong to the Piton de la Fournaise series since it includes differentiated rocks, more alkalic in bulk composition than typical Piton de la Fournaise lavas (Albarède et al. 1997; Luais 2004). Instead, it would be considered as the upper part of its basement, possibly related to the old Les Alizés volcano according to Merle et al. (2010). Yet, such an interpretation does not take into account the radial distribution of the dykes intruded during this period, which suggests a volcanic centre around the Plaine des Sables, i.e. the locus of the ancient Piton de la Fournaise (Bachèlery and Mairine 1990). Thus, whether the *Pintades* unit belongs to the Alizés volcano or corresponds to an initial alkalic phase of Piton de la Fournaise is still debated. The magma composition changed at around 400 ka when the *Pintades* unit was overlain by the basaltic *Olivine Lavas* unit (Luais et al. 2004; Merle et al. 2010). The numerous dykes cropping out in the Rivière des Remparts and Rivière Langevin exhibit a radial distribution similar to that of the *Pintades* unit, suggesting a similar location of the volcanic centre during both phases. Morphological reconstructions determined from the remnant massif built during this period, west of the Rivière des Remparts, suggest that the edifice had reached an elevation of around 2700 m asl (Gayer et al. 2014). Note that Chevallier and Bachèlery (1981) proposed two coeval volcanic centres for this old phase, one located north of the Plaine des Sables and the other one close to the current location of the summit of Piton de la Fournaise. The Ancient Piton de la Fournaise then experienced alternating episodes of volcanic destruction and reconstruction, destructive episodes of the edifice being characterised by giant landslides and summit caldera collapses (Merle et al. 2010) (Fig. 4.5).

The oldest recognized landslide occurred at about 290 ka. This landslide resulted in the systematic stripping of the *Olivine Lavas* unit over a large area from the Rivière des Remparts to the Rivière Langevin canyons. At about 250 ka, a first caldera collapse forming the 8.5-km-wide Rivière des Remparts caldera would have affected the volcano (Merle et al. 2010). This event was followed by the development of the Mahavel volcano whose deposits correspond to (1) west dipping interstratified scoria and lava flows in the Bras de Mahavel, tentatively interpreted as evidencing the building of a main scoria cone at this time (Bachèlery and Mairine 1990), and (2) a pile of regular lava flows with a gentle dip toward the south that form the base on the Grand Coude plateau (Fig. 4.2). The Mahavel volcano was then cut at around 150 ka by the second recognised caldera collapse of Piton de la Fournaise. Unlike the Rivière des Remparts caldera, the western limit of this collapse structure, named the Morne Langevin caldera after Bachèlery and Mairine (1990), is well visible in the scarps of the Rivière des Remparts (see Fig. 4.4a in Merle et al. 2010). It bounds to the south and to the west the Plaine des Remparts plateau (Fig. 4.5). This caldera was entirely filled around 80–100 ka ago by a >200-m-thick series of thick lava units that flowed over the western caldera scarps. Lava flows invaded a paleo-depression located between the current course of the Rivière des Remparts and the Rivière Langevin, which was resulting either from erosion (Bachèlery and Mairine 1990; Mairine and Bachèlery 1997) or from a southward directed landslide (Merle et al. 2010). Whatever the origin of this depression, geochronological dating indicates that the Ancient Piton de la Fournaise constitutes the basement of the modern one east of the Rivière Basse Vallée (Gillot and Nativel 1989; Gillot et al. 1994) (Fig. 4.2).

Finally, the evolution of the Ancient Piton de la Fournaise indicates that after each event of caldera collapse, the lava flows remained initially confined to the caldera depression, allowing deep canyons to be eroded on the external slopes of the volcano. These canyons were later filled up when lavas finally overflowed the caldera rim and were able to flow down the external slopes. This excavating/infilling process occurred twice after the formation of the Rivière des Remparts and Morne Langevin calderas (Merle et al. 2010).

4.4.2 The Recent Shield of Piton de la Fournaise

For the Recent Shield, the main eruptive centre was located at the same place of the today's Central Cone, but Letourneur et al. (2008) suggest the possibility of the occurrence of an intermediate location before activity attained the present one. The age of the eastward migration of the eruptive centre remains controversial, as no decisive arguments have been found for an indisputable chronology (Merle et al. 2010). Bachèlery and Mairine (1990) propose that the eastward migration could have taken place at about 150 ka, concurrently with the formation of the Morne Langevin caldera. A second possibility is that it occurred after the Plaine des Sables caldera formed at about 65 and/or 40 ka (Bachèlery and Mairine 1990; Gillot and Nativel 1989; Merle et al. 2010; Staudacher and Allègre 1993).

A new edifice built east of the ancient volcanic centre after the collapse of the Plaine des Sables. The formation of the most recent 8 km-wide caldera, termed the Enclos Fouqué caldera, stopped its activity. This event marked the most recent major volcano-tectonic event in the history of Piton de la Fournaise whose timing and dynamics remains unclear (see discussion in Sect. 4.3.1). Caldera formation has been initially associated with the emplacement of the Bellecombe Ash Member, a sequence of ash deposits cropping along the western area of the Enclos Fouqué caldera (Bachèlery 1981; Mohamed-Abchir 1996; Ort et al. 2014). An age of 4745 ± 130 yr BP is usually assumed as an older limit for this caldera collapse event and the beginning of the Bellecombe paroxysmal explosive events (Mohamed-Abchir 1996; Staudacher and Allègre 1993). This age corresponds to a radiocarbon dating of the uppermost lava flows of the Bellecombe scarp (Bachèlery 1981).

Two ages are available for the lowermost ashes of the Bellecombe sequence, one in a gully inside the Plaine des Sables (4175 ± 145 yr BP; Mohamed-Abchir 1996) and another on the Langevin cliff (4880 ± 35 yr BP; Morandi et al. 2016, this book, Chap. 8). This age range is broadly consistent with that provided by cosmogenic dating, which indicate ages older than 3340 ± 1012 yr for the lava flows constituting the lavas topping the western rim of Enclos Fouqué caldera (Staudacher and Allègre 1993). However, the recent reappraisal of the existing radiocarbon data, enriched by new dating, suggests that Bellecombe ashes groups several explosive events, whose age ranges between 4880 ± 35 yr BP and 2340 ± 30 yr BP (Morandi et al. 2016, this book, Chap. 8).

Whatever the origin and the age of the depression formed by the Grandes Pentes and the Grand Brûlé, the U-shaped structure bounds most of the recent volcanic activity (Villeneuve and Bachèlery 2006; Michon et al. 2013) and plays a key role in the current sedimentation on the submarine slopes (Ollier et al. 1998; Saint-Ange et al. 2013).

The concentration of volcanic activity in the upper part of the caldera led to the construction of the 400-m-high steep Central Cone inside the Enclos Fouqué (Fig. 4.2). Its internal structure has been partially exposed down to about 350 m below the summit by the 2007 caldera collapse (Fig. 4.6). Five main types of units have been recognized: piles of thin lava flows, a few and thin tephra beds, pit crater infills, intrusive bodies and localized hydrothermal haloes (Peltier et al. 2012; Michon et al. 2013). A large portion of the N, E and SE caldera walls exposes uniform piles of thin lava flows (Fig. 4.6a). In two areas, accumulations of thick horizontal lava flows have been interpreted as infilling of summit pit craters. To the south, a 150-200-m-diameter filled crater, initially named Dolomieu (Bory de Saint-Vincent 1804) likely formed during the 1791 large eruption (Peltier et al. 2012; Michon et al. 2013). To the west, a much larger pit crater, the Pre-Bory pit crater, with an estimated diameter of about 800 m, cut the pile of thin lava flows forming most of the volcano upper part (Fig. 4.6b). Its vertical extent is greater that the 200 m exposed in the wall. A dense network of subvertical dykes, generally less than one meter in width, is observed in the walls of the Dolomieu summit caldera, more specifically in the N, SW and SE scarps. The sectors of highest concentration of dykes correspond to the proximal zone of the summit N25-30 and N120 rift (Michon et al. 2013, 2016, this book, Chap. 7).

A recent reappraisal of historical reports combined with new geological observations suggests that the historical activity of the Central Cone was characterized by a lava lake activity centred on the current inactive Bory crater (Michon et al. 2013). This lava lake fed the largest recent pahoehoe lava field, the Enclos Fouqué lava field (named CLEF in Lénat et al. 2001b), between 1733 and 1750 CE. The volcanic activity then shifted to the eastern part of the volcano summit with the development of a long-lasting summit eruption focused on the Mamelon Central, which ended with the large 1791 explosive eruption. A new lava lake was observed in 1841 in the eastern part of the current Dolomieu crater (Maillard 1862). Again, this summit activity continued until the 1860 paroxysmal explosive eruption (Michon et al. 2013). Since then, Piton de la Fournaise entered a phase of frequent, short-lived eruptions where pit crater collapses recurrently affected the Dolomieu crater.

The 2007 caldera collapse has led several authors (Michon et al. 2007; Gailler et al. 2009; Peltier et al. 2009; Staudacher et al. 2009; Lénat et al. 2012) to suggest that, as a consequence of the subsidence, a cylinder of faulted and fractured rocks must exist between the surface and the top of the magma reservoir. Indeed, the interpretations of the geophysical and visual observations during the collapse converge toward a model of a piston-like subsidence in one or several magma pockets being emptied by a voluminous lateral eruption. Lénat et al. (2012), noting that episodes of collapse are recurrent in the summit area in historical times, infer that the collapse column constitutes a major lithological heterogeneity with a significantly lower strength and higher permeability than that of the surrounding rocks.

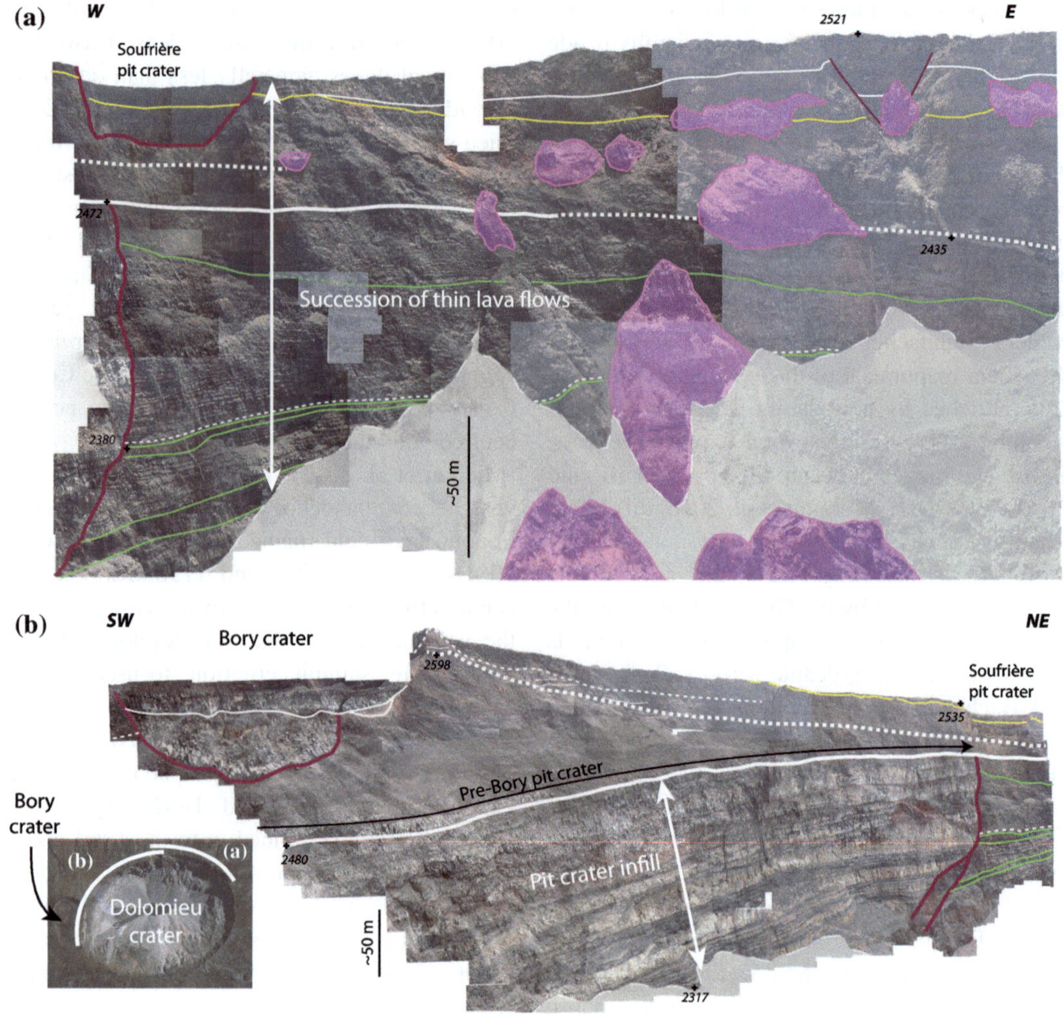

Fig. 4.6 Inner structure of the Central Cone exposed in the Enclos Fouqué summit caldera (modified after Michon et al. 2013). The 2007 caldera walls reveal that the central cone is mostly built by thin lava flows (**a**). The cone experienced a large pit crater collapse (Pre-Bory), which was subsequently filled by thick lava units (**b**). *Purple lines* Boundaries of collapse structures. *White lines* limits of the main litho-stratigraphical units. *Green* and *yellow lines* secondary units. *Pink* slide blocks and remaining terraces. *Grey* screes and talus. Selected point elevations are shown in metres

4.5 Synthesis

This review on the geology and morphostructural evolution of Piton de la Fournaise summarizes the intense debate that exists since the 80's to explain the origin of the main volcano-tectonic structures. Concerning the collapse structures, most of the controversies are focused on the involved processes, i.e., collapse calderas versus lateral landslides. A second and major issue is the timing and dynamics of caldera formation and its link with explosive phreatomagmatic activity like the one recorded in Bellecombe ashes. Another critical point resides in the location of the emission centre for the old alkaline series forming the base of the Piton de la Fournaise volcano. Do these plagioclase-rich early lavas correspond to the final activity of the oldest volcano forming La Réunion (Les Alizés volcano) or to an initial phase of Piton de la Fournaise? Indubitably, further works are required to potentially answer these issues, lateral versus vertical collapse and shift of the magma composition, which are shared with other volcanoes worldwide.

References

Albarède F, Luais B, Fitton G, Semet M, Kaminski E, Upton BGJ, Bachèlery P, Cheminée JL (1997) The geochemical regimes of Piton de la Fournaise volcano (Réunion) during the last 530,000 years. J Petrol 38:171–201

Babonneau N, Delacourt C, Cancouët R, Sisavath E, Bachèlery P, Mazuel A, Jorry SJ, Deschamps A, Ammann J, Villeneuve N (2013) Direct sediment transfer from land to deep-sea: insights from new shallow-marine multibeam data at La Réunion Island. Mar Geol 346: 47–57. doi:10.1016/j.margeo.2013.08.006

Babonneau N, Villeneuve N, Mazuel A, Bachèlery P (2016) Erosion and volcaniclastic sedimentation at Piton de la Fournaise: from source to deep marine environment. In: Bachèlery P, Lénat JF, Di Muro A, Michon L (eds) Active volcanoes of the Southwest Indian Ocean: Piton de la Fournaise and Karthala. Active Volcanoes of the World. Springer, Berlin

Bachèlery P (1981) Le Piton de la Fournaise (Ile de la Réunion). Etude volcanologique, structurale et pétrologique. PhD thesis 215 pp., Univ Clermont-Ferrand II, Clermont-Ferrand

Bachèlery P, Lénat JF (1993) Le Piton de la Fournaise. Mémoires de la Société Géologique de France. Nouvelle Série 163:221–229

Bachèlery P, Mairine P (1990) Evolution volcano-structurale du Piton de la Fournaise depuis 0.53 Ma. In: Lénat JF (ed) Le volcanisme de la Réunion–Monographie. Centre de Recherche en Volcanologie. Clermont-Ferrand, pp 213–242

Bachèlery P, Villeneuve N (2013) Hot spots and large igneous provinces.In: Shroder J, Owen L (eds) Treatise on geomorphology. Academic Press, San Diego, 5, pp 193–233

Bory de Saint-Vincent JBG (1804) Voyage dans les quatre principales iles des mers d'Afrique, fait par ordre du gouvernement, pendant les années neuf et dix de la République (1801 et 1802). Buisson (4 volumes)

Chevallier L, Bachèlery P (1981) Evolution structurale du volcan actif du Piton de la Fournaise, Ile de la Réunion —Océan indien occidental. Bull Volcanol 44:723–741

Cochonat P, Lénat JF, Bachèlery P, Boivin P, Cornaglia B, Deniel C, Labazuy P, Ledrezen E, Lipman P, Ollier G, Savoye B, Vincent P, Voisset M (1990) Gravity events as a primary process in the construction of a submarine volcano-sedimentary system (Fournaise volcano, Reunion Island). C. Rendus Acad Sci Serie II 311:679–686

Duffield WA, Stieltjes L, Varet J (1982) Huge landslide blocks in the growth of Piton de la Fournaise, La Reunion, and Kilauea Volcano, Hawaii. J Volcanol Geoth Res 12:147–160

Gailler LS, Lénat JF (2010) Three-dimensional structure of the submarine flanks of La Réunion inferred from geophysical data. J Geophys Res 115(B12). doi: 10. 1029/2009JB007193

Gailler L, Lénat JF, Lambert M, Levieux G, Villeneuve N, Froger JL (2009) Gravity structure of Piton de la Fournaise volcano and inferred mass transfer during the 2007 crisis. J Volcanol Geotherm Res 184:31–48. doi:10.1016/j.jvolgeores.2009.01.024

Garcin M, Poisson B, Pouget R (2005) High rates of geomorphological processes in a tropical area: the Remparts River case study (Réunion Island, Indian Ocean). Geomorphology 67:335–350. doi:10.1016/j. geomorph.2004.11.002

Gayer E, Lopez P, Michon L (2014) DEM-based model for reconstructing volcano's morphology from primary volcanic landforms. Geophys Res Abstracts 16: EGU2014-10701

Gillot PY, Nativel P (1989) Eruptive history of the Piton de la Fournaise volcano, Réunion Island, Indian Ocean. J Volcanol Geotherm Res 36:53–65

Gillot PY, Lefèvre JC, Nativel PE (1994) Model for the structural evolution of the volcanoes of Réunion Island. Earth Planet Sci Lett 122:291–302

Hürlimann M, Marti J, Ledesma A (2004) Morphological and geological aspects related to large slope failures on oceanic islands: the huge La Orotava landslides on Tenerife, Canary Islands. Geomorphology 62:143–158

Labazuy P (1996) Recurrent landslides events on the submarine flank of Piton de la Fournaise volcano (Réunion Island). In: McGuire WJ, Jones AP, Neuberg J (eds) Volcano instability on the earth and other planets. vol 110, Geol Soc Spec Publ, pp 293–30

Le Friant A, Lebas E, Clément V, Boudon G, Deplus C, de Voogd B, Bachèlery P (2011) A new model for the evolution of La Réunion volcanic complex from complete marine geophysical surveys. Geophys Res Lett 38:L09312. doi:10.1029/2011GL047489

Lénat J, Gibert-Malengreau B, Galdeano A (2001a) A new model for the evolution of the volcanic island of Reunion (Indian Ocean). J Geoph Res-Solid Earth 106:8645–8663

Lénat JF, Bachèlery P, Desmulier F (2001b) Genèse du champ de lave de l'Enclos Fouqué; une éruption d'envergure exceptionnelle du Piton de la Fournaise (Réunion) au 18ème siècle. Bull Soc géol Fr 172:177–188

Lénat JF, Boivin P, Deniel C, Gillot PY, Bachèlery P (2009) Age and nature of deposits on the submarine flanks of Piton de la Fournaise (Reunion Island). J Volcanol Geotherm Res 184:199–207. doi:10.1016/ j.jvolgeores.2009.01.013

Lénat JF, Bachèlery P, Peltier A (2012a) The interplay between collapse structures, hydrothermal systems and magma intrusions: the case of the central area of Piton de la Fournaise volcano. Bull Volc 7:407–421. doi:10.1007/s00445-011-0535-3

Lénat JF, Bachèlery P, Merle O (2012b) Anatomy of Piton de la Fournaise volcano (La Réunion, Indian Ocean). Bull Volcanol 74:1945–1961. doi:10.1007/s00445-012-0640-y

Lénat JF, Vincent PM, Bachèlery P (1989) The off-shore continuation of an active basaltic volcano: Piton de la Fournaise (Réunion Island, Indian Ocean); structural and geomorphological interpretation from sea beam mapping. J Volcanol Geoth Res 36:1–36.

Letourneur L, Peltier A, Staudacher T, Gudmundsson A (2008) The effects of rock heterogeneities on dyke paths and asymmetric grounddeformation: the example of Piton de la Fournaise (Réunion Island). J Volcanol Geotherm Res 173:289–302

Luais B (2004) Temporal changes in Nd isotopic composition of Piton de la Fournaise magmatism (Reunion Island, Indian Ocean). Geochem Geophys Geosyst 5:Q01008. doi:10.1029/2002GC000502

MacDonald GA (1972) Volcanoes. Prentice-Hall, Englewood Cliffs 510 pp

Maillard L (1862) Notes sur l'île de la Réunion (Bourbon). Lanée, Paris 344 pp

Mairine P, Bachèlery P (1997) Un grand épisode érosionnel dans l'histoire ancienne du Piton de la Fournaise. C Rendus Acad Sci Paris 325:243–249

Malahoff A (1987) Geology of the summit of Loihi submarine volcano. US Geol Surv Prof Paper 1350:133–144

Mark RK, Moore J (1987) Slopes of the Hawaiian ridge. US Geol Surv Prof Pap 1350:101–107

Merle O, Borgia A (1996) Scaled experiments of volcanic spreading. J Geophys Res 101:13805–13817

Merle O, Lénat J (2003) Hybrid collapse mechanism at Piton de la Fournaise volcano, Reunion Island, Indian Ocean. J Geophys Res-Solid Earth 108(B3):2166. doi:10.1029/2002JB002014

Merle O, Mairine P, Michon L, Bachèlery P, Smietana M (2010) Calderas, landslides and paleo-canyons on Piton de la Fournaise volcano (La Réunion Island, Indian Ocean). J Volcanol Geotherm Res 189:131–142. doi:10.1016/j.jvolgeores.2009.11.001

Michon L, Saint-Ange F (2008) Morphology of Piton de la Fournaise basaltic shield volcano (La Réunion Island): characterization and implication in the volcano evolution. J Geophys Res-Solid Earth 113:B03203. doi:10.1029/2005JB004118

Michon L, Staudacher T, Ferrazzini V, Bachelery P, Marti J (2007) April 2007 collapse of Piton de la Fournaise: a new example of caldera formation. Geophys Res Lett 34:L21301. doi:10.1029/2007GL031248

Michon L, Di Muro A, Villeneuve N, Saint-Marc C, Fadda P, Manta F (2013) Explosive activity of the summit cone of Piton de la Fournaise volcano (La Réunion island): a historical and geological review. J Volcanol Geotherm Res 264:117–133. doi:10.1016/j.jvolgeores.2013.06.012

Michon L, Ferrazzini V, Di Muro A (2016) Magma paths at Piton de la Fournaise volcano. In: Bachèlery P, Lénat JF, Di Muro A, Michon L (eds) Active volcanoes of the Southwest Indian Ocean: Piton de la Fournaise and Karthala. Active Volcanoes of the World. Springer, Berlin

Mohamed-Abchir A (1996) Les Cendres de Bellecombe: un évènement majeur dans le passé récent du Piton de la Fournaise, Ile de la Réunion. PhD thesis, Univ Paris VII 248 pp

Moore JG (1964) Giant submarine landslides on the Hawaiian ridge. US Geol Surv Prof Pap 501:D95–D98

Moore JG, Mark MK (1992) Morphology of the Island of Hawaii. GSA Today 2:257–262

Morandi A, Di Muro A, Principe C, Michon L, Leroi G, Norelli F, Bachèlery P (2016) Pre-historic explosive activity at Piton de la Fournaise volcano. In: Bachèlery P, Lénat JF, Di Muro A, Michon L (eds) Active volcanoes of the Southwest Indian Ocean: Piton de la Fournaise and Karthala. Active Volcanoes of the World. Springer, Berlin

Oehler JF, Labazuy P, Lénat JF (2004) Recurrence of major flank landslides during the last 2-Ma-history of Reunion Island. Bull Volcanol 66:585–598. doi:10.1007/s00445-004-0341-2

Oehler JF, Lénat JF, Labazuy P (2008) Growth and collapse of the Reunion Island volcanoes. Bull Volcanol 70:717–742. doi:10.1007/s00445-007-0163-0

Ollier G, Cochonat P, Lénat JF, Labazuy P (1998) Deep-sea volcaniclastic sedimentary systems: an example from La Fournaise volcano, Reunion Island, Indian Ocean. Sedimentology 45:293–330

Ort M, Di Muro A, Michon L, Bachèlery P (2014) The Bellecombe ash: indications of explosive eruptions at Piton de la Fournaise, Reunion Island. IAVCEI - 5IMC conference. Queretaro, Mexico

Peltier A, Staudacher T, Bachèlery P, Cayol V (2009) Formation of the April 2007 caldera collapse at Piton de La Fournaise volcano: Insights from GPS data. J Volcanol Geotherm Res 184:152–163. doi:10.1016/j.jvolgeores.2008.09.009

Peltier A, Massin F, Bachèlery P, Finizola A (2012) Insights on internal structures and edification of basaltic shield volcanoes: new field investigations on Piton de La Fournaise. Bull Volc 74:1881–1897. doi:10.1007/s00445-012-0636-7

Rowland SK (1996) Slopes, lava flow volumes, and vent distributions on Volcan Fernandina, Galapagos Islands. J Geophys Res 101:27657–27672

Rowland S, Garbeil H (2000) Slopes of oceanic basalt volcanoes. In: Mouginis-Mark PJ, Crisp JA, Fink JH (eds) Remote sensing of active volcanism. AGU Geophys Monogr Ser 116, pp 223–247

Saint-Ange F, Savoye B, Michon L, Bachèlery P, Deplus C, de Voogd B, Dyment J, Le Drezen E, Voisset M, Le Friant A, Boudon G (2011) A volcaniclastic deep-sea fan off La Réunion Island (Indian Ocean): gradualism versus catastrophism. Geology 39:271–274. doi:10.1130/g31478.1

Saint-Ange F, Bachèlery P, Babonneau N, Michon L, Jorry S (2013) Volcaniclastic sedimentation on the submarine slopes of a basaltic hotspot volcano: Piton de la Fournaise volcano (La Réunion Island, Indian Ocean). Mar Geol 337:35–52. doi:10.1016/j.margeo.2013.01.004

Sisavath E, Babonneau N, Saint-Ange F, Bachèlery P, Jorry S, Deplus C, de Voogd B, Savoye B (2011) Morphology and sedimentary architecture of a modern volcaniclastic turbidite system: the Cilaos fan, offshore La Réunion Island. Mar Geol 288:1–17. doi:10.1016/j.margeo.2011.06.011

Sisavath E, Mazuel A, Jorry S, Babonneau N, Bachèlery P, de Voogd B, Salpin M, Emmanuel L, Beaufort L, Toucanne S (2012) Processes controlling a volcaniclastic turbiditic system during the last climatic cycle: example of the Cilaos deep-sea fan, offshore La Reunion Island. Sedim Geol 281:180–193. doi:10.1016/j.sedgeo.2012.09.010

Smietana M (2011) Pétrologie, géochronologie (K-Ar) et géochimie élémentaire et isotopique (Sr, Nd, Hf, Pb) de laves anciennes de La Réunion: Implications sur la construction de l'édifice volcanique. Unpublished PhD thesis, Univ. La Réunion pp. 364

Staudacher T, Allègre CJ (1993) Ages of the second caldera of Piton de la Fournaise volcano (Réunion) determined by cosmic ray produced ^{3}He and ^{21}Ne. Earth Planet Sci Lett 119:395–404

Staudacher T, Ferrazzini V, Peltier A, Kowalski P, Boissier P, Catherine P, Lauret F, Massin F (2009) The April 2007 eruption and the Dolomieu crater collapse, two major events at Piton de la Fournaise (La Réunion Island, Indian Ocean). J Volcanol Geotherm Res 184:126–137. doi:10.1016/j.jvolgeores.2008.11.005

Tort A, Finizola A (2005) Structural survey of misti volcanic cone (southern Peru) combining elliptical fourier function analysis of the volcano morphology and self-potential measurements. J Volcanol Geotherm Res 141:283–297

Villeneuve N, Bachèlery P (2006) Revue de la typologie des éruptions au Piton de La Fournaise, processus et risques volcaniques associés. Cybergeo Eur J Geog (http://cybergeo.revues.org/2536)

Walker GPL (1986) Koolau dike complex, Oahu: intensity and origin of a sheeted-dike complex high in a Hawaiian volcanic edifice. Geology 14:310–313

Groundwater Resources on Active Basaltic Volcanoes: Conceptual Models from La Réunion Island and Grande Comore

5

Jean-Lambert Join, Jean-Luc Folio, Anli Bourhane
and Jean-Christophe Comte

Abstract

La Réunion Island and Grande Comore are oceanic islands entirely formed by young and still active volcanoes, namely Piton de la Fournaise and Karthala. These geological features are responsible for a particular type of hydrosystem and present a unique challenge regarding the prospection and management of water resources. Young volcanic terrains are composed of lava flows with exceptional hydraulic conductivities. On both islands, despite high levels of precipitation, surface water resources are scarce due to high infiltration rates. Water infiltrates deep into the ground and reaches the water table at an elevation close to sea level. Access to water resources is particularly challenging on these islands due to: (i) the relative absence of perennial surface water resources, (ii) the widespread marine contamination in littoral areas, (iii) the excessive depth of the groundwater table due to high elevations. We present the main results obtained from twenty years of multidisciplinary investigation programs dedicated to groundwater resources development in La Réunion Island. This approach includes geological and hydrogeological investigations, geophysical prospections and geochemical analysis. Results have led us to propose an improved conceptual model describing groundwater resources and to implement a numerical model of groundwater flow applied to Piton de la Fournaise volcano. In the interior of the island, the presence of a piezometric dome reaching a height of over 1200 m above sea level has been revealed and is of major interest for groundwater development plans.

J.-L. Join (✉) · J.-L. Folio · A. Bourhane
Laboratoire Géosciences Réunion, Université de La Réunion, Institut de Physique du Globe de Paris, Sorbonne Paris Cité, CNRS, 97744 Saint-Denis, France
e-mail: join@univ-reunion.fr

J.-C. Comte
School of Geosciences, University of Aberdeen, Aberdeen, Scotland, UK

5.1 Introduction

Despite high precipitation rates, young or still active volcanic islands, which are mainly composed of stacks of lava flows, present difficulties

P. Bachèlery et al. (eds.), *Active Volcanoes of the Southwest Indian Ocean*,
Active Volcanoes of the World, DOI 10.1007/978-3-642-31395-0_5

for the access of water resources. High hydraulic conductivities promote deep infiltration and the depth of the water table increases dramatically with the island's elevation. As a consequence, inland prospection of groundwater is expensive and uncertain. Near the coast, wells are drilled until they reach the water table, which is close to sea level and at permanent risk of saltwater contamination. In addition, hydrogeologists commonly have to deal with the lack of direct observations or monitoring points. The major challenge today is to define conceptual groundwater models for inland aquifers (Izquierdo 2014).

A number of key hydrogeological studies have described aquifers in active volcanic islands, such as the Teide volcano (Ecker 1976; Custodio 1978), Karthala in the Comoros archipelago (Savin et al. 2001), the island of Mauritius (Join et al. 2000), Pico Island in the Azores archipelago (Cruz and Silva 2001) and the Hawaiian archipelago (Izuka and Gingerich 2003). Several conceptual models are used to describe volcanic island aquifers (Cruz and Silva 2001). They may be classified in two categories: (i) Hawaiian models (Peterson 1972; MacDonald et al. 1983), and (ii) Canary Islands models (Custodio 1974, 1978; Custodio and Saenz de Oiza 1973; Falkland and Custodio 1991).

Hawaiian models (Izuka et Gingerich 2003; Lau et Mink 2006), distinguish between the "basal groundwater" and the "high-elevation groundwater saturated zones". The basal groundwater is characterized by a lens-shaped water body floating on saltwater with a flat water table. This aquifer is composed of a thick accumulation of basaltic lava flows with high hydraulic conductivity. The high-level saturated zones are found at elevations much higher than the basal groundwater. These bodies are described as "perched" or "dyke impounded".

The Canary Islands models consider a progressive decrease of hydraulic conductivity with depth, which controls the general hydrogeological behaviour of active shield volcanoes. The basal groundwater is assumed to extend continuously from coastal areas to the caldera and rift zones. In the mountainous interior parts of the volcanoes, the water table has the shape of a high elevation dome (Custodio et al. 1988). This is because hydraulic conductivities are assumed to be smaller as the average depth of the basal groundwater below the land surface increases. In addition, the recharge is usually higher in these areas. In this model, perched or dike-impounded water bodies can be found locally but it is assumed that the main streams or springs in the high elevation zones are connected to the basal groundwater.

Those two categories of conceptual models were synthesized by Join et al. (2005) and are clearly different with regard to hydraulic conductivity distribution and the level of hydraulic continuity between coastal and mountainous areas (Fig. 5.1). Hawaiian models consider a binary distribution of hydraulic conductivities (very permeable/impervious) and a hydraulic discontinuity between basal groundwater and high-level saturated zones. In contrast, Canary Islands models assume a smoother hydraulic conductivity distribution that enables better hydraulic continuity. Custodio (1989) noted that the widely-used classification for Hawaiian models is not suitable for the case of the Canarian archipelago, especially the relationship between the high-level saturated zones and the basal aquifer that consists in a huge and relatively sudden head loss. More recently, Izuka and Gingerich (in 1998 then 2003) suggested that a new conceptual model of groundwater occurrence in shield-volcano islands is needed to explain the conditions observed in Kauai Island (Hawaii). They described a hydrogeological feature called "fully saturated vertically extensive fresh water body" which is similar to the continuous basal groundwater proposed in the Canary Islands models. In fact, the Canary Islands and Hawaiian models are not antagonistic but complementary and influenced by the age of the geological materials of volcanic islands, their degree of erosion, the relative position of young basaltic lava flows and the type of volcanism (Join et al. 2005).

Taking into account the hydrological similarities between the islands of Grande Comore and La Réunion, we propose here to use La Réunion and the two massifs of Piton des Neiges and

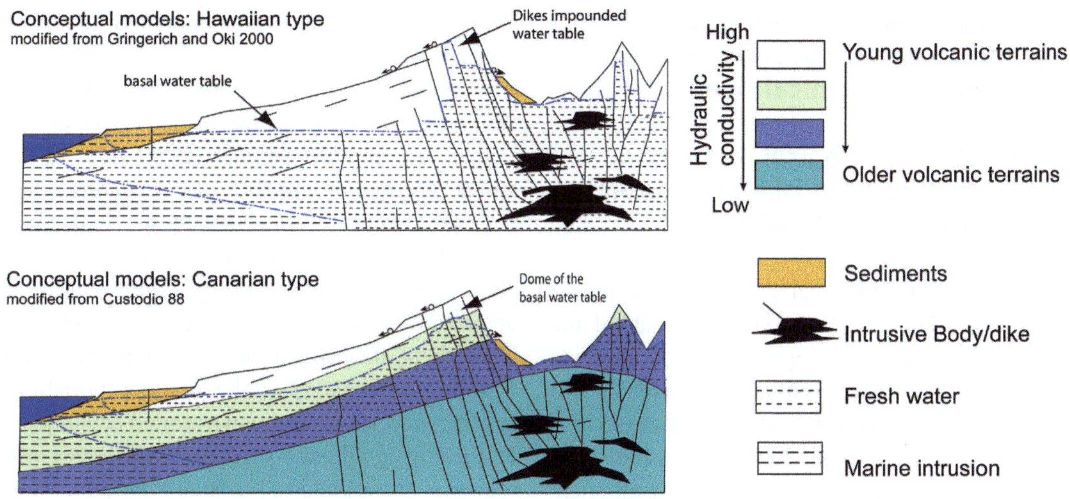

Fig. 5.1 Hydrogeological conceptual models of young oceanic volcanoes (modified from Join et al. 2005)

Piton de la Fournaise as a pilot site for the understanding of the groundwater settings of these two young volcanic islands. The main hydrogeological analysis results obtained in La Réunion will be set out below. These results have enabled a conceptual model to be defined and a simulation of flows path to be generated. Finally, the potential applications of this simulation are expounded in terms of prospection of groundwater resources on this type of volcano.

5.2 Climatology of La Réunion: The Case of Piton de la Fournaise

Precipitations fluctuate from a massif to another due mainly to its situation to tradewinds. However in both La Réunion and Grande Comore islands, the climate is inter-tropical with two seasons: a hot rainy season from December to April and a temperate dry season from May to November. In La Réunion the windward eastern slopes are subjected to high precipitations, with between 2 and 12 m of rainfall per year. Conversely, the western leeward side receives only between 0.5 and 2 m of rainfall per year. In La Grande Comore, annual rainfall varies from less than 1.5 m in the north-east to more than 4 m in the south-west of the island. Despite the abundance of the precipitations, there are no permanent rivers on the flanks of these volcanoes. In La Réunion Island, permanent streams only exist within the deep erosion cirques that are absent in Grande Comore. However, in both islands the increase of precipitations with elevation promotes the existence of small springs in higher-altitude areas.

Within La Réunion Island, the massif of Piton de la Fournaise is the rainiest area, whose water budget is given by Barcelo and Coudray (1996). Mean annual rainfall is 6200 mm. Evaporation and runoff are each close to 1000 mm, which is approximately the uncertainty regarding estimated rainfall. Recharge would therefore be around 4200 mm/year. However, a very contrasted distribution of rainfall over the volcano is observed, with more than 12,000 mm/year on the eastern side of the massif.

On the whole massif and despite abundant rainfalls, only three rivers feature perennial flow reaching the sea; they are located in the deepest incised valleys of the massif. On the volcano flanks, runoff occurs only during the most intense rainfall events. Barcelo (1996) has shown that these runoff events are associated with high infiltration in the river bed that contributes to recharging the aquifers.

5.3 Groundwater Bodies: Typology, Geological, Hydrodynamical and Geochemical Aspects

5.3.1 Hydrogeological Analogy Between the Islands of Grande Comore and La Réunion

The first deep geophysical surveys performed on the flanks of the Karthala volcano enabled the analogy of the hydrogeological conditions between the Karthala massif and the better-known context of La Réunion Island to be discussed for the first time (Savin et al. 2001). Later, Join et al. (2005) revised the conceptual hydrogeological model of La Réunion Island and proposed its application to other volcanic islands such as Grande Comore. According to these works, the hydrogeological context of La Réunion can be divided into two hydrogeological domains: the littoral domain and the altitude domain. These two categories are artificial and result from historic uncertainty relative to inland continuity of the well-known volcanic coastal aquifer.

5.3.2 The Coastal Area

The geographical zone corresponding to the coastal area is defined arbitrarily as the zone of prospection of the basal groundwater by drilling. In practice, this zone is limited to the area whose elevation is below 300 m. At higher elevations, survey soundings become scarce and access to hydrogeological information becomes more difficult. In the littoral domain, the analysis tools stem from geological and hydrodynamic data derived from interpolations of soundings and hydraulic tests.

Since the eighties, thanks to a large number of deep wells drilled within the framework of a development programme of the Water Research Department undertaken in La Réunion Island as well as the United Nations Development Programme in Grande Comore, the hydrogeological conditions in both islands' coastal areas have been better known than inland. In most cases, the piezometric level measured in the wells is close to the mean sea level (the mean hydraulic head is under 1 m). This confirms the very low hydraulic gradient of the coastal groundwater. Moreover, the water level in the wells and groundwater salinity are strongly influenced by tidal fluctuations. The marine impact is, however, greater in Grande Comore where the mean tidal range reaches 3 m (1.2 m in La Réunion Island). In both cases, it is tricky to determine both upgradient and downgradient hydraulic conditions. Upgradient, the estimation of fluxes is even more complicated because of the non-conformability between superficial and underground catchments. At the downgradient boundary, the interface with seawater corresponds to a head condition. Nevertheless, control of the outflows remains uncertain.

5.3.3 The Inland Domain

The inland domain can be defined from the point of view of the problem of groundwater accessibility by the classical hydrogeological investigation methods based on drilling. In la Réunion island, the rare drilling surveys performed at the centre of the island show that the water table depth of the base aquifer can reach several hundred meters (typically >300 m up more than 3 km from the shoreline). Therefore, beyond an elevation of 300 m, drilling investments are exceptional and groundwater exploration requires different investigation methods. The distinction between altitude and littoral domains is thus based on two types of methodological approaches of investigation.

In the high elevation zones of the two volcanic massifs of La Réunion island, the existence of aquifer systems is evidenced by more than 400 springs which appear mostly in the wettest part of the island or in the erosion morphology and mark the points of groundwater emergence. This abundance of springs in La Réunion contrasts with Grande Comore, which lacks such erosion. This difference is related to the much wetter climate of La Réunion.

Field work carried out on the Piton des Neiges massif contributed to defining the main conditions

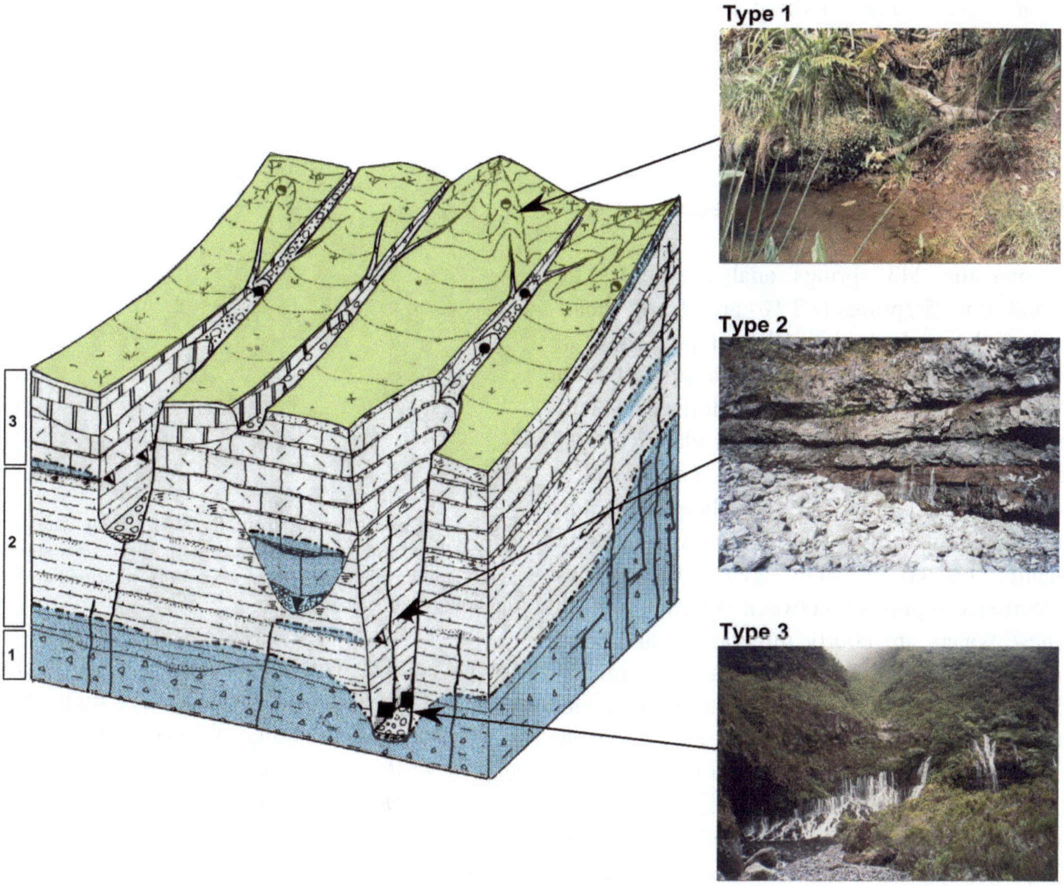

Fig. 5.2 Hydrogeological conceptual model of inland aquifers. Groundwater: *type 1* springs from superficial aquifer, *type 2* springs from intermediate aquifer, *type 3* springs from deep aquifer. Geology: from older unit (*1*) to younger unit (*3*) volcanic terrains

of aquifer deposits in altitude zones (Join and Coudray 1993). A hydrogeological conceptual model of the springs and their reservoirs was proposed with a view to describing the different aquifer systems occurring in the inland zone (Fig. 5.2). Three types of aquifer systems were differentiated: this classification was later adapted to the Piton de la Fournaise.

- Type 1 represents the superficial aquifers developed in weathered horizons facilitated by the presence of pyroclastic accumulations at the surface. In the wettest zones, a weakly incised hydrological network participates in the groundwater drainage of these small aquifers. The variability of the instantaneous outflows confirms the absence of reserve.
- Type 2 represents the perched aquifers observed in the cliff walls of erosion canyons and cirques. They are supported by thin impervious horizons associated with paleo-valleys or paleo-soils. Like type 1, these aquifers are characterized by high depletion rates.
- Type 3 represents the deep saturated zone (the basal aquifer) within the oldest volcanic units of the massif. In La Réunion, this deepest saturated zone yields most of the emergences observed as well as the biggest rivers in the two volcanic massifs. Springs appear preferentially at the bottom of the erosional amphitheatre of the largest rivers such as the Rivière des Remparts, the Rivière de l'Est and the Rivière Langevin (Coudray et al. 1990). They are also

observed in the massif of Piton des Neiges (Atlas Hydrogéologique de la Réunion, coll. 1986). The outflow values are remarkably sustained during the driest periods.

5.3.4 Hydrochemical Aspects

Among the 243 springs analysed across the island, only 5 springs (≈2 %) are clearly identified as being hydrothermal in type. They are essentially located in the three cirques of Piton des Neiges. They can be characterized by temperatures higher than 25 °C and electrical conductivities higher than 200 μS cm^{-1}, with a predominance of the SO_4–Na chemical feature (Daesslé et al. 1988). The remaining springs are mainly located in high areas. The highest occurrence is situated between 900 and 1300 m. These waters are poorly mineralized and water mineralization does not appear to be a sufficient criterion to differentiate the three types of inland aquifers. Join et al. (1997) have shown that the electric conductivity of the inland springs ranges from 17 to 143 μS cm^{-1}, while that of the basal aquifer springs varies only between 52 and 64 μS cm^{-1} and is below the mean value of the springs from superficial and perched aquifers (71 μS cm^{-1}). The waters are characterized by HCO_3–Mg or HCO_3–Ca chemical facies, but sodium can constitute the secondary facies and become preponderant in some springs within the cirques. By classifying samples on a Piper diagram (Fig. 5.3), one can identify the differences between superficial springs (related to presence of ando soils) which are characterised by chloride (and also nitrates), and the basal aquifer springs (issuing from the oldest volcanic rocks) which are characterised in part by sodium.

Silica is always present in high quantities [the mean value is 27 mg/L (Daesslé et al. 1988)]. Because of the high solubility of silica in the soils, high concentrations can be reached after a short time in the subsurface (Nicolini et al. 1988).

In the basal aquifer, Grünberger (1989) showed the preponderant influence of the marine environment on the groundwater chemistry of the littoral aquifer. This has been confirmed by Daesslé and Duchamps (1989) who have studied the hydrochemical variability of 73 boreholes drilled mainly in the coastal area. A modal analysis is given as representative of the mean groundwater mineralization in the basal aquifer (Table 5.1). As an example, we present the analysis provided from the drill hole of a saltwater-contaminated site (Well "La Saline").

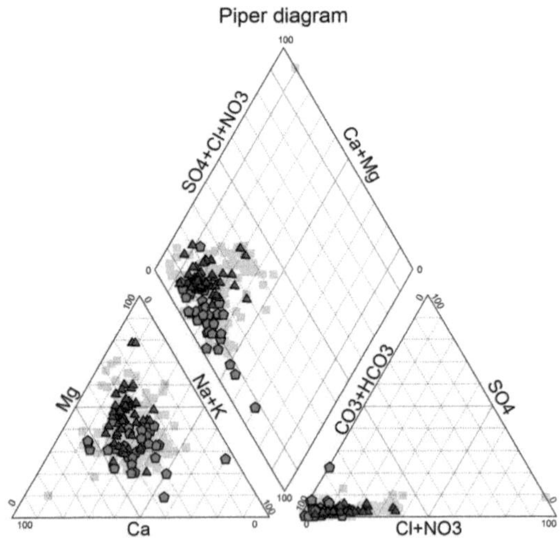

Fig. 5.3 Piper diagram from the 243 spring data set. *Green dots* basal aquifer characteristics, *red dots* superficial aquifer characteristics, *grey dots* intermediate or unknown characteristics

The preponderant role of marine contamination is well highlighted; the less mineralized waters define an initial chemical feature that is HCO_3–Mg or HCO_3–Ca in type, that changes to Cl–Na type. The mean electrical conductivity is 120 μS cm^{-1}, corresponding to a mineralization of 100–120 mg/L (BRGM, Atlas hydrogéologique de La Réunion, *coll.* 1986).

5.3.5 Continuity Between Littoral Basal Groundwater and Deep Groundwater of High Elevation Zones

The chloride and sodium contents of 240 springs from Join et al. (1997) are plotted on a graph of Cl (mg/L) as a function of Na (mg/L). Springs plotted on this graph have first been differentiated

Table 5.1 Chemical analysis of the basal groundwater

	Cond. (μS)	Ca^{2+} (mg/L)	Mg^{2+} (mg/L)	Na^{+} (mg/L)	K^{+} (mg/L)	HCO_3^{-} (mg/L)	Cl^{-} (mg/L)	NO_3^{-} (mg/L)	SO_4^{2-} (mg/L)
Modal analysis (73 samples)	275	13	9	11	2.1	115	7	1.5	0.5
Forage La Saline	635	18.5	19	91	4.9	107	145	24	10.3

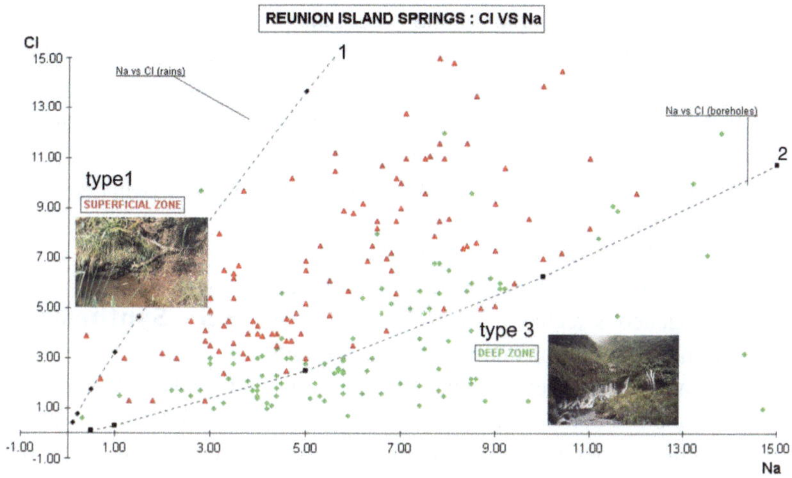

Fig. 5.4 La Réunion Island springs: chloride versus sodium (modified from Join et al. 1997). *Green dots* basal aquifer characteristics, *red dots* superficial aquifer characteristics; *1* curve of best fit defined in rainwater; *2* curve of best fit defined in boreholes

by their structural situation. Points represented by green filled symbols represent the basal springs and the remainder represent the superficial springs.

Data from poorly mineralized boreholes and rainwater are indicated in the same graph by two adjustment curves of chlorides (mg/L) as a function of sodium (mg/L). These relationships are indicated by dashed curves in Fig. 5.4.

The graph shows that the Na/CI ratio of the spring waters is well correlated with the structural distinctions between the superficial and basal aquifers. In fact, it increases from the superficial toward the basal aquifers. Variations range from low values such as the Na/Cl ratios typical of rainwater in the superficial aquifers, to the high ratios representative of basal aquifers (cf. rains and boreholes dashed curve Cl = f(Na) in Fig. 5.3). This process is independent of the total mineralization of the water. Join et al. (1997) suggest that in the superficial zone, aquifers are associated with an atmospheric source of chloride ions related to marine spray. During percolation, Na/Cl ratios increase progressively by the dissolving of sodium from volcanic rocks. The similarity Na/Cl of the deepest springs and coastal boreholes supports the hypothesis of inland continuity of the basal aquifer.

5.4 Hydrogeological Modelling of Piton de la Fournaise

Groundwater continuity between the coastal aquifer toward the inland deepest aquifers remains a debated question. In order to document this issue, a new hydrogeological model was proposed by using the 3D Femwater finite element code. This approach has been undertaken on the Piton de la Fournaise, considered as a perfect representative of young oceanic volcanoes

In order to describe the aquifer geometry, a 3D geological model of Piton de la Fournaise was built from geometrical hypotheses established from the former previous geophysical surveys (Courteaud et al. 1997; Descloitres et al. 1997; Lénat et al. 2000) as well as new geological investigations carried out on the lesser known sectors. More than a hundred EM soundings have provided reliable information about the geological structures, the existence of a substratum and the estimation of the water table's elevation (Folio et al. 2000; Join et al. 2005).

The model was built to represent the general structure of the volcano as defined by four main volcanic units from the oldest stage (Unit I) to historic and present activity (Unit IV) used as hydrogeological layers.

The boundary conditions compute the average annual groundwater recharge spatially distributed into homogeneous recharge zones, and constant head boundary conditions corresponding to both the sea level and the level of permanent rivers, which define the borders of the modelled area.

Hydraulic conductivities are distributed based on field survey results including the investigation of the oldest volcanic units of Piton de la Fournaise. They were carried out on outcrops in the deepest valleys of Piton de la Fournaise. The main results are provided in Table 5.2.

The results of numerical simulations of groundwater flow have provided hydraulic heads that reach 1800 m a.s.l in the vicinity of the volcano summit (Fig. 5.5). The simulated groundwater outflows are consistent with the discharge rates of springs and resurgences observed within the deepest valleys or along the shoreline.

These results have provided hydrodynamic evidence that supports the hypothesis of a hydraulic continuity from coastal areas to inland areas characterized by high water table. It improves the conceptual model previously proposed on the massif of Piton des Neiges. These results support the presence of a central groundwater dome, similar to those observed in the Canary or Azores islands, where volcanic terrains are older. The shape of the piezometric surface suggests that drilling infiltration galleries may be a relevant technique to access inland groundwater (Pennober et al. 2004). This have been recently validated by deep underground works within the Piton des Neiges volcano where a 18 km long tunnel was drilled for surface water transfer and have been interrupted by groundwater discharge up to 1 m^3/s (Bret et al. 2000).

5.5 Synthesis

Twenty years of hydrogeological investigations in La Réunion Island including both Piton des Neiges and Piton de la Fournaise massif have provided a refined conceptual model of groundwater resources in active oceanic volcanoes. Results are supported by various approaches from field geological survey of springs to hydrochemical characterisation of groundwater and flow path modelling. This model reflects the main issues of groundwater resources management in La Réunion. We assume that it could be applicable to the island of Grande Comore as another typical young and active oceanic volcano. In both islands, the low groundwater hydraulic gradients in the littoral zone are accessible by drilling, but the proximity of the sea makes this resource highly vulnerable to saline contamination. In the higher-altitude areas,

Table 5.2 Evaluation of hydraulic conductivity anisotropic (Kx, Ky, Kz) and isotropic equivalent (k) in the oldest units of Piton de la Fournaise (from Folio 2001)

Parameters	Unit I	Unit II	Unit III and IV
Kx, Ky (m/s)	$5 \times 10^{-5} - 8 \times 10^{-5}$	$10^{-4} - 3.10^{-4}$	10^{-2} (average value from pumping test)
Kz (m/s)	$4 \times 10^{-7} - 4 \times 10^{-5}$	2.10^{-4}	
Isotropic K model values	10^{-6}	10^{-4}	10^{-2}

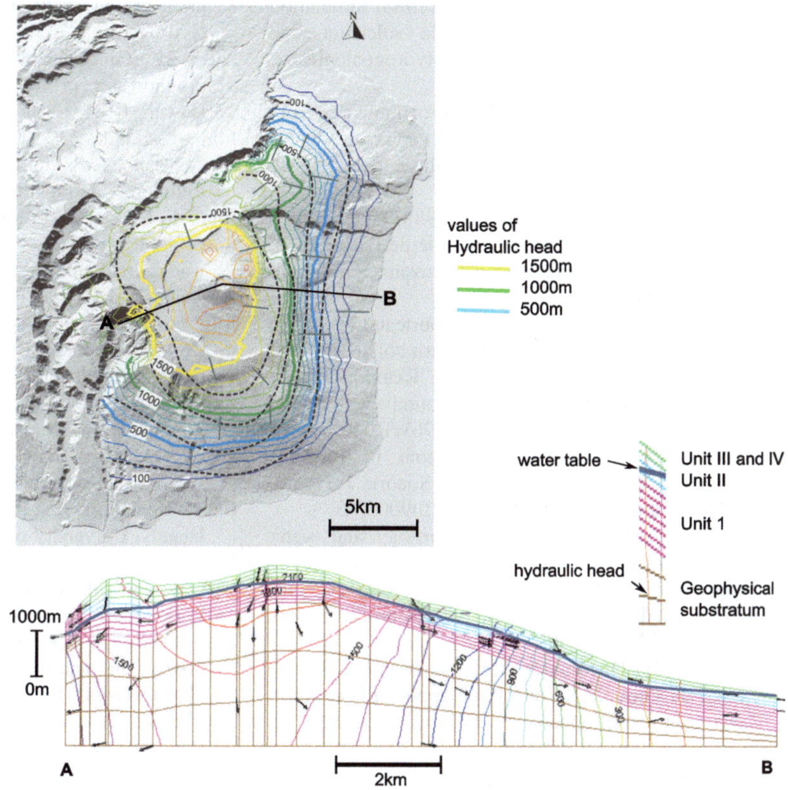

Fig. 5.5 Groundwater table in the Piton de la Fournaise volcano (modified from Folio 2000)

despite greater hydraulic gradients, the excessive depth of the water table leads to unacceptable operating costs. The implementation of horizontal infiltration galleries within elevations comprised between around 600 and 900 m constitutes an interesting technical alternative for exploiting the inland aquifer.

References

Barcelo A (1996) Analyse des mécanismes hydrologiques en domaine volcanique insulaire tropical à relief jeune. Apports à la connaissance du bilan hydrique. Massif du Piton de la Fournaise (île de la Réunion). Montpellier 2

Barcelo A, Coudray J (1996) Nouvelle carte des isohyètes annuelles et des maxima pluviométriques sur le massif du Piton de la Fournaise (Ile de la Réunion). Rev Sci Eau 9:457. doi:10.7202/705262ar

Bret L, Join J-L, Coudray J (2000) Investigation on deep water resources by tunnels within the Piton des Neiges volcano, Réunion Island. In: Proceeding XXX IAH congress on groundwater, p 4

BRGM (1986) Atlas hydrogéologique de La Réunion. BRGM, Réunion Island

Coudray J, Mairine P, Nicolini E, Clerc JM (1990) Approche hydrogéologique. In: Lénat JF (ed) Le volcanisme de l'île de la Réunion. Centre de Recherches Volcanologiques, Clermont-Ferrand

Courteaud M, Ritz M, Robineau B et al (1997) New geological and hydrogeological implications of the resistivity distribution inferred from audiomagnetotellurics over La Fournaise young shield volcano (Réunion Island). J Hydrol 203:93–100

Cruz VJ, Silva OM (2001) Hydrogeologic framework of Pico Island, Azores, Portugal. Hydrogeol J 9:177–189. doi:10.1007/s100400000106

Custodio E (1974) Datos sobre la Hidráulica de las galerías de captación de agua subterránea en el Macizo de Famara, Lanzarote (Islas Canarias, España), pp 235–284

Custodio E (1978) Geohidrología de terrenos e islas volcánicas. Centro de estudios hidrograficos, Universidad politécnica de Barcelona

Custodio E (1989) Strict aquifer control rules versus unrestricted groundwater exploitation: comments on economic consequences. Dev Water Sci 39:381–395

Custodio E, Saenz de Oiza J (1973) Hydrology of Famara's volcanic block supply galleries, Lanzarote Island, Spain. Conv. Int. Acque Sotterranee, Atti [serial online] 2:487–494

Custodio E, Lopez Garcia L, Amigo E (1988) Simulation par modèle mathématique de l'Île volcanique de Ténériffe : Canaries, Espagne. Hydrogéologie 2: 153–167

Daesslé M, Duchamps JM (1989) Hydrochimie des eaux souterraines de l'île de La Réunion : eaux de sources et eaux de forages. Bilan et synthèse début 1989. BRGM

Daesslé M, Join J-L, Duchamps JM (1988) Hydrochimie des émergences et des nappes de l'île de La Réunion. Compte-rendu d'avancement des travaux. Decembre 1988. BRGM

Descloitres M, Ritz M, Robineau B, Courteaud M (1997) Electrical structure beneath the eastern collapsed flank of Piton de la Fournaise volcano, Réunion Island: Implications for the quest for groundwater. Water Resour Res 33:13–19. doi:10.1029/96WR02673

Ecker A (1976) Groundwater behaviour in Tenerife, Volcanic Island (Canary Islands, Spain). J Hydrol 28:73–86. doi:10.1016/0022-1694(76)90053-6

Falkland A, Custodio E (1991) Hydrology and water resources of small Islands: a practical guide. In: A contribution to the international hydrological programme, IHP-III, Project 4.6. [s.l.]: Unesco, p 460. ISBN: 9789231027536

Folio J-L (2001) Distribution de la perméabilité dans le massif du Piton de la Fournaise : apport à la connaissance du fonctionnement hydrogéologique d'un volcan-bouclier. La Réunion

Folio J-L, Join J-L, Robineau B et al (2000) Combined electromagnetic prospecting and ground water modelling to study shield volcano hydrogeology: Piton de La Fournaise volcano case study (Réunion Island). In: Sililo et al (eds) Ground water past achiev. Future Chall., Balkema, Rotterdam, pp 385–389

Grünberger O (1989) Etude géochimique et isotopique de l'infiltration sous climat tropical contrasté, massif du Piton des Neiges, Ile de la Réunion. Université Paris Sud - Paris XI

Izquierdo T (2014) Conceptual hydrogeological model and aquifer system classification of a small Volcanic Island (La Gomera; Canary Islands). CATENA 114:119–128. doi:10.1016/j.catena.2013.11.006

Izuka SK, Gingerich SB (1998) Estimation of the depth to the fresh-water/salt-water interface from vertical head gradients in wells in coastal and island aquifers. Hydrogeol J 6:365–373

Izuka SK, Gingerich SB (2003) A thick lens of fresh groundwater in the southern Lihue Basin, Kauai, Hawaii, USA. Hydrogeol J 11:240–248

Join J-L, Coudray J (1993) Caractérisation géostructurale des nappes d'altitude en milieu insulaire, Ile de la Réunion. Géodinamica Acta 243–254

Join J-L, Coudray J, Longworth K (1997) Using principal components analysis and Na/Cl ratios to trace groundwater circulation in a Volcanic Island: the example of Réunion. J Hydrol 190:1–18

Join J-L, Robineau B, Courteaud M, Ritz M (2000) CSAMT mapping of a deep hydrogeological structure in Mauritius Island. Proc. IAH Congr Groundw 30

Join J-L, Folio J-L, Robineau B (2005) Aquifers and groundwater within active shield volcanoes. Evolution of conceptual models in the Piton de la Fournaise volcano. J Volcanol Geotherm Res 147:187–201

Lau L-KS, Mink JF (2006) Hydrology of the Hawaiian Islands. University of Hawaii Press, Hawaii

Lénat JF, Fitterman D, Jackson DB, Labazuy P (2000) Geoelectrical structure of the central zone of Piton de la Fournaise volcano (Réunion). Bull Volc 62:75–89

Macdonald GA, Abbott AT, Peterson FL (1983) Volcanoes in the Sea: the geology of Hawaii. University of Hawaii Press, Hawaii

Nicolini E, Jusserand C, Coudray J (1988) Approche de l'hydrogéologie du massif du Piton de La Fournaise, chimie et isotopes des eaux superficielles et souterraines. Bull Soc Geol Fr 6–7

Pennober G, Odon O, Join J-L, Folio J-L (2004) Approche par analyse spatiale de la faisabilité de captage d'eau souterraine sur le Massif du Piton de la Fournaise (île de La Réunion - France - Océan Indien). Cybergeo Eur J Geogr. doi:10.4000/cybergeo.3298

Peterson FL (1972) Water development on tropic Volcanic Islands—type example: Hawaiia. Ground Water 10:18–23. doi:10.1111/j.1745-6584.1972.tb03586.x

Savin C, Ritz M, Join JL, Bachelery P (2001) Hydrothermal system mapped by CSAMT on Karthala volcano, Grande Comore Island, Indian Ocean. J Appl Geophys 48:143–152. doi:10.1016/S0926-9851(01)00078-7

Erosion and Volcaniclastic Sedimentation at Piton de la Fournaise: From Source to Deep Marine Environment

6

Nathalie Babonneau, Nicolas Villeneuve, Aude Mazuel and Patrick Bachèlery

Abstract

With very high rainfall, steep slopes and young reliefs, La Réunion Island is renowned for its highest erosion rates worldwide. The erosion mainly occurs by rockfalls and landslides in the steep sides of the river canyons. Most sediment is transported from inland to the sea via rivers and streams. River channels are formed by incision into the basaltic pile of lava and breccia constituting the original volcanic shield. Erosion and sediment products in the largest rivers are mainly driven by the recurrence of tropical cyclones. Cyclonic conditions induce heavy rainfalls and torrential floods, causing land erosion and hyperconcentrated sediment loads in the main river mouths. The absence of coastal platform induces no sediment storage in shallow marine environment. Erosion processes onland and sediment transport to the coast, via river systems, induce direct sediment feeding of the upper submarine slope and canyon heads. The morphology of the submarine flanks of Piton de la Fournaise is rough and steep. Different submarine features were identified as debris avalanches deposits, erosive canyons and volcanic constructions. This chapter proposes a synthesis of morphological and sedimentological data on

N. Babonneau (✉) · A. Mazuel
UMR CNRS 6538 Domaines Océaniques, Université de Brest, UBO, Institut Universitaire Européen de la Mer, Place Copernic, 29280 Plouzané, France
e-mail: nathalie.babonneau@univ-brest.fr

N. Villeneuve
Observatoire Volcanologique du Piton de la Fournaise (OVPF), Institut de Physique du Globe de Paris, Sorbonne Paris-Cité, UMR 7154 CNRS, Université Paris Diderot, 75238 Paris, France

P. Bachèlery
Laboratoire Magmas et Volcans, UMR CNRS-IRD 6524, Observatoire de Physique du Globe de Clermont-Ferrand, Université Blaise Pascal, 5, Rue Kessler, 63038 Clermont-Ferrand, France

P. Bachèlery et al. (eds.), *Active Volcanoes of the Southwest Indian Ocean*, Active Volcanoes of the World, DOI 10.1007/978-3-642-31395-0_6

Piton de la Fournaise volcano, onland and offshore, at different scales. It describes the relationships between the main erosional structures onland and major submarine sedimentary systems. Combining onland erosion, sediment transport on the submarine slopes and sediment deposition in the deep-sea allow a better understanding of the transfer of volcaniclastic material on oceanic shield-volcanoes. The morphology of onland and submarine slopes of Piton de la Fournaise suggests two domains, distinguished by the maturity of land-to-sea sediment transfers, directly related to the maturity of hydrological networks onland. The first domain corresponds to limited land-to-sea transfer, with no hydrographic network, as observed on the volcanic rift-zones. The coastal area forms a rocky promontory and only coastal erosion of recent fragmented volcanic rocks, produces sediments in the upper submarine slope. Submarine instabilities, scarps, and local erosional canyons are superficial and discontinuous on the slope. The second domain corresponds to high-efficient land-to-sea sediment transfer, characterised by a direct connection between river mouths and submarine canyon heads (as for "Rivière des Remparts"). The submarine sedimentary systems are well-developed with continuous canyons, feeding deep-sea fans at the base of the slope. These volcaniclastic deep-sea fans are characterised by sandy lobe accumulation fed by small channels and are typically interpreted as the result of turbidite processes and deposits.

Keywords
Erosion · Gravity processes · Turbidites · Volcanic geomorphology · La Réunion Island

6.1 Introduction

Large oceanic shield-volcanoes as Hawai'i, La Réunion or Canary Islands are associated with high magma production rates and rapid growing. Although reaching high altitudes, these huge edifices are largely submarine, and their evolution through time leads to the production of large quantities of volcaniclastic material to the deep-sea. Numerous studies have evidenced that dismantling processes, as landslides, slumps, and submarine gravity flows, can transport sedimentary material over areas larger than the volcanic edifice itself, spreading over the abyssal plain (e.g. McGuire 1996).

The major role of flank landslides in the evolution of Hawaiian Islands or La Réunion Island has long been recognized (Moore 1964; Lipman et al. 1988; Moore et al. 1989; Lénat et al. 1989; Cochonat et al. 1990; Bachèlery et Villeneuve 2013; Lénat 2016b, Chap. 3). Further studies on other oceanic islands confirmed that most shield-volcanoes experience considerable modifications of their shape due to large-scale gravitational displacement of the island's flanks, mainly by massive slumps and huge landslides in debris avalanches (Chadwick and Howard 1991; Holcomb and Searle 1991; Moore et al. 1994; Filmer et al. 1994; Bachèlery et al. 1996; Carracedo 1999; Naumann and Geist 2000; Clouard et al. 2001; Krastel et al. 2001; Mitchell et al. 2002, 2003; Masson et al. 2002; Clément et al. 2003; Hildenbrand et al. 2004; Oehler et al. 2008; Le Friant et al. 2011).

If large-scale catastrophic slope-failure and slumps can be seen as the dominant edifice-modifying process, volcaniclastic sedimentation is mainly related to other processes such as submarine eruptions, brecciation of lava flows entering into the ocean, hyaloclastites formation, and mobilization of material by sediment gravity flows. Sub-aerial volcanism may also supply volcaniclastic material through fallout of tephras from explosive eruptions, pyroclastic density currents and lahars reaching the sea, and, may be the most prevalent, the fluvial supply of eroded sub-aerial volcanic materials.

Volcanic Islands, affected by a wet and hot tropical climate, are the site of significant erosion, combining the effects of intense chemical weathering and strong mechanical denudation. In addition to climate, factors as topography, tectonic or lithology are identified as key parameters, but their relative weight is generally poorly constrained. With very high rainfall, steep slopes and young reliefs, La Réunion Island is renowned for its highest values of chemical and mechanical erosion rates worldwide (Louvat and Allègre 1997; Rad et al. 2007). The mechanical erosion mainly occurs by rockfalls and landslides in the steep sides of the river canyons. Most sediment is transported from inland to the sea via rivers and streams (locally called "ravines"). River channels are formed by incision into the basaltic pile of lava and breccia constituting the original volcanic shield. They result from (1) differential resistance of materials, (2) regressive erosion mechanisms and (3) hydrogeological erosion mechanisms (Fèvre 2005).

The aim of this chapter is to describe the strong relationships established between the main erosional structures onland and major submarine sedimentary systems. It proposes a synthesis of morphological and sedimentological data on Piton de la Fournaise volcano, onland and offshore, at different scales. Combining onland erosion, sediment transport on the submarine slopes and sediment deposition in the deep-sea allow to fully understand the transfer of volcaniclastic material on shield-volcanoes.

6.2 Geological and Climatic Settings

6.2.1 Geological Settings of La Réunion Island

The geodynamical setting in which the island of La Réunion has grown, is already widely described in the Chap. 2 of this volume (Lénat 2016a). The emerged part of La Réunion Island only corresponds to 4 % of the entire volcanic edifice, which is characterized by a diameter of about 240 km at its base and rises up to more than 7000 m above the abyssal plain (Fig. 6.1). The sub-aerial morphology of the island is dominated by two large juxtaposed basaltic shield-volcanoes: Piton des Neiges and Piton de la Fournaise.

The Piton des Neiges (3070 m above sea level) occupies the north-western part of the island. Its activity started more than 2.4 My ago (McDougall 1971; Quidelleur et al. 2010; Smietana 2011) and stopped about 22 ka (Delibrias et al. 1986; Salvany et al. 2012). The inner part of the volcano is deeply incised by valleys and three major depressions called the "cirques" of Mafate, Salazie and Cilaos, formed by both volcano-tectonic processes and intense tropical erosion (Arnaud 2005; Oehler et al. 2004; Salvany et al. 2012).

The active Piton de la Fournaise is located in the south-eastern part of the island. Its morphology is characterized by concentric caldera rims open eastward to the sea and three deep valleys cutting its flanks. The volcano mostly produces basaltic lava flows and fountains, with sometimes phreatic and phreatomagmatic explosions. Its activity is considered to have started about 450 ka on the relics of an older volcano named "Les Alizés" (Gillot and Nativel 1989; Merle et al. 2010; Lénat et al. 2012; Michon et al. 2016, Chap. 4). Since its initiation, the Piton de la Fournaise edifice underwent successive stages of building alternating with caldera collapses and intense erosion stages (Mairine and Bachèlery 1997; Merle et al. 2010). The presence of widespread debris avalanches deposits on the

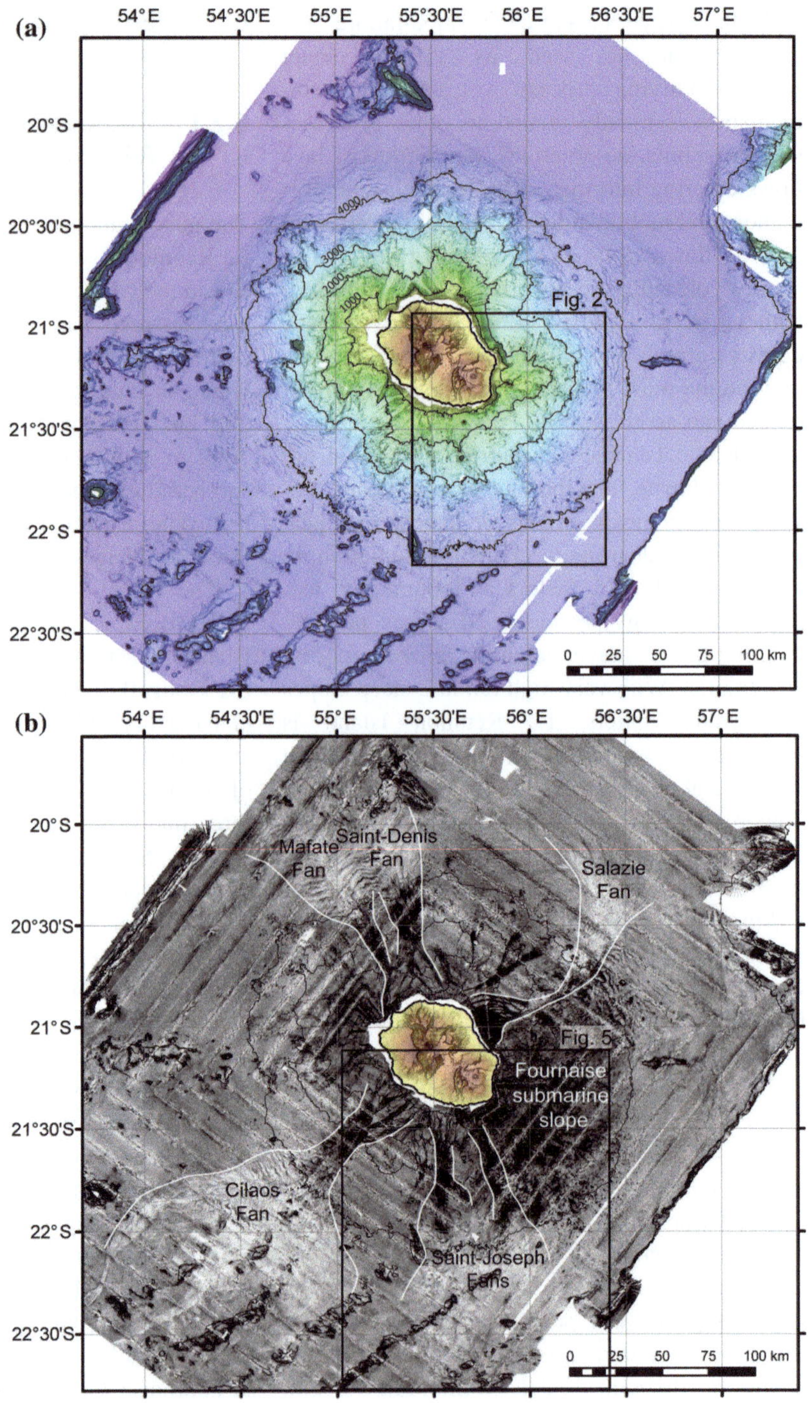

Fig. 6.1 **a** Topographic and bathymetric map of La Réunion Island (ERODER and FOREVER data). **b** Reflectivity map around La Réunion Island (ERODER and FOREVER data), modified after Sisavath et al. (2011)

submarine flank attested that major flank landslides occurred during the recent evolution of the edifice but the ages of these landslides remains poorly constrained (Bachèlery et al. 1996; Oelher et al. 2008; Lénat et al. 2009; Le Friant et al. 2011).

6.2.2 Climatic Settings of La Réunion Island and Erosion Conditions

La Réunion Island is characterized by a sub-tropical climate characterised by the alternation of two seasons: a hot and wet austral summer and a dry and cold austral winter. Trade winds from the East induce variable precipitation regimes in time and space; the windward (eastern) side being dominantly wet (more than 5000 mm/year of rain) while dryer conditions are prevailing on the leeward (western) side (Robert 2001). Erosion and sediment products in the largest rivers are mainly driven by the recurrence of tropical cyclones.

Cyclonic conditions induce heavy rainfalls and torrential floods, causing land erosion and hyperconcentrated sediment loads in the main river mouths (Bret et al. 2003; Fèvre 2005; Garcin et al. 2005). The rainfalls and the basaltic nature of the island induce high erosion rates, in the same order as active orogenic areas (Gaillardet et al. 1997). They are estimated about 0.47–3.34 m ka^{-1} (Louvat and Allègre 1997; Rad et al. 2007). The sediments are transported seaward by hyperconcentrated flows throughout a dense hydrographic network, which is constituted by more than 750 temporal gullies and rivers (only a twenty of them are perennial).

6.3 The Submarine Flanks of La Réunion Volcanic Edifice

The MD-32 (1982), FOURNAISE 1 (1984) and FOURNAISE 2 (1988) surveys around La Réunion Island first revealed the morphology of the submarine flanks of the volcanic edifice (Averous 1983; Cochonat et al. 1990; Lénat et al. 1989). The chaotic morphology of most surfaces is an obvious feature of the submarine slopes. Thus, many studies were focused on mass-wasting processes like fast-running debris avalanches and submarine slope instabilities (Lénat and Labazuy 1990; Lénat et al. 2009, 2001; Oehler et al. 2004, 2008). Oehler et al. (2008) listed 47 collapse events grouped into four main bulges around the island, and separated by large submarine canyons. Oceanographic cruises FOREVER and ERODER (from 2006 to 2011) offer a new dataset to map the superficial structure of the submarine slopes and the surrounding abyssal plain. In the light of these data, Le Friant et al. (2011) confirmed that the submarine flanks of Piton de la Fournaise volcano are characterised by debris avalanche deposits extending on the abyssal plain. They were induced by catastrophic large flank-collapse events. But, they proposed an alternative interpretation for the submarine flanks of the Piton des Neiges, suggesting they are mainly affected by slow deformation and spreading, leading to secondary submarine slope instabilities (Le Friant et al. 2011).

The bathymetric map acquired during FOREVER and ERODER cruises showed the detailed morphology of canyon networks incising the submarine flanks (Fig. 6.1a). The largest canyon systems are directly linked to the sub-aerial hydrographical network and more particularly to rivers draining the "cirques" (Saint-Ange 2009). They fed six large volcaniclastic deep-sea fans, which were highlighted by the analysis of backscatter data (Fig. 6.1b). They are named Mafate and Saint-Denis fans at North of the island, Salazie fan at the North-East, Saint-Joseph and Saint-Pierre fans at the South and Cilaos fan at the South-West. These deep-sea fans are characterised by low reflectivity acoustic facies of the sea floor, extending over hundred kilometres on the abyssal plain (Saint-Ange et al. 2011; Sisavath et al. 2011). Sedimentological studies, focused on piston cores collected at the edge of the Cilaos fan, showed that the Cilaos fan is composed of successions of sandy accumulations corresponding to recent turbidite deposits (Saint-Ange et al. 2011; Sisavath et al. 2011). The canyon is directly connected to a torrential river ("Rivière Saint-Etienne"), that floods during tropical cyclone, with limited sediment storage at the coast (Babonneau et al. 2013; Saint-Ange et al. 2011). This configuration suggests the continuity of hyperconcentrated fluvial flows to hyperpycnal flows in the submarine canyon (Babonneau et al. 2013). These submarine gravity currents, similar to turbidity currents, flow down to the base of the slope feeding the fan with volcaniclastic sandy sediments. In that case, the sediment transfer from land (erosion in the drainage area) to deep sea (turbidite accumulation) is direct (Babonneau et al. 2013). Similar morphology and processes were observed in the other deep-sea fans around La Réunion Island (Mazuel 2014).

6.4 Onland Morphological Features of Piton de la Fournaise

The Piton de la Fournaise volcano is composed of several different geomorphological landscapes (Fig. 6.2 and Chap. 4 of this volume).

The "Enclos Fouqué - Grand Brûlé" structure is the large U-shaped depression, open to sea on the eastern part of the island, where the modern volcanic activity is concentrated. This area is bounded by two short and diffuse rift-zones, trending in NE and SE directions, which can be prolonged toward the submarine domain. The rift-zones are broad convex surfaces extending from 2000 m in altitude down to 1000 m below sea level.

The "Enclos Fouqué" is the upper part of this large depression. It includes the present summit of the Piton de la Fournaise volcano (central cone in Fig. 6.2). The "Grand Brûlé" corresponds to the lower part of the structure including the slope domain extending to the coast. The transition between "Enclos Fouqué" area and "Grand Brûlé" area is characterised by steep slope, named "Grandes Pentes", and by a convex structure at the North, named "Plaine des Osmondes". The "Plaine des Osmondes" is now considered as a partly filled paleo-valley from arguments such as the convex shape of the scarp, but also by the presence of thick levels of alluvial deposits drilled by boreholes located southeast of the scarp near the seashore (Kieffer 1990; Courteaud 1996).

The whole "Enclos Fouqué - Grand Brûlé" depression might result from combined processes of caldera collapses and flank destabilisation (Bachèlery and Michon 2010). Actually, both the processes involved in the "Enclos Fouqué - Grand Brûlé" formation and the age of this structure are badly constrained. A close relation was proposed between the phreatic to phreatomagmatic "Bellecombe Ash Unit" deposit and the formation of the Enclos Fouqué caldera (Bachèlery 1981; Mohamed-Abchir 1996). If such a link exists, the recent revisions of the interpretation and age of deposition of the "Bellecombe Ash Unit" (Morandi et al. 2016, Chap. 8) allow to place the most recent phases of collapse of the "Enclos Fouqué" caldera between 4745 and 2140 years BP. However, this does not provide any relevant information about the origin and age of the "Grand Brûlé" depression, its submarine extensions and the "Plaine des Osmodes" paleo-valley, or even relationships that may exist between these structures and the "Enclos Fouqué" caldera.

The areas located on the outskirts of the rift zones are characterised by deep river incisions (Fig. 6.2). In the South, the "Rivière des Remparts" and "Rivière Langevin" and, to a lesser extent, the "Ravine Basse Vallée" deeply incise the topography. In the North, the "Rivière de l'Est" is the only river incising the volcano.

The "Rivière des Remparts" is the largest river with a drainage area about 65 km^2, 24 km long and a vertical incision locally reaching 1200 m (Villeneuve and Bachèlery 2014) (Fig. 6.3).

Mairine and Bachèlery (1997), Merle et al. (2008, 2010) clearly show thanks to the study of lahars interbedded between lava filling and thanks to outcrops datation that the geological histories of "Rivière des Remparts" and "Rivière Langevin" are closely related. It reminds us that growth and destruction phases of volcanoes are, at geologic time scale, near to be simultaneous (Thouret 1999). Merle et al. (2010) show that the evolution of the upper part of the two rivers is related to (1) the structural evolution of the Piton

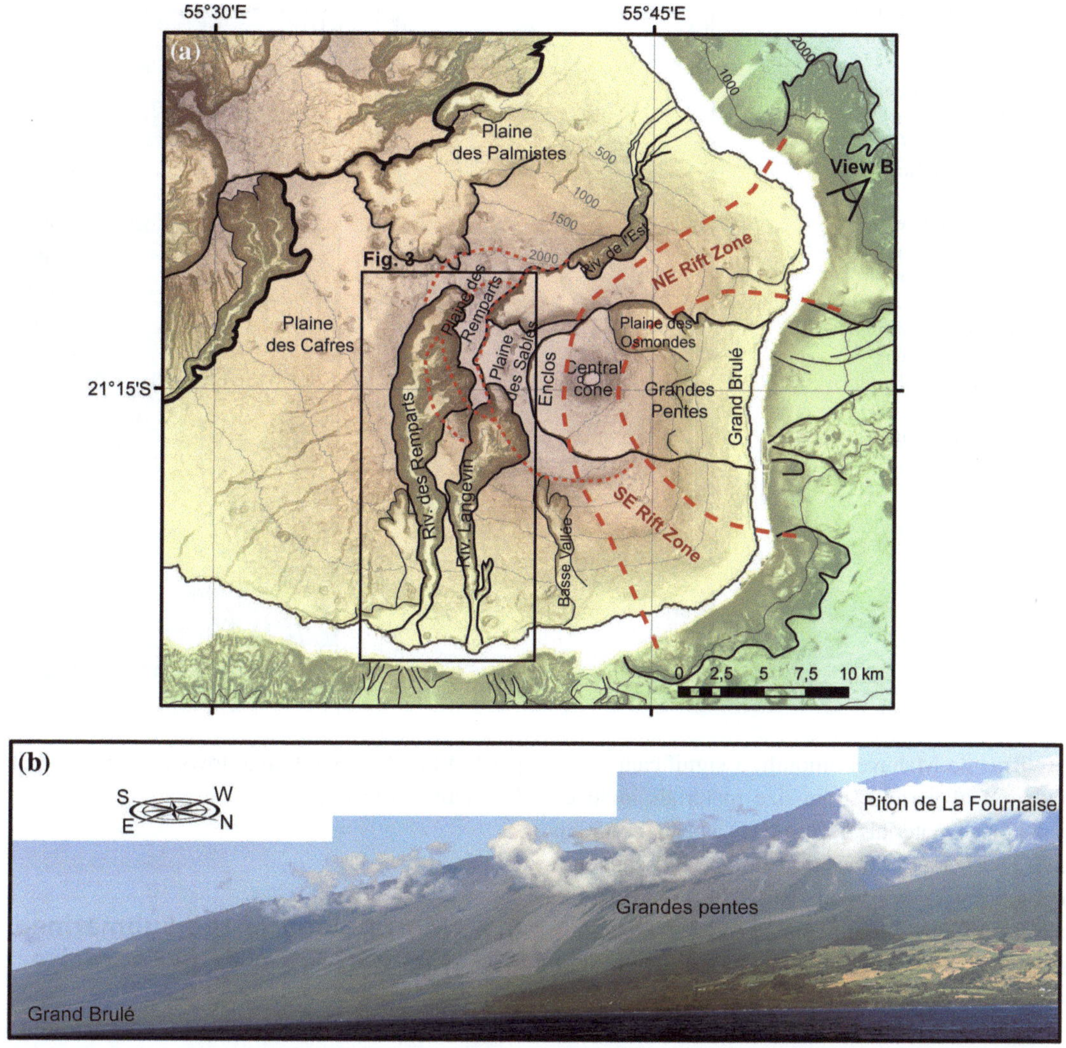

Fig. 6.2 **a** Topographic data of La Réunion Island (resolution of the grid at 25 m) and bathymetric data of the coastal area (ERODER and FOREVER multibeam bathymetric data, 2006), and interpretation of main morphological structures in *black lines* and geological structures in *red* (after Merle et al. 2010). Detailed relief is underlined by slope gradient (high slope values in *dark*). Topographic contours every 500 m and isobaths every 1000 m. **b** Mosaic of "Grand Brulé" area

de la Fournaise and (2) the infill with new thick lava flows. The succession of these episodes occurred several times between 530 and 40 ka BP (Merle et al. 2010), explaining horizontal thick lava piles, up to 1000 m thick, like the Morne Langevin unit (Fig. 6.3).

The upstream part of the "Rivière des Remparts", as below the "Morne Langevin", a dip discordance between two volcanic series (Mahavel and Morne Langevin series) is responsible of a significant gravity instability on the right side of the "Rivière des Remparts" in an affluent head valley called the "Bras de Mahavel". Gravity instabilities are responsible of many major landslides in the riverbed.

The upstream riverbed is composed by recent lava flows erupted during the "Cratère Commerson" eruption, estimated at 1890 ± 75 years BP by radiocarbon dating (Tanguy et al. 2011). In this section, the river is a typical torrential bed (Fig. 6.4).

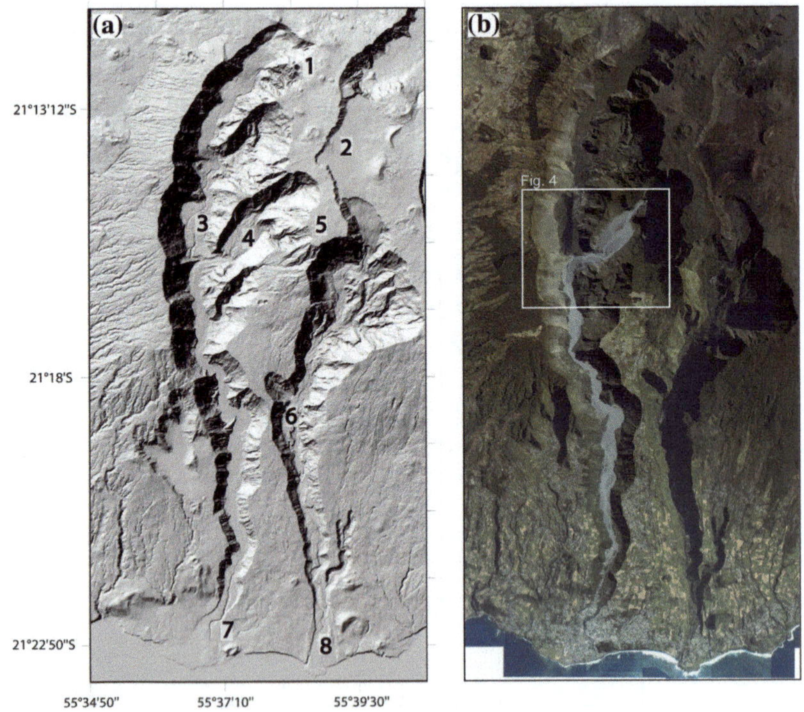

Fig. 6.3 **a** Extraction of 5 m DEM processed in 2008 by Institut Géographique National (*1* Cratère Commerson, *2* Plaine des Sables, *3* Rivière des Remparts, *4* Bras de Mahavel, *5* Morne Langevin, *6* Rivière Langevin, *7* City of Saint Joseph, *8* Village of Langevin). **b** 2012 aerial photography of Rivière des Remparts and Rivière Langevin (IGN)—the *black box* corresponds to the location of Fig. 6.4a

At "Bras de Mahavel" mouth, a significant accumulation of various grain-size materials includes very coarse materials, forming (1) a barrier of 50 m high and (2) a net transition in the landscape and topography. Downstream "Bras de Mahavel", the bed is only made of gravels, masking the bedrock associated to "Cratère Commerson" event. The topography is a gentle and regular slope. The river runs under the gravel. The resurgence is finally visible 2 km from the coast.

The most external domain of Piton de la Fournaise is "Les Plaines", mainly including the "Plaine des Cafres" and the "Plaine des Palmistes", in the saddle between Piton des Neiges and Piton de la Fournaise massifs. The surface of the "Plaine des Cafres" is a relatively smooth south-eastward slope with narrow and superficial gully incisions. The "Plaine des Palmistes" is a large depression bounded by very steep slopes and open to the sea toward the North-East. The "Plaine des Palmistes" probably corresponds to an ancient erosional depression dug during the last 400 ka. "Plaine des Palmistes" shows a smooth morphology in the inner part because it is partly filled by recent lava flows from Piton de la Fournaise volcano.

6.5 Morphology of the Submarine Flank of Piton de la Fournaise

The morphology of the submarine flanks of La Réunion Island is irregular, rough and steep. Different submarine features were identified: coastal shelf, debris avalanches and sedimentary deposits, erosive canyons, volcanic constructions and seamounts. The submarine flanks of Piton de la Fournaise are divided in several morphological areas (Fig. 6.5): (1) the southern flank dominated by canyon incisions connected to the coast (including the "Canyons de Saint-Pierre" and "Canyon de Saint-Joseph", (2) the SE rift-zone submarine spur and its lower slope, (3) the eastern flank dominated by a E-W oriented bulge called "Plateau Est" and (4) the NE rift-zone submarine spur and the Salazie canyon complex at North.

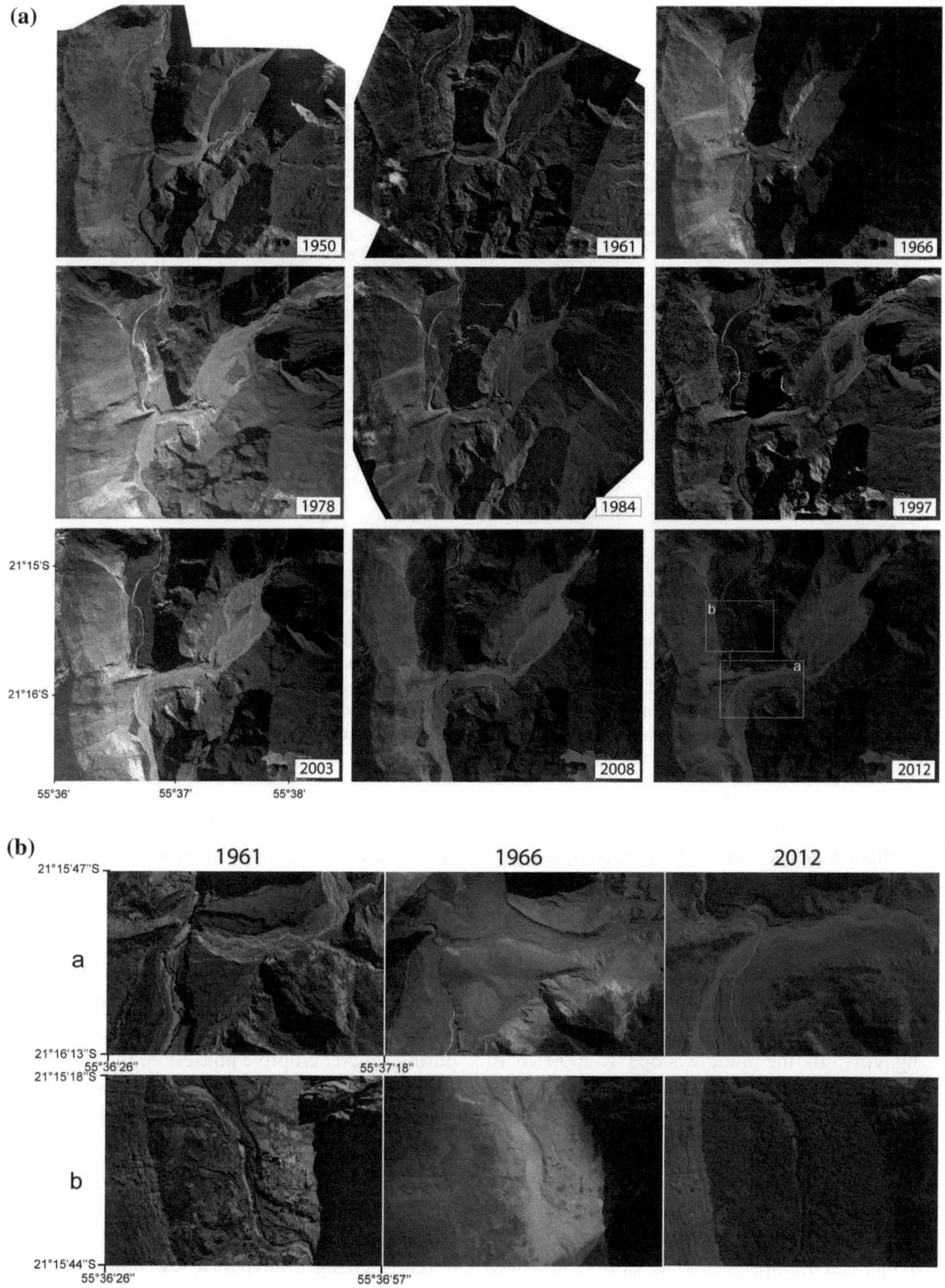

Fig. 6.4 **a** Time series of aerial photographs focused on the "Bras de Mahavel" between 1950 and 2012, see location in Fig. 6.3b. **b** Zooms of two sectors in 1961, 1966 and 2012

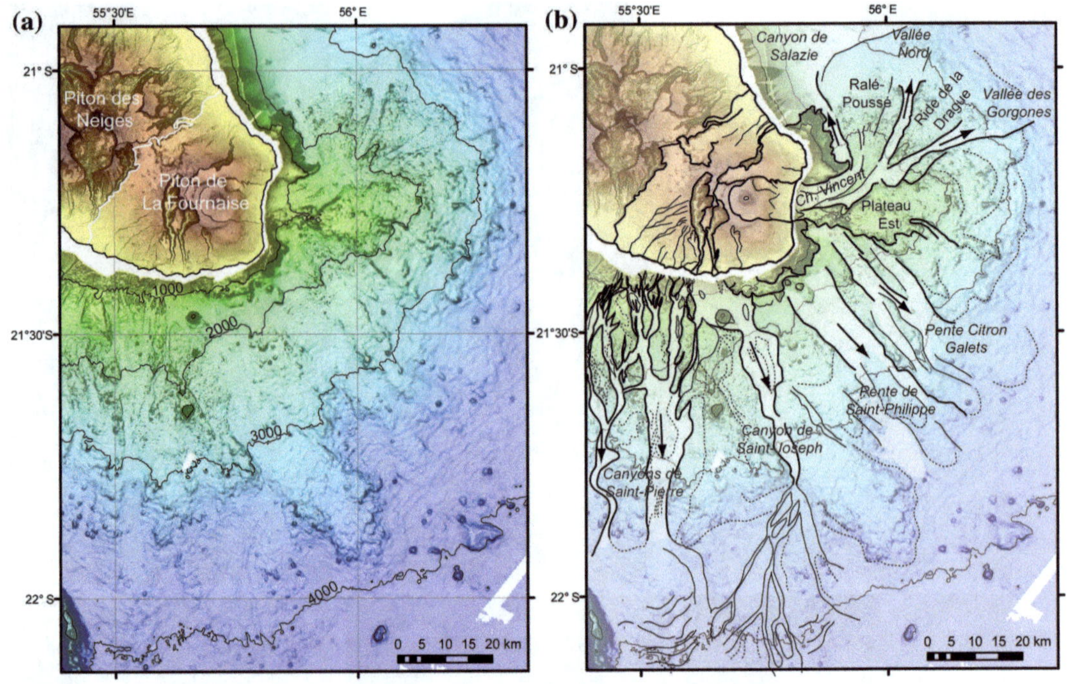

Fig. 6.5 **a** Bathymetric map of the southeastern submarine flank of La Réunion Island (ERODER and FOREVER multibeam bathymetric data, 2006), and topographic data of La Réunion Island. Detailed relief is underlined by slope gradient (high slope values in *dark*). Isobaths every 1000 m. **b** Morphological interpretation of the main sedimentary and volcanic structures of the submarine flank of the Piton de La Fournaise, modified after Saint-Ange et al. (2013) for the eastern submarine flank and modified after Mazuel (2014) for the southern submarine flank. Submarine canyons and valleys are underlined by *light shades*

The southern flank is characterised by a rough and chaotic surface incised by a complex network of canyons (Fig. 6.5). Most of canyon heads are connected to the coast in the prolongation of river and ravine mouths. The largest canyon head, characterised by a wide smooth incision, is located in front of the two main rivers: the "Rivière des Remparts" and "Rivière Langevin". In the middle slope (>2000 m of water depth), the canyon networks are divided into three main canyons: the two "Saint-Pierre" canyons in the western part and the "Saint-Joseph" canyon in the eastern part. Minor ravines incising the western side of the Piton de la Fournaise volcano and the western part of "Rivière des Remparts" mouth dominantly feed the "Saint-Pierre" canyons. "Rivière Langevin", "Basse Vallée" and the eastern part of "Rivière des Remparts" feed the "Saint-Joseph" canyon. The two canyons networks are separated by a higher submarine area, where several huge volcanic edifices have been identified between 1500 and 2500 m of water depth. Among these volcanic edifices, the most prominent is the "Cone Eliane", about 700 m high, with a summit reaching the water depth of 800 m and a base diameter of about 3000 m. The "Cone Eliane" was probably built by the accumulation of lava during several eruptive phases. At the base of the slope, the canyons are prolonged by small-ramified channels feeding depositional areas.

The SE rift-zone submarine spur (Figs. 6.2 and 6.5) is the extension of the onland volcanic building of the SE rift-zone but its morphology strongly changes beyond a few hundred meters deep. It is composed of at least three morphological spurs extending to 5 km from the coast and forming abrupt rocky features down to 1200 m of water depth. Dredging carried out on the slope of these reliefs has allowed to bring a wide variety of aerial origin samples (presence of oxidized scoria). The lower part of the submarine slope is irregular, with wide valley-features separated by rough bulges composed of large blocks ("Pente de Saint-Philippe" in Fig. 6.5). The widest valley-feature (10 km wide) begins at the base of rift-zone spur and becomes diffuse under 3000 m of water depth.

In the eastern submarine flank of Piton de la Fournaise, debris avalanche deposits are clearly identified in bathymetric data (Oehler et al. 2008). They form an E-W oriented bulge ("Plateau Est" in Fig. 6.5), characterised by a rough and chaotic surface composed of blocks with size from tens to hundreds of meters (Figs. 6.5 and 6.6). The surface of the debris avalanche deposits is incised by wide and elongated depressions with abrupt walls forming submarine valleys (Oehler et al. 2008; Saint-Ange et al. 2013). They initiate at the different water depths between the coast and 2000 m (light shades in Fig. 6.5b). These valleys are short, about 15–30 km long. Their width varies from 1 to 5 km. Among these valleys, only the "Chenal Vincent" is connected to the coast. The other valleys, starting deeper than 1000 m of water depth, are interpreted as the scars of submarine secondary instabilities (Oelher et al. 2008), as the "Pente Citron Galets" in the southern part and the "Vallée Nord" and "Vallée des Gorgones" in the northern part (Fig. 6.5b). The "Chenal Vincent" has morphological characteristics similar to canyons directly connected to hydrographical network onland. The channel floor is smooth and regular (Oehler et al. 2008; Saint-Ange et al. 2013). It turns to north getting around the toe of the NE rift-zone spur and joins the southern part of the Salazie canyon complex, which constitute the largest and the most mature canyon complex around the island (Fig. 6.1b).

The morphological analysis of the submarine prolongation of the NE rift zone shows that rugged eroded terrains constitute the lower flanks, from 1000 to 2000 m in depth, while the upper part is smoother and less eroded. K–Ar dating recently obtained (Smietana 2011) confirms that this morphological difference is determined by the relative age of the formations. While rocks older than 3 Ma are found in the basal area, the upper formations are younger than 0.45 Ma.

North of the NE rift-zone, the upper submarine slope is smooth with very steep slopes along the coast (Fig. 6.5). This smooth morphology corresponds to the wide canyon complex of Salazie. The canyon heads are continuous along 30 km of the eastern coast of La Réunion Island (Fig. 6.1). Several river watersheds feed them. The largest river is the "Rivière du Mât" fed by the "cirque" of Salazie incising the Piton des Neiges volcano. Sediments coming from the Piton de la Fournaise volcano, mainly by the river draining the "Plaine des Palmistes" and the "Rivière de l'Est", also feed the southern part of the Salazie canyon. In the lower slope, the wide canyon network of Salazie focuses in one main valley feeding the Salazie deep-sea fan, extending at the base of the slope (Fig. 6.1).

6.6 Onland Sediment Production and Transport from the High Relief to the Coastal Areas

6.6.1 Lava Flows Fragmentation

The volcanic activity is characterised by frequent and small-volume eruptions (Peltier et al. 2008). Larger eruptions occur occasionally, reaching the coastal and submarine domains. The hot lava flow entrance into the sea causes instantaneous fragmentation. This thermal effect produces fine black sand to be transported by currents and/or deposited either on the coast or on the submarine slope.

If the lava flow flux is high, it induces the building of coastal volcanic platform (Saint-Ange et al. 2013). Erosion of these platforms by wave action produces coarse-grained sediments (from sand to blocks) feeding the

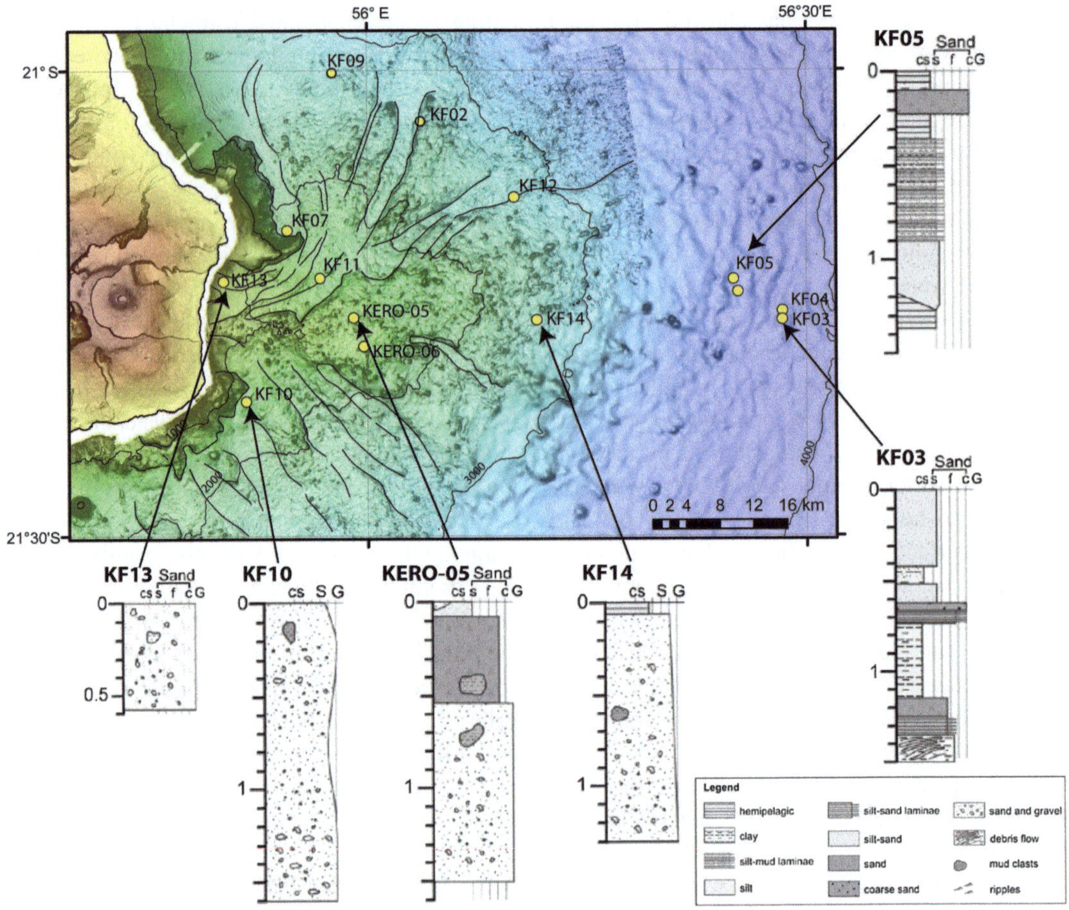

Fig. 6.6 Bathymetric map of the eastern submarine flank of La Réunion Island (ERODER and FOREVER multibeam bathymetric data, 2006), and topographic data of La Réunion Island. Location of piston cores collected during FOURNAISE 2 cruise and ERODER 1 cruise and sedimentological logs of selected cores showing the main volcaniclastic sedimentary facies on the eastern submarine flank of Piton de la Fournaise volcano (modified after Saint-Ange et al. 2013)

upper submarine slope as potential sediment source of turbidity currents (Ollier et al. 1998; Saint-Ange et al. 2013).

6.6.2 Landslides

For La Réunion rivers, considered as "young", major landslides feed solid transports. Transitional period as talus are short. For example, the Bras de Mahavel, affluent of "Rivière des Remparts" (Fig. 6.3), has many frequent landslides and produce most of the sediment load of the downstream river. Mairine and Bachèlery (1997), Garcin et al. (2005) and Kemp (2010) consider that the largest landslide never observed at Mahavel occured in may 1965 (Fig. 6.4). This event produced about 50 million cubic meters of rock, fell from the head into the valley below. The extent of the detached material measured nearly 5000 m long and 700 m wide. This 1965 landslide caused a natural dam at the confluence with "Rivière des Remparts", creating a lake with a water capacity of 14 million cubic meters.

Study of aerial photography (Fig. 6.4a, b) clearly shows three main points:

- The 50 m high barrier visible on the right bank at the outlet of the "Bras de Mahavel" predates the May 1965 event. This morphology is visible in the 1950 and 1961 aerial photography. As shows the 1966 zoom (Fig. 6.4ba), this topographic high was flooded by the products of the May 1965 landslide. The whole area of confluence between "Bras de Mahavel" and "Rivière des Remparts" is smoothed by these episode products. Between 1966 and 1978 erosion processes have (i) re-shape the river bed and (ii) changed the morphology of the confluence area.

- The lake sediment deposits are visible on more than one kilometre upstream. This area is currently completely occupied by an anthropic forest (Fig. 6.4bb).
- Dense vegetation is visible all along "Bras de Mahavel" in 1950 and 1961 aerial photography. After 1966, this "forest" is not visible apart for a small patch in 1997. It is no longer visible in 2003 aerial photography and never till reappear. We can concluded that (1) events that occurred before 1966 dates back several decades or more, (2) the destabilization that occurred in 1965 was followed by many significant events.

For instance, in 1996 and 2001, additional landslides in the "Bras de Mahavel" occurred. Landslide volumes were estimated respectively to 5 million and 1 million cubic meters (Kemp 2010).

Rain seasons, storms and cyclones play an important role in the triggering mechanisms of the landslide but also in the transport processes of materials available. 1965, 1996 and 2001 events was directly related to the cyclones Freda (01/1965), Bonita (01/1996) and Edwige (02/1996) and Ando (01/2001).

In term of transport, storm rainfalls allow the sediments remobilising. The Piton de la Fournaise edifice is particularly prone to heavy rains; the material is transported downstream, as hyper-concentrated flows on short distances to the sea. For example the 7th of January 1966, local weather monitoring network has measured 1825 mm of rain fell in 24 h for the Foc-Foc station located at less than 10 km of Mahavel deposits. This measure is still a world record for 24 h measurements.

6.6.3 Quantitative Approach of Erosional Processes

Erosion is highly active on La Réunion Island, so several authors have recently published articles focused on different quantitative approaches. Kemp (2010) thanks to the comparison of two 1 m resolution DEM processed during two stereophotogrammetry campaigns over la "Rivière des Remparts" (1st April 2005 and 21st January 2007) calculated an average erosion in the transfer compartment of 4217.9 t/ha/year. For the same area, Garcin et al. (2005) calculated a sediments supply of about 6400 t/ha/year for the 1997–2002 time interval. The significant value difference could be due to: (i) the difference in methodology, (ii) climate variation and (iii) human impacts. Garcin et al. (2005) measurements include the Cyclone Dina (January 2002) impact, and Kemp (2010) measurements include the tropical storm Diwa (March 2006). In January 2002 and in March 2006, Dina and Diwa respectively precipitated nearly 1500 and 2900 mm of rain at the "Cratère Commerson" network station, near the upper catchments of "Rivière des Remparts" (Meteo-France, 2010). Considering the human impact, an internal report of BRGM (2008) shows that the need for aggregates in La Réunion, is increasing with time and the industrial extraction in "Rivière des Remparts" also increases. The value given by the BRGM (2008) is about half of Kemp (2010) erosion measurements.

Louvat and Allègre (1997) have give erosion rates estimations over long time periods by considering geochemistry and suspended sediment concentrations. Their results applied on several rivers including "Rivière Saint Etienne" (for the "Bras de Cilaos" part); "Rivière de l'Est" and la "Rivière Langevin". Unfortunately, they didn't provide any results for La "Rivière des Remparts". Results clearly show that dramatic events are much more efficient even if chemical erosion rate are between 0.63 and 1.69 t/ha/year and mechanical rock weathering of between 12 and 91 t/ha/year.

6.7 Sedimentary Deposits and Processes on the Submarine Flanks

6.7.1 Eastern Submarine Flank

A first sedimentological study of the eastern submarine flank of Piton de la Fournaise was done by Ollier et al. (1998), based on bathymetry, deep-towed side-scan sonar images, submarine photographs and piston cores, collected during the FOURNAISE 2 cruises in 1988. The dataset was completed between 2006 and 2008 during the FOREVER and ERODER cruises with new bathymetry and backscatter imagery and new piston cores. Saint-Ange et al. (2013) provided a sedimentological facies model showing a wide range of gravity flows and erosional features.

In the eastern submarine flank of Piton de la Fournaise, all cores are dominated by coarse volcaniclastic sediments from silt and fine sand to gravels and pebbles (Ollier et al. 1998; Saint-Ange et al. 2013). Figure 6.6 illustrates examples of six sedimentological logs of cores collected at different depths from 600 to 3800 m. Core KF13 is the most proximal core, located inside the main channel. It mainly consists of gravels and sparse coarse sands. The core is less than 0.6 m long and does not show vertical facies organisation (Fig. 6.6). This deposit likely represents the basal layers of high-density gravity flow (Ollier et al. 1998).

Cores KF10, KERO-05, and KF14 are located at the top of the bathymetric bulge formed by debris avalanche deposits. They are dominated by bad-sorted volcaniclastic debris of grain-size ranging from silt and fine sand to coarse gravels (Fig. 6.6). The poor sorting, the absence of grading and the presence of mud clasts suggest that these sedimentary deposits result from local granular flow events, related to small-scale landslides (Saint-Ange et al. 2013).

Cores KF03 and KF05 are the most distal cores. They are located in the lower part of the submarine slope. They consist of successions of thin sandy beds alternating with silty-clayed hemipelagic deposits (Fig. 6.6). The sandy beds are various in thickness and grain-size. They locally show normal-grading and laminations, typical from turbidite deposits (Saint-Ange et al. 2013). Saint-Ange et al. (2013) described different gravity flow processes in this distal area: from the debris flow and the high-density turbulent flows to the low-density turbidity currents.

6.7.2 Southern Submarine Flank

No sedimentological data (core) was directly collected in the southern submarine flank of Piton de la Fournaise. But the study of the sedimentary processes was initiated by the interpretation of the reflectivity facies of the backscatter image, acquired during FOREVER and ERODER surveys (Fig. 6.1b). Figure 6.7 shows a zoom of the backscatter image and its interpretation concerning the sedimentary systems observed at the south of the island.

Canyon networks, described in the morphology of the submarine slope, are characterised by high reflectivity area (in black in Fig. 6.7a). At the base of the submarine slope, canyon incision decreases and divides into several small channels feeding large low-reflectivity areas (in white in Fig. 6.7a). These low-reflectivity areas are characterised by lobated shape extending to more 80 km over the abyssal plain. By analogy with the Cilaos deep-sea fans described at the South-West of La Réunion Island in the same data set (Saint-Ange et al. 2011; Sisavath et al. 2011), these low-reflectivity areas correspond to superficial sediment accumulations typical from sandy turbidite deposits.

In the prolongation of the "Saint-Pierre" and "Saint-Joseph" canyons, we respectively distinguish two main deep-sea fans: the "Saint-Pierre" deep-sea fan in the western part and the "Saint-Joseph" fan in the eastern part. They are

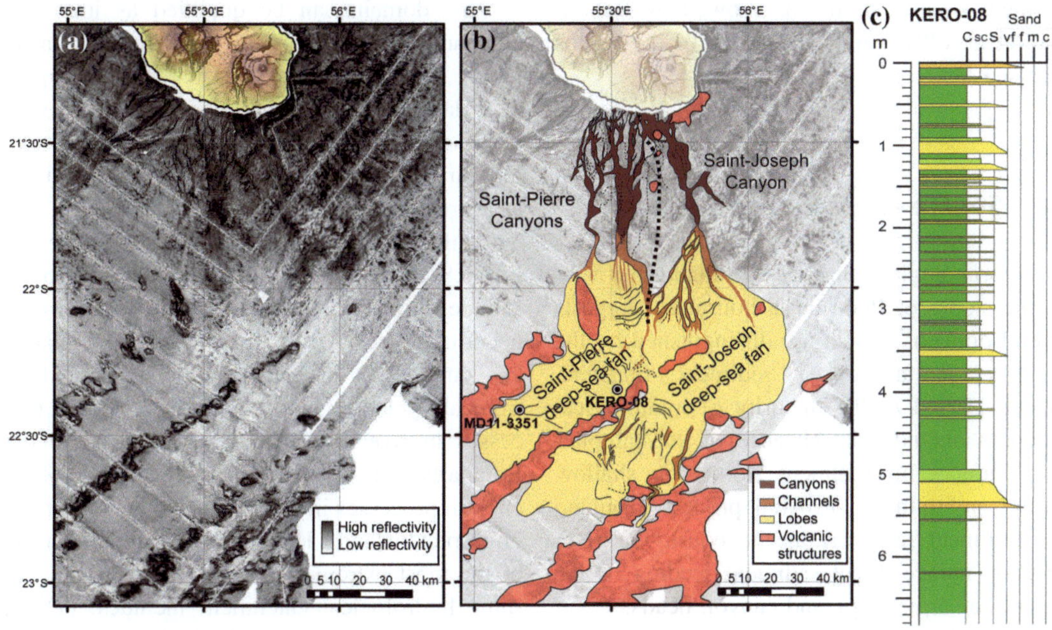

Fig. 6.7 **a** Backscatter image (high reflectivity in *black*, low reflectivity in *white*) from ERODER and FOREVER data (2006), focused at the southwest of La Réunion Island (see location in Fig. 6.1). **b** Interpretation of the submarine morphology and the acoustic facies at the south of La Réunion Island, showing the large extension of turbidite systems constituting Saint-Joseph fans. **c** Sedimentological log of the KERO-08 piston core showing the succession of sandy and silty turbidite deposits

partly coalescent at the base of the slope and morphologically constrained by volcanic ridges forming bathymetric high in the abyssal plain (Fig. 6.7).

Two cores were collected in the distal part of the "Saint-Pierre" fan: KERO-08 and MD11-3351. Figure 6.7c illustrates the sedimentological log of core KERO-08, collected at 4126 m. This core is 6.59 m long and is composed of a succession of sandy and silty turbidites alternating with hemipelagic muddy layers (Sisavath et al. 2012). The turbidite ranges from 2 to 30 cm thick and show normal grading and horizontal laminations. The dark sandy layers are composed of volcaniclastic grains and locally show white laminations concentrating bioclastic debris (foraminifera and other bioclasts) (Sisavath et al. 2012). Sedimentary facies of core KERO-08 confirm that the main processes transporting volcaniclastic sediment to the abyssal plain at the south of La Réunion Island are turbidity currents flowing from the canyons along the southern submarine flank of Piton de la Fournaise. The organisation of sedimentary systems, including incised canyons along the slope, turbidite channels and distal lobes, corresponds to a typical organisation of sandy turbidite systems, as those described on continental margins (Reading and Richards 1994; Piper and Normark 2009).

Formation of incised canyons implies a high recurrence of turbidity currents, triggering at the canyon heads. Turbidity currents, transporting sandy sediments down to the distal end of the deep-sea fan, are high-density currents in the upper slope. They maintained high transport capacity of large volume of sand along hundred kilometres. This type of current behaviour is favoured by the steep submarine slope inducing high-velocity turbidity currents, and large volume of mobilized sediment. This suggests frequent renewal of sediment supply with high sandy content, consistent with direct fluvial sediment supply, similar as showed by Babonneau et al. (2013) for the feeding of the Cilaos deep-sea fan.

6.8 Source to Sink Interpretation

The morphological analysis of onland and submarine slopes of Piton de la Fournaise suggests several domains, which are distinguished by the maturity of land-to-sea sediment transfers, directly related to the maturity of hydrological networks onland. The absence of coastal platform induces no sediment storage in shallow marine environment. Erosion processes onland and sediment transport to the coast, via river systems, induce direct sediment feeding of the upper submarine slope and canyon heads.

The first domain (type 1 in Fig. 6.8) corresponds to limited land-to-sea transfer, with no hydrographic network, as observed on the rift-zones (especially on the SE rift-zone). The coastal area forms a rocky promontory and only coastal erosion, produced by wave surf on recent fragmented volcanic rocks, produces sediments in the upper submarine slope. This sediment production, added to the instable structure of debris avalanches deposits forming the steep submarine slope, induced secondary gravity instabilities, which could be initiated at different water depths. Many submarine erosional scars or secondary landslide scars are visible in the morphology at different depths. They are characterised by various sizes and shapes, and feed short and low-incised valleys (green in Fig. 6.8).

Different types of gravity process can be initiated related to the size of the initial gravity instability, the local slope and the sediment source (type and size of destabilised material), as it is observed in cores collected in the eastern submarine flank of Piton de la Fournaise volcano. Based on core descriptions and sedimentary features observed on the seafloor, Saint-Ange et al. (2013) described different gravity flow processes: debris flow, high-density turbulent flows and low-density turbidity currents.

This domain can be qualified as immature submarine slope concerning the sediment transfer to the deep ocean: no direct sediment supply from land, various instabilities in terms of size, process, depth triggering, short distance of sediment transport, and immature submarine incision networks.

At the opposite, the second domain (Type 2 in Fig. 6.8) corresponds to high-efficient land-to-sea sediment transfer, characterised by a direct connection between river mouths and submarine canyon heads. Torrential floods, generated by tropical cyclones, induce intense erosional processes in river watersheds, as in "Rivière des Remparts". This river is a perfect example of the erosion processes in active volcanic environment on tropical islands. Here gathered all the features of the most extreme cases we can imagine. The same hydrological catchment includes: (1) the extremes orographic and cyclonic rainfalls, (2) significant slopes, (3) a lithology characterized by a major discordence in the "Bras de Mahavel", lahars, debris flows and lava flows, (4) a large amount of landslides deposits that feds the river alluvium which are easily transported by the river.

All these associated characters gives to this river very strong erosion rates. Then large amounts of sediments are transported to the coast, in hyper-concentrated flows. The absence of coastal platform favours the continuity from fluvial flows to submarine gravity processes, such as hyperpycnal flows as suggested by Saint-Ange et al. (2011) for the Cilaos sedimentary system. It probably exists several triggering mechanisms of submarine flows at the canyons heads (hyperpycnal flows, coastal erosion by storm waves, or local gravity instabilities) as shown by Babonneau et al. (2013). In all cases, sediments mobilised for the initiation of submarine gravity flow are granular sediments providing by rivers with high sand content, inducing the development of turbidity currents. Sand-rich turbidity currents, generated in steep slope, are erosional and incise the submarine slope forming deep and long canyons. They can reach high velocities, flow toward long distances and favour the development of canyons and

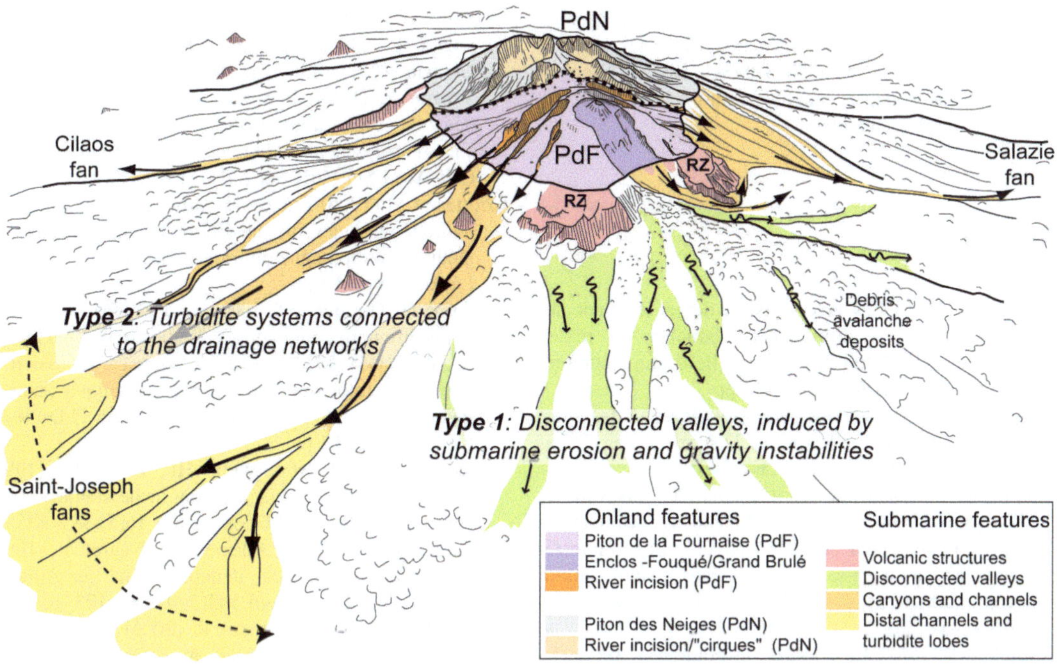

Fig. 6.8 3D schematic view from the southeast of La Réunion volcanic edifice, illustrating the land-to-sea sediment transfers around Piton de la Fournaise

channels. The high occurrence of turbidity currents generated in the same canyon heads during long time intervals (tens to hundreds ka) leads to the development of complex sedimentary architecture as channel-levee systems and distal lobe complexes. The "Saint-Joseph" and "Saint-Pierre" turbidite systems show simple architecture with canyons, channels and lobes (Figs. 6.7 and 6.8) indicating relatively young and immature deep-sea fans, comparing to deep-sea fans development on continental margins or to their neighbouring Cilaos fan (Sisavath et al. 2011).

The submarine slope of the eastern flank of Piton de la Fournaise volcano is marked by the presence of debris avalanches deposits and by the incision of the "Chenal Vincent" in the continuity of the "Enclos Fouqué" and "Grand Brûlé" structures (Fig. 6.5). Secondary gravity instabilities affect the top of debris avalanches deposits, as attested by erosional scars and irregular valley incision ("Vallée Nord", "Vallée des Gorgones" and "Pente Citron Galets" in Fig. 6.5). The gravity processes are probably similar to those described for the type 1 (Fig. 6.8), without connection with onland structures. The "Chenal Vincent" shows a good connection with the onland structures. Its morphology corresponds to an incised canyon, as those connected to river mouths. "Chenal Vincent" corresponds to an older canyon incision related to the Osmondes paleo-valley. Saint-Ange et al. (2013) described the land-to-sea transition and showed two recent lava-flow deltas in the proximal part down to 900 m of water depth. The recent infill by lava flow shows another process of sediment production, by fragmentation erosion and dismantling of coastal and submarine lava flows.

Acknowledgments This chapter is a morphological synthesis based on sub-aerial and submarine data. The marine data (bathymetry, backscatter images and cores) have been acquired during oceanographic cruises FOREVER and ERODER between 2006 and 2011, in collaboration with many partners. Authors thank all the collaborators and particularly the FOREVER survey's PI C. Deplus (IPGP), S. Jorry (IFREMER) and C. Delacourt (Univ. Brest) for their collaboration and project enterprise at sea and onland. We are also grateful to the PhD students contributing to the projects concerning the erosional and sedimentary processes in La Réunion Rivers

(J. Kemp) and on the submarines slopes of La Réunion edifice (E. Sisavath, F. Saint-Ange). These works were financially supported by CNRS-INSU, Région Réunion and DGA. This paper benefited from the constructive comments of JL Schneider.

References

Arnaud N (2005) Les processus de démantèlement des volcans. Le cas d'un volcan bouclier en milieu océanique : le Piton des Neiges (Ile de La Réunion). Université de La Réunion, 422 pp

Averous P (1983) Esquisse géomorphologique des atterages de lîle de La Réunion. Document des Terres Australes et Antarctiques Françaises

Babonneau N, Delacourt C, Cancouet R, Sisavath E, Bachèlery P, Deschamps A, Mazuel A, Ammann J, Jorry SJ, Villeneuve N (2013) Direct sediment transfer from land to deep-sea: Insights into shallow multibeam bathymetry at La Réunion Island. Mar Geol 346:47–57

Bachèlery P (1981) Le Piton de la Fournaise (Ile de La Réunion). Etude volcanologique, structurale et pétrologique, PhD thesis, Université Blaise Pascal, Clermont-Ferrand, France, 215 pp

Bachèlery P, Michon L (2010) The Enclos caldera: how did it form? In: IAVCEI third workshop on collapse calderas, La Réunion, pp 105–108

Bachèlery P, Villeneuve N (2013) Hot spots and large igneous provinces. In: Shroder J, Owen LA (eds) Treatise on geomorphology, vol 5. Academic Press, San Diego, Tectonic Geomorphology, pp 193–233

Bachelèry P, Labazuy P, Lenat JF (1996) Avalanches de débris sous-marines et subaériennes à La Réunion. C R Acad Sci Paris Série IIa 323:475–482

Bret L, Fevre Y, Join JL, Robineau B, Bachèlery P (2003) Deposits related to degradation processes on Piton des Neiges Volcano (Reunion Island): overview and geological hazard. J Volcanol Geoth Res 123(1–2):25–41

Carracedo JC (1999) Growth, structure, instability and collapse of Canarian volcanoes and comparisons with Hawaiian volcanoes. J Volcanol Geoth Res 94(1–4):1–19. doi:10.1016/s0377-0273(99)00095-5

Chadwick WW Jr, Howard KA (1991) The pattern of circumferential and radial eruptive fissures on the volcanoes of Fernandina and Isabela islands, Galapagos. Bull Volcanol 53(4):259–275. doi:10.1007/BF00414523

Clément J-P, Legendre C, Caroff M, Guillou H, Cotten J, Bollinger C, Guille G (2003) Epiclastic deposits and "horseshoe-shaped"calderas in Tahiti (Society Islands) and Ua Huka (Marquesas Archipelago), French Polynesia. J Volcanol Geoth Res 120:87–101

Clouard V, Bonneville A, Gillot P-Y (2001) A giant landslide on the southern flank of Tahiti Island, French Polynesia. Geophys Res Lett 28:2253–2256

Cochonat P, Lénat J-F, Bachèlery P, Boivin P, Cornaglia B, Deniel C, Labazuy P, Le Drezen E, Lipman PW, Ollier G, Savoye B, Vincent PM, Voisset M (1990) Importance des dépôts gravitaires dans la mise d'un système volcano-sédimentaire sous-marin (volcan de la Fournaise, Ile de la Réunion). C R Acad Sci Paris Série II 311:679–686

Courteaud M (1996) Etudes des structures géologiques et hydrogéologiques du massif de la Fournaise par la méthode audiomagnétollurique. PhD thesis, Univ. de La Réunion, 212 pp

Delibrias G, Guillier MT, Labeyrie J (1986) Gif natural radiocarbon measurements X. Radiocarbon 28(1):9–68

Fèvre Y (2005) Mécanismes et vitesses d'érosion à l'échelle géologique sur une île volcanique jeune à relief élevé – La Réunion (Océan Indien). PhD thesis, La Réunion, 217 pp

Filmer PE, McNutt MK, Webb HF, Dixon DJ (1994) Volcanism and archipelagic aprons in the Marquesas and Hawaiian Islands. Mar Geophys Res 16(5):385–406

Gaillardet J, Dupré B, Allègre CJ, Négrel P (1997) Chemical and physical denudation in the Amazon River Basin. Chem Geol 142(3–4):141–173

Garcin M, Poisson B, Pouget R (2005) High rates of geomorphological processes in a tropical area: the Remparts River case study (Réunion Island, Indian Ocean). Geomorphology 67(3–4):335–350

Gillot PY, Nativel P (1989) Eruptive history of the Piton de la Fournaise volcano, Réunion Island, Indian Ocean. J Volcanol Geoth Res 36:53–65

Hildenbrand A, Gillot P-Y, Le Roy I (2004) Volcano-tectonic and geochemical evolution of an oceanic intra-plate volcano: Tahiti-Nui (French Polynesia). Earth Planet Sci Lett 217:349–365

Holcomb RT, Searle RC (1991) Large landslides from oceanic volcanoes. Mar Geotechnol 10:19–32

Kemp JN (2010) Développement de méthodologies basées sur une combinaison de télédétection optique et radar pour la cartographie et la quantification d'érosion. Thèse de doctorat, La Réunion, Université de La Réunion, France, 221 pp

Kieffer G (1990) Recent dynamic and structural evolution ("phase IV") of the Piton des Neiges (Reunion island, Indian Ocean). In: Lénat JF (ed) Le Volcanisme de la Réunion, Monogr. Cent. De Rech. Volcanol.: Clermont-Ferrand, France, pp 163–185

Krastel S, Schmincke HU, Jacobs CL, Rihm R, Le Bas TM, Alibés B (2001) Submarine landslides around the Canary Islands. J Geoph Res 106(B3):3977–3997

Le Friant A, Lebas E, Clément V, Boudon G, Deplus C, de Voogd B, Bachèlery P (2011) A new model for the evolution of La Réunion volcanic complex from complete marine geophysical surveys. Geophys Res Lett 38:L09312

Lénat JF (2016a) Geodynamic setting of La Réunion. In: Bachèlery P, Lénat JF, Di Muro A, Michon L (eds) Active volcanoes of the Southwest Indian Ocean: Piton de la Fournaise and Karthala. Active Volcanoes of the World. Springer, Berlin

Lénat J-F (2016b) Construction of La Réunion Island. In: Bachèlery P, Lénat J-F, Di Muro A, Michon L (eds) Active volcanoes of the Southwest Indian Ocean: Piton de la Fournaise and Karthala. Active volcanoes of the World. Springer, Berlin

Lénat JF, Labazuy P (1990) Morphologies et structures sous-marines de la Reunion. In: Lénat JF (ed) Le volcanisme de la Réunion, Monographie. Centre Rech. Volcanol., Clermont-Ferrand, France, pp 43–74

Lénat JF, Vincent PM, Bachèlery P (1989) Sea-Beam mapping of the off-shore continuation of an active basaltic volcano: Piton de la Fournaise (Reunion Island, Indian Ocean). Structural and geomorphological interpretation. J Volcano Geoth Res 36(1–3):1–36

Lénat JF, Boivin P, Deniel C, Gillot P-Y, Bachèlery P, Fournaise2 Team (2009) Age and nature of deposits on the submarine flanks of Piton de la Fournaise (Réunion Island). J Volcanol Geotherm Res 184:199–207

Lénat JF, Bachèlery P, Merle O (2012) Anatomy of Piton de la Fournaise volcano (La Réunion, Indian Ocean). Bull Volc 74:1945–1961. doi:10.1007/s00445-012-0640-y

Lipman PW, Normark WR, Moore JG, Wilson JB, Gutmacher CE (1988) The giant submarine Alika Debris Slide, Mauna Loa, Hawaii. J Geophys Res Solid Earth 93:4279–4299

Louvat P, Allegre CJ (1997) Present denudation rates on the island of Reunion determined by river geochemistry: basalt weathering and mass budget between chemical and mechanical erosions. Geochim Cosmochim Acta 61(17):3645–3669

Mairine P, Bachèlery P (1997) Major erosional period in the building of Piton de la Fournaise (Réunion Island). C R Acad Sci 325:243–249

Masson DG, Watts AB, Gee MJR, Urgeles R, Mitchell NC, Le Bas TP, Canals M (2002) Slope failures on the flanks of the western Canary Islands. Earth-Sci Rev 57:1–35

Mazuel A (2014) Processus d'érosion et de transferts sédimentaires actifs sur les flancs sous-marins de l'édifice volcanique de l'île de La Réunion (Océan Indien), du littoral à l'océan profond. Thèse de doctorat de l'Université de Bretagne Occidentale, 474 pp

McDougall I (1971) The geochronology and evolution of the young volcanic island of Réunion, Indian Ocean. Geochim Cosmochim Acta 35:261–288

McGuire WJ (1996) Volcano instability: a review of contemporary themes. Geol Soc Lond Spec. Publ 110:1–23

Merle O, Michon L, Bachèlery P (2008) Caldera rim collapse: a hidden volcanic hazard. J Volcanol Geoth Res 177:525–530

Merle O, Mairine P, Michon L, Bachèlery P, Smietana M (2010) Calderas, landslides and paleo-canyons on Piton de la Fournaise volcano (La Réunion Island, Indian Ocean). J Volcanol Geoth Res 189:131–142

Michon L, Lénat JF, Bachèlery P, Di Muro A (2016) Geology and morphostructural evolution of Piton de la Fournaise. In: Bachèlery P, Lénat JF, Di Muro A, Michon L (eds), Active volcanoes of the Southwest Indian Ocean: Piton de la Fournaise and Karthala. Active Volcanoes of the World. Springer, Berlin

Mitchell NC, Masson DG, Watts AB, Gee MJ, Urgeles R (2002) The morphology of the submarine flanks of volcanic ocean islands: a comparative study of the Canary and Hawaiian hotspot islands. J Volcanol Geotherm Res 115:83–107

Mitchell NC, Dade WB, Masson DG (2003) Erosion of the submarine flanks of the Canary Islands. J Geophys Res-Earth Surf 108. doi:10.1029/2002JF000003F1-6002

Mohamed-Abchir A (1996) Les Cendres de Bellecombe (CB) : un événement explosif majeur dans le passé récent du Piton de la Fournaise, Ile de la Réunion. Thèse de doctorat, Paris, Université de Paris VII, France, 248 pp

Moore JG (1964) Giant submarine landslides on the Hawaiian ridge. US Geol Surv Prof Pap 501-D:95–98

Moore JG, Clague DA, Holcomb RT, Lipman PW, Normark WR, Torresan ME (1989) Prodigious submarine landslides on the Hawaiian Ridge. J Geophys Res Solid Earth 94:17465–17484

Moore JG, Normark WR, Holcomb RT (1994) Giant Hawaiian landslides. Annu Rev Earth Planet Sci 22:119–144

Morandi A, Di Muro A, Principe C, Michon L, Leroi G, Norelli F, Bachèlery P (2016) Pre-historic explosive activity at Piton de la Fournaise volcano. In: Bachèlery P, Lénat JF, Di Muro A, Michon L (eds) Active volcanoes of the Southwest Indian Ocean: Piton de la Fournaise and Karthala. Active Volcanoes of the World. Springer, Berlin

Naumann T, Geist D (2000) Physical volcanology and structural development of Cerro Azul Volcano, Isabela Island, Galapagos: implications for the development of Galapagos-type shield volcanoes. Bull Volc 61:497–514

Oehler J-F, Labazuy P, Lénat J-F (2004) Recurrence of major flank landslides during the last 2 Ma-history of Reunion Island. Bull Volcanol 66:585–598

Oehler J-F, Lenat JF, Labazuy P (2008) Growth and collapse of the Reunion Island volcanoes. Bull Volc 70(6):717–742

Ollier G, Cochonat P, Lénat JF, Labazuy P (1998) Deep-sea volcaniclastic sedimentary systems: an example from La Fournaise volcano, Réunion Island, Indian Ocean. Sedimentology 45:293–330

Peltier A, Famin V, Bachèlery P, Cayol V, Fukushima Y, Staudacher T (2008) Cyclic magma storages and transfers at Piton de la Fournaise volcano (La Réunion hotspot) inferred from deformation and geochemical data. Earth Planet Sci Lett 270:180–188

Piper DJW, Normark WR (2009) Processes that initiate turbidity currents and their influence on turbidites: a marine geology perspective. J Sediment Res 79:262–347. doi:10.2110/jsr.2009.046

Quidelleur X, Holt JW, Salvany T, Bouquerel H (2010) New K-Ar ages from La Montagne massif, Reunion

Island (Indian Ocean), supporting two geomagnetic events in the time period 2.2–2.0 Ma. Geophys J Int 182:699–710

Rad SD, Allegre CJ, Louvat P (2007) Hidden erosion on volcanic islands. Earth Planet Sci Lett 262:109–124

Reading HG, Richard MT (1994) The classification of deep-water siliciclastic depositional systems by grain size and feeder systems. AAPG Bull 78:792–822

Robert R (2001) Pluviométrie à l'île de La Réunion: des travaux de J. Defos du Rau (1960) à nos jours. L'information géographique 1:53–59

Saint-Ange F (2009) La sédimentation volcaniclastique en contexte de point chaud (île de La Réunion, Océan Indien). PhD Université de La Réunion, France, 279 pp

Saint-Ange F, Savoye B, Michon L, Bachelèry P, Deplus C, De Voogd B, Dyment J, Le Drezen E, Voisset M, Le Friant A, Boudon G (2011) A volcaniclastic deep-sea fan off La Réunion Island (Indian Ocean): gradualism versus catastrophism. Geology 39:271–274. doi:10.1130/G31478.1

Saint-Ange F, Bachèlery P, Babonneau N, Michon L, Jorry S (2013) Volcanoclastic sedimentation on the submarine slopes of a basaltic hotspot volcano: Piton de la Fournaise volcano (La Réunion Island, Indian Ocean). Mar Geol 337:35–52

Salvany T, Lahitte P, Nativel P, Gillot P-Y (2012) Geomorphic evolution of the Piton des Neiges volcano (Reunion Island, Indian Ocean): competition between volcanic construction and erosion since 1.4 Ma. Geomorphology 136:132–147

Sisavath E, Babonneau N, Saint-Ange F, Bachèlery P, Jorry S, Deplus C, De Voogd B, Savoye B (2011) Morphology and sedimentary architecture of a modern volcaniclastic turbidite system: the Cilaos fan, offshore La Réunion Island. Mar Geol 288:1–17. doi:10.1016/j.margeo.2011.06.011

Sisavath E, Mazuel A, Jorry S, Babonneau N, Bachèlery P, De Voogd B, Salpin M, Emmanuel L, Beaufort L, Toucanne S (2012) Processes controlling a volcaniclastic turbiditic system during the last climatic cycle: example of the Cilaos deep-sea fan, offshore La Réunion Island. Sed Geol 281:180–193

Smietana M (2011) Pétrologie, géochronologie (K–Ar) et géochimie élémentaire et isotopique (Sr, Nd, Hf, Pb) de laves anciennes de La Réunion: Implications sur la construction de l'édifice volcanique. PhD thesis, University of La Réunion, 362 pp

Tanguy JC, Bachèlery P, LeGoff M (2011) Archeomagnetism of Piton de la Fournaise: Bearing on volcanic activity at La Réunion Island and geomagnetic secular variation in Southern Indian Ocean. Earth Planet Sci Lett 303(3–4):361–368

Thouret J-C (1999) Volcanic geomorphology—an overview. Earth-Sci Rev 47:95–131

Villeneuve N, Bachèlery P, Kemp J (2014) La Réunion Island: a typical Example of basaltic shield volcano with rapid evolution. In: Fort M, André M-F (eds) Lanscapes and landforms of France. World Geomorphological Landscapes, Springer-Science

7 Magma Paths at Piton de la Fournaise Volcano

Laurent Michon, Valérie Ferrazzini and Andrea Di Muro

Abstract

Several patterns of magma paths have been proposed since the 1980s for Piton de la Fournaise volcano. Given the significant differences, which are presented here, we propose a reappraisal of the magma intrusion paths using a 17-years-long database of volcano-tectonic seismic events and a detailed mapping of the scoria cones. At the edifice scale, the magma propagates along two N120 trending rift zones. They are wide, linear, spotted by small to large scoria cones and related lava flows and involving magma resulting from high-pressure fractionation of ol ± cpx and presents an eruption periodicity of around 200 years over the last 30 kiloyears. The upper plumbing system originates at the base of the edifice below the Enclos Fouqué caldera. It feeds frequent (1 eruption every 9 months on average), short-lived summit and distal (flank) eruptions along summit and outer rift zones, respectively. Summit rift zones are short and present an orthogonal pattern restricted to the central active cone of Piton de la Fournaise whereas outer rift zones extend from inside the Enclos Fouqué caldera to the NE and SE volcano flanks. To sum up, rift zones of Piton de la Fournaise present strong geometrical and dynamical differences. On the one hand, the lower plumbing system feeds rift zones showing striking similarities to those developed in Hawaii during the alkaline postshield

L. Michon (✉)
Laboratoire Géosciences Réunion, Université de La Réunion, Institut de Physique du Globe de Paris, Sorbonne Paris Cité, UMR 7154 CNRS, F-97744 Saint Denis, France
e-mail: laurent.michon@univ-reunion.fr

V. Ferrazzini · A. Di Muro
Observatoire Volcanologique du Piton de la Fournaise, Institut de Physique du Globe de Paris, Sorbonne Paris Cité, UMR 7154 CNRS, F-75005 Paris, France

P. Bachèlery et al. (eds.), *Active Volcanoes of the Southwest Indian Ocean*,
Active Volcanoes of the World, DOI 10.1007/978-3-642-31395-0_7

stage. On the other hand, the rift zones connected to upper plumbing system can be compared the rift system of Mount Etna, whose dynamics is know to be linked with the lateral displacement of the east flank.

Keywords
Piton de la Fournaise · Plumbing system · Rift zones · Magma paths

7.1 Introduction

On ocean basaltic volcanoes, magma transfer from the source to the surface is generally assumed to occur along sub-vertical ascent paths through the mantle lithosphere, the oceanic crust and the volcanic edifice, eventually followed by lateral propagation at shallower levels along preferential intrusion paths named rift zones (e.g. MacDonald 1972). At the surface, volcanic rift zones are outlined by a concentration of scoria cones and eruptive fissures, which results from their overall history of activity. Their wide range of geometry, from radial or circumferential diffuse rift zones (e.g. Galapagos; Chadwick and Howard 1991) to narrow km-long rift zones (e.g. Hawaiian shields; Walker 1999) is known as dependent, among other parameters, on the dynamics of the plumbing system, the gravitational deformation of the volcanic edifice and the influence of the crustal inheritance (Fiske and Jackson 1972; MacDonald 1972; Dieterich 1988; Chadwick and Dieterich 1995; Walker 1999; Walter and Troll 2003; Michon et al. 2007; Chaput et al. 2014a). For the world's most active volcanoes, such as Kilauea and Etna, the rift zones are well identified and provide the opportunity of searching/studying the interrelation between rift zone development, flank displacement and plumbing system dynamics (Dieterich 1988; Walter et al. 2005; Montgomery-Brown et al. 2011). By contrast, the geometry of the rift zones of Piton de la Fournaise (La Réunion hotspot) is still debated and the link between their development and the dynamics of the plumbing system still unclear.

This paper aims first at reviewing the spatial distribution of preferential intrusion paths proposed for Piton de la Fournaise during the last decades using different methodological approaches and, second, at determining the geometry of the plumbing system and its link with the rift zones.

7.2 Review of the Intrusion Paths at Piton de la Fournaise

7.2.1 Magma Paths at the Edifice Scale

Two different approaches have been conducted to determine the rift zone geometry at the edifice scale, taking into account either the concentration of the scoria cones (Villeneuve and Bachèlery 2006) or the distribution and the morphometrical parameters of the cinder cones(Bonali et al. 2011). Despite the use of the same 25-m-step digital elevation model of the French Institut Géographique National (IGN), each methodology has produced contrasting results. Only three rift zones were identified from the volcanic cone concentration: two arched rift zones connected to the volcano's summit, prolonged in the NE and SE directions toward the ocean, and a N120 rift zone located between the summits of the Piton des Neiges and Piton de la Fournaise volcanoes (Fig. 7.1a; Villeneuve and Bachèlery 2006). An additional minor direction of intrusion,

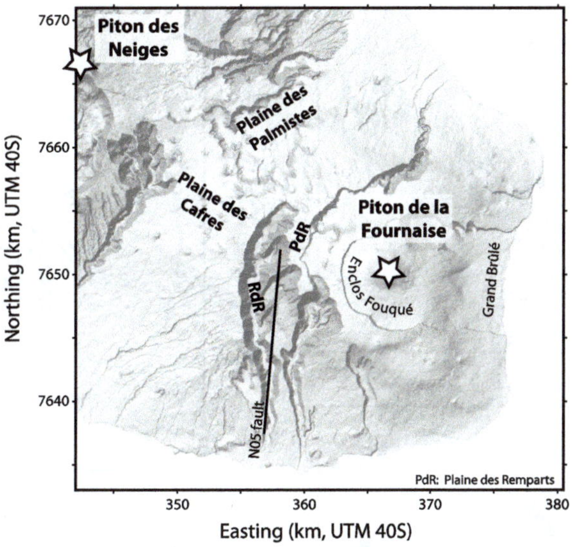

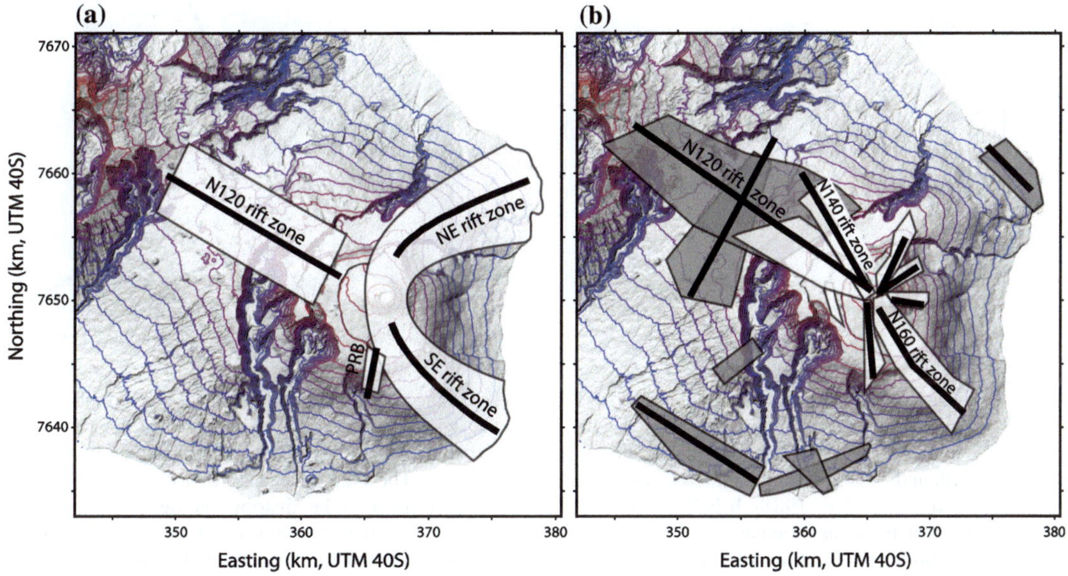

Fig. 7.1 Rift zone distribution determined for Piton de la Fournaise from **a** the concentration of the scoria cones (Bachèlery 1981; Villeneuve and Bachèlery 2006) and **b** the distribution and asymmetry of the scoria cones (Bonali et al. 2011). *White* and *grey zones* account for recent and old rift zones, whose distinction is based on the degradation stage of the scoria cones. PRB: Puys Ramond Baril volcanic alignment. *PdR* Plaine des Remparts. *RdR* Rivière des Remparts

corresponding to the Puys Ramond alignment, extends south of the Enclos Fouqué caldera. Integrating the degradation stage of the scoria cones to the morphometric analysis, Bonali et al. (2011) distinguished active and inactive rift zones. It appears that the intrusion paths identified on the volcano flanks, and most of the N120 rift zone lying between Piton des Neiges and Piton de la Fournaise, were considered as inactive and related to the volcano's ancient history (Fig. 7.1b; Bonali et al. 2011). All the active rift zones were connected to the summit of Piton de la Fournaise.

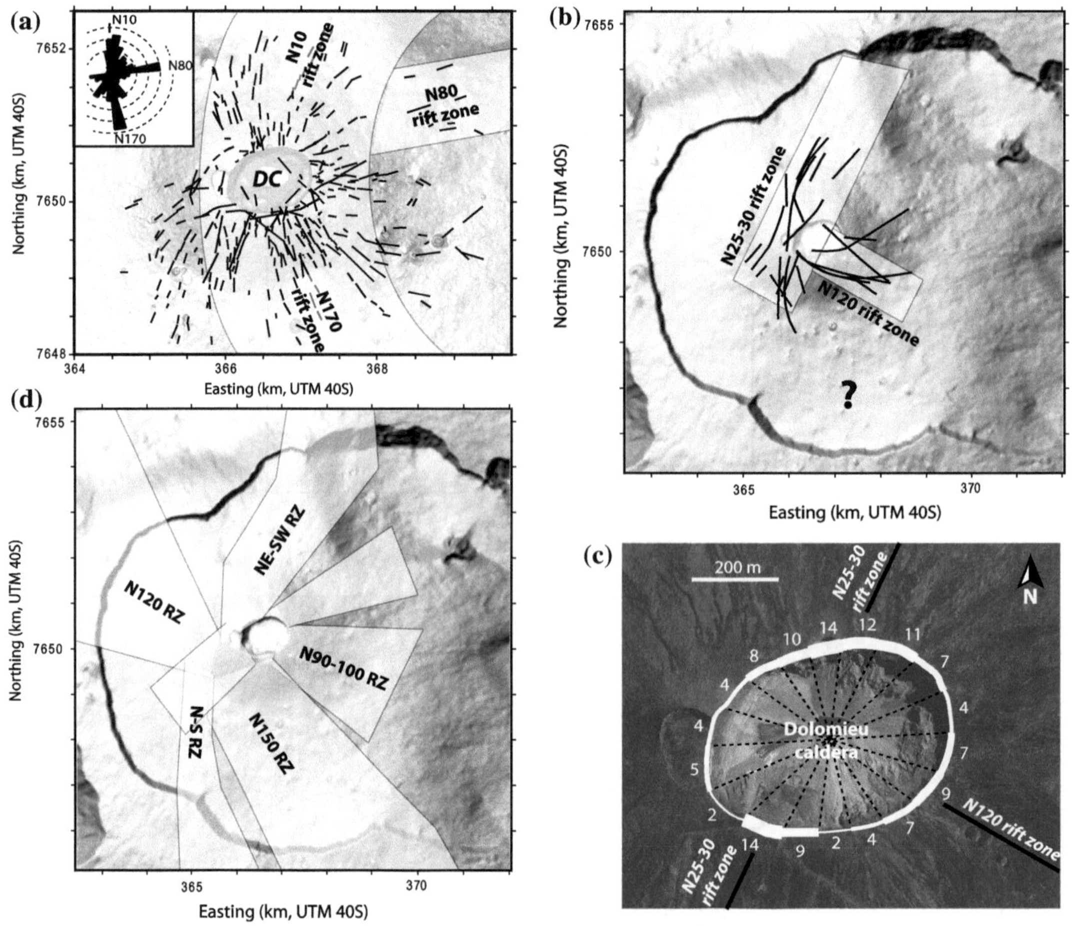

Fig. 7.2 **a** Distribution and main orientations of the eruptive fissures opened in the Central Cone of Piton de la Fournaise, enabling to determine two main N10 and N170 rift zones North and South of the Dolomieu caldera (DC), respectively (after Bachèlery 1981). **b** N25-30 and N120 rift zones inferred from the orientation of recent dykes (*black lines*) intruding the Central Cone between 1981 and 2006 (after Michon et al. 2009). **c** Distribution of radial dykes in the walls of the Dolomieu caldera. Numbers refer to the dyke population detected in each of the 18 sectors. The thickness of the *white line* is proportional to the dyke population. The N25-30 and N120 rift zones previously determined from dyke distribution outside the caldera (Michon et al. 2009) are indicated for comparison. **d** Rift zone distribution determined from the scoria cone morphometry (after Fig. 12 in Bonali et al. 2011)

7.2.2 Magma Paths Centred on the Enclos Fouqué Caldera

Seismological data indicate that above 7.5 km below sea level (b.s.l.), the plumbing system is almost exclusively restricted to the Enclos Fouqué caldera (Nercessian et al. 1996; Aki and Ferrazzini 2000; Battaglia et al. 2005; Prôno et al. 2009). The magmatic activity related to this part of the volcanic system progressively led to the building, on the floor of the Enclos Fouqué caldera, of the 400-m-high Central Cone capped by the Bory crater and the Dolomieu caldera (Fig. 7.2a). This cone and to a lesser extent the Enclos Fouqué caldera are cut by hundreds of eruptive fissures that formed during the recent history of Piton de la Fournaise.

Considering the eruptive fissures orientation, Bachèlery (1981), Carter et al. (2007) proposed that the main shallow intrusion paths fed by the summit plumbing system are organized in two main N10 and N170 rift zones located North and South of the summit craters, respectively, and in a secondary N80 rift zone on the cone's eastern flank (Fig. 7.2a). A contrasting geometry characterized by two perpendicular N25-30 and N120 rift zones has been suggested from the distribution of the dykes intruded in the Central Cone between 1981 and 2006 (Fig. 7.2b; Michon et al. 2009). This pattern is supported by the dyke distribution in the wall of the Dolomieu caldera (Fig. 7.2c; Michon et al. 2013). Finally, a much more complex rift zone distribution was recently determined from the scoria cone morphometry (Fig. 7.2d; Bonali et al. 2011). Magmas ascending through the summit plumbing system could propagate along seven radial rift zones. Noteworthily, this rift zone pattern partly corresponds to the aggregate of preferential intrusion paths described in previous works (Bachèlery 1981; Michon et al. 2007). Given the important differences in the rift zone geometries (N10/N170 arcuate rift zones, N25-30/N120 perpendicular rift zones or seven radial rift zones), it is worth wondering (1) the cause of these differences and (2) the link between the rift zones identified inside the Enclos Fouqué caldera and the NE and SE rift zones propagating outside the Enclos Fouqué.

The question arises whether these rift zones connect to a single or to different magma reservoirs. Lénat and Bachèlery (1990) explained the main characteristics of the summit eruptions, i.e. short-lived, small volume of erupted magma and a short duration between the initiation of dyke injection and the opening of eruptive fissures, as resulting from relatively small magma pockets scattered between the volcano's summit and sea level (2.6 km below volcano's summit) (Fig. 7.3). The draining of these shallow magma reservoirs during distal eruptions was thought to trigger summit collapses (pit crater) inside the Dolomieu crater like in 1986 and 2002 (Hirn et al. 1991; Longpré et al. 2007). Beside this shallow system, two distinct magma-rich zones have been

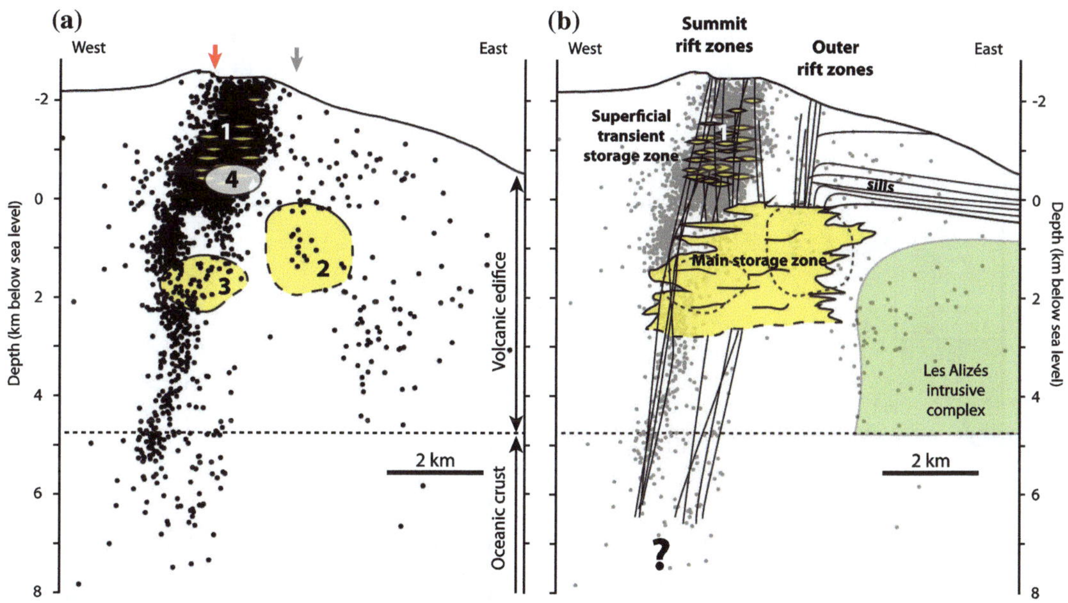

Fig. 7.3 **a** Seismic events related to the summit and distal eruptions between 1998 and 2010. Structures labelled 1, 2, 3 and 4 account for small magma pockets related to summit eruptions (Lénat and Bachèlery 1990), a low velocity zone inferred from Vp and Vs waves (Nercessian et al. 1996), a low velocity zone deduced from a Vp-wave tomography (Prôno et al. 2009) and the magma reservoir inferred from geodetic data (Peltier et al. 2008), respectively. *Red* and *grey arrows* indicate the locus of the pre-eruptive inflation related to summit and distal eruptions. **b** Interpretative summit plumbing system below the Enclos Fouqué caldera (see text for explanation)

evidenced from seismic wave propagation. These low velocity bodies are located between 1–2 km depth bsl below the Bory crater (Prôno et al. 2009) or 0–2 km depth bsl, 1 km east of Dolomieu crater (Nercessian et al. 1996) (Fig. 7.3). Importantly, the identification of these two zones was made from two different sets of data, at two different periods, with two different networks of seismic stations. It cannot be excluded that each zone is a part of a larger structure or the same low velocity zone located at different places. Assuming a unique structure, this zone situated between sea level and 2 km bsl is supposed to be the locus of a transient magma storage, which feeds the most voluminous distal eruptions like in 1977, 1986 and 2007 (Nercessian et al. 1996; Prôno et al. 2009). Finally, a quite different picture has been proposed from the inversion of geodetic data (Peltier et al. 2008, 2009). According to Peltier et al. (2009), all the eruptions, except for that of 1998, are fed by a single magma reservoir, located above sea level, below the Dolomieu crater and regularly recharged by deeper sources (Fig. 7.3). However, this interpretation does not consider nor explain the intense seismicity that occurs in the proposed magma reservoir.

7.3 Reappraisal of the Magma Paths

In order to decipher the rift zone geometry and their current activity or inactivity, an integrated analysis of (1) the recent (<30 kiloyears) volcanic cones distribution, (2) the volcano-tectonic events recorded by the Piton de la Fournaise Volcano Observatory between 1996 and 2013, and (3) the available eruption datings (Bachèlery 1981; Tanguy et al. 2011; Morandi et al. 2016) has been carried out. The density map calculated from a total of 644 scoria cones first shows a N–S elongated maximum cone concentration restricted to the Enclos Fouqué caldera, the site of most of the recent eruptions (Fig. 7.4a). This area is superimposed to a dense seismicity developed below the summit from the surface to about 7.5 km below sea level, tracking the magma propagation paths along the summit plumbing system (Fig. 7.5a; Aki and Ferrazzini 2000; Battaglia et al. 2005). The evolution of the April 1977, March 1986 and March 1998 eruptions, starting inside the Enclos Fouqué caldera and propagating along the NE and SE rift zones clearly indicates that these rift zones, which are

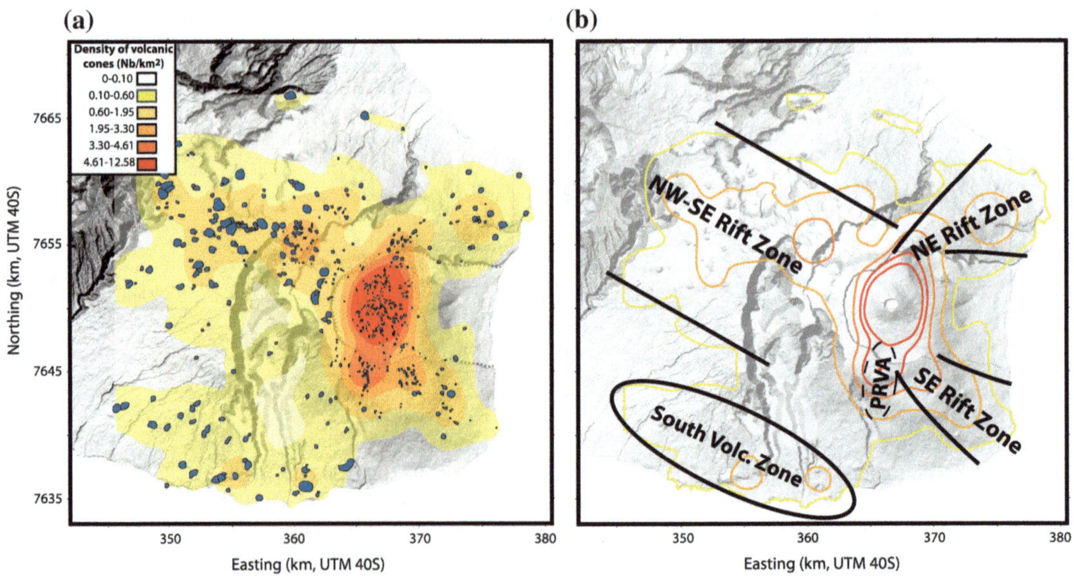

Fig. 7.4 **a** Density map of the scoria cones (*blue*) on the massif of Piton de la Fournaise younger than 60 kiloyears. **b** Preferential intrusion paths proposed in this study from the density map

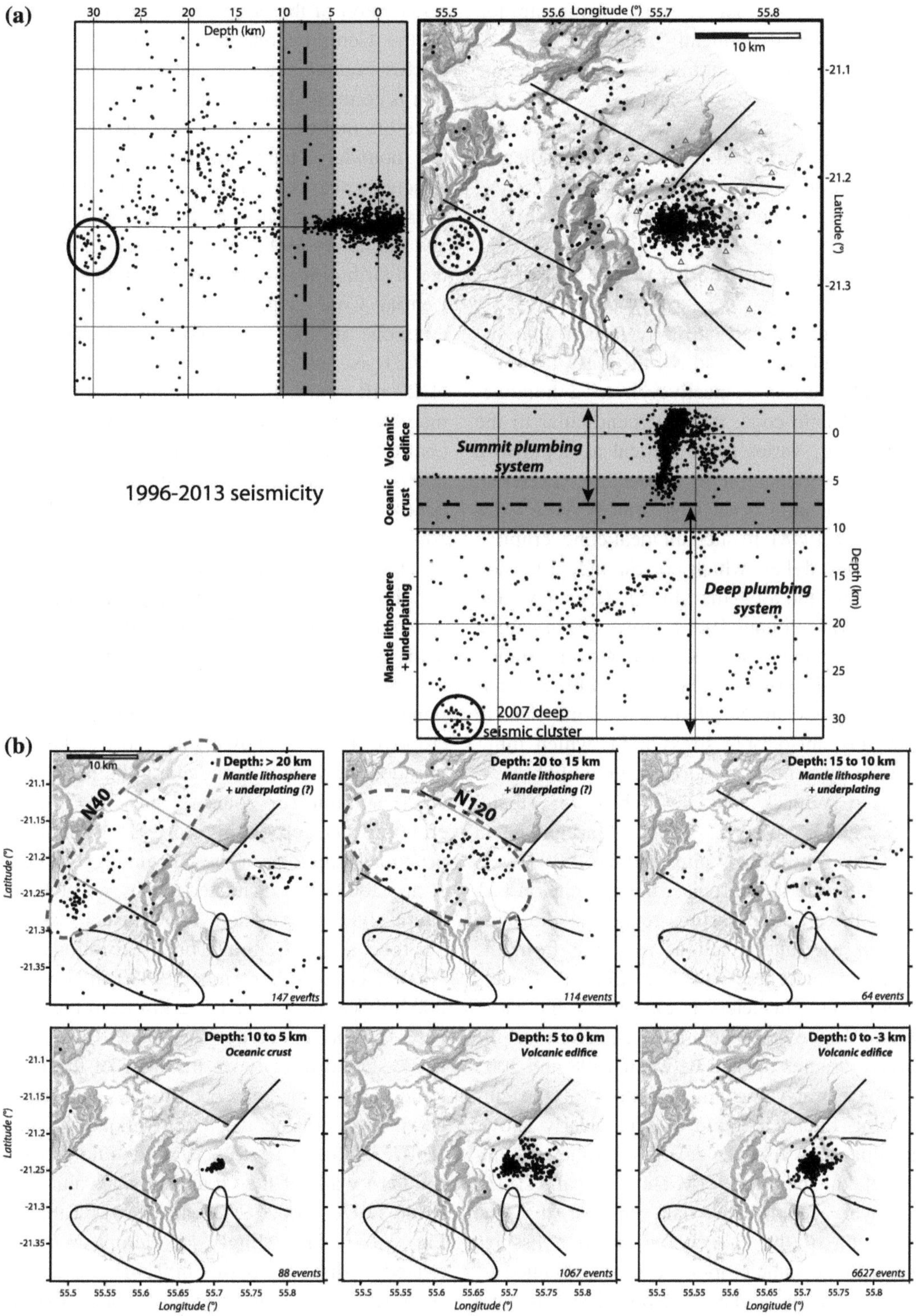

Fig. 7.5 **a** Volcano-tectonic seismic hypocenters (*black dots*) localized between 1996 and 2013 by the Piton de la Fournaise volcano observatory (OVPF/IPGP) on the massif of Piton de la Fournaise. Location is determined by a manual picking of the arrival times at different stations of the network. Depth of the top and the base of the oceanic crust is from Fontaine et al. (2012). The occurrence of underplating between 10 and 15 km bsl has been determined by Gallart et al. (1999). **b** Distribution of the seismicity at different depths. The *solid lines* account for the preferential intrusion paths inferred from the scoria cones distribution (Fig. 7.4)

underlined by a larger cone concentration (Fig. 7.4a), are dynamically linked to the shallower part of the plumbing system feeding the Central Cone.

At the edifice scale, two further areas of cone concentration can be identified, corresponding to the NW–SE rift zone and the South Volcanic Zone, both trending in the N120 direction (Fig. 7.4b).

7.3.1 The Summit Plumbing System

Magma intrusions are strongly sensitive to the stress distribution (σ_3 and σ_1, the least and main principal stresses, respectively). In a homogeneous volcanic cone, σ_3 is perpendicular to the slope and σ_1 varies from vertical at depth to a downslope orientation in subsurface (Dieterich 1988; Cayol and Cornet 1998; van Wyk de Vries and Matela 1998). In such an ideal cone, eruptive fissures and dykes, which correspond to tension cracks are radially distributed (Delaney and Pollard 1981; Pollard 1987). However, in nature, additional effects such as magma overpressure, rheological layering, volcano instabilities, faulting activity and structural inheritance may lead to a strongly anisotropic stress field, which influence the dyke intrusion along preferential orientations (Chadwick and Dieterich 1995; Tibaldi 1996, 2003; Walker 1999; Marinoni and Gudmundsson 2000; Walter and Troll 2003).

At Piton de la Fournaise, the en échelon pattern of the eruptive fissures related to the same eruption was interpreted as resulting from the eastern displacement of the Central Cone's eastern flank (Bachèlery et al. 1983; Lénat et al. 1989a). However a detailed inspection of the eruptive fissures opened between 1998 and 2005 in the Central Cone and the floor of the surrounding Enclos Fouqué caldera clearly shows that almost all fissures develop parallel or subparallel to the slope (Michon et al. 2007). Moreover, geophysical data suggest that the orientation of the en échelon eruptive fissures related to the same eruption can be explained by a single feeding dyke at depth (Battaglia and Bachèlery 2003; Fukushima et al. 2005; Peltier et al. 2005). Thus, the radial to subradial distribution of the eruptive fissures on the flanks of the Central Cone and on the floor of the Enclos Fouqué suggests that the fissure orientation is controlled by downslope σ_1 rather than by dyke orientation (Michon et al. 2007). Note that the boundary faults of the summit collapse structures also control the concentric distribution of the eruptive fissures in the summit area (Lénat and Bachèlery 1990). We therefore consider that the distribution of the eruptions intruding the Central Cone, e.g. the summit and proximal eruptions in the terminology of Peltier et al. (2008), is best explained by the N25-30 and N120 rift zones. In such a configuration, a few magma intrusions in the cone's eastern flank are controlled by N65 trending faults (Michon et al. 2009). What is the link between these summit rift zones and the much broader NE and SE rift zones on the volcano flank?

The location of the eruptions, their dynamics and the related deformation pattern for the period spanning between 1972 and 2013 have been taken into account to answer this question. During this time span, Piton de la Fournaise experienced 81 eruptions. The distribution of the eruptive fissures shows that most dykes (58) intruded the Central Cone along the N25-30 and N120 summit rift zones (Table 7.1). Summit eruptions, corresponding to both summit and proximal eruptions in the terminology of Peltier et al. (2008), are characterized by no or limited lateral dyke propagation, which usually stops at the base of the Central Cone. Exceptionally, in November 1987, September 1988 and November 2002, dykes propagated 1–2 km from the cone's base. Twenty-two eruptions, named distal eruptions, started directly on the floor of the Enclos Fouqué caldera and occasionally propagated downslope outside the Enclos Fouqué caldera (1977, 1986 and 1998) or along the northern and southern walls of the Grandes Pentes and Grand Brûlé (1976, 2002, 2005 and 2007; Table 7.1; Fig. 7.6a). Interestingly, the eruptive fissures related to distal eruptions in Table 7.1, are concentrated in three areas located South, East and North of the Central Cone. They define three preferential intrusion paths named outer rift

Table 7.1 Eruptions of Piton de la Fournaise since 1972

Eruption timing		Eruption type	Rift zone	Volume (10^6 m^3)
Start (dd/mm/yy)	End (dd/mm/yy)			
09/06/72	11/06/72	Summit	N25-30 SRZ	0.27
25/07/72	17/08/72	Distal	East RZ	2.9
07/09/72	26/09/72	Summit	N25-30 SRZ	4
10/10/72	10/12/72	Distal	South-East RZ	9.3
10/05/73	28/05/73	Summit	N120 SRZ	1.6
04/11/75	18/11/75	Summit	N120 SRZ	1.4
18/12/75	25/01/76	Summit	N120 SRZ	0.03
12/01/76	06/04/76	Distal	South-East RZ	11.6
02/11/76	03/11/76	Summit	N25-30 SRZ	0.46
24/03/77	24/03/77	Distal	South-East RZ	0.06
05/04/77	16/04/77	Distal	North-East RZ	20
24/10/77	17/11/77	Summit	N120 SRZ	23
28/05/79	29/05/79	Summit	N25-30 SRZ	0.2
13/07/79	14/07/79	Summit	N25-30 SRZ	0.3
03/02/81	24/02/81	Summit	N25-30 SRZ	3
24/02/81	31/03/81	Summit	N25-30 SRZ	4
01/04/81	05/05/81	Distal	East RZ	5
04/12/83	18/01/84	Summit	N25-30 SRZ	8
18/01/84	18/02/84	Summit	N25-30 SRZ	9
14/06/85	15/06/85	Summit	N25-30 SRZ	1
05/08/85	01/09/85	Summit	N25-30 SRZ	7
06/09/85	10/10/85	Summit	N120 SRZ	14
01/12/85	03/12/85	Summit	N25-30 SRZ	0.7
29/12/85	07/02/86	Summit	N120 SRZ	7
19/03/86	05/04/86	Distal	South-East RZ	13.5
13/07/86	13/07/86	Summit	N120 SRZ	0.28
12/11/86	13/11/86	Summit	N120 SRZ	0.27
26/11/86	27/11/86	Summit	N120 SRZ	0.24
06/12/86	07/01/87	Summit	N120 SRZ	2
07/01/87	10/02/87	Distal	East RZ	10
10/06/87	29/06/87	Summit	N120 SRZ	1
19/07/87	20/07/87	Distal	East and South RZ	1
06/11/87	08/11/87	Summit	N25-30 SRZ	1.6
30/11/87	01/01/88	Distal	South-East RZ	10
07/02/88	02/04/88	Distal	South-East RZ	8
18/05/88	01/08/88	Summit	N25-30 SRZ	7
31/08/88	12/09/88	Summit	N25-30 SRZ	4
14/12/88	29/12/88	Distal	North-East RZ	4

(continued)

Table 7.1 (continued)

Eruption timing		Eruption type	Rift zone	Volume (10^6 m^3)
Start (dd/mm/yy)	End (dd/mm/yy)			
18/01/90	19/01/90	Summit	N120 SRZ	0.97
18/04/90	08/05/90	Distal	South-East RZ	8
19/07/91	20/07/91	Summit	N120 SRZ	5
27/08/92	23/09/92	Summit	N120 SRZ	5.5
09/03/98	21/09/98	Summit	N25-30 SRZ	60
11/03/98	01/04/98	Other	NW-SE RZ	0.75
08/08/98	15/09/98	Distal	North-East RZ	0.9
19/07/99	31/07/99	Summit	N120 SRZ	1.3
28/09/99	23/10/99	Summit	N25-30 SRZ	1.4
13/02/00	03/03/00	Summit	N25-30 SRZ	8.2
23/06/00	30/07/00	Distal	South-East RZ	10
12/10/00	13/11/00	Summit	N120 SRZ	9
27/03/01	04/04/01	Summit	N120 SRZ	4.8
11/06/01	09/07/01	Summit	N120 SRZ	9.5
05/01/02	16/01/02	Distal	North-East RZ	13
16/11/02	03/12/02	Summit	N120 SRZ	8
30/05/03	30/05/03	Summit	N25-30 SRZ	0.16
04/06/03	09/06/03	Summit	N25-30 SRZ	0.72
12/06/03	15/06/03	Summit	N25-30 SRZ	0.59
22/06/03	07/07/03	Summit	N25-30 SRZ	0.69
22/08/03	27/08/03	Summit	N25-30 SRZ	6.2
30/09/03	01/10/03	Summit	N25-30 SRZ	1
07/12/03	25/12/03	Summit	N120 SRZ	1.2
08/01/04	08/01/04	Distal	North-East RZ	1.9
02/05/04	18/05/04	Summit	N25-30 SRZ	16.9
12/08/04	16/10/04	Summit	N120 SRZ	20
17/02/05	26/02/05	Distal	North-East RZ	19
04/10/05	17/10/05	Summit	N25-30 SRZ	2
29/11/05	29/11/05	Summit	N25-30 SRZ	1
26/12/05	18/01/06	Distal	North-East RZ	20
20/07/06	14/08/06	Summit	N25-30 SRZ	2
30/08/06	01/01/07	Summit	N120 SRZ	20
18/02/07	19/02/07	Summit	N120 SRZ	1
30/03/07	31/03/07	Distal	South-East RZ	0.6
02/04/07	01/05/07	Distal	South-East RZ	240
21/09/08	02/10/08	Summit	N25-30 SRZ	1
27/11/08	28/11/08	Summit	N25-30 SRZ	0.1
14/12/08	04/02/09	Summit	N25-30 SRZ	1.5

(continued)

Table 7.1 (continued)

Eruption timing		Eruption type	Rift zone	Volume (10^6 m^3)
Start (dd/mm/yy)	End (dd/mm/yy)			
05/11/09	06/11/09	Summit	N120 SRZ	0.14
14/12/09	15/12/09	Summit	N25-30 SRZ	0.16
02/01/10	12/01/10	Summit	N25-30 SRZ	1.2
14/10/10	31/10/10	Distal	South-East RZ	2.7
09/12/10	10/12/10	Summit	N25-30 SRZ	0.53

Emitted volumes from Peltier et al. (2009), Roult et al. (2012)

zones, that are spatially disconnected from the summit rift zones; the South-East, East and North-East rift zones that gather 12, 4 and 7 distal eruptions respectively (Fig. 7.6). These rift zones roughly correspond to the NE and SE rift zones and the N80 intrusion path identified by Bachèlery (1981). The important new observation brought by the present work is that the outer rift zones are solely fed by distal eruptions rather than by all eruption types. According to Peltier et al. (2009), between 1998 and 2007, proximal and distal eruptions were fed by two distinct vertical pathways, located below the western and eastern part of the Dolomieu caldera, respectively, and connected to a single magma reservoir lying at about 2300 m depth below the Dolomieu caldera. The inclinometers data recorded during the eruptions of 1983, 1984, 1985, 1986, 1988, 1990 and 1992 also suggest two different inflation sources located below the western part of Dolomieu and below the eastern flank of the Central Cone for the summit and distal eruptions, respectively (Fig. 7.3; Delorme et al. 1989; Lénat et al. 1989a, b; Toutain et al. 1992; Delorme 1994). We consequently propose that magma intrusions feeding the summit and distal eruptions not only follow different vertical pathways but also propagate laterally along two distinct systems of rift zones, the summit and outer rift zones (Fig. 7.6b). Considering the geophysical data (Nercessian et al. 1996; Prôno et al. 2009), the dynamics of the summit and distal eruptions (Lénat and Bachèlery 1990) and the observations made in the incised Piton des Neiges volcano (Famin and Michon 2010; Chaput et al. 2014a), we tentatively propose the existence of a single main magma storage below sea level (Fig. 7.3b). This magmatic zone feeds a superficial storage system below the Central Cone that is involved during the summit eruptions. Magma injections from the eastern part of the main magmatic system occasionally occur, forming sill intrusions that act as detachment levels capable to trigger lateral flank displacement (Michon and Famin 2010; Chaput et al. 2014b) and subsequent dyke intrusions along the outer rift zones.

Finally, among the 81 eruptions, the March–April 1998 Hudson eruption did not follow the summit or outer rift zones. Its magma chemical composition, similar to that of the eccentric cones of the NW–SE rift zone (Pietruszka et al. 2009; Lénat et al. 2012), and its location south-west of the Central Cone suggest that the magma intrusion was not controlled by the summit or distal rift zones but instead followed the NW–SE rift zone.

7.3.2 The NW–SE Rift Zone, the South Volcanic Zone and the Deep Plumbing System

The scoria cone density map clearly shows that eruptions occurred along the NW–SE rift zone and the South Volcanic Zone, both trend in the N120 direction (Fig. 7.4b). The NW–SE rift zone, 15 km in width, is composed of about 160 scoria cones and pit craters scattered between the Enclos Fouqué caldera in the east and the eastern slopes of Piton des Neiges in the west. The volcanic activity started at least 29 kiloyears ago (McDougall 1971) and continued until recently (140 yr BP for Piton Rampe 14), given an

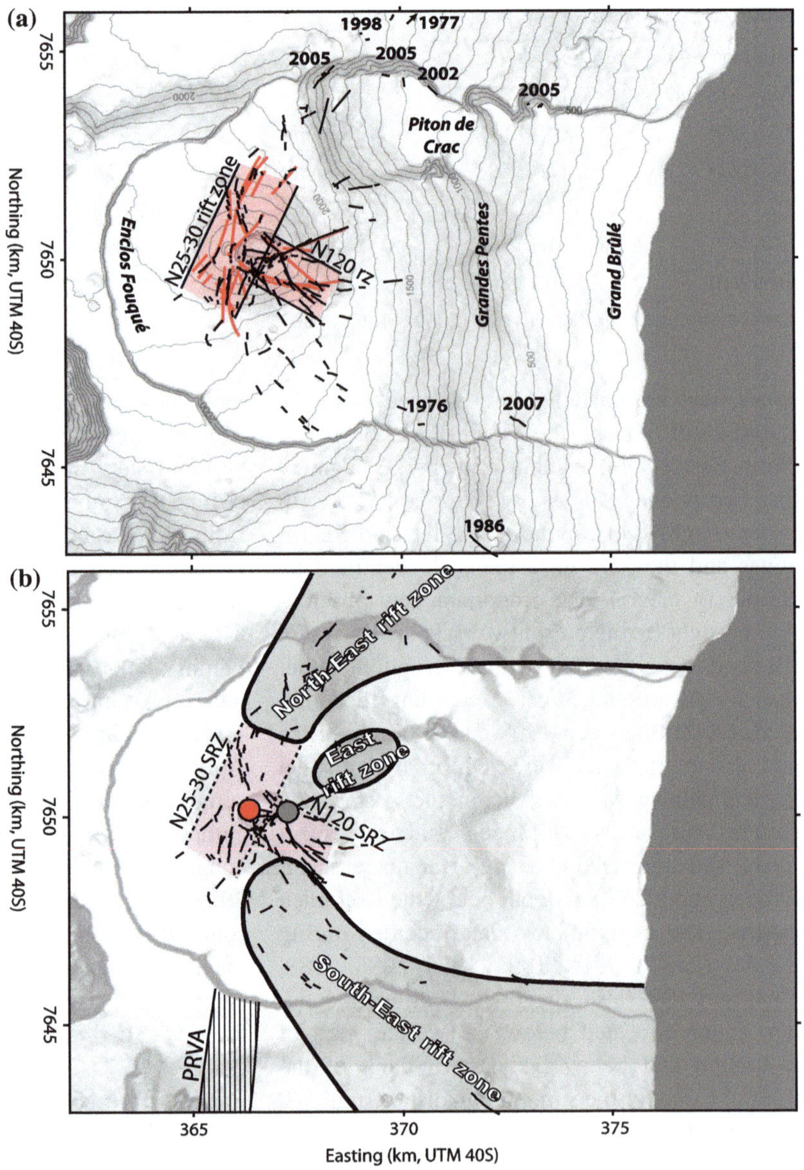

Fig. 7.6 **a** Eruptive fissures formed between 1972 and 2010 (*black*), summit dykes for the period between 1981 and 2007 (*red*, after Michon et al. 2009) and summit rift zones channeling the summit eruptions. Dates indicate the eruptive vents opened outside the Enclos Fouqué caldera or close to the northern and southern walls of the Grandes Pentes. **b** Distribution of the outer (*grey*) and summit (*light red*) rift zones related to the distal and summit eruptions, respectively. *Red* and *grey circles* indicate the location of the edifice inflation triggered by magma intrusions before the summit and distal eruptions, respectively

average eruption periodicity of around 200 years (Morandi et al. 2016; Chap. 8). Interestingly, the NW–SE rift zone is superimposed to a deep seismicity concentrated in a N120 trending zone ranging mostly between 15 and 20 km bsl, i.e. in the mantle lithosphere, just below the oceanic crust (Fig. 7.5). This seismicity is particularly active north of the Plaine des Remparts where the cone density is maximum. It is also worth noting that below 20 km bsl, the seismicity is organized in an orthogonal direction joining Le Tampon, the Plaine des Cafres and the Plaine des Palmistes (Fig. 7.5b).

A link between the current deep seismicity and the pre-historical scoria cones may exist only if (1) the volcano-tectonic events are associated with magma transfer instead of purely tectonic (a-magmatic) processes and (2) magmas involved in the eruptions along the NW–SE rift zone originate from at least 15 km in depth. Such a relationship is supported by the following geophysical and geochemical data. First, during the April 2007 major eruption, the seismicity progressively migrated down to 7 km bsl. The migration continued after the eruption along the NW–SE rift zone, down to about 30 km below sea level (Fig. 7.5a; Massin 2009). A similar downward migration of the seismicity was recorded during the Eyjafjallajökul 2010 eruption, where it was interpreted as resulting from a downward-propagating decompression of the plumbing system (Tarasewicz et al. 2012). Second, magma involved in the eruptions along the NW–SE rift zone indicates a high-pressure (>0.4 GPa) fractionation of clinopyroxene ± olivine (Kornprobst et al. 1984; Albarède et al. 1997; Famin et al. 2009; Lénat et al. 2012), which is consistent with the depth of the earthquakes located below the rift zone (depth >15 km). Third, high-pressure fluid inclusions (P up to 400–500 MPa) are common olivine carried by magmas erupted along the N120 rift zone (Piton des Cailles; Chisny; Bureau et al. 1998). Combining these geochemical and geophysical data, we tentatively interpret the deep zone of dense seismicity below the Plaine des Cafres and Plaine des Palmistes, in the west, and the NW–SE rift zone toward the east as the deep plumbing system of Piton de la Fournaise. This zone of magma transfer could feed recurrent eruptions and subsequently form a deeply rooted NW–SE rift zone characterized by a low eruption rate.

Contrary to the NW–SE rift zone, the 20-km-long South Volcanic Zone is not associated with recent seismic events. The wide range of degradation stages of the large volcanic cones and the related lava flows, from smooth—recent—to incised—old—morphologies, suggests a recurrent volcanic activity along this intrusion path. Yet, the lack of chronological data combined with the location of the volcanic zone in apparent disconnection with the active volcanic area makes the origin and the dynamics of this volcanic zone enigmatic.

An additional, short vent alignment, the Puys Ramond-Baril volcanic alignment, occurs south of the Enclos Fouqué by the density map (Fig. 7.4b). This narrow, 10-km-long area is composed of several spatter and cinder conesthat are in average significantly smaller than those of the NW–SE rift zone and the South Volcanic Zone, but similar to those found in the Enlos Fouqué and along the SE and NE rift zones. Radiocarbon datings and the occurrence of dykes and buried cones in the southern wall of the Enclos Fouqué caldera indicate a long-lasting activity, which developed until recently (355 BP; Bachèlery 1981; Tanguy et al. 2011; Morandi et al. 2016, Chap. 8). Despite its location close to the summit plumbing system, the link between this intrusion part and the summit activity and/or the SE rift zone remains unclear.

7.4 Comparison with Other Volcanoes

Our work clearly shows that different types of rift zone developed on Piton de la Fournaise since at least 30 kiloyears. On the one hand, the 120 N trending NW–SE rift zone and the South Volcanic Zone present strong similarities with the rift developed in the post shield phase in Mauritius and Hawaii alkaline postshield rift zones in Hawaii (Clague 1987; Moore et al. 2011). In all these cases, they correspond to wide, linear rift zones with eruptions with a secular frequency involving magmas fractionating at minimum depths of 15–20 km (Roeder 1965; Clague 1987; Albarède et al. 1997; Lénat et al. 2012). On the other hand, the arched rift zones geometry of the NE and SE outer rift zones resembles to that of the NE and S rift zones of Mount Etna (Kieffer 1975). Following the model proposed by Swanson et al. (1976) for Kilauea, Bachèlery (1981), Duffield et al. (1982), Lénat et al. (1989b) postulated that the recurrent dyke intrusions along the NE and SE rift zones forcefully push the east flank and subsequently triggered the lateral collapse of the eastern flank of Piton de la Fournaise. However, determining whether or not dyking along the outer rift zones of Piton de la Fournaise actively pushes the east flank or if it passively follows the upper boundaries of a sliding sectors, like at Mount Etna (Acocella et al. 2003; Neri et al. 2005) requires at least a thorough inspection of the rift zone and eruption dynamics, which is beyond the scope of this review. Recent work carried out on Piton des Neiges volcano, the inactive twin of Piton de la Fournaise, suggests that rift zone development and flank landslides are linked by a common process, sill injections below the volcano flanks that act as detachment (Famin and Michon 2010; Chaput et al. 2014a).

References

Acocella V, Behncke B, Neri M, D'Amico S (2003) Link between major flank slip and 2002–2003 eruption at Mt. Etna (Italy). Geophys Res Lett 30(24). doi:10.1029/2003GL018642

Aki K, Ferrazzini V (2000) Seismic monitoring and modeling of an active volcano for prediction. J Geophys Res 105(B7):16617–16640

Albarède F, Luais B, Fitton G, Semet M, Kaminski E, Upton B, Bachèlery P, Cheminée JL (1997) The geochemical regimes of Piton de la Fournaise volcano (Réunion) during the last 530,000 years. J Petrol 38 (2):171–201

Bachèlery P (1981) Le Piton de la Fournaise (Ile de la Réunion). Etude volcanologique, structurale et pétrologique. Unpublished Ph.D. thesis, pp 215, Univ. Clermont-Ferrand II, Clermont-Ferrand, France

Bachèlery P, Chevallier L, Gratier JP (1983) Caractères structuraux des éruptions historiques du Piton de la Fournaise (Ile de la Réunion). C R Acad Sci Paris 296:1345–1350

Battaglia J, Bachèlery P (2003) Dynamic dyke propagation deduced from tilt variations preceding the March 9, 1998, eruption of the Piton de la Fournaise volcano. J Volcanol Geoth Res 120:289–310

Battaglia J, Ferrazzini V, Staudacher T, Aki K, Cheminée JL (2005) Pre-eruptive migration of earthquakes at the Piton de la Fournaise volcano (Réunion Island). Geophys J Int 161(2):549–558. doi:10.1111/j.1365-246X.2005.02606.x

Bonali FL, Corazzato C, Tibaldi A (2011) Identifying rift zones on volcanoes: an example from La Réunion island. Indian Ocean Bull Volcanol 73(3):347–366. doi:10.1007/s00445-010-0416-1

Bureau H, Pineau F, Metrich N, Semet M, Javoy M (1998) A melt and fluid inclusion study of the gas phase at Piton de la Fournaise volcano (Reunion Island). Chem Geol 147:115–130

Carter A, Wyk de Vries B, Kelfoun K, Bachèlery P, Briole P (2007) Pits, rifts and slumps: the summit structure of Piton de la Fournaise. Bull Volcanol 69 (7):741–756. doi:10.1007/s00445-006-0103-4

Cayol V, Cornet FH (1998) Three-dimensionnal modeling of the 1983–1984 eruption at Piton de la Fournaise volcano, Réunion Island. J Geophys Res 103:18025–18037

Chadwick WW, Howard KA (1991) The pattern of circumferential and radial eruptive fissures on the volcanoes of Fernandina and Isabela islands. Galapagos Bull Volcanol 53(4):259–275

Chadwick W, Dieterich J (1995) Mechanical modeling of circumferential and radial dike intrusion on galapagos volcanoes. J Volcanol Geoth Res 66:37–52

Chaput M, Famin V, Michon L (2014a) Deformation of basaltic shield volcanoes under co-intrusive stress permutations. J Geophys Res Solid Earth 119. doi:10.1002/2013JB010623

Chaput M, Pinel V, Famin V, Michon L (2014b) Cointrusive shear displacement by sill intrusion in a detachment: a numerical approach. Geophys Res Lett 41. doi:10.1002/2013GL058813

Clague DA (1987) Hawaiian xenolith populations, magma supply rates, and development of magma chambers. Bull Volcanol 49:577–587

Delaney PT, Pollard DD (1981) Deformation of host rocks and flow of magma during growth of minette dikes and -bearing intrusions near Ship Rock, New Mexico. US Geol Surv Prof Pap 1202:1–61

Delorme H (1994) Apport des déformations à la compréhension des mécanismes éruptifs: le Piton de la Fournaise. Unpublished HDR Thesis pp. 615. University Paris VII

Delorme H, Bachèlery P, Blum PA, Cheminée JL, Delarue JF, Delmond JC, Him A, Lépine JC, Vincent PM, Zlotnicki J (1989) March 1986 eruptive episodes at Piton de la Fournaise volcano (Reunion Island). J Volcanol Geoth Res 36:199–208

Dieterich JH (1988) Growth and persistence of Hawaiian volcanic rift zones. J Geophys Res 93(B5):4258–4270

Duffield WA, Stieljes L, Varet J (1982) Huge landslide blocks in the growth of Piton de la Fournaise, La Réunion, and Kilauea Volcano, Hawaii. J Volcanol Geoth Res 12:147–160

Famin V, Michon L (2010) Volcano destabilization by magma injections in a detachment. Geology 38 (3):219–222

Famin V, Welsch B, Okumura S, Bachèlery P, Nakashima S (2009) Three differentiation stages of a single magma at Piton de la Fournaise volcano (Reunion hot spot). Geochem Geophys Geosyst 10 (1). doi:10.1029/2008GC002015

Fiske RS, Jackson ED (1972) Orientation and growth of Hawaiian volcanic rifts: the effect of regional structure and gravitational stress. Proc Roy Soc London Series A 329:299–320

Fontaine FR, Barruol G, Tkalcic H, Haugmard M (2012) Mapping crustal structure variation beneath the Piton de la Fournaise volcano. Abstract T31B-2590 presented at 2012 Fall Meeting. AGU. San Francisco, California, pp 3–7

Fukushima Y, Cayol V, Durand P (2005) Finding realistic dike models from interferometric synthetic aperture radar data: the February 2000 eruption at Piton de la Fournaise. J Geophys Res Solid Earth 110(B3). doi:10.1029/2004JB003268

Gallart J, Driad L, Charvis P, Sapin M, Hirn A, Diaz J, de Voogd B, Sachpazi M (1999) Perturbation to the lithosphere along the hotspot track of La Réunion from an offshore-onshore seismic transect. J Geophys Res 104(B2):2895–2908

Hirn A, Lépine J, Sapin M, Delorme H (1991) Episodes of pit-crater collapse documented by seismology at Piton de la Fournaise. J Volcanol Geoth Res 47:89–104

Kieffer G (1975) Sur l'existence d'une 'rift zone' à l'Etna (Sicile). C R Acad Sci Paris 280:263–266

Kornprobst J, Boivin P, Lénat JF, Bachèlery P, Bonneville A, Dupont P, Lecointre J, Seidel JL, Thomas P, Vincent PM (1984) Le Piton de la Fournaise, île de la Réunion. Colloque Prévision et Surveillance des Eruptions Volcaniques. (C.N.R.S.–I.N.A.G.) Clermont-Ferrand, pp 75–82

Lénat JF, Bachèlery P, Bonneville A, Hirn A (1989a) The beginning of the 1985–1987 eruptive cycle at Piton de la Fournaise (La Réunion); New insights in the magmatic and volcano-tectonic systems. J Volcanol Geoth Res 36:209–232

Lénat JF, Vincent P, Bachèlery P (1989b) The off-shore continuation of an active basaltic volcano: piton de la Fournaise (Reunion Island, Indian Ocean): structural and geomorphological interpretation from Sea Beam mapping. J Volcanol Geoth Res 36:1–36

Lénat JF, Bachèlery P (1990) Structure and dynamics of the central zone of Piton de la Fournaise volcano. In: Lénat JF (ed) Le volcanisme de la Réunion, Monographie. Cent. De Rech. Volcanol, Clermont Ferrand, pp 257–296

Lénat JF, Bachèlery P, Merle O (2012) Anatomy of Piton de la Fournaise volcano (La Réunion, Indian Ocean). Bull Volcanol 74:1945–1961. doi:10.1007/s00445-012-0640-y

Longpré MA, Staudacher T, Stix J (2007) The November 2002 eruption at Piton de la Fournaise volcano, La Réunion Island: ground deformation, seismicity, and pit crater collapse. Bull Volcanol 69(5):511–525. doi:10.1007/s00445-006-0087-0

MacDonald GA (1972) Volcanoes. Prentice-Hall Inc Englewood Cliffs, New Jersey, 510 pp

McDougall I (1971) The geochronology and evolution of the young volcanic island of Réunion (Indian Ocean). Geochim Cosmochim Acta 35:261–288

Marinoni LB, Gudmundsson A (2000) Dykes, faults and palaeostresses in the Teno and Anaga massifs of Tenerife (Canary Islands). J Volcanol Geoth Res 103 (1):83–103

Massin F (2009) Transferts et Stockages Magmatiques au Piton de la Fournaise. unpublished Ph.D. thesis, Université de la Réunion, St Denis, France, pp 176

Michon L, Saint-Ange F, Bachèlery P, Villeneuve N, Staudacher T (2007) Role of the structural inheritance of the oceanic lithosphere in the magmato-tectonic evolution of Piton de la Fournaise volcano (La Reunion Island). J Geophys Res Solid Earth 112: B04205. doi:10.1029/2006JB004598

Michon L, Cayol V, Letourneur L, Peltier A, Villeneuve N, Staudacher T (2009) Edifice growth, deformation and rift zone development in basaltic setting: insights from Piton de la Fournaise shield volcano (Reunion Island). J Volcanol Geoth Res 184:14–30. doi:10.1016/j.jvolgeores.2008.11.002

Michon L, Di Muro A, Villeneuve N, Saint-Marc C, Fadda P, Manta F (2013) Explosive activity of the summit cone of Piton de la Fournaise volcano (La Réunion island): a historical and geological review. J Volcanol Geoth Res 263:117–133. doi:10.1016/j.jvolgeores.2013.06.012

Montgomery-Brown EK, Sinnett DK, Larson KM, Poland MP, Segall P, Miklius A (2011) Spatiotemporal evolution of dike opening and décollement slip at Kīlauea Volcano, Hawaii. J Geophys Res 116(B3). doi:10.1029/2010JB007762

Moore J, White WM, Paul D, Duncan RA, Abouchami W, Galer SJG (2011) Evolution of shield-building and rejuvenescent volcanism of Mauritius. J Volcanol Geoth Res 207(1–2):47–66. doi:10.1016/j.jvolgeores.2011.07.005

Morandi A, Di Muro A, Principe C, Michon L, Leroi G, Norelli F, Bachèlery P (2016) Pre-historic explosive activity at Piton de la Fournaise volcano. In: Bachèlery P, Lénat JF, Di Muro A, Michon L (eds) Active volcanoes of the Southwest Indian Ocean: Piton de la Fournaise and Karthala. Active Volcanoes of the World. Springer, Berlin

Nercessian A, Hirn A, Lépine JC, Sapin M (1996) Internal structure of Piton de la Fournaise volcano from seismic wave propagation and earthquake distribution. J Volcanol Geoth Res 70(3–4):123–143

Neri M, Acocella V, Behncke B, Maiolino V, Ursino A, Velardita R (2005) Contrasting triggering mechanisms of the 2001 and 2002–2003 eruptions of Mount Etna (Italy). J Volcanol Geoth Res 144(1–4):235–255. doi:10.1016/j.jvolgeores.2004.11.025

Peltier A, Ferrazzini V, Staudacher T, Bachèlery P (2005) Imaging the dynamics of dyke propagation prior to the 2000–2003 flank eruptions at Piton de la Fournaise, Reunion Island. Geophys Res Lett 32(22). doi:10.1029/2005GL023720

Peltier A, Famin V, Bachèlery P, Cayol V, Fukushima Y, Staudacher T (2008) Cyclic magma storages and transfers at Piton de la Fournaise volcano (La Réunion hotspot) inferred from deformation and geochemical data. Earth Planet Sci Lett 270(3–4):180–188. doi:10.1016/j.epsl.2008.02.042

Peltier A, Bachèlery P, Staudacher T (2009) Magma transport and storage at Piton de la Fournaise (La Réunion) between 1972 and 2007: a review of geophysical and geochemical data. J Volcanol Geoth Res 184(1–2):93–108. doi:10.1016/j.jvolgeores.2008.12.008

Pietruszka AJ, Hauri EH, Blichert-Toft J (2009) Crustal contamination of mantle-derived magmas within Piton de la Fournaise, Réunion Island. J Petrol 50:661–684

Pollard DD (1987) Elementary fracture mechanics applied to the structural interpretation of dykes. In: Halls HC, Fahrig WF (eds) Mafic dyke swarms. Geological Association of Canada Special Paper, vol 34, pp 5–24

Prôno E, Battaglia J, Monteiller V, Got JL, Ferrazzini V (2009) P-wave velocity structure of Piton de la Fournaise volcano deduced from seismic data recorded between 1996 and 1999. J Volcanol Geoth Res 184(1–2):49–62. doi:10.1016/j.jvolgeores.2008.12.009

Roedder E (1965) Liquid CO_2 inclusions in olivine-bearing nodules and phenocrystsfrom basalts. Am Min 50:1746–1782

Roult G, Peltier A, Taisne B, Staudacher T, Ferrazzini V, Di Muro A, team TO, (2012) A new comprehensive classification of the Piton de la Fournaise activity spanning the 1985–2010 period. Search and analysis of short-term precursors from a broad-band seismological station. J Volcanol Geoth Res 241–242: 78–104. doi:10.1016/j.jvolgeores.2012.06.012

Swanson DA, Duffield WA, Fiske RS (1976) Displacement of the south flank of Kilauea volcano: the result of forceful intrusion of magma into the rift zones. US Geol Surv Prof Pap 963:39 p

Tanguy JC, Bachèlery P, Le Goff M (2011) Archeomagnetism of Piton de la Fournaise: bearing on volcanic activity at La Réunion Island and geomagnetic secular variation in Southern Indian Ocean. Earth Planet Sci Lett 303(3–4):361–368. doi:10.1016/j.epsl.2011.01.019

Tarasewicz J, White RS, Woods AW, Brandsdóttir B, Gudmundsson MT (2012) Magma mobilization by downward-propagating decompression of the Eyjafjallajökull volcanic plumbing system. Geophys Res Lett 39(19). doi:10.1029/2012GL053518

Tibaldi A (1996) Mutual influence of diking and collapses at Stromboli volcano, Aeolian Arc, Italy. Geol Soc London Spec Publ 110:55–63

Tibaldi A (2003) Influence of volcanic cone morphology on dikes, Stromboli, Italy. J Volcanol Geoth Res 126:79–95

Toutain JP, Bachèlery P, Blum PA, Cheminée JL, Delorme H, Fontaine L, Kowalski P, Taochy P (1992) Real time monitoring of vertical ground deformations during eruptions at Piton de la Fournaise. Geophys Res Lett 19(6):553–556

van Wyk de Vries B, Matela R (1998) Styles of volcano-induced deformation: numerical models of substratum flexure, spreading and extrusion. J Volcanol Geoth Res 81(1–2):1–18

Villeneuve N, Bachèlery P (2006) Revue de la typologie des éruptions au Piton de la Fournaise, processus et risques volcaniques associés. Cybergeo: European J Geography. http://cybergeo.revues.org/2536)

Walker GPL (1999) Volcanic rift zones and their intrusion swarms. J Volcanol Geoth Res 94(1–4):21–34

Walter TR, Troll VR (2003) Experiments on rift zone evolution in unstable volcanic edifices. J Volcanol Geoth Res 127(1–2):107–120. doi:10.1016/S0377-0273(03)00181-1

Walter TR, Troll VR, Cailleau B, Belousov A, Schmincke HU, Amelung F, Bogaard P (2005) Rift zone reorganization through flank instability in ocean island volcanoes: an example from Tenerife, Canary Islands. Bull Volcanol 67(4):281–291. doi:10.1007/s00445-004-0352-z

Pre-historic (<5 kiloyear) Explosive Activity at Piton de la Fournaise Volcano

8

Andrea Morandi, Andrea Di Muro, Claudia Principe, Laurent Michon, Gabrielle Leroi, Francesco Norelli and Patrick Bachèlery

Abstract

The characterization of the recent (<5 kiloyears) explosive activity and the research of violent paroxystic events over Piton de la Fournaise edifice has been performed through a drilling and excavation campaign supported by the integration of new radiocarbon ages to previous chronologic data. Fine grained "Bellecombe" phreatomagmatic ashes represent the product of the most violent explosive Piton de la Fournaise activity inside the investigated period. This activity results from a series of eruptions occurred over a time span much longer than previously thought. Anyhow, it represents the most traceable horizon (up to 13 km W-NW from the central cone) among the studied deposits and no other pyroclastic blanket exhibits a similar regional dispersion. The lack of a continuous lapilli/ash horizon from the proximal to

A. Morandi
Department of Earth Sciences,
University of Florence, Florence, Italy

A. Di Muro · G. Leroi
Institut de Physique du Globe OVPF, Université Paris Diderot, Sorbonne Paris Cité, CNRS UMR-7154, 75005 Paris, France

C. Principe (✉) · F. Norelli
Istituto di Geoscienze e Georisorse—CNR,
via G. Moruzzi 1, 56124 Pisa, Italy
e-mail: c.principe@igg.cnr.it

L. Michon
Laboratoire Géosciences Réunion Université de La Réunion, Institut de Physique du Globe de Paris, Sorbonne Paris Cité CNRS, 97744 Saint-Denis, France

P. Bachèlery
Laboratoire Magmas et Volcans UMR, Observatoire de Physique du Globe de Clermont-Ferrand. Université Blaise Pascal, CNRS-IRD 6524 5, rue Kessler, 63038 Clermont-Ferrand, France

P. Bachèlery et al. (eds.), *Active Volcanoes of the Southwest Indian Ocean*,
Active Volcanoes of the World, DOI 10.1007/978-3-642-31395-0_8

distal areas points out that the lapilli cover on the volcano flank result of several local blankets linked to as many Hawaiian to Strombolian emission centers. As highlighted by new radiocarbon ages this kind of activity persisted along the N120 rift zone up to very recent times and ended probably just before the island's colonization, leaving no trace in historical records. The maximum expected magmatic event (Chisny-type eruption) has therefore to be related to intense Hawaiian fountaining. Its dangerousness is restricted to a relatively brief distance from the source and a regional deposition is strongly unlikely. The hazard represented by this type of activity resides in the possible positioning of the vent close to inhabited areas and in the possible occurrence of repeated events with short but unpredictable time interval from one to the other. Forecasting the location of the future eccentric eruptions is thus of paramount importance to minimize the potential impact of mild explosive eruptions on the inhabitants and the infrastructures.

8.1 Introduction

Piton de la Fournaise (PdF) is one of the most active basaltic intra-plate volcanoes. Even if the number of identified eruptive events depends on the completeness of written records (Lénat 2016, Chap. 1), at least 238 eruptions have been recognized since the second half of the 17th century, when permanent settlement on the island began (Ludden 1977; Bachèlery 1981; Lénat and Bachèlery 1988; Stieltjes and Moutou 1989; Roult et al. 2012; Di Muro et al. 2012; Michon et al. 2013). Since the establishment of permanent monitoring networks, coinciding with the foundation of Piton de la Fournaise Observatory (OVPF/IPGP) in 1979, there have been 61 instrumentally recorded eruptions. During the last 100 year, the average eruptive rate has been of about 1 eruption every 10 months and the longest eruptive pauses have occurred in recent times (1966–1972; 1992–1998) and have lasted up to 6 years. During recorded history, most of the eruptive vents were located close to the PdF central cone and more generally inside the horseshoe-shaped Enclos Fouqué depression, which encloses it. Only rare lateral eruptive fissures opened outside the Enclos Fouqué, along the NE and SE rift zones (Michon et al. 2016, Chap. 7; Staudacher et al. 2016, Chap. 9). Since human settlement on the island on 1638 CE, eruptive events have not been recorded along the third and largest NW rift zone materialized by a N 120° trending cone alignment. However, the first settlements on La Réunion were located on the opposite coast of the island (NW coast) in respect to PdF. The corollary is that written records cover only a very short period at La Réunion, reliable written reports on PdF activity dating only back to 1754 CE (Michon et al. 2013). Therefore, it is of paramount importance to assess volcanic hazard over a period long enough to be representative of the variability in eruptive dynamics of PdF.

Nowadays, the island accounts for more than 830,000 permanent residents and ca. 470,000 tourists per year. Densely populated cities and villages are widespread on the whole active volcanic massif and progressively develop towards the most active volcanic zone. In recent times, effusive eruptions have periodically affected important infrastructure like the eastern national road, buildings and farmlands. In 1977 and 1986, the volcanic activity produced the opening of lateral eruptive fissures emitting lava that flowed through inhabited areas (Piton Sainte-Rose and Tremblet, respectively) up to the sea-shore (Delorme et al. 1989; Kieffer et al. 1977).

Recent review of historic reports has revealed that weak to mild phreatic and phreatomagmatic ash plumes (the most violent corresponding to the 1860 CE eruption) repeatedly punctuated the dominantly effusive activity of the central cone, spreading fine ash on the volcano edifice and sometimes over most of La Réunion Island (Bachèlery 1981; Villeneuve and Bachèlery 2006; Peltier et al. 2012; Michon et al. 2013). Dispersal of Pele's hairs over the whole island can also occur, as in 1939 and in 2007 (Michon et al. 2013). More violent explosive phreatic to phreatomagmatic eruptions of PdF have traditionally been related to major old caldera forming events contributing to the formation of the Enclos Fouqué caldera. The Bellecombe Ash unit would account for such paroxysmal events around 4175–3200 BP (Bachèlery 1981; Mohamed-Abchir 1996). This chapter reviews the dynamics of pre-historic eruptions of PdF, whose age is comprised between the Bellecombe Ash unit and the recorded history. Magmatic explosive activity, like violent strombolian or subplinian events, cannot be a priori discarded even for a basaltic volcano in an intraplate environment. Even if effusive or mild explosive dynamics of Strombolian and Hawaiian types characterize the basaltic volcanic activity, it has been since long recognized that mafic magmas can produce violent explosive events, with ash columns up to 30 km high and their products spreading over an area of tens to more than 1000 km^2 (MacDonald 1972; Arrighi et al. 2001; Houghton et al. 2004; Sable et al. 2006, 2009; Pioli et al. 2008; Andronico et al. 2009; Wong and Larsen 2010). The possible occurrence of such kind of events has thus motivated the recent re-assessment of volcanic hazard at La Réunion (Di Muro et al. 2012). In this chapter, we address this issue by integrating previous and new stratigraphic and chronologic data obtained on stratigraphic sections and shallow drillings (<10 m).

8.2 Wind Distribution Over La Réunion Island and Microstratigraphic Drilling Strategy

Wind direction and strength play an important role on the regional distribution of pyroclastic products dispersed by vertical plumes carrying ash and lapilli. This is particularly true for the finest particles that can be carried tens to hundreds of kilometers away from the eruptive source. The main synoptic winds over La Réunion Island are the Trade winds that blow northwestwards dominating the air circulation at ground level (Trenberth et al. 1990). Local breezes effects become more influent mostly in downwind and protected areas, where the surface winds are weaker (Lesouëf et al. 2011).

The vertical winds distribution over La Réunion shows a marked inversion of their direction at an altitude of about 3000–4000 m asl (Taupin et al. 1999), depending on the season. The dispersion of pyroclasts is therefore strongly controlled by the height of the eruptive column. Examples are given by the ash and aerosol plumes generated during the April 2007 eruption, when an initial westward spreading of small ash and gas plumes was followed by a clockwise rotation toward N and NE sectors as the aerosol plume exceeded 4 km height (Tulet and Villeneuve 2011). Ash dispersal during recent phreatomagmatic explosive eruptions (e.g. 1860 CE) confirms this trend and suggests even the possibility of a splitting of the eruptive columns (Hugoulin 1860; Villeneuve and Bachèlery 2006). In agreement with this winds pattern, only sparse low amounts of basaltic shards, possibly attributed to PdF explosive events, were identified during the last 30 k year in a deep-sea marine core located south of the volcano edifice (Fretzdorff et al. 2000; Sisavath et al. 2012). Consequently, we performed in 2011 new microstratigraphic drillings and surface outcrops descriptions along a N-NW axis, between the proximal area (Plaine des Sables, about 5 km from the central cone) and the distal area (Grand Étang, 18 km NW from the summit vents), passing through the N120 rift zone. As a whole, 30 outcrops were described and 5 drillings were performed (Fig. 8.1).

The investigation of the spatial distribution of pyroclastic deposits must take carefully into account their potential preservation over the territory. This is particularly true in the tropical environment of La Réunion where a steep, pronounced topography is associated with extreme weather conditions (up to 6 m of rain per year

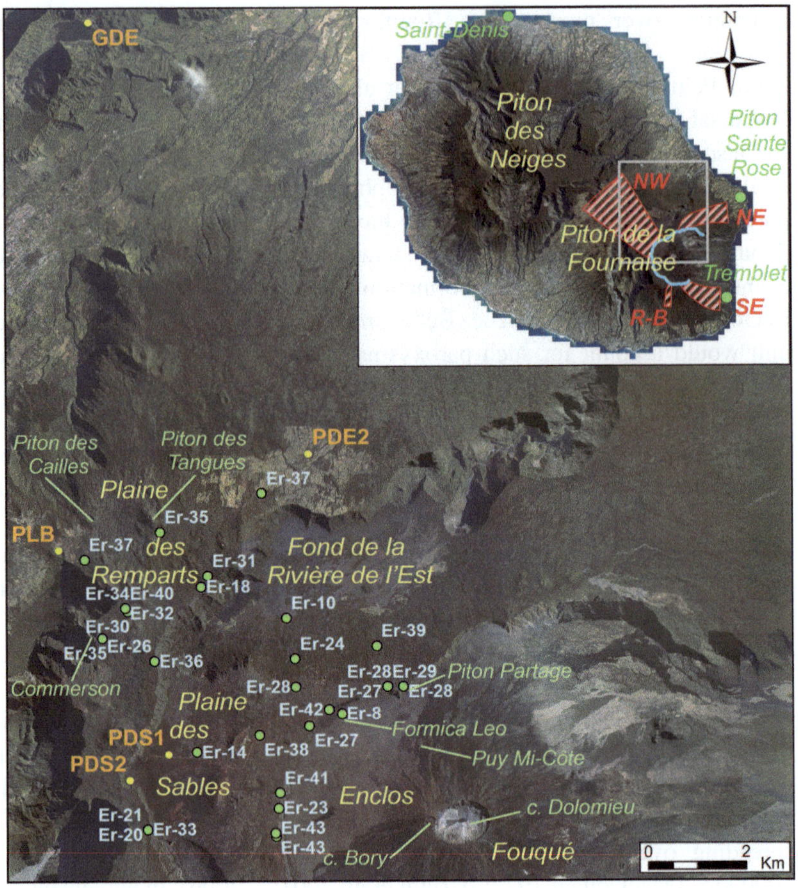

Fig. 8.1 Drilling locations (*yellow dots*) and dated eruptive events, inside stratigraphic sections, younger than 5000 year (*green dots*) on the NW rift. In the *upper right corner*: the Enclos Fouqué rim (*light blue*) and the three major rift zones (*red areas*) intersecting the PdF massif are shown. NW, SE and NE stand respectively for NW, SE and NE rift; R-B is the Ramond-Baril eruptive axis, corresponding to a N-S scoria cones alignment

and wind speed up to 250 km/h; Météo France source). Such a natural background produces a high rate of removal and reworking especially on the incoherent and easy-erodible pyroclastic sediments. The preservation potential of loose lapilli fallouts or thin ash layers (typical of small eruptions or distal outcrops) is therefore very low. As an example, the 1–2 cm-thick ash deposit from the 1986 Dolomieu phreatic explosion on the slopes of the summit cone of PdF was completely removed in less than one week, after two rainy days. In order to overcome these severe constraints, we selected low-energy sedimentation environments as drilling sites. This condition is better achieved in the endorheic volcanic lakes where the sedimentation is a quite sporadic event (Principe et al. 1997, 2007). In these setting, the deposition of tephra is limited to primary fallout deposits and the reworking is restricted to the bank's material. At La Réunion, this kind of sampling point is represented by intracrateric ponds on the top of strombolian cones spotted over the territory. On the oter hand, we have to stress that the volume of these permanent ponds/lakes is highly sensitive to the alternation of extreme pluviometric conditions and dryer periods. In other terms the sedimentation conditions are never optimal. To reach the selected locations, we used a portable drilling system (AF topcore 40 modified; Fig. 8.2a, b), able to produce 10 m long continuous and undisturbed micro-cores (Principe et al. 1997). This method is able to avoid the loss of information even for very thin layers, due to the presence of a textile sheet around the core (Fig. 8.2c). Wherever possible, the condensed tephra sequences, which sediment on the lakes shores, are drilled in order to reach the maximum length of tephra record with the minimum coring depth (Principe et al. 2007).

Fig. 8.2 AF topcore 40 modified coring system and drilling operation. **a** The infission is performed by a jackhammer thrust, while **b** the extraction by a mechanical jack system. **c** Thin ash beds preserved in the PDS2 core

8.3 Drilled Cores and Outcrop Description

The performed cores have been opened in laboratory for description and sampling (Fig. 8.2c). Macroscopical description has been performed and tephra layers have been characterized by means of their color, texture, granulometry, componentry and mineralogical assemblage. Thanks to the broad spreading of the grey, fine-grained Bellecombe Ash falls and to their peculiar mineral assemblage, i.e. abundant pyroxene + olivine crystals together with gabbro fragments and rare quartz crystals, they represented a perfect stratigraphic marker to identify tephra sequences younger than ca. 4000 year in the proximal-medial areas of the volcano.

In the Grand Étang distal drilling (GDE; Table 8.2) two units of fine ashes overly the proximal facies of a scoria cone, they contain a variable amount of granule to pebble-size clasts, the lower unit is characterized by the presence of organic matter. In the middle sector, Piton de l'Eau core (PDE2; Table 8.3) shows, within the scoria cone edifice (represented by the lower poorly sorted breccia), the presence of fine-grained deposits related to a reddish ashy paleosol containing organic matter and an upper stratigraphic succession of ashes which includes the Bellecombe Ash unit. Piton dans le Bout core (PLB; Table 8.4) is instead made of six units, of approximately alternating lapilli and ash beds. Bellecombe Ash unit underlies the uppermost black lapilli bed. We here stress that the "Bellecombe ash unit" group a sequence of events spanning about 2 millennia. The cpx-gabbro rich "Bellecombe ashes" cropping out in the Piton de l'eau—Piton dans le Bout area has been dated at 3735 ± 90 and 3715 ± 30 BP (Table 8.1). Younger ages in the range 3110 ± 30 BP–3060 ± 35 have been obtained for the uppermost beds of the "Bellecombe ash unit" along the Langevin plateau (Table 8.1). Notably, the ages of the lowermost beds in the "Bellecombe unit" cropping out in the proximal-medial area are somewhat older (4880 ± 35 BP; 4175 ± 145 BP; Table 8.1) than the uppermost beds, which attain the medial and distal sectors.

In the proximal sector, the two cores of the Plaine des Sables (PDS2 and PDS1; Tables 8.5 and 8.6) show, above the Bellecombe Ash unit, the presence of a black, glassy, and largely reworked lapilli sequence, which is intercalated with thin lava flows when approaching the Piton Chisny cone. In Tables 8.2, 8.3, 8.4, 8.5 and 8.6

Table 8.1 Dated eruptive event at Piton de La Fournaise volcano

Eruption ID	Location	Unit name	Unit type	Note on the sample	Coordinates		Dating method	Age (BP)	±	Median probability (A.D.)	Calibrated min (A.D) 2σ	Calibrated max (A.D.) 2σ
					x	y						
Er-1	RZNE	1998 lava flow	Lava flow							1998	1998	1998
Er-2	RZSE	1986 lava flow	Lava flow							1986	1986	1986
Er-3	RZNE	1977 lava flow	Lava flow							1977	1977	1977
REU140212-10/9	Central activity	Partage cliff	Ash beds	Historical eruptions; B layer; multiple fragments	366,212.7	7,653,718.8	14C	120	30	1831	1939	1679
REU140212-S1-2b	Central activity	Partage cliff	Ash beds	Historical eruptions; B1 layer; top of historical sequence; single fragment	365,582.4	7,653,224.7	14C	130	30	1821	1941	1675
Er-4	RZSE	1800 lava flow	Lava flow							1802	1802	1802
Er-10	N120	Piton Rampe 14	Strombolian cone	Below the lava flow	363,034.827	7,654,507.466	14C	140	90	1791	nd	1528
AND 02 / REU140721 / S1	N120	Trous Blancs area	Modern soil	Uppermost layer rich in roots and organic matter; multiple fragments	355,338.72	7,655,654.69	14C	145	30	1801	1948	1668
REU031014-3	N120	Rivière des Remparts					14C	150	30	1785	1950 (nd)	1667
Er-5	RZSE	1776 lava flow	Lava flow							1776	1776	1776
Er-6	RZSE	1774 lava flow	Lava flow							1776	1774	1774
REU140212-S4-2	Central activity	Partage cliff	Ash beds	Historical eruptions; B layer; top of historical sequence; multiple fragments	364,086.9	7,653,098.6	14C	160	30	1773	1950 (nd)	1664
REU140521-BB1	RZNE	Piton Moka (near Anses des Cascades)	Ash beds	In lapilli fall above the lahar reworking Bellecombe breccia; modern agricultural activity? multiple fragments	376,943.4	7,656,622.4	14C	175	30	1770	1950 (nd)	1657
AND 08 / REU140728-2a / S1	N120	Top of Trous Blancs site	Modern soil		355,672.20	7,654,589.84	14C	195	30	1770	1950 (nd)	1648
Er-7	RZSE	Puy Raymond	Strombolian cone	Inside ashes. 1 m below the lava flow	365,883.571	7,645,619.532	14C	45-115	75		nd	nd

(continued)

Table 8.1 (continued)

Eruption ID	Location	Unit name	Unit type	Note on the sample	Coordinates		Dating method	Age (BP)	±	Median probability (A.D.)	Calibrated min (A.D) 2σ	Calibrated max (A.D.) 2σ
					x	y						
Er-8	EF (lava above the Formica leo cone)	CLEF lava field	Lava field	Pahoehoe lava	364,160.653	7,652,535.955	Archeomagnetic			1750	ca. 1750	
Er-8	EF	CLEF lava field	Lava field	Pahoehoe lava			Historical sources			1740	1730	1750
Er-9	RZNE	1708 lava flow	Lava flow							1708	1708	1708
Er-8	EF (lava close to Bellecombe steps)	CLEF lava field	Lava field	Sample RE-90-28; pahohehoe	364,160.653	7652535.955	Cosmogenic	250	800	1700	nd	nd
REU140212-S1-2a	Central activity	Partage cliff	Ash beds	Historical eruptions; B1 layer; top of historical sequence; multiple fragments	365,582.4	7,653,224.7	14C	255	30	1650	1950 (nd)	1521
Er-11	RZNE	Piton Nelson—Piton de Bois Blancs	Strombolian cone	Near Bois Blanc village	374,006.015	7,654,760.755	14C	260	80	1643	nd	1450
REU140320-S5-4	Central activity	Partage cliff	Ash beds	Historical eruptions; multiple fragments	366,109.7	7,653,616.5	14C	340	30	1559	1639	1470
Er-12	RZSE	Piton Taipoul	Strombolian cone	Below the scoriae of the cone and above an ash bed	366,430.576	7,645,458.601	14C	355	75	1549	1794	1427
Er-13	RZSE	Mare Longue lava flow	Lava flow	Below the lava flow			14C	364	25	1517	1632	1451
Er-15	N120	Piton Chisny—lava with dunite enclaves near Chisny cone	Polyphase cone	Near Chisny cone; phase 2 in Bachèlery 1981			Archeomagnetic			1500	1600	1400
Er-16	RZNE	Piton Indivis - Bois Blanc	Strombolian cone		376,033.1	7,655,634.376	14C	420	80	1500	1649	1327
Er-14	N120	Piton Chisny—Plaine des Sables black scoria fall	Multiple cones	Section 20; Plaine des Sables. near Piton Chisny	361,198.937	7,651,746.67	14C	381	26	1493	1630	1445
Er-17	RZSE	Ravine d'Ango lava	Strombolian cone	In conglomerate below the lava (the soil is not burned)	374,511.652	7,638,425.555	14C	455	70	1457	1635	1318
Er-18	N120	Petit Cratère	Strombolian cone	Below the lava flow	361,293.907	7,655,150.227	14C	470	75	1443	1634	1305
Er-19	RZSE	Le Baril 1 (Brulé du Baril)	Strombolian cone	Basal scoriae below the lava flow (Ravine	365,841.284	7,638,409.566	14C	575	75	1358	1444	1281

(continued)

Table 8.1 (continued)

Eruption ID	Location	Unit name	Unit type	Note on the sample	Coordinates		Dating method	Age (BP)	±	Median probability (A.D.)	Calibrated min (A.D) 2σ	Calibrated max (A.D.) 2σ
					x	y						
				Perote river) and above a soil								
Er-20	N120	Langevin plateau fall—bed 4	Scoriae bed	Section 1—Caldera des Sables cliff	360,176.998	7,650,157.619	14C	620	30	1349	1399	1292
Er-21	N120	Langevin plateau fall—bed 5	Scoriae bed	Section 1—Caldera des Sables cliff	360,176.998	7,650,157.619	14C	868	30	1177	1251	1046
Er-22	N120	"Piton Chisny lavas" in Langevin valley	Lava flow	Below aa lava flow in the Langevin valley	359,323.842	7,640,947.249	14C	1105	60	925	1022	775
Er-24	N120	Piton sous le Gite	Strombolian cone	Below the lava flow	363,211.109	7,653,672.45	14C	1465	75	578	675	420
Er-23	Central activity?	Ashes on western rim of Enclos Fouqué (near Chisny)	Ash beds		362,860.285	7,650,591.453	14C	1450	250	576	1122	50
Er-25	N120	Tephra on Les Trous Blancs	Scoria bed	Sample REU 1301-243 Trou Blanc	356,229.000	7,654,133.000	14C	1495	30	573	641	434
Er-26	N120	Commerson crater	Scoria bed		359,258.678	7,654,094.665	14C	1890	55	119	249	−20
Er-27	Central activity	Partage Tephra (Unit 3; CC2 layer)	Scoria bed	Section 6 (Partage cliff)	365,088.054	7,653,096.733	14C	2140	35	−179	−52	−356
Er-27	Central activity	Ashes on NW cliff of Enclos Fouqué (Bellecombe cliff)	Ash bed	Ashes draping the lower third of the Enclos Fouqué cliff	363495.684	7652284.643	14C	2140	80	−184	1	−381
Er-28	Central activity	Ashes near the "Gite du Volcan" ("Cendres de Bellecombe" in B'81 and V&B'82)	Ash beds	Sampled near the Gites du Volcan	363,223.126	7,653,091.586	14C	2320	90	−403	−178	−755
Er-28	Central activity	Partage Tephra (Unit 2; CC1 layer)	Ash bed	Section 6 (Partage cliff)	365,088.054	7,653,096.733	14C	2340	30	−402	−366	−506
Er-28	Central activity	Ashes near the "Gite du Volcan"	Ash beds	Upper part of an ash bed. below a bed of scoriae	365,412.448	7,653,098.027	14C	2390	85	−521	−235	−784
Er-28?	N120	Piton des Tangues tephra (Below Red Tephra Unit)	Lapilli bed?	Ash or lapilli bed above the Upper Bellecombe ash			14C	2665	615	−841	608	−2345
Er-29	Central activity	Partage Tephra (Unit1; CC0 layer)	Ash bed	Section 6 (Partage cliff)	365,088.054	7,653,096.733	14C	2855	35	−1021	−919	−1121
REU140825-PDT2	N120	Near Piton des Tangues parking	Ash beds	Black soil below red tephra lapillis covering Bellecombe falls	360,408.8	7,656,179.8	14C	2920	30	−1115	−1020	−1211

(continued)

Table 8.1 (continued)

Eruption ID	Location	Unit name	Unit type	Note on the sample	Coordinates		Dating method	Age (BP)	±	Median probability (A.D.)	Calibrated min (A.D) 2σ	Calibrated max (A.D.) 2σ
					x	y						
Er-30	Central activity	Upper unit of Bellecombe ash	Ash bed	Sample REU 1303-07 A1	359,775.000	7,654.645.000	14C	3060	35	−1326	−1228	−1411
Er-31	N120	Piton Sauvetage	Strombolian cone	Ash bed	361,424.89	7,655.382.022	14C	3085	75	−1335	−1127	−1501
Er-32	Central activity	Upper unit of Bellecombe ash	Ash bed	Sample REU 1303-07 A2 2A	359,728.000	7,654.714.000	14C	3095	30	−1350	−1280	−1428
Er-33	N120	Langevin plateau fall- bed 6	Ash and lapilli bed	Section 1—Caldera des Sables cliff	360,176.998	7,650.157.619	14C	3100	30	−1355	−1284	−1431
Er-34	Central activity	Upper unit of Bellecombe ash	Ash bed	Sample REU 1303-07 A2 2C/1 km west of Piton des Basaltes	359,728.000	7,654.714.000	14C	3110	30	−1377	−1288	−1437
Er-41	Central activity	Lava3 of W Enclos Fouqué rim	Lava flow	Sample RE-90-15b (below Bellecombe ash); oceanite	362,889.735	7,650.907.916	Cosmogenic	3340	1012	−1390		
REU140822-2	N120	0.5 km from Nez de Bœuf. above Bellecombe ash fall	Ash beds	Soil above Bellecombe and below a thin lapilli layer	357,317	7,655.847	14C	3130	30	−1407	−1300	−1494
Er-41	Central activity	Lava3 of W Enclos Fouqué rim	Lava flow	Sample RE-90-15a (below Bellecombe ash); oceanite	362,889.735	7,650.907.916	Cosmogenic	3420	820	−1470		
Er-35	N120	Piton des Tangues ashes ; Upper Bellecombe ash (U3a)?	Ash bed	Cpx-gabbro rich ash bed (above Piton des Tangues cone; Below Red Tephra Unit)	360,442.092	7,656.281.636	14C	3205	75	−1483	−1288	−1656
Er-35	N120	Ashes near commerson (below red tephra unit)	Ash bed	Fine ash bed near commerson	359,258.678	7,654.094.665	14C	3220	50	−1495	−1411	−1615
Er-42	Central activity	Lava2 of NW Enclos Fouqué rim	Lava flow	Sample RE-90-31 (below partage unit)	363,890.37	7652624.595	Cosmogenic	3500	1150	−1550		
Er-36	N120	Piton des Basaltes	Strombolian cone	Scoriae bed	360,326.603	7,653,622.523	14C	3305	55	−1584	−1453	−1735
Er-37	Central activity	Bellecombe ash; upper bellecombe ash (U3a?)	Ash bed	Sample REU 1303-07 A6/About 1 km east of Piton dans l'bout	358,895.000	7,655,721.000	14C	3715	30	−2098	−2029	−2200

(continued)

Table 8.1 (continued)

Eruption ID	Location	Unit name	Unit type	Note on the sample	Coordinates		Dating method	Age (BP)	±	Median probability (A.D.)	Calibrated min (A.D) 2σ	Calibrated max (A.D.) 2σ
					x	y						
Er-37	Central activity	Piton des Tangues ashes; upper Bellecombe ash (U3a)?	Ash beds	Cpx-gabbro rich ash bed (above Piton des Tangues lava; below red Tephra unit)	362,534.552	7,657,077.969	14C	3735	90	−2148	−1920	−2456
Er-43	Central activity	Laval of W Enclos Fouqué rim	Lava flow	Sample RE-90-24a	362,798.548	7,650,082.1	Cosmogenic	4310	821	−2360		
Er-43	Central activity	Laval of W Enclos Fouqué rim	Lava flow	Sample RE-90-25	362,806.183	7,650,013.384	Cosmogenic	4550	1270	−2600		
AND 02/ REU140721/ S3b	N120	Trous Blanc site (S3b ash below Bellecombe and above Trous Blanc eruption)	Ash beds	Charcoals at the base of S3B yellow layer (unit just below Bellecombe fall)	355,338.72	7,655,654.69	14C	4115	30	−2693	−2577	−2865
Er-38	Central activity	Lower Bellecombe ash Unit (U1b)	Ash beds	Fossil wood (unburned)	362,477.784	7,652,096.79	14C	4175	145	−2746	−2307	−3316
AND 04/ REU140723-1/ S3b	N120	Piton Sec (S3b ash below Bellecombe)	Ash beds	Charcoals at the base of S3B yellow layer (unit just below Bellecombe fall)	355,201.53	7,655,305.20	14C	4400	30	−3013	−2917	−3261
AND 08/ REU140728-2a/ bglaBR	N120	Trous Blancs site; top of the Trous Blancs Breccia	Ash beds	Charcoals at the base of L2(?) orange red layer (unit just below Bellecombe fall)	355,672.20	7,654,589.84	14C	4685	30	−3447	−3370	−3625
AND 12/ REU140801-2	N120	At the feet of Nez de Bœuf spatter (below Bellecombe)	Ash beds	Soi below thin lapilli; layer just below Bellecombe fall	356,562.22	7,654,352.30	14C	4695	35	−3454	−3370	−3629
Er-39	Central activity	Lava2 of NW Enclos Fouqué rim	Lava flow	Basal scoriae below the uppermost lava flow near Bellecombe (below Partage unit)	364,873.000	7,653,932.000	14C	4745	130	−3515	−3100	−3794
Er-40	Central activity	Second unit of Bellecombe ash	Ash bed	Sample REU 1303-07 A2 2b/1 km west of Piton des Basaltes	359,728.000	7,654,714.000	14C	4880	35	−3669	−3540	−3758
Er-43	Central activity	Laval of W Enclos Fouqué rim	Lava flow	Sample RE-90-24b; oceanite	362,798.548	7,650,082.1	Cosmogenic	5130	690	−3180		
AND 02/ REU140721/ S3a	N120	Trous Blancs area	Ash beds	S3a ashes above the Trous Blanc eruption	355338.72	7655654.69	14C	5400	35	−4277	−4077	−4341

(continued)

Table 8.1 (continued)

Eruption ID	Location	Unit name	Unit type	Note on the sample	Coordinates		Dating method	Age (BP)	±	Median probability (A.D.)	Calibrated min (A.D) 2σ	Calibrated max (A.D.) 2σ
					x	y						
REU140825-PDT1	N120	Piton des Tangues parking	Ash beds	Thick red lahar below Bellecombe falls	360,408.8	7,656,179.8	14C	5570	35	−4406	−4346	−4460
Er-44	Piton des Neiges?	Grand Etang deep drilling	Conglomerate	Fossil wood (unburned) in conglomerates at the base of the Puy de l'Etang ashes	359,436.931	7,666,718.692	14C	5730	110	−4584	−4352	−4823
Er-45	N120	"Les Trous Blancs" lava field	Lava flow		356,050.106	7,654,462.838	Stratigraphic	6500				
Er-46	N120	Ashes along the path to "Piton des Tangues"	Ash beds	Section 12; layer 5 (above the "Cendres de Notre Dame" unit)	360,408.937	7,656,181.043	14C	6965	38	−5847	−5745	−5975
Er-47	N120	Piton textor	Ash beds		358,199.373	7,656,293.751	14C	7125	125	−5999	−5740	−6231
Er-48	N120	Ashes between Piton textor and Piton des Cailles	Ash beds		358,729.735	7,656,540.14	14C	7930	90	−6837	−6611	−7060
AND 04/ REU140723-1/ S3a	N120	Trous Blancs area (near Piton sec; abobe Trous Blancs eruption)	Ash beds	S3a ash layer	355,201.53	7,655,305.20	14C	7985	40	−6915	−6709	−7051
Er-49	N120	Piton dans le bout	Strombolian cone		358,474.994	7,655,901.2	14C	9355	190	−8658	−8254	−9223
AND 06/ REU140723-3/ S3b	N120	Near Piton Sec (S3b ashes above the Trous Blanc eruption)	Ash beds		355,352.97	7,655,202.68	14C	9430	45	−8712	−8572	−8823
Er-50	Piton des la Fournaise?	Grand Etang deep drilling	Conglomerate	Fossil wood in conglomerate below the deep lavas (−40 m) met by the drill	359,436.931	7,666,718.692	14C	10,650	110	−10,644	−10,238	−10,801
REU031014-1BC	Piton des Neiges	Bras des Lianes					14C	12,040	60	−11,942	−11,802	−12,111
Er-51	Central activity	Notre Dame de la Paix -2	Ash beds		353,792.779	7,646,998.349	14C	15,200	1300	−16,378	−12,873	−19,716
LAS 17/ REU140718-1/ Part B	N120	Bois Court lava flow	Lava flow		348,715.72	7,655,411.65	14C	17,780	110	−19,582	−19,222	−19,908

(continued)

Table 8.1 (continued)

Eruption ID	Location	Unit name	Unit type	Note on the sample	Coordinates x	Coordinates y	Dating method	Age (BP)	±	Median probability (A.D.)	Calibrated min (A.D) 2σ	Calibrated max (A.D.) 2σ
LAS 17/ REU140718-1/ Part A	N120	Bois Court lava flow	Lava flow		348,715.72	7,655,411.65	14C	17,840	100	−19,664	−19,347	−19,948
LAS 17/ REU140718-1/ Part C	N120	Bois Court lava flow	Lava flow		348,715.72	7,655,411.65	14C	17,970	100	−19,817	−19,507	−20,114
Er-52	Central activity	"Cendres de Notre Dame" Unit	Ash beds				Stratigraphic	19,000–37,000				
Er-53	N120	Bois court	Strombolian cone	Altered ash layer. below the lava flow	348,644.795	7,655,086.233	14C	19,185	265	−21,168	−20,540	−21,776
Er-54	Central activity	Plaine des Sables caldera fault	Caldera fault	Sample RE-90-07; surface of the caldera fault	360,074.551	7,650,976.811	Cosmogenic	23,800	2000	−21,850	−21,478	−30,115
Er-55	N120	Piton Manuel—Piton des Cabris	Strombolian cone		351,186.326	7,653,685.743	K/Ar	29,000		−27,050		
Er-56	Central activity	Notre Dame de la Paix -1	Ash beds		354,230.713	7,647,126.81	14C	37,200	1500	−39,467	−36,591	−42,016
Er-57	N120 SVZ	Piton Rouge (St. Joseph)	Strombolian cone		356,019.22	7,638,466.955	K/Ar	57,000		−55,050		
Er-58	Central activity	Summit lava of the Plaine des Ramparts	Lava flow	Sample RE-90-06; last lavas of the caldeira	359,640.634	7,651,410.728	Cosmogenic	62,000	4000	−60,050		
Er-59	Central activity?	Submarine tephra (S17-656 drill)	Ash bed	Sample R4-1; basaltic ash			18O	67,000	5000	−65,050		

Eruption ID	Location	Unit name	Unit type	Note on the sample	Calibrated probability distribution (A.D.)–Max probability 2σ			Calibrated probability distribution (A.D.) –Secondary probability 2σ			Calibrated probability distribution (A.D.)–Tertiary probability 2σ			Reference
					Min age	Max age	% prob.	Min age	Max age	% prob.	Min age	Max age	% prob.	
Er-1	RZNE	1998 lava flow	Lava flow											Michon et al. (2013)
Er-2	RZSE	1986 lava flow	Lava flow											Michon et al. (2013)
Er-3	RZNE	1977 lava flow	Lava flow											Michon et al. (2013)
REU140212-10/9	Central activity	Partage cliff	Ash beds	Historical eruptions; B layer; multiple fragments	1939	1801	65.9	1764	1679	34.1				This chapter
REU140212-S1-2b	Central activity	Partage cliff	Ash beds	Historical eruptions; B1 layer; top of historical sequence; single fragment	1894	1799	44.4	1777	1675	39.9	1941	1905	15.7	This chapter

(continued)

Table 8.1 (continued)

Eruption ID	Location	Unit name	Unit type	Note on the sample	Calibrated probability distribution (A.D.)–Max probability 2σ			Calibrated probability distribution (A.D.) –Secondary probability 2σ			Calibrated probability distribution (A.D.)–Tertiary probability 2σ			Reference
					Min age	Max age	% prob.	Min age	Max age	% prob.	Min age	Max age	% prob.	
Er-4	RZSE	1800 lava flow	Lava flow											Michon et al. (2013)
Er-10	N120	Piton Rampe 14	Strombolian cone	Below the lava flow										–
AND 02 / REU140721 / S1	N120	Trous Blancs area	Modern soil	Uppermost layer rich in roots and organic matter; multiple fragments	1890	1797	35	1781	1716	30.7	1711	1668	17.2	This chapter
REU031014-3	N120	Rivière des Remparts			1783	1717	33.2	1889	1831	19.5	nd	1910	18.1	This chapter
Er-5	RZSE	1776 lava flow	Lava flow											Michon et al. (2013)
Er-6	RZSE	1774 lava flow	Lava flow											Michon et al. (2013)
REU140212-S4-2	Central activity	Partage cliff	Ash beds	Historical eruptions; B layer; top of historical sequence; multiple fragments	1820	1719	50.1	nd	1913	18.4	1706	1664	17.8	This chapter
REU140521-BB1	RZNE	Piton Moka (near Anses des Cascades)	Ash beds	In lapilli fall above the lahar reworking Bellecombe breccia; modern agricultural activity? multiple fragments	1815	1725	55.6	1696	1657	19.4	nd	1917	18.7	This chapter
AND 08 / REU140728-2a / S1	N120	Top of Trous Blancs site	Modern soil		1811	1728	57.6	1691	1648	25.2	nd	1922	17.2	This chapter
Er-7	RZSE	Puy Raymond	Strombolian cone	Inside ashes. 1 m below the lava flow										–
Er-8	EF (lava above the Formica leo cone)	CLEF lava field	Lava field	Pahoehoe lava										Tanguy et al. (2011)
Er-8	EF	CLEF lava field	Lava field	Pahoehoe lava										Michon et al. (2013)
Er-9	RZNE	1708 lava flow	Lava flow											Michon et al. (2013)
Er-8	EF (lava close to Bellecombe steps)	CLEF lava field	Lava field	Sample RE-90-28; pahohehoe										–

(continued)

Table 8.1 (continued)

Eruption ID	Location	Unit name	Unit type	Note on the sample	Calibrated probability distribution (A.D.)–Max probability 2σ			Calibrated probability distribution (A.D.) –Secondary probability 2σ			Calibrated probability distribution (A.D.)–Tertiary probability 2σ			Reference
					Min age	Max age	% prob.	Min age	Max age	% prob.	Min age	Max age	% prob.	
REU140212-S1-2a	Central activity	Partage cliff	Ash beds	Historical eruptions; B1 layer; top of historical sequence; multiple fragments	1675	1621	59.1	1579	1521	21.3	1799	1777	16.4	This chapter
Er-11	RZNE	Piton Nelson—Piton de Bois Blancs	Strombolian cone	Near Bois Blanc village										Bachèlery (1981)
REU140320-S5-4	Central activity	Partage cliff	Ash beds	Historical eruptions; multiple fragments	1639	1470	100							This chapter
Er-12	RZSE	Piton Taipoul	Strombolian cone	Below the scoriae of the cone and above an ash bed										Tanguy et al. (2011)
Er-13	RZSE	Mare Longue lava flow	Lava flow	Below the lava flow										Straberg (comm pers)
Er-15	N120	Piton Chisny—lava with dunite enclaves near Chisny cone	Polyphase cone	Near Chisny cone; phase 2 in Bachèlery (1981)										Tanguy et al. (2011)
Er-16	RZNE	Piton Indivis—Bois Blanc	Strombolian cone											Bachèlery (1981), map, Tanguy et al. (2011)
Er-14	N120	Piton Chisny—Plaine des Sables black scoria fall	Multiple cones	Section 20; Plaine des Sables. near Piton Chisny										Di Muro et al. (2012)
Er-17	RZSE	Ravine d'Ango lava	Strombolian cone	In conglomerate below the lava (the soil is not burned)										Tanguy et al. (2011)
Er-18	N120	Petit Cratère	Strombolian cone	Below the lava flow	1529	1384	69.8	1634	1544	17	1364	1305	13.2	Tanguy et al. (2011)
Er-19	RZSE	Le Baril 1 (Brulé du Baril)	Strombolian cone	Basal scoriae below the lava flow (Ravine Perote river) and above a soil	1444	1281	100							Tanguy et al. (2011)
Er-20	N120	Langevin plateau fall—bed 4	Scoriae bed	Section 1—Caldera des Sables cliff	1399	1292	100							Di Muro et al. (2012)

(continued)

Table 8.1 (continued)

Eruption ID	Location	Unit name	Unit type	Note on the sample	Calibrated probability distribution (A.D.)−Max probability 2σ			Calibrated probability distribution (A.D.) −Secondary probability 2σ			Calibrated probability distribution (A.D.)−Tertiary probability 2σ			Reference
					Min age	Max age	% prob.	Min age	Max age	% prob.	Min age	Max age	% prob.	
Er-21	N120	Langevin plateau fall—bed 5	Scoriae bed	Section 1—Caldera des Sables cliff	1251	1148	79.5	1090	1046	16.1	1139	112	4.4	Di Muro et al. (2012)
Er-22	N120	"Piton Chisny lavas" in Langevin valley	Lava flow	Below aa lava flow in the Langevin valley	1022	775	100							Albarède et al. (1997), pg. 174, Tanguy et al. (2011)
Er-24	N120	Piton sous le Gite	Strombolian cone	Below the lava flow	675	420	100							–
Er-23	Central activity?	Ashes on western rim of Enclos Fouqué (near Chisny)	Ash beds		1047	50	98.9	1122	1088	0.7				Bachèlery (1981), pg. 137
Er-25	N120	Tephra on Les Trous Blancs	Scoria bed	Sample REU 1301-243 Trou Blanc	641	534	93.8	487	470	3.2	453	434	2.9	This chapter
Er-26	N120	Commerson crater	Scoria bed		249	−1	99.5	−12	−20	0.5				Bachèlery (1981), Tanguy et al. (2011)
Er-27	Central activity	Partage Tephra (Unit 3; CC2 layer)	Scoria bed	Section 6 (Partage cliff)	−54	−231	78.2	−292	−354	21.8				Di Muro et al. (2012)
Er-27	Central activity	Ashes on NW cliff of Enclos Fouqué (Bellecombe cliff)	Ash bed	Ashes draping the lower third of the Enclos Fouqué cliff	1	−380	100							–
Er-28	Central activity	Ashes near the "Gite du Volcan" ("Cendres de Bellecombe" in B'81 and V&B'82)	Ash beds	Sampled near the Gites du Volcan	−178	−597	86	−680	−755	8.7	−607	−671	5.4	Bachèlery (1981)
Er-28	Central activity	Partage Tephra (Unit 2; CC1 layer)	Ash bed	Section 6 (Partage cliff)	−366	−490	99.2	−501	−506	0.8				Di Muro et al. (2012)
Er-28	Central activity	Ashes near the "Gite du Volcan"	Ash beds	Upper part of an ash bed. below a bed of scoriae	−357	−784	97.5	−255	−283	1.9	−235	−246	0.6	–
Er-28?	N120	Piton des Tangues tephra (Below Red Tephra Unit)	Lapilli bed?	Ash or lapilli bed above the Upper Bellecombe ash	608	−2345	100							Abchir (1996), pg. 86
Er-29	Central activity	Partage Tephra (Unit1; CC0 layer)	Ash bed	Section 6 (Partage cliff)	−919	−1121	100							Di Muro et al. (2012)

(continued)

Table 8.1 (continued)

Eruption ID	Location	Unit name	Unit type	Note on the sample	Calibrated probability distribution (A.D.)–Max probability 2σ			Calibrated probability distribution (A.D.) –Secondary probability 2σ			Calibrated probability distribution (A.D.)–Tertiary probability 2σ			Reference
					Min age	Max age	% prob.	Min age	Max age	% prob.	Min age	Max age	% prob.	
REU140825-PDT2	N120	Near Piton des Tangues parking	Ash beds	Black soil below red tephra lapillis covering Bellecombe falls	−1020	−1211	100							This chapter
Er-30	Central activity	Upper unit of Bellecombe ash	Ash bed	Sample REU 1303-07 A1	−1228	−1411	100							This chapter
Er-31	N120	Piton Sauvetage	Strombolian cone	Ash bed	−1152	−1501	97.7	−1127	−1150	2.3				Bachèlery (1981)
Er-32	Central activity	Upper unit of Bellecombe ash	Ash bed	Sample REU 1303-07 A2 2A	−1280	−1428	100							This chapter
Er-33	N120	Langevin plateau fall- bed 6	Ash and lapilli bed	Section 1—Caldera des Sables cliff	−1284	−1431	100							Di Muro et al. (2012)
Er-34	Central activity	Upper unit of Bellecombe ash	Ash bed	Sample REU 1303-07 A2 2C/1 km west of Piton des Basaltes	−1288	−1437	100							This chapter
Er-41	Central activity	Lava3 of W Enclos Fouqué rim	Lava flow	Sample RE-90-15b (below Bellecombe ash); oceanite										–
REU140822-2	N120	0.5 km from Nez de Bœuf. above Bellecombe ash fall	Ash beds	Soil above Bellecombe and below a thin lapilli layer	−1371	−1457	69.8	−1300	−1359	27	−1477	−1494	3.2	This chapter
Er-41	Central activity	Lava3 of W Enclos Fouqué rim	Lava flow	Sample RE-90-15a (below Bellecombe ash); oceanite										–
Er-35	N120	Piton des Tangues ashes ; Upper Bellecombe ash (U3a)?	Ash bed	Cpx-gabbro rich ash bed (above Piton des Tangues cone; Below Red Tephra Unit)	−1288	−1643	99.9	−1655	−1656	0.1				–
Er-35	N120	Ashes near commerson (below red tephra unit)	Ash bed	Fine ash bed near commerson	−1411	−1615	100							Bachèlery (1981)
Er-42	Central activity	Lava2 of NW Enclos Fouqué rim	Lava flow	Sample RE-90-31 (below partage unit)										–
Er-36	N120	Piton des Basaltes	Strombolian cone	Scoriae bed	−1492	−1694	93.6	−1453	−1482	4	−1716	−1735	2.3	–
Er-37	Central activity	Bellecombe ash; upper bellecombe ash (U3a?)	Ash bed	Sample REU 1303-07 A6/About 1 km east of Piton dans l'bout	−2029	−2155	81.5	−2157	−2200	18.5				This chapter
Er-37	Central activity	Piton des Tangues ashes; upper Bellecombe ash (U3a)?	Ash beds	Cpx-gabbro rich ash bed (above Piton des Tangues lava; below red Tephra unit)	−1920	−2369	94.2	−2417	−2456	2.9	−2373	−2408	2.9	–
Er-43	Central activity	Lava1 of W Enclos Fouqué rim	Lava flow	Sample RE-90-24a										–

(continued)

Table 8.1 (continued)

Eruption ID	Location	Unit name	Unit type	Note on the sample	Calibrated probability distribution (A.D.)–Max probability 2σ			Calibrated probability distribution (A.D.) –Secondary probability 2σ			Calibrated probability distribution (A.D.)–Tertiary probability 2σ			Reference
					Min age	Max age	% prob.	Min age	Max age	% prob.	Min age	Max age	% prob.	
Er-43	Central activity	Lava1 of W Enclos Fouqué rim	Lava flow	Sample RE-90-25										–
AND 02/ REU140721/ S3b	N120	Trous Blanc site (S3b ash below Bellecombe and above Trous Blanc eruption)	Ash beds	Charcoals at the base of S3B yellow layer (unit just below Bellecombe fall)	−2577	−2764	73.5	−2804	−2865	26.2	−2770	−2773	0.3	This chapter
Er-38	Central activity	Lower Bellecombe ash Unit (U1b)	Ash beds	Fossil wood (unburned)	−2336	−3110	97.9	−3237	−3266	0.8	−3293	−3316	0.6	Abchir (1996), pg. 86
AND 04/ REU140723-1/ S3b	N120	Piton Sec (S3b ash below Bellecombe)	Ash beds	Charcoals at the base of S3B yellow layer (unit just below Bellecombe fall)	−2917	−3098	99.1	−3254	−3261	0.9				This chapter
AND 08/ REU140728-2a/ bglaBR	N120	Trous Blancs site; top of the Trous Blancs Breccia	Ash beds	Charcoals at the base of L2(?) orange red layer (unit just below Bellecombe fall)	−3370	−3477	68.9	−3482	−3525	23.7	−3600	−3625	7.4	This chapter
AND 12/ REU140801-2	N120	At the feet of Nez de Bœuf spatter (below Bellecombe)	Ash beds	Soi below thin lapilli; layer just below Bellecombe fall	−3370	−3476	60.7	−3483	−3532	23.4	−3582	−3629	15.9	This chapter
Er-39	Central activity	Lava2 of NW Enclos Fouqué rim	Lava flow	Basal scoriae below the uppermost lava flow near Bellecombe (below Partage unit)	−3263	−3794	92.3	−3100	−3246	7.7				Bachèlery (1981)
Er-40	Central activity	Second unit of Bellecombe ash	Ash bed	Sample REU 1303-07 A2 2b/1 km west of Piton des Basaltes	−3634	−3713	97.2	−3540	−3556	1.6	−3743	−3758	1.3	This chapter
Er-43	Central activity	Lava1 of W Enclos Fouqué rim	Lava flow	Sample RE-90-24b; oceanite										–
AND 02/ REU140721/ S3a	N120	Trous Blancs area	Ash beds	S3a ashes above the Trous Blanc eruption	−4227	−4341	89.5	−4167	−4202	7.8	−4077	−4095	2	This chapter
REU140825-PDT1	N120	Piton des Tangues parking	Ash beds	Thick red lahar below Bellecombe falls	−4346	−4460	100							This chapter
Er-44	Piton des Neiges?	Grand Etang deep drilling	Conglomerate	Fossil wood (unburned) in conglomerates at the base of the Puy de l'Etang ashes	−4352	−4800	99.7	−4819	−4823	0.3				Banton (1985)

(continued)

Table 8.1 (continued)

Eruption ID	Location	Unit name	Unit type	Note on the sample	Calibrated probability distribution (A.D.)–Max probability 2σ			Calibrated probability distribution (A.D.) –Secondary probability 2σ			Calibrated probability distribution (A.D.)–Tertiary probability 2σ			Reference
					Min age	Max age	% prob.	Min age	Max age	% prob.	Min age	Max age	% prob.	
Er-45	N120	"Les Trous Blancs" lava field	Lava flow											Villeneuve and Bachèlery (2006)
Er-46	N120	Ashes along the path to "Piton des Tangues"	Ash beds	Section 12; layer 5 (above the "Cendres de Notre Dame" unit)	−5745	−5918	93.4	−5950	−5975	6.6				Di Muro et al. (2012)
Er-47	N120	Piton textor	Ash beds		−5740	−6231	100							Bachèlery (1981)
Er-48	N120	Ashes between Piton textor and Piton des Cailles	Ash beds		−6611	−7060	100							Bachèlery (1981)
AND 04/ REU140723-1/ S3a	N120	Trous Blancs area (near Piton sec; abobe Trous Blancs eruption)	Ash beds	S3a ash layer	−6752	−7051	98.7	−6709	−6720	1.3				This chapter
Er-49	N120	Piton dans le bout	Strombolian cone		−8254	−9223	100							Bachèlery (1981)
AND 06/ REU140723-3/ S3b	N120	near Piton Sec (S3b ashes above the Trous Blanc eruption)	Ash beds		−8597	−8823	99.2	−8572	−8565	0.8				This chapter
Er-50	Piton des la Fournaise?	Grand Etang deep drilling	Conglomerate	Fossil wood in conglomerate below the deep lavas (−40 m) met by the drill	−10,427	−10,801	96.2	−10,287	−10,339	2	−10,350	−10,396	1.5	Banton (1985)
REU031014-1BC	Piton des Neiges	Bras des Lianes			−11,802	−12,111	100							This chapter
Er-51	Central activity	Notre Dame de la Paix -2	Ash beds		−12,873	−19,716	100							–
LAS 17/ REU140718-1/ Part B	N120	Bois Court lava flow	Lava flow		−19,222	−19,908	100							This chapter
LAS 17/ REU140718-1/ Part A	N120	Bois Court lava flow	Lava flow		−19,347	−19,948	100							This chapter

(continued)

Table 8.1 (continued)

Eruption ID	Location	Unit name	Unit type	Note on the sample	Calibrated probability distribution (A.D.)–Max probability 2σ			Calibrated probability distribution (A.D.) –Secondary probability 2σ			Calibrated probability distribution (A.D.)–Tertiary probability 2σ			Reference
					Min age	Max age	% prob.	Min age	Max age	% prob.	Min age	Max age	% prob.	
LAS 17/ REU140718-1/ Part C	N120	Bois Court lava flow	Lava flow		−19,507	−20,114	100							This chapter
Er-52	Central activity	"Cendres de Notre Dame" Unit	Ash beds											
Er-53	N120	Bois court	Strombolian cone	Altered ash layer. below the lava flow	−20,540	−21,776	100							–
Er-54	Central activity	Plaine des Sables caldera fault	Caldera fault	Sample RE-90-07; surface of the caldera fault										–
Er-55	N120	Piton Manuel—Piton des Cabris	Strombolian cone											–
Er-56	Central activity	Notre Dame de la Paix -1	Ash beds		−36,591	−42,016	100							–
Er-57	N120 SVZ	Piton Rouge (St. Joseph)	Strombolian cone											–
Er-58	Central activity	Summit lava of the Plaine des Ramparts	Lava flow	Sample RE-90-06; last lavas of the caldeira										–
Er-59	Central activity?	Submarine tephra (S17-656 drill)	Ash bed	Sample R4-1; basaltic ash										Fretzdorff et al. (2000)

Abbreviation: *EF* Enclos Fouqué; *CLEF* Enclos Fouqué Lava Field; *RZSE* SE rift zone; *RZNE* NE rift zone; *N120* NW rift zone

Table 8.2 GDE core description and coarse ash (710 < φ < 1000 μm) optic microscope analysis

CORE: **GDE**
DEPTH: **623 cm**

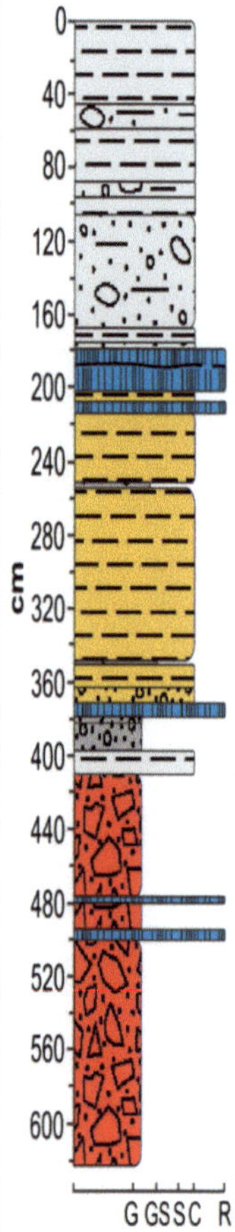

cm	Description
	Fine ashes
0-46	fine dark grayish (0-14 cm) to reddish (14-36 cm) and yellowish (36-46 cm) brown ashes. Coarse sand fraction increase toward the base
46-59	poorly sorted layer of lithics (ϕ=5 mm) and altered scoriae (ϕ=2-3 cm) within an abundant mostly fine matrix
59-87	fine ash. A 6 cm thick cobble at 65 cm
87-96	poorly sorted lapilli and gravels within a fine matrix
96-106	fine ash with few coarse ash
106-168	fine matrix with very poorly sorted scoriae and gravels
168-176	fine ash
176-179	pebbles and scoriae within fine ash
	Soil
179-200	vesicular dark lavas
200-207	light brown clay
207-214	lava
214-347	yellowish brown clay with some very altered scoriae and rounded gravel (ϕ=4-10 mm). At 251-255 cm black lapilli (ϕ=5 mm) within a coarse matrix, at 320-321 black lapilli (ϕ=5 mm) within a fine matrix
	Proximal Scoria
347-350	altered lapilli (ϕ=5 mm; ϕmax< 1 cm) surrounded by an orange (at the top) to dark matrix
350-362	yellowish brown clay with some very altered scoriae and rounded gravel.
362-371	lapilli and lithics surrounded by a fine to coarse matrix
371-378	lava
378-381	lapilli and gravel surrounded by a brown poorly sorted matrix
381-397	very poorly sorted lapilli (ϕmax< 2.5 cm) within an orange (381-387 cm) to dark (387-397 cm) matrix
397-409	fine reddish light brown ash
409-623	very poorly sorted scoriae (ϕ< 5 cm) with some to scarce matrix. Lavas at 476 and 494 cm

Coarse ash

cm	ol	pl	px	qz	mt
3					
10					
21					
34					
41					
53	X				
63	X				
78	x				
88					
92					
105					
123					
166					
170					
205					
230					
275					
330					
349					
353					
384					
404					

cm	v	μxx		xx	ag
			lth		
3					X
10					X
21					X
34					X
41					X
53					X
63					X
78					X
88					X
92					X
105					X
123					X
166					X
170					X
205					X
230					X
275					X
330					X
349					X
353					X
384					X
404					X

Legend Stratigraphic log: *G* gravel; *GS* gravel and sand; *S* sand; *C* Clay; *R* coherent rock. Microscopic analysis: *ol* olivine; *pl* plagioclase; *px* pyroxene; *qz* quartz; *mt* magnetite; *v* glass; *μxx* microcrystalline; *xx* crystals; *ag* fine sand aggregates; *lth* lithics/lava fragments (as part of μxx). Quantities: *x* abundant/frequent; *x* some; *x* few; *tr* trace

Table 8.3 PDE2core description and coarse ash (710 < φ < 1000 μm) optic microscope analysis

CORE: **PDE2**
DEPTH: **395 cm**

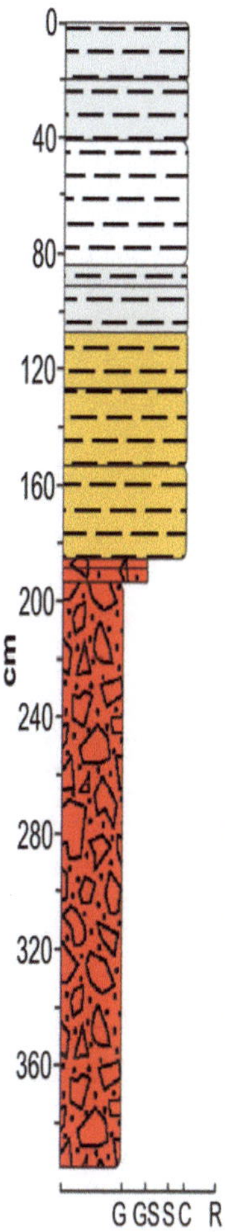

cm	
	Fine Ashes
0-20	dark fine ash with many roots
20-27	dark gray fine ash with few achnelith lapilli $\varnothing_{max}$< 1 cm
27-41	reddish dark gray fine ash with few achnelith lapilli $\varnothing_{max}$< 5 mm
	Bellecombe Ashes
41-84	grayish brown fine ash with few granule, sometimes forms aggregates
	Fine Ashes
84-91	brown fine ash
91-107	dark brown fine ash
	Soil
107-127	reddish clay with few coarse sand and altered scoriae $\varnothing_{max}$< 1 cm
127-153	light brown ash with a flat and rounded 2 cm thick scoria,rare granule
153-185	reddish brown ash with few granule (at the top) and some altered scoria ($\varnothing_{m}$= 1 cm; $\varnothing_{max}$< 4 cm). A 3 cm thick lava fragment at 155 cm
	Proximal deposit
185-189	dark altered scoria ($\varnothing_{m}$= 1-2 cm; $\varnothing_{max}$< 4 cm) and matrix
189-194	matrix sustained yellowish brown layers with altered lapilli $\varnothing_{max}$< 1 cm
194-395	very poorly sorted scoria layer $\varnothing_{max}$< 4 cm. A yellowish matrix decrease from the top to base

Coarse ash					
cm	ol	pl	px	qz	mt
23					
35	X				
49	tr	tr	tr	tr	tr
80	X				
88	tr		tr		tr
97	X				

cm	v	μxx	μxx	xx	ag
			lth		
23	x				X
35	x	x	x	tr	X
49	tr	tr	x	x	X
80		x	X	x	X
88		x	x	x	X
97		X	x	x	X

Legend Stratigraphic log: *G* gravel; *GS* gravel and sand; *S* sand; *C* Clay; *R* coherent rock. Microscopic analysis: *ol* olivine; *pl* plagioclase; *px* pyroxene; *qz* quartz; *mt* magnetite; *v* glass; *μxx* microcrystalline; *xx* crystals; *ag* fine sand aggregates; *lth* lithics/lava fragments (as part of μxx). Quantities: *x* abundant/frequent; *x* some; *x* few; *tr* trace

Table 8.4 PLB core description and coarse ash (710 < φ < 1000 μm) optic microscope analysis

CORE: **PLB**
DEPTH: **348 cm**

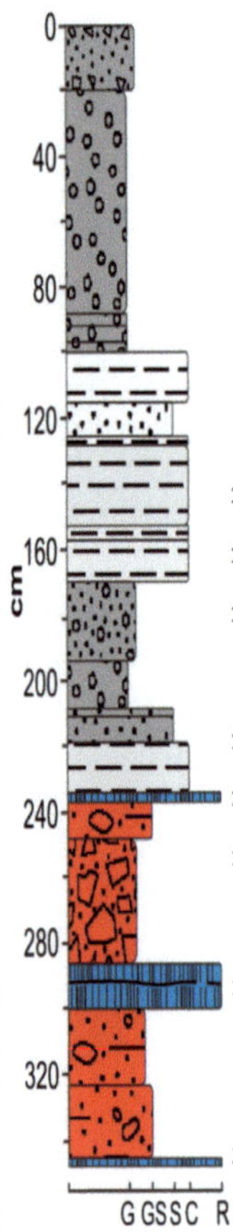

cm	Description
	Black fine lapilli sequence
0-20	fresh black lapilli ($Ø_m$= 2-5 mm) and ash
20-88	well sorted, fresh, fine lapilli ($Ø_m$= 2-5 mm)
88-92	fine reddish ash-coated lapilli ($Ø_m$= 2-5 mm)
92-97	lapilli ($Ø_m$= 5 mm; $Ø_{max}$ <1 cm)
97-100	very fine lapilli (Ø< 2 mm)
	Bellecombe and pyroxene ashes
100-115	dark fresh fine ash
115-126	olivine-bearing coarse ashes
126-128	dark to brown fine ash with few granule, olivine and pyroxene
	Fine Ashes
128-153	brown ash
153-157	dark brown ash
157-170	reddish brown with some granule toward the base
	Lapilli sequence
170-209	dark red fine lapilli ($Ø_m$= 5 mm, $Ø_{max}$< 8 mm) very altered. Reverse grading and well sorting with just few coarse matrix at the top
209-211	coarse (but finer than the preceding) reddish brown ash
211-219	coarse dark ash; 1 cm thick scoria at the top
	Fine Ashes
219-234	reddish brown fine ash
	Proximal deposit
234-237	lava
237-248	fine brown ash with very altered red scoriae (Ø< 1 cm)
248-286	Poorly sorted dark scoriae ($Ø_m$= 1-2 cm) with a reddish fine to coarse ashes . A 2 cm thick lava fragment at the roof
286-300	lava
300-323	poorly sorted altered scoriae within abundant brown reddish matrix
323-346	poorly sorted rusty red scoriae and matrix hardened by alteration
346-348	lava

Coarse ash

cm	ol	pl	px	qz	mt
2					
11			tr	tr	tr
20	X				
40	X				
65					
86	X				
89	X				
111	X	x	tr		
117	X	x	tr	tr	
123	X	x	tr	tr	tr
127	X				
133	X				
146	X				
155					
164	X		tr		
180	X				

cm	v	μxx		xx	ag
			lth		
2	X				
11	X	tr	x	tr	
20	X	tr		tr	
40	X	tr		tr	
65	X				
86	X	tr		tr	
89	X	tr		tr	
111	tr	X	x	x	
117		X	x	X	x
123		X	X	x	x
127		x	X	x	X
133	tr	x		tr	
146	tr	X	tr	tr	
155	tr	x		tr	
164	x			x	X
180	X			x	x

Legend Stratigraphic log: *G* gravel; *GS* gravel and sand; *S* sand; *C* Clay; *R* coherent rock. Microscopic analysis: *ol* olivine; *pl* plagioclase; *px* pyroxene; *qz* quartz; *mt* magnetite; *v* glass; *μxx* microcrystalline; *xx* crystals; *ag* fine sand aggregates; *lth* lithics/lava fragments (as part of μxx). Quantities: *x* abundant/frequent; *x* some; *x* few; *tr* trace

Table 8.5 PDS2 core description and coarse ash (710 < φ < 1000 μm) optic microscope analysis

CORE: **PDS2**
DEPTH: **243 cm**

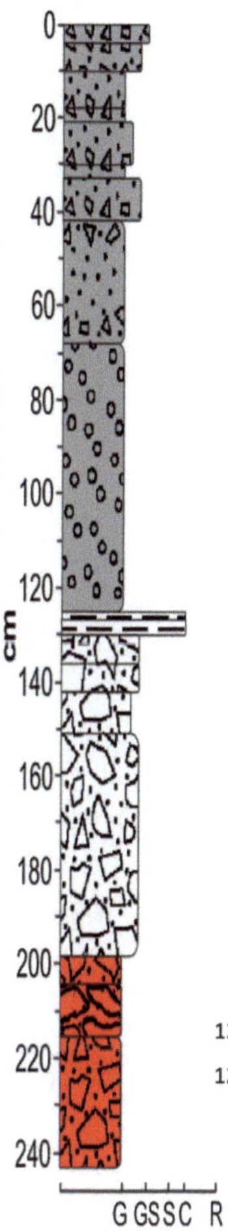

cm	Description
	Reworked lapilli and ashes
0-4	black fine lapilli ($Ø_m$= 2 mm; $Ø_{max}$= 5 mm) with abundant dark matrix. Rare red lapilli
4-10	black fine lapilli ($Ø_m$= 5mm) with yellowish ash
10-18	black lapilli ($Ø_m$= 5 mm; $Ø_{max}$= 8 mm) more granular at the base. Scarce matrix. Rare red lapilli
18-21	well sorted lapilli ($Ø_m$= 5 mm)
21-30	very fine lapilli. Coarse ash fraction increase toward the base. Rare red lapilli. Some Olivine crystals in the lower part
30-33	fine lapilli without matrix
33-42	normally graded lapilli bed (from $Ø_m$= 2 mm to $Ø_m$= 5- 8 mm)
42-68	black fine lapilli ($Ø_m$= 2-5 mm). Coarse ash augmentation toward the top. Some olivine crystals. Rare red lapilli
	Black Lapilli
68-125	black well sorted ($Ø_m$= 5-10 mm; $Ø_{max}$< 2 cm), achnelith and vesicular lapilli. Slight reverse grading. A 3 cm thick olivine bearing lava fragment at the base. The lapilli layer has a very thin (< 1mm) coarse ash layers at the base
	Bellecombe ashes
125-126.5	brown olivine-bearing fine ash
126.5-130	grayish brown fine ash with few granule
130-198	Coarse olivine-bearing scoriaceous and partially altered lava fragments ($Ø_{max}$< 5 cm) surrounded by fine grayish ash that forms aggregates. Sometimes the gray matrix is replaced by a fine red-brown ash (132-133 and 136-142 cm) or by a reddish dark coarse ash (142-151 cm)
	Pre-Bellecombe deposits
198-204	dark scoriaceous lava fragments ($Ø_{max}$< 2 cm) and some coarse ash
204-215	poorly sorted red lapilli ($Ø_{max}$< 1 cm) and spatter bombs ($Ø_{max}$< 4 cm). Scarce matrix
215-243	poorly sorted dark, scoriaceous lava fragments ($Ø_{max}$=1-3 cm) and few red matrix and lapilli which decrease towards the base

Coarse ash

cm	ol	pl	px	qz	mt
7	X				
14	X				
20	X				
22	X				
26	X				
32	X				
41	X				
47	X		tr		
67	X				
72	X				
99	X				
124	X				
126	X	x	tr		x
128	X				
135	X	x		tr	tr
140	X		tr		tr
148	X				
163	X	x		tr	
176	X	x	tr	tr	
194	X	x			
201	X				
211	X				

cm	v	μxx	μxx lth	xx	ag
7	X			x	
14	X	tr		x	
20	X			x	
22	X	tr		x	
26	X	tr		x	
32	X	tr		x	
41	X	tr		x	
47	X	tr		x	
67	X	tr		x	
72	X			x	
99	X			x	
124	X	x	x	x	
126	x	x	x	X	X
128			x		X
135	x	x	x	X	
140	x	x	tr	x	
148	x	x	tr	x	
163	tr	X	X	X	
176	tr	X	X	X	
194		X	X	X	
201	tr	X		x	
211	X	x		X	

Legend Stratigraphic log: *G* gravel; *GS* gravel and sand; *S* sand; *C* Clay; *R* coherent rock. Microscopic analysis: *ol* olivine; *pl* plagioclase; *px* pyroxene; *qz* quartz; *mt* magnetite; *v* glass; *μxx* microcrystalline; *xx* crystals; *ag* fine sand aggregates; *lth* lithics/lava fragments (as part of μxx). Quantities: *x* abundant/frequent; *x* some; *x* few; *tr* trace

Table 8.6 PDS1 core description and coarse ash (710 < φ < 1000 μm) optic microscope analysis

CORE: **PDS1**
DEPTH: **222 cm**

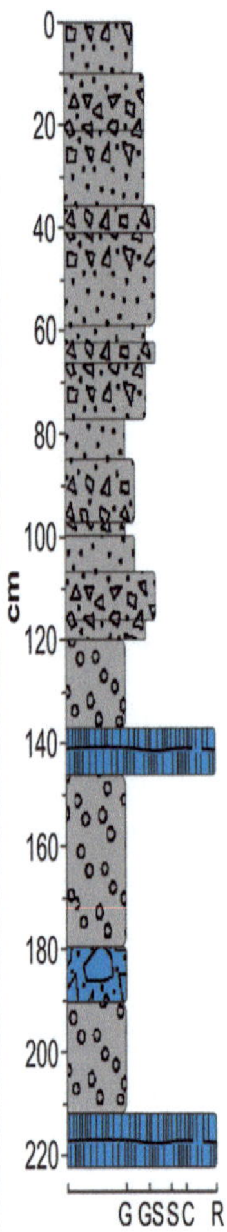

cm	**Reworked lapilli and ashes**
0-10	fresh black lapilli (Øm= 2-5 mm; Ømax< 8 mm) and few coarse ash . Slightly normal grading. Some oxidized red lapilli
10-36	fresh black lapilli ($Ø_m$= 2-5 mm; $Ø_{max}$< 8mm) and coarse ash . Slightly normal grading. Some oxidized red lapilli
36-41	Fine lapilli ($Ø_m$= 2-5 mm) and abundant yellowish matrix. Coarse olivine-bearing ash and granule at the roof
41-66	fine lapilli ($Ø_m$= 2-5 mm but coarser at the roof and at the base $Ø_m$= 5-10 mm). Abundant coarse matrix, decreasing at the base. Some red lapilli
66-97	fine lapilli with reverse grading (Øm= 5 mm to 2 mm at the base). Matrix in the superior part. Some red lapilli
97-100	well sorted lapilli ($Ø_m$= 5 mm). Some red lapilli
100-120	normal graded lapilli layer ($Ø_{max}$< 5 mm to 10 mm at the base) with decreasing ash toward the base
	Black lapilli and lavas
120-137	well sorted black, fresh and vesicular lapilli (ϕ= 5-8 mm; ϕmax< 1.5 cm)
137-146	very vesicular fresh olivine basalt; slightly reddish oxidation; few ol crystal
146-180	massive well sorted black fresh and vesicular lapilli; ϕ=5-8 mm $ϕ_{max}$< 1.5 cm
180-190	vesicular basalt; slightly reddish oxidation
190-212	massive well sorted black, fresh and vesicular lapilli (ϕ=8-10; mm $ϕ_{max}$< 1.5 cm)
212-222	vesicular basalt; slightly reddish oxidation

Coarse ash					
cm	ol	pl	px	qz	mt
12	X				
28	X				
36	X				
39	X				
43	X				
54	X				
64	X				
70	X				
77	X				
80	X				
88	X				
98	X				
102	X				
116	X				
122	X				
132	X				
151	X				
162					
174	X				
192	X				
202	X				
211	X				

cm	v	µxx		xx	ag
			lth		
12	X	tr		x	
28	X	x		x	
36	X	tr		x	
39	X	tr		x	x
43	X	x	tr	x	
54	X			x	
64	X			x	
70	X			x	
77	X			x	
80	X			x	
88	X	tr	tr	x	
98	X		tr	x	
102	X			x	
116	X			x	
122	X			x	
132	X	tr	x	tr	
151	X			x	
162	X	x			
174	X	tr		x	
192	x	x	x	x	
202	x	X	X	x	
211	x	X	X	x	

Legend Stratigraphic log: *G* gravel; *GS* gravel and sand; *S* sand; *C* = Clay; *R* coherent rock. Microscopic analysis: *ol* olivine; *pl* plagioclase; *px* pyroxene; *qz* quartz; *mt* magnetite; *v* glass; *µxx* microcrystalline; *xx* crystals; *ag* fine sand aggregates; *lth* lithics/lava fragments (as part of µxx). Quantities: *x* abundant/frequent; *x* some; *x* few; *tr* trace a synthesis of all macroscopically and microscopically performed observations performed on drilling cores is presented.

In order to better constrain the explosive activity and the stratigraphic correlations in the N120 rift zone, the excavation of about 30 tephra sections has been performed in the proximal and middle sector of the volcano. It leads to the identification of several, spatially limited, lapilli deposits, and of three major tephra sequences, respectively in the Piton de Partage, Plaine des Sables, and Plaine des Remparts sectors (Fig. 8.1). These three sequences are described in the following.

8.3.1 The Partage Tephra

The "Partage Tephra" (Fig. 8.3) is here defined for the first time. It consists in a <1.5 m thick sequence of ash, lapilli and scoria beds that drape almost continuously the northern wall of the Enclos Fouqué caldera. The sequence attains the maximum thickness close to the northern caldera rim, thins out quickly when moving westward in the Plaine des Sables and is not observed in the Fond de la Rivière de l'Est, neither in more distal areas, as on the Plaine des Remparts plateau. The tephra sequence underlies the modern soil and has been divided in three units (U1; U2; U3 from base to top), on the basis of grain size, componentry and ^{14}C age. This tephra sequence was initially included in the "Bellecombe ash unit", but it results to be younger. Thin, usually <2 cm-thick, laterally continuous charcoal-rich layers are widespread inside the "Partage Tephra" sequence. The lowermost layer (CC0: 2855 ± 35 BP) occurs in the middle of Unit 1, while the others, CC1 and CC2, occur respectively at the base of Unit 2 (CC1: 2340 ± 30 BP) and of Unit 3 (CC2: 2140 ± 35 BP) (Table 8.1).

Near the Enclos Fouqué cliff, Unit 1 (U1) overlies a thin, usually <0.5 m thick bed of matrix-supported breccia containing scattered angular decimeter-sized lithic clasts ("Bellecombe breccia" sensu Bachèlery 1981). In more distal areas, Unit 1 often overlies older lavas. In

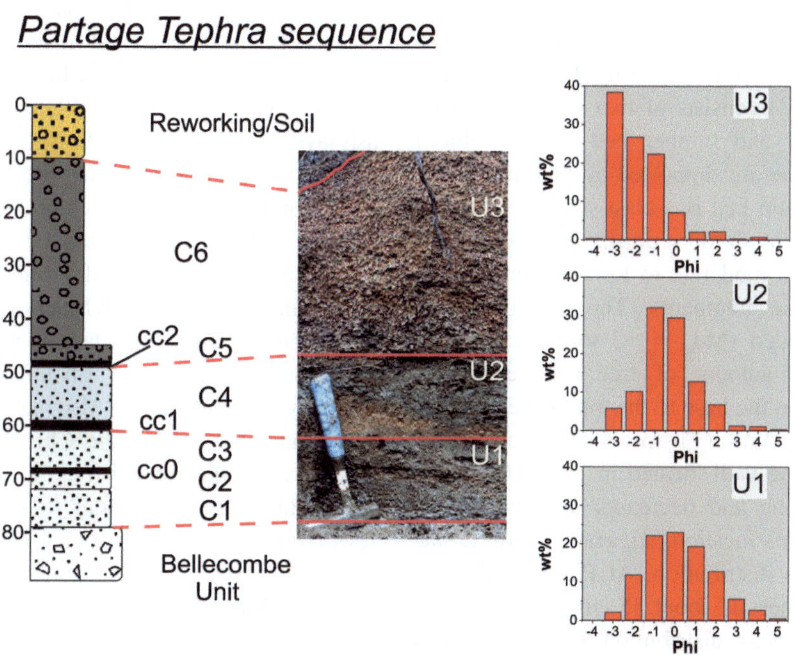

Fig. 8.3 "Partage Tephra" sequence. *C1, C2, C3* fine ash (U1); *C4* coarse ash (U2); *C5* lapilli and ash, *C6* well-sorted lapilli bed (U3); *cc0, cc1, cc2* charcoal-rich layers (respectively dated to 2855, 2340 and 2140 year BP; Table 8.1). Average grain size distribution for each Unit is plotted as histogram. The matrix-supported breccia underlying the "Partage Tephra" sequence is attributed to the "Bellecombe Unit"

the most proximal outcrops, Unit 1 groups at least four cm thin ash layers, whose maximum global thickness (≈30 cm) is observed close to Piton de Partage site. Piton de Partage scoria cone lies below Bellecombe breccia and "Partage" ashes. U1 ashes are brown to pink colored, fine grained ($Md_{\varphi} < -1.5$ φ) and poorly sorted ($\sigma_{\varphi} = 1.5-2$ φ) (Fig. 8.3). The ash layers are composed of minor highly fluidal glassy juvenile clasts in a volumetrically dominant matrix formed by variably altered lava fragments, possible small gabbro clasts, olivine crystals and rare euhedral quartz and magnetite (mostly in the lowermost layers). Similarities (in terms of grain size and componentry) between Unit 1 of the "Partage Tephra" and the (older) "Bellecombe Ash unit" suggest the possibility of an analogous mechanism at the origin of these deposits.

Unit 2 (U2) is a grey homogeneous lapilli and ash bed ($Md_{\varphi} = -1$ φ), moderately sorted ($\sigma_{\varphi} = 1-1.5$ φ), with a high amount of dense clasts (mm- to cm-sized grey lava fragments) (Fig. 8.3). Coarse fragments are dispersed in a fine-grained ash matrix containing olivine crystals and rare fluidal glassy fragments, whose amount increases towards the top of the bed. Bed thickness changes between 6 and 12 cm along the outcrop area.

Unit 3 (U3) consists of two scoria-rich beds. In the lower and thinner bed (<5 cm thick), reddish scoriae are supported by an ashy matrix. The upper main bed is a clast-supported deposit of mm- to cm-sized, reddish, variably altered scoriae with a moderate to low amount of crystals and lithic fragments. The bed is weakly normally graded ($Md_{\varphi} = -3$ to -2 φ; $\sigma_{\varphi} = 1-2$ φ). The amount of free olivine crystals increases from the base to the top of the bed. The maximum thickness (1 m) is observed on the Enclos Fouqué wall located in front of the Formica Leo cone and decreases quickly moving away from this location, suggesting a quite narrow isopachs distribution. At Piton de Partage, the U3 thickness is about 35 cm.

The thickness distribution of the "Partage Tephra" sequence strongly supports the emission from one or more sources located close and (at least for U1 and U2 eruptions) inside the Enclos Fouqué. Eruptive dynamics possibly evolve from phreatomagmatic in the lowermost units (U1 and U2) to magmatic in the upper one (U3). U2 bed drapes the inner cliffs of Enclos Fouqué and gives a minimum age to the formation of this depression.

8.3.2 The Plaine des Sables Black Tephra

The black pyroclastic blanket forming the surface of large part of the Plaine des Sables consists mainly of very fresh and vesicular, fine to medium achnelith and black lapillis, which cover the areas surrounding between Piton Chisny and Piton Haüy cones. We have referred to this deposit as "Black Tephra" unit. It shows a maximum thickness of about 2.5 m in the area between Piton Chisny and Demi-Piton cones, near the NW foot of Piton Chisny edifice (Fig. 8.4). The deposit is absent on the Enclos Fouqué rim, while it is still present, although less than 10–20 cm thick, on the west edge of the Plain des Sables. In the proximal area, the "Black Tephra" sequence is a crudely stratified, well-sorted lapilli bed. Locally, a thin and coarse ash layer can be observed at its base (S13 and S20) that vanishes quickly when moving outside the main dispersal axis. In PDS2 core the basal ash layer is almost absent. In the proximal facies, the S20 section shows the presence of some spatter bomb layers. The distal facies (S3 section) is instead characterized by a basal coarse ash, a middle well sorted lapilli body, and a lapilli and ash top layer. Whereas not overlain by lavas, the deposit is often largely reworked (as in PDS1 and PDS2). In the ash fraction, olivine is the only mineral phase. Moving northwards in the Plaine des Sables, the Chisny black lapilli layer overlies the yellowish lapilli layer related to Piton Haüy activity. A dm to m thick bed of reworked ash and lapillis occur between them.

8.3.3 The "Red Tephra"

In the Plaine des Remparts area, the most widespread deposit consists in a bed of reddish glassy fluidal scoriae, here named "Red Tephra" unit,

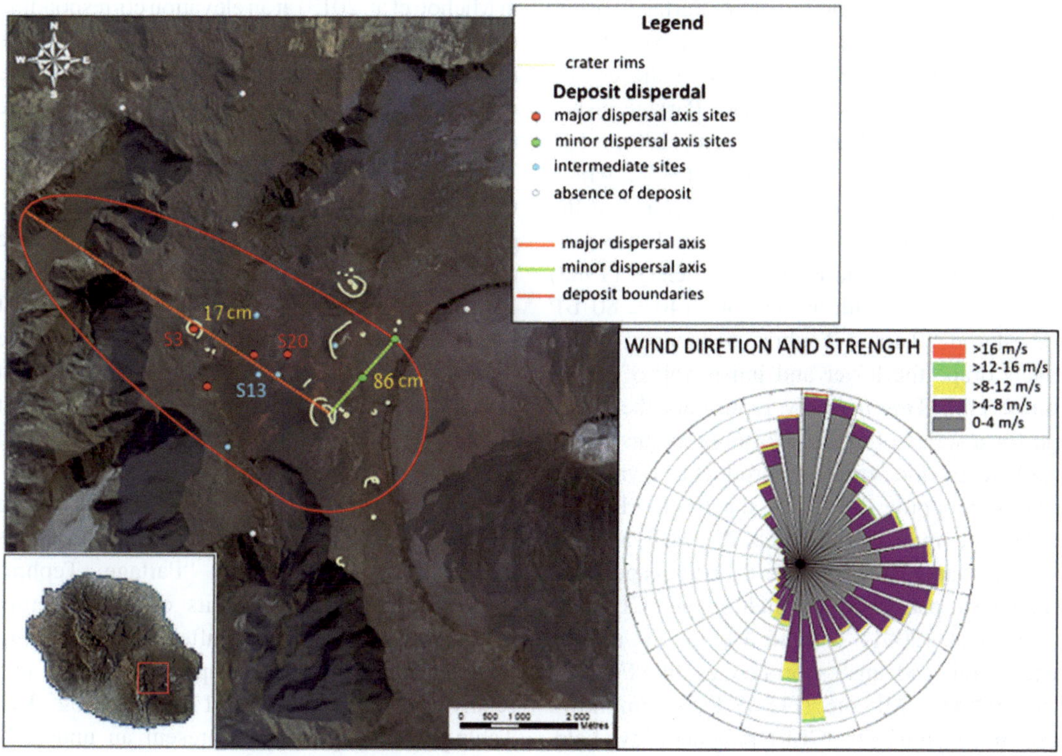

Fig. 8.4 "Black Tephra" unit maximum dispersal. *S*3, *S*13 and *S*20 are the sites whose logs are shown in Fig. 7.5. Wind strength and direction records at Pas de Bellecombe between 01/09/2007 and 15/03/2011 are reported in the down right corner (source Meteo France)

cropping out between Piton de l'Eau cone and Commerson crater. Maximum thickness (up to 2.5 m) is observed close to Piton des Cailles. Near Piton des Tangues, the fall bed is separated by a decimeter-thick black soil from the underlying Bellecombe Ash unit. The Red Tephra is restricted to the middle-distal sectors and is not observed in the proximal area of Plaine des Sables. As a whole, this unit has not been characterized yet and the source vent is unknown.

8.4 Frequency of PdF Explosive Activity During the Last 4000 Years

Available ages of PdF pre-historic volcanic events, together with new ^{14}C dating (Centre de Datation par le Radiocarbone, University Lyon 1), are here reviewed in order to identify the main periods of recent explosive volcanic activity (Table 8.1).

The main tephra identified all around the Enclos Fouqué is the "Bellecombe Ash unit". It corresponds to a complex sequence of ash, crystal and lithic-rich deposits (Bachèlery 1981; Mohamed-Abchir 1996). This deposit is thought to be emplaced during violent phreatic/phreatomagmatic events related to the flashing of an extensive hydrothermal system during the collapse of the Enclos Fouqué caldera (Bachèlery 1981; Mohamed-Abchir 1996; Fontaine et al. 2002; Upton et al. 2000).

The ages of the caldera formation and related Bellecombe Ash unit are poorly constrained. They are likely comprised in a time span ranging between the emplacement of the lavas forming the uppermost part of the Enclos Fouqué walls that underly Bellecombe and Partage ashes (4745 ± 130 BP, 3340 ± 1012; respectively Er-39 and Er-41 in Table 8.1) and that of the crystal and lithic-rich Bellecombe and Partage ashes themselves (several ages between 4175 ± 145 and 2320 ± 90 BP; respectively Er-38, which matches the Bellecombe Ash type-section at the Petite Carrière outcrop, and Er-28 in Table 8.1). Interestingly, Vergniolle and Bachèlery (1982) report that ashes with an age of 2140 ± 80 BP (U3 of the "Partage Tephra" sequence, in this paper) drape the lower and inner wall of Bellecombe cliff. This possibly represents the upper age for a series of caldera-forming events.

Most of the eruptive products, lavas and tephras, cropping out inside the Enclos Fouqué caldera are much younger than 2320 BP, actually younger than 1750 CE (Table 8.1), final age of the widespread pāhoehoe lava field named CLEF by Lénat et al. (2001). Notable exceptions are the central cone and small scoria cones, like Formica Leo and Puy Mi-côte (Fig. 8.1), of unknown age, but covered by the CLEF pāhoehoe lava field.

Recent and detailed stratigraphic analysis of the 340 m high walls of the summit PdF caldera (Peltier et al. 2012; Michon et al. 2013) failed in identifying deposits that could be related to major summit explosive events. Michon et al. (2013) highlight that the summit cone evolution is mainly made by the alternation of crater collapse, lava infilling and outpouring episodes, suggesting a dominant effusive magmatic style punctuated by mild to weak explosive events unable to leave significant and durable traces on the territory. Moreover, the span of time covered by these processes is probably short (Michon et al. 2013). For instance, caldera filling appears to be relatively rapid. The ca. 300 m-deep, 600 m-across 1931 summit collapse structure was filled in 2006, just before the major 2007 summit collapse event. This recent and dominantly effusive activity of PdF was preceded by an important strombolian to hawaiian activity, whose remnants is a basal cone made of red to ochre scoriae visible at the bottom of the present Dolomieu crater (Michon et al. 2013). Both hawaiian, strombolian and possible phreatomagmatic deposit have been recognized in this lower sequence (Unit 1a and 1b in Peltier et al. 2012; U1a in Michon et al. 2013) at an elevation corresponding to the Enclos Fouqué caldera filling. The deposits related to this explosive phase of unknown age have not been identified outside the Enclos Fouqué (Peltier et al. 2012; Michon et al. 2013).

Weak to mild explosive eruptions related to the central activity have left very little stratigraphic record, except near the summit cone (Michon et al. 2013). For instance, fine ash and Pele's hairs of the 2007 caldera forming event have been quickly eroded away and only little traces are found inside the Bory crater, just 6 years after the event. The deposits of the most intense recent explosive eruptions (1961 and 1860) are only found near the volcano summit.

A notable exception is represented by the "Partage Tephra" sequence. Our new datings clearly indicate that the "Partage Tephra" sequence groups the deposits of several eruptions, which have occurred after a main event of breccia formation. These eruptions occurred in a relatively long period (−1124 to −52 BC; Table 8.1) and possibly represent an unusually violent series of events occurring soon after (and partially associated with) the last phase of Enclos Fouqué caldera collapse. These deposits give thus a lower limit for the age of the PdF central cone.

Along the N120° rift zone, the "Red Tephra" unit seems to pertain to a similar age interval. In this area, two ^{14}C ages (3735 ± 90 and 3205 ± 75 BP; respectively Er-37 and Er-35 in Table 8.1) are reported bracketing the underlying Bellecombe cpx-gabbro rich ash. Very similar ages (3715 ± 30 and 3220 ± 50 BP) were also found 1 km East from Piton dans le Bout (Er-37, Table 8.1) and in an ash bed underlying the Red Tephra near Commerson crater (Er-35, Table 8.1). The widespread presence in this sector, up to the Caldera des Sables rim (Er-30 to Er-36 in Table 8.1), of ash beds belonging to the "Bellecombe Ash unit" provide hence a upper limit for the age of the Red Tephra.

All tephra units identified in the period succeeding the "Partage" and "Red" tephra sequences have a very limited degree of dispersion. For instance, Er-21 and Er-20 new radiocarbon ages, respectively dated to 868 ± 30 and 620 ± 30 BP, identify two relatively recent and very localized events occurred close to Langevin area, inasmuch their deposits are absents both within the near Plaine des Sables (PDS2 core) and Pas des Sables (section S3; Fig. 8.4).

The most widespread lapilli bed of the Piton de la Fournaise's recent history is the "Black Tephra" unit, which marks the beginning of Chisny cone activity and is here dated for the first time at 381 ± 26 BP (Er-14 Table 8.1). It likely represents the most explosive magmatic event of a series of eccentric eruptions that have occurred along the three rift zones (NW, SE and NE) in a short and recent period (1400–1600 CE; Er-10 to Er-18 Table 8.1; Tanguy et al. 2011). Interestingly, this series of closely spaced violent eruptions immediately preceded the first permanent human settlements on the island; consequently, without leaving any trace in the historical record.

8.5 Concluding Discussion

The characterization of the explosive activity and the research of violent paroxystic events over PdF massif has been performed through a drilling and excavation campaign supported by the integration of new radiocarbon ages to previous chronologic data. The use of a light shallow and continuous drilling system has proved to be a suitable methodology to execute surveys in an impervious territory, such in La Réunion, and to realize a detailed micro-stratigraphic study of the tephra emitted by the volcano over a relative recent period (last 5 kiloyear).

The stratigraphic study carried out on the cores confirms the presence of the "Bellecombe Ash unit" in a relatively superficial position, over a wide area ranging from the proximal area, near the Bellecombe caldera cliff and the Plaine des Sables, to at least the Plaine des Remparts, the middle-distal sector. The observation made at the microscopic scale does not allow us to extend this correlation to the most distal Grand Étang drill, located at 18 km from the PdF central cone, suggesting either the lacking of this deposit in the more distal areas or an age of the cones of Grand Étang younger than 3 kiloyear. Actually, the uncertainty about the age of the pyroclastic cones which close the Grand Etang lake basin (younger than 5730 ± 110 year BP; Banton 1985) does not permit to constrain its stratigraphic position with respect to -Bellecombe ashes. Occurrence of "Bellecombe" layers (<20 cm total thickness) on the Trous Blancs area ca. 13 km W-NW from the central cone suggest that this is the main dispersion axis of these ashes.

The review of the Bellecombe Ash unit shows that only few ash fallout beds attain the medial area, with a thickness quickly decreasing when moving away from the Enclos Fouqué caldera. These decimeter-thick beds correspond to the upper part of the type-section described by Mohamed-Abchir (1996) in proximal areas. The age interval (3735–3060 BP) of the upper beds results significantly younger than that of the lower beds (4880–4175 BP) and of the lava flow underlying the ashes near the rim of the Enclos Fouqué caldera (4745 BP) (Bachèlery 1981; Staudacher and Allègre 1993). The "Partage Tephra" sequence, here described for the first time, can thus represent a sequence of events marking the end of the "Bellecombe ash unit" and puts an upper limit for the Enclos Fouqué caldera collapse events (2340 year BP for Unit 2). Therefore the Bellecombe Ash unit groups a sequence of eruptive events, which span a much wider time-period than previously thought. Anyhow, it represents the most traceable horizon (Fig. 8.5) among the studied deposits and no other pyroclastic blanket exhibits a similar regional dispersion in the last 5 kiloyear.

A second, more limited, correlation has been found in the drilling cores. It involves the "Black Tephra" lapilli bed drilled in the Plaine des Sables (PDS1 and PDS2, Fig. 8.5). The "Black Tephra" originates and is mainly distributed westward (Fig. 8.4) within the Plaine des Sables from possibly linear (fracture) source structure. The deposit dispersal exhibits a W elongation, in agreement with the Easterlies direction at Pas de Bellecombe (inset in Fig. 8.4, ground level records by Météo France). Such fact highlights that an eventual plume generated during the eruption was confined below the Trade winds inversion zone, at about 4 km a.s.l., and thus cannot exceeds the height of about 2 km

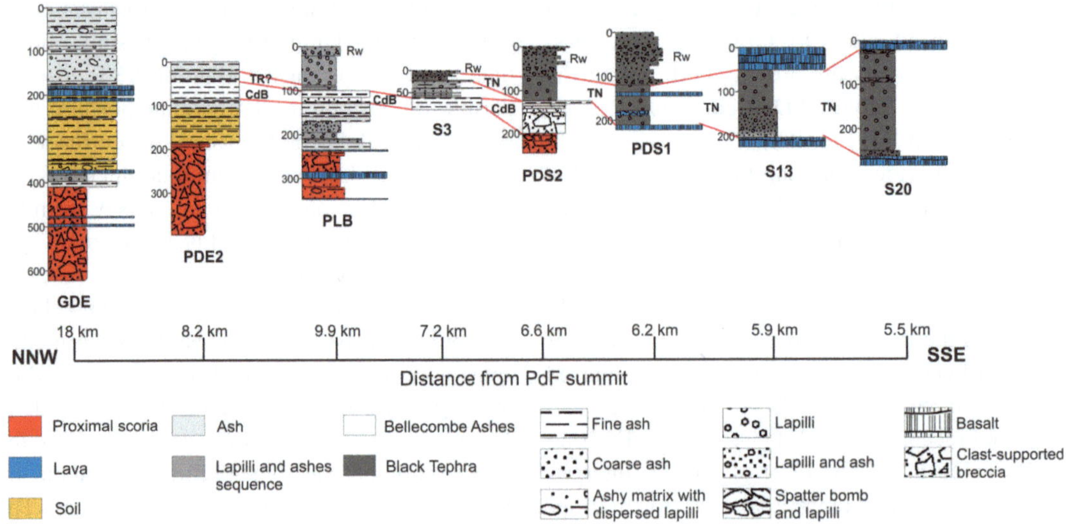

Fig. 8.5 Stratigraphic correlations from the distal to the proximal sector at NW of Piton de la Fournaise volcano. *CdB* Bellecombe Ashes; *TN* Black Tephra; *TR?* Red Tephra possible correlation; *Rw* reworked lapilli and ash

inasmuch the tephra source is located at an altitude of 2300 m a.s.l. The Black Tephra maximum dispersal seems to vanish quickly, within about 3.7 km in the downwind direction and over an area of less than 13 km^2. This deposit presents the typical features of the hawaiian activity. In particular it seems to be related to a lava fountaining generated through a fissural source. This activity results to be very young (381 ± 26 BP); calibrated ages (1447–1628 CE) suggest this unusual event occurred shortly before the beginning of permanent settlement on La Réunion island.

A fountaining activity of similar intensity to that of the Chisny event is possibly represented by the Red Tephra unit dispersed in the medial area. This deposit could be represented in the PLB core (Plaine des Remparts West) upper black lapilli sequence at least by the basal red ash-coated lapilli (Table 8.4) and in the PDE2 core (Plaine des Remparts East) by the reddish ash layer overlying the Bellecombe Ash unit (Table 8.3). However this correlation still remains a hypothesis to verify cause of the scarce characterization for the middle-distal sector deposits. The age of this event (younger than 3200 BP) and its stratigraphic position suggests that it occurred near the end of the Bellecombe ash unit emplacement.

Besides the Bellecombe Ash unit, the lack of regionally distributed lapilli blankets suggests that no violent strombolian or more intense eruptions (subplinian to plinian) have occurred in the recent story of the volcano's activity (<5 kiloyear). In particular, the distribution of the "Partage Tephra" sequence highlights that phreatomagmatic events can impact the proximal areas outside the Enclos Fouqué caldera, but with a limited extension. Moreover, the lack of a continuous lapilli/ash horizon from the proximal to distal areas points out that the lapilli cover that overlay the Plaine des Sables and the Plaine des Remparts is the result of several local blankets linked to several hawaiian to strombolian emission centers. As highlighted by new radiocarbon ages obtained on tephra lying on the Langevin Plateau and Plaine des Sables (respectively 868, 620 and 381 year BP in Table 8.1) this kind of activity persisted in the N120° rift up to very recent times and ended probably just before the island's colonization, leaving no trace in historical records. The eccentric activity persisted instead along the NE and SE rift zones but with a dominantly effusive style (Michon et al. 2013).

The maximum expected magmatic events have therefore to be related to intense hawaiian fountaining. The Black Tephra unit, emitted by a source area located near the Piton Chisny cone, represents a very recent and intense manifestation of this typology. The low amount of fine ashes and the NW elongation of its deposit, in fact, suggest the absence of the well-developed "cineritic" plume, which characterizes violent strombolian eruptions. Therefore, the implication for the risk assessment, and consequently for risk maps production and organization, must consider that lava fountaining, even if more vigorous than the normal, is the maximum magmatic explosive activity produced by Piton de la Fournaise in the last 5000 years. It implies that its dangerousness is restricted to a relatively brief distance from the source and that a regional deposition is strongly unlikely. The hazard represented by this type of activity resides in the possible positioning of the vent close to inhabited areas and in the possible occurrence of repeated events with short but unpredictable time interval from one to the other. Forecasting the location of the future eccentric eruptions is thus of paramount importance to minimize the potential impact of mild eruptions on the inhabitants and the structures.

Acknowledgments The Préfecture, Protection Civile and DEAL of the La Réunion and the MEEDDAT (Ministère de l'Ecologie, du Développement Durable et de l'Energie) are acknowledged for promoting, supporting and partially financing this project on volcanic hazard assessment at Piton de la Fournaise (convention n°11_037 and 1321; 2011–2014). A. Morandi was supported by a grant of the Department of Earth Sciences of the University of Florence (entitled to prof. Orlando Vaselli).

References

Andronico D, Cristaldi A, Del Carlo P, Taddeucci J (2009) Shifting styles of basaltic explosive activity during the 2002–03 eruption of Mt. Etna. Italy. J Volcanol Geotherm Res 180:110–122

Arrighi S, Principe C, Rosi M (2001) Violent strombolian and subplinian eruptions at Vesuvius during post-1631 activity. Bull Volcanol 63:126–150

Bachèlery P (1981) Le Piton de la Fournaise (Ile de la Réunion). Etude Volcanologique, Structurale et Pétrologique. Thèse de l'Université de Clermont Ferrand, p. 217

Banton O (1985) Rapport de thèse: Etude hydrogéologique d'un complexe alluvial en pays volcanique, sous climat tropical, site du Grand Étang—Ile de la Réunion. Université des sciences et techniques du Languedoc de Montpellier et Université française de l'Océan Indien, Atelier duplication—USTL, p 237

Delorme H, Bachèlery P, Blum PA, Cheminée J-L, Delarue J, Delmond J, Hirn A, Lepine J, Vincent P, Zlotnicki J (1989) March 1986 eruptive episodes at Piton de la Fournaise volcano (Réunion Island). J Volcanol Geotherm Res 36:199–208

Di Muro and project team (2012) Evaluation de l'alea volcanique a La Réunion. Rapport final-année I°. Projet BRGM/IPGP, p 80

Fontaine FJ, Rabinowicz M, Boulègue Jouniaux L (2002) Constraints on hydrothermal processes on basaltic edifices: inferences on the conditions leading to hydrovolcanic eruptions at Piton de la Fournaise, Réunion Island, Indian Ocean. Earth Planet Sci Lett 200:1–14

Fretzdorff S, Paterne M, Stoffers P, Ivanova E (2000) Explosive activity of the Reunion Island volcanoes through the past 260,000 years as recorded in deep-sea sediments. Bull Volcanol. doi:10.1007/s004450000095

Houghton BF, Wilson CJN, Del Carlo P, Coltelli M, Sable JE, Carey RJ (2004) The influence of conduit processes on changes in style of basaltic Plinian eruptions: Tarawera 1886 and Etna 122 BC. J Volcanol Geotherm Res 137:1–14

Hugoulin F (1860) Dernière éruption du volcan de l'île de la Réunion (19 mars 1860). Revue algérienne et coloniale 2:483–487

Kieffer G, Tricot B, Vincent PM, (1977) Une éruption inhabituelle (Avril 1977) du Piton de La Fournaise (Ile de La Réunion): ses enseignements volcanologiques et structuraux. C R Acad Sci Paris, 285(D): 957–960

Lesouëf D, Gheusi F, Delmas R, Escobar J (2011) Numerical simulations of local circulations and pollution transport over Reunion Island. Ann Geophys 29:53–69

Lénat J -L (2016) A brief history of the observation of Piton de la Fournaise central area. In: Bachèlery P, Lénat J-F, Di Muro A, Michon L (eds) Active volcanoes of the Southwest Indian Ocean: Piton de la Fournaise and Karthala. Active Volcanoes of the World. Springer, Berlin

Lénat JF, Bachèlery P (1988) Dynamics of magma transfer at Piton de la Fournaise volcano (Réunion Island, Indian Ocean). In: Chi-Yu et Scarpa (eds) Earth evolution sciences special issue "Modeling of Volcanic Processes". Friedr. Vieweg and Sohn, Brauschweig/Wiesbaden, pp 57–72

Lénat JF, Bachèlery P, Desmulier F (2001) Genèse du champ de lave de l'Enclos Fouqué: une éruption d'envergure exceptionnelle du Piton de la Fournaise (Réunion) au 18 ème siècle. Bull Soc Géol Fr 2:177–188

Ludden JN (1977) Eruptive patterns for the volcano Piton de la Fournaise, Reunion Island. J Volcanol Geotherm Res 2:385–395

Macdonald GA (1972) Volcanoes. Prentice-Hall inc., Englewood Cliffs, New Jersey. p 510

Michon L, Di Muro A, Villeneuve N, Saint-Marc C, Fadda P, Manta F (2013) Explosive activity of the summit cone of Piton de la Fournaise volcano (La Réunion island): a historical and geological review. J Volcanol Geotherm Res 263:117–133

Michon L, Ferrazzini V, Di Muro A (2016) Magma paths at Piton de la Fournaise volcano. In: Bachèlery P, Lénat J-F, Di Muro A, Michon L (eds) Active volcanoes of the Southwest Indian Ocean: Piton de la Fournaise and Karthala. Active Volcanoes of the World. Springer, Berlin

Mohamed-Abchir MA (1996) Les Cendres de Bellecombe: Un événement majeur dans le passé récent du Piton de la Fournaise. Thèse d'université, Univ. de Paris VII, Paris, Ile de la Réunion

Peltier A, Massin F, Bachèlery P, Finizola A (2012) Internal structures and building of basaltic shield volcanoes: the example of the Piton de La Fournaise terminal cone (La Réunion). Bull Volcanol. doi:10.1007/s00445-012-0636-7

Pioli L, Erlund E, Johnson E, Cashman K, Wallace P, Rosi M, Delgado Granados H (2008) Explosive dynamics of violent strombolian eruptions: the eruption of Parícutin volcano 1943–1952 (Mexico). Earth Planet Sci Lett 271:359–368

Principe C, Malfatti A, Rosi M, Ambrosio M, Fagioli MT (1997) Metodologia innovativa di carotaggio microstratigrafico: esempio di applicazione alla tefrostratigrafia di prodotti vulcanici distali. Geol Tec Ambient 4 (97):39–50

Principe C, Malfatti A, Ambrosio M, Fagioli MT, Rosi M, Ceccanti B, Arrighi S, Innamorati D (2007) Finding distal Vesuvius tephra at the borders of Lago Grande di Monticchio. In: AF shallow coring system micro-cores. Atti Soc Tosc Sci Nat Mem Serie A, 112:189–197

Roult G, Peltier A, Taisne B, Staudacher T, Ferrazzini V, Di Muro A, OVPF team (2012) A new comprehensive classification of the Piton de la Fournaise activity spanning the 1985–2010 period. Search and analysis of short-term precursors from a broad-band seismological station. J Volcanol Geotherm Res 241:78–104

Sable JE, Houghton BF, Del Carlo P, Coltelli M (2006) Changing conditions of magma ascent and fragmentation during the Etna 122 BC basaltic Plinian eruption: evidence from clast microtextures. J Volcanol Geotherm Res 158:333–354

Sable JE, Houghton BF, Wilson CJN, Carey R (2009) Eruption mechanisms during the climax of the Tarawera 1886 basaltic Plinian inferred from microtextural characteristics of deposits. In: Larsen G, Rowland SK, Self S, Hoskuldsson A (eds) Thordarson T. Studies in Volcanology, The Legacy of George Walker, pp 129–154

Sisavath E, Mazuel A, Jorry SJ, Babonneau N, Bachèlery P, de Voogd B, Salpin M, Emmanuel L, Beaufort L, Toucanne S (2012) Processes controlling a volcaniclastic turbiditic system during the last climatic cycle: example of the Cilaos deep-sea fan, offshore La Réunion Island. Sediment Geol 281:180–193. doi:10.1016/j.sedgeo.2012.09.010

Staudacher T, Allègre CJ (1993) Ages of the second caldera of Piton de la Fournaise volcano (Réunion) determined by cosmic ray produced 3He and 21Ne. Earth Planet Sci Lett 119:395–404

Staudacher T, Peltier A, Ferrazzini V, Di Muro A, Boissier P, Catherine P, Kowalski P, Lauret F (2016) Fifteen years of intense (1998–2013) at Piton de La Fournaise volcano: a review. In: Bachèlery P, Lénat J-F, Di Muro A, Michon L (eds) Active volcanoes of the Southwest Indian Ocean: Piton de la Fournaise and Karthala. Active Volcanoes of the World. Springer, Berlin

Stieltjes L, Moutou P (1989) A statistical and probabilistic study of the historic activity of the Piton de la Fournaise, Réunion Island, Indian Ocean. J Volcanol Geotherm Res 36:67–86

Tanguy JC, Bachèlery P, Le Goff M (2011) Archeomagnetism of Piton de la Fournaise: bearing on volcanic activity at La Réunion Island and geomagnetic secular variation in Southern Indian Ocean. Earth Planet Sci Lett 303:361–368. doi:10.1016/j.epsl.2011.01.019

Taupin FG, Bessafi M, Baldy S, Brémaud PJ (1999) Tropospheric ozone above the southwestern Indian Ocean is strongly linked to dynamical conditions prevailing in the tropics. J Geophys Res 104:8057–8066

Tulet P, Villeneuve N (2011) Large scale modeling of the transport, chemical transformation and mass budget of the sulfur emitted during the eruption of April 2007 by the Piton de la Fournaise. Atmos Chem Phys 11:4533–4546

Trenberth KE, Large WG, Olson JG (1990) The mean annual cycle in global ocean wind stress. J Phys Oceanogr 20:1742–1760

Upton BGJ, Semet MP, Joron JL (2000) clasts in the, Piton de la Fournaise, Réunion Island, and their bearing on cumulative processes in the petrogenesis of the Réunion lavas. J Volcanol Geotherm Res 104:297–318

Villeneuve N, Bachèlery P (2006) Revue de la typologie des éruptions au Piton de La Fournaise, processus et risques volcaniques associés. Cybergéo: revue européenne de Géographie, n° 336

Wong LJ, Larsen JF (2010) The middle scoria sequence: a Holocene Subplinian and of Okmok volcano, Alaska. Bull Volcanol 72:17–31. doi:10.1007/s00445-009-0301-y

9 Fifteen Years of Intense Eruptive Activity (1998–2013) at Piton de la Fournaise Volcano: A Review

Thomas Staudacher, Aline Peltier, Valérie Ferrazzini, Andrea Di Muro, Patrice Boissier, Philippe Catherine, Philippe Kowalski, Frederic Lauret and Jacques Lebreton

Abstract

Piton de la Fournaise (La Réunion Island) is amongst the most active volcanoes in the world, having had about 130 eruptions during the last century. Over the last 50 years it has produced 970 Mm^3 of magma, at a mean eruption rate of 19.4 Mm^3 per year, of which 489 Mm^3 had been erupted over the last 15 years (at a rate of 32.7 Mm^3 per year). This indicates a relative increase in its output. In this paper we present a detailed description of the high volcanic activity spanning the 1998–2013 period (eruptive precursors, location, characteristics and volume of emitted lavas, …). This period of high volcanic activity has been the subject of numerous studies that have led to significant advances in the knowledge of the shape and dynamics of the shallow plumbing system of the volcano.

9.1 Introduction

The island of La Réunion is located at 21°S and 55°30′E in the western Indian Ocean. It represents the active end member of the 65 Ma old Deccan Trap hotspot chain (Duncan 1981; Morgan 1981; see also Lénat 2016a, Chap. 2). The island is composed of two volcanoes, Piton des Neiges (3071 m) and Piton de la Fournaise (2632 m) (Fig. 9.1). Lénat (2016b) in Chap. 3 develops the aspects specific to the construction of the island.

Piton des Neiges is the main massif. Activity began 5 Ma ago on the 4000 m deep sea floor (Gillot et al. 1994). The oldest sub-areal rocks are found on the NW side of the island and have an age of 2.1 Ma (McDougall 1971). Large landslides (Oehler 2005) and high erosion rates have characterized the evolution of the Piton des Neiges massif, leaving behind three "cirques" and steep-sided river valleys with scarps of up to 1000 m in height. The latest known eruptions of Piton des Neiges occurred about 12,000 years

T. Staudacher (✉) · A. Peltier · V. Ferrazzini · A. Di Muro · P. Boissier · P. Catherine · P. Kowalski · F. Lauret · J. Lebreton
Observatoire Volcanologique du Piton de la Fournaise, Institut de Physique du Globe de Paris, CNRS, Sorbonne Paris Cité, Univ. Paris Diderot, CNRS, F-97418 La Plaine des Cafres, La Réunion, France
e-mail: thomasstaudacher1950@gmail.com

P. Bachèlery et al. (eds.), *Active Volcanoes of the Southwest Indian Ocean*, Active Volcanoes of the World, DOI 10.1007/978-3-642-31395-0_9

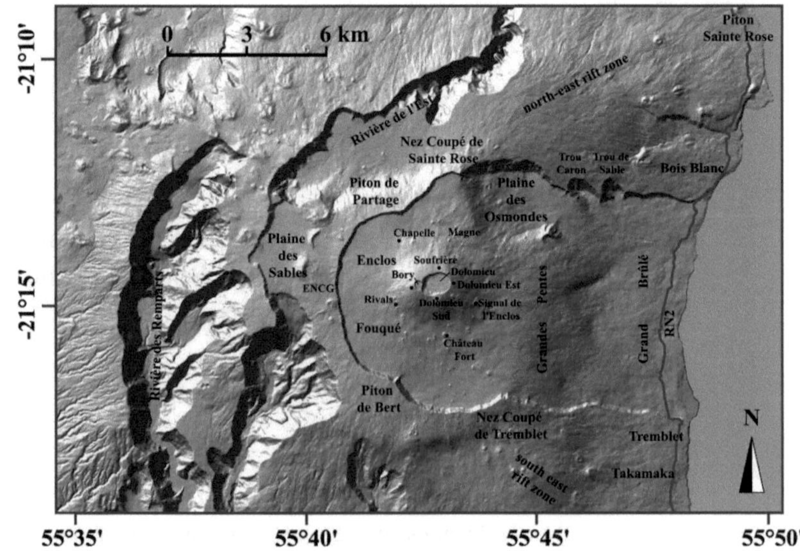

Fig. 9.1 Digital Elevation Model of the Piton de la Fournaise massif. Locations and names of areas mentioned in the text are reported

ago (Deniel et al. 1992). The volcanic massif has a diameter of about 220 km at the sea floor and its total height is ~7000 m. The active volcano, Piton de la Fournaise (PdF), grew on the SE flank of the Piton des Neiges massif. It has a sub areal diameter of about 40 km and an elevation of 2632 m above sea level (Fig. 9.1).

The first eruptions of Piton de la Fournaise have occurred about 500,000 years ago (McDougall 1971; Gillot and Nativel 1989). The massif is characterized by three large calderas (Fig. 9.1). The first, ~150,000 years old, is highly eroded and roughly delimited to the west by the Rivière des Remparts. The second one, ~24 000 years old (Gillot and Nativel 1989; Staudacher and Allègre 1993), is delimited by Plaine des Sables and Rivière de l'Est. The current active terminal cone is located inside of the third caldera, the ~4500 years old Enclos Fouqué caldera (Bachèlery 1981). This caldera has a size of 8 × 13 km and is open to the east to the ocean. Two rift zones extend to the north-east (NE rift) and south-east (SE rift) of the central cone (Fig. 9.1). Piton de la Fournaise currently hosts two craters at its summit, the 300 m wide Bory crater to the west and the 1 km wide Dolomieu crater to the east. In 2007, the Dolomieu crater floor collapsed during the largest known low-elevation flank eruption, and formed a km-scale caldera (Staudacher et al. 2009).

Piton de la Fournaise is an extremely active volcano, with 93 documented eruptions since 1950 (Fig. 9.2). Since 1950, 970 Mm^3 of lava has been erupted, 489 Mm^3 of which was erupted during the last 15 years. Most of the eruptions between 1950 and 2010 were confined inside of the Enclos Fouqué caldera. Only three eruptions occurred outside of the caldera ("hors Enclos"). In April 1977 an eruption "hors Enclos" happened along the north-east rift zone and inundated the village of Piton Sainte Rose with lava. A further eruption in March 1986, along the south-east rift zone, occurred in the Le Tremblet and the Takamaka area. Finally, a small "hors Enclos" event occurred during the March 1998 eruption, when a vent opened to the north of PdF above the village of Bois Blanc. Since 1950, eight eruptions have reached the ocean (Fig. 9.3), five of which occurred during the last 15 years. The period was also marked by two hiatus, each lasting ~6 years; the first lasting from May 1966 to June 1972, and the second from October 1992 to March 1998 (Fig. 9.2). The 1992–1998 hiatus ended with a 6.5 month long eruption, which was then followed by 33 eruptions over 13 years.

Here we present a review of the last 15 years (1998–2013) of activity at Piton de la Fournaise, including eruption precursors, eruptive activity, the sites and the evolution of each eruption. As part of the discussion, we present an overview of

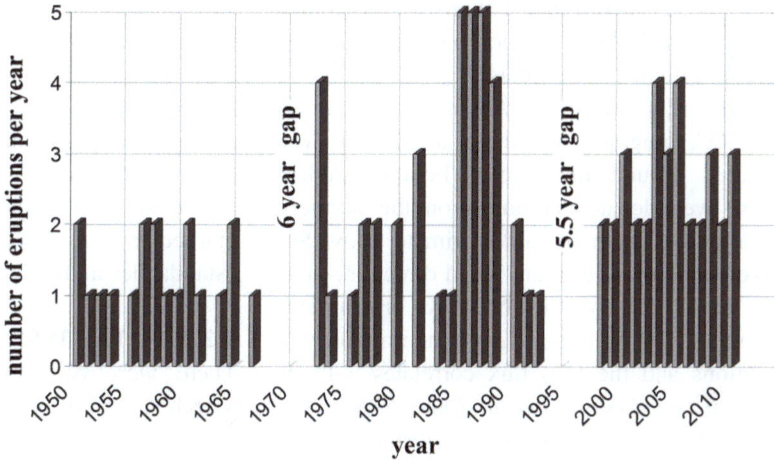

Fig. 9.2 Number of eruption versus time at Piton de la Fournaise between 1950 and 2013

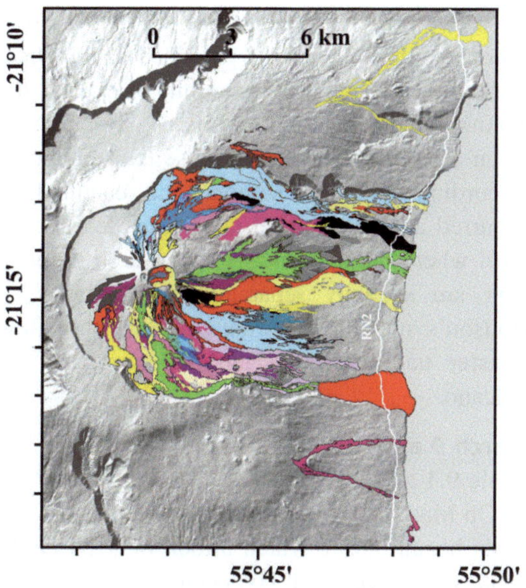

Fig. 9.3 Lava flow map of 1950–2010. Only three eruptions occurred outside of the U-shaped Enclos Fouqué caldera, open to the ocean, in 1977, 1986 and 1998

the current plumbing system (its geometry, location, dynamics) which has fed recent activity.

9.2 OVPF Networks

Despite the high number of eruptions, the volcanological observatory at La Réunion Island only established in 1979. The decision of the creation of an observatory followed the 1977 eruption. This occurred outside of the Enclos Fouqué caldera and led to the destruction of several houses in the village of Piton Sainte Rose. After this event, the local authorities became conscious of the necessity to monitor the volcano and to attempt to forecast eruptions. Observatoire Volcanologique du Piton de la Fournaise (OVPF), which belongs to the Institut de Physique du Globe de Paris (IPGP), became operational in 1980 with the installation of the first seismic and deformation networks. Today monitoring of Piton de la Fournaise is assured by a dense seismic network of more than 40 seismic stations (Lemarchand et al. 2013). The deformation network includes Global Navigation Satellite System (GNSS) receivers, tiltmeters and extensometers. Twenty-three GNSS receivers cover mainly the Enclos Fouqué area, but also the region outside of this caldera (Staudacher and Peltier 2016). Seven tiltmeter stations are located in the Enclos Fouqué: three on the summit, three at the base of the central cone and one 4 km to the south-east. Each station has two Blum-type pendulum silica tiltmeters (Blum 1963) or two electrolytic SPECTRON tiltmeters. These are fixed to the rock and are respectively orientated radial and tangential to the summit craters. Finally, four 3-component extensometers are located over deep, several cm-wide fractures to measure the extension, shear and vertical components every minute (Peltier et al. 2006). Staudacher and Peltier (2016, Chap 15) and

Peltier et al. (2016, Chap. 17) provide a complete review of the GNSS, tiltmeters and extensometers data for Piton de la Fournaise.

In late 2007, three DOAS instruments of the NOVAC program (Galle et al. 2010) were installed around the Enclos Fouqué caldera. These were located at points on the northern, western and southern caldera rim to allow SO_2 fluxes to be tracked in any wind direction. Since the installation of the NOVAC instruments, recorded SO_2 emissions are synchronous with eruptions and the SO_2 flux correlates with the effusion rate (Garofalo et al. 2009; Allard et al. 2014). A multi-gas analyzer was installed on the summit in 2011 and CO_2 flux has been measured continuously since 2012 (Di Muro et al. 2015). For all networks, data are transmitted to the OVPF in real time by radio or WiFi.

9.3 Activity from 1998 to 2013

After the August 27, 1992 eruption, which lasted 27.2 days, Piton de la Fournaise entered in a phase of eruptive and seismic rest that lasted 5½ years (Fig. 9.2). Seismic activity [mainly volcano-tectonic (VT) events] beneath PdF at that time declined from a mean of 3.3 events per day in 1993 to 0.6 events per day in the first six months of 1997 (Fig. 9.4). This trend was only interrupted by some deep seismic events in September 1996 below Plaine des Sables and in October 1996 close to Chapelle. In addition a seismic crisis of 138 events lasting 104 min occurred in November 26, 1996. A further, smaller seismic crisis occurred in August 23, 1997. From mid-1997, seismicity beneath PdF began to increase significantly, reaching a frequency of about 3 events per day by late 1997, and 4.2 events per day by early 1998. No significant ground deformation was recorded during this period (Staudacher and Peltier 2016).

9.3.1 Eruption Activity: 1998–2006

Following this increase in seismicity, the 6-year long hiatus ended in March 1998. There followed 34 eruptions through 2013, which we split into two phases: 1998–2006 and 2007–2013, separated by the voluminous 2007 eruption (and the related eruptions in February and March), which caused the Dolomieu collapse. The 1998–2006 phase is marked by a global long-term inflation of the summit, while the 2007–2013 phase is marked by a global deflation of the massif (Staudacher and Peltier 2016).

9.3.1.1 The 1998 Eruptions

There were two eruptions simultaneously in March 1998 (Kapor and Hudson vents). The chemical composition and the partial pressures deduced from melt inclusion studies indicated different depths for the magma associated to these two eruptions (Bureau et al. 1999). However, Salaün et al. (2010) proposed cannibalism of olivine-rich cumulate xenoliths to explain the distinct geochemical compositions, which could mean a direct link between these two eruptions. According to these authors, the cannibalism occurred at shallow level (above sea level), at a level where the two dikes feeding each eruptive vent are spatially separated. The scarcity of Hudson's magma is attributed to their shallow transfer path in rarely intruded zones of the volcano.

March 9 and March 12 event (No. 1 and 2 in Table 9.1 and Fig. 9.5a)

On March 8, 1998, a strong 35-h long seismic crisis began during which a total of 3135 earthquakes were recorded. Hypocenters were located first at 6–5 km below sea level, moving upwards along a narrow path, up to 1.5 km below sea level at an average velocity of 1.6 m min^{-1} (Battaglia et al. 2005). Between 1.5 km below sea level and sea level the velocity was slightly lower at 1.2 m min^{-1}. Above sea level, the dyke migrated to the surface in less than one hour, and the eruption began on March 9, at 11:05 (all hours are given in GMT = local time −4 h and as hh:mm) from an *en echelon* fissure located between 2460 and 2070 m (Table 9.1). Lava fountains of about 50 m in height were observed along the fissures. Primary sources became focused at two vents: Piton Kapor

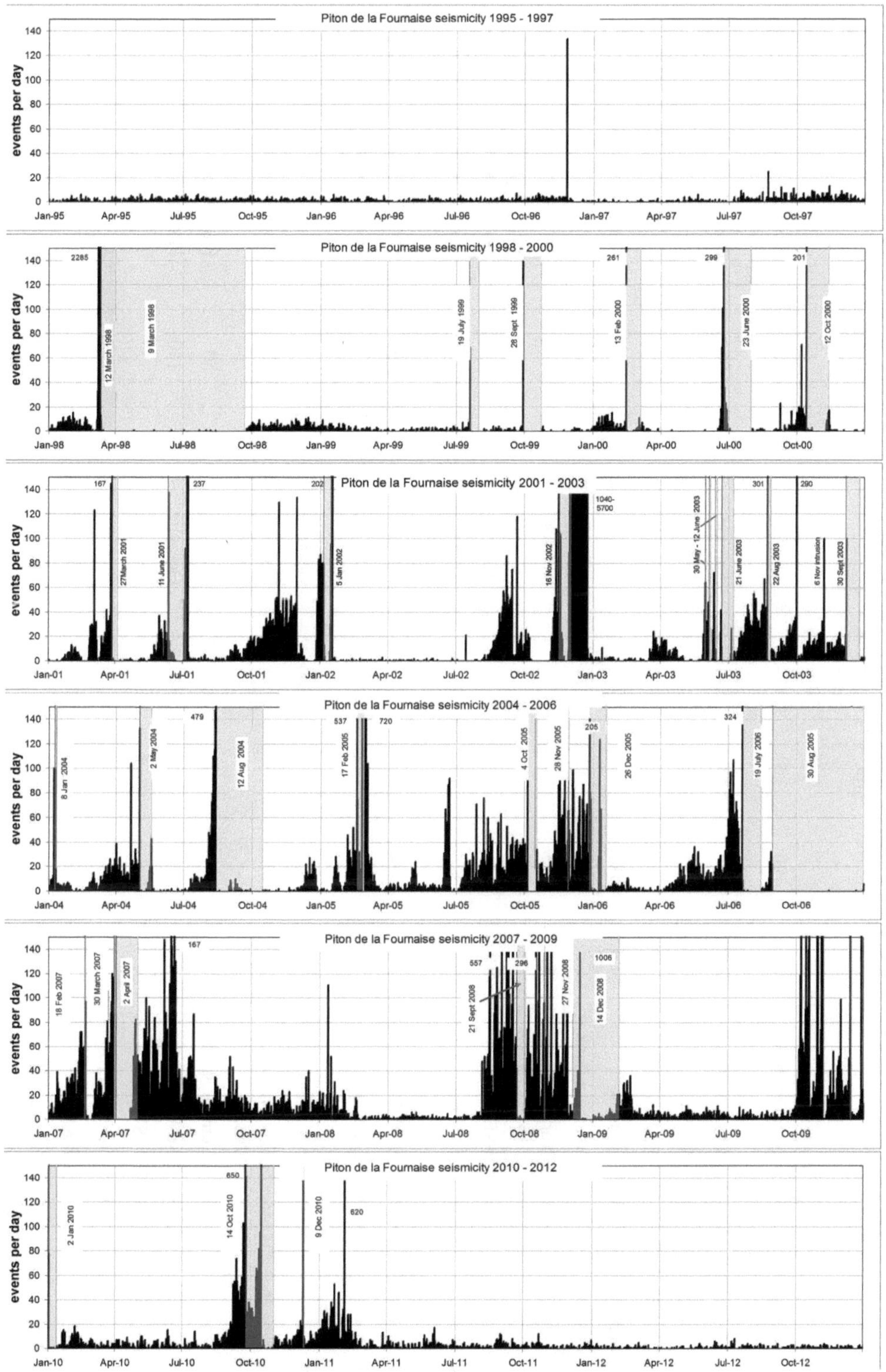

Fig. 9.4 Piton de la Fournaise daily seismicity from 1995 to 2012. Dates on the plots indicate the beginning of eruptions, *light grey* areas indicate the period of longer lasting eruptive events. Numbers denote daily seismic events that exceed the scale. Present-day seismicity (2013) is extremely low, and with an average of 1.1 events per day is comparable to 2012

Table 9.1 Characteristics of eruptions spanning the 1998-2010 period (after Roult et al. 2012)

No.	Date	Start time	Site	Volume (Mm^3)	Duration (days)	MOR ($m^3\ s^{-1}$)	Max effusion rate ($m^3\ s^{-1}$)	Lava type	Crater diameter (m)	Crater name
1	1998, March 9	11:05	Northern flank	60	196	3.5	15–30	Aphyric	120, 100	Piton Kapor, Piton M. and K. Krafft
2	1998, March 11	22:45	Western flank	0.75	21	0.4	5–10	Aphyric	60	Crater Hudson
3	1999, 19 July	14:50	Summit	1.3	11.9	1.3		Aphyric		–
3′	1999, 24 July	nd	Eastern flank					Aphyric		–
4	1999, Sept 28	07:58	Summit	1.4	24.5	0.7	70	Aphyric		Crater Dupavillon
4′	1999, Oct 11	nd	Southern flank					Aphyric		–
5	2000, Feb 13	20:18	Northern flank	8.2	19.9	4.8		Aphyric	130	Piton Célimène, Crater Legros
6	2000, June 23	14:00	ESE flank	10	37	3.1		Aphyric	120	Piton Pârvédi
7	2000, Oct 12	01:05	ESE flank	9	32.7	3.2	40–60	Aphyric ⇒ olivine bearing	170	Piton Morgabim
8	2001, March 27	09:20	SSE flank	4.8	7.6	7.3		Aphyric	160	Piton Tourkal
9	2001, June 11	09:50	SE flank	9.5	26	4.2		Aphyric ⇒ olivine basalt	200	Piton Madoré
10	2002, Jan 5	19:00	NE flank, PdO[a]	13	10.7	14	100	Oceanite		–
11	2002, Nov 16	00:56	Eastern flank, distal	8	19.5	4.8		Aphyric ⇒ olivine basalt	180	Piton Guanyin
12	2003, May 30	08:00	Summit WSW	2.1	38.3	0.7	0.5–20	Aphyric	50	Piton Kaf
13	2003, Aug 22	16:50	Summit, N flank	6.2	5.1	14.1	34	Oceanite	140	Piton Payankë
14	2003, Sept 30	19:40	WSW flank	1	0.5	24.5		Aphyric		–
15	2003, Dec 7	11:20	Summit ESE	1.2	18.2	0.8		Aphyric		–
16	2004, Jan 8	22:45	PdO[a]	1.9	1.1	19.4		Olivine basalt		–
17	2004, May 2,	15:35	Southern flank	16.9	15.9	12.3		Aphyric		–
18	2004, Aug 12	22:40	Summit, E flank	20	62.2	3.7		Aphyric	70	Piton Kala

(continued)

Table 9.1 (continued)

No.	Date	Start time	Site	Volume (Mm^3)	Duration (days)	MOR ($m^3\ s^{-1}$)	Max effusion rate ($m^3\ s^{-1}$)	Lava type	Crater diameter (m)	Crater name
19	2005, Feb 17	16:35	PdO[a]	19	9.1	24.3		Aphyric ⇒ oceanite		–
20	2005, Oct 4	10:35	Summit WSW	2	12.5	1.9		Aphyric	40	Piton Kafrine
21	2005, Nov 29	02:30	Summit, N flank	1	0.4	30.9		Aphyric		–
22	2005, Dec 26	13:10	PdO[a]	20	23.3	9.9		Olivine basalt ⇒ oceanite		–
23	2006, July 20	00:06	SW + S flank	2	25.6	0.9		Aphyric	50	Piton la Paix
24	2006, Aug 30	07:42	Summit ESE, flank	20	123.6	1.9		Aphyric	75, 70	Pitons la Wouandzani, Moinama,
									45, 65	Pongal, Moinache
25	2007, Feb 18	12:35	Summit	1	1.4	8.6		Aphyric		–
26	2007, March 30	19:00	Southern flank	0.6	0.3	20.8		Aphyric		–
26′	2007, April 2	06:00	Southern Grand Brûlé	240	29.6	93.9	300	Oceanite	400 × 250	Piton Tremblet
27	2008, Sept 21	11:38	Summit SW	1	10.5	1.1		nd		–
28	2008, Nov 27	07:55	Summit SW	0.1	1	1		nd		–
29	2008, Dec 14	22:55	Summit E&N	1.5	51.8	0.3		nd		–
30	2009, Nov 05	17:00	SSE flank	0.14	0.3	5.4		Aphyric		–
31	2009, Dec 14	14:40	Southern flank	0.16	0.7	2.8		Aphyric		–
32	2010, Jan 02	10:20	Western Dolomieu	1.2	9.6	5		nd		–
33	2010, Oct 14	15:20	Southern flank	2.7	16.4	1.96		Aphyric		–
34	2010, Dec 09	18:15	Northern flank	0.53	0.6	9.7		Aphyric		

[a]*PdO* Plaine des Osmondes; *nd* not determined. Lava flow volumes are the volumes estimated from the field observations and mapping; they do not represent DRE

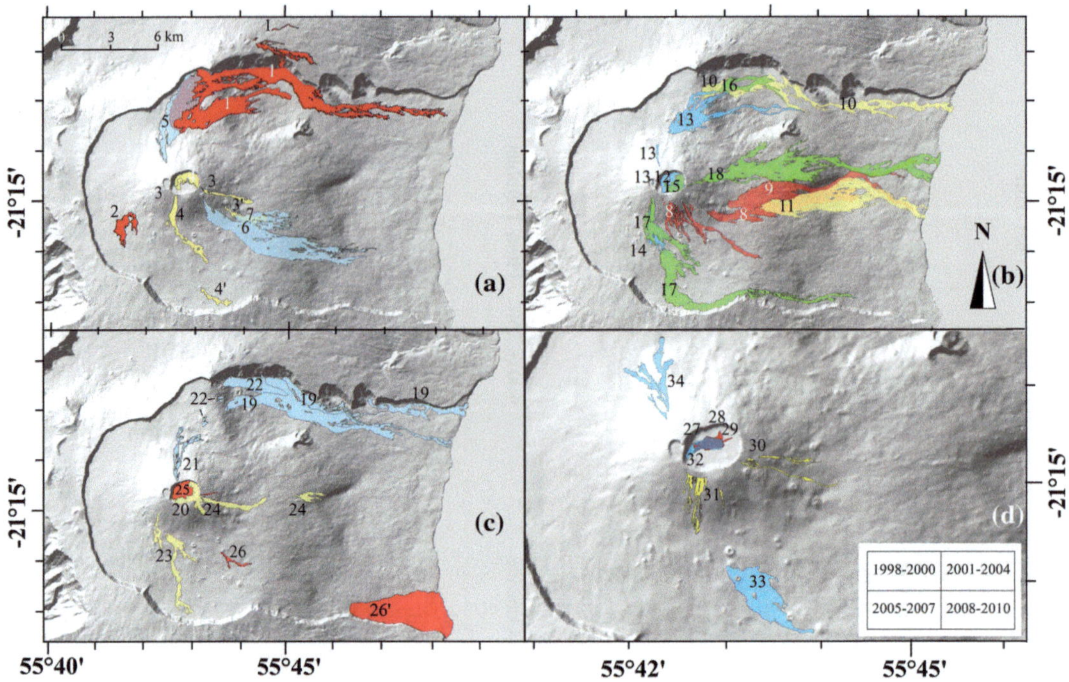

Fig. 9.5 Lava flow maps for four periods, **a** 1998–2000, **b** 2001–2004, **c** 2005–2007 and **d** 2008–2010. Numbers stand for the eruptions mentioned in the text and in Table 9.1

(55°42′47″, −21°13′40″, 2170 m) and Crater M. & K. Krafft (55°43′02″, −21°13′26″, 2071 m). The latter vent was active until March 31, while the former vent was active until September 21. The lava flow from the first vent flow around the second vent which was 560 m downstream and the two lava flows merged together, forming a single lava flow. The first lava flow extended 4.4 km onto Plaine des Osmondes. After this effusion was of tube-fed pahoehoe, which covered 6 km^2 in the northern Enclos Fouqué caldera, Plaine des Osmondes and moved 11 km down the northern Grand Brûlé.

In spite the fact that the volcanic-tectonic seismicity usually stops after the beginning of eruptions at Piton de la Fournaise (Hirn et al. 1991) several seismic events were recorded beneath the western flank during the first days of eruption.

On March 11 at 22:45, a vent, Crater Hudson, opened on the western flank of Piton de la Fournaise (55°41′44″, -21°15′11″, 2190 m). This vent fed a 1.5 km long aa lava flow. It remained active for three weeks until April 1. Initial flow rates of 5–10 $m^3\ s^{-1}$ were estimated.

As the eruption at Piton Kapor went on, the dike steadily extended down the north-east rift zone. On August 8, three small fissures opened between 4 and 5 km NNE of Piton Kapor; outside of the Enclos Fouqué caldera. Only short lava flows of highly degassed magma, 2 km long, formed here; with the eruption lasting 32 days. Part of the lava flow felt back into Plaine des Osmondes.

After 6.5 months of activity, the eruption finally stopped, while the lava flow front was just 5 m from the main road. The Piton Kapor and M. & K. Krafft eruptions erupted 61 Mm^3 of basalt and the lava flows covered 7.9 km^2 (Staudacher and Villeneuve 1998; Villeneuve 2000). The initial effusion flow rate was estimated in the field to be 15–30 $m^3\ s^{-1}$. Given a duration of 196 days, the mean output rate (MOR) was 3.5 $m^3\ s^{-1}$. The Hudson lava flow covered 0.44 km^2 and produced 0.75 Mm^3 of basalt. Its initial flow rate was estimated to be 5–10 $m^3\ s^{-1}$ (Villeneuve 2000) and its MOR was 0.4 $m^3\ s^{-1}$. The emitted lava was aphyric (Bureau et al. 1999; Vlastélic and Staudacher 2005).

9.3.1.2 The 1999 Eruptions

After the 1998 eruption, seismicity resumed, attaining a daily mean of 4–5 events by late-1998. Seismicity then decreased until mid-June 1999 reaching a mean value of 1 event per day (Fig. 9.4). Two short eruptions occurred in 1999, both beginning with a short eruptive phase within the Dolomieu crater, followed by a flank event some days after.

July 19 (No. 3 and 3′)
After a slight increase in seismicity during mid-July (with a magnitude 2 earthquake at sea level occurring below the summit on July 14), a seismic crisis comprising 68 events of low magnitude (<0.5) started beneath the summit on July 19 at 14:18 (Fig. 9.4). At 14:20, tiltmeter variations of 180 and 140 μradians were observed at Soufrière and Bory stations, respectively, and then at 14:30 at Dolomieu Sud (50 μradians; see Fig. 9.1 for the station locations). At 14:50, eruption tremor was recorded, centered on the summit. Simultaneously, lava fountaining reaching heights of up to 100 m was observed in the eastern Dolomieu crater, where a 500 m long fissure cut the crater, orientated N110. The main vent was located at the eastern edge of Dolomieu (55°43′9″, −21°14′43″, 2480 m) and aa lava gathered inside of Dolomieu, where it covered about 30 % of the 0.5 km^2 floor. A second aa lava flow from the same vent (outside of the Dolomieu crater) moved down 2 km on the eastern flank of PdF (northern flow). The intensity of tremor declined rapidly and most of the lava flows were already stagnating 40 h after the beginning of the eruption.

On July 24, the tremor amplitude increased and a new vent opened outside of Dolomieu crater at 55°43′41″, −21°15′8″, 2030 m, producing exclusively tube-fed pahoehoe lava flows that extended 1.7 km to reach 1650 m elevation (southern flow). The eruption ended on July 31 at 11:59 after numerous gas-piston events, which were caused by gas accumulation within the lava column, followed by series of pulses of gas release. The erupted lava volumes were estimated at 0.3 Mm^3 inside Dolomieu, 0.3 Mm^3 for the northern flow, and 0.7 Mm^3 for the southern flow. The initial effusion rate was estimated at about 10 $m^3\ s^{-1}$, whereas the MOR was 1.3 $m^3\ s^{-1}$. The lava was exclusively aphyric composition.

September 28 (No. 4 and 4′)
Seismicity during the August and September remained low with at an average of 1 event per day (Fig. 9.4). On September 28, after a summit inflation event of 40 μradians, a seismic crisis of 180 events began at 06:37. At 07:10 the summit (Soufrière) tiltmeter indicated inflation of 15 μradians, followed by deflation of 60 μradians. The Bory tiltmeter indicated inflation of 140 μradians at 07:30, followed by deflation.

At 07:58 the opening of a fissure in the Dolomieu crater was observed directly by observatory staff. After a strong whistling release of gas, lava fountains of 40–50 m in height erupted at the south-western edge of the Dolomieu crater floor (55°42′43″, −21°14′48″, 2460 m). Next, a N140 trending fissure opened, from which white smokes first emanated before to release lava flows. The fissure reached the foot of the western wall of the Dolomieu crater, just beneath the Bory crater, where another lava fountain appeared. Ten minutes later four *en echelon* fissures opened on the southern flank of Piton de la Fournaise, that fed an aa flow.

On September 30, the tremor suddenly decreased and emission of lava was low. The tremor increased again on October 8 and fumaroles appeared in the southern part of Enclos Fouqué (55°42′50″, -21°16′30″, 1890 m). On October 11, a new vent opened (55°43′12″, −21° 16′46″, 1840 m), almost 4 km from the summit and 730 m from the southern Enclos Fouqué caldera rampart. This erupted highly degassed magma and a 1.5 km long pahoehoe type lava flow formed. The eruption ended on October 23.

The lava flow field inside the Dolomieu crater and on the southern flank had volumes of 0.5 and

0.6 Mm^3 respectively, while that of the southern Enclos lava flow had a volume of 0.3 Mm^3. A total area of 0.75 km^2 was thus covered by 1.4 Mm^3 of lava. The initial effusion rate during the first hour of the eruption on September 28 was estimated at 70 $m^3\ s^{-1}$, while the MOR was only 0.7 $m^3\ s^{-1}$. The lava was exclusively aphyric.

9.3.1.3 The 2000 Eruptions

In 2000, three eruptions occurred on the flanks of the central Piton de la Fournaise cone (Fig. 9.5a): one on the northern flank, and two on the east-south-east flank.

February 13 (No. 5)

Seismicity beneath Piton de la Fournaise increased in January 2000, reaching up to 15 events per day (Fig. 9.4). On February 13, a seismic crisis began at 19:14 with 261 earthquakes in 64 min. Immediately an intrusion was recorded by the tiltmeter network. At first summit inflation was recorded at 19:17, followed at 19:40 by migration of a dike to N320 with a velocity of 1 $m\ s^{-1}$. Eruption tremor was recorded at 20:18, and five *en echelon* fissures opened on the northern flank of Piton de la Fournaise. These extended between the elevations of 2470 and 2250 m. No direct observations were possible in the first days of this eruption due to heavy rain (200 mm per day) associated with a cyclone that was passing close to La Réunion. The primary vent focused at 2250 m (55°42′34″, −21°13′48″). A 2.8 km long aa lava flow went to the north, followed by eruption of numerous pahoehoe flows, which covered 1.8 km^2 and attained a volume of 8.2 Mm^3. The eruption ended on March 4 after hundreds of gas-piston events. The lava was exclusively aphyric and the MOR was 4.8 $m^3\ s^{-1}$.

June 23 (No. 6)

In mid-April the extensometer station at Magne began to record a fracture opening at a rate of 0.0059 mm day^{-1}. The fracture at Château Fort then began to open in June at a rate of 0.0089 mm day^{-1}. At the same time the tiltmeter network recorded inflation of the summit by up to 80 μradians at the southern edge of the Dolomieu crater. From mid-June, seismicity increased up to 100 events per day (Fig. 9.4). On June 22 a first small seismic crisis was recorded. The next day a seismic crisis of 270 events preceded the eruption which began at 14:00 on the east-south-eastern flank of PdF. Six fissures *en echelon* were located between the 2080 and 1820 m elevations. Effusion became focused on the lower part of the fissure system at 1830 m elevation (55°44′1″, −21°15′32″). The eruption lasted 37 days and the flow was 4.8 km long. It produced 10 Mm^3 of lava and covered 2.2 km^2. The MOR was 3.1 $m^3\ s^{-1}$ and the lava was aphyric.

October 12 (No. 7)

Immediately after the June eruption, the Dolomieu Sud and Château Fort tiltmeters recorded tilt variations of up to 50 μradians. At the same time extensometer at Château Fort recorded fracture opening with a rate of 0.003 mm day^{-1}. Seismic activity ranged between 10 and 20 small VT events per day (with magnitudes <1). On October 6 a first seismic crisis of 57 events occurred, followed on October 12 by a second crisis of 201 events at 00:09. This preceded the onset of the eruption at 01:05. Three vents became active on the east-south-eastern flank of PdF between the 2320 and 1980 m elevations. The main eruptive vent (55°43′38″, −21°15′22″, 2000 m) was located 700 m above the June 23 flow field whose lava flows were partly buried by the new lava. Initially aa flows were erupted, but later flows were of pahoehoe. The initial effusion rate was estimated in the field to be between 40 and 60 $m^3\ s^{-1}$. On October 9 a small phreatic explosion formed a second crater on the active fissure, some tens of meters above the upper-most crater. The phreatic crater was eventually consumed by the initial crater, which increased in size during the eruption to create a 100 × 75 m crater. The eruption ended on November 13 at 18:45 after a strong increase of activity. The lava flow field covered 4.1 km^2 and had a volume 9 Mm^3, which was erupted at a

MOR of 3.2 $m^3\ s^{-1}$. Initially aphyric lava was erupted, but some olivine phenocrysts appeared towards the end of the eruption.

9.3.1.4 The 2001 Eruptions

Two eruptions occurred in 2001. Both began high on the volcano where numerous *en echelon* fissures extended from the south of the summit, running down counterclockwise around the edifice to the eastern flank. The main activity was focused between the 1800 and 1900 m elevations (Fig. 9.5b).

March 27 (No. 8)
The March 27 eruption was preceded by three months of continuous fracture opening, as recorded by the Château Fort and Magne extensometers at the base of the terminal cone. Here fractures attained maximum widths of 0.3 and 0.35 mm respectively. Simultaneously, the Dolomieu Sud and Soufrière tiltmeters indicated summit inflation of 110 μradians. Seismicity was located at ~500 m above sea level, and varied between 0 and up to 120 events per day during the four weeks prior to the eruption (Fig. 9.4). On March 27 at 08:55 a 25 min-long seismic crisis of 120 events began. This followed by the eruption tremor at 09:20. Five *en echelon* fissures opened between 2545 and 1940 m elevations oriented east-south-east. The upper vents were active only for a few hours, and the primary eruptive activity became focused at 55°43′37″, −21°15′13″ and 2050 m, slightly below the October 2000 eruption site and close to crater Signal de l'Enclos. The eruption ended on April 4 at 00:20. The lava flow field covered 2.5 km^2 and had a volume of 4.8 Mm^3, which was erupted at a MOR of 7.3 $m^3\ s^{-1}$. The lava was exclusively aphyric.

June 11 (No. 9)
Shortly after the end of the March eruption, the extensometer and tiltmeter networks indicated new inflation of the volcano (up to 0.35 mm total opening of the monitored fractures and 150 μradians of inflation respectively). From May 23, seismicity increased to 10–40 events per day (Fig. 9.4). On May 30 a small seismic crisis occurred. Finally, on June 11 a seismic crisis of 125 events preceded the appearance of eruption tremor at 09:50. Seven *en echelon* fissures opened from the summit (at 2480 m elevation), down the eastern flank to the 1800 m elevation, where a pyroclastic cone formed (55°44′8″, −21° 15′5″). The lava flows descended the eastern flank, and on July 6 two lava flows overrun the main road, having attained a length of 7.3 km (Fig. 9.5b). The eruption ended on July 7, after an unusually high level seismic activity of up to 173 events per day. The lava flow field covered 3.6 km^2 with a volume of 9.5 Mm^3 and a MOR of 4.2 $m^3\ s^{-1}$. The lava was of aphyric type at the beginning of eruption, which olivine rich lava being erupted later. Enhanced SO_2 concentrations in the air were measured in the northern part if the island (Bhugwant et al. 2002), about 60 km from the eruption site.

9.3.1.5 The 2002 Eruptions

Two eruptions occurred during 2002; the first was located above, and in Plaine des Osmondes; the second on the eastern PdF flank. The latter eruption was followed by formation of a pit crater in the Dolomieu crater (Fig. 9.5b).

January 5 (No. 10)
After two months of low seismicity (1.7 events per day), seismicity increased up to 50 events per day between early-September to late-November 2001 (Fig. 9.4). Two seismic crises occurred in November 2001. In mid-December seismicity was low again (between 0 and 4 events per day), but increased up to 60 events per day during late-December. In the same period, tiltmeters indicated inflation of up to +120 μradians, and the Château Fort extensometer recorded up to 0.32 mm fracture opening (Peltier et al. 2006). On January 5, 2002 a strong seismic crisis began at 12:37. Initially located beneath the summit, after 40 min the seismic hypocenters moved to the north-east at an average velocity of 9 m min^{-1}. Eruption tremor was recorded at 19:00 and simultaneously four eruptive vents opened about 3 km north-east of the summit, above Plaine des Osmondes (55°43′33″, −21°12′51″, 1860 m).

Only low fountaining was observed from Piton de Partage. An aa lava flow covered a distance of 4.5 km length. On January 13 at 11:59 a new vent opened at the northern base of Plaine des Osmondes rampart, 3 km to the east of the first vent (55°45′15″, −21°12′34″, 1090 m). It released highly degassed magma. Presumably the dike progressed down the NE rift zone, before being trapped by pre-existing fractures along the scarp of the caldera. The eruption tremor intensity increased dramatically on January 13 when the lava flow cut the main road and reached the ocean, having covered a distance of 6.6 km. On January 16 the effusion rate was estimated in the field at 100 m^3 s^{-1}. Shortly (5 min) afterwards the eruption tremor broke down and the eruption ended at 12:00. A total of 13 Mm^3 of lava was erupted, covering an area of 2.8 km^2. A small bench of 15 ha was formed at the coast. The MOR was 14 m^3 s^{-1}. The initial lava was aphyric, while olivine rich to oceanite (see 9.4.2.4) lava erupted from the second vent at the end, during the most intense stage of the eruption.

November 16 (No. 11)
Since early-August all networks showed signs of unrest. The tiltmeters indicated summit inflation of up to 130 μradians, and the extensometers at Magne and Château Fort recorded continuous fracture opening at a rate of 0.06 and 0.035 mm day^{-1}, respectively. At the same time the permanent GNSS station "SNEG" (Soufrière) showed horizontal displacements to the north-east and vertical displacements of about 0.6 mm day^{-1}. Daily seismicity increased to up to 60 events per day (Fig. 9.4). On September 14 and 20 two seismic crises of 55 and 40 events occurred. On November 12, a further seismic crisis of 75 events occurred and ground deformation recorded by the Magne and Château Fort extensometers continued, although at slower rates of about 0.02 mm day^{-1}. On November 15, a final seismic crisis occurred with more than 1000 events. This was followed, the next morning at 00:56, by the appearance of eruption tremor located on the eastern flank above Grandes Pentes (Vigouroux et al. 2009). Four vents opened between the 1760 and 1570 m elevations. The eruption became focused at the 1600 m elevation (55°44′46″, −21°15′8″). An aa lava flow cut the main road on November 30 and reached the ocean some hours later, having extended 6.2 km. On December 5 eruption tremor disappeared. The final days of the eruption were accompanied by strong seismicity beneath the summit, which continued after the end of eruption and was related to pit crater formation in the Dolomieu. The November 16 eruption produced 8 Mm^3 of lava which covered an area of 3.6 km^2 at a MOR of 4.8 m^3 s^{-1}. Whereas the initial lava was aphyric, the eruption finished with olivine rich lava.

December 23 Pit Crater Formation
At the end of the November–December eruption, seismicity beneath Dolomieu crater increased from about 300 events on December 1 to 5700 events on December 22 (Longpre et al. 2006; Fig. 9.4). On December 23 at 06:02 a final seismic event of magnitude 3 terminated this period of heightened seismicity. A helicopter over flight on the same day revealed that a 120 m wide and 10 m deep pit crater had opened in the south-western part of the Dolomieu crater (55° 42′45″, −21°14′51″, 2480 m). This event was related to the preceding eruption at low elevation, which created empty space between the magma reservoir and the Dolomieu crater floor (Longpre et al. 2006). Simultaneous with pit crater formation, the Dolomieu crater contracted by about 10 cm north-south and 5 cm west-east.

9.3.1.6 The 2003 Eruptions

There were four eruptions in 2003: two summit eruptions, contained inside the Dolomieu crater, and two flank eruptions, on the northern and the south-western flanks (Fig. 9.5b).

May 30 (No. 12)
By March 2003 the Château Fort extensometer had recorded a total fracture opening of 0.22 mm. Rapid static GNSS campaigns made between January and April 2003 confirmed inflation of the

summit, and indicated Dolomieu crater expansion of 5 cm. Seismicity in May was at about 20 events per day (Fig. 9.4). On May 30 at 08:00, after a short seismic crisis of 37 events in 22 min, eruption tremor was recorded on the summit zone. After two hours, eruptive activity ceased, but tremor continued for another six hours; decreasing continuously. The eruption vent (55° 42′41″, −21°14′50″, 2500 m) was located on the south-western floor of Dolomieu, 110 m west of the December 2002 pit crater. A small 200 × 400 m lava flow field formed in the Dolomieu comprising 0.16 Mm^3 of aphyric material. The initial effusion rate was 20 $m^3\ s^{-1}$ and the MOR was 5.6 $m^3\ s^{-1}$. Rhythmic degassing at the vent continued for the next few days, but no visible lava output was observed (Coppola et al. 2005).

On June 4 at 07:20 a new lava flow was erupted from a second vent 60 m to the north of the first vent (55°42′41″, −21°14′52″, 2500 m). Lava fountains of 15–20 m in height were observed. The eruptive activity ended after a small phreatic explosion on June 6 at 17:20. The resulting lava flow field covered about 400 × 600 m with a volume of 0.72 Mm^3 and a MOR 3.4 $m^3\ s^{-1}$. However the extensometer network (Magne and Château Fort) indicated ongoing fissure opening. On June 12, the June 4 vent reactivated at 23:08, remaining active for 50 h and covering 80 % of the crater floor with 380 × 600 m of pahoehoe and aa. Activity ceased on June 15 on 02:00, having erupted 0.6 Mm^3 of lava at a MOR of 3.2 $m^3\ s^{-1}$. Over the following days, the extensometers continued to record fissure opening, indicating ongoing inflation.

A final eruptive phase began on June 21 at 19:30 from the same vent. It lasted for 16 days with eruption tremor disappearing on July 7. Again, a large part of the surface in the western Dolomieu was surfaced by aa-lava, with pahoehoe extruding from the frontal flow zones. This final eruptive phase produced 0.69 Mm^3 of lava at a MOR 0.5 $m^3\ s^{-1}$.

The four eruptive episodes of May to July 2003 emplaced a 420 × 640 m large lava flow field with a thickness of up to 16.5 m, as measured by rapid static GNSS. The total emitted volume was 2.2 Mm^3, which was erupted at a MOR of 0.7 $m^3\ s^{-1}$. The erupted lava was exclusively aphyric. A detailed analysis of the effusion rates and emplacement mechanisms during this eruption are given in Coppola et al. (2005).

August 22 (No. 13)

Five days after the end of the May–July eruption, seismicity reappeared reaching 20–60 events per day by mid-July (Fig. 9.4). On August 22, a seismic crisis of 300 events occurred. The seismicity, located between 200 and 800 m asl, was accompanied by summit inflation of up to 30 cm (Peltier et al. 2007). Three fissures opened. The first, within the Bory crater at 16:50, produced a small (160 m long) pahoehoe flow of aphyric lava. At 17:20 a second vent opened at the 2475 m elevation to feed a small (820 m long) aphyric aa flow on the northern flank. A final vent opened at 19:35 between the 2210 and 2150 m elevations. Primary activity became focused on the eastern flank of the 1998 Piton Kapor crater at 55°42′41″, −21°14′50″ and 2160 m elevation. Lava fountains of 30–40 m in height were observed. Pahoehoe flows of oceanites were effused from a 100 m wide and 30 m high crater. The intensity of effusion increased on August 27, when large blocks of several hundred of kg in weight were projected from the eruption cone to land several dozens of meters away. At 18:50 the activity ended abruptly, crashing to nothing in less than one minute. A total area of 1.7 km^2 in the Enclos Fouqué was covered, inundating parts of the 1998 lava flow. A few tiny lava flows reached Plaine des Osmondes, which was crossed by one of them. These flows attained a maximum length of 5.2 km. A total of 6.2 Mm^3 of pahoehoe lava was erupted, at a MOR of 14.1 $m^3\ s^{-1}$, although pulses of 34 $m^3\ s^{-1}$ were recorded (Coppola et al. 2007).

September 30 (No. 14)

After the August eruption, seismicity increased to up to 35 events per day (Fig. 9.4), even if the intensity of the earthquakes remained small. The tiltmeter network indicated summit inflation of 25–57 μradians. On September 30, at 18:25, a 65 min-long seismic crisis began with 290 events. Two vents opened between the 2370 and 2210 m elevations on the south-south-western flank of Piton de la Fournaise. Activity lasted 12 h, produced 1 Mm^3 of aa lava, covering an area of 0.51 km^2 at a MOR of 25 $m^3\ s^{-1}$. Because of the short duration of the eruption, no significant pyroclastic cone formed. After the eruption Soufrière, Dolomieu Sud and Château Fort tiltmeters continued to indicate summit inflation.

November 6 Intrusion

Immediately after the September 30 eruption, seismicity gradually increased over the next few weeks, increasing from 1 to 30 events per day (Fig. 9.4). The extensometer at Château Fort indicated fracture opening of 0.13 mm and the Dolomieu Sud tiltmeter revealed 50 μradians of inflation. Finally, on November 6, a seismic crisis of 100 events was recorded. The GNSS network recorded a summit inflation of 30 cm, centered at 55°43′15″, −21°14′49″ on the upper south-eastern flank of PdF. Seismic data indicated that an intrusion stalled between 500 and 1500 m below the summit, probably representing an aborted eruption (this is commonly observed on Kilauea, see Dzurisin et al. 1984). From GNSS data and surface deformation, the volume of the intrusion was estimated to 1.2 Mm^3.

December 7 (No. 15)

From mid-November the tiltmeter network recorded summit inflation of 18, 132 and 32 μradians at Dolomieu Sud, Château Fort and Chapelle, respectively. The Dolomieu Sud extensometer recorded a total fracture opening of 0.23 mm. During the first days of December an increase in seismicity beneath the summit was recorded (Fig. 9.4). On December 7, a 51 min-long seismic crisis began. This was followed, at 11:20, by the onset of the eruption tremor. A 200 m long fissure orientated N45°, opened on the south-eastern floor of Dolomieu (55°43′5″, −21°14′44″, 2500 m). This emitted mostly aphyric pahoehoe lava which covered part of the May–July lava flow field. A second 250 m long fracture opened on the "southern plateau" (Dolomieu Sud), but only released gas and a total of about 1 m^3 of lava. The eruption continued until December 25 and covered 0.27 km^2, erupting a volume of 1.2 Mm^3 at a MOR of 0.76 $m^3\ s^{-1}$. No pyroclastic cone was formed.

9.3.1.7 The 2004 Eruptions

Three eruptions took place in 2004: a distal one to the north, a flank eruption to the south and a summit-to-flank eruption to the east (Fig. 9.5b).

January 8 (No. 16)

Immediately after the December 2003 eruption, seismicity increased from 0 to 13 events per day (Fig. 9.4). The deformation network recorded summit inflation and fracture opening. On January 7, after a 30 min-long seismic crisis (of hundreds of events), seismicity diminished to about 20–50 earthquakes per hour. During 41 h, hypocenters of these events shifted slowly along the north-east rift zone. This, in agreement with the deformation network, indicated migration of a dike to the north east. At 22:45 on January 8, eruption tremor was recorded close to Nez Coupé de Sainte Rose (~55°44′, −21°12′44″). A 600 m long fissure opened between 1530 and 1400 m elevations, above Plaine des Osmondes. This emitted, over 27 h, 1.9 Mm^3 of olivine-rich aa lava. Moderately high lava fountains were observed the next morning, and no eruption cone was formed. The lava flow covered a length of 2.5 km and an area of 0.6 km^2, at a MOR of 19.5 $m^3\ s^{-1}$.

May 2 (No. 17)

After two months of low seismicity from mid-January to mid-March, seismicity increased to reach an average value of 15 events per day between March 10 and March 30 (Fig. 9.5). Two seismic crises occurred on March 31 and April 20, comprising 39 and 104 events, respectively. At the same time, the permanent GNSS station at Soufrière shifted 9 cm to the north-east and rose by 2 cm. On May 2 at 15:04 a third crisis of 31 earthquakes preceded the eruption, which began at 15:35. Two fissures opened. The first one was located on the SSW flank, just below Bory crater, where no lava fountaining was observed. The second fissure opened to the southern Enclos Fouqué, 2.4 km from summit (55°42′42″, −21° 15′58″, 2010 m). During the first days two eruptive cones formed on this fissure 50 m apart. These then merged in one single cone. A 9.4 km-long aphyric lava flow followed the southern scarp of the caldera and reached the 450 m elevation. On May 17, lava fountains of 20–30 m in height were observed. The eruption ended on May 18. It produced a volume of lava of 16.9 Mm^3 and covered an area of 3.3 km^2, at a MOR of 12.3 $m^3\ s^{-1}$.

August 12 (No. 18)
Between end-May to early-July seismicity was low with an average of 0.5 events per day (Fig. 9.4). From early-July seismicity increased from <10 events to >100 event per day. During this high seismic period, the GNSS network recorded a north-south expansion of the Dolomieu crater by 4 cm and uplift of ~5 cm. On August 12 at 22:40 eruption tremor appeared. Three fissures, orientated west-east, were active in the Dolomieu crater. A fourth 450 m long fissure opened at 2180 m elevation at which a pyroclastic cone formed (55°43′37″, −21°14′ 33″). Over the next few days this fissure extended 1.5 km down to 1850 m elevation. The summit fissure rapidly shut down, while the pyroclastic cone at 2180 m elevation released gas. The lower part of the fissure at 1850 m elevation emitted large volumes of gas poor pahoehoe basalt. The 9.4 km long lava flow reached the ocean on August 25 and formed a 300 m deep and 680 m wide bench. Two littoral hornitos were fed by a lava tube and lava was emitted directly into the ocean. This formed, together with inflowing sea water, strong explosions and LIMU o Pele and phreatomagmatic particles. The eruption ended on October 14, having covered 4.8 km^2. It produced 20 Mm^3 at a MOR of 3.7 $m^3\ s^{-1}$; and the lava was aphyric. The bench at the ocean entry was rapidly eroded by waves and its area was reduced to about 1/3 of its original size within 5 years.

Rapid static GNSS measurements were recorded before and after the eruption on eight stainless steel rods, which survived the eruption. These were located inside of the Dolomieu crater, some meters to tens of meters from the eruption fissures, and on the eastern flank. These measurements allowed us to determine that the width of the eruption fissure was 44 cm.

9.3.1.8 The 2005 Eruptions

Four eruptions occurred in 2005, two being distal and close to the northern scarp of the Enclos Fouqué caldera. A third one was a summit eruption, and the fourth was a summit-to-flank eruption (Fig. 9.5c).

February 17 (No. 19)
In early December 2004, the extensometer at Soufrière recorded fracture opening at a rate of 0.015 mm day^{-1}. At the same time seismicity increased simultaneously (with an average of 16 events per day). Both were interrupted on December 26 for twenty days, when the Piton de la Fournaise was shacken by the passage of the wave front from the magnitude 9.3 earthquake in Sumatra (distance ~6000 km). Seismicity and ground deformation increased again from February 5 when up to 50 events per day were recorded. While the Dolomieu crater expanded 4 cm in north-south direction, the Bory station to the west shifted 1.3 cm to the west, testifying to ongoing inflation of the summit. On February 17 at 09:20, numerous low energy earthquakes were measured beneath the summit and, at 13:35, a seismic crisis began. As the crisis weakened, the seismic events shifted to the north-east. A total of 537 seismic events were recorded. At 16:35 eruption tremor was recorded. The eruptive fissure was located between the Enclos Fouqué wall and Plaine des Osmondes (~55°43′44″, −21°13′ 1″, 1720 m) about 500 m from the scarp. No significant eruption cone formed. The lava flow crossed Plaine des Osmondes and extended 6.8 km down to Grand Brûlé reaching an elevation of 250 m. At that moment, still numerous seismic events were observed. On February 22, eruption tremor increased, and a second vent opened at the base of the caldera scarp in Plaine

des Osmondes (55°44′30″, −21°12′23″, 1250 m). It fed a 7.3 km long flow of degassed pahoehoe, which cut the main road on February 24, before reaching the ocean. In the evening of February 24, intense earthquakes were recorded and at 19:00 two new fissures opened in the Trou de Caron (55°46′8″, −21°13′5″, 720 m) and the Trou de Sable (55°46′37″, −21°12′49″). While the first fissure released only gas, the second, located at an elevation of 540 m, released degassed magma, forming an aa and pahoehoe lava flow field, which extended to 4 km to reach the ocean on February 25 at 03:45. On February 26 the eruption ceased. Seismicity continued at 720 events per day and lasted further two weeks but at a decreasing frequency. The eruption covered an area of 4.2 Mm^2, erupting 19 Mm^3 of lava at a MOR of 24.3 $m^3\ s^{-1}$, beginning with aphyric type lava and ending with oceanite.

October 4 (No. 20)

Since early-June the extensometer of Soufrière indicated fracture opening at a rate of 0.012 mm day^{-1}. The permanent GNSS network recorded summit inflation of 2–3 cm, and extension of Dolomieu by 8.4 cm north-south (at a rate of 0.06 cm day^{-1}). On October 4, at 09:30 a seismic crisis began beneath the summit with 112 low energy events. At 10:35 seismic tremor began to be recorded. Three fissures opened on the Dolomieu crater floor, close to the 2002 pit-crater and the 2003 summit eruption site, and a small pyroclastic cone formed at 55°42′42″, −21°14′50″ (2500 m). Lava fountains up to 40 m in height were observed. The eruption ended on October 17. The erupted volume was 2 Mm^3 of aphyric lava and the lava flow field covered 66 % of the Dolomieu crater floor at a MOR of 1.9 $m^3\ s^{-1}$.

November 29 (No. 21)

Immediately after the end of the October eruption, summit inflation resumed at the same rate as before, and seismicity reappeared with an average of 35 events per day (Fig. 9.5). On November 29 at 02:00 a new seismic crisis began and at 02:30 eruption tremor began to be recorded. Two small fissures opened on the Dolomieu crater floor on its western border and fed small lava flows. Fountains 40 m in height were observed and a small pyroclastic cone formed. The main fissure was located on the northern flank of Piton de la Fournaise at 2510 m elevation (~55°42′42", −21°14′18"). It erupted pahoehoe lava in the upper part and aa in the lower part. No eruption cone formed. The lava flow extended 2 km down the northern flank to run around Piton Kapor and Piton Payankë cones. The eruption lasted 10 h and produced 1 Mm^3 of aphyric lava at a MOR of 31 $m^3\ s^{-1}$.

December 26 (No. 22)

The last eruption was followed by high, but variable levels of seismicity with 7–100 events per day being recorded at an average value of 43 events per day (Fig. 9.4). Deformation also continued at the same rate as before (Fig. 9.6). On December 26, after a short seismic crisis, a small fissure opened at 13:10 at the 2020 m elevation

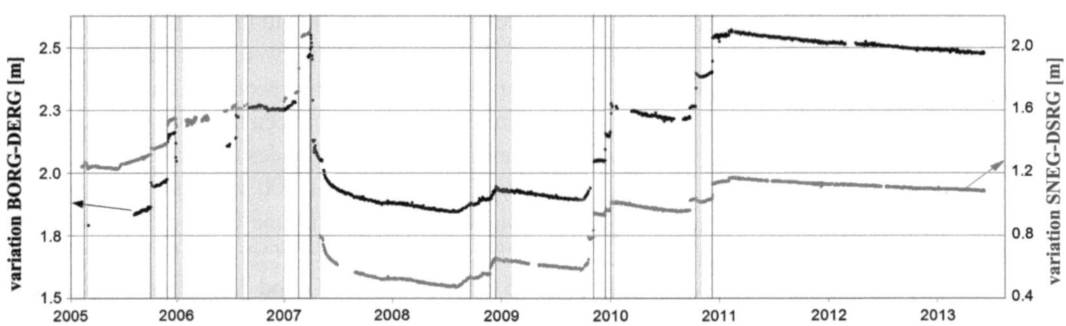

Fig. 9.6 Example of long-term deformation at Piton de la Fournaise. The *gray* and *black curves* stand for the baseline variations between SNEG (Soufrière) and DSRG (Dolomieu Sud) and BORG (Bory) and DERG (Dolomieu Est) respectively, measured by GNSS. *Vertical lines* and *gray areas* stand for short and longer lasting eruptions

(55°43′16″, −21°13′16″), mid-way between the summit and Nez Coupé de Sainte Rose (NCR). Tremor lasted only 3 h. At 16:30 a second small fissure opened (55°43′34″, −21°12′49″) at the 1840 m elevation, 700 m from NCR. It emitted lava for one hour. The main fissure opened at 18:30. It ran up the northern Enclos Fouqué scarp (55°43′49″, −21°12′24″, 1630 m). Obliquely orientated lava fountains with heights of 50–70 m erupted from mid-way up the scarp. Several lava flows crossed Plaine des Osmondes and flowed down the Grand Brûlé area, stopping about 2 km above the main road, having attained a length of 6.5 km. The eruption ended on January 18, at 20:00. The erupted lava was aphyric in the beginning of the eruption, turning to oceanite by the end of eruption. The erupted volume was 20 Mm^3 and the MOR was 10 $m^3\ s^{-1}$.

9.3.1.9 The 2006 Eruptions

Two eruptions occurred in 2006 (Fig. 9.5c), the second of which was the most voluminous and longest duration summit eruption at Piton de la Fournaise since the inception of the OVPF.

July 20 (No. 23)
From mid-January to mid-April 2006 the level of seismic activity was low, with an average of only two events per day (Fig. 9.4). At the end of April, however, it increased to 10 to >100 events per day. The increase was accompanied by continuous ground deformation. The permanent GNSS network recorded a general expansion of the Dolomieu north-south and east-west axis by 7.2 and 9.3 cm, respectively (Fig. 9.6) and an uplift of 4–7 cm. The Château Fort extensometer indicated fracture opening at a rate of 0.002 mm day^{-1}. On July 19, 22:20 a seismic crisis with 324 events preceded the eruption, which began on July 20 at 00:06. Two fissures opened. The first fissure between the 2410 and 2320 m elevations on the south-western flank between Bory crater and Rivals fed a 1.4 km long aa lava flow. The second fissure orientated N170° at the southern base of Piton de la Fournaise formed a pyroclastic cone at the 2140 m elevation (55°42′30″, −21°15′30″). Two aa lava flows extended 3.1 km to the south and south-east and covered 0.7 km^2. The eruption continued until August 14, erupting 2 Mm^3 of aphyric lava at a MOR of 0.9 $m^3\ s^{-1}$.

August 30 (No. 24)
After the July eruption, only a small amount of ground deformation was observed by the GNSS and the tiltmeter networks. The level of seismic activity ranged between 0 and 32 events per day (Fig. 9.4). On August 30, after a small crisis of 22 events, eruption tremor, focused on the summit was recorded. At 07:42 a fissure opened on the south-east floor of Dolomieu (55°43′5″, −21° 14′46″, 2490 m), about 50 m from the crater rim. A small lava flow extended 2.5 km down the east-flank of Piton de la Fournaise reaching the 1600 m elevation. Eruptive activity then became focused within the Dolomieu crater, where 75 % of the crater floor was inundated by new lava flows. An 80 m wide and 50 m high eruption cone formed, containing a lava lake. On October 10, a second vent appeared on the eastern floor of Dolomieu crater (55°43′, −21°14′47″). In the following weeks, two further eruptive cones formed between the first and the second one. Lava flows entirely covered the Dolomieu crater floor with a thickness of up to 30 m. On November 27, the Dolomieu crater overflowed to the east, forming a 570 m long pahoehoe lava flow. On December 9, after a small increase in the tremor intensity, a further fissure (55°45′13″, −21°14′44″) opened on the eastern flank, 3.5 km from summit, at the 1380 m elevation. The erupted, highly degassed magma fed a 1 km long lava flow that reached the 800 m elevation. The eruption ended on December 31 after a paroxysmal phase in the Dolomieu crater. The eruption

produced 20 Mm^3 of aphyric lava at a MOR of 1.9 $m^3 s^{-1}$.

9.3.2 The 2007–2010 Activity

9.3.2.1 The 2007 Eruptions and Dolomieu Collapse

Three eruptions occurred in 2007 (Fig. 9.5c): a summit, a summit–flank and a distal eruption. The latter eruption represented the most voluminous and intense eruption recorded for at least two centuries.

February 18 (No. 25)
Seismicity beneath Piton de la Fournaise and inflation of the massif resumed immediately after the end of the August–December 2006 eruption. Between January and mid-February, seismicity gradually increased to 60 events per day, while summit inflation continued. On February 18, after a seismic crisis of 25 min, a summit eruption began at 12:35. A straight fracture running west-east across the floor of the Dolomieu crater erupted less than 1 Mm^3 of aphyric lava during the 34 h-long eruptive event. The fissure extended on to the east-flank of Piton de la Fournaise. This fed and a small 500 m long lava flow with a volume of <0.03 Mm^3. The MOR for both intra-crater and flank lavas was 8.6 $m^3 s^{-1}$. On the Dolomieu floor spatter row along the fissure formed.

March 30 (No. 26)
After the February eruption, seismicity continued, with tens up to > 100 events per day (Fig. 9.4) being recorded. Summit inflation also continued, as recorded by GNSS and the extensometer at Château Fort. On March 30, after a 160 min-long seismic crisis, GNSS data and seismic hypocenters indicated an intrusion beneath the eastern flank of Piton de la Fournaise extending to the north, and then to the south (Peltier et al. 2009a, Staudacher et al. 2009). An eruptive fissure opened at 19:00 at 1900 m on the east-south-east flank (55°43′37″, −21°15′50″), which fed a 1 km long aphyric lava flow. Ten to twenty m high fountains were observed and spatter rows along the fissure, no pyroclastic cone formed. Eruption ended the next morning after 7 h of activity. The erupted volume was 0.6 Mm^3, at a MOR of 21 $m^3 s^{-1}$.

April 2007 (No. 26′)
Seismicity beneath Piton de la Fournaise continued with hundreds of events per day being recorded (Fig. 9.4), and eruptive activity soon resumed; the April eruption is in fact considered as a continuation of the March 30 eruption. On April 2 at 06:00, eruption tremor began. A 1 km long fissure, oriented N130°, opened in the southern Grand Brûlé. The effusion became focused at 570 m elevation, only 300 m from the foot of the southern Enclos Fouqué scarp (55°46′27″, −21°16′56″). The danger of a propagation of the eruptive fissures beyond the scarp was thus a real one. Such activity outside of the Enclos Fouqué caldera would threaten population and housing. At 15:15 lava flows crossed the main road to reach the ocean at 21:25. Lava fountains, up to 200 m high, were observed, and effusion rates of up to 300 $m^3 s^{-1}$ were estimated in the field (Staudacher et al. 2009). The eruption covered 3.6 km^2, buried 1.4 km length of the main road, and formed a lava flow field with a thickness of up to 60 m. The erupted volume, on land, was determined to be 90 Mm^3, and 0.45 km^2 new land was formed at the coast (Staudacher et al. 2009). The volume lost to the ocean was estimated to be 150 Mm^3 (Bachèlery et al. 2010). The total erupted volume of 240 Mm^3 ranked this eruption as the most voluminous eruption on Ile de La Réunion since at least two centuries. It formed a 500 × 250 m large eruption cone and the MOR was 94 $Mm^3 s^{-1}$. The eruption began with aphyric type lava, with the lava in the final stages being extremely rich in olivine (>50 %, Vlastélic et al. 2009). The eruption buried several hectares of primeval forest, and the radiation heat of the flows (which were 100's of meter wide) ignited trees at the borders of the Enclos Fouqué caldera. Abundant acid rain (rich in H_2SO_4 and HCl; Bhugwant et al. 2009), formed from the volcanic gases and from the lava flow running into the

ocean, damaged vegetation and metal roofs of buildings within a radius of about 1 km and represented severe health hazards (Viane et al. 2009).

April 5 Dolomieu collapse

Hundreds per day of large seismic events were recorded during the April 2 eruption. This seismicity, combined with the low elevation of the eruption, let us to suspect at that time a pit crater was forming in the Dolomieu (being similar to those after the November–December 2002 eruption). On April 5, eruption tremor and seismicity increased. The GNSS network observed summit deflation of 4 cm and Dolomieu contraction of 23 cm (Peltier et al. 2009a; Staudacher and Peltier 2016). At 20:48, a 4.8 M_S seismic event was followed by sudden summit inflation of 10 cm and Dolomieu east-west expansion of 30 cm. This marked the first collapse of the Dolomieu crater floor (Urai et al. 2007; Michon et al. 2007; Staudacher et al. 2009). Forty-eight rhythmic seismic shocks and summit deformation cycles (Fontaine et al. 2014; Staudacher and Peltier 2016), accompanied by formation of ash clouds, testified to the ongoing collapse of the Dolomieu crater. Most of the crater collapse occurred over 36 h from April 5, 20:48:30 to April 7. The final funnel shaped caldera is was about 1000 × 800 m in dimension and 340 m deep. Rims were unstable, and the walls were sub-vertical to the north-west, and 35°–40° steep to the east and south-east.

On April 6, just after the collapse, seismicity hitherto located above sea level moved downward to 8 km below sea level. These deep earthquakes were much less numerous and of smaller magnitude than those of the 1998 March migration, but borrow the same path in reverse (Battaglia et al. 2005). They may have been triggered by the depressurisation of the shallow levels of the volcano (Pinel and Jaupart 2005) but their focal mechanisms can not be resolved due to their size and low depth (Massin 2009).

The eruption of a large volume emptied the magma reservoir, removing about 50 % of its volume. This was calculated using a magma reservoir of 300 Mm^3 (Albarède F. 1993; Sigmarsson et al. 2005; Peltier et al. 2007) and a DRE volume of the April eruption of 0.6 × 240 Mm^3 [porosity of 40 %, (see Krafft and Gérentre 1977; Villeneuve 2000)]. The eruption at a low elevation created empty space in the magma reservoir. This destabilised the rock column between the shallow magma reservoir and the Dolomieu crater and favoured the collapse of the Dolomieu crater floor (Michon et al. 2007; Peltier et al. 2009a; Staudacher et al. 2009).

After the Dolomieu collapse, the central volcanic edifice deflated (Staudacher and Peltier 2016). Seismic activity over the following months was high due to stress readjustment following the collapse event (Fig. 9.4). Permanent GNSS and rapid static GNSS measurements showed that the Dolomieu crater had contracted in the first year after the collapse between 1.2 and 2.6 m in east-west and north-south directions, respectively, and that its border had subsided between 35 and 85 cm (Peltier et al. 2010).

9.3.2.2 The 2008 Eruptions

After the April 2007 eruption, the volume of the magma reservoir was reduced. The likelihood of voluminous eruptions over the next few years thus became lower. Nevertheless, three small eruptive phases occurred in 2008. These were all located in the now inaccessible Dolomieu crater (Fig. 9.5d), so that no lava samples other than Pele's hairs on the crater rim were collected (Di Muro et al. 2015). For the first time, SO_2 measurements using the new permanent DOAS network were performed (Garofalo et al. 2009; Allard et al. 2014).

September 21 (No. 27)

Level of seismic activity between March and July 2008 was low at 1.5 events per day. In August seismic activity increased, with tens of events per day, and several seismic crises, being recorded. Simultaneously, summit west-east extension by 2.8 cm was recorded by the GNSS network. On September 12 tremor was recorded for 10 h, and enhanced gas emission occurred in the western Dolomieu crater. The NOVAC network recorded a SO_2 peak at the

Piton de Bert station. On September 21, after a final seismic crisis, a fissure opened on the south-western flank of the Dolomieu crater (55° 42′40″, −21°14′38″, 2340 m). The eruption ended on October 2, having formed a 300 × 160 m lava flow. The bottom of the Dolomieu crater rose by 50 m due to fill by lava erupted onto the crater floor. The total erupted volume was estimated to be 1 Mm^3, at a MOR of 1.1 m^3 s^{-1} (Staudacher 2010).

November 27 (No. 28)
After the end of the September eruption, seismic activity remained at high levels (Fig. 9.4). Several seismic crises were recorded accompanied on October 20 and 31, by ground deformation of up to 2.5 cm. On November 27 an eruptive episode of 26 h occurred from the same vent as the September 21 eruption. Only 0.1 Mm^3 of lava was erupted, at a MOR of 1.1 m^3 s^{-1} (Staudacher 2010).

December 14 (No. 29)
Over the following weeks, seismicity continued at high levels with about 40 events per day being recorded (Fig. 9.4). The Dolomieu crater expanded by 4 cm over 2½ weeks. On December 14 a 17-hour-long seismic crisis of >1000 events was recorded. At 22:55 eruption tremor was recorded. Two vents were located mid-way up the northern and north-eastern wall of the Dolomieu crater. The main vent was at 55°42′57″, −21°14′31″ and at an elevation of 2335 m. The eruption lasted 52 day and erupted aa lava. It ended on April 4, 2009. The emitted volume was 1.5 Mm^3 with at a MOR of 0.3 m^3 s^{-1}. The September, November and December 2008 eruptions raised the Dolomieu crater floor by 75 m (Staudacher 2010) so that its depth was reduced to 265 m.

9.3.2.3 The 2009 Eruptions

In 2009, Piton de la Fournaise witnessed two short-duration low-volume summit eruptions. Both had their sources on the Piton de la Fournaise flanks close to the Dolomieu crater rim (Fig. 9.5d).

November 5 (No. 30)
Seismicity between March and September remained at low levels with around 3 events per day being recorded (Fig. 9.4). On June 2009 a swarm of VT events were recorded at about 4 km b.s.l. In early-October the Dolomieu crater began to expand radially at a rate of 0.2–0.3 cm per day. On October 18 at 16:20 a seismic crisis occurred. This was associated with a north-south orientated extension of the Dolomieu by 14.8 cm (Fig. 9.6), and was followed at 18:10 by tremor located beneath the Dolomieu crater. Tremor ended after a duration of 6 h. Visual observations during next morning revealed release of steam and sulfur deposition on the south-western Dolomieu crater wall (55°42′48″, −21°14′49″, 2400 m). However there was no evidence of lava eruption. Three other seismic crises occurred in October.

On November 5 at 17:00 tremor was recorded at a location on the east-south-eastern flank of PdF. Two eruption fissures were opened. The first fissure had a length of 180 m and ran almost parallel to the border of the Dolomieu crater at about 150 m distance from the rim (~55°43′15″, −21°14′46″, 2430 m), forming a 1.8 km long flow field. The second fissure of 60 m length was radially orientated and was located 350 m distance from the Dolomieu rim (~55°43′24″, −21° 14′45″, 2350 m). An aa lava flow, 1.1 km long, extended from the second fissure. The first fissure emitted pahoehoe lava and up to 5 m high spatter cones were formed. On the second fissure less than 2 m high spatter cones exist. The eruption lasted only 7 h and produced 0.14 Mm^3 of aphyric lava at a MOR of 5.6 m^3 s^{-1}.

December 14 (No. 31)
Only 5 days after the November eruption seismicity increased again, maintaining an average of 26 events per day though the month. The GNSS network however recorded no ground deformation. A new seismic crisis began on December 14. At 14:40 the successive opening

of three fissures were observed by one of the OVPF webcams. The fissures were located on the southern PdF flank, and run parallel to the Dolomieu crater from east to west, 120 m from its rim (~55°42′45″, −21°14′58″, 2490 m). Lava effusion there lasted until 16:10, and when emission from the first vent ended, a fourth radial fissure opened (at 16:23) 330 m from the Dolomieu rim (~55°42′57″, −21°15′04″, 2430 m). The eruption tremor ended at 06:30 next day. On the first fissures no sputtering was observed and pahoehoe lava was emitted, extending to a length of 1.1 km. The fourth fissure formed a <2 m high sputter cone and emitted aa lava, extending to a length of 380 m. The eruption emitted 0.16 Mm^3 of aphyric lava at a MOR of 2.8 $m^3\ s^{-1}$.

9.3.2.4 The 2010 Eruptions

Three eruptions occurred in 2010 (Fig. 9.5d). The first was in the Dolomieu crater, the second was located close to Château Fort in the south and the third was located on the northern PdF flank.

January 2 (No. 32)
Seismicity remained low between 15 and 26 December 2009 (Fig. 9.5). On December 29, a first seismic crisis occurred, 188 events were recorded. On January 2, 2010 after a second crisis (78 events), an eruption began at 10:20 on the western crater wall of the Dolomieu crater (55°42′36″, −21°14′44″, 2490 m), only meters from the September 21 and November 27, 2008 fissures. This may explained the absence of pre-eruptive ground deformation measured by the GNSS network (Fig. 9.6). Because the eruption began on the Dolomieu crater wall, several large rock-falls occurred from the wall, forming ash plumes. The eruption ended on January 12 and emitted 1.2 Mm^3 at a MOR of 1.4 $m^3\ s^{-1}$. Only lapilli, Pele's hair and ash related to the rocks-fall had been recovered within the adjacent Bory crater. The eruption formed an aa lava flow in the Dolomieu crater, however pahoehoe emerged at its border. It raised the Dolomieu crater floor by 25 m, reducing the depth to 240 m.

October 14 (No. 33)
Between January and the end of August seismicity was low (average of 3.6 events per day). Between September and early-October, seismic activity increased to between 20 and 100 events per day, with a seismic crisis occurring on September 23 (650 events, Fig. 9.4). On October 14, after another seismic crisis (870 events being recorded), the hypocenters, which had initially been centered beneath the Dolomieu, moved to the south and became increasingly superficial. The tiltmeter network indicated an inflation of the summit, followed by a migration of the inflation center to the south-flank. The GNSS network recorded a movement of the Château Fort receiver by 27 cm to the east and 5.5 cm to the north, accompanied by an uplift of 31 cm. Finally, at 02:00, a 170 m long fissure opened, orientated N170, forming three pyroclastic cones (between 15 and 30 m high) close to the Château Fort crater. The main eruptive cone was located at 55°43′2″, −21°15′52″ and the 1990 m elevation. A 2 km-long aphyric lava flow extending to the south-east was emplaced. The eruption lasted 16 days, covered an area of 0.67 km^2 and erupted a volume of 2.7 Mm^3 at a MOR of 1.9 $m^3\ s^{-1}$. The flow field was mainly aa lava.

December 9 (No. 34)
Seismicity remained at an average level of 5 events per day until the end of November. It increased in December to 15 events per day. The GNSS network recorded a dilatation at the Dolomieu crater by 2.2 cm (Fig. 9.6). On December 9 a seismic crisis of 159 events occurred at 15:11. At 18:15 eruption tremor was recorded. The eruption was located on the northern flank of PdF where a 430 m-long fissure opened, orientated north. The main activity established at 55°42′24″, −21°14′9″ and at

2390 m elevation, where a 10 m-high spatter cone formed. The eruption lasted 14.5 h and erupted an aphyric lava with a thickness of 0.5–2 m. The aa lava flow had a length of 1.5 km and a volume of 0.53 Mm^3, having been erupted at a MOR of 9.7 $m^3\ s^{-1}$.

9.3.3 Post 2011 Repose Period

The February 2, 2011 seismic crisis

Between mid-December 2010 and early February 2011 the Dolomieu crater enlarged by 1.8 cm in north-south direction and by 1.2 cm in the east-west direction (Fig. 9.6). Seismicity in January was characterized by 22 events per day. On February 2 a new seismic crisis of 620 events occurred. This was accompanied by 2 cm of Dolomieu west-east extension, but no eruption followed.

Activity 2011–2013

The average seismicity between February and December 2011 declined from 8.7 events per day in February to 1.5 events in December (Fig. 9.4). In 2012 and 2013 seismicity remained low with average of 0.7 events per day in 2012 and 1.1 event per day in 2013. In 2011 (after the seismic crisis), 2012 and 2013, the GNSS network recorded a 4.8, 3.2 and 2.7 cm contraction of the Dolomieu crater in west-east direction, respectively, at an average rate of $\sim 1 \times 10^{-2}$ cm day^{-1} (Fig. 9.6). No seismic crises, intrusive or eruptive events were recorded.

9.3.4 Instability of the Eastern Flank

Until 2009, only the terminal cone of the volcano was monitored by the permanent GNSS networks. Interferometric Synthetic Aperture Radar (InSAR) monitoring allowed to access to the deformation occurring farther away. During the

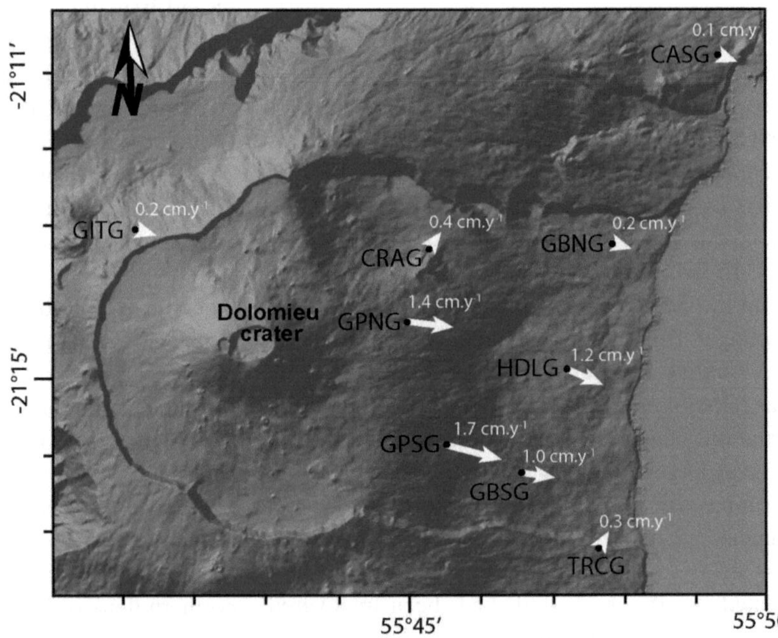

Fig. 9.7 Velocity of the eastern flank motion recorded by GNSS. *Arrows* represent the mean velocity recorded by each GNSS station since their implementation in 2009–2010, corrected for the plate velocity

course of the April 2007 eruption, a large motion of up to 1.4 m of the eastern flank toward the sea has been detected by InSAR (see Augier et al. 2008 and Clarke et al. 2013 for more detail concerning the April 2007 large instability). Following this large flank motion and in the framework of the UnderVolc project (Brenguier et al. 2012), six new permanent GNSS stations had been installed on the volcano's eastern flank, at elevations comprised between 136 and 1414 m (Fig. 9.7).

Since their installation, these GNSS stations recorded continuous and ongoing displacements of the eastern flank toward the east (from 0.2 to 1.7 cm $year^{-1}$, values corrected from the plate velocity), accompanied by a subsidence (from 0.6 to 3.5 cm $year^{-1}$). These displacements are quite constant since the first records about 4 years ago (Fig. 9.7; Staudacher and Peltier 2016, Chap. 15).

Such a preferential long-term motion of the eastern flank was already proposed by several authors to explain the asymmetrical deformation pattern observed during dyke injections (Briole et al. 1998; Froger et al. 2004; Fukushima et al. 2005; Peltier et al. 2007, 2009b), but was never recorded and quantified before. Preferential flank motions are common in volcanic area, and are similarly observed at Kilauea and Etna (e.g. Swanson et al. 1976; Puglisi and Bonforte 2004). On these volcanoes, as observed at Piton de la Fournaise, the seaward-facing flanks are unstable, with steady motion (resulting from gravity processes presumably accommodated along a basal fault plane) punctuated by episodic movement that is sometimes associated with magmatic activity.

9.4 General Characteristics of Recent Eruptions at Piton de la Fournaise

A period of 15 years is short in the life of a volcano, nevertheless, the 34 eruptions of Piton de la Fournaise since 1998 allow us to give a general view on the of the present-day Piton de la Fournaise, in context of its behavior during pre-eruptive and eruptive periods, and the characteristics of eruptions: precursors, frequency, location, emitted lava volumes and covered areas and in particular hazard for inhabitants (see below).

9.4.1 Pre-eruptive, Eruptive and Post-eruptive Behavior

9.4.1.1 Characteristics of the Pre-eruptive Period

Seismicity

All eruptions at Piton de la Fournaise recorded between 1998 and 2010 were preceded by increases in the number of seismic events per day and seismic crises, as well as by inflation and extension of the terminal cone.

Typically the number of volcano tectonic events increases during the weeks, or months, before an eruption due to pressurization by influx of magma supply from a deeper magma reservoir or due to bubble formation in the shallow magma reservoir and interaction with the hydrothermal system. The pre-eruptive long-term frequency of seismic events range between 5 and 50 events per day, and sometimes higher as before the March 2001 and February 2005 eruptions, when up to 130 events per day were recorded (Fig. 9.4). Seismic crises may comprise ten's (e.g. July 1999, August 2006) to more than thousand (e.g. March 1998, December 2008) of seismic events. Roult et al. (2012) distinguishes four phases in a pre-eruptive seismic crisis at PdF. The first is characterized by a rapid increase in the frequency of seismic events with at least two volcano-tectonic (VT) events per minute being recorded. The second phase corresponds to a swarm of more than six events per minute which have larger magnitudes, than those of the first phase. Strong ground deformation is generally observed during this phase. This is followed by a seismically quiet/low intensity sequence with a small number of VT events and corresponds to a lateral migration of the magma (Fig. 9.8: 17 February 2005, 14:10–16:35). The fourth phase corresponds to the onset of the eruptive tremor. This is associated with the opening of eruptive fissures (Fig. 9.8: 17 February 2005 at 16:35). Summit or proximal eruptions (i.e. on the flanks or at the base of the terminal cone) are mostly

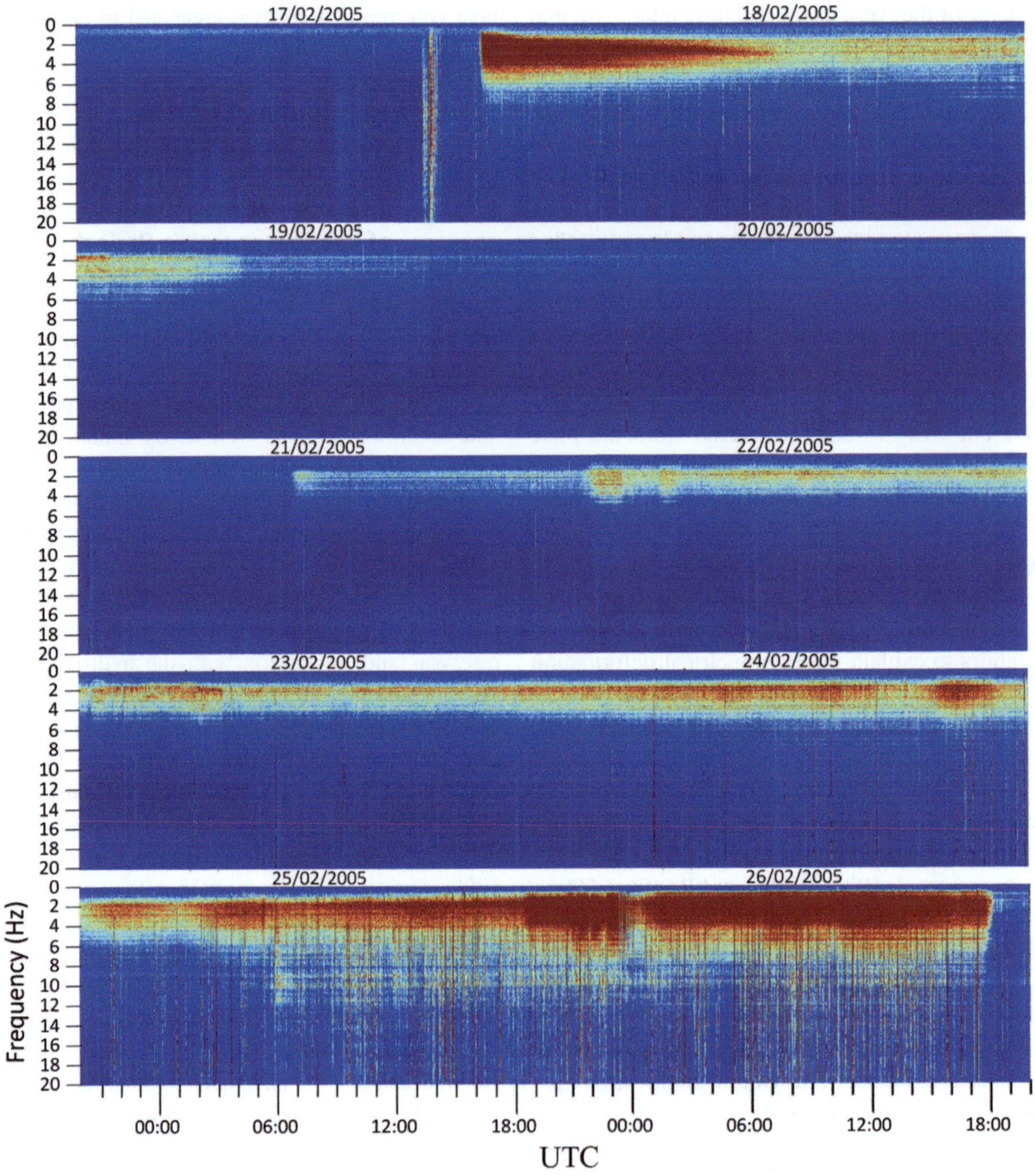

Fig. 9.8 Spectrogram of the February 17–26, 2005 distal eruption, described in Sect. 9.3.1.8. It comprises a seismic crisis (high frequencies), an initial opening on February 17 (characterised by strong tremor in the beginning, slowing down with time), followed by the opening of secondary vents at low elevation on February 22 and 24, seismicity during eruption and a final strong activity shutting down suddenly on February 26

characterized by short-duration seismic crises (typically lasting 20 min–2 h). Long-duration crises (of several hours in duration) are often associated with distal eruptions (i.e. more than 4 km away from the summit, Aki and Ferrazzini 2000; Schmid et al. 2012).

Since the Dolomieu collapse in 2007, numerous seismic crises without eruptions have occurred (Roult et al. 2012). They were often accompanied by inflation of the summit cone and likely correspond to magma intrusions into the highly fractured volcanic edifice. However the shallow magma chamber is not sufficiently over pressurised to allow magma arrival at the surface and filled up shallow empty spaces left in the collapsed zone.

Ground Deformation

Between 1998 and 2011, pre-eruptive ground deformation were observed before most eruptions at Piton de la Fournaise. Between 1998 and 2003 the extensometer network at Château Fort and Magne recorded systematic fracture opening at the base of PdF at a rate of 0.0004–0.0083 mm day^{-1}, with an average rate of 0.0035 mm day^{-1} (Peltier et al. 2006, 2016). Since 2004, deformation at Piton de la Fournaise has been recorded by the GNSS network (Staudacher and Peltier 2016). Seventeen eruptions out of nineteen were preceded by ground deformation. The most precise measure are baseline variations between the summit stations. These recorded a systematic expansion of the Dolomieu crater at a rate of 0.3–1.2 mm day^{-1} over six days (January 2010) to three months (May 2004; October 2005) prior to eruptions.

Two exceptions occurred. The November 2008 eruption was not preceded by ground deformation, despite the fact that seismicity was high (with an average of 60 events per day, with peaks of up to 280 events per day). This eruption started from the same vent as for the September 2008 eruption. The magma probably used the same pathway without causing detectable deformation. The second exception was the December 2009 eruption. This eruption occurred in a region where large—ten's of centimeter to meter—fracture tangential to the Dolomieu rim are present. It might be possible that the dike used such fractures and caused no detectable ground deformation.

9.4.1.2 Eruptive and Post Eruptive Period

Seismicity

Eruption tremor at Piton de la Fournaise is associated with arrival of the dike at (or near) the surface and the emission of gas and lava (e.g. Fig. 9.8, February 17, 16:35). During the last 15 years, with only two exceptions, tremor has begun with eruptions. Two exceptions were recorded on September 12, 2008 and October 18, 2009, when tremor was recorded without emission of lava. In both cases, only gas emission was observed in the Dolomieu crater escarpments, forming sulfur deposits.

Most eruptions begin with high tremor levels. These levels are maintained for few hours and correspond to the phase of continuous lava fountaining (Fig. 9.8: February 17, 16:35). After this tremor intensity decreases gradually until the end of the eruption. With the beginning of eruptions, seismicity under the summit usually stops. However from June 2000, in the course of both, proximal and distal eruptions (never in the case of summit eruptions) earthquakes have been recorded, mostly above sea level under the summit craters and about 2 km below sea level under the east flank. This seismicity has been noticed first by Hirn et al. during the 1985–86 eruptive cycle (Hirn et al. 1991). The eruption and the seismic activities ended usually in phase, as if the system went from an opened to a closed state. During the distal eruptions of January 2002 and February 2005 in Plaine des Osmondes (Fig. 9.8: from February 24 to February 26), numerous seismic events occurred under the summit, but also 2 km below sea level under the north-east flank. They were felt by the inhabitants of Bois Blanc before the opening of fissures at low elevations in the northern Grand Brûlé (No. 10 and No. 19 in Figs. 9.5b and c, respectively). The origin of these earthquakes is still unknown, but earthquake mechanisms below the summit correspond mostly to movement of normal faults in the north-western part of the

Dolomieu crater that appears to reflect the Dolomieu crater subsidence forcing movement on the fault (Massin 2009).

An exception to this pattern occurred following the November 2002 eruption. In the last days of this eruption, seismicity increased dramatically above sea level. No deep events were recorded. It continued after the end of eruption on December 5, at rates of up to 5700 events per day before the formation of a pit crater in the Dolomieu crater on December 23 (Longpré et al. 2006). The characteristics of the seismic crisis was similar to the one recorded during the 1986 pit crater formation in the Dolomieu crater. For that pit crater collapse Hirn et al. (1991) described similar increase of number of events, waveforms of the events, and the final collapse with a seismic event of magnitude about 3.

In 2007 the collapse of the Dolomieu crater during the April eruption was also preceded by hundreds of seismic events, but occurred in the course of the eruption and was characterised by a rhythmic nature, similar to that described on Fernandina by Filson et al. (1973) and Michon et al. (2011). The main shock was nearly 2 order of magnitude greater that for the 1986 and 2002 pit craters formation.

Ground Deformation

Strongest ground deformation was observed during dike propagation to the surface. The intensity of the ground deformation depends on the proximity of the dike to the GNSS stations or geodetic points. In August 2004, for example we recorded deformation in the Dolomieu crater and on the eastern flank of up to 44 cm. These measurements were made at several meters distance of the eruptive fissure; with values decreasing with distance. In October 2010, the dike passed about 200 m west of the Château Fort permanent GNSS station and ground deformation of several ten's of cm were recorded. After eruptions, the edifice usually deflates slightly, as for example after the eruptions between May 2004 and October 2005. In other cases, the inter-eruptive period was too short so that the deflation was absent or truncated by the next eruption, as for example in November 2005, February 2007 and November 2008 (Staudacher and Peltier 2016).

9.4.2 Characteristic of Eruptions

9.4.2.1 Location

All eruptions since 1998 have begun within the Enclos Fouqué caldera. The U-shape structure of the caldera, open to the ocean to the east, funnels the lava flows in that direction. This avoids any damage to the inhabited regions on the volcano flanks. Nevertheless, long lasting eruptions, or those at low elevation, might overrun the main road and reach the ocean. The main road was cut in June 2001, January 2002, November 2002, August 2004, February 2005 and April 2007, and lava flows reached the ocean in January 2002, November 2002, August 2004, February 2005 and April 2007.

In 1998, during the March to September eruption, three small vents opened outside of the Enclos Fouqué caldera, but they were located at high elevations (1680–1590 m); far away from inhabited areas. Activity was short and the emitted volume of lava was low. By contrast, in 1977 and 1986, hors Enclos eruptions damaged houses in the village of Piton de Sainte Rose and Le Tremblet, respectively. Eruptions outside of the Enclos Fouqué caldera occurred in 1776 and 1800 along the south-east rift zone, and in 1708 and 1774 on the north-east rift zone (Bachèlery 1981). Since 1800 more than 200 eruptions have been recorded, only seven (<4 %) of them occurring outside of the Enclos Fouqué caldera.

9.4.2.2 Duration and Erupted Volume

The duration of eruptions at Piton de la Fournaise is extremely variable (Table 9.1). The shortest recorded duration since 1998 lasted less than 10 h (March 2007, November 2005). The longest lasted 196 days, being the March 1998 eruption (Table 9.1). The average duration was 25 days.

Eruptions produced between 0.14 and 240 Mm^3 of lava and the MOR was 14×10^6 m^3, while the average production rate was 1.03 m^3 s^{-1}. The total emitted volume was

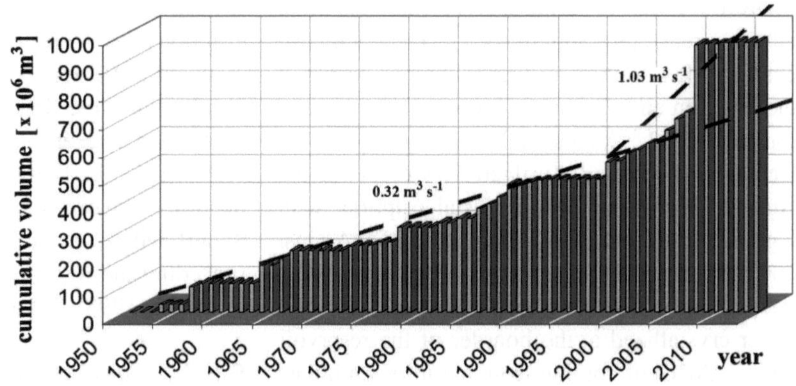

Fig. 9.9 Cumulative volume of lava emitted by Piton de la Fournaise since 1950. The average slope between 1950 and 2000 is 0.32 m^3 s^{-1}, while it is 1.03 m^3 s^{-1} during the last decade

489 Mm^3 (293 Mm^3 DRE) partitioned between 34 eruptions. This DRE volume is close to the shallow magma reservoir volume of ~300 Mm^3 inferred from geochemical or geodetic data (Sigmarsson et al. 2005; Peltier et al. 2008).

If we consider the period between 1950 and 1998, the erupted volume was 481 Mm^3 (DRE), corresponding to a rate of 0.32 m^3 s^{-1}. Piton de la Fournaise thus erupted as much volume in the last 15 years than in the preceding 48 years. Figure 9.9 gives the cumulative volume plot for lava erupted since 1950. This illustrates an increase of the emitted material with time. The total emitted volume in the last 60 years was 970 Mm^3 (580 Mm^3 DRE), with an average production rate of only 0.5 m^3 s^{-1}. This compares with 0.32 m^3 s^{-1} before 1998 and 1.03 m^3 s^{-1} after 1998.

There is no correlation between duration and emitted volume of magma. For example, while the April 2007 eruption produced 240 Mm^3 in one month, the December 2008 eruption produced just 1.5 Mm^3 in 2 months (Table 9.1). The thickness of the lava-flows range between 0.5 and 60 m. The average thickness of lava-flows range between 1 and 25 m, with the maxima corresponding to the April 2007 eruption. The overall average thickness is about 5.2 m. If we exclude the eruptions within the funnel shaped Dolomieu crater and April 2007, during which lavas became entrapped and pond, the average value reduces to 3.6 m.

9.4.2.3 Lava Areas

The surface areas covered by lava flows during one eruption ranged between 0.1 (March 2007, November 2009) and 7.3 km^2 (March 1998), with an average value of 0.79 km^2. The total surface covered by lava flows in the last 15 years is determined to be about 40 km^2, a surface that corresponds to ~38 % of the Enclos Fouqué.

9.4.2.4 Lava Compositions

Since 1998 Piton de la Fournaise has erupted aphyric lava and olivine bearing lava (Table 9.1). Aphyric lavas contain less than 7 wt% of MgO. Picrites contain >8 wt% of MgO and are also termed oceanites at La Réunion if their olivine contents exceeds 20 %. The April 2007 eruption ended with an olivine content of >50 % (Vlastélic et al. 2013).

Most eruptions at PdF produce aphyric lava. This is particularly true for summit and proximal eruptions, even though oceanitic (or pycritic) lava had been erupted there in the past (witnesses of such summit oceanites lavas are still visible on the east border of the Dolomieu crater). Some olivine phenocrysts appeared at the end of the "proximal" October 2000 eruption, which finished with a small phreatic explosion and its highest effusion rate of 40–60 m^3 s^{-1}. Other olivine bearing lava flows erupted during distal eruptions, as in November 2002 and January 2004. Oceanitic lavas are often emitted during eruptions which were characterized by high

MOR (>19 m^3 s^{-1}) and/or very high maximal effusion rates (30–300 m^3 s^{-1}). They are also emitted when high paroxysmal activity occurs at the eruption end. Oceanitic lavas were emitted in January 2002, August 2003, February 2005, December 2005 and April 2007. Except for the August 2003 event, all were distal eruptions with high MOR (14–94 m^3 s^{-1}; January 2002, August 2003, February 2005, April 2007). These MOR rates would allow the incorporation of olivines, either crystallized at the boarder of the reservoir (Welsh 2010) or coming from a dunitic cumulate body (Albarède et al. 1997).

9.5 Discussion

Since the installation of the Volcanological Observatory in the early 1980s, the large number of eruptions has allowed accumulation of a large database for pre-eruptive, eruptive and post-eruptive activity. This have allowed us to have a better understanding of the eruptive phenomenon, the behavior of the volcano and the dynamics of the current shallow magma plumbing system of Piton de la Fournaise.

9.5.1 Current View of the Shallow Magma Plumbing System

Geodetic and seismic data suggest the presence of a shallow magma reservoir located around 2500 ± 1000 m below the summit craters (Battaglia et al. 2005; Peltier et al. 2008, 2009b; Massin et al. 2011; Prôno et al. 2009). The shape of the shallow plumbing system (single reservoir or network of dykes and sills) under Piton de la Fournaise remains a matter of strong debate (e.g. Lénat and Bachèlery 1990; Peltier et al. 2009b). All eruptions of the last 15 years have been fed by magmas rising into this system. This led to a two stage dyke propagation sequence; the first stage corresponds to a vertical dyke propagation (at a rate of ~2 m s^{-1}), centered below the summit crater, followed by lateral propagation toward one of the rift-zones in the case of lateral eruptions (at a rate of 0.2–0.8 m s^{-1}) (Toutain et al. 1992; Peltier et al. 2005). Summit eruptions involve only the first stage.

A deeper reservoir is thought to exist around 7500 m depth (Battaglia et al. 2005). The connection between the two reservoirs has been underlined by an ascending seismicity of 35 h preceding the resumption of the eruptive activity in March 1998 (Battaglia et al. 2005).

9.5.2 Conclusion: Recent Activity of Piton de la Fournaise

After a six year-long hiatus, renewal of eruptive activity in March 1998 marked the beginning of a high intensity eruptive period at Piton de la Fournaise that ended in 2010. In March 1998, refilling of the shallow magma reservoir was highlighted by an ascending seismic swarm (Battaglia et al. 2005). Between 2000 and 2007, deformation networks recorded quasi-continuous summit inflation during inter-eruptive periods, signs of a gradual magma pressure build-up in the shallow reservoir and a long-term magma refilling (Peltier et al. 2009b; Staudacher and Peltier 2016). This quasi-continuous refilling of the shallow plumbing system is the origin of the high eruption frequency observed between 2000 and 2007. This activity followed eruptive cycles, with successive summit/proximal eruptions ended by a distal one, more than 4 km from the summit (Peltier et al. 2008, 2009b). These eruptive cycles (of less than 1.5 years in duration) have been linked to a non-linearity in the stress state of the edifice (Got et al. 2013). This allows the edifice to modulate magma transfer from a constant input rate to discrete, cyclical production of magma at the surface.

The eruption of 2007, which emitted more lava than any other eruption since at least 2 centuries largely drained the shallow plumbing system. This led to the collapse of the Dolomieu crater (Michon et al. 2007; Staudacher et al. 2009) and was at the origin of a change in the eruptive activity in the following years. The activity that followed (2008–2010) involved eight low-volume eruptions, the majority of which occurred in or close to the Dolomieu crater. Over the same period there were 18 intrusive events, compared with two intrusive events between 1998 and 2007 (Roult et al. 2012). The deformation network only recorded summit deflation during pre-eruptive periods. No detectable long-term magma refilling was observed, but only post-collapse relaxation. All these changes can be explained by major stress perturbations of the shallow edifice and major changes in the shallow plumbing system, and its refilling, after the voluminous 2007 eruption (Peltier et al. 2010; Roult et al. 2012).

Since the end of 2010 through 2013, no eruption has occurred. Deformation data recorded an ongoing summit contraction, and a continuous motion of the eastern flank toward the ocean. In the absence of a decrease of the motion with time, this long-term motion is most probably linked with an ongoing eastern flank instability that could be associated with gravity processes, rather than by relaxation of the flank movement that occurred in April 2007. As proposed at Kilauea, the steady eastern flank motion could be slip along a basal fault plane (e.g., Denlinger and Okubo 1995; Got et al. 2008; Peltier et al. 2015b). The presence of a décollement plane at sea level has been mentioned in case of Piton de la Fournaise (Hirn et al. 1991; Battaglia et al. 2005) but because the seismicity is rare on this flank more robust constraints are still being needed.

Acknowledgements The authors wish to thank the Institut National des Sciences de l'Univers, the Conseil Général de La Réunion, le Conseil Régional de La Réunion, the «prefecture de la Réunion» and in particular the FEDER funding (Fonds européen de développement regional) for their precious financial help to support the OVPF and allowed the equipment of Piton de la Fournaise by a dense and powerful network. We are also grateful to the Gendarmerie for logistical help during eruptions and we thank the members of the IPGP and all others, which we can not all mention, for their help during the long nights of survey in front of the computer screens at the observatory waiting the eruptions to come out and during the field work in order to get the best possible lava samples and observations. Paule Annick Davoine and Cecile Saint-Marc are warmly thanked for their great help in mapping the lava flows. We are particularly grateful to the anonymous reviewer who made the manuscript readable.

References

Aki K, Ferrazzini V (2000) Seismic monitoring and modelling of an active volcano for prediction. J Geophys Res 105(B7):16617–16640

Albarède F (1993) Residence time analysis of geochemical fluctuations in volcanic series. Geochem Cosmochemica Acta 57:615–621

Albarède F, Luais B, Fitton G, Semet M, Kaminski E, Upton BGJ, Bachèlery P, Cheminée JL (1997) The geochemical regimes of Piton de la Fournaise volcano (Reunion) during the last 530000 years. J Petrol 38 (2):171–201

Allard P, Di Muro A, Spina A, Burton M, Tamburello G, Métrich N, Staudacher T, Michel A (2014) First measurement of magma degassing composition and budget during an eruption of Piton de la Fournaise hot spot volcano, La Réunion island, EPSL (submitted)

Augier A, Froger JL, Cayol V, Fukushima Y, Tinard P, Souriot T, Mora O, Staudacher T, Durand P, Fruneau B, Villeneuve N (2008) The April 2007 eruption at Piton de la Fournaise, Réunion Island, imaged with ENVISAT-ASAR and ALOS-PALSAR data, USEReST workshop, Napoli, Italy

Bachèlery P (1981) Le Piton de la Fournaise (Ile de la Réunion). Etude volcanologique, structurale et pétrologique, PhD thesis, University of Clermont-Ferrand II, France, 215 pp

Bachèlery P, Saint-Auge F, Villeneuve N, Savoye B, Normand A, Le Drezen E, Barrere A, Quod JP, Deplus C (2010) Huge lava flows into the sea and caldera collapse, April 2007, Piton de la Fournaise volcano, Workshop on collapse Calderas—La Réunion, pp 73–74

Battaglia J, Ferrazzini V, Staudacher T, Aki K, Cheminée JL (2005) Pre-eruptive migration of earthquakes at Piton de la Fournaise volcano (Réunion Island). Geophys J Int 161:549–558

Bhugwant C, Siéja B, Perron L, Rivière E, Staudacher T (2002) Impact régional du dioxyde de soufre d'origine volcanique induit par l'éruption du Piton de la Fournaise (Ile de la Réunion) en juin-juillet 2001, Pollution Atmosphérique, vol 44(176)

Bhugwant C, Siéja B, Bessafi M, Staudacher T, Ecornier J (2009) Atmospheric sulphur dioxide measurements during 2005 and 2007 eruptions of the Piton de la Fournaise volcano: implications for human health and environmental changes. J Volcano Geotherm Res 184 (1–2):208–224

Blum PA (1963) Contribution à l'étude des variations de la verticale en un lieu. Ann Géophys 19:215–243

Brenguier F, Kowalski P, Staudacher T, Ferrazzini V, Lauret F, Boissier B, Catherine P, Lemarchand A,

Pequegnat C, Merix O, Pardo C, Peltier A, Tait S, Shapiro NM, Campillo M, Di Muro A (2012) First results from the undervol High resolution seismic and gps network deployed on Piton de la Fournaise Volcano. Seismolog Res Lett vol. 83(7). doi: 10.1785/gssrl.83.1.97

Briole P, Bachèlery P, Mc Guire B, Moss J, Ruegg JC, Sabourault Ph (1998) Deformation of Piton de la Fournaise: evolution of the monitoring techniques and knowledge acquired in the last five years. In Casale R et al (eds) The European laboratory volcanoes. Second Workshop Santorini, Greece

Bureau H, Métrich N, Semet MP, Staudacher T (1999) Fluid-magma decoupling in a hot-spot volcano. Geophys Res Lett 3501–3504

Clarke D, Brenguier F, Froger JL, Shapiro NM, Peltier A, Staudacher T (2013) Timing of a large volcanic flank movement at Piton de la Fournaise volcano using noise-based seismic monitoring and ground deformation measurements. Geophys J Int. doi:10.1093/gji/ggt276

Coppola D, Staudacher T, Cigolini C (2005) The May-July eruption at Piton de la Fournaise (La Réunion): volume, effusion rate and emplacement mechanisms inferred by thermal imaging and GPS survey. In: Manga M, Ventura G (eds) Kynematics and dynamics of lava flows, Geological Society of America Special Paper, vol 396, pp 103–124

Coppola D, Staudacher T, Cigolini C (2007) Field thermal monitoring during the August 2003 eruption at Piton de la Fournaise (La Réunion). J Geophys Res 112:B05215. doi:10.1029/2006JB004659

Deniel C, Kieffer G, Lecoindre J (1992) New ^{230}Th-^{238}U and ^{14}C age determinations from Piton des Neiges volcano, Reunion—a revised chronology for the Differentiated Series. J Volcanol Geotherm Res 51:253–267

Denlinger RP, Okubo P (1995) Structure of the mobile south flank of Kilauea Volcano, Hawaii. J Geophys Res 100(B12):24499–24507

Di Muro A, Staudacher T, Ferrazzini V, Métrich M, Villemant B, Besson P, Garofalo C (2015) Shallow magma storage at Piton de la Fournaise Volcano after the 2007 summit caldera collapse tracked in Pele's hairs. In: Carey R, Cayol V, Poland M, Weis D (eds) Hawaiian volcanoes: from source to surface. Geophysical monograph series, vol 208, pp 189–212

Duncan RA (1981) Hotspots in the southern oceans—an absolute frame of reference for motion of the Gondwana continents. Tectono-Physics 74:29–24

Dzurisin DR, Koyanagi Y, English TT (1984) Magma supply and storage at Kilauea Volcano, Hawaii, 1956–1983. J Volcanol Geotherm Res 21(3–4):177–206

Filson J, Simkin T, Leu LK (1973) Seimicity of a caldera collapse: Galapagos Islands 1968. J Geophys Res 78:8591–8622

Fontaine FR, Roult G, Michon L, Barruol G, Di Muro A (2014) The 2007 eruptions and caldera collapse of the Piton de la Fournaise volcano (La Réunion Island) from tilt analysis at a single very broadband seismic station. Geophys Res Lett 41:2803–2811. doi:10.1002/2014GL059691

Froger JL, Fukushima Y, Briole P, Staudacher T, Souriot T, Villeneuve N (2004) The deformation field of the August 2003 eruption at Piton de la Fournaise, Reunion Island, mapped by ASAR interferometry. Geophys Res Lett 31:L14601. doi:10.1029/2004GL020479

Fukushima Y, Cayol V, Durand P (2005) Finding realistic dike models from interferometric synthetic aperture radar data: the February 2000 eruption at Piton de la Fournaise. J Geophys Res 110:B03206. doi:10.1029/2004JB003268

Galle B, Johansson M, Rivera C, Zhang Y, Kihlman M, Kern C, Lehmann T, Platt U, Arellano S, Hidalgo S (2010) Network for observation of volcanic and atmospheric change (NOVAC)—a global network for volcanic gas monitoring: network layout and instrument description. J Geophys Res 115:D05304. doi:10.1029/2009JD011823

Garofalo K, Staudacher T, Ferrazzini V, Kowalski P, Boissier P, Dupond A, Peltier A, Villemant B, Boudon G (2009) Eruptive SO2-plume measurements at Piton de la Fournaise (Ile de La Réunion) by stationary NOVAC scanning MAX-DOAS instruments, abstract, EGU Vienna April 2009

Gillot PY, Nativel P (1989) Eruption history of the Piton de la Fournaise volcano, Reunion Island, Indian Ocean. J Volcanol Geotherm Res 36:53–55

Gillot PY, Lefèvre JC, Nativel P (1994) Model for the structural evolution of the volcanoes of Réunion Island. Earth Planet Sci Lett 122:291–302

Got JL, Monteiller V, Monteux J, Hassani R, Okubo P (2008) Deformation and rupture of the oceanic crust may control growth of Hawaiian volcanoes. Nature 451:453–456

Got JL, Peltier A, Staudacher T, Kowalski P, Boissier P (2013) edifice strength and magma transfer modulation at Piton de la Fournaise volcano. J Geophys Res 118:5040–5057. doi:10.1002/jgrb.50350

Hirn A, Lépine JC, Sapin M, Delorme H (1991) Episodes of pit-crater collapse documented by seismology at Piton de la Fournaise. J Volcanol Geotherm Res 47:89–104

Krafft M, Gérente A (1977) L'activité du Piton de la Fournaise entre novembre 1975 et avril 1976 (île de la Réunion, Océan Indien). C.R Académie des Sciences, tome 284:2091–2094

Lemarchand A, Tait S, Beauducel F, Bouin MP, Brenguier F, de Chabalier JB, Clouard V, Di Muro A, Ferrazzini V, Shapiro N, Staudacher T, Anglade A, Kowalski P, Saurel JM (2013) Significant breakthroughs in monitoring networks of the volcanological and seismological French observatories, AGU abstract, San Francisco

Lénat J-F (2016a) Geodynamic setting of La Réunion Island. In: Bachèlery P, Lénat J-F, Di Muro A, Michon L (ed), Active volcanoes of the Southwest Indian Ocean: Piton de la Fournaise and Karthala.

Active Volcanoes of the World (Chap. 2). Springer, Berlin
Lénat J-F (2016b) The construction of La Réunion Island. In: Bachèlery P, Lénat J-F, Di Muro A, Michon L (eds) Active volcanoes of the Southwest Indian Ocean: Piton de la Fournaise and Karthala. Active volcanoes of the World (Chap. 3). Springer, Berlin
Lénat JF, Bachèlery P (1990) Structure et fonctionnement de la zone centrale du Piton de la Fournaise. In: Lénat JF (ed) Le volcanisme de La Réunion. Centre de Recherches Volcanologiques, Clermont-Ferrand, pp 257–296
Longpre MA, Staudacher T, Stix J (2006) The November 2002 eruption at Piton de la Fournaise volcano, La Reunion Island: ground deformation, seismicity, and pit crater collapse. Bull Volcanol. doi:10.1007/s00445-006-0087-0
Massin F (2009) Transferts et Stockages Magmatiques au Piton de la Fournaise. PhD thesis, Université de La Réunion, France
Massin F, Ferrazzini V, Bachèlery P, Nercessian A, Duputel Z, Staudacher T (2011) Structures and evolution of the plumbing system of Piton de la Fournaise volcano inferred from clustering of 2007 eruptive cycle seismicity. J Volcanol Geotherm Res 202:96–106
McDougall I (1971) The geochronology and evolution of the younger volcanic island of Reunion, Indian Ocean. Geochim Cosmochim Acta 36:261–288
Michon L, Staudacher T, Ferrazzini V, Bachèlery P, Marti J (2007) April 2007 collapse of Piton de la Fournaise: a new example of caldera formation. Geophys Res Lett 34(L21301). doi:10.1029/2007GL031248
Michon L, Massin F, Famin V, Ferrazzini V, Roult G (2011) Basaltic calderas: collapse dynamics, edifice deformation, and variations of magma withdrawal. J Geophys Res 116:B03209. doi:10.1029/2010JB007636
Morgan WJ (1981) Hotspot tracks and the opening of the Atlantic and Indian Oceans. In: Emiliani C (ed) The sea. Wiley, New York, pp 443–487
Oehler JF (2005) Les déstabilisations de flanc des volcans de l'Ile de La Réunion (Océan Indien): Mise en évidence, implications et origines, PhD thesis Université Blaise Pascal, Clermont Ferrand II
Peltier A, Ferrazzini V, Staudacher T, Bachelery P (2005) Imaging the dynamics of dyke propagation prior to the 2000–2003 flank eruptions at Piton de la Fournaise, Reunion Island. Geophys Res Lett 32. doi:10.1029/2005GL023720
Peltier A, Staudacher T, Catherine P, Ricard LP, Kowalski P, Bachèlery P (2006) Subtle precursors of volcanic eruptions at Piton de la Fournaise detected by extensometers. Geophys Res Lett 33:L06315. doi:10.1029/2005GL025495
Peltier A, Staudacher T, Bachèlery P (2007) Constraints on magma transfers and structures involved in the 2003 activity at Piton de la Fournaise from displacement data. J Geophys Res 112(B03207). doi:10.1029/2006JB004379
Peltier A, Famin V, Bachèlery P, Cayol V, Fukushima Y, Staudacher T (2008) Cyclic magma storages and transfers at Piton de la Fournaise volcano (La Réunion hotspot) inferred from deformation and geochemical data. Earth Planet Sci Lett 270:180–188
Peltier A, Staudacher T, Bachèlery P, Cayol V (2009a) Formation of the April 2007 caldera collapse at Piton de la Fournaise volcano: insights from GPS data. J Volcanol Geotherm Res 184:152–163
Peltier A, Bachèlery P, Staudacher T (2009b) Magma transfer and storage at Piton de la Fournaise (La Réunion Island) between 1972 and 2007: a review of geophysical and geochemical data. J Volcanol Geotherm Res 184(1–2):93–108
Peltier A, Staudacher T, Bachèlery P (2010) New behaviour of the Piton de la Fournaise volcano feeding system (La Réunion Island) deduced from GPS data: influence of the 2007 Dolomieu caldera collapse. J Volcanol Geotherm Res 192:48–56
Peltier A, Poland M, Staudacher T (2015) Are Piton de la Fournaise (La Réunion) and Kīlauea (Hawai'i) really "analog volcanoes"? In: Carey R, Cayol V, Poland M, Weis D (eds) Hawaiian volcanoes: from source to surface. Geophysical monograph series, vol 208, pp 507–531
Peltier A, Beauducel F, Staudacher T, Catherine P, Kowalski P (2016) Contribution of Tiltmeters and Extensometers to Monitor Piton de la Fournaise Activity, In: Bachèlery P, Lénat JF, Di Muro A, Michon L (eds) Active volcanoes of the Southwest Indian Ocean: Piton de la Fournaise and Karthala. Active Volcanoes of the World. Springer, Berlin
Pinel V, Jaupart C (2005) Some consequences of volcanic edifice destruction for eruption conditions. J Volcanol Geotherm Res 145:68–80
Prôno E, Battaglia J, Monteiller V, Got JL, Ferrazzini V (2009) P-wave velocity structure of Piton de la Fournaise volcano deduced from seismic data recorded between 1996 and 1999. J Volcanol Geotherm Res 184:49–62
Puglisi G, Bonforte A (2004) Dynamics of Mount Etna Volcano inferred from static and kinematic GPS measurements. J Geophys Res 109 (B11404). doi:10.1029/2003JB002878
Roult G, Peltier A, Taisne B, Staudacher T, Ferrazzini V, Di Muro A, The OVPF Group (2012) A new comprehensive classification of the Piton de la Fournaise eruptions spanning the 1986–2011 period. Search and analysis of eruption precursors from a broad-band seismological station. J Volcanol Geotherm Res 241–242:78–104
Salaün A, Villemant B, Semet MP, Staudacher T (2010) Cannibalism of olivine-rich cumulate xenoliths during the 1998 eruption of Piton de la Fournaise (La Réunion hotspot): Implications for the generation of magma diversity. J Volcanol Geotherm Res 198:187–204

Schmid A, Grasso JR, Clarke D, Ferrazzini V, Bachèlery P, Staudacher T (2012) Eruption forerunners from multiparameter monitoring and application for eruptions time predictability (Piton de la Fournaise). J Geophys Res 117:B11203. doi:10.1029/2012JB009167

Sigmarsson O, Condamines M, Bachèlery P (2005) Magma residence time beneath the Piton de la Fournaise volcano, Reunion Island, from U-series disequilibria. Earth Planet Sci Lett 234:223–234

Staudacher T (2010) Field observations of the 2008 summit eruptions at Piton de la Fournaise (Ile de La Réunion) and implications on the 2007 Dolomieu collapse. J Volcanol Geotherm Res 198:60–68

Staudacher T, Allègre CJ (1993) Ages of the second caldera of Piton de la Fournaise volcano (Réunion) determined by cosmic ray produced ^{3}He and ^{21}Ne. Earth Planet Sci Lett 119:395–404

Staudacher T, Peltier A (2016) Ground deformation at Piton de la Fournaise, a review from 20 years of GNSS monitoring, In: Bachèlery P, Lénat JF, Di Muro A, Michon L (eds) Active volcanoes of the Southwest Indian Ocean: Piton de la Fournaise and Karthala. Active Volcanoes of the World. Springer, Berlin

Staudacher T, Villeneuve N (1998) Piton de la Fournaise, chronique de l'éruption du 9 et 12 mars 1998, imprimerie La N.I.D. St Denis La Réunion, 20 pp

Staudacher T, Ferrazzini V, Peltier A, Kowalski P, Boissier P, Catherine P, Lauret L, Massin F (2009) The April 2007 eruption and the Dolomieu crater collapse, two major events at Piton de la Fournaise. J Volcano Geotherm Res 184(1–2):126–137

Swanson DA, Duffield WA, Fiske RS (1976) Displacement of the south flank of Kilauea Volcano: the result of forceful intrusion of magma into the rift zones. US Geol Surv Prof Pap 963: 39 pp

Toutain JP, Bachelery P, Blum PA, Cheminee JL, Delorme H, Fontaine L, Kowalski P, Taochy P (1992) Real time monitoring of vertical ground deformations during eruptions at Piton de la Fournaise, Geophys Res Lett 19(6)

Urai M, Geshi N, Staudacher T (2007) Size and volume evaluation of the caldera collapse on Piton de la Fournaise volcano during the April 2007 eruption using ASTER stereo imagery. Geophys Res Lett doi:10.1029/2007GL031551

Viane C, Bhugwant C, Sieja B, Staudacher T, Demoly (2009) Etude comparative des émissions de gaz volcanique du Piton de la Fournaise et des hospitalisations pour asthme de la population de 2005 à 2007. Revue française d'allergologie 49:346–351

Vigouroux N, Williams-Jones AE, Wallace P, Staudacher T (2009) The November 2002 eruption of Piton de la Fournaise, Réunion: tracking the pre-eruptive thermal evolution of magma using melt inclusions. Bull Volcanol 71:1077–1089. doi:10.1007/s00445-009-0287-5

Villeneuve N (2000) Apports multi-sources à une meilleure compréhension de la mise en place des coulées de lave et des risques associés au Piton de la Fournaise : Géomorphologie quantitative en terrain volcanique, thesis Institut de Physique du Globe de Paris

Vlastélic I, Staudacher T (2005) Rapid change of lava composition from 1998 through 2002 at Piton de la Fournaise (Réunion Island) inferred from Pb isotopes and trace elements: evidence for variable crustal contamination. J Petrol 46:79–107

Vlastelic I, Deniel C, Bosq C, Telouk P, Boivin P, Bachelery P, Famin V, Staudacher T (2009) Pb isotope geochemistry of Piton de la Fournaise historical lavas. J Volcanol Geoth Res 184(1–2):63–78

Vlastélic I, Menard G, Gannoun A, Piro JL, Staudacher T, Famin V (2013) Magma degassing during the April 2007 collapse of Piton de la Fournaise: the record of semi-volatile trace elements (Li, B, Cu, In, Sn, Cd, Re, Tl, Bi). J Volcanol Geotherm Res 254:94–107

Welsh B (2010) Signification des Océanites dans le fonctionnement du Piton de la Fournaise, Ile de La Réunion, PhD Thesis, University of La Réunion, 303 pp

Petrological and Experimental Constraints on the Evolution of Piton de la Fournaise Magmas

10

Michel Pichavant, Yann Brugier and Andrea Di Muro

Abstract

This chapter outlines essential petrological features of Piton de la Fournaise magmas. The main characteristics of the different magma types, and of their components (mineral and glass phases) are detailed. The available geophysical and petrological informations on the structure of the feeding system are combined and the magmatic conditions (temperature, volatile concentrations, redox state) summarized. Experimental modelling of the magmatic evolution is limited by the presently available database which concerns volatile-free compositions mostly from Piton des Neiges.

10.1 Piton de la Fournaise Magmas

10.1.1 Magma Types

Lavas and volcanic deposits of Piton de la Fournaise are distributed in 4 different compositional groups (Fig. 10.1; Table 10.1), which correspond to the activity inside, but also outside, the present-day Enclos Fouqué collapse caldera.

The Steady State Basalt (SSB) group is the most abundant in the recent activity of the volcano. SSB lavas are transitional basalts with a narrow range of compositions, from MgO = 5–8 wt% and CaO/Al_2O_3 = 0.75–0.85 (e.g., Albarède et al. 1997; Famin et al. 2009; Villemant et al. 2009; Salaün et al. 2010; Fig. 10.1). The SSB group is the point of convergence of 2 evolutive trends, one involving the Picrite group (MgO = 8–28 wt%, CaO/Al_2O_3 = 0.75–0.85) and the other the Differentiated Lavas group (MgO = 2–6 wt%, CaO/Al_2O_3 = 0.2–0.8, Fig. 10.1).

Picritic lavas, locally designated as oceanites (Lacroix 1936), are characteristic of the volcanic activity of La Réunion Island, being found at Piton de la Fournaise but also at Piton des Neiges. They contain millimetric to centimetric crystals of olivine embedded in a microcrystalline dark matrix. Much has been written on the petrology of La Réunion oceanites and their significance is still not completely clear. However, oceanite whole-rock compositions clearly plot on olivine control lines (e.g., Salaün et al.

M. Pichavant (✉) · Y. Brugier
ISTO, Orléans, France
e-mail: pichavan@cnrs-orleans.fr

A. Di Muro
IPGP-OVPF, La Réunion, France

P. Bachèlery et al. (eds.), *Active Volcanoes of the Southwest Indian Ocean*,
Active Volcanoes of the World, DOI 10.1007/978-3-642-31395-0_10

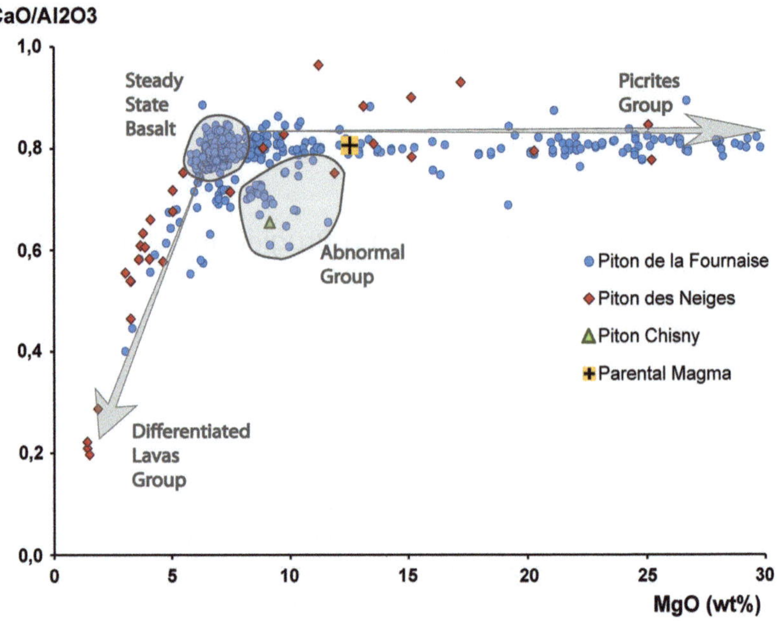

Fig. 10.1 CaO/Al_2O_3 versus MgO plot for Piton de la Fournaise and Piton des Neiges magmas. Data from Albarède et al. (1997), Boivin and Bachèlery (2009), Brugier et al. (2015), Bureau et al. (1999), Fisk et al. (1988), Peltier (2007), Salaün et al. (2010), Sobolev and Nikogosian (1994), Upton and Wadsworth (1972), Vigouroux et al. (2009), Villemant et al. (2009), Vlastélic et al. (2005, 2007), Welsch et al. (2013). The 4 different magma groups and one representative composition of parental magma (Villemant et al. 2009) are shown

2010; Fig. 9.1) suggesting that their chemistry reflects the proportion of olivine crystals present in the rock. In this context, an important observation is that the composition of matrix glasses in oceanites is nearly constant and similar to SSB bulk rocks (e.g., Welsch et al. 2009, see below). Most workers now consider that oceanites, instead of representing the product of crystallization of a picritic magmatic liquid at depth, result from a general mechanism of shallow incorporation of olivine crystals into a basaltic parental magma.

The Differentiated Lavas group, with a compositional evolution marked by strongly decreasing CaO and increasing Al_2O_3 at decreasing MgO, has been defined mainly from differentiated lavas erupted near the end of activity of Piton des Neiges (Upton and Wadsworth 1966). Differentiated alkaline lavas are found during the early stages of activity of the Piton de la Fournaise (Albarède et al. 1997) but are uncommon among recent products. This group includes chemically evolved terms such as hawaiites, mugearites and benmoreites and, in the recent activity, rare trachytic pumices (Vlastélic et al. 2011). Recent work (Smietana 2011) has suggested that these early differentiated alkaline lavas (the Pintades unit) in fact do not belong to the Piton de la Fournaise series but rather should be associated with the activity of the older Les Alizés volcano which forms the basement of the Piton de la Fournaise (Lénat et al. 2012).

The 4th lava group, the Abnormal Group (AbG) is compositionally different from the 3 others. It comprises lavas with MgO = 8–11 wt% and CaO/Al_2O_3 = 0.60–0.80, and shows enrichments in Fe, Mg, Ti, Na, K and a depletion in Ca relative to SSB basalts. AbG-type compositions are rarely observed inside the Enclos Fouqué caldera (e.g., Hudson eruption, early 1998 activity) being mostly found at eccentric vents or at recent vents west of the central cone (Lénat et al. 2012).

For example, the basaltic cones outside the Enclos Fouqué (Piton Chisny, Piton Haüy) have

Table 10.1 Representative major and trace element compositions of Piton de la Fournaise and Piton des Neiges magmas

	Piton de la Fournaise				Piton des Neiges				Parental magma
	SSB	Picrite	AbG	Diff.	SSB	Picrite	AbG	Diff.	
SiO_2 (wt %)	49.37	44.13	46.70	52.91	48.55	45.03	44.97	58.74	48.39
Al_2O_3	14.40	7.47	13.83	15.99	13.18	9.35	12.57	18.62	12.06
FeO	–	13.36	–	10.10	–	12.13	–	–	1.83
Fe_2O_3	11.98	–	13.20	–	12.72	–	12.84	6.71	10.09
MnO	0.17	0.19	0.18	0.19	0.18	0.17	0.17	0.17	0.19
MgO	6.79	25.30	9.14	3.04	9.74	20.28	11.89	1.40	12.52
CaO	11.39	6.10	9.05	6.41	10.91	7.43	9.44	3.90	9.72
Na_2O	2.71	1.35	2.90	5.04	2.44	1.62	2.61	5.97	2.26
K_2O	0.71	0.38	0.91	2.36	0.62	0.48	0.68	3.14	0.55
TiO_2	2.64	1.36	2.68	2.50	2.35	1.77	2.32	1.02	2.33
P_2O_5	0.32	0.15	0.39	0.92	0.28	0.29	0.25	0.39	0.06
L.O.I.	*−0.79*	–	*−0.39*	–	–	*−0.37*	–	*−0.44*	–
Sum	99.69	99.79	98.59	99.46	100.97	98.92	97.74	99.62	100.00
Ba (ppm)	119.10	69.50	166.10	430.20	144.00	74.00	170.00	648.00	102.00
Be	1.07		1.29						
Cd	0.16		0.14						
Ce	39.78	23.10	50.41	125.90	36.00	20.30	27.00	125.00	31.70
Co	39.68		50.93			76.00			59.00
Cr	230.60	1712.00	411.00	0.00	472.00		721.00	7.00	691.00
Cs	0.24	0.13	0.33	0.36		0.19			
Cu	125.40	66.30	73.38	11.40	98.00		75.00	8.00	
Dy	5.16	2.96	5.38	10.88	5.20	1.50			5.23
Er	2.47	1.36	2.50	5.48	2.60				2.66
Eu	2.08	1.05	2.27	4.15	1.90	1.63			1.64
Ga	21.41		20.71						
Gd	6.14	3.36	6.53	12.67	5.80				5.96
Ge	1.25		1.16						
Hf	4.40	2.55	5.10	11.72		3.00			4.24
Ho	0.94	0.54	0.96	2.03					0.93
La	17.17	9.90	22.19	56.30	18.00	12.99	14.00	50.00	14.20
Lu	0.31	0.17	0.31	0.63	0.31	0.18			0.29
Mo	1.13		1.38						
Nb	15.81	13.20	21.01	72.70	20.00	18.00	19.00	68.00	
Nd	23.99	13.50	28.76	65.10	20.00		17.00	45.00	18.80
Ni	91.93	1012.80	213.20	2.10	244.00	900.00	376.00	5.00	404.00
Pb	1.56	1.15	2.71	4.85					
Pr	5.22	2.96	6.40	15.71					4.64

(continued)

Table 10.1 (continued)

	Piton de la Fournaise				Piton des Neiges				Parental magma
	SSB	Picrite	AbG	Diff.	SSB	Picrite	AbG	Diff.	
Rb	16.00	9.50	22.08	59.70	14.00	13.00	19.00	100.00	12.80
Sc	33.42	19.60	26.05	13.20	35.00	21.00	29.00	3.00	27.00
Sm	5.96	3.07	6.62	13.29	5.60	1.10			5.38
Sn	3.84		1.81						
Sr	320.50	185.00	367.50	448.00	324.00	247.00	309.00	531.00	324.00
Ta	1.44		1.94			1.13			1.90
Tb	0.93	0.51	0.97	1.92		0.61			0.92
Th	1.82	0.96	2.40	7.50	3.00	1.44	3.00	10.00	1.62
Tm	0.33		0.34						
U	0.47	0.32	0.60	1.53		0.32			0.44
V	280.60	180.90	218.70	95.30	321.00		272.00	0.00	
W	0.32		0.40						
Y	25.80	14.60	26.52	57.80	26.00	20.00	25.00	28.00	
Yb	2.10	1.10	2.13	4.36	2.10	1.47			2.24
Zn	105.90	107.10	119.50	128.30	100.00		100.00	89.00	
Zr	154.60	99.50	183.10	493.10	168.00	129.00	145.00	543.00	
References	Brugier (2012)	Albarède et al. (1997)	Brugier (2012)	Albarède et al. (1997)	Fisk et al. (1988)	Sobolev and Nikogosian (1994)	Fisk et al. (1988)	Fisk et al. (1988)	Villemant et al. 2009

Data from Albarède et al. (1997), Brugier (2012), Fisk et al. (1988), Sobolev and Nikogosian (1994), Villemant et al. (2009)

whole-rock compositions typical of the AbG group, with relatively low CaO/Al_2O_3 at high MgO (Fig. 10.1). The AbG group has mixed geochemical features, some (e.g., high MgO) suggesting a primitive character and others (e.g., low CaO/Al_2O_3, high K_2O) indicating a differentiated evolution similar to that shown by the Differentiated Lavas group (Fig. 10.1). Nevertheless, it seems clear from Fig. 10.1 that the AbG group of lavas does not evolve directly from SSB-like parental liquids.

Piton de la Fournaise basalts range from tholeiitic to alkaline. The He, Nd and Sr isotopic homogeneity of La Réunion lavas has been emphasized (e.g., Fisk et al. 1988; Graham et al. 1990). More recent studies have deciphered small, but systematic, changes in trace element abundances and Pb, Sr, Nd, Hf isotope ratios (Albarède et al. 1997; Vlastélic et al. 2005, 2007, 2009; Bosch et al. 2008; Pietruszka et al. 2009). This geochemical variability has been interpreted to reflect melting of different source components intrinsic to La Réunion mantle plume, and superimposed crustal processes including magma contamination and fractionation. Clinopyroxene (+olivine) fractionation ("wehrlitic trend") has been proposed to generate AbG compositions (Kornprobst et al. 1979, 1984) and to drive tholeiitic basalts in the field of alkaline basalts (Albarède et al. 1997). Evidence for an important petrogenetic role of clinopyroxene is globally clear from both major and trace elements variations (Figs. 10.1 and 10.2). However, clinopyroxene fractionation is commonly cryptic. For example, the early lavas from Piton Chisny bear a strong geochemical imprint of clinopyroxene

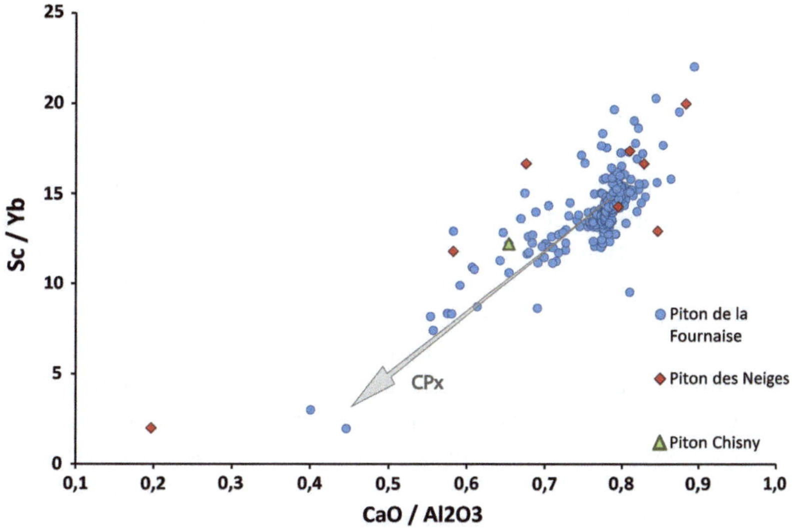

Fig. 10.2 Sc/Yb versus CaO/Al_2O_3 plot emphasizing the importance of clinopyroxene fractionation. Data from Albarède et al. (1997), Brugier et al. (2015), Fisk et al. (1988), Peltier (2007), Salaün et al. (2010), Sobolev and Nikogosian (1994)

fractionation (low Ca and Sc/Yb, Figs. 10.1 and 10.2), yet they contain olivine as the sole silicate phenocryst phase, clinopyroxene being found only in the ultramafic cumulate nodules (Upton and Wadsworth 1972). The genesis of more differentiated lavas such as those belonging to the prehistoric activity (Rempart series) has been modeled by crystal fractionation of a gabbroic assemblage ("gabbroic trend") dominated by clinopyroxene (30–70 %), plagioclase (20–40 %) and minor olivine (10 %, Albarède et al. 1997).

10.1.2 Magma Components

Olivine, clinopyroxene, plagioclase and Cr-spinel are found systematically in Piton de la Fournaise products. Ti-rich magnetite is also present, together with sulfides. Ilmenite and amphibole are rare (Albarède et al. 1997). In general, the four main crystalline phases are found in a variety of textural types, occurring as phenocrysts, microphenocrysts and microlites (Salaün et al. 2010; Welsch et al. 2009, 2013). However, the characteristics of the mineral phase assemblage are specific of the lava group.

SSB lavas host olivine, clinopyroxene, plagioclase and spinel (the latter mainly as inclusions in olivine) in variable but generally low proportions. For example, phenocryst contents range from 0 to 10 vol.% in 1985 lavas (Boivin and Bachèlery 2009, see also Salaün et al. 2010 for 1998 products). Textural evidence suggests that, in SSB, clinopyroxene and plagioclase crystallize together with olivine, this lava group being often designated as "cotectic". The olivine + clinopyroxene + plagioclase + spinel four phase assemblage is also typical of the AbG Group. Compared to SSB lavas erupted contemporaneously with AbG lavas during 1998, the latter (Hudson vent) are richer in olivine and poorer in clinopyroxene and plagioclase than the former (Kapor vent, see Salaün et al. 2010). Early AbG lavas from Piton Chisny are crystal-poor (<2 vol.%) and have only phenocrysts of olivine and Cr-spinel. Oceanite lavas from the Picrite group contain macrocrysts of olivine, mesocrysts of either olivine or an assemblage of clinopyroxene and plagioclase, and microcrysts of olivine + clinopyroxene + plagioclase + spinel ± Ti-magnetite (Welsch et al. 2009). The proportion of olivine macrocrysts in oceanite can reach ~50 vol.% (e.g., Salaün et al. 2010). Piton de la Fournaise basalts generally contain small amounts of plagioclase (but see Boivin and Bachèlery 2009 and Salaün et al. 2010 for exceptions). However, plagioclase is the dominant phase in the Differentiated Lava group from Piton des Neiges and early Piton de la

Fournaise (feldspar phyric hawaiites, mugearites and benmoreites, Upton and Wadsworth 1966; Albarède et al. 1997).

Olivine compositions at Piton de la Fournaise are rarely highly forsteritic. Histograms of Fo contents yield frequency maxima at 86 (early activity, Albarède et al. 1997), 81 and 84 (1977–1998 activity, Boivin and Bachèlery 2009), 81 and 84 (1998 activity, Villemant et al. 2009), 84 (2005 oceanite, Welsch et al. 2009), 85–86 (Chisny lavas, Brugier 2012) and 83 and 88 for both the Chisny ultramafic cumulate nodules (Brugier et al. 2015) and the dunite nodules analyzed by Sobolev and Nikogosian (1994). Bureau et al. (1999) found Fo contents as high as 89 in AbG lavas erupted from the Hudson vent, in agreement with the Fo88 compositions found by Salaün et al. (2010). This supports 89 as the maximum Fo content for Piton de la Fournaise olivines (Fisk et al. 1988; Albarède et al. 1997). However, the dataset of Sobolev and Nikogosian (1994) contains more forsteritic olivine compositions, up to Fo90.6 in an oceanite. Fretzdorff and Haase (2002) found olivines with Fo contents up to 90.6 in a suite of picritic submarine lavas dredged in the northeastern flank of the Réunion island. Normal core-rim zonation is commonly observed in olivine crystals, for example from Fo84 cores to Fo80 rims in oceanite (Welsch et al. 2009, see also Salaün et al. 2010). Olivine microphenocrysts and microlites have Fo contents lower than phenocrysts (or macrocrysts, Welsch et al. 2009). Texturally, the occurrence of deformed olivines has led to the proposition that many crystals at Piton de la Fournaise are xenocrysts derived from the mechanical disaggregation of earlier cumulates (e.g., Albarède et al. 1997). However, Welsch et al. (2013) have recently shown that the olivine macrocrysts from oceanite mostly originate from dendritic growth and ripening in suspension within the melt; since very few olivine macrocrysts have experienced intracrystalline deformation, they cannot be considered as xenocrysts.

Clinopyroxenecompositions are mostly augite, some extending into the fields of diopside and salite in the En-Fs-Wo diagram (Boivin and Bachèlery 2009; Welsch et al. 2009). Boivin and Bachèlery (2009) distinguished two pyroxene compositional groups, one with an average composition En49Fs9Wo42 (Mg-augitic) and the other En48Fs12Wo40 (augitic). They suggested that the former group, with higher Al/Ti ratios, crystallized at higher pressures than the latter. Fisk et al. (1988) mentioned the occurrence of aluminous augite crystals in lavas carrying Fo89 olivine. In the same way, Fretzdorff and Haase (2002) described Al-rich augites coexisting with Fo-rich olivines in the picritic submarine lavas. Clinopyroxene Mg/(Mg + Fe_t) is rather homogeneous, either for a given period of activity, within a single eruption, or between magma groups; core to rim variations are generally not very strong. However, clinopyroxene microlites may show more evolved (i.e., less calcic) compositions than phenocrysts (Boivin and Bachèlery 2009; Salaün et al. 2010).

Plagioclase in recent (1977–1998) Piton de la Fournaise products ranges from An45 to An80 (Boivin and Bachèlery 2009). Phenocrysts and microlites are only slightly compositionally different: the former have a frequency maximum at An70 (total range: An58–80) and the latter at An65 (total range: An45–76). The 1998 products have similar characteristics (Salaün et al. 2010). In contrast, plagioclase compositions were more calcic during the early activity of the volcano (frequency maximum at An84 in the Rempart series, Albarède et al. 1997).

Spinels (both inclusions in olivine and isolated crystals dispersed in the matrix) are mostly chromites with Cr/(Cr + Al) around 70 and Mg/(Mg + Fe_t) between 50 and 65. They have TiO_2 contents between 2 and 3 wt%, increasing with decreasing Mg/(Mg + Fe_t), and Fe^{3+}/(Fe^{3+}+ Al + Cr) of 10–15 % (Brugier 2012). Spinel compositions at Piton de la Fournaise do not show systematic variations with magma types (Sobolev and Nikogosian 1994; Albarède et al. 1997; Welsch et al. 2009; Brugier 2012). However, lavas from Piton Chisny contain unusual Mg-rich (Mg/(Mg + Fe_t) = 60–70) and Cr-poor (Cr/(Cr + Al) = 30) crystals, besides the normal compositions (Brugier 2012).

Data on matrix glasses concern the SSB and Picrite groups and, to a lesser extent, the AbG group. One difficulty resides in the microcrystallinity of the samples, especially those having cooled in air, because groundmass crystallization limits the size of the glass pools amenable to electron microprobe analysis. This has motivated the use of unconventional rapidly quenched glassy materials such as Pélé's hairs and tears to document changes in matrix glass compositions (Villemant et al. 2009). In first approximation, the composition of the matrix glass in most Piton de la Fournaise products is similar to SSB bulk rocks (MgO = 5.5–6 wt%, CaO/Al_2O_3 = 0.79–0.82), i.e., it is in equilibrium with olivines in the range Fo75–80 (Welsch et al. 2009). However, systematic compositional variations have been recently discovered. Matrix glasses with MgO contents as high as 8–9 wt% have been analyzed in 2007 products (Villemant et al. 2009). The 1998–2007 matrix glasses define a unique evolution trend starting from these high-MgO glasses and ending at ~5.5 wt% MgO, which corresponds to the evolved Hudson and Kapor matrix glasses (Villemant et al. 2009; Salaün et al. 2010). This evolution mimics that shown by whole rocks during the same period, if olivine-rich oceanites and AbG lavas are excepted. For more differentiated rocks, interstitial glasses from the Bellecombe cumulate xenoliths (Upton et al. 2000) and aphyric magmas from the early Piton de la Fournaise (Albarède et al. 1997) pursue the 1998–2007 evolution trend toward lower MgO (Salaün et al. 2010).

Melt inclusions are common in olivine crystals from Piton de la Fournaise. Primary inclusions are either glassy (ejected lapilli) or recrystallized (lava samples). Secondary inclusions are mainly fluid inclusions. These occur along healed fractures in olivine and are associated with glass and crystals in variable proportions (Sobolev and Nikogosian 1994; Bureau et al. 1998a; Famin et al. 2009). Compositions of primary glassy inclusions (recalculated for post-entrapment crystallization of olivine) range from 8 to 10 wt% MgO (Bureau et al. 1998a), 8–11.5 wt% MgO (Famin et al. 2009) to up to ~14 wt% MgO (Sobolev and Nikogosian 1994). Less magnesian compositions (5.5–7 wt% MgO) are reported by Villemant et al. (2009) for inclusions in 1998 and 2007 products. In 2005 oceanite, Welsch et al. (2009) emphasized the large variability of glass inclusion compositions in terms of K_2O, P_2O_5, K_2O/TiO_2 and CaO/Al_2O_3, especially when compared to the tightly grouped matrix glass compositions.

H_2O and CO_2 concentrations in glass inclusions are generally <1 wt% and <500 ppm (Bureau et al. 1998b; Famin et al. 2009). However, certain samples (Piton de Caille, Hudson) have H_2O and CO_2 concentrations largely exceeding these values, with H_2O up to ~1.6 wt % and CO_2 up to ~2800 ppm (Bureau et al. 1998b; 1999). The secondary fluid inclusions contain essentially pure CO_2 with a large range of densities, from 0 to 0.8 g/cm^3 (Bureau et al. 1998a; Famin et al. 2009).

There seems to exist overall agreement on the order of crystallization of silicate phases in Piton de la Fournaise magmas, excluding rocks from the Differentiated series. Albarède et al. (1997) suggested progressive crystallization in the order olivine-clinopyroxene-plagioclase. This sequence is consistent with data on the Chisny ultramafic cumulate nodules (Upton and Wadsworth 1972). These are dominated by dunites made of olivine and Cr-spinels; wehrlitic nodules are less abundant. In these, clinopyroxene is present in low amounts and the Fo content of olivine is lower than in dunitic nodules (83 vs. 87, Brugier et al. 2015). A minor proportion of gabbroic nodules also exists (Bachèlery, personal communication, 2013). Overall, these observations leave little doubt on the order of crystallization of the Chisny magmas. The fact that clinopyroxene precedes plagioclase in the crystallization sequence of Piton de la Fournaise basalts is also supported by recent textural studies and numerical runs using the MELTS software (Famin et al. 2009; Welsch et al. 2009, 2013).

10.2 Structure of the Magma Feeding System and Magmatic Conditions

10.2.1 Combined Geophysical and Petrological Constraints on the Structure of the Feeding System

The dense monitoring network on La Réunion island (OVPF/IPGP), coupled with the high eruption frequency, allows magma ascent at Piton de la Fournaise to be followed in real time. This provides constraints on the overall structure of the feeding system and three zones of magma storage are presently defined (see also the description in Michon et al. 2016, Chap. 7). For the *shallow* magma storage zone, models essentially divide into two groups. Lénat and Bachèlery (1990) suggested the presence of an array of sills and dykes between 0.5 and 1.5 km beneath the Dolomieu crater whereas Nercessian et al. (1996) and Peltier et al. (2007) favoured the existence of a single reservoir of about 0.3 km^3 at about 2.5 km depth. Whereas the structure of the magma chamber is still debated, it is worth stressing that seismic and deformation constraints essentially agree with petrological data on the depth of shallow magma storage. Most melt-fluid saturation pressures (calculated from H_2O and CO_2 concentrations in glass inclusions) are <50 MPa ($< \sim 1.9$ km for a crustal density of 2.7 g/cm^3, Bureau et al. 1998b; Famin et al. 2009). For the fluid inclusions, entrapment pressures are either of a maximum of ~ 70 MPa (2.6 km, Famin et al. 2009) or <95 MPa (~ 3.6 km, Bureau et al. 1998b). The seismicity that preceeded the 1998 eruption, and also during 2005 and 2007, provide evidence that the magmas stored near the surface have ascended from a second reservoir located at *intermediate* crustal depths (Battaglia et al. 2005; Peltier et al. 2009). The root of the 1998 seismic swarms is located at about 200 MPa (7.6 km depth), which corresponds to the top of the oceanic crust beneath La Réunion island (Gallard et al. 1999). At greater depths, the intermediate velocity layer described in the southwestern part of La Réunion has been interpreted as underplating of the base of the crust by mantle magmas, followed by their crystallization and differentiation (Gallard et al. 1999). This layer has thus all attributes of a *deep* magma storage zone. It is located at 320–400 MPa (12–15 km depth), a pressure range consistent with melt-fluid saturation pressures computed for the Piton de Caille (190–420 MPa, Bureau et al. 1998b) and Hudson (<420 MPa, Bureau et al. 1999) glass inclusions. The CO_2-rich inclusions from the Piton de Caille and Chisny ultramafic nodules also yield high pressures, up to 500 and 350 MPa respectively (Bureau et al. 1998a).

10.2.2 Magma Temperature

Information on temperatures of Piton de la Fournaise magmas comes from different methods. First, optical pyrometry and thermocouple measurements in the field provide direct data on lava temperatures, from 872 °C and up to 1170 °C with a frequency maximum between 1100 and 1150 °C. The pyrometry is systematically higher than the thermocouple data because active parts of lava flows can be measured with that method (Boivin and Bachèlery 2009). Second, magma temperatures have been computed from olivine-liquid equilibrium. When applied to olivine phenocryst cores and their host bulk lava, grouped values are obtained, close to 1200 °C for the 1977–1998 products (Boivin and Bachèlery 2009). As a variant of this method, the MgO content of the melt (from the analysis of glass inclusions) at equilibrium with olivine yields temperatures (e.g., Helz and Thornber 1987) between 1170 and 1215 °C (Bureau et al. 1998a, b) and between 1192 and 1236 °C (Famin et al. 2009). It should be noted however that results with methods based on glass inclusion compositions depend on procedures adopted to correct for post-entrapment olivine crystallization. Third, thermometric measurements on melt inclusions have yielded homogeneization temperatures from

1170 to about 1215 °C (Bureau et al. 1998a, b). A markedly higher temperature range (1235–1345 °C) has been reported by Sobolev and Nikogosian (1994).

Except the field measurements which apparently give minimum values, the other methods define a consistent temperature range for the Piton de la Fournaise magmas, broadly centred on 1200 °C. The data of Sobolev and Nikogosian (1994) reach very high temperatures, but are consistent with their proposal of the existence of picritic (up to ~14 wt% MgO) melt inclusions in Piton de la Fournaise olivines. It is emphasized that the available temperature estimates concern mainly the SSB and Picrite lava groups. This 1200 °C value should best be viewed as an "average" or "steady-state" temperature for PTF lavas since magma temperatures would be expected to vary along with differentiation and depth of the tapped magma storage zone.

10.2.3 Oxygen Fugacity

Oxygen fugacities between NNO+0.7 to NNO-1.8 have been computed from the olivine-liquid equilibrium (Boivin and Bachèlery 2009). Most values are however <NNO, and so Piton de la Fournaise magmas evolve under moderately reducing conditions (Fig. 10.3). Similar results (from NNO+0.3 to NNO-1.8) were obtained (Brugier et al. 2015) from the analysis of FeO and Fe_2O_3 in 5 bulk rocks (2 SSB and 3 AbG), using the calibration of Kress and Carmichael (1991). Slightly more oxidizing log fO_2 values (from NNO+1 to NNO-0.5) have been obtained (Brugier et al. 2015) by taking the Fe2+/Fe3+ of spinels to calculate the FeO/Fe_2O_3 of magmatic liquids (Maurel and Maurel 1982; Danyuschevsky and Sobolev 1994; Kress and Carmichael 1991). In the same way, log fO_2 between NNO-0.3 and NNO-0.9 are reported by Bureau et al. (1998b) from spinel-melt equilibria calculated with the expression of Maurel and Maurel (1984). Finally, oxygen fugacities from NNO-0.4 to NNO-0.8 (Brugier et al. 2015) were calculated from the olivine-spinel oxybarometric expression of Ballhaus et al. (1991).

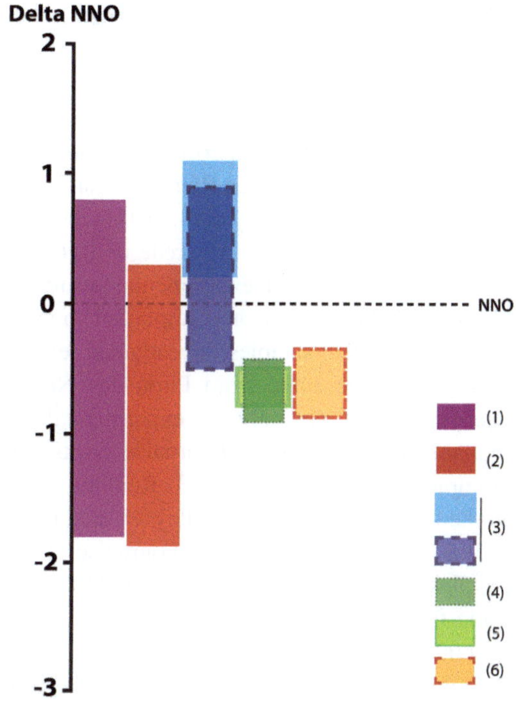

Fig. 10.3 Compilation of log fO_2 data (shown as deviations from the NNO buffer) obtained with different methods emphasizing the moderately reducing redox state of Piton de la Fournaise magmas. (1) olivine-liquid equilibrium (Boivin and Bachèlery 2009); (2) bulk rock analyses of FeO and Fe_2O_3 (Brugier et al. 2015); (3) calculated FeO and Fe_2O_3 of magmatic liquids based on Kress and Carmichael (1991) for two compositional groups of spinels (Brugier et al. 2015); (4) spinel-melt equilibria (Bureau et al. 1998b); (5) olivine-spinel oxybarometry (Brugier et al. 2015); (6) log fO_2 results for mantle source region obtained with V/Sc ratios in Piton de la Fournaise magmas, based on Lee et al. (2005) (Brugier et al. 2015). See also text

We conclude from these data that the fO_2 range of the Piton de la Fournaise magmas is reasonably well defined, around NNO-0.5. This makes the average Fe_2O_3/FeO of basaltic liquids at Piton de la Fournaise slightly higher than commonly assumed (the proportion of Fe_2O_3 is 18 % and not 15 %, e.g., Fisk et al. 1988).

10.3 Experimental Constraints

10.3.1 General

There are only a few experimental studies on magmatic products from La Réunion island. This includes the work of Tilley et al. (1971) and Fisk et al. (1988). In addition, Albarède et al. (1997) mentioned unpublished experimental results by Mesnil and Baker. First, it is important to stress that most of the samples investigated experimentally to date come from Piton des Neiges. There are in fact relatively few experimental data available on Piton de la Fournaise rocks. For example, on the total of 13 Réunion lavas worked experimentally by Tilley et al. (1971), only 6 come from Piton de la Fournaise. The 6 samples investigated by Fisk et al. (1988) all come from Piton des Neiges. Second, no experimental information is as yet available on La Réunion magmas in presence of volatile components, either H_2O or CO_2.

10.3.2 Results

Tilley et al. (1971) performed melting experiments under unbuffered redox conditions at 1 atm., either in argon or in air above 1230 °C. Liquidus temperatures ranged from 1522 °C (oceanite, about 27 wt% MgO) down to 1185 °C (hawaiite, 4.8 wt% MgO), decreasing progressively with decreasing bulk rock MgO. For rocks with bulk MgO higher than ~7 wt%, the order of crystallization is olivine, plagioclase and clinopyroxene. Olivine and plagioclase crystallized together on the liquidus of a 5 wt% MgO basalt whereas, for the hawaiite sample investigated, the liquidus phase was plagioclase.

Fisk et al. (1988) basically extended the work of Tilley et al. (1971). Some compositions used previously by Tilley et al. (1971) were re-investigated at 1 atm.; high pressure experiments were conducted mostly at 250–1000 MPa in gas vessel, and a few at 1500 and 2000 MPa in piston-cylinder. No volatile components were added to the starting materials. Redox conditions were mostly close to the iron-wustite buffer (Fisk et al. 1988), i.e., several log units more reducing than Piton de la Fournaise magmas. However, some 1 atm. experiments were also conducted closer to the natural fO_2 range. Although 6 rocks from 8.8 to nearly 27 wt% MgO were studied, the data mostly bear on 4 compositions that cover a more restricted range (8–13 wt% MgO). A high degree of internal consistency appears in results for the different samples and several general conclusions can be drawn (Fig. 10.4; see also Albarède et al. 1997, Fig. 21).

1. As expected, liquidus temperatures increase both with pressure (below 1000 MPa the gradient is about 4 °C/100 MPa) and, at a given pressure, with the MgO content of the starting material. For example, the 8 wt% MgO rock has liquidus temperatures from 1240 (1 atm.) to 1280 °C (1000 MPa) and the 13 wt% MgO rock from 1330 to 1370 °C.
2. The liquidus phase is always olivine at low pressures. Upon increasing pressure, it is replaced on the liquidus by clinopyroxene, this transition occurring at 800–1000 MPa for compositions with bulk MgO < 11 wt%, and at ~1800 MPa for the composition with 13 wt% MgO.

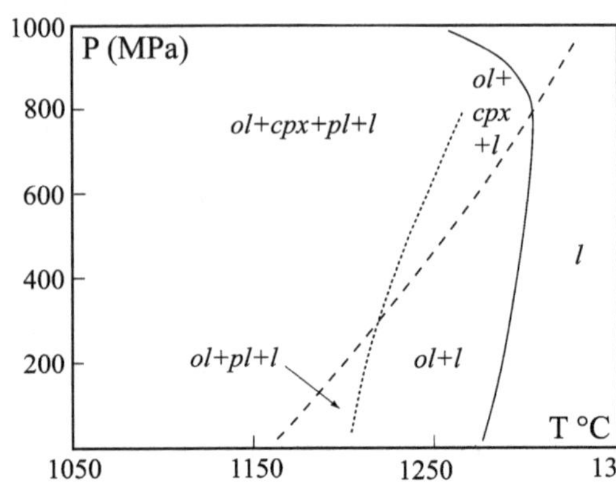

Fig. 10.4 Summary phase diagram for Piton de la Fournaise magmas based on the experimental data of Fisk et al. (1988). The diagram is constructed from experimental data mainly on a 11 wt% MgO oceanite from Piton des Neiges. *L* liquid; *ol* olivine; *cpx* clinopyroxene; *pl* plagioclase

3. The order of crystallization is (a) olivine-clinopyroxene-plagioclase below 800–1000 MPa (or below 1800 MPa for the 13 wt% MgO composition), and (b) olivine-plagioclase-clinopyroxene below 250–500 MPa, the latter sequence being in agreement with the experiments of Tilley et al. (1971).
4. Spinel is absent from most runs, being restricted to the few 1 atm. experiments performed at fO_2 around the natural range.
5. Orthopyroxene never occurs as an experimental phase.

Although compositions of experimental phases were not detailed, Fisk et al. (1988) give Fo, Mg/(Mg + Fe_t) and An contents of experimental olivine, clinopyroxene and plagioclase, respectively. Fo88 is the maximum olivine Fo content, consistent with the proposition that Piton de la Fournaise olivines have Fo < 89 (see above). However, the 27 wt% MgO composition was not studied at near-liquidus temperatures where it would have presumably crystallized olivines with Fo > 88. Clinopyroxenes have Mg/(Mg + Fe_t) equal or higher than olivine. The investigated sample with 11 wt% MgO (which is quite Ca-rich, CaO ~13 wt%) yields clinopyroxene Mg/(Mg + Fe_t) and plagioclase An contents up to 86 and 90 respectively (the highest of the experimental database), for a maximum olivine Fo content of 86.

10.3.3 Application to Piton de la Fournaise Magmas

Experiments by Fisk et al. (1988) have provided a framework for interpreting the genesis and evolution of Piton de la Fournaise magmas. In particular, these experimental data have been used to define pressures of magma fractionation (Albarède et al. 1997). The Fisk et al. (1988) data imply minimum pressures of 250–500 MPa for clinopyroxene to crystallize before plagioclase. Fractionation of olivine together with clinopyroxene would also require a minimum of 250–500 MPa from their data. Still higher crystallization pressures (a minimum of 800–1000 MPa for compositions <11 wt% MgO) would be necessary for fractionation of clinopyroxene alone.

Assuming that all Piton de la Fournaise primitive magmas crystallize in the sequence olivine-clinopyroxene-plagioclase (see above and Albarède et al. 1997; Welsch et al. 2009), then the Fisk et al. (1988) experimental results would lead to the conclusion that magma fractionation takes place for the most part at the base of the crust or in the mantle. The crust-mantle boundary beneath La Réunion is located at 12–15 km depths (Gallard et al. 1999; Lénat et al. 2012) or 320–400 MPa. Thus, the 250–500 MPa minimum pressures required by the Fisk et al. (1988) data makes it possible that crystallization of either olivine + clinopyroxene or olivine + clinopyroxene + plagioclase takes place at the base of the crust. Fractionation of a clinopyroxene-rich, olivine-poor assemblage was proposed by Albarède et al. (1997) to occur at mantle pressures (800–1000 MPa), consistent with the pressure (900 MPa) determined for the crystallization of aluminous augites (Fretzdorff and Haase 2002). Therefore, the Fisk et al. (1988) data implies relatively deep magma fractionation, in a pressure range difficult to reconcile with the presence of shallow crustal storage zones beneath Piton de la Fournaise (see above). It is worth noting that both melt and fluid inclusions are consistent with fluid-melt saturation and fluid entrapment at crustal pressures, and that petrological evidence for crystallization pressures above 400–500 MPa (Bureau et al. 1998b, 1999) is presently lacking.

These difficulties lead to an alternative interpretation of the Fisk et al. (1988) data, reminding that their experiments were performed under volatile-free conditions (see above). Recently, numerical runs using the MELTS software (Famin et al. 2009; Welsch et al. 2009) have been performed for a parental Piton de la Fournaise magma composition (see below), in presence of 1 wt% H_2O and 0.1 wt% CO_2, and for a fO_2 set at the FMQ buffer (about 0.7 log unit below NNO). They show that clinopyroxene crystallizes second after olivine at 100 MPa

(Welsch et al. 2009), i.e., at pressures much lower than the 250–500 MPa minimum pressure range found by Fisk et al. (1988).

These results open the possibility that magma fractionation at Piton de la Fournaise takes place at levels much shallower than previously thought. We conclude that there is a need for experiments on Piton de la Fournaise magmas in presence of volatile components, H_2O and CO_2. Determination of (1) the pressure of the transition from clinopyroxene to olivine as the liquidus phase and (2) the lower pressure limit for clinopyroxene to crystallize before plagioclase, both in presence of H_2O and CO_2, represent critical experimental objectives for constraining the structure of the feeding system at Piton de la Fournaise.

The lack of orthopyroxene (Fisk et al. 1988) implies that none of the rocks experimentally studied is saturated on its liquidus with an olivine-orthopyroxene phase assemblage. Therefore, none can be interpreted as a melt directly produced from either a peridotitic or harzburgitic source. It follows that Piton de la Fournaise melts must have suffered differentiation following extraction from their mantle source. On this basis, several studies have attempted to reconstruct the composition of Réunion parental magmas (Ludden 1978; Fisk et al. 1988; Sobolev and Nikogosian 1994; Fretzdorff and Haase 2002; Villemant et al. 2009; Dasgupta et al. 2010). In detail, results differ between studies, with substantial dispersions especially for SiO_2 (45–48.5 wt%), Al_2O_3 (10–12.5 wt%), MgO (12.5–20 wt%) and CaO (8–10 wt%). One representative composition of parental magma is shown in Table 10.1. The observation that the evolutionary trend for the AbG group intersects the olivine control line in Fig. 10.1 at about 11 wt% MgO indicates that 11 wt% MgO is actually a minimum for the parental magmas. Taking 89 as the maximum Fo in olivine (see above), an olivine-melt Fe–Mg partition coefficient of 0.306 (Fisk et al. 1988), an average total FeO of 11.74 wt% (Fisk et al. 1988) and a $Fe_2O_3/(Fe_2O_3 + FeO)$ proportion of 18 % (corresponding to NNO-0.5, see above), parental melts must have a maximum of 13.7 wt% MgO. Higher melt MgO concentrations are obtained by taking more Fo-rich olivine compositions (15–20 wt% MgO, Sobolev and Nikogosian 1994; 17 wt% MgO, Fretzdorff and Haase 2002). Therefore, there is agreement on the picritic nature of the Réunion parental melts. However, all reconstructions so far have assumed fractionation of only olivine at depth. This assumption may be correct if differentiation takes place for pressures and melt compositions where olivine is the sole liquidus phase. According to Fisk et al. (1988), this occurs at <800–900 MPa for melt MgO contents <11 wt%, and at <1800 MPa for 13 wt% MgO. However, clinopyroxene may precede/join olivine on the liquidus. Wehrlitic differentiation, either at crustal (Kornprobst et al. 1979, 1984) or mantle (Albarède et al. 1997; Fretzdorff and Haase 2002) depths, necessarily involves clinopyroxene fractionation. In this case, the reconstruction of parental Piton de la Fournaise magmas becomes more difficult because the composition of clinopyroxene must be known for reverse fractionation calculations to be reliably performed.

Acknowledgements The authors thank Didier Laporte for his constructive review. The project about the Piton de la Fournaise at Orléans is supported by the ANR DEGAZMAG. Ida Di Carlo and Michel Fialin provided assistance with electron microprobe analyses at Orléans and Paris.

References

Albarède F, Luais B, Fitton G, Semet M, Kaminski E, Upton BGJ, Bachèlery P, Cheminee JL (1997) The geochemical regimes of Piton de la Fournaise volcano (Réunion) during the last 530,000 years. J Petrol 38:171–201

Ballhaus C, Berry RF, Green DH (1991) High pressure experimental calibration of the olivine-orthopyroxene-spinel oxygen barometer: implications for the oxidation state of the upper mantle. Contrib Mineral Petrol 107:27–40

Battaglia J, Ferrazzini V, Staudacher T, Aki K, Cheminée JL (2005) Pre-eruptive migration of earthquakes at the Piton de la Fournaise Volcano (Réunion Island). Geophys J Int 161:549–558

Boivin P, Bachèlery P (2009) Petrology of 1977 to 1998 eruptions of Piton de la Fournaise, La Réunion island. J Volcanol Geotherm Res 184:109–125

Bosch D, Blichert-Toft J, Moynier F, Nelson BK, Télouk P, Gillot P-Y, Albarède F (2008) Pb, Hf and Nd isotope compositions of the two Réunion volcanoes

(Indian Ocean): a tale of two small-scale mantle blobs. Earth Planet Sci Lett 265:748–768

Brugier Y (2012) Pétrologie naturelle et expérimentale de laves du Piton de la Fournaise, La Réunion. Msc thesis, Université d'Orléans, 76 p

Brugier Y-A. Pichavant M. Di Muro A. (2015) Experimental determination of phase equilibria of a basalt from Piton de la Fournaise (La Réunion island): 1atm data and high pressure results in presence of volatiles. European Geosciences Union, General Assembly 2015

Bureau H, Métrich N, Pineau F, Semet M (1998a) Magma-conduit interaction at Piton de la Fournaise volcano (Réunion island): a melt and fluid inclusion study. J Volcanol Geotherm Res 84:39–60

Bureau H, Pineau F, Métrich N, Semet M, Javoy M (1998b) A melt and fluid inclusion study of the gas phase at Piton de la Fournaise volcano (Réunion island). Chem Geol 147:115–130

Bureau H, Métrich N, Semet M, Staudacher T (1999) Fluid-magma decoupling in a hot-spot volcano. Geophys Res Lett 26:3501–3504

Danyuschevsky LV, Sobolev AV (1994) Ferric-ferrous ratio and oxygen fugacity calculations for primitive mantle-derived melts: calibration of an empirical technique. Mineral Petrol 57:229–241

Dasgupta R, Jackson MG, Lee C-TA (2010) Major element chemistry of ocean island basalts—conditions of melting and heterogeneity of mantle source. Earth Planet Sci Lett 289:377–392

Famin V, Welsch B, Okumura S, Bachèlery P, Nakashima S (2009) Three differentiation stages of a single magma at Piton de la Fournaise volcano (Réunion hot spot). Geochem Geophys Geosyst 10. doi:10.1029/2008GC002015

Fisk MR, Upton BGJ, Ford CE, White WM (1988) Geochemical and experimental study of the genesis of magmas of Réunion island, Indian Ocea. J Geophys Res 93:4933–4950

Fretzdorff S, Haase KM (2002) Geochemistry and petrology of lavas from the submarine flanks of Réunion Island (western Indian Ocean): implications for magma genesis and the mantle source. Mineral Petrol 75:153–184

Gallard J, Driad L, Charvis Ph, Sapin M, Hirn A, Diaz J, de Voogd B, Sachpazi M (1999) Perturbation to the lithosphere along the hotspot track of La Réunion from an offshore-onshore seismic transect. J Geophys Res 104:2895–2908

Graham D, Lupton J, Albarède F, Condomines M (1990) Extreme temporal homogeneity of helium isotopes at Piton de la Fournaise, Réunion Island. Nature 347:545–548

Helz RT, Thornber CR (1987) Geothermometry of Kilauea Iki lava lake, Hawaii. Bull Volcanol 49:651–668

Kornprobst J, Boivin P, Bachèlery P (1979) L'alimentation des éruptions récentes du Piton de la Fournaise (Ile de la Réunion, Océan Indien): degré d'évolution et niveau de ségrégation des laves émises. CR Acad Sci 288:1691–1694

Kornprobst J, Boivin P, Lénat JF, Bachèlery P, Bonneville A, Dupont P, Lecointre J, Seidel JL, Thomas P, Vincent PM (1984) Le Piton de la Fournaise, île de la Réunion. Colloque Prévision et Surveillance des Eruptions Volcaniques. (C.N.R.S.–I.N.A.G.) Clermont-Ferrand, pp 75–82

Kress VC, Carmichael ISE (1991) The compressibility of silicate liquids containing Fe_2O_3 and the effect of composition, temperature, oxygen fugacity and pressure on their redox states. Contrib Mineral Petrol 108:82–92

Lacroix A (1936) Le volcan actif de l'île de la Réunion et ses produits. Gauthier et Villard, 297 p

Lee CTA, Leeman WP, Canil D, Li ZXA (2005) Similar V/Sc systematics in MORB and arc basalts: implications for the oxygen fugacities of their mantle source regions. J Petrol 46:2313–2336

Lénat JF, Bachèlery P (1990) Structure et fonctionnement de la zone centrale du Piton de la Fournaise. In: Lénat JF (ed) Le volcanisme de La Réunion. Centre de Recherches Volcanologiques, Clermont-Ferrand, pp 257–296

Lénat JF, Bachèlery P, Merle O (2012) Anatomy of Piton de la Fournaise volcano (La Réunion, Indian Ocean). Bull Volcanol 74:1945–1961

Ludden J (1978) Magmatic evolution of the basaltic shield volcanoes of Réunion island. J Volcanol Geotherm Res 4:171–198

Maurel C, Maurel P (1982) Etude expérimentale de l'équilibre Fe_2+–Fe_3+ dans les spinelles chromifères et les liquides silicatés basiques coexistant à 1 atm. CR Acad Sci 295:209–212

Maurel C, Maurel P (1984) Etude expérimentale de la distribution du fer ferrique entre spinelle chromifère et liquide silicaté basique. Bull Mineral 107:25–33

Michon L, Ferrazzini V, Di Muro A (2016) Magma paths at Piton de la Fournaise volcano. In: Bachèlery P, Lénat J-F, Di Muro A, Michon L (eds) Active volcanoes of the Southwest Indian Ocean: Piton de la Fournaise and Karthala. Active Volcanoes of the World (Chap. 7). Springer, Berlin, Heidelberg

Nercessian A, Hirn A, Lépine J-C, Sapin M (1996) Internal structure of Piton de la Fournaise volcano from seismic wave propagation and earthquakes distribution. J Volcanol Geotherm Res 70:123–143

Peltier A (2007) Suivi, modélisation et évolution des processus d'injections magmatiques au Piton de La Fournaise. PhD thesis, Université de La Réunion, 365 p

Peltier A, Staudacher T, Bachèlery P (2007) Constraints on magma transfers and structures involved in the 2003 activity at Piton de la Fournaise from displacement data. J Geophys Res 112. doi:10.1029/2006JB004379

Peltier A, Bachèlery P, Staudacher T (2009) Magma transport and storage at Piton de la Fournaise (La Réunion) between 1972 and 2007: a review of geophysical and geochemical data. J Volcanol Geotherm Res 184:93–108

Pietruszka AJ, Hauri EH, Blichert-Toft J (2009) Crustal contamination of mantle-derived magmas within Piton de la Fournaise volcano, Réunion Island. J Petrol 50:611–684

Salaün A, Villemant B, Semet M, Staudacher T (2010) Cannibalism of olivine-rich cumulate xenoliths during the 1998 eruption of Piton de la Fournaise (La Réunion hotspot): implications for the generation of magma diversity. J Volcanol Geotherm Res 198:187–204

Smietana, M (2011) Pétrologie, géochronologie (K–Ar) et géochimie élémentaire et isotopique (Sr, Nd, Hf, Pb) des laves anciennes de La Réunion. Implications sur la construction de l'édifice volcanique. PhD thesis, Université de La Réunion, 362 p

Sobolev AV, Nikogosian IK (1994) Petrology of long-lived plume magmatism: Hawaii, Pacific and Réunion islands, Indian Ocean. Petrology 2:111–144

Tilley CE, Thompson RN, Wadsworth WJ, Upton BGJ (1971) Melting relations of some lavas of Réunion island, Indian Ocean. Mineral Mag 38:344–352

Upton BGJ, Wadsworth WJ (1966) The basalts of Réunion island, Indian Ocean. Bull Volcanol 29:7–23

Upton BGJ, Wadsworth WJ (1972) Peridotitic and gabbroic rocks associated with the shield-forming lavas of Réunion. Contrib Mineral Petrol 35:139–158

Upton BGJ, Semet M, Joron J-L (2000) Cumulate clasts in the Bellecombe Ash Member, Piton de la Fournaise, Réunion Island, and their bearing on cumulative processes in the petrogenesis of the Réunion lavas. J Volcanol Geotherm Res 104:297–318

Vigouroux N, Williams-Jones AE, Wallace P, Staudacher T (2009) The November 2002 eruption of Piton de la Fournaise, Réunion: tracking the pre-eruptive thermal evolution of magma using melt inclusions. Bull Volcanol 71:1077–1089

Villemant B, Salaün A, Staudacher T (2009) Evidence for a homogeneous primary magma at Piton de la Fournaise (La Réunion): a geochemical study of matrix glass, melt inclusions and Pélé's hairs of the 1998–2008 eruptive activity. J Volcanol Geotherm Res 184:79–92

Vlastélic I, Staudacher T, Semet M (2005) Rapid change of lava composition from 1998 to 2002 at Piton de La Fournaise (Réunion) inferred from Pb isotopes and trace elements: evidence for variable crustal contamination. J Petrol 46:79–107

Vlastélic I, Peltier A, Staudacher T (2007) Short-term (1998–2006) fluctuations of Pb isotopes at Piton de la Fournaise volcano (Réunion Island): origins and constrains on the size and shape of the magma reservoir. Chem Geol 244:202–220

Vlastélic I, Deniel C, Bosq C, Télouk P, Boivin P, Bachèlery P, Famin V, Staudacher T (2009) Pb isotope geochemistry of Piton de la Fournaise historical lavas. J Volcanol Geotherm Res 184:63–78

Vlastélic I, Staudacher T, Bachèlery P, Télouk P, Neuville D, Benbakkar M (2011) Lithium isotope fractionation during magma degassing: Constraints from silicic differentiates and natural gas condensates from Piton de la Fournaise volcano (Réunion Island). Chem Geol 284:26–34

Welsch B, Faure F, Bachèlery P, Famin V (2009) Microcrysts record transient convection at Piton de la Fournaise volcano (La Réunion hotspot). J Petrol 50:2287–2305

Welsch B, Faure F, Famin V, Baronnet A, Bachèlery P (2013) Dendritic crystallization: a single process for all the textures of olivine in basalts? J Petrol 54:539–574

11 A Review of the Recent Geochemical Evolution of Piton de la Fournaise Volcano (1927–2010)

Ivan Vlastélic and Aaron J. Pietruszka

Abstract

Between 1927 and 2010, more than one hundred eruptions of Piton de la Fournaise produced ~1 km^3 of lava, and the volcano's summit collapsed twice (in 1931 and 2007). These lavas display, respectively, 20 and 65 % of the Sr–Nd and the Pb isotope ranges reported for La Réunion volcanoes over their known eruptive record (3.8 Ma). Variations in major and trace element concentrations and Sr–Pb isotopes do not define a temporal trend at the scale of the century, but display systematic short-term cyclic fluctuations. The positive correlation between $^{87}Sr/^{86}Sr$ and ratios of trace elements that are more versus less incompatible during partial melting of the mantle (e.g., Nd/Sm, La/Sm) probably results from the sampling of small-scale heterogeneities within the plume source. Changes in the degree of melting and/or crystallization are debated, but these appear ultimately linked to source properties. Lead isotopes do not co-vary with Sr isotopes, in part because of the partitioning of Pb into dense metallic phases that are preferentially sampled during high-flux eruptions. Taken together, Sr–Nd–Pb–Os–Th isotopes do not support contamination of magma with genetically unrelated components, such as the underlying Indian oceanic crust, mantle lithosphere, seawater, or seawater-altered lavas. Yet, in some rare cases (e.g. the 1998 Hudson eruption), the compositional patterns suggest that the parental magma assimilated older volcanic products within the edifice, such as crystal cumulates and/or interstitial differentiated melts. The geochemical fluctuations over the 1927–2010 time period constrain the residence time of magma in the

I. Vlastélic (✉)
Laboratoire Magmas et Volcans, Clermont Université, Université Blaise Pascal, CNRS UMR 6524, IRD R 163, Aubière, France
e-mail: I.Vlastelic@opgc.univ-bpclermont.fr

A.J. Pietruszka
U.S. Geological Survey, Denver Federal Center, Denver, CO 80225, USA

P. Bachèlery et al. (eds.), *Active Volcanoes of the Southwest Indian Ocean*, Active Volcanoes of the World, DOI 10.1007/978-3-642-31395-0_11

shallow reservoir to 10–30 years and its size to 0.1–0.3 km^3. The magma residence time during the course of the long-lived 1998 eruption is estimated to be an order of magnitude shorter, but the reservoir was probably of similar size. Instead, the shorter magma residence for the 1998 eruption was probably due to a higher magma flux.

11.1 Introduction

During the last century, more than one hundred eruptions of Piton de la Fournaise volcano produced about one cubic kilometer of lava (Stieltjes and Moutou 1989; Peltier et al. 2009). A regular sampling of the eruptions allowed the identification of cyclic variations in lava petrography, chemistry, and isotopic composition. One of the most striking features is that these rapid compositional changes encompass a significant, if not a dominant, part of the total compositional range known at Piton de la Fournaise. For instance, historical eruptions produced the major types of lava erupted at this volcano over its entire known compositional record (poorly phyric transitional basalts, olivine-rich basalts, abnormal basalts enriched in both MgO and alkali elements and differentiated basalts; see Sect. 11.4 and Chap. 10). The same is true for magma source tracers, with Pb isotope variations in recent lavas accounting for more than half of the range reported for the last 530 ka. The temporal geochemical variations may have multiple causes, depending on which compositional parameter is considered. The most commonly cited explanations include mixing of melts in a periodically refilled magma chamber (Albarède 1993; Sigmarsson et al. 2005), magma differentiation, or entrainment and/or assimilation of cumulative phases (Albarède and Tamagnan 1988; Boivin and Bachèlery 2009; Villemant et al. 2009; Famin et al. 2009; Salaün et al. 2010), and contamination of plume-derived magmas in the crust or lithospheric mantle and/or sampling small-scale mantle heterogeneities (Vlastélic et al. 2005, 2007; Pietruszka et al. 2009).

This review starts in 1927, which marks an increase in sampling density as well as in the precision with which sample eruption dates are known. We do not aim to provide an exhaustive report of published results and conclusions. Instead, we focus on the most commonly debated issues, including (1) the relative influence of heterogeneity in La Réunion mantle plume source versus contamination of plume-derived magma during ascent through the lithosphere, (2) the roles of magma differentiation and entrainment of cumulative phases, (3) the trace element record of melting and crystallization processes, and (4) the size and shape of the plumbing system. Lastly, the recent geochemical evolution of Piton de la Fournaise is briefly compared to Kilauea, another frequently active ocean-island volcano on the Island of Hawaii.

11.2 Eruptive History and Samples

At least 106 eruptions have occurred between 1927 and 2010 at Piton de la Fournaise (Stieltjes and Moutou 1989; Bachèlery 1999; Peltier et al. 2009, and reports from the volcano observatory). During this period, the volcano remained rarely inactive for more than 3 years (1966–1972 and 1992–1998). The 1927–2010 period is marked at its beginning and end by major collapses of the Dolomieu crater summit in 1931 and 2007. Second-order collapses (forming pit craters) also occurred in March 1986 and November 2002. The two major collapses were ascribed to massive lateral withdrawal of magma from the shallow reservoir (Lénat and Bachèlery 1990; Staudacher et al. 2009) during the voluminous eruptions of July 1931 (130×10^6 m^3) and April 2007 (240×10^6 m^3). Between the two major collapses, eruptions of generally modest size ($<30 \times 10^6$ m^3) and duration (<2 months)

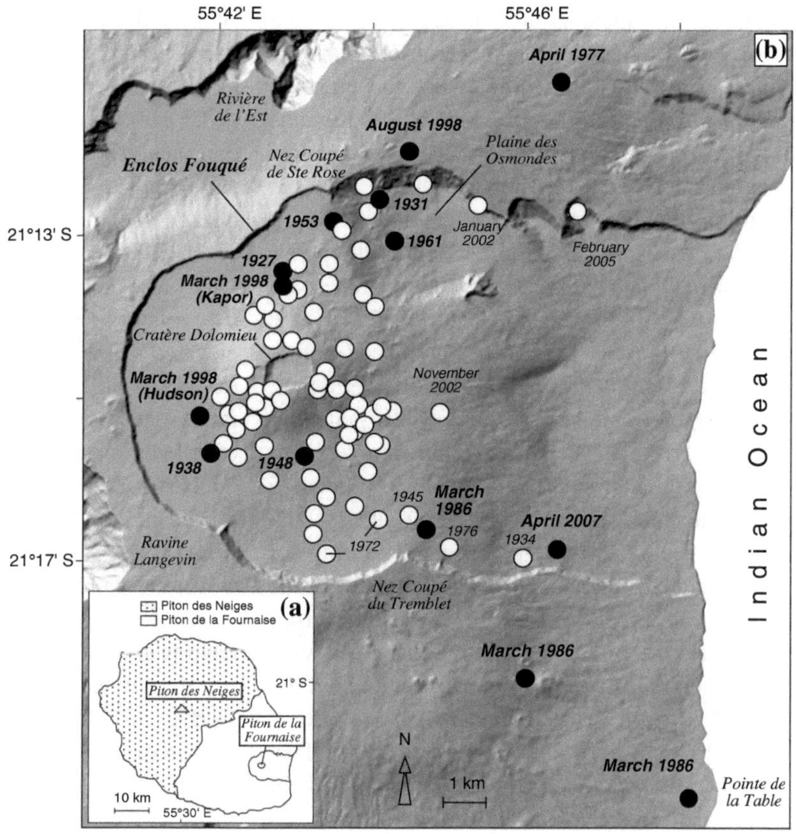

Fig. 11.1 a Map of La Réunion Island showing the location of Piton des Neiges and Piton de la Fournaise. **b** Map of Piton de la Fournaise showing the locations of the 1927–2010 eruptions that have been studied. Eruptive vents within the Dolomieu crater are not shown. Major eruptions discussed in text (*black circles*) and remarkable vents are labeled

covered the floor of the Dolomieu and Bory craters, or propagated along the northeast and southeast flanks (Fig. 11.1). A lava lake was observed within the Bory crater in 1937 and 1942. Peltier et al. (2009) who studied the most recent activity (1972–2007) sorted eruptions into three groups. Most of the eruptions (68 %) started near the top of the central cone at 2500 m, or within the summit crater, before the erupted fissures propagated down slope on the flank of the cone and the activity stabilized near its base (between 1800 and 2000 m). The remaining events occurred entirely within the summit crater (21 %), or at low altitude, more than 4 km from the summit (11 %). All eruptions occurred inside the Enclos Fouqué caldera, with the exception of the April 1977 and March 1998 eruptions (which propagated along the Northeast rift zone), and the March 1986 eruption (which propagated along the Southeast rift zone) (Fig. 11.1). However, it is possible that some vents (e.g., in January 2002) located at the bottom of the cliff of the Enclos Fouqué might have been fed by magma that originated from outside the boundaries of the Enclos Fouqué at depth (Villeneuve and Bachèlery 2006). Besides the July 1931 and April 2007 events, relatively voluminous eruptions (60×10^6 m^3) occurred in 1961 and 1998. The March 1998 eruption (Kapor vent) also stands out because of its long duration (six and a half months). High-flux eruptions producing olivine-rich lava occurred more or less regularly between 1927 and 1977 (in 1927, 1931, 1938, 1948, 1953, 1961, 1977) and frequently between 2002 and 2007 (in 2002, 2005, 2006 and 2007), but such eruptions did not occur between 1977 and 2002. The rate of magma production (0.32 m^3/s for the whole period) seems to have increased during the 20th century and has been particularly high (>0.6 m^3/s) since 1998 (Stieltjes and Moutou 1989; Bachelery 1999; Villeneuve and Bachèlery 2006; Vlastélic et al. 2009), although the apparent low productivity before 1961 could reflect less detailed observations of volcanic activity. For comparison, the eruption rate of Piton de la Fournaise since 1998 remains 6–10 times lower than the 3.5–6 m^3/s range of magma supply rates estimated for the Puu Oo eruption of Kilauea volcano (Poland et al. 2012).

The density of sampling as well as the precision with which the sample eruption dates are known have improved considerably during the 1927–2010 period. Lava flows emplaced before 1975 were sampled several years after eruption, and only one sample per eruption has generally been analyzed. In these cases, the sample eruption dates are not precisely constrained (Albarède and Tamagnan 1988; Albarède et al. 1997a). Since 1975, lavas were collected during or shortly after eruption, so the eruption dates of the samples are precisely known. These later eruptions were often sampled continuously, allowing the resolution of compositional variations within individual events (e.g., Bachèlery 1981; Vlastélic et al. 2005; Boivin and Bachèlery 2009). Historical lavas have been extensively analyzed for major and trace element concentrations and Sr–Nd–Pb radiogenic isotopes (Bachèlery 1981; Albarède and Tamagnan 1988; Albarède et al. 1997a; Boivin and Bachèlery 2009; Villemant et al. 2009; Salaün et al. 2010; Luais 2004; Bosch et al. 2008; Pietruszka et al. 2009; Vlastélic et al. 2005, 2007, 2009, 2013).

11.3 Source Heterogeneity Versus Contamination Processes

La Réunion lavas are noteworthy for showing considerably less isotopic variation compared to lavas produced by other major hotspots (Fisk et al. 1988; Graham et al. 1990; Albarède et al. 1997a; Fretzdorff and Haase 2002). For example, the $^{206}Pb/^{204}Pb$, $^{87}Sr/^{86}Sr$ and εNd ranges of sub-aerial lavas (18.76–19.08, 0.70398–0.70436 and +3.8 to +4.5, respectively) represent less than 7 % of the isotopic variability reported in oceanic basalts worldwide (Fig. 11.2). As suggested by Luais (2004), this view is potentially biased because the sample set on which it is based includes dominantly young lavas (<0.5 Ma) from Piton de la Fournaise. However, the analysis of additional older (2.1 Ma) sub-aerial lavas from Piton des Neiges did not change the picture (Bosch et al. 2008; Nauret et al. 2007). Only recently, the study of old (>3.3 Ma) submarine lavas from the Northeast rift zone of Piton de Fournaise has revealed elevated $^{87}Sr/^{86}Sr$ (up to 0.70478) and low εNd (down to +2.8) signatures (but no Pb isotope ratios outside the known range) in a sub-group of alkali basalts, providing the first firm evidence for coupled variations of Sr and Nd isotopes in La Réunion lavas (Smietana 2011; Fig. 11.2).

Furthermore, the Sr and Pb isotopic variations in historical lavas from Piton de la Fournaise are significant ($0.70407 < {}^{87}Sr/^{86}Sr < 0.70424$ based on the most recent high-precision data, and $18.87 < {}^{206}Pb/^{204}Pb < 18.94$ excluding the short-term dips to highly unradiogenic Pb isotope ratios) and display clear temporal trends (Fig. 11.3). However, the absence of a simple correlation between Sr and Pb isotope ratios (not shown) and the apparent constancy of Nd isotope ratios (Fig. 11.3) (+3.9 < εNd < +4.3, excluding data from Luais (2004), which were collected in the early days of MC-ICPMS and may be analytically problematic) are not fully understood. For instance, it is not clear whether Nd isotopes are truly constant in the youngest Piton de la Fournaise lavas or whether they vary systematically [within analytical error of ±0.15 ε units (2σ)] with Sr isotopes. Based on the larger Sr–Nd isotope trend that exists at the scale of the entire edifice (Smietana 2011), εNd variations of only ~0.3 ε units are expected from the historical range of $^{87}Sr/^{86}Sr$ (0.00014), which is similar to the observed range (0.4 ε units). Therefore, the possibility that Sr and Nd isotopes co-vary in younger lavas cannot be excluded.

The question arises as to whether the short-term isotopic variations reflect fine-scale compositional heterogeneity of the mantle source or contamination of plume-derived magmas on

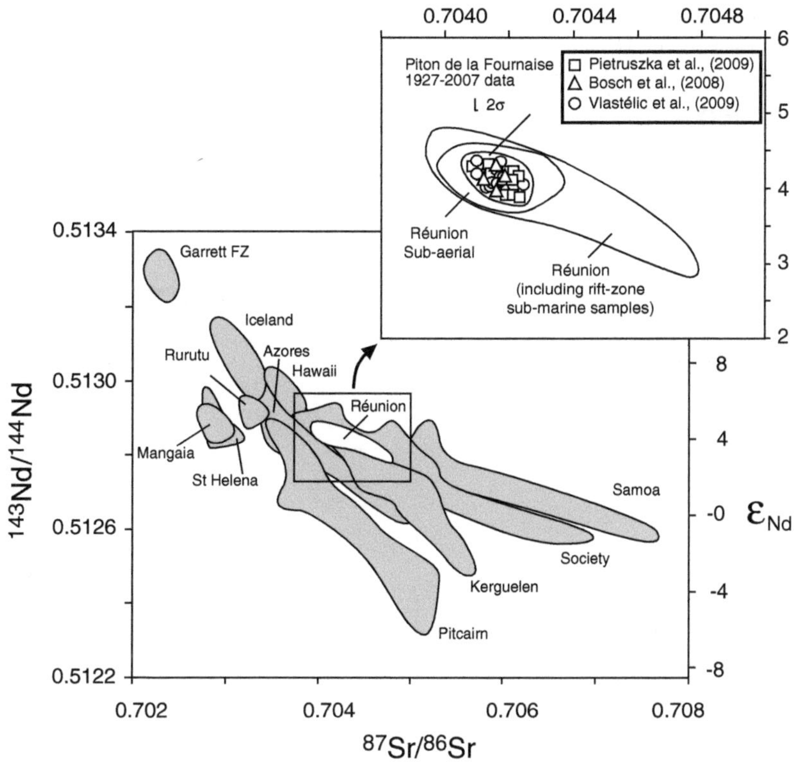

Fig. 11.2 Sr–Nd isotope plot showing ranges for La Réunion compared to other hotspot volcanoes. The *inset* shows the Sr–Nd isotope range of historical lavas of Piton de la Fournaise (Bosch et al. 2008; Pietruszka et al. 2009; Vlastélic et al. 2009), by comparison with the range of sub-aerial lavas from Piton de la Fournaise and Piton des Neiges (Fisk et al. 1988; Bosch et al. 2008) and the total range of La Réunion including sub-marine rift zone samples (Fretzdorff and Haase 2002; Smietana 2011; Bachèlery, unpublished data). The isotopic fields of major hotspot volcanoes are drawn using the GEOROC database (http://georoc.mpch-mainz.gwdg.de/georoc/)

their way to the surface (Albarède and Tamagnan 1988; Albarède et al. 1997a; Vlastélic et al. 2005, 2007, 2009). Potential contaminants of La Réunion magmas include Indian Ocean crust or lithospheric mantle, Indian Ocean sediments, and/or seawater-related components (e.g., brines or seawater-altered basalts), or old cumulates and/or lavas making up the interior of the volcanic edifice. The contribution of the Indian Ocean lithosphere was quantitatively tested using a mixing model in Sr versus Th isotope space, where the compositional contrast between plume melts and the contaminants is relatively large (Pietruszka et al. 2009). The Sr–Th isotope mixing trends can potentially be explained by assimilation of either unaltered Indian oceanic crust (in bulk) or a low-degree partial melt of the crystalline lower oceanic crust, but unreasonably large amounts of assimilation are required in both cases (>25 % bulk crust or >12 % of a 0.5 % partial melt), which makes these scenarios unlikely. This conclusion was supported by the relatively small variations of Os isotopes in lavas from Piton de la Fournaise, which are inconsistent with assimilation of aged crustal material enriched in radiogenic Os (Schiano et al. 2012).

Marine ferromanganese deposits lying atop the Indian oceanic crust are potential contaminants because they have high concentrations of Sr, Nd and Pb (in the range of 10^2–10^3 ppm), and continental-type isotopic compositions with high $^{87}Sr/^{86}Sr$ and $^{207}Pb/^{204}Pb$ and low εNd (Albarède et al. 1997b; Vlastélic et al. 2001). However, no isotopic trends consistent with the involvement

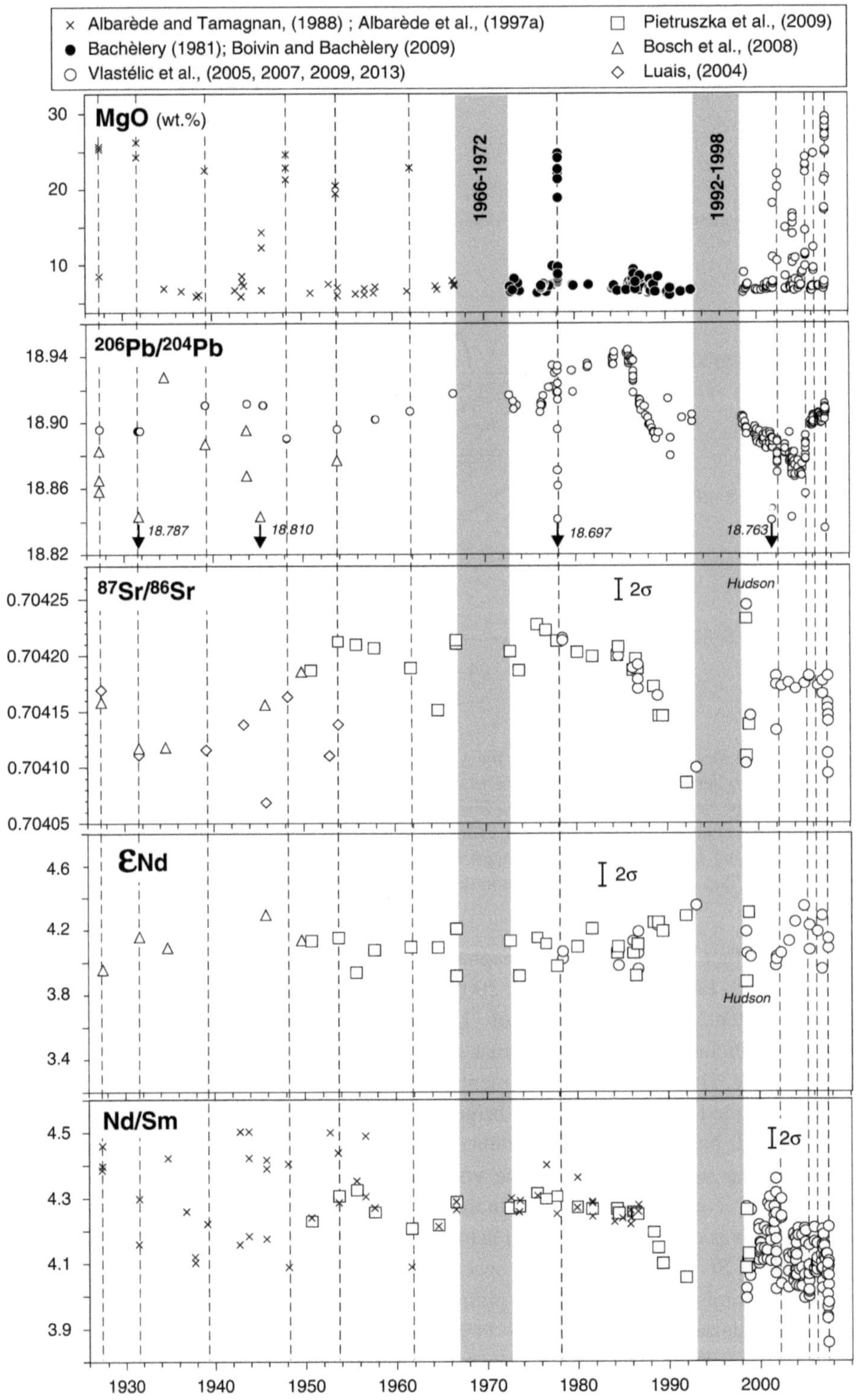

Fig. 11.3 Overview of the geochemical variations in historical lavas. *Vertical grey bands* indicate the longest periods of inactivity. *Dashed lines* indicate olivine-rich eruptions. *Arrows* indicate samples that plot off the diagram. External 2σ error is smaller than symbols [MgO, $^{206}Pb/^{204}Pb$, isotope-dilution Nd/Sm ratios of Pietruszka et al. (2009)] or as indicated. The error of Nd/Sm ratios determined by ICPMS (Vlastélic et al. 2005, 2007, 2013) is less than 1.5 % because of the strong correlation between Nd and Sm errors

of such components are observed in recent lavas. Hydrothermal deposits formed near Indian spreading ridges (presumably now lying beneath the volcanic edifice) have both unradiogenic (mantle-like) Pb and unradiogenic (continental-like) Nd. Their contribution was envisioned to explain the unradiogenic Pb anomalies observed in some lavas from Piton de la Fournaise (Vlastélic et al. 2007), but the absence of coupled unradiogenic Nd signatures is inconsistent with this idea (Vlastélic et al. 2009).

Unlike ferromanganese deposits, seawater has a very small amount of Nd (5×10^{-6} ppm) and Pb (3.5×10^{-5} ppm), and its addition to magmas (through assimilation of brines or salts) is expected to influence the Sr, but not the Nd and Pb isotopic composition of contaminated lavas. Thus, contamination with a seawater-derived component could explain the Sr isotope variations at apparently constant Nd isotope ratios, but another process would be required to control the Pb isotope variations. However, the small negative Sr–Nd co-variation observed between 1977 and 1998 (Vlastélic et al. 2009) is inconsistent with the idea that the first-order Sr isotope variations are related to contamination with a seawater-derived component. Contamination of the magma with either modern or ancient seawater (or seawater-altered materials) may also be independently ruled out by the observation of equilibrium (^{234}U)/(^{238}U) and nearly constant (^{230}Th)/(^{232}Th) activity ratios in historical Piton de la Fournaise lavas (Pietruszka et al. 2009).

Conversely, assimilation of ancient lavas or cumulates within La Réunion volcanic edifice may be an important control on the compositional variations of some historical lavas from Piton de la Fournaise. However, modeling based on high-precision ratios of incompatible trace elements suggests that bulk assimilation of older volcanic products is of only minor to moderate importance for most Piton de la Fournaise lavas erupted between 1950 and 1998 (Pietruszka et al. 2009). In rare cases, the effects of crustal contamination were probably more extreme. For example, anomalous trace element ratios (such as low Th/U) in the 1964 and 1998 Hudson lavas (as well as three other lavas out of the 27 analyzed) can be explained if their parental magmas assimilated small amounts (~3–6 %) of low degree (<1 %) partial melts of young, hydrothermally altered gabbroic and wehrlite cumulates within the volcanic edifice (Pietruszka et al. 2009). This interpretation is consistent with the evidence for an active hydrothermal system beneath the summit crater, in particular since its collapse in April 2007 (Lénat et al. 2000, 2012a).

Nevertheless, the overall range in Sr isotope and incompatible trace element ratios of Piton de la Fournaise lavas is probably controlled by mantle processes. The co-variations between ^{87}Sr/^{86}Sr and Nd/Sm observed between 1950 and 2007 (Fig. 11.4) probably result from the sampling of small-scale heterogeneities within the plume source without significant variations in the degree of partial melting (Pietruszka et al. 2009). The existence of systematic ^{87}Sr/^{86}Sr variations through time (Fig. 11.3) suggests that the heterogeneities are not randomly distributed in the magma source and/or are not randomly sampled. For instance, successive temporal trends of decreasing ^{87}Sr/^{86}Sr and ratios of trace elements that are more versus less incompatible occurred between 1953 and 1991 (1953–1964, 1966–1973, and 1975–1991), while an inverse trend occurred between 1998 and 2000. The 1953–1991 trends were interpreted as individual melting events within the mantle, during each of which the contribution of melts derived from a less fertile, chemically depleted lithology progressively increased (Pietruszka et al. 2009). On the other hand, the 1998–2000 trend was first interpreted as mixing between an old magma (similar as that erupted in 1992) with a newly injected magma (the 1998 Hudson magma), with an increasing contribution of the latter though time (M. Semet, personal communication). The peculiar geochemical signature of the Hudson magma (including elevated ^{87}Sr/^{86}Sr and enrichment in both incompatible and compatible elements) has been subsequently debated (see Sect. 11.3). Whatever the exact origin of the Hudson magma, the simultaneous eruptions in 1998 at the two vents ~4 km apart (Hudson and Kapor) of two types of magmas with different ^{87}Sr/^{86}Sr and incompatible element ratios that

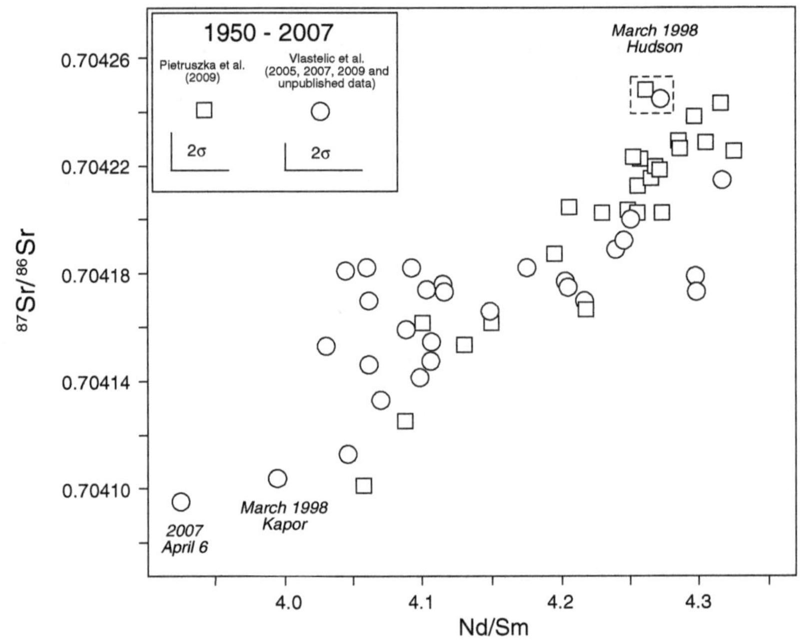

Fig. 11.4 Plot of the $^{87}Sr/^{86}Sr$ versus Nd/Sm ratios of Piton de la Fournaise lavas. Only recent data with similar precision are shown (Pietruszka et al. 2009; Vlastélic et al. 2005, 2007, 2009 and unpublished data). The measured $^{87}Sr/^{86}Sr$ ratios are not re-normalized, as they refer to NBS987 Sr values [0.710243 ± 10 (LMV) and 0.710255 ± 17 (SDSU)] that are identical within errors. The locations of the two vents (Kapor and Hudson) that were simultaneously active in March 1998 are outlined with *dashed lines*

span the historical range (Fig. 11.4) indicates that the plumbing system contains melts of various ages and isotopic compositions that are poorly mixed or physically isolated from one another. The production of such heterogeneous melts over a short period of time requires partial melting of small-scale compositional heterogeneities in La Réunion plume. The question remains whether the heterogeneities are intrinsic to the plume (or rising blobs) or whether they are entrained from the ambient upper mantle [see discussion by Bosch et al. (2008) and Pietruszka et al. (2009)].

The absence of an overall correlation between Sr and Pb isotope ratios (Fig. 11.3) indicates that at least some of the Pb isotope variation results from a process other than the mantle source control on the Sr and trace elements ratios that was previously discussed. The time-integrated ratios involving Pb (U/Pb and Th/Pb) show the most variation in lavas from Piton de la Fournaise, whereas the time-integrated Th/U ratio ($^{208}Pb^*/^{206}Pb^*$) is essentially constant (Vlastélic et al. 2009). This observation indicates that the time-integrated behavior of elemental Pb is the dominant cause of the Pb isotopic variations. The affinity of Pb for sulfides is generally invoked to account for its anomalous behavior in the mantle (Hart and Gaetani 2006). Recent eruptions of olivine-rich lavas, and in particular the April 2007 eruption, revealed a direct link between the unradiogenic Pb signatures and period of high magma flux (Vlastélic et al. 2007, 2009). One possible explanation for the unradiogenic Pb signatures is the entrainment of dense, Pb-rich sulfide phases that may accumulate in the plutonic roots of the volcano (Upton et al. 2000; Collins et al. 2012). Hart and Gaetani (2006) speculated that plume material stripped of its S near the sulphide solidus tends to assimilate or equilibrate with genetically unrelated sulfides from the lithosphere or the crust. Such sulphides were recently shown to have extremely unradiogenic signatures ($^{206}Pb/^{204}Pb$ < 17) (Burton et al. 2012). Neither sulfide phases nor their unradiogenic Os isotope fingerprint (Schiano et al. 2012) have yet been identified in Piton de la Fournaise lavas. However, thin, Fe-rich coatings with unradiogenic Pb signatures ($^{206}Pb/^{204}Pb$ down to 17.6) were found in quenched lava samples (Vlastélic 2014). Without further constraints, we suggest that the Fe-rich coatings originate from the entrainment of old mantle sulfides that destabilize during lava eruption. In addition, it is possible that the smooth temporal fluctuation of Pb isotopes (Fig. 11.3) reflects to some extent changes in the ability of the magma feeding system to sample these dense unradiogenic phases. Further work is needed to confirm this hypothesis, and determine the portion of the Pb isotopic variations that might be related to the heterogeneous source components reflected in the Sr isotope and incompatible trace element ratios (e.g., Pietruszka et al. 2009).

11.4 Petrogenetic Processes

Over the past 0.5 Ma, Piton de la Fournaise has produced four main types of lava, including, in order of decreasing frequency in historical eruptions (see also Pichavant et al. 2016, Chap. 10): (1) Nearly aphyric basalts with transitional compositions between the alkali and tholeiitic fields. Albarède et al. (1997a) suggested that the minor compositional variability (5–8 wt% MgO) of this common group reflects the buffering of major elements in a cumulate-rich plumbing system with a low fraction of percolating melt. This group is either referred to as "cotectic" or "steady state" basalts. (2) Olivine-rich basalts (named picrites or oceanites) with MgO contents up to 30 wt%. These lavas are thought to form when high-flux magmas disrupt and entrain cumulates of olivine xenocrysts (genetically unrelated) or antecrysts (genetically related) (Albarède and Tamagnan 1988; Famin et al. 2009). (3) Aphyric basalts, referred to as "abnormal", with high MgO and alkali abundances and low CaO and SiO_2 concentrations compared to steady state basalts. The origin of these signatures, first ascribed to high-pressure differentiation, is a matter of debate (Salaün et al. 2010). (4) Differentiated lavas (1–5 wt% MgO) that occur mainly in a single 395–527 ka outcrop, which were first ascribed to the early activity of Piton de la Fournaise before being associated with the late stage activity of the pre-existing Les Alizés volcano (Smietana 2011). Most of the differentiated lavas are plagioclase-rich. Albarède et al. (1997a) noted that Piton de la Fournaise lavas evolved from "mildly alkalic" to "mildly tholeiitic" during the growth history of the volcano. Based on the negative correlation between CaO/Al_2O_3 and Sc/Yb, Albarède et al. (1997a) emphasized the importance of clinopyroxene fractionation, and, following the original idea of O'Hara (1968), proposed that the alkalinity of Piton de la Fournaise lavas is related to the fractionation of clinopyroxene at high pressure (>9 kb or ~30 km). In this scenario, the decrease in alkalinity during the evolution of Piton de la Fournaise may have resulted from a gradual decrease in the depth of crystallization over time (Albarède et al. 1997a).

All lava types, with the exception of the plagioclase-rich lavas, have occurred in historical eruptions. This short-term petrologic diversity has been the subject of debate, especially with respect to the respective contributions of magma differentiation and crystal entrainment or assimilation. As proposed for the long-term evolution of the volcano (Albarède et al. 1997a), changes in crystallization depth and corresponding pyroxene/olivine ratio of the crystallizing mineral assemblage are thought to control the major element variability in recent lavas (Bureau et al. 1999; Vlastélic et al. 2007; Boivin and Bachèlery 2009). Famin et al. (2009) refined this picture with a study of melt inclusions from the 2005 eruptions, and proposed three differentiation levels: high-pressure (>7.5 km) crystallization of pyroxene and plagioclase, low-pressure (<2.5 km) crystallization of olivine, and low-pressure crystallization of pyroxene and plagioclase near the surface (<0.6 km). Conversely, Villemant et al. (2009) argued that the highly variable entrainment of phenocrysts in

Piton de la Fournaise magmas hampers the interpretation of bulk rock compositions. Using an approach based only on recent glassy samples (matrix glasses and Pélé's hair), they found that the major and trace element abundances in the 1998–2008 melts define a single differentiation trend. Using the MELTS code, they modeled this trend as a single "low-pressure" (<26 km) crystal fractionation process, where the composition of the crystallizing mineral assemblage changes as the temperature decreases. They distinguish a high temperature step (>1160 °C) where only olivine crystallizes from a low temperature step (<1130 °C) where clinopyroxene and plagioclase (but no olivine) crystallize.

The case of the 1998 Hudson lava is at the heart of this debate, since this lava is the only historical representative of the so-called "abnormal group" that is potentially derived from high-pressure differentiation. The enrichment of MgO and depletion of CaO in this lava, first interpreted as reflecting high-pressure pyroxene fractionation (Bureau et al. 1999; Boivin and Bachèlery 2009; Famin et al. 2009) could instead reflect assimilation of olivine-rich cumulates (Salaün et al. 2010). Similarly, the enrichment in highly incompatible elements with respect to less incompatible elements in this lava, first interpreted as a fingerprint of a relatively low degree of partial melting in the plume source (Vlastélic et al. 2005), could instead result from contamination of the parental magma with a low-degree partial melt of gabbro and wehrlite cumulates (Pietruszka et al. 2009) or with a small amount of highly differentiated melt (Salaün et al. 2010). The major and trace element signatures of the Hudson lava may be reconciled if the parental melt assimilated olivine-rich xenoliths containing highly differentiated interstitial melts (Salaün et al. 2010). Evidence for the existence for such evolved melts are rare (e.g., Upton and Wadsworth 1971), but the entrainment of a small volume of trachytic melt during the paroxysmal phase of the April 2007 eruption suggests that small pockets of differentiated magma exist within the volcanic edifice (Vlastélic et al. 2011).

Another implication of the model of Villemant et al. (2009) is that late stage crystallization may significantly fractionate (by ~20 %) incompatible element ratios such as La/Yb or La/Sm, even though such ratios were previously thought to record dominantly melting processes and/or source properties (Albarède and Tamagnan 1988). The reason for this is that Albarède and Tamagnan (1988) considered that the fraction of melt crystallizing and the difference in solid/liquid partition coefficients (D) between rare earth elements cannot exceed 30 % and 0.3, respectively, whereas Villemant et al. (2009) inferred (from glass data) higher values up to 35 % and 0.63 ($D_{Yb} - D_{La}$). Albarède and Tamagnan (1988) identified antithetic variations of Ce/Yb and CaO/Al_2O_3 over 17 years cycles, which they explain in a scenario where lower degrees of melting produce smaller volumes of magma that undergo larger degrees of fractionation in a magma chamber or the melt-transport conduits. If incompatible element ratios fractionate significantly during crystallization (Villemant et al. 2009), such coupled variations in major and trace elements could result from magma differentiation alone.

Vlastélic et al. (2005, 2007) noted that La/Sm ratios are lower (on average) during the eruptive events producing olivine-rich lavas compared to the bracketing eruptions producing nearly aphyric lavas. Yet, it is clear that olivine addition alone cannot significantly modify incompatible element ratios because all of these elements are expected to be highly incompatible in olivine. Entrainment of dunitic cumulates with very little interstitial plagioclase such as that recovered from Piton Chisny (Upton et al. 2000) may have significant influence on the La/Sm ratio, but only if addition exceeds 80 %, which is a much larger amount of olivine than observed in any lavas from Piton de la Fournaise. The monitoring of the April 2007 caldera-forming eruption provided new clues about the processes that may lead to changes in the incompatible element ratios of the lavas (Fig. 11.5). This eruption, the most voluminous of the last century (230×10^6 m^3), started on April 2 along the Southeast rift zone (Fig. 11.1). The rate of magma production increased significanly from 55 to >200 m^3/s (Coppola et al. 2009) until the summit Dolomieu crater collapsed on the night of the 5–6 April (Staudacher et al. 2009). These rates are one

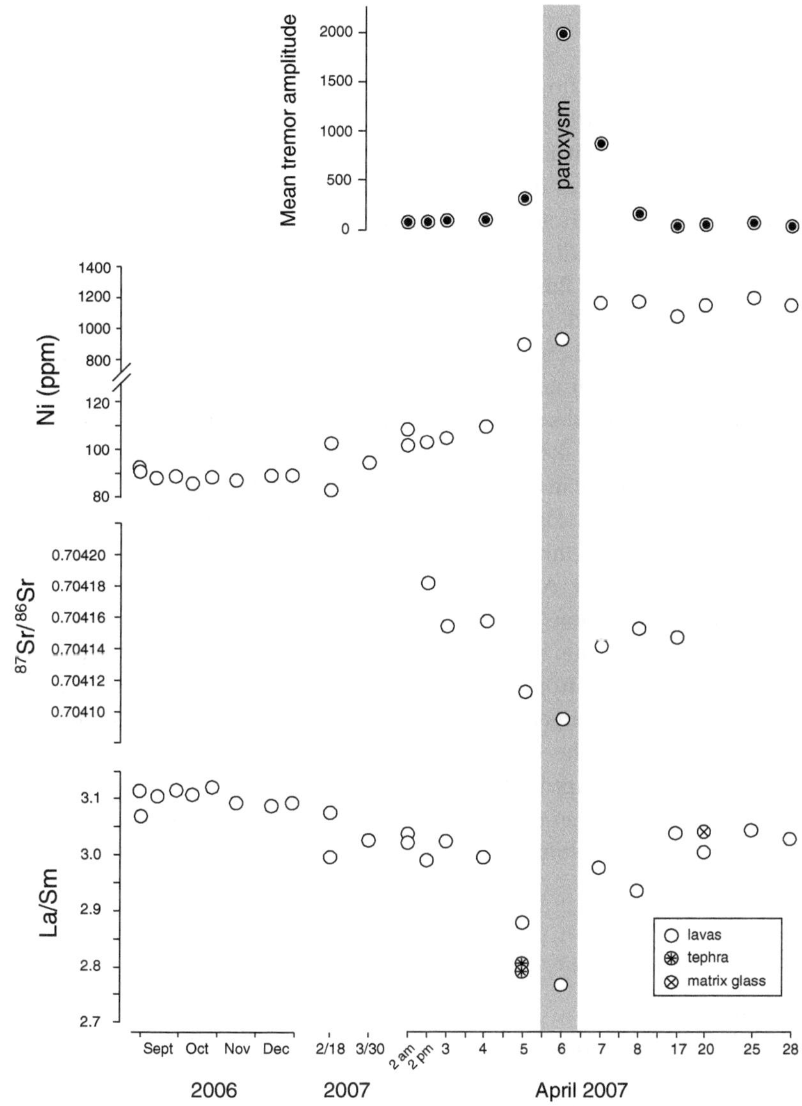

Fig. 11.5 Temporal variations of Ni concentrations, $^{87}Sr/^{86}Sr$ and La/Sm ratios for Piton de la Fournaise lavas from September 2006 to April 2007 (Vlastélic et al. 2013 and unpublished Sr isotope data). Ni greater than 140 ppm is assumed to be a proxy for olivine accumulation (MgO was not measured in all samples). Different time scales are used to emphasize the rapid chemical variations occurring during the very first days of the April 2007 eruption. *Inset* shows the variation in eruption rate for the series of lavas erupted during April 2007, as inferred from the daily average of tremor amplitude (root mean square) calculated in the 0.5–1.5 Hz band at Bory station (Battaglia, personal communication). This figure shows that the ratios of trace elements that are more versus less incompatible (e.g., La/Sm) and $^{87}Sr/^{86}Sr$ ratios are both the lowest during eruption paroxysm

order of magnitude higher than those estimated for typical eruptions (0.5–20 m^3/s) (Coppola et al. 2009). Following the paroxysmal phase, the eruption continued at an elevated rate until May 1. The amount of olivine crystals in the lavas increased rapidly between April 4 and 5 and remained elevated (≥50 wt%) until the end of the eruption. The CaO/Al_2O_3 ratio remains constant at ~0.809. In contrast, the La/Sm ratio in whole-rock samples decreased until April 6th (eruption paroxysm) reaching the lowest value (~2.8) measured in historical lavas, and increased subsequently towards values that typified lavas from 2006 and earlier in 2007 (Vlastélic et al. 2013). A detailed temporal record of glass composition does not exist, but available data on glassy samples indicate that the low La/Sm ratios coincide with an unusually MgO-rich melt (8.6 wt%; Villemant et al. 2009). The connection between magma flux and trace element ratios could be explained if fast rising magmas undergo limited crystal fractionation. Using a primary melt composition and D values from Villemant et al. (2009), the decrease of the La/Sm ratio from 3.1 to 2.8 during the April 2007 paroxysm corresponds to a decrease in the extent of crystallization from 35 to 12 %. However, the constant CaO/Al_2O_3 ratios during this period are inconsistent with this interpretation. Alternatively, it is possible that the La/Sm variations reflect variations in the degree of partial melting. For instance, the high-flux magmas might result from high rates of melt production, and thus, a larger degree of melting at depth. In any case, the co-variation between La/Sm and $^{87}Sr/^{86}Sr$ in April 2007 lavas (Fig. 11.5) indicates that the rapid variation of trace ratios during eruption paroxysm is ultimately linked to sources properties.

11.5 The Size and Shape of the Plumbing System

The magma feeding system of Piton de la Fournaise has been characterized on the basis of a wide range of observations, including earthquake locations, compositional variation and trapping pressure of olivine melt inclusions, eruption volumes and durations (e.g., Nercessian et al. 1996; Battaglia et al. 2005; Famin et al. 2009; Aki and Ferrazzini 2001). Altogether, the data support the existence of a deep magma reservoir at the crust-mantle interface (7.5 km below the summit) and a shallower reservoir near sea level (2.5 km below the summit), whereas an array of sills and dykes make up the shallowest part of the plumbing system (0.5–1 km below the surface; see also Michon et al. 2016, Chap. 7). The size of the storage system is less well constrained. From the summit inflation that preceded the December 2005 eruption, Peltier et al. (2008) modeled the shallow magma reservoir as an ellipsoidal body about 500 m wide (i.e., $\sim 65 \times 10^6$ m^3). The withdrawal of this reservoir during the April 2007 eruption produced 230×10^6 m^3 of lava suggesting either that its size has been underestimated, or that that the shallow reservoir was continuously recharged during the eruption.

The geochemical fluctuations of Piton de la Fournaise lavas can be regarded either as magma mixing within a periodically refilled reservoir, or as tapping distinct parts of a complex plumbing system including magmas of various ages and histories. In the former case, magma residence time (τ) and size of the magma reservoir can be estimated using an approach based on the damping of the source signal as magmas pass through the storage system (Albarède 1993). A simplified expression of τ from the model of Albarède (1993), assuming the elements or isotopes are highly incompatible in the crystallizing mineral assemblage (Pietruszka and Garcia 1999a), is given by:

$$\tau = \frac{C_2^{in}}{C_2^{res}} \frac{R^{in} - R^{res}}{dR/dt}$$

where "in" and "res" refer to the input magma and initial magma in the reservoir, respectively. The variable R is the ratio of two incompatible elements or isotopes, C_2 is the concentration of the normalizing species, and dR/dt is the temporal change of R. The volume of the reservoir is $\tau \times Q$, where Q is either the flux of magma entering (magma supply rate) or leaving (lava eruption rate) a reservoir of constant volume. This approach suffers from uncertainties on input parameters, in particular with respect to the composition of newly injected magma (R^{in}), which is implicitly never sampled. Albarède (1993) modeled the variations of Ce/Yb between 19 and 22 observed during the 1930–1986 period by injection of a differentiated magma with an assumed Ce/Yb ratio in the range of 30–50 into a less differentiated magma with Ce/Yb of 19. He estimated a mean magma residence time in the range of 10–30 years, and, using a magma supply rate of 0.32 m^3/s (Stieltjes and Moutou 1989), inferred a reservoir size of 0.1–0.3 km^3. The eruption in April 2007 of trachytic melt with Ce/Yb ranging from 34 to 46 (Salaün et al. 2010; Vlastélic et al. 2011) suggests this estimation is based on a realistic assumption. Following a similar approach, Sigmarsson et al. (2005) modeled the geochemical fluctuations of ($^{226}Ra/^{230}Th$) in Piton de la Fournaise lavas. They assumed that the increase in ($^{226}Ra/^{230}Th$) from 1.22 in 1964 to 1.33 in 1989 was due to the injection in 1961 of a deep magma with ($^{226}Ra/^{230}Th$) of 1.40 (sampled by olivine-rich eruptions only) into a magma reservoir with ($^{226}Ra/^{230}Th$) of 1.20. They inferred a magma residence time of 25 years in good agreement with the previous approach. Using a slightly higher magma production rate of 0.44 m^3/s (taking into account intrusions) they estimated a reservoir size of 0.35 km^3.

The model of Albarède (1993) can be applied to the long-lived (six and a half months) 1998 eruption, which shows a remarkable compositional evolution that most likely reflects magma mixing. Assuming the addition of a Hudson-type magma (with $^{87}Sr/^{86}Sr$ of 0.70425) to a 1992-type magma (with $^{87}Sr/^{86}Sr$ of 0.70410), then the compositional evolution during the eruption (0.70410–0.70415) is consistent with a residence time of only ~1.5 years. Using the well constrained lava output rate of 3.5 m^3/s (Vlastélic et al. 2005), a magma chamber size of ~0.2 km^3 is inferred (assuming the Sr concentration of the input magma is identical to the erupted lava at the start of the eruption). This inferred reservoir size is consistent with previous estimates, although the residence time is much shorter due to the fact that the eruption rate was ~10 times higher than the long-term average used for previous estimates.

Alternately, the geochemical fluctuations of Piton de la Fournaise lavas might reflect changes in the magma feeding system and its ability to sample deep or eccentric magma reservoirs. This scenario might account for the wider range in lava chemistry during periods with frequent eruptions of olivine-rich lavas (e.g., 2002–2007). Beyond the direct influence of olivine accumulation on lava chemistry, the geochemical record of the genetic processes associated with olivine-rich lavas includes (1) more primitive composition of cumulative olivine (Fo85-83 vs. Fo80 in olivine phenocrysts from "steady state" basalts), (2) low ratios of trace elements that are more versus less incompatible (probably due to larger melting degree and/or lower crystallization extent of high-flux magmas), (3) unradiogenic Pb isotope ratios (ascribed to the entrainment of dense Pb-rich phases from the mantle), and (4) high ($^{226}Ra/^{230}Th$) (a possible fingerprint of deep magmas) (Peltier 2007; Welsch 2010; Albarède and Tamagnan 1988; Vlastélic et al. 2007, 2009, 2013; Sigmarsson et al. 2005). While the distinctive geochemical signatures of the olivine-rich lavas are not yet fully understood, they may be related to the continuous recharge of the shallow magma reservoir by magmas rising from the crust-mantle interface (Peltier et al. 2009).

11.6 Brief Comparison with Kilauea Volcano

The isotopic variations of historical lavas from Piton de la Fournaise are best explained in the framework of a heterogeneous mantle source within La Réunion plume, rather than by variable contamination of magmas with genetically unrelated components such as Indian oceanic crust or mantle lithosphere, oceanic sediments or hydrothermal deposits, seawater, or seawater-altered materials. Yet, the complexity of the temporal geochemical trends, the response of isotopic ratios to rapid changes in eruptive behavior, and the relatively small isotopic range at Piton de la Fournaise all suggests that the heterogeneity of the mantle source is not only small, but also blurred by the process of melt generation, transport, and storage. This picture is especially true for periods of frequent eruption of olivine-rich lavas, which suggests that some of the isotopic heterogeneity (i.e., Pb isotopes) originates from mantle-derived phases carried up from the plutonic roots of the volcano.

Piton de la Fournaise is comparable to Kilauea volcano, Hawaii, in the sense that both are active intraplate basaltic shield volcanoes sampling long-lived mantle plumes. However, Piton de la Fournaise and Kilauea are different because the former rests on a slow-moving oceanic plate, lacks well-developed rift zones, is less regularly active and shows its own eruptive pattern (e.g., increasing lava flow rate with increasing eruption volume; Aki et al. 2001). The complex geochemical signal of Piton de la Fournaise also contrasts with the temporal compositional trends shown by historical eruptions of Kilauea volcano, Hawaii. Although low $\delta^{18}O$ values have been found in some Kilauea lavas and olivines (Garcia et al. 1998, 2008), crustal contamination has not obscured the mantle-derived signatures of melting variations and source heterogeneity at Kilauea. Olivine-rich lavas at Kilauea generally have similar isotope and trace element ratios as contemporaneous aphyric lavas (Pietruszka and Garcia 1999b; Garcia et al. 2003), except for the differences in element abundances due to the accumulation of olivine (i.e., there is no evidence of trapped melt, evolved residual liquids, or unusual phases other than olivine). Over the last 1000 years (Marske et al. 2007) and during the historical period (Pietruszka and Garcia 1999b; Marske et al. 2008), Kilauea lavas display a much larger range in ratios of highly to moderately incompatible element than Piton de la Fournaise lavas (e.g., La/Yb varies by more than a factor of ~2 at Kilauea). These variations in trace element ratios are most likely related to a factor of ~2 range in the degree of partial melting of a heterogeneous source (Pietruszka and Garcia 1999a; Pietruszka et al. 2013). This picture contrasts with the relative lack of variations in the degree of partial melting at Piton de la Fournaise, and the relatively homogenous source tapped by this volcano. This behavior at Kilauea is probably related to the presence of variable amounts of recycled oceanic crust within the Hawaiian plume (Pietruszka et al. 2013), which presumably is less important at La Réunion (Vlastélic et al. 2006).

Unlike Piton de la Fournaise lavas, Kilauea summit lavas are dominantly olivine controlled (Wright 1971; Garcia et al. 2003), whereas many rift lavas (and rare summit lavas from periods of low eruption rate) may have differentiated beyond olivine control (cpx ± plag) in the relatively shallow parts of the rift zone (Wright and Fiske 1971). Variations in CaO/Al_2O_3 ratios in olivine-controlled Kilauea lavas probably do not result from change in crystallization depth, but are thought to be controlled by changes in the composition of the parental magmas delivered to the volcano from the mantle (Garcia et al. 2003). The origin of CaO/Al_2O_3 variations is controversial, but they might be related to the presence of recycled oceanic crust in the mantle source (e.g., Jackson et al. 2012). In general, nearly aphyric summit lavas from Piton de la Fournaise tend to be significantly more evolved than their equivalent at Kilauea [$5.5 < MgO < 8.6$ vs. $6.7 < MgO < 10.1$ wt % (Garcia et al. 2003)]. The relatively large extent of crystallization at Piton de la Fournaise, between 10 and 35 % according to Villemant et al. (2009), suggests that crystals accumulate at a high rate within the magma feeding system. The amount of crystals estimated to have accumulated within the volcano over the last century (0.10–0.35 km^3) could thus match the size of the shallow magma reservoir. Although cumulative crystals are regularly removed from the plumbing system during high-flux eruptions, the plutonic system of Piton de la Fournaise (or of the pre-existing volcano) has developed over time (probably aided by the very slow motion of the underlying oceanic plate), so that Piton de la Fournaise is described today as a relatively small volcano resting on a huge intrusive complex (Lénat et al. 2012b). In such a framework, it is difficult to escape the idea that the plutonic basement of Piton de la Fournaise has a major influence on magma path and storage.

11.7 Synthesis

Historical lavas of Piton de la Fournaise display short-term, generally coupled variations in $^{87}Sr/^{86}Sr$ and incompatible element ratios that are most likely related to small-scale compositional heterogeneities in the mantle source. The measured $^{143}Nd/^{144}Nd$ ratios vary only slightly outside of analytical error (±0.15 ε units), but such small variations are consistent with those expected from the historical $^{87}Sr/^{86}Sr$ range and the Sr–Nd isotope relationship existing at the scale of the entire edifice (Smietana 2011). Lead isotopes do not co-vary with Sr isotopes, in part because of the partitioning of Pb into dense metallic phases that are variably sampled, mostly depending on magma flux.

There is no evidence for contamination of magmas by genetically unrelated components. Conversely, interaction of magmas with previous volcanic products of Piton de la Fournaise is likely. The plutonic roots of the volcano seem to be extensively developed and exert a strong control on magma path and storage. In some cases, partial melting of cumulates or entrainment of differentiated melts trapped in cumulates interstices or in isolated pockets play a key role in controlling lava chemistry. The overall homogeneity of the erupted lavas, together with the relative low magma production rate, suggest that the mantle source and melting processes at Piton de la Fournaise are fundamentally different from those at Kilauea Volcano.

Acknowledgements This paper benefited from the constructive comments of E. Todd and M. Coombs from USGS. The studied samples were collected by G. Kieffer, P.M. Vincent, P. Boivin, P. Bachèlery, F. Albarède, T. Staudacher, and several members from the Piton de la Fournaise observatory and La Réunion University. D. Auclair, and C. Bosq are thanked for the new Sr isotope data on the April 2007 eruption. This work was financed by the French Government Laboratory of Excellence initiative n°ANR-10-LABX-0006, the Région Auvergne and the European Regional Development Fund. This is Laboratory of Excellence ClerVolc contribution number 118.

References

Aki K, Ferrazzini V (2001) Comparison of Mount Etna, Kilauea, and Piton de la Fournaise by a quantitative modeling of their eruption histories. J Geophys Res 106:4091–4102

Albarède F (1993) Residence time analysis of geochemical fluctuations in volcanic series. Geochim Cosmochim Acta 57:615–621

Albarède F, Tamagnan V (1988) Modelling the recent geochemical evolution of the Piton de la Fournaise volcano, Réunion island, 1931–1986. J Petrol 29:997–1030

Albarède F, Luais B, Fitton G, Semet M, Kaminski E, Upton BGJ, Bachèlery P, Cheminée JL (1997a) The geochemical regimes of Piton de la Fournaise volcano (Réunion) during the last 530 000 years. J Petrol 38:171–201

Albarède F, Goldstein SL, Dautel D (1997b) The neodymium isotopic composition of manganese nodules from the Southern and Indian oceans, the global oceanic neodymium budget, and their bearing on deep ocean circulation. Geochim Cosmochim Acta 61:1277–1291

Bachèlery P (1981) Le Piton de la Fournaise (Ile de la Réunion). Etude volcanologique, structurale et pétrologique. Thèse, Université de Clermont-Ferrand, France

Bachèlery P (1999) Le fonctionnement des volcans boucliers. Exemple des volcans de La Réunion et de la Grande Comore. Mémoire d'Habilitation à Diriger les Recherches, Université de Clermont-Ferrand, France

Battaglia J, Ferrazzini V, Staudacher T, Aki K, Cheminée J-L (2005) Pre-eruptive migration of earthquakes at the Piton de la Fournaise volcano (Réunion Island). Geophys J Int 161:549–558

Boivin P, Bachèlery P (2009) Petrology of 1977 to 1998 eruptions of Piton de la Fournaise, La Réunion Island. J Volcanol Geotherm Res 184:105–109

Bosch D, Blichert-Toft J, Moynier F, Nelson BK, Télouk P, Gillot P-Y, Albarède F (2008) Pb, Hf and Nd isotope compositions of the two Réunion volcanoes (Indian Ocean): a tale of two small-scale mantle "blobs"? Earth Planet Sci Lett 265:748–768

Bureau H, Métrich N, Semet M, Staudacher T (1999) Fluid-magma decoupling in a hot-spot volcano. Geophys Res Lett 23:3501–3504

Burton KW, Cenki-Tok B, Mokadem F, Harvey J, Gannoun A, Alard O, Parkinson IJ (2012) Unradiogenic Pb in Earth's upper mantle. Nature Geosci 5:570–573

Collins SJ, Maclennan J, Pyle DM, Barnes SJ, Upton BGJ (2012) Two phases of sulphide saturation in Réunion magma: evidence from cumulates. Earth Planet Sci Lett 337–338:104–113

Coppola D, Piscopo D, Staudacher T, Cigolini C (2009) Lava discharge rate and effusive pattern at Piton de la Fournaise from MODIS data. J Volcanol Geotherm Res 184:174–192

Famin V, Welsch B, Okumura S, Bachèlery P, Nakashima S (2009) Three differentiation stages of a single magma at Piton de la Fournaise volcano (Reunion hotspot). Geochem Geophys Geosyst 10: Q01007. doi:10.1029/2008GC002015

Fisk MR, Upton BGJ, White WM (1988) Geochemical and experimental study of the genesis of magmas of Réunion Island, Indian Ocean. J Geophys Res 93:4933–4950

Fretzdorff S, Haase KM (2002) Geochemistry and petrology of lavas from the submarine flanks of Réunion Island (western Indian Ocean): implications for magma genesis and mantle source. Mineral Petrol 75:153–184

Garcia MO, Ito E, Eiler JM, Pietruszka AJ (1998) Crustal contamination of Kilauea volcano magmas revealed by oxygen isotope analyses of glass and olivine from Puu Oo eruption lavas. J Petrol 39:803–817

Garcia MO, Pietruszka AJ, Rhodes JM (2003) A petrologic perspective of Kilauea volcano's summit magma reservoir. J Petrol 44:2313–2339

Garcia MO, Ito E, Eiler JM (2008) Oxygen isotope evidence for chemical interaction of Kilauea historical magmas with basement rocks. J Petrol 49:757–769

Graham D, Lupton J, Albarède F, Condomines M (1990) Extreme temporal homogeneity of helium isotopes at Piton de la Fournaise, Réunion Island. Nature 347:545–548

Hart S, Gaetani GA (2006) Mantle Pb paradoxes: the sulfide solution. Contrib Mineral Petrol 152:295–308

Jackson MG, Weis D, Huang S (2012) Major element variations in Hawaiian shield lavas: source features and perspectives from global ocean island basalt (OIB) systematics. Geochem Geophys Geosyst 13: Q09009. doi:10.1029/2012GC004268

Lénat JF, Bachèlery P (1990) Structure et fonctionnement de la zone centrale du Piton de la Fournaise. In: Lénat JF (ed) Le volcanisme de la Réunion—Monographie. Centre de Recherche volcanologique. Université de Clermont-Ferrand, pp 257–296

Lénat JF, Fitterman D, Jackson DB, Labazuy P (2000) Geoelectrical structure of the central zone of Piton de la Fournaise volcano (Réunion). Bull Volcanol 62:75–89

Lénat JF, Bachèlery P, Peltier A (2012a) The interplay between collapse structures, hydrothermal systems, and magma intrusions: the case of the central area of Piton de la Fournaise volcano. Bull Volcanol 74:407–421

Lénat JF, Bachèlery P, Merle O (2012b) Anatomy of Piton de la Fournaise volcano (La Réunion, Indian Ocean). Bull Volcanol 74:1945–1961

Luais B (2004) Temporal changes in Nd isotopic composition of Piton de la Fournaise magmatism (Réunion Island, Indian Ocean). Geochem Geophys Geosyst 5:Q01008. doi:10.1029/2002GC000502

Marske JP, Pietruszka AJ, Weis D, Garcia MO, Rhodes JM (2007) Rapid passage of a small-scale mantle heterogeneity through the melting regions of Kilauea and Mauna Loa Volcanoes. Earth Planet Sci Lett 259:34–50

Marske JP, Garcia MO, Pietruszka AJ, Rhodes JM, Norman MD (2008) Geochemical variations during Kilauea's Puu Oo eruption reveal a fine-scale mixture of mantle heterogeneities within the Hawaiian plume. J Petrol 49:1297–1318

Michon L, Ferrazzini V, Di Muro A (2016) Magma paths at Piton de la Fournaise volcano. In: Bachèlery P, Lénat J-F, Di Muro A, Michon L (eds) Active volcanoes of the Southwest Indian Ocean: Piton de la Fournaise and Karthala. Active volcanoes of the World. Springer, Berlin

Nauret F, Arigot M, Louvat P, Moreira M (2007) Is the isotope composition of reunion plume really homogeneous? Geochim Cosmochim Acta 71:A707

Nercessian A, Hirn A, Lépine J-C, Sapin M (1996) Internal structure of Piton de la Fournaise volcano from seismic wave propagation and earthquake distribution. J Volcanol Geotherm Res 70:123–143

O'Hara MJ (1968) The bearing of phase equilibria studies on synthetic and natural systems on the origin and evolution of basic and ultrabasic rocks. Earth Sci Rev 4:69–133

Peltier A (2007) Suivi, modélisation et évolution des processus d'injection magmatique au Piton de la Fournaise (Réunion) à partir d'une analyse croisée des données de déformation, géochimiques et structurales. Thèse, Université de la Réunion, France

Peltier A, Famin V, Bachèlery P, Cayol V, Fukushima Y, Staudacher T (2008) Cyclic magma storages and transfers at Piton de la Fournaise volcano (La Réunion hotspot) inferred from deformation and geochemical data. Earth Planet Sci Lett 270:180–188

Peltier A, Bachèlery P, Staudacher T (2009) Magma transport and storage at Piton de la Fournaise (La Réunion) between 1972 and 2007: a review of geophysical and geochemical data. J Volcanol Geotherm Res 184:93–108

Pichavant M, Brugier Y, Di Muro A (2016) Petrological and experimental constraints. In: Bachèlery P, Lénat J-F, Di Muro A, Michon L (eds) Active volcanoes of the Southwest Indian Ocean: Piton de la Fournaise and Karthala. Active volcanoes of the World. Springer, Berlin

Pietruszka AJ, Garcia MO (1999a) The size and shape of Kilauea volcano's summit magma storage reservoir: a geochemical probe. Earth Planet Sci Lett 167:311–320

Pietruszka AJ, Garcia MO (1999b) A rapid fluctuation in the mantle source and melting history of Kilauea volcano inferred from the geochemistry of its historical summit lavas (1790–1982). J Petrol 40:1321–1342

Pietruszka AJ, Hauri EH, Blichert-Toft J (2009) Crustal contamination of mantle-derived magmas within Piton de la Fournaise volcano, Réunion Island. J Petrol 4:661–684

Pietruszka AJ, Norman MD, Garcia MO, Marske JP, Burns DH (2013) Chemical heterogeneity in the Hawaiian mantle plume from the alteration and dehydration of recycled oceanic crust. Earth Planet Sci Lett 361:298–309

Poland MP, Miklius A, Sutton AJ, Thornber CR (2012) A mantle-driven surge in magma supply to Kilauea volcano during 2003–2007. Nat Geosci 5:295–300

Salaün A, Villemant B, Semet MP, Staudacher T (2010) Cannibalism of olivine-rich cumulate xenoliths during the 1998 eruption of Piton de la Fournaise (La Réunion hotspot): implication for the generation of magma diversity. J Volcanol Geotherm Res 198:187–204

Schiano P, David K, Vlastélic I, Gannoun A, Klein M, Nauret F, Bonnand P (2012) Osmium isotope systematics of historical lavas from Piton de la Fournaise (Réunion Island, Indian Ocean). Contrib Mineral Petrol 164:805–820

Sigmarsson O, Condomines M, Bachèlery P (2005) Magma residence time beneath the Piton de la Fournaise volcano, Réunion Island, from U-series disequilibria. Earth Planet Sci Lett 234:223–234

Smietana M (2011) Pétrologie, geochronology (K-Ar) et géochimie élémentaire et isotopique (Sr, Nd, Hf, Pb) de laves anciennes de la Réunion: Implications sur la construction de l'édifice volcanique. Thèse, Université de la Réunion, France

Staudacher T, Ferrazzini V, Peltier A, Kowalski P, Boissier P, Catherine P, Lauret F, Massin F (2009) The April 2007 eruption and the Dolomieu crater collapse, two major events at Piton de la Fournaise (La Réunion Island, Indian Ocean). J Volcanol Geotherm Res 184:126–137

Stieltjes L, Moutou P (1989) A statistical and probabilistic study of the historic activity of Piton de la Fournaise, Réunion Island, Indian Ocean. J Volcanol Geotherm Res 36:67–86

Upton BGJ, Wadsworth WJ (1971) Rhyodacite glass in Réunion basalt. Mineral Mag 38:152–159

Upton BGJ, Semet MP, Joron J-L (2000) Cumulate clasts in the Bellecombe Ash Member, Piton de la Fournaise, Réunion Island, and their bearing on cumulative processes in the petrogenesis of the Réunion lavas. J Volcanol Geotherm Res 104:297–318

Villemant B, Salaün A, Staudacher T (2009) Evidence for a homogeneous primary magma at Piton de la Fournaise (La Réunion): a geochemical study of matrix glass, melt inclusions and Pélé's hairs of the 1998–2008 eruptive activity. J Volcanol Geotherm Res 184:79–92

Villeneuve N, Bachèlery P (2006) Revue de la typologie des éruptions au Piton de La Fournaise, processus et risques volcaniques associés. Cybergeo: Eur J Geogr Environ Nat Paysage, article 336. http://cybergeo.revues.org/2536. doi:10.4000/cybergeo.2536

Vlastélic I, Staudacher T (2014) Iron-oxide coatings in Quenched Lavas from Piton de la Fournaise Volcano (Réunion Island). In: 2014 Goldschmidt conference, 8–13 June 2014, Sacramento, USA

Vlastélic I, Abouchami W, Galer SJG, Hofmann AW (2001) Geographic control on Pb isotope distribution and sources in Indian Ocean Fe-Mn deposits. Geochim Cosmochim Acta 65:4303–4319

Vlastélic I, Staudacher T, Semet M (2005) Rapid change of lava composition from 1998 to 2002 at Piton de la Fournaise (Réunion) inferred from Pb isotopes and trace elements: evidence for variable crustal contamination. J Petrol 46:79–107

Vlastélic I, Lewin E, Staudacher T (2006) Th/U and other geochemical evidence for the Réunion plume sampling a less differentiated mantle domain. Earth Planet Sci Lett 248:364–378

Vlastélic I, Peltier A, Staudacher T (2007) Short-term (1998–2006) fluctuations of Pb isotopes at Piton de la Fournaise volcano (Réunion Island): origins and constraints on the size and shape of the magma reservoir. Chem Geol 244:202–220

Vlastélic I, Deniel C, Bosq C, Télouk P, Boivin P, Bachèlery P, Famin V, Staudacher T (2009) Pb isotope geochemistry of Piton de la Fournaise historical lavas. J Volcanol Geotherm Res 184:63–78

Vlastélic I, Staudacher T, Bachèlery P, Télouk P, Neuville D, Benbakkar M (2011) Lithium isotope fractionation during magma degassing: constraints from silicic differentiates and natural gas condensates from Piton de la Fournaise volcano (Réunion Island). Chem Geol 284:26–34

Vlastélic I, Menard G, Gannoun A, Piro J-L, Staudacher T, Famin V (2013) Magma degassing during the April 2007 collapse of Piton de la Fournaise: the record of semi-volatile trace elements (Li, B, Cu, In, Sn, Cd, Re, Tl, Bi). J Volcanol Geotherm Res 254:94–107

Welsch B (2010) Signification des Océanites dans le fonctionnement du Piton de la Fournaise, Île de la Réunion. Thèse, Université de la Réunion, France

Wright TL (1971) Chemistry of Kilauea and Mauna Loa in space and time. US Geol Surv Prof Pap 735:1–40

Wright TL, Fiske RS (1971) Origin of differentiated and hybrid lavas of Kilauea Volcano. Hawaii J Petrol 12:1–65

Magma Degassing at Piton de la Fournaise Volcano

12

Andrea Di Muro, Nicole Métrich, Patrick Allard, Alessandro Aiuppa, Mike Burton, Bo Galle and Thomas Staudacher

Abstract

Since about 1860 AD, magmatic gas release at Piton de la Fournaise volcano is very weak during intra-eruptive phases and essentially occurs during the relatively short-lived eruptions. Recent gas measurements performed during an eruption in October 2010, combined with detailed review of melt and fluid inclusion composition in magmas erupted over the past 50 kyrs, indicate that most PdF eruptions extrude magmas having variably degassed at shallow depth ($P < 1$ kbar). The average low gas flux results from the low volume of both the magma recharges (from $P > 1$ kbar) and the shallow magmas, whose fluids are efficiently scrubbed by the hydrothermal system and the water table. Quantification of SO_2 fluxes permits to track syn-eruptive magma ascent at shallow level (above sea level). Deeper exsolution of CO_2 (below sea level) provides a potential long-term geochemical precursor for the detection of new magma recharges and volcano unrest.

A. Di Muro (✉) · T. Staudacher
Observatoire Volcanologique du Piton de la Fournaise (OVPF), 14 RN3, 97418 La Plaine des Cafres, France
e-mail: dimuro@ipgp.fr

A. Di Muro · N. Métrich · P. Allard · T. Staudacher
Institut de Physique du Globe de Paris, Sorbonne Paris Cité, Université Paris Diderot, UMR 7154 CNRS, 75005 Paris, France

A. Aiuppa
University of Palermo, Palermo, Italy

A. Aiuppa
INGV Palermo, Palermo, Italy

M. Burton
INGV Pisa, Pisa, Italy

B. Galle
Chalmers University, Gothenburg, Sweden

P. Bachèlery et al. (eds.), *Active Volcanoes of the Southwest Indian Ocean*,
Active Volcanoes of the World, DOI 10.1007/978-3-642-31395-0_12

12.1 Introduction

Although Piton de la Fournaise volcano (PdF) is a very active basaltic volcano (3 eruptions and 22 Mm^3 of bulk lava emitted per year, on average, over the past 28 years; Roult et al. 2012), its gas emissions have long remained poorly documented. This lack of data results from two singularities at this volcano: (i) its lack of hydrothermal manifestations (e.g. hot springs) and the weakness of its intracrateric gas emanations during inter-eruptive periods; and (ii) a difficult access to magmatic gas emissions during its short-lived and often remote eruptions. Geophysical soundings indicate the presence of a hydrothermal system at depth, with hydrothermal fluids rising below the central cone and along the three rift zones that cut the edifice (Malengreau et al. 1994; Barde-Cabusson et al. 2012; Lénat et al. 2012). However, no persistent hydrothermal manifestation and modest CO_2 flux soil anomalies affect these main volcano-tectonic structures (Toutain et al. 2002; Liuzzo et al. 2014).

Intense and continuous degassing of sulphur-rich gases was only described by Maillard (1862) reporting his discovery in 1852 of a lava lake persistently active at the volcano summit (Cratère Brulant). After the disappearance of lava lake activity following the collapse of volcano summit in 1860, Vélain (1878) explored the summit cone soon after the 1874 eruption and described only low-temperature fumaroles (47.3 °C) composed essentially by water vapour, carbon dioxide and traces of chloridric acid. Since then, only weak fumaroles, with temperatures close or lower to that of boiling water at ~2600 m a.s.l (90 °C) have been reported during intra-eruptive phases inside the summit craters (Marty et al. 1993; Baubron and Toutain 1993; Staudacher 2010; Di Muro et al. 2012). Marty et al. (1993) found a mantle $\delta^{13}C$ signature of −4.1 per mil (‰) for CO_2 in summit emanations at 90 °C, despite their severe air contamination (90 %), demonstrating a weak but sustained leakage of magma-derived carbon dioxide through the central volcano conduit during non-eruptive periods. The most abundant gas release at PdF occurs during the eruptions and quickly vanishes once these end (Lénat et al. 1989; Allard et al. 2011a, b). Unfortunately, many eruptions are of very short duration and happen in remote areas or inaccessible sites (e.g. inner cliffs of Dolomieu crater or eastern volcano flank), preventing the possibility of gas sampling.

Since September 2007, three scanning MAX-DOAS stations have been installed on PdF within the frame of NOVAC European project in order to remotely monitor the volcanic SO_2 emissions. The scanning DOAS are located along the rim of the Enclos Fouqué caldera, between 3 and 4 km away from the Piton de la Fournaise central cone. The instruments measure the SO_2 optical depth during daylight between 7 a.m. and 6 p.m. local time, acquiring data every 15 min on average during favourable weather conditions, and, for a given plume height, the SO_2 flux is determined from the geometry of the network and the wind speed. Data acquired during eruptions and intrusions occurred since September 2008 have shown that increasing SO_2 release begins in broad coincidence with the onset of seismic tremor, tracking short-term magma ascent towards the surface (Garofalo et al. 2009) and broadly correlates with lava extrusion rate (Hibert et al. 2015). Since 2011, the summit crater emanations have been monitored using a MultiGaS sensor device. It was found that PdF intracrateric fumaroles are steam-dominated (average H_2O/CO_2 molar ratio: 250) and contain little H_2S as the only sulphur species (average C/S molar ratio: 120). This continuous monitoring reveals that the edifice contraction and the drop in seismicity recorded since the beginning of the pause in eruptive activity in February 2011 was paralleled by decreasing H_2O/CO_2 ratio and increasing CO_2/H_2S ratio, likely due to a declining rate of H_2O and H_2S residual emissions (Di Muro et al. 2012).

On the whole, monitoring data indicate that Piton de la Fournaise volcano is characterized by very weak gas fluxes during intra-eruptive periods (Di Muro et al. 2012). This behaviour clearly contrasts with that of other persistently active basaltic volcanoes such as Etna, Kilauea and Stromboli.

12.2 Review of Volcanic Gas Analyses at Piton de la Fournaise

The first available data on gases released by PdF basalts were obtained by extracting residual volatiles from 1943, 1952, 1959 and 1960 lavas (Chaigneau 1968). Volatiles were measured by mass spectrometry and selective absorption methods after their extraction at 1000 °C and dehydration on P_2O_5 (phosphorous pentoxide). The anhydrous volatile phase (0.15–0.22 wt% of the total sample) turned out to be very rich in reduced species (CO and H_2) and composed of CO_2, CO, H_2, HCl, SO_2, N_2 and hydrocarbons by order of abundance (Table 12.1). Most importantly, Chaigneau suggested that PdF volatile composition was relatively constant during a given eruption, but that the CO_2/CO molar ratio could significantly change (from 0.86 to 2.12) from one eruption to the other. Krafft and Chaigneau (1980) actually identified temporal changes in the eruptive gas phase by looking at residual volatiles extracted from a set of aphyric bombs collected during the long lasting November 1975 to April 1976 PdF eruption (Table 12.1). The residual volatile content was again in the range 0.11–0.22 wt%, with the following anhydrous molar proportions: CO_2 (31–61 %), CO (16–34 %), H_2 (1–17 %), HCl (10–19 %), SO_2 (<4 %), N_2 (0.4–2.3 %) and hydrocarbons (0.7–2.1 %). Interestingly, regular sampling of bombs permitted to identify a progressive increase in the CO_2/CO ratio (0.9–3.8) and a decrease in H_2 and hydrocarbons from the beginning to the end of the eruption. HCl was highest at the beginning and at the end of the eruptive sequence. Krafft and Chaigneau (1980) attributed such an evolution of the residual volatile composition to the involvement of deeper-derived fluids at the beginning of the eruption and to an increasing input of external fluids at the end of it.

High temperature gases (apparent equilibrium temperature: 1098 °C) were sampled for the first time from cracks on top of lava tubes, near eruptive vents, at the beginning of the December 4th, 1983–January 18th, 1984 PdF eruption (Lénat et al. 1989). Gases were collected in evacuated Pyrex vials containing P_2O_5 and connected to a quartz tube insulated with asbestos. Gas samples were analyzed by mass spectrometry (Table 12.1). Most samples turned out to be highly contaminated by air and affected by sulphur precipitation in the vials. Nevertheless, these analyses showed that the dominant volatile component was vapour water (88 mol%), followed by CO_2 then SO_2. The authors could not discriminate whether the relatively low proportion of carbon dioxide (C/S: 1.4; H/C: 30) reflected an advanced degassed stage of the outflowing lavas or previous shallow degassing of an older magma batch emplaced during the large 1977 eruption (Lénat et al. 1989). High temperature (850 °C) sublimates were collected in silica tube inserted into cracks on the roof of lava tubes during the 1985–1986 eruptions (Toutain et al. 1990). These sublimates were found to be rich in chlorides (NaCl, KCl) and fluorides but devoid of sulphides, sulphates and native sulphur, strongly suggesting previous outgassing of sulphur at the eruptive vents upstream of the lava flow.

In August 1998 Evans and Staudacher (2001) analysed cooler gases (T < 130 °C) from cracks in the crust of the active lava field and in the soil surrounding the new Kapor crater. Gases were drained out with a stainless steel pipe and were analyzed after cooling using a "MultiGAS" apparatus (IR multi-analyzer for CO_2, O_2 and CH_4 + electrochemical sensors for CO and H_2S). Helium was subsequently measured in the lab by mass spectrometry. Peaks in H_2S and CO on concentration profiles permitted to identify and locate hidden fractures and the buried lava tube. Increasing concentrations of CO_2, CO and CH_4 with increasing distance from the vent indicated that the combustion of the vegetation overridden by the lava flow was an important source of external volatiles.

The 2007 eruptive sequence was the largest event at PdF in the past hundred years. It ended a 9 year-long period (1998–2007) of continuous edifice inflation and intense eruptive activity and generated (i) the greatest lava output, (ii) the

Table 12.1 Composition of gases emitted by Piton de la Fournaise Volcano

Author sample	K&C80	K&C80	K&C80	K&C80	K&C80	K&C80	K&C80	Len89	Ev&St01	Ev&St01	Ev&St01	Ev&St01	Ev&St01	Ev&St01	Ev&St01	Ev&St01
	15	16	17	18	19	20	22	Except	South-Kap	South-Kap	South-Kap	South-Kap	South-Kap	South-Kap	South-Kap	South-Kap
	(0 °C-1 bar)	(0 °C-1 bar)	(0 °C-1 bar)	(0 °C-1 bar)	(0 °C-1 bar)	(0 °C-1 bar)	(0 °C-1 bar)	N_2, O_2, Ar	8/4/1998	8/4/1998	8/4/1998	8/4/1998	8/4/1998	8/4/1998	8/4/1998	8/4/1998
Unit	vol.%	vol.%	vol.%	vol.%	vol.%	vol.%	vol.%	wt% of sample	wt%	wt%	wt%	wt%	wt%	wt%	wt%	wt%
H_2O	nd	nd	nd	nd	nd	nd	nd	74.1								
CO_2	38.67	36.21	30.71	50.15	59.67	58.76	61.56	11.6								0.7
CO	31.56	33.82	32.76	33.64	22.50	22.04	16.20	0.4	0.0004	0.0004	0	0.0003	0	0.0002	0.0004	0.0005
H_2	14.48	12.47	17.08	4.69	0.99	2.74	1.05									
HCl	13.30	14.60	17.90	9.80	10.00	11.50	19.00	nd								
SO_2	0.17			0.54	3.78	3.01	0.00	12.6								
H_2S								0.03	0	0	0.0001	0.0002	0.0001	0	0.0004	0.0009
CH_4	0.17	0.09	0.16	0.18	0.27	0.18	0.41	0.01	0.1	0.2	0	0.1	0	2	1	0
C_2H_6	0.09	0.26	0.33	0.09	0.09	0.44	0.08									
C_3H_8	0.17	0.17	0.16	0.18	0.18	0.18	0.24									
C_2H_4		0.09	0.41	0.00	0.09	0.35	0.41									
C_3H_6	0.09	0.17	0.00	0.18	0.09	0.09	0.24									
C_4H_8							0.16									
C_2H_2						0.09	0.08									
Benzene	0.09	0.09	0.08	0.09	0.09		0.08									
Toluene																
N_2	1.21	2.05	0.41	0.45	2.25	0.62	0.41	1.2								
HePPB									5						18	16
O_2									18.7	18.7	18.7	18.9	18.7	18.7	18.6	18.2
Total	100	100	100	100	100	100	100	100								
Tot. Gas wt%	0.11	0.14	0.13	0.13	0.13	0.22	0.22									
Molar																
CO_2/CO	1.23	1.07	0.94	1.49	2.65	2.67	3.80	18.28								890.91
S/C	0.002			0.006	0.046	0.037		0.71								
H/C								29.59								

(continued)

Table 12.1 (continued)

Author sample	Ev&St01	Ev&St01	Ev&St01	Ev&St01	Ev&St01	Ev&St01	Ev&St01	Ev&St01	Ev&St01	Ev&St01	Ev&St01	Ev&St01	Ev&St01	Ev&St01	Ev&St01	Ev&St01	Ev&St01
	South-Kap	South-Kap	South-Kap	South-Kap	North-Kap	North-Kap	North-Kap	North-Kap	North-Kap	North-Kap	North-Kap	North-Kap	North-Kap	North-Kap	LavaFlow-lower	LavaFlow-lower	LavaFlow-lower
	8/4/1998	8/4/1998	8/4/1998	8/4/1998	8/6/1998	8/6/1998	8/6/1998	8/6/1998	8/6/1998	8/6/1998	8/6/1998	8/6/1998	8/6/1998	8/6/1998	August-1998	August-1998	August-1998
Unit	wt%	wt%	wt%	wt%	wt%	wt%	wt%	wt%	wt%	wt%	wt%	wt%	wt%	wt%	wt%	wt%	wt%
H_2O																	
CO_2	0.35					0.1	0.1	0.2	0.5	0.7	0.6	0.5	0.2				
CO	0.00045	0.0002	0	0.0002	0	0.0002	0.0005	0.0003	0.0008	0.001	0.0007	0.0006	0.0002	0	0.0102	0.029	0.0188
H_2																	
HCl																	
SO_2																	
H_2S	0.00065	0.0005	0	0.0001	0	0	0	0.0002	0.0003	0.0003	0.0003	0	0.0001	0	0	0.0003	0.0003
CH_4	1.5	0	2	2	0.2	0.1		0.2	0.3	0.3	0.2	0.1		0	2.3	2.8	0.1
C_2H_6																	
C_3H_8																	
C_2H_4																	
C_3H_6																	
C_4H_8																	
C_2H_2																	
Benzene																	
Toluene																	
N_2																	
HePPB	17				12				28					19	4	6	11
O_2	18.4	18.6	18.4	18.5	17.5	16.4	16.9	16.1	14.3	13.4	14.9	14.1	15.2	16.8	17.6	15.2	14.1
Total																	
Tot. Gas wt%																	
Molar																	
CO_2/CO	494.95					318.18	127.27	424.24	397.73	445.45	545.45	530.30	636.36				
S/C																	
H/C																	

(continued)

Table 12.1 (continued)

	South-Kap	South-Kap	South-Kap	South-Kap	South-Kap	South-Kap	South-Kap	South-Kap	South-Kap	South-Kap	South-Kap	South-Kap	North-Kap
	8/4/1998	8/4/1998	8/4/1998	8/4/1998	8/4/1998	8/4/1998	8/4/1998	8/4/1998	8/4/1998	8/4/1998	8/4/1998	8/4/1998	8/6/1998
H_2O													
CO_2 %								0.7	0.35				
COPPM	4	4	0	3	0	2	4	5	4.5	2	0	2	
H_2													
HCl													
SO_2													
H_2SPPM			1	2	1		4	9	6.5	5		1	
CH_4 %	0.1	0.2	0	0.1	0	2	1	0	1.5	0	2	2	0.2
C_2H_6													
C_3H_8													
C_2H_4													
C_3H_6													
C_4H_8													
C_2H_2													
Benzene													
Toluene													
N_2													
HePPB	5						18	16	17				12
O_2	18.7	18.7	18.7	18.9	18.7	18.7	18.6	18.2	18.4	18.6	18.4	18.5	17.5

(continued)

Table 12.1 (continued)

	North-Kap	North-Kap	North-Kap	North-Kap	North-Kap	North-Kap	North-Kap	North-Kap	North-Kap	LavaFlow-lower	LavaFlow-lower	LavaFlow-lower
	8/6/1998	8/6/1998	8/6/1998	8/6/1998	8/6/1998	8/6/1998	8/6/1998	8/5/1998	8/6/1998	August-1998	August-1998	August-1998
H_2O												
CO_2 %	0.1	0.1	0.2	0.5	0.7	0.6	0.5	0.2				
COPPM	2	5	3	8	10	7	6	2		102	290	188
H_2												
HCl												
SO_2												
H_2SPPM			2	3	3	3		1		0	3	3
CH_4 %	0.1		0.2	0.3	0.3	0.2	0.1		0	2.3	2.8	0.1
C_2H_6												
C_3H_8												
C_2H_4												
C_3H_6												
C_4H_8												
C_2H_2												
Benzene												
Toluene												
N_2												
HePPB				28					19	4	6	11
O_2	16.4	16.9	16.1	14.3	13.4	14.9	14.1	15.2	16.8	17.6	15.2	14.1

most violent lava fountaining activity (>200 m high), (iii) the largest SO_2 release ever measured (≥230 kt) at PdF (Tulet and Villeneuve 2011; Roult et al. 2012). The bulk magma volume (~0.24 or ~0.15 km^3 DRE) involved in the 31 days of eruption was similar to that emitted during the entire 1998–2006 period (~0.25; ~0.16 km^3 DRE; Roult et al. 2012). Space-borne monitoring of the 2007 eruption plume revealed a huge SO_2 release, although diverging from 230 kt (Tulet and Villeneuve 2011) to 935 ± 244 kt (Gouhier and Coppola 2011). While the lower value is compatible with the syn-eruptive degassing of 0.15 km^3 (DRE) olivine-rich basalt, the upper value was ascribed to the degassing of a shallow sill intrusion below the eastern volcano flank (Massin et al. 2011), to a deep magma recharge before April climactic event (Di Muro et al. 2014b) and/or to a sudden flashing of S-rich PdF hydrothermal system upon collapse of the Dolomieu summit crater (Gouhier and Coppola 2011). This latter interpretation, however, is unlikely since flashing of SO_4-bearing hydrothermal system is unable to generate SO_2. Huge magma withdrawal during the 2007 event was followed since then by (i) a trend of continuous deflation centered on PdF summit edifice (Chap. 14), (ii) small ($<3 \times 10^6$ m^3) summit eruptions in 2008–2010 that produced relatively evolved magmas (MgO: 7.1 ± 0.6 wt%) (Chap. 10) and (iii) a long lasting rest in volcanic activity recorded between February 2011 and June 2014 by the Piton de la Fournaise Volcanological Observatory (OVPF/IPGP).

Viane et al. (2009) showed that SO_2 emissions during PdF eruptions, including the 2007 one, do not result in a broad increase in hospitalizations for asthma, but can have a local health impact depending on ambient SO_2 concentrations and meteorological conditions.

During about 20 years (1983–2004), geochemical monitoring of PdF was performed by continuous measurements of radon in soil variations. The network successfully permitted to detect a strong radon anomaly preceding the 1996 seismic crisis (Monnin and Seidel 1998). The network was dismantled in 2004 because of the difficulty in filtering the environmental effects on the radon signal and its non-systematic correlation with volcanic activity (Seidel et al. 1988; Segovia et al. 1997; Barillon et al. 1993; Basséguy et al. 2006). After the major 2007 eruption, a new effort has been made to modernize the OVPF geochemical monitoring network.

The very first complete analysis of high temperature magmatic gas emissions from PdF was realized by Allard et al. (2011a, b) during the October 2010 eruption. The compositions and fluxes of gases released from three active fissure vents ejecting aphyric magma were determined by combining both remote sensing tools (OP-FTIR absorption spectroscopy, DOAS spectroscopy for SO_2 flux) and in situ techniques (MultiGaS sensor, filter pack plume sampling). In particular, the high frequency (~1–4 s) of OP-FTIR measurements allowed the first cross-correlation of temporal changes in gas composition with seismic tremor intensity and the magma extrusion rate, the latter being deduced by space-borne thermal (MODIS) data. Like 1983 gases, the 2010 gases resulted to be water-rich (95–98 mol%; average molar ratio H_2O/CO_2: 52–70) and have a low C/S (1.2–2.9). Their low Cl/F (0.9–1.8) and high Cl/Br (700–1000) and S/Cl (average: 15–17) molar ratios are typical of Hawaiian-type hot spot basaltic magmatism (Allard et al. 2011a). Both SO_2 flux and lava extrusion rate decreased synchronously and progressively during the first week of the eruption. High-frequency correlations between CO_2/HCl and seismic tremor suggested either the involvement of a previously accumulated CO_2-rich gas phase or the occurrence of differential CO_2 bubbling across the feeding dyke. On the contrary, sulphur, chlorine and fluorine were interpreted to track a syn-eruptive (closed system) degassing during magma ascent and decompression. The measured excess in H_2O flux opens the possibility of entrainment of external steam from meteoric groundwater during dyke propagation. Evolution from closed to open system degassing near the end of the eruption was inferred to produce the observed strong decoupling between seismic tremor (increasing) and lava extrusion rate (decreasing). At PdF, the critical influence of magma ascent rate on the evolution from closed to open system degassing regime has also been evidenced by cross correlating thermal and seismic data (Coppola et al. 2009).

12.3 Compositions of Olivine and of Their Melt and Fluid Inclusions at Piton de la Fournaise

Vélain (1878) performed in its report of the 1874 eruption a first detailed description of glass inclusions and aqueous fluid inclusions in olivine, pyroxene and plagioclase in PdF lavas, only 16 years after Sorby's (1858) classic paper.

Major and volatile element compositions of melt inclusions and glassy matrices have been quantified for the different magma types which were extruded during the last 50 kys at PdF and are here reviewed (see also Pichavant et al. 2016, Chap. 10). These include (i) the Steady State basalts (SSB) and the olivine cumulative "picrites" (oceanites) where olivine, in association with Cr-spinel, is the dominant phase (Albarède et al. 1997; Boivin and Bachèlery 2009), (ii) the cotectic basalts (ol + cpx + pl) typical of the PdF summit activity which display the higher modal fraction of plagioclase (Boivin and Bachèlery 2009; Di Muro et al. 2014a), (iii) the rarely extruded alkali basalts in which the clinopyroxene is present as phenocrysts (Upton et al. 2000). Hence olivine is ubiquitous that explains why the vast majority of PdF melt and all fluid inclusions have been analysed in this mineral phase (Bureau 1996, 1998a, b, 1999; Famin et al. 2009; Vigouroux et al. 2009; Villemant et al. 2009; Welsch et al. 2009; Di Muro et al. 2014b). In contrast, only a few analyses exist for MIs in pyroxene (Bureau 1996) and spinel (Bureau 1996; Clocchiatti et al. 1979) and none for MIs in plagioclase.

12.3.1 The Olivine Crystals

In most eruptive products, olivine is compositionally homogeneous and contains one or several primary glass inclusions (often <150–100 μm) with a relatively small bubble. Secondary MIs aligned along healed secondary shear fractures in association with abundant CO_2-rich fluid inclusions and sulphide globules are frequently observed. Systematic mineralogical studies of the lava emplaced during the last 100 years reveal that the molar Fo amount in olivine did not exceed 85.5 (Mg# 0.855), with the notable exception of Hudson 1998 eruption (up to Fo_{88}). Actually, magnesian olivine (Fo > 87) is quite uncommon at PdF (Albarède et al. 1997; Boivin and Bachèlery 2009; Di Muro et al. 2014a). It was described in old picrites of Piton des Neiges ($Fo_{90.6}$; Sobolev and Nikogosian 1994) and in the submarine products of Catherine cone on the noth-eastern PdF flank ($Fo_{90.6}$; Fretzdorff and Haase 2002). It is also typical of Piton des Neiges dunite cumulates (Sobolev and Nikogosian 1994) which result of olivine accumulation during the basalt crystallization (Albarède et al. 1997). The average composition of olivine in PdF dunite is more evolved and corresponds to that of olivine carried by sub-aerial Pdf basalts.

PdF basalts display two olivine modes (Fo_{84} and Fo_{82}; Fig. 12.1a). The Fo_{84} mode is best represented in crystal-rich oceanites, while the Fo_{82} mode prevails in aphyric products. Some aphyric products (e.g. November 2009) may display a bimodal distribution (Fig. 12.1a). In recent eruptive products, the most magnesian olivines have only been identified in magmas erupted during eruptions associated with deeper than usual seismicity (Hudson 1998 eruption;

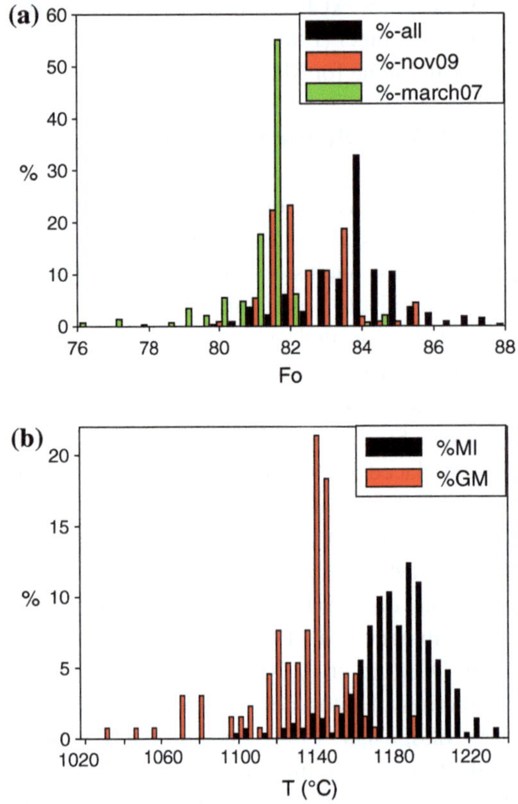

Fig. 12.1 **a** Distribution of olivine compositions analyzed for their MIs. **b** Distribution of melt temperature in primary melt inclusions (up to 1240 °C) and glassy ground masses (only seldom above 1170 °C) of Piton de la Fournaise volcano

Bureau et al. 1999). Aphyric magmas can thus contain olivine with both lower (e.g. March 2007 and November 2009) or higher (e.g. Hudson 1998) Fo contents than the average.

12.3.2 Experimental Olivine Geothermometry

The experimental temperature measurement of the melt inclusion homogenization under heating stage with very fast heating rates indicates that the olivines Fo_{87-81} of PdF crystallized in a temperature range between 1215 and 1170 °C (Bureau et al. 1998b; Massare et al. 2002). These results are in good agreement with the temperature range calculated using the geothermometer of Helz and Thornber (1987) calibrated for olivine from the 1959 Kilauea Iki eruption of Hawai'i (Fig. 12.1b). The relatively low temperatures (<1170–1190 °C) computed for the glassy groundmasses are fully consistent with the lava emission temperatures (1170–870 °C) which have been measured in situ (Boivin and Bachèlery 2009) or remotely (Staudacher 2010) (Fig. 12.1b).

12.3.3 Major Element Compositions of Melt Inclusions and Matrices

Isolated "primary" melt inclusions, entrapped during crystal growth, have been analysed in mm-sized olivine phenocrysts, whose compositions range from Fo_{78} to Fo_{88} and average at $Fo_{83.6 \pm 1.8}$ (N = 310). The MgO content of these inclusions (after correction for post-entrapment crystallization) does not exceed 10.8 wt% (Fig. 12.2a). Average MgO of MIs is 8.3 ± 1.1 wt% a value closely similar to the highest concentrations occasionally measured in Pelé hairs (MgO 8.7 wt %; Villemant et al. 2009; Di Muro et al. 2014a) and corresponds to the expected melt MgO content in equilibrium with Fo_{84} olivines. These MgO concentrations are lower than that estimated (15 wt%) for PdF parental magma (Sobolev and Nikogosian 1994; Albarède et al. 1997; Fretzdorff and Haase 2002; Villemant et al. 2009) (Fig. 12.2a).

However, we stress that primary magma compositions and temperatures have been estimated assuming that olivine is the only liquidus phase together with Cr-spinel at high pressure, and excluding/neglecting the possible influence of early clinopyroxene crystallization.

These melt inclusions span a relatively narrow range in CaO/Al_2O_3 (0.8 ± 0.1 wt%) and K_2O contents (0.7 ± 0.1 wt%), consistent with the dominant fractionation of olivine and the late stage crystallization of clinopyroxene and plagioclase (Chap. 9 and Fig. 12.2a). This homogeneity contrasts with the chemical variability of the secondary melt inclusions aligned along healed fractures. These latter are enriched in incompatible elements and have average K_2O contents of 2.2 ± 2.6 wt% (K_2O can attain up to 9.8 wt%) at CaO/Al_2O_3 of 0.7 ± 0.3 wt% (Fig. 12.2b). Their content in incompatible elements is higher than

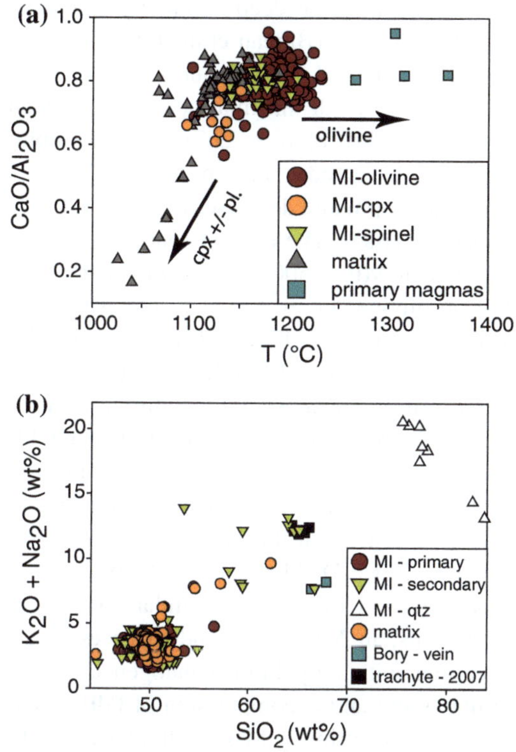

Fig. 12.2 a Evolution of CaO/Al_2O_3 ratio in parental basaltic melts, melt inclusions and groundmasses at Piton de la Fournaise as a function of melt temperature (directly proportional to MgO content). **b** Silica and total alkali contents in primary and secondary melt inclusions and evolved melts of Piton de la Fournaise and Piton des Neiges. The 2007 trachyte melt and most secondary melt inclusion plot on a similar enrichment trend

that of most glassy groundmasses, which have average K_2O of 0.8 ± 0.2 wt% and CaO/Al_2O_3 of 0.8 ± 0.1 wt%, values very close to those of primary MIs (Fig. 12.2). Secondary MIs plot along a trend of alkali enrichment towards highly differentiated melts (Fig. 12.2b). These melts have been documented as thin veins in lavas filling the Bory crater (Upton and Wadsworth 1971) and as small volume trachyte pumices erupted during the 2007 eruption (Vlastélic et al. 2011). At PdF, silica-alkali rich melts have been interpreted either as the ultimate products of magma differentiation (Clocchiatti and Nativel 1984; Vlastélic et al. 2011) or as the product of partial re-melting of older intrusives (gabbros or dunites; Bureau et al. 1998b; Pietruszka et al. 2009; Salaun et al. 2010). The variability of PdF lava geochemical spectrum, and possibly of dissolved volatile contents, could be partly related to mixing/assimilation processes between mafic and highly evolved melts (Bureau et al. 1998b; Pietruszka et al. 2009; Salaun et al. 2010; Di Muro et al. 2014a).

12.3.4 Volatile Content of "Primary" Melt Inclusions and Matrices

We present here a general overview of the CO_2, H_2O, S, Cl and F amounts dissolved in PdF magmas as inferred from the analysis of "primary" melt inclusions hosted in olivine. In order to be able to compare the published data sets on H_2O, which was systematically obtained by FTIR, we have recalculated all analyses using a single IR molar absorptivity of 63 L mol^{-1} cm^{-1} (Mercier et al. 2010) and a melt density of 2700 kg/m^3. All reported analyses have been corrected for post-entrapment crystallization following the procedure of Bureau et al. (1998a). Hence the database is self-consistent. It gathers about 1000 analyses, of which 40 % include the determination of the main volatile species (H_2O, CO_2, S, Cl, F). About 10 % of MIs were analysed for volatiles only. Most analyses include only some of the volatile species. For instance, only a few data exist for fluorine and no data at all about F content in glasses with low ($K_2O < 0.5$) or high ($K_2O > 1$ wt%) alkali contents. No data exist for H_2O or CO_2 content in MIs hosted in pyroxene, spinel or plagioclase. Very few data are available for H_2O and CO_2 contents in groundmasses.

Water. The H_2O content measured in 171 primary, isolated MIs ranges from 0.1 to 2.0 wt% and averages at 0.8 ± 0.3 wt%, against 0.2 ± 0.1 wt% in matrix glasses (Fig. 12.3a). The highest values of 1.7–2 wt% (1998 Hudson eruption; Bureau et al. 1999) and 1.8 wt% (November 2002 eruption; Vigouroux et al. 2009) have been found in a very few MIs. These high water concentrations require further confirmation. An pre-eruptive H_2O content in present-day PdF basaltic melts of 1.3 wt% has been estimated by Di Muro et al. (2014b) by cross-correlating H_2O and Ce concentrations in 2007 MIs. H_2O content correlates only weakly with

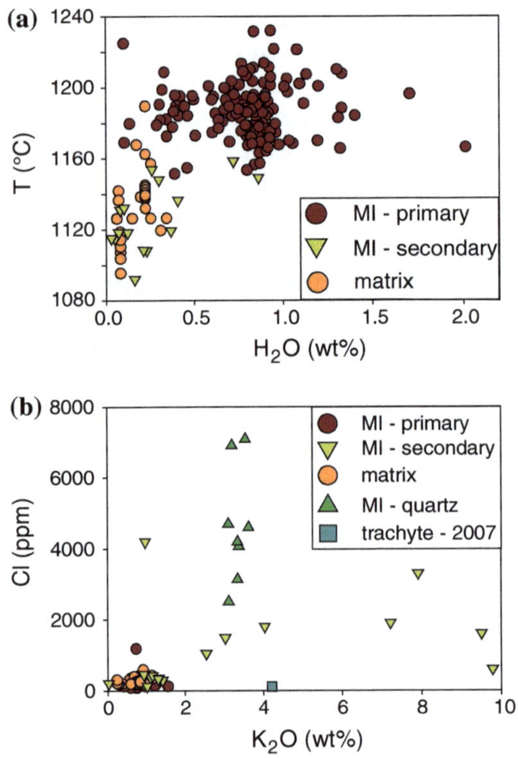

Fig. 12.3 **a** H_2O contents in primary and secondary melt inclusions, and glassy matrices of PdF eruptive products. Note that secondary MIs and matrices share apparent similar temperature. This is partly and artefact because secondary MIs cannot be corrected for post-entrapment crystallization. **b** Cl and K_2O in primary and secondary melts inclusions, glassy matrices and evolved melts of Piton de la Fournaise and Piton des Neiges. Secondary melt inclusions of PdF and evolved residual melts of PdN plot on very distinct enrichment trends. Note the very low halogen content in 2007 trachyte

K_2O, MgO or CaO/Al_2O_3 ratio suggesting only limited magma crystallization during deep magma ascent and degassing. Plagioclase crystallization initiates in relatively evolved melts having 0.8–0.1 wt% of H_2O, and is enhanced at low concentrations (H_2O < 0.4 wt%) as observed in PdF low-MgO cotectic basalts (Di Muro et al. 2014a).

Carbon. Dissolved carbon content in primary MIs of PdF magmas, expressed as CO_2, spans a significant range (4363–25 ppm; n: 95). Average CO_2 content in MIs is relatively low (450 ppm). The highest concentrations (2000–4360 ppm) have been measured in primary MIs in Fo_{83-87} olivines carried by 1998 Hudson lava, consistently with the deep seismicity associated with this eruption (Bureau et al. 1999).

Sulphur. Dissolved sulphur content (S) in primary MIs of PdF magmas spans a significant range (2200–120 ppm; n: 263).Average sulphur content dissolved in primary MIs is 1050 ± 330 ppm and decreases to 250 ± 170 ppm in matrix glasses. The highest concentrations are recorded in MIs (1500–2200 ppm) in the 1998 Hudson olivine (Fo ~ 88) and, interestingly, in those of pyroxenes (970–2200 ppm) of the Bellecombe ash (Bureau 1996; Bureau et al. 1998b). PdF magmas are at sulphide saturation and S does not correlate with other major elements like K_2O, CaO, FeO or FeO_T. Early and high pressure sulphide immiscibility could be responsible for the low S content of the most PdF basalts (Collins et al. 2012; Di Muro et al. 2014b).

Halogens. Cl/F ratio is similar, within the error limits, in MIs (0.54) and matrices (0.48), consistent with only limited halogen degassing during ascent and decompression of PdF basalts. Cl contents of primary MIs in olivines are in the range 551–94 ppm (n: 216). Average chlorine content in primary MIs of olivines is 260 ± 60 ppm. Again, the high-pressure MIs of 1998 Hudson eruption record the highest average Cl concentrations (350 ± 40 ppm). MIs in clinopyroxene have slightly higher average Cl contents (317 ± 65) with respect to olivines. Fast quenched Pele's hairs erupted in 2008 contain average residual Cl contents of 230 ± 20 ppm (Di Muro et al. 2014a). In primary MIs, Chlorine concentration does not correlate with MgO and exhibits weak positive correlation with K_2O and slight reverse correlation with CaO/Al_2O_3. Secondary MIs are often enriched in Cl (average 1160 ppm; maximum content is 4200 ppm in Piton des Cailles olivines; Bureau 1996).

Average fluorine content is 500 ± 70 ppm in primary MIs (n: 34) and 540 ± 160 in 2008 Pele's hairs. Fluorine is slightly more variable in cooler (low MgO) and high K_2O glasses, possibly because of the competition between late stage degassing and crystallization. The rare trachyte melt erupted during the 2007 PdF eruption has very low Cl content (120 ppm) and very high Cl/F (1.2) ratio (Vlastélic et al. 2011).

As discussed in detail by Bureau et al. (1998b), both mantle source heterogeneity and shallow contamination by evolved melts can potentially produce variable F/K_2O and Na_2O/K_2O ratios in PdF melts.

12.3.5 CO_2-Rich Fluid Inclusions Trapped in Healed Fractures

Fluid inclusions confirm that an exsolved CO_2-rich fluid phase coexists with the PdF basaltic melts over a very large range of pressure and depth. High density CO_2 fluid inclusions are reported from products erupted along the N120 rift zone (Chisny: up to 427 MPa; Piton des Cailles up to 503 MPa; Hudson: up to 389 MPa) (Bureau et al. 1998a, b). In magmas emitted closer to the central cone, fluid density is lower than 0.29 g/cm^3 (P_{CO_2}) < 146 MPa) and most fluid inclusions record trapping pressure (P_{CO_2}) lower than 73 MPa (ca. 2750 m below summit; Bureau et al. 1998a, b; Famin et al. 2009). This range is consistent with the low H_2O–CO_2 contents of most PdF MIs, which require saturation pressures lower than 100 MPa (Fig. 12.4). We stress that these P_{CO_2} values represent the last stage of magma re-equilibration. These fluid inclusions are formed by H_2O and CO_2 vapours with a high estimated H_2O/CO_2 molar proportion estimated at 3/5 (Sobolev et al. 1983), again consistent with low pressure entrapment (Table 12.2).

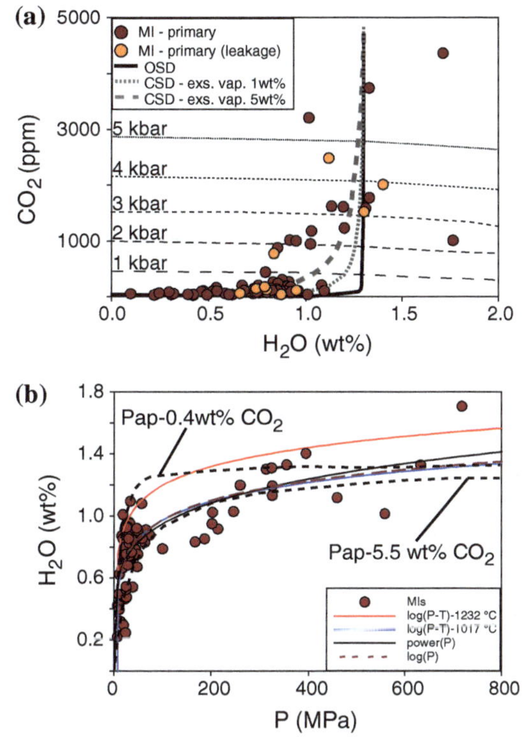

Fig. 12.4 **a** Dissolved H_2O and CO_2 in PdF primary glass inclusions and matrices. Trapping saturation pressures and degassing paths calculated using the Newman and Lowenstern (2002) saturation model. The curves show the CO_2–H_2O evolution during open system degassing (*OSD*) and closed system degassing (*CSD*) and variable amount of CO_2 in the bulk system. Available data indicate significant deviation from degassing curves at P < 3 kbar. Melt inclusions possibly affected by post-entrapment water-loss have been identified on the basis of their shift from expected sulphur degassing curves. **b** Fitting of dissolved water contents in PdF primary melt inclusions and matrices using several semi-empirical equations. For comparison, thermodynamic modelling of H_2O solubility as a function of pressure and variable total CO_2 content in the system is shown (Papale et al. 2006; Pap-lines in the figure)

12.4 Modelling Magma Degassing at Piton de la Fournaise

Dissolved water and carbon contents provide a minimum estimate of the melt saturation pressures (Fig. 12.4a). Olivine-hosted melt inclusions track magma crystallization and degassing over a pressure range similar to that estimated by fluid inclusions (≤5 kbar). Newman and Lowenstern's model predicts maximum trapping pressure of 7.2 kbar for the melt inclusions with the highest volatile content. However, we stress that this range is outside the calibration domain of the model. Lower pressures (no more than 5.1 kbar) are obtained when using the Papale et al. (2006) model (Table 12.2). Papale's model predicts pressures about 40 % lower than Newman's model in the 1–7 kbar range. The discrepancy becomes much smaller in the low pressure domain (<0.5 kbar) recorded by most PdF melt inclusions. In order to permit an easier comparison with datasets on other volcanoes, fitting and data description are performed following Newman and Lowenstern pressure data.

Table 12.2 Representative melt inclusion composition of PdF magmas trapped at decreasing saturation pressures (5–0.1 kbar)

Eruption	Hudson	Hudson	Hudson	Hudson	Hudson	Hudson	P. Manapany	November	03 April	25 April
Age	1998	1998	1998	1998	1998	1998	50 kyrs	2009	2007	2007
H_2O (wt%)	1.7	1.3	1.1	1.4	2.0	0.9	0.9	0.8	1.1	0.3
CO_2 (ppm)	4363	3741	2480	2009	1771	878	306	178	101	56
SiO_2	50.4	50.4	49.8	50.3	50.0	50.5	49.9	49.2	48.4	49.1
TiO_2	2.7	2.6	2.5	2.6	3.0	2.5	2.3	2.8	2.6	2.6
Al_2O_3	14.3	14.2	14.4	14.0	15.0	14.1	14.8	14.3	13.6	13.9
Fe_2O_3	1.3	1.3	1.3	1.2	1.1	1.3	1.2	1.1	1.4	1.3
FeO	7.9	8.1	8.1	8.9	8.5	8.2	9.1	9.9	10.7	9.8
MnO	0.1	0.1	0.1	0.2	0.2	0.2	0.2	0.2	0.2	0.2
MgO	9.1	8.7	8.8	8.5	7.6	8.8	8.1	8.2	9.4	9.2
CaO	11.4	11.6	11.9	11.1	11.3	11.4	10.9	10.6	10.6	10.6
Na_2O	2.2	2.4	2.5	2.5	2.6	2.6	2.8	2.9	2.5	2.6
K_2O	0.6	0.6	0.6	0.7	0.8	0.5	0.7	0.9	0.7	0.7
Total	100.0	100.0	100.0	100.0	100.0	100.0	100.0	100.0	100.0	100.0
Fo	87.4	86.4	86.7	84.7	83.9	86.4	83.6	82.8	83.6	84.6
x	0.061	0.078	0.067	0.061	0.026	0.065	0.039	0.061	0.057	0.085
T (Helz)	1196	1188	1191	1184	1166	1191	1177	1179	1202	1199
Pressure (Mpa) N&L	716	632	458	394	377	187	73	44	34	13
Pressure (Mpa) Pap96	506	452	315	300	266	132	51	34	31	10
Diff%	42	40	45	31	42	42	44	30	8	29
Gas phase (Papale)										
H_2O (wt%)	4.8	3.4	3.4	4.6	8.8	4.7	10.4	11.9	22.0	12.0
CO_2 (wt%)	95.3	96.6	96.6	95.4	91.2	95.3	89.6	88.1	78.0	88.0
Molar H_2O/CO_2	0.18	0.13	0.13	0.18	0.35	0.18	0.42	0.49	1.03	0.50

In order to model degassing trends at PdF, the definition of the initial dissolved volatile contents is required. At PdF this is a highly challenging task because of (i) the very limited number of MIs recording high pressure, (ii) the scattering in high pressure data and (iii) the potential effect of H_2O leakage and/or CO_2 migration in the bubble, and (iv) the dependence of barometric estimations on adopted model. In order to constrain the possible initial concentration of dissolved volatiles at high pressure, we ha fitted a population of 130 MI analyses. Fitting of dissolved volatile concentrations as a function of pressure (P, in MPa) using a simple power equation like with a = 0.394 and b = 0.19

$$H_2O(wt\%) = a * P^b \quad (12.1a)$$

matches closed system degassing trends with high exsolved CO_2 contents in the system, but results in relatively poor fitting ($R^2_{H_2O} = 0.576$; $R^2_S = 0.612$) of the dataset of PdF melt inclusions (Fig. 12.4b). In order to improve the quality of the fitting, we introduced the effect of temperature (T, in °C) and composition (X, in wt%). We tested 2 groups of regression equations in order to reproduce the concentration of volatiles dissolved in the PdF silicate melt.

$$Vol(wt\%) = a * P^b + c * \exp(d * T) + e * \exp(f * X) + g \quad (12.1b)$$

$$Vol(wt\%) = a * \ln(P) + b * \exp(c * T) + d * \exp(e * X) + f \quad (12.2)$$

The simple power equation (Eq. 12.1a) has the advantage of permitting to fit the dataset at its lower pressure end (P < 10 MPa), and predicts a moderate but continuous increase in volatile concentration with pressure (Fig. 12.4b). The more complex equations like Eq. 12.1b (power) or Eq. 12.2 (logarithmic) produce a sensible increase in the quality of fitting, result in flatter trends at high pressure (Fig. 12.4b), but are discontinuous at very low pressure (P < 10 MPa). All equations respect the assumptions of normality and constant variance of the fitted population, but Eq. 12.1b (power) permits a stricter adherence to the assumptions underlying non linear regression. Nevertheless, in the following we adopted Eq. 12.2 because it gives larger weight to water-rich MIs for a given pressure range and reduces the potential influence of MIs having suffered post-entrapment leakage. We stress that all tested equations produce quite similar results at low (eruptive) temperature, but Eq. 12.2 permits to better describe the ascent and degassing of high temperature water-rich melts (Fig. 12.4b; Table 12.3). In any case, more data are required in the high pressure range (P > 3 kbar) in order to better constraining the regression equation for PdF melts.

Water and sulphur. Modelling of H_2O and S release during magma ascent at PdF has usually assumed a strong dependence on pressure only (Bureau et al. 1998a, b; Famin et al. 2009). An equation like volatile% = a * In(P) produces only poor fitting for H_2O (R^2: 0.587; Fig. 12.4b), somewhat better for S (R^2: 0.738). In both cases, better fitting of available melt inclusion data is obtained by integrating the effect of temperature T (estimated from the MgO content of MIs).

Table 12.3 Fitting parameters of the equations describing the content (wt%) of volatiles dissolved in Piton de la Fournaise basalts (X_{melt}) as a function of pressure (P, MPa) and temperature (T, °C)

Equation	$X_{melt}(wt\%) = a * (In)P + b * \exp(c * T) + d$				R^2
H_2O	a = 0.1759	b = 4.019E−31	c = 0.0556	d = 0.1640	0.672
S	a = 0.0265	b = 70.43	c = 8.21E−6	d = −71.12	0.788
Cl	a = 0.0015	b = 0.0	c = 0.0	d = 0.0206	0.160
F	a = −0.0011	b = 0.0	c = 0.0	d = 0.0552	0.003
Equation	$X_{melt}(wt\%) = a * P^\wedge 2 + b * P + c$				R^2
CO_2	a = 3.019E−7	b = 0.0004	c = −0.0005		0.999

The dataset is reproduced in this case with a moderate coefficient of determination (R^2: 0.672) for water and a R^2: 0.788 is obtained for S. Error on fitting is 0.15 wt% for H_2O and 250 ppm for S. In the case of water, best fitting (R^2: 0.862) of the extremes of the dataset (very high and very low water content) can be obtained by integrating total alkalinity (X: Na_2O + K_2O in Eqs. 12.1b and 12.2) to express the dependence from melt composition. We do not observe improvement of the fitting for S when parameters like FeO_T or total alkalis are integrated in the equations.

Comparison between measured and fitted dissolved volatile contents facilitates the identification of MIs potentially affected by post-entrapment water loss (e.g. sulphur rich inclusions with S > 1350 ppm at very low P < 44 MPa) and S loss because of sulphides precipitation. For instance, those inclusions plot on the trend deviating from degassing curves at P < 3 kbar (Fig. 12.4a). This approach helps filtering filter the MIs database for spurious data, but we stress that MIs having suffered post-entrapment re-equilibration represent only a very small group, equally distributed among the studied eruptions.

Carbon. Dissolved CO_2 contents in PdF magmas decreases with decreasing pressure and is best modelled using a polynomial equation.

$$CO_2(\mathrm{wt\%}) = a * P^2 + b * P + c \qquad (12.3)$$

This equation permits an excellent fit (R^2: 0.999) and a very low error on the estimate (22 ppm) (Table 12.3). Much worse fit is obtained using logarithmic (R^2: 0.636) or exponential (R^2: 0.649) decay laws as a function of P. Even if low temperature melts contain very low CO_2 (<300 ppm), we did not introduce a temperature term in the fitting equation in order not to induce overfitting. MIs can be potentially affected by post-entrapment migration of CO_2 into one or more bubbles. The very low dispersion of MIs data about the fitting curve for CO_2 and the much higher dispersion observed for water suggest that the magma ascent rate is high enough to preserve pristine variability in water and CO_2 contents.

Halogens. Chlorine. As a whole, dissolved Cl concentration changes little with decompression: degassing up to a very low pressure can be fitted using the simplest form of Eq. 12.2

$$\mathrm{Cl}(\mathrm{wt\%}) = a * \ln(P) + b \qquad (12.2b)$$

Because of the high scattering of available data, regression is always very poor ($R^2 < 0.3$) independently on the adopted equation. Adding a temperature term does not improve fitting significantly. Introducing the effect of composition (X: total alkalis) produces the best fitting and would suggest some minor chlorine degassing already at a pressure of about 60 MPa. Unfortunately, lack of data in the pressure range 67–164 MPa does not permit of better constraining the fitting equation. Comparison between measurements and fitting suggest that the average concentration at high pressure (720 MPa) is 310 ppm, even if only very few data are available. Interestingly, MIs data are very scattered in the P range 374–165 MPa, where two groups can be discriminated. A first MI group has Cl contents similar or slightly higher (330–390 ppm) than observed at higher pressure and a second one contains appreciably lower Cl (190–170 ppm). Comparable scattering is observed in the lowest pressure range P < 65 MPa. Possible explanations for this scattering in halogen concentrations are (i) source heterogeneity and/or (ii) (more likely) mixing between uprising deep magmas and shallower volatile-poor partial melts.

The same kind of logarithmic equation can be used to describe fluorine behaviour. We note however that F dataset is the smallest available on PdF volatiles and data exist only for the low pressure range <48 MPa. In this pressure range, average F is almost constant (522 ppm), some scattering occurs at very low pressure, best documented in groundmasses of 2008 products (1150–240 ppm; average 539 ppm). This scattering is consistent with highly incompatible behaviour of F, late stage concentration in the melt and eventual final degassing during cooling and strong melt crystallization as documented by the abundance in F-rich minerals in the fumarolic sublimates (Toutain et al. 1990; Vlastélic et al. 2011, 2013) (Fig. 12.5).

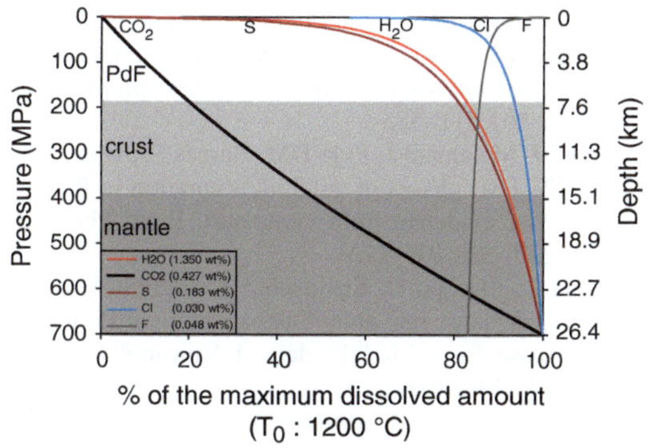

Fig. 12.5 Closed system Degassing curves calculated for PdF melt at 1200 °C as a function of pressure. Note that magmas propagating in the PdF volcanic edifice have lost most of the initial carbon amount. On the contrary, S and H_2O degassing occurs mostly during magma ascent inside the volcanic edifice. Halogen degassing is confined to very shallow depth

12.5 Conclusions

Petrological studies suggest that melting of mantle peridotite producing PdF basaltic magmas begins very deep, in the 30–40 kbar pressure range (100–130 km) and in a temperature range of 1650–1500 °C (Sobolev and Nikogosian 1994; Fretzdorff and Haase 2002) (Chap. 9). The absence of significant thinning of oceanic lithosphere (90–100 km) possibly induces low degree of source melting and correlates with higher water contents in PdF basaltic melts, with respect to other hot-spot contexts like Hawaii (Edmonds et al. 2013 and references therein).

Ultramafic melt compositions (15–20 wt% MgO) in PdF magmas have not been documented yet. Such a feature may result from the thickness of the volcanic pile which achieves nearly 7000 m and the relatively low magma production rates. Maximum temperature tracked in PdF melt inclusions is about 1230 °C. The evolved composition of most PdF olivines and of their MIs testify of magma storage and partial crystallization over a similar depth range (20 km to the surface) to that tracked by hypocenter distribution below PdF volcanic edifice (Chap. 6). Melt and fluid inclusions suggest that a major step of magma crystallization may occur near the mantle-crust boundary at a depth of about 12 km (ca. 3 kbar), consistently with the identification of an underplating layer at this depth (Gallart et al. 1999).

Fluid inclusions suggest that deep magma storage and crystallization is associated with early CO_2 exsolution and possible accumulation. Deep CO_2 accumulation may represent a potential trigger for periodic ascent of deep melts towards shallower levels. Deep exsolution and potential degassing of carbon provides a potential long-term precursor of new magma input in the plumbing system of Piton de la Fournaise. At relatively high pressure, scattering in halogen and water contents may suggest that other processes (slight source heterogeneity and mixing between uprising magmas and volatile poor liquids produced by partial melting of intrusives) operate to produce some noise in the MIs message. Most PdF melt and fluid inclusions record very shallow pressures (<0.5 kbar), with low carbon content and track sulphur and water degassing at shallow depth during magma ascent inside the PdF volcanic edifice. MIs composition and the few available data on syn-eruptive gases together with the low gas fluxes in intraeruptive phases are consistent with magma residing most of its time below sea level, at a pressure exceeding the S–H_2O exsolution threshold (Fukushima et al. 2010; Massin et al. 2011). Their ascent above sea level and eventual eruption is potentially triggered by ascent of small volumes of deeper melts. Low gas fluxes at the surface of PdF arise from both the small volume of variably degassed magma residing at shallow depth and the scrubbing effect of gases by the shallow hydrothermal system and the water table.

References

Albarède F, Luais B, Fitton G, Semet M, Kaminski E, Upton BGJ, Bachèlery P, Cheminée JL (1997) The geochemical regimes of Piton de la Fournaise volcano (Réunion) during the last 530,000 years. J Petrol 38:171–201

Allard P, La Spina A, Tamburello G, Aiuppa A, Coquet A, Brenguier F, Coppola D, Di Muro A, Burton M, Staudacher T (2011a) First cross-correlated measurements of magma dynamics and degassing during a dyke eruption at Piton de la Fournaise hot spot volcano, Réunion Island. AGU Fall Meeting, San Francisco, V22A-04

Allard P, La Spina A, Tamburello G, Aiuppa A, Di Muro A, Burton M, Staudacher T (2011b) First measurements of magmatic gas composition and fluxes during an eruption (October 2010) of Piton de la Fournaise hot spot volcano, La Réunion Island. 11th workshop of the commission on the chemistry of volcanic gases (CCVG)—IAVCEI, Kamchatka, Petropavlovsk, Russia

Barde-Cabusson S, Finizola A, Peltier A, Chaput M, Taquet N, Dumont S, Duputel Z, Guy A, Mathieu I, Saumet S, Sobardère F, Vieille M (2012) Structural control of collapse events inferred by self-potential mapping on the Piton de la Fournaise volcano (La Réunion Island). J Volcanol Geotherm Res 209–210:9–18

Barillon R, Violette S, Nicolini E, Klein D, Chambaudet A, Carbonnel JP, Heath MJ, Merefield J (1993) Continuous measurements of radon content in groundwater on the volcanic site of Piton de la Fournaise (Island of Réunion, France). Nucl Tracks Radiat Meas 22:277–280

Basséguy S, Nicolini E, Staudacher T, Klein D, Delcher E, Bachelery P (2006) Parameters influencing radon content in groundwater of La Réunion Island. EGU General Assembly, vol 8, 01148

Baubron JC, Toutain JP (1993) Etude de l'emission du 222Rn et du CO_2 en relation avec l'activité volcanique et sismique sur les volcans de la Réunion. Partie 1: prospection des zones de fuite et surveillance en continu d'une source hydrothermale. Projet 91 D 207

Boivin P, Bachèlery P (2009) Petrology of 1977 to 1998 eruptions of Piton de la Fournaise, La Réunion Island. J Volcanol Geotherm Res 184:109–125

Bureau H (1996) Les elements volatils associés aux magmas du Piton de la Fournaise: une approche par l'étude des inclusions fluids et vitreuses. Thèse

Bureau H, Pineau F, Métrich N, Semet PM, Javoy M (1998a) A melt and fluid inclusion study of the gas phase at Piton de la Fournaise volcano (Reunion Island). Chem Geol 147:115–130

Bureau H, Métrich N, Pineau F, Semet MP (1998b) Magma-conduit interaction at Piton de la Fournaise volcano (Réunion Island): a melt and fluid inclusion study. J Volcanol Geotherm Res 84:39–60

Bureau H, Métrich N, Semet MP, Staudacher T (1999) Fluid-magma decoupling in a hot-spot volcano. Geophys Res Lett 26:3501–3504

Chaigneau M (1968) Sur les gaz occlus dans les laves émises par le Piton de la Fournaise (Ile de la Réunion). Bull Volcanol 32:515–520

Clocchiatti R, Nativel P (1984) Etude minéralogique et découverte de reliquats magmatiques silicosodiques hydratés dans les phénocristaux de quartz de la syénite de Cilaos (Ile de la Réunion). C R Acad Sc Paris, t. 299, Série II, n°8

Clocchiatti R, Havette A, Nativel P (1979) Relations pétrogénétiques entre les basaltes transitionnels et les océanites du Piton de la Fournaise (Ile de la Réunion, océan Indien) à partir e la composition chimique des inclusions vitreuses des olivines et des spinelles. Bull Minér 102:511–525

Collins SJ, Mclennan J, Pyle DM, Barnes SJ, Upton BGJ (2012) Two phases of sulphide saturation in Réunion magmas: evidence from cumulates. Earth Planet Sci Lett 337–338:104–113

Coppola D, Piscopo D, Staudacher T, Cigolini C (2009) Lava discharge rate and effusive pattern at Piton de la Fournaise from MODIS data. J Volcanol Geotherm Res 184:174–192

Di Muro A, Aiuppa A, Burton M, Métrich N, Allard P, Fougereoux T, Giudice G, Guida R (2012) Intra-eruptive gas emissions and shallow magma storage after the 2007 summit caldera collapse of Piton de la Fournaise, Réunion Island. EGU Meeting, Vienna

Di Muro A, Staudacher T, Ferrazzini V, Métrich N, Besson P, Villemant B, Garofalo K (2014a) Shallow magma storage at Piton de la Fournaise volcano after the 2007 summit caldera collapse tracked in Pele's hairs, in Hawaiian volcanoes: from source to surface. Geophysicsal Monograph Series, AGU, Washington, D.C. (Accepted in press)

Di Muro A, Métrich N, Vergani D, Rosi M, Armienti P, Fougeroux T, Deloule E, Arienzo I, Civetta L (2014b) The shallow plumbing system of Piton de la Fournaise Volcano (La Réunion Island, Indian Ocean) revealed by the major 2007 caldera forming eruption. J Petrol 55:1287–1315

Edmonds M, Sides IR, Swanson DA, Werner C, Martin RS, Mather TA, Herd RA, Jones RL, Mead MI, Sawyer G, Roberts TJ, Sutton AJ, Elias T (2013) Magma storage, transport and degassing during the 2008–2010 summit eruption at Kilauea Volcano, Hawai'i. Geochim Cosmochim Acta 123:284–301

Evans BM, Staudacher T (2001) In situ measurements of gas discharges across fissures associated with lava flows at Réunion Island. J Volcanol Geotherm Res 106:255–263

Famin V, Welsch B, Okumura S, Bachèlery P, Nakashima S (2009) Three differentiation stages of a single magma at Piton de la Fournaise (Réunion hotspot). Geoch Geoph Geos 10:Q01007. doi:10.1029/2008GC002015

Fretzdorff S, Haase KM (2002) Geochemistry and petrology of lavas from the submarine flanks of Réunion Island (western Indian Ocean): implications for magma genesis and the mantle source. Mineral Petrol 75:153–184

Fukushima Y, Cayol V, Durand P, Massonnet D (2010) Evolution of magma conduits during the 1998–2000 eruptions of Piton de la Fournaise volcano, Réunion Island. J Geoph Res, Solid Earth, 115(B10)

Gallart J, Driad L, Charvis P, Sapin M, Hirn A, Diaz J, de Voogd B, Sachpazi M (1999) Perturbation to the lithosphere along the hotspot track of La Réunion from an offshore-onshore seismic transect. J Geoph Res 104:2895–2908

Garofalo K, Staudacher T, Ferrazzini V, Kowalski P, Boissier P, Dupont A, Peltier A, Villemant B, Boudon G (2009) Eruptive measurements at Piton de la Fournaise (Ile de la Réunion) by stationary NOVAC scanning MAX-DOAS instruments. EGU General Assembly, p 12113, 19–24 April, Vienna

Gouhier M, Coppola D (2011) Satellite-based evidence for a large hydrothermal system at Piton de la Fournaise volcano (Reunion Island). Geoph Res Lett 38:L02302. doi:10.1029/2010GL046183

Helz RT, Thornber CR (1987) Geothermometry of Kilauea Iki lava lake, Hawaii. Bull Volcanol 49:651–668

Hibert C, Mangeney A, Polacci M, Di Muro A, Vergniolle S, Ferrazzini V, Taisne B, Burton M, Dewez T, Grandjean G, Dupont A, Staudacher T, Brenguier F, Shapiro NM, Kowalski P, Boissier P, Catherine P, Lauret F (2014) Multidisciplinary monitoring of the January 2010 eruption of Piton de la Fournaise volcano, La Réunion island. J Geoph Res (Under review)

Krafft M, Chaigneau M (1980) Les gaz occlus dans les bombes volcaniques de l'activité du Piton de la Fournaise en 1975-1976 (Ile de la Réunion). Bull Volcanol 43:225–232

Lénat JF, Bachèlery P, Bonneville A, Tarits P, Cheminée JL, Delorme H (1989) The December 4, 1983 to February 18, 1984 eruption of Piton de la Fournaise (La Réunion, Indian Ocean): description and interpretation. J Volcanol Geotherm Res 36:87–112

Lénat EF, Bachelery PB, Merle O (2012) Anatomy of Piton de la Fournaise volcano (La Réunion, Indian Ocean). Bull Volcanol 74:1945–1961

Liuzzo M, Giudice G, Di Muro A, Ferrazzini V, Michon L (2014) New observational evidence of CO_2 degassing anomalies on the Piton de la Founaise and the relationship between seismotectonic structures and CO_2 flux from the soil. EGU, Vienna

Maillard L (1862) Notes sur l'île de la Réunion (Bourbon), Lanée editeur

Malengreau B, Lénat JF, Bonneville A (1994) Cartography and temporal observation of self-potential (SP) anomalies at Piton de la Fournaise. Bull Soc Geol France 165:221–232

Marty B, Meynier V, Nicolini E, Griesshaber E, Toutain JP (1993) Geochemistry of gas emanations: a case study of the Réunion hot spot, Indian ocean. App Geochem 8:141–152

Massare D, Métrich N, Clocchiatti R (2002) High temperature experiments on silicate melt inclusions in olivine at 1 atm: inferences on temperatures of homogeneization and H_2O concentrations. Chem Geol 183:87–98

Massin F, Ferrazzini V, Bachèlery P, Nercessian A, Duputel Z, Staudacher T (2011) Structures and evolution of the plumbing system of Piton de la Fournaise volcano inferred from clustering of 2007 eruptive cycle seismicity. J Volcanol Geotherm Res 202:96–106

Mercier M, Di Muro A, Métrich N, Giordano D, Belhadj O, Mandeville CW (2010) Spectroscopic analysis (FTIR, Raman) of water in mafic and intermediate glasses and glass inclusions. Geochim Cosmochim Acta 74:5641–5656

Monnin M, Seidel JL (1998) An automatic radon probe for earth science studies. J Appl Geoph 39:209–220

Newman S, Lowenstern JB (2002) VOLATILECALC: a silicate melt-H_2O–CO_2 solution model written in Visual Basic for Excel. Comp Geosci 28:597–604

Papale P, Moretti R, Barbato D (2006) The compositional dependence of the saturation surface of $H_2O + CO_2$ fluids in silicate melts. Chem Geol 229:78–95

Pichavant M, Brugier Y, Di Muro A (2016) Petrological and experimental constraints. In: Bachèlery P, Lénat J-F, Di Muro A, Michon L (eds) Active volcanoes of the southwest Indian Ocean: Piton de la Fournaise and Karthala. Active volcanoes of the world. Springer, Berlin

Pietruszka AJ, Hauri EH, Blichert-Toft J (2009) Crustal contamination of mantle-derived magmas within Piton de la Fournaise, Réunion Island. J Petrol 50:661–684

Roult G, Peltier A, Staudacher T, Ferrazzini V, Taisne B, Di Muro A, The OVPF Team (2012) A comprehensive classification of the Piton de la Fournaise eruptions (La Réunion Island) spanning the 1986–2010 period. Search for eruption precursors from the broad-band GEOSCOPE RER station analysis and interpretation in terms of volcanic processes. J Volcanol Geotherm Res 241:78–104

Salaun A, Villemant B, Semet M, Staudacher T (2010) Cannibalism of olivine-rich cumulate xenoliths during the 1998 eruption of Piton de la Fournaise (La Réunion hotspot): implications for the generation of magma diversity. J Volcanol Geotherm Res 198:187–204

Segovia N, Mena M, Monnin M, Pena P, Seidel JL, Tamez E (1997) Radon-in-soil variations related to volcanic activity. Radiat Meas 28:745–750

Seidel JL, Bonneville A, Lenat JF (1988) Radon measurements related to Piton de la Fournaise (Réunion) from 1983 to 1987. C R Acad Sci Paris 306:89–92

Sobolev AV, Nikogosian IK (1994) Petrology of long-lived mantle plume magmatism: Hawaii, Pacific, and Réunion Island, Indian Ocean. Petrol 2:111–144

Sobolev AV, Clocchiatti R, Dhamelincourt P (1983) Variations of the temperature, melt composition and water pressure during olivine crystallization in oceanitic rocks from the Piton de la Fournaise volcano (Réunion Island, 1966 eruption). C R Acad Sci Paris II:275–280

Sorby HC (1858) On the microscopical structure of the crystals indicating the origin of minerals and rocks. Quat J Geol Soc London 14:453–500

Staudacher T (2010) Field observations of the 2008 summit eruption at Piton de la Fournaise (Ile de La Réunion) and implications for the 2007 Dolomieu collapse. J Volcanol Geotherm Res 191:60–68

Toutain JP, Aloupogiannis P, Delorme H, Person A, Blanc P, Robaye G (1990) Vapor deposition of trace elements from degassed basaltic lava, Piton de la Fournaise volcano, Réunion Island. J Volcanol Geotherm Res 40:257–268

Toutain JP, Baubron JC, François L (2002) Runoff control of soil degassing at an active volcano. The case of Piton de la Fournaise, Réunion Island. Earth Planet Sci Lett 19:83–94

Tulet P, Villeneuve N (2011) Large scale modeling of the transport, the chemical transformation of the sulphur emitted during the April 2007 eruption of Piton de la Fournaise. Atm Chem Phys 11:4533–4546

Upton BG, Wadsworth WJ (1971) Rhyodacite glass in Réunion basalt. Min Mag 38:152–159

Upton BGJ, Semet MP, Joron JL (2000) Cumulate clasts in the Bellecombe ash member, Piton de la Fournaise, Réunion Island, and their bearing on cumulative processes in the petrogenesis of the Réunion lavas. J Volcanol Geotherm Res 104:297–318

Vélain C (1878) Description géologique de la presqu'île d'Aden, de l'île de la Réunion, des îles Saint-Paul et Amsterdam. Hennuyer, Paris (356 p)

Viane C, Bhugwant C, Sieja B, Staudacher T, Demoly P (2009) Etude comparative des émissions de gaz volcanique du Piton de la Fournaise et des hospitalisations pour asthme de la population réunionnaise de 2005 à 2007. Revue Française d'Allergologie 49:346–351

Vigouroux N, Williams-Jones AE, Wallace P, Staudacher T (2009) The November 2002 eruption of Piton de la Fournaise: tracking the pre-eruptive thermal evolution of magma using melt inclusions. Bull Volcanol 71:1077–1089

Villemant B, Salaün A, Staudacher T (2009) Evidence for a homogeneous primary magma at Piton de la Fournaise (La Réunion): A geochemical study of matrix glass, melt inclusions and Pélé's hairs of the 1998–2008 eruptive activity. J Volcanol Geotherm Res 184:79–92

Vlastélic I, Staudacher T, Bachèlery P, Télouk P, Neuville D, Benbakkar M (2011) Lithium isotope fractionation during magma degassing: constraints from silicic differentiates and natural condensates from Piton de la Fournaise volcano (Réunion Island). Chem Geol 284:26–34

Vlastélic I, Menard G, Gannoun A, Piro JL, Staudacher T, Famin V (2013) Magma degassing during the April 2007 collapse of Piton de la Fournaise: the record of semi-volatile trace elements (Li, B, Cu, In, Sn, Cd, Re, Tl, Bi). J Volcanol Geotherm Res 254:94–107

Welsch B, Faure F, Bachelery P, Famin V (2009) Microcrysts record transient convection at Piton de la Fournaise Volcano (La Réunion Hotspot). J Petrol 50:2287–2305

Seismic Monitoring at Piton de la Fournaise

13

Jean Battaglia, Florent Brenguier and Geneviève Roult

Abstract

The Piton de la Fournaise Volcanological Observatory ("Observatoire Volcanologique du Piton de la Fournaise", OVPF) started operating at the end of 1979. The first monitored eruption by the OVPF was that of 1981 (Bachèlery et al. in Nature 297(5865):395–397, 1982; Chevallier et al. in L'éruption du mois de février 1981 au Piton de la Fournaise (Ile de la Réunion, Océan Indien). Phénoménologie et remarques structurales, 293 (II):187–190, 1981). Since then, the surveillance networks, and particularly the seismic network, have been dramatically improved. In addition, various seismic experiments have been carried out in parallel with the monitoring of the volcano. This chapter is aimed at summarizing the main results obtained within both the framework of the monitoring and that of research experiments.

13.1 Seismic Monitoring Network

Since the installation of the observatory, the seismic network has evolved from a short period network recording on a triggered basis, to a modern broadband and short period network recording continuously. The main points in the evolution of the network during the last 15 years can be summarized as follows (Fig. 13.1).

- In 1999 the network included 19 short period stations installed mostly near the summit and around the volcano, with only three stations having horizontal components. The acquisition system recorded only triggered data

J. Battaglia (✉)
Laboratoire Magmas et Volcans, Université Blaise Pascal—CNRS—IRD, OPGC, 5 rue Kessler, 63038 Clermont-Ferrand, France
e-mail: J.Battaglia@opgc.univ-bpclermont.fr

F. Brenguier
Institut des Sciences de la Terre, University of Grenoble, Grenoble, France

G. Roult
Equipe de Sismologie, Institut de Physique du Globe de Paris-Sorbonne Paris Cité, UMR CNRS 7154, Université Paris Diderot, 1 rue Jussieu, 75238 Paris Cedex 05, France

P. Bachèlery et al. (eds.), *Active Volcanoes of the Southwest Indian Ocean*, Active Volcanoes of the World, DOI 10.1007/978-3-642-31395-0_13

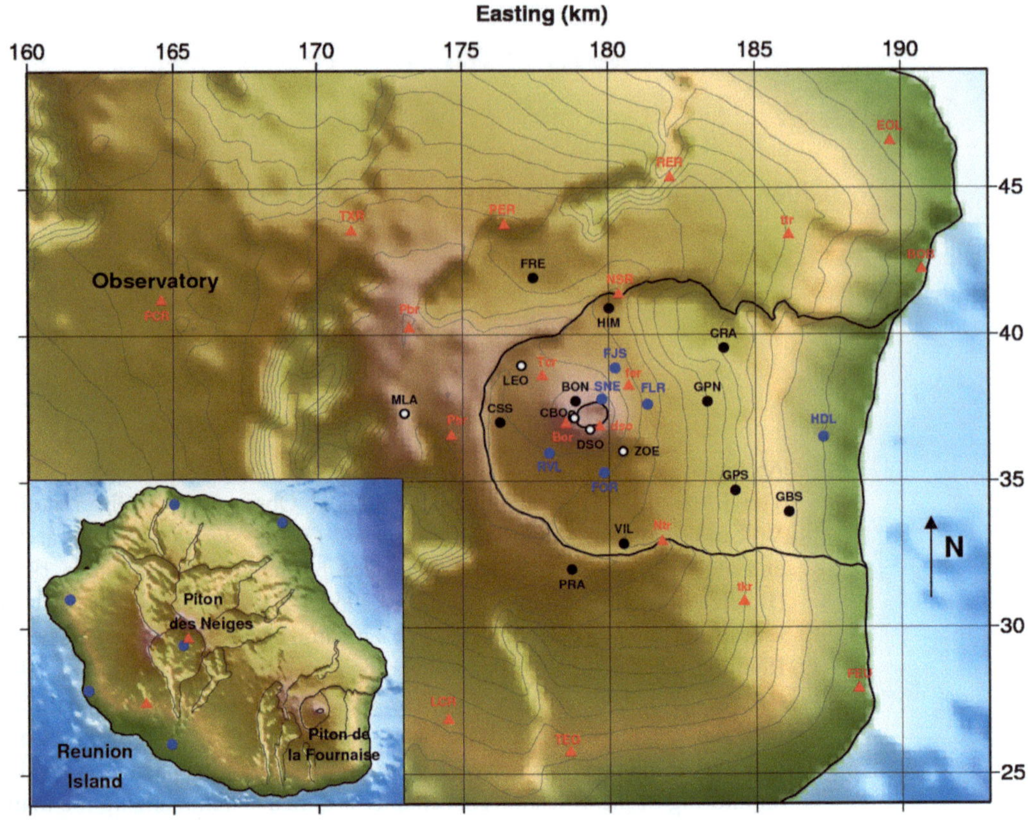

Fig. 13.1 Shaded relief view of Piton de la Fournaise. Short period seismic stations are shown as *red triangles*, *blue filled circles* are broadband stations installed between 2007 and 2009. *Filled and empty black circles* are stations installed during the Undervolc project, the *filled black circles* are stations which became permanent after the end of the project

digitally with a 100 Hz sampling frequency, all stations being recorded for each event. At the end of 1999, the system evolved toward continuous recording with storage on CD-ROMs.

- After the 2007 collapse of the Dolomieu crater, the network started including broadband 3-component stations, with up to 6 stations by the middle of 2009.
- Starting in October 2009, the temporary project Undervolc (Brenguier et al. 2012) led to the installation of 15 broadband stations located on the volcano, with 4 stations situated in the Grandes Pentes on the eastern flank.
- In 2011, at the end of the Undervolc project, the temporary broadband stations were removed, with 10 station sites being progressively replaced with new broadband sensors and digitizers.
- In 2012, the network, which had up until then focused mainly on monitoring Piton de la Fournaise, with only 2 stations in the area of Piton des Neiges, was extended to incorporate 8 stations spread over the western part of the island.

At the time of writing, in 2014, the monitoring network of the volcano (Fig. 13.1) includes a total of 16 short period sensors radio-transmitted to the observatory and 16 broadband digital sensors transmitted by Wifi or Freewave to the observatory. Additionally the western part of the island is monitored by 7 stations. All data are continuously stored in MiniSEED format and available for monitoring and research in near real-time.

13.2 Overview of Piton de la Fournaise Seismicity

The seismic monitoring network records a wide variety of signals which can be classified into three categories: (1) events related to the volcano, (2) those related to the island without any direct relation with the volcanic activity (local tectonic earthquakes, rockfalls), and finally (3) events originating from outside the island (teleseismic waves, regional earthquakes, T-phases). Volcanic seismicity mostly includes volcano-tectonic (VT) events and tremor during eruptions. Rare long period (LP) events are observed.

LP events are rare in comparison to the number at other active volcanoes. Aki and Ferrazzini (2000) interpreted LP events observed during the years preceding the March 1998 eruption as being associated with lateral movements of magma to the volcano's rift zones. Zecevic et al. (2013) identified only 15 LPs during intrusive or pre-eruptive swarms between November 2009 and January 2011 and showed that their sources are located a few hundred meters below the Dolomieu crater. Tremor at Piton de la Fournaise is mostly eruption tremor observed in conjunction with eruptions. It starts with the beginning of eruptive activity and, when no field observations are available, the beginning time of an eruption is defined as being the time the tremor amplitude increases. This tremor has a relatively broad spectrum and should be referred to as "non-harmonic tremor". The decrease and disappearance of eruption tremor defines the end of an eruption episode. Exceptionally, similar broadband tremor has been observed without any surface activity in 2008 and 2009. A second type of tremor, which may be named pre-eruptive tremor, was observed before and during the beginning of the March 1998 eruption. Its spectrum was narrower than that of typical eruption tremor and peaked around 1.0 Hz. Pre-eruptive tremors are, however, not commonly observed prior to eruptions at Piton de la Fournaise.

The most common events related to the volcano are the high frequency VT earthquakes. They have the same characteristics as classic tectonic events and are routinely located at the observatory by picking their arrival times. They occur in swarms often shortly preceding eruptions, as well as on a daily basis as background seismicity. Except for a couple of large non-tectonic events which occurred during the Dolomieu collapse and a small number of deep ($Z < 15$ km b.s.l.) events below the island, all earthquakes recorded below the volcano have magnitudes lower than 2.5. Figure 13.2 shows hypocenters for VT events recorded between 1996 and 2011, corresponding to both background seismicity and events occurring during swarms. These hypocenters outline the main zones of activity observed at Piton de la Fournaise since the installation of the observatory in 1980:

- A zone located right below the summit and above sea level in which most (80–90 %) of the events occur. Most of them are found between 0 and 1 km a.s.l., with a peak about 0.8 km a.s.l. This cluster has tended to be slightly deeper since 2007, but this effect could be due to improved 3-component station coverage. Most of the seismicity occurring during pre-eruptive swarms is between 0 and 1 km a.s.l, with only few shallower events shortly before the opening of the fissures.
- A second zone of seismicity is found linking sea level to a depth of about 6 km. This deeper seismicity marks a conduit along which the earthquakes migrated upward during the March 1998 pre-eruptive swarm (Battaglia et al. 2005c) when this zone was activated for the first time. Later on, activity occurred along this path during the 2007 eruption and Dolomieu collapse (Massin et al. 2011) and occasionally a few earthquakes have been observed, mostly at 4–6 km b.s.l.
- A third zone is found below the eastern flank, with a rather diffuse seismicity. Unfortunately these events are poorly constrained because of the lack of stations in the Grandes Pentes before 2009.

Finally we note the presence of a deeper seismicity, often located west of the volcano, at

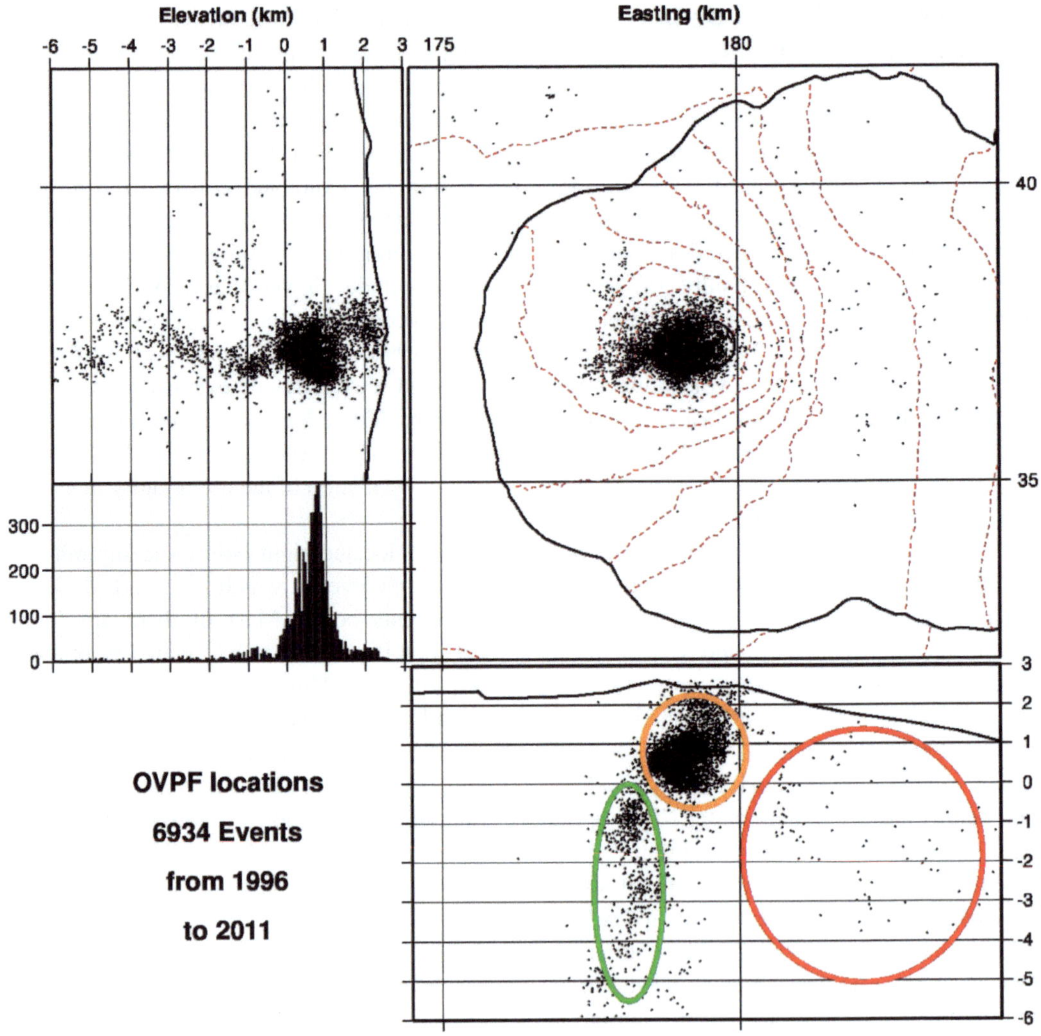

Fig. 13.2 The 6934 events located between 1996 and 2011 define 3 main zones of activity: (1) a zone above sea level and below the summit including 80–90 % of the seismicity, (2) a seismic zone describing a vertical conduit (1998, 2007), and (3) a diffuse and poorly constrained seismic zone below the Eastern Flank

depths up to 15–20 km. The relation between those events and the eruptive activity of the volcano remains, however, unclear.

As shown above, a significant part of the VT seismicity occurs during swarms in which the rate of earthquakes increases, sometimes merging into a quasi continuous pattern. Such swarms systematically precede eruptions. These pre-eruptive swarms are often short-lived at Piton de la Fournaise, commonly lasting 45–90 min. Aki and Ferrazzini (2000) showed that the duration of these seismic crises is longer for eruptions occurring at lower elevations, farther from the summit. Looking at the temporal evolution of the average seismic amplitude (Fig. 13.3) during the swarms we note that there is commonly a peak in the amplitude followed by a period of relative quiescence and finally the appearance of eruption tremor with the opening of the eruptive fissures. Some swarms do not lead to eruptions, and we term them intrusions; in most cases they correspond to aborted eruptions in which dike propagation did not reach the surface.

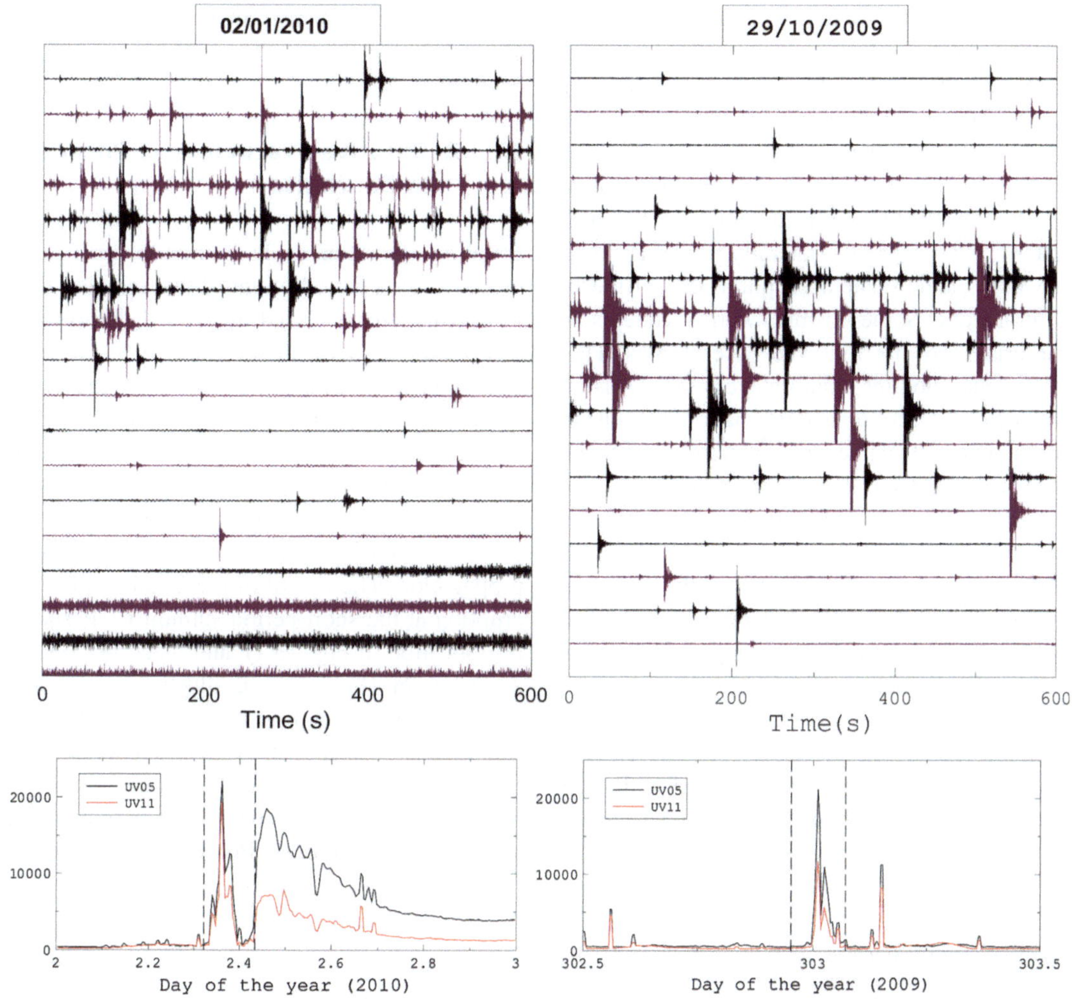

Fig. 13.3 Upper plots: examples of a pre-eruptive swarm (*left*) and an intrusion (*right*). Lower plots show seismic amplitude (RMS) for over a 1-day period. For the pre-eruptive swarm the *first dashed vertical line* corresponds to the approximate beginning of the crisis and the *second* to the beginning of the eruption. For the intrusion the *lines* correspond to the approximate beginning and end of the crisis

13.3 Compilation of the Piton de la Fournaise Activity Spanning the 1985–2010 Period

A comprehensive database of 83 volcanic events (54 eruptions, 11 intrusions, 15 seismic crises and 3 pit crater/collapse episodes) during the December 1985–December 2010 period, has been established using the OVPF data (Roult et al. 2012). The main features were compiled (eruption type, eruption timing, site, fissure locations and elevation, surface and volume of lava flows, timing and characteristics of seismic crises and seismic swarms etc.).

13.3.1 Types of Volcanic Events at Piton de la Fournaise

At Piton de la Fournaise, the recent eruptions occurred either inside the summit craters, on the flanks of the Central Cone, or along the rift zones

(Michon et al. 2016, Chap. 7). The beginning and duration of the volcanic events since December 1985 are plotted in Fig. 13.4, with an indication of the family type. We distinguish three families of eruptions, according to the location of the eruptive fissures: (1) eruptions inside the summit craters, (2) eruptions that begin inside or near the summit craters and propagate laterally towards lower elevations, and (3) lateral eruptions i.e. outside the Central Cone. Some seismic swarms, called seismic crises, are not followed by eruptions. They can be classified into four categories: (1) seismic crises without detectable deformation, (2) seismic crises with detectable deformation, (3) seismic crises without detectable deformation followed by tremor, and (4) seismic crises with detectable deformation followed by tremor. The first category is in grey on Fig. 13.4 and the last three categories in yellow. The seismic crises without detectable deformation (15 during the study period) can be interpreted either in terms of deep intrusion, or in terms of brittle failures most likely associated with hydrothermal fluid. The occurrence of seismic crises not associated with eruptions (with or without deformation) clearly increased after 2007 (Fig. 13.4). This would be linked to stress changes in the volcanic edifice after the 2007 caldera collapse (Peltier et al. 2010; Roult et al. 2012).

13.3.2 Eruptive Precursors at Piton de la Fournaise

Eruptions at Piton de la Fournaise are usually preceded by a period of unrest that may last days to months, characterized by ground deformation and a slow increase in seismic activity, in most cases composed of shallow events located below the central cone, from 0 to 2300 m above sea level (e.g. Battaglia and Bachèlery 2003; Massin et al. 2011; Peltier et al. 2009a). The onset of a seismic swarm marks the acceleration of the process accompanying the shallow magma propagation. The record of the October 14, 2010 eruption at the short-period summit BOR station (see location in Fig. 13.1), and its corresponding spectrogram (Fig. 13.5a), illustrates one such typical crisis. Four distinct phases can be identified. Phase 1 is characterized by an acceleration of the seismic activity, with a large number of VT events, more than one event per minute and sometimes overlapping. Phase 2 (swarm phase)

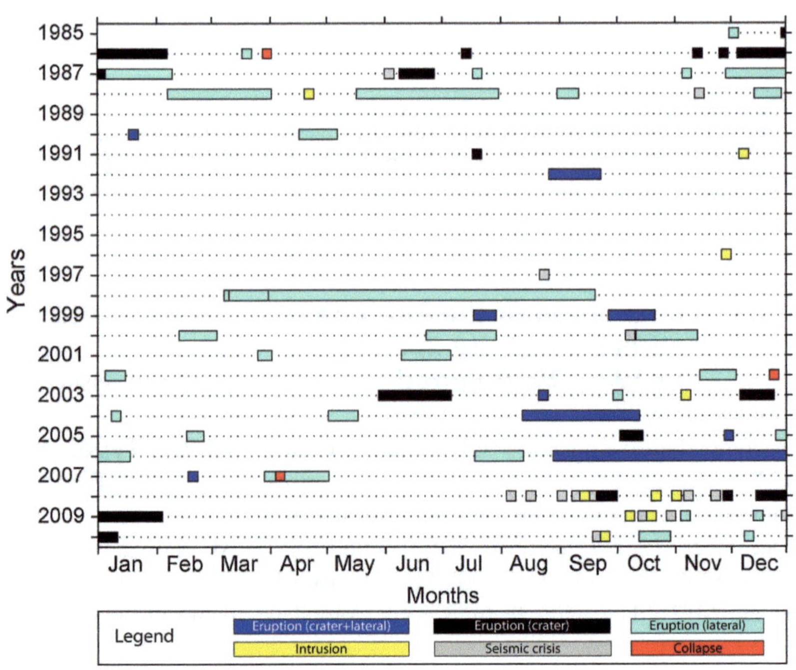

Fig. 13.4 Distribution and duration of volcanic events in the December 1985–December 2010 period. Onset times and durations, with indication of the family type (see *legend*). Note the increase in frequency of intrusions and seismic crises after the major 2007 caldera-forming eruption

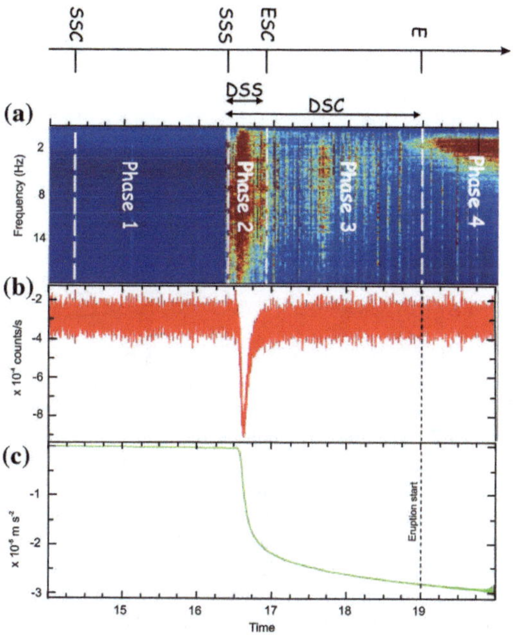

Fig. 13.5 Illustration of a typical crisis recorded at BOR and RER stations using the October 14, 2010 eruption. **a** Spectrogram at the short-period BOR station, **b** ground velocity (in counts) at the RER station, and **c** acceleration trace (in m s^{-2} after a low-pass filtering at 100 s and removal of the transfer function) at the RER station. *Vertical dashed line* eruption start

is characterized by numerous merged events (more than six events per minute with larger magnitude than in phase 1). This phase is not always present. Phase 3 corresponds to a seismically quiet/low intensity sequence with a small number of VT events. Phase 4 corresponds to the building and stabilization of eruptive tremor, in most cases associated with the opening of eruptive fissures in the surface. A clear correlation is observed between the duration of the seismic crises and the distance from the summit for eruptions occurring at distances of less than 2500 m (except possibly for the post-2007 events) (Aki and Ferrazzini 2000; Peltier et al. 2005; Roult et al. 2012). This correlation disappears at larger distances.

13.3.3 Short-Term Precursors from the RER Broad-Band Seismic Station

A broad-band GEOSCOPE station is located at RER (Fig. 13.1), 8.5 km north of the summit crater (Roult et al. 2010), in a 4.7 km-long tunnel, at 834 m in elevation. The high sensitivity of the station allows precise identification of transient signals related to Piton de la Fournaise activity (see Fig. 13.5). These unusual long period seismic signals, observed on the horizontal components before the eruptions, are commonly interpreted as tilt signals induced by the inflation and/or deflation of the volcano, and as a result of pressure variations within the volcanic edifice (e.g. Battaglia et al. 2000). Here, they are considered as tilt variations related to magma transfer inside the shallow part of the volcano edifice. The database presented in Roult et al. (2012) includes the parameters of the associated signals recorded by the RER station.

The RER transients can be sorted into four classes, depending on their waveform (Class 1+, Class 1−, Class 2+, Class 2−; the + and − refers to the sign of the first transient signal). Examples of transients are given in Fig. 13.6. Signals of Class 1 exhibit two successive impulses with opposite senses, whereas the Class 2 signals are simpler and exhibit a single decrease. Most Class 1− and Class 2− transients precede eruptions north of the summit, when dikes propagate towards the RER station. Conversely, Class 2+ and Class 1+ transients are recorded during eruptions located south of the summit. The dominant transient classes are Class 2− and Class 1+. Class 2− transients are often associated with lateral eruptive fissures, whereas most of the Class 1+ transients are associated with summit or near summit eruptions. The possible mechanisms explaining these transient signals are discussed in Roult et al. (2012). Note that two other classes may be defined. Class 0 events correspond to the

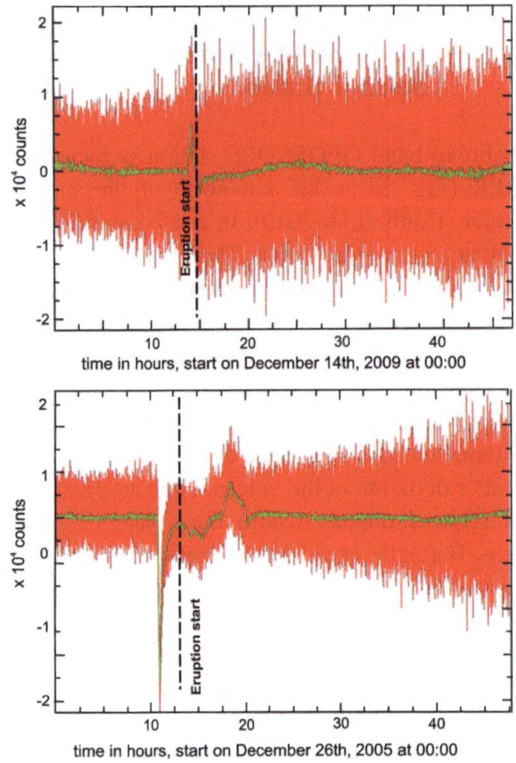

Fig. 13.6 Examples of transients observed at the RER station on the N-S component. The *red traces* correspond to raw data (ground velocity in counts), the *green* ones to the same traces after a low-pass filtering at 100 s. The *black vertical dashed lines* indicate the eruption onset time. *Top* Transient of Class 1+ before the December 14, 2009 eruption. *Bottom* Transient of Class 2− before the December 26, 2005 eruption

absence of a signal and Class 3 correspond to pit-craters or caldera collapses (Roult et al. 2012).

13.3.4 Acceleration at the Very Broad-Band RER Station During Eruptions and Intrusions. Tools for Prediction

The acceleration (after a low pass filtering at 100 s and removal of the instrumental response) from the RER north-south component has been

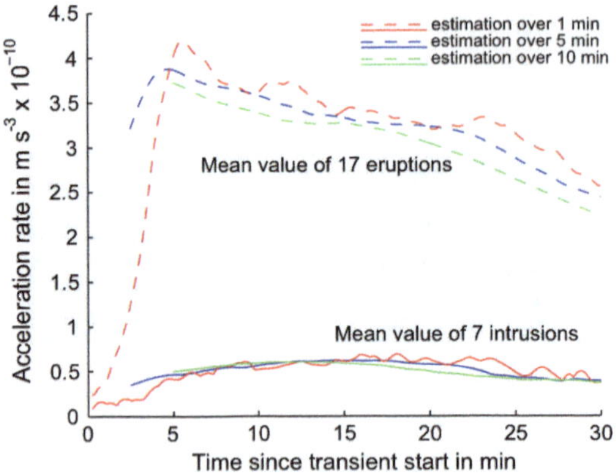

Fig. 13.7 Absolute acceleration rate (in m s^{-3}) of RER N-S long period during intrusions and eruptions (2005–2010). 3 different sliding time windows (1, 5 and 10 min) of the absolute acceleration are shown for 17 eruptions (*dashed lines*) and 7 intrusions (*solid lines*). After 7 min from the onset of the seismic crisis, the difference between the crises leading to intrusion or to eruption becomes significant (maximum ratio of 7). After 12 min the ratio decreases slightly (5). The unusual and exceptional March 30, 2007 eruption is not considered here. On the basis of these observations, we suggest that if, after 7 min from the beginning of the crisis, the acceleration rate is larger than a threshold value (2×10^{-10} m s^{-3}) an eruption will most likely follow the seismic crisis

calculated for 17 eruptions and 7 intrusions, spanning the 2005–2010 period. For each event, the analysis brackets the start of the eruption or of the seismic swarm for intrusions (4 days before and 8 days after). A clear difference is found between intrusions (low acceleration rate) and eruptions (higher rate). The absolute acceleration rate (Fig. 13.7) can thus be used to differentiate intrusions and eruptions. The average maximum values for intrusions and eruptions are ~4.96 × 10–11 m s^{-3} and ~2.86 × 10–10 m s^{-3}, respectively. This large ratio (~6) offers the possibility of anticipating if an ongoing crisis will terminate in an intrusion or an eruption, if the acceleration is calculated in real time. However, this criterion requires further refinements, in particular in the cases of small volume eruptions or very shallow intrusions.

13.4 The March 1998 Pre-eruptive Earthquake Migration

Volcanic eruptions are nearly always preceded by an increase of seismic activity and since they are caused by the migration of magma towards the surface, we may expect such a process to be accompanied by simultaneously migrating earthquakes. However, well documented cases of such phenomena are rare. Horizontal pre-eruptive migrations of earthquakes have been observed in rift zones (Brandsdóttir and Einarsson 1979; Koyanagi et al. 1988; Rubin and Gillard 1998) and rare vertical migrations have been reported (Ukawa and Tsukahara 1996; Umakoshi et al. 2001). More commonly, discontinuously shallowing seismicity has been observed prior to the onset of eruptions (Harlow et al. 1996; Zobin et al. 2002). Most of the time, pre-eruptive seismicity is diffuse or not directly related to magma propagation. At Piton de la Fournaise, the absence of deeper seismicity before 1998 has been interpreted (Lénat and Bachèlery 1990) as being due to the absence of deep magma transfers, the frequent small eruptions thus being caused by the emptying of shallow magma pockets.

After an unusually long period of dormancy, which lasted five and a half years, the volcano started erupting on March 9, 1998, at 11:05 GMT on the northern flank of the summit cone (Kapor and Krafft vents in Fig. 13.8). It was preceded by a pre-eruptive crisis which included more than 3100 volcano-tectonic events and lasted for about 35 h, what is unusually long for this volcano. Battaglia et al. (2005a) showed that the seismicity during this crisis was characterized by a monotonic upward migration of the hypocenters from 5 to 6 km b.s.l. to the surface, leading to the onset of the eruption.

Manually picked arrival times and the software HYPO71 (Lee and Lahr 1975) have been used (Battaglia et al. 2005c) to determine the hypocenters for 583 sufficiently large events, using a 1-D velocity model composed of 8 horizontal layers which has been developed for routine hypocenter determination at the OVPF. The first group of events directly related to the precursory seismic crisis started on March 6 and 7, with about 20 earthquakes mainly clustered around two levels: the first group around 5–6 km b.s.l., and the second between 1 and 2 km b.s.l., situated under station Tcr. According to results from a reflection-refraction seismic survey (Gallart et al. 1999), the starting depth of the crisis (D1 in Fig. 13.8) corresponds approximately to the top of the pre-existing oceanic crust upon which the volcanic edifice is built.

Around 00:00 on March 8, the seismicity increased and remained relatively steady during the remaining part of the day. The depth to the hypocenters slowly moved towards the surface with a steady migration velocity of about 1.59 ± 0.06 m/min (0.095 km/h). Around 1.5 km b.s.l. a discontinuity (D2) was observed in the migration, with the presence of a low seismicity zone. In space, discontinuity D2 also corresponds to the depth at which persistent seismicity was observed below station Tcr during March 6, 7 and 8, separate from the main path of migration. Above discontinuity D2, we observed a region of very high seismicity (histogram in Fig. 13.8) through which the vertical migration continued at a slightly lower mean velocity of 1.25 ± 0.09 m/min (0.075 km/h) until 1 h before the beginning of lava outflow. The region of high seismicity became active during the first peak observed in the number of events (Fig. 13.9) at about 0:00 on March 9. The activity was first focused in the southern part of the cluster of events, and then during the second peak of activity, starting at about 3:00, the activity shifted back to the central part of the cluster and propagated northeastwards towards sea level. Assuming that the seismicity reflects magma movements, we may interpret the above shift as magma searching for an exit first to the south and then to the north. The first path was oriented in the direction of the Hudson vent, where a second eruption started only three days later, and the second towards the Kapor-Krafft area.

The seismic activity below sea level outlines an almost continuous vertical path which could be interpreted as a magmatic conduit. Earthquakes appear to be mostly generated at the tip

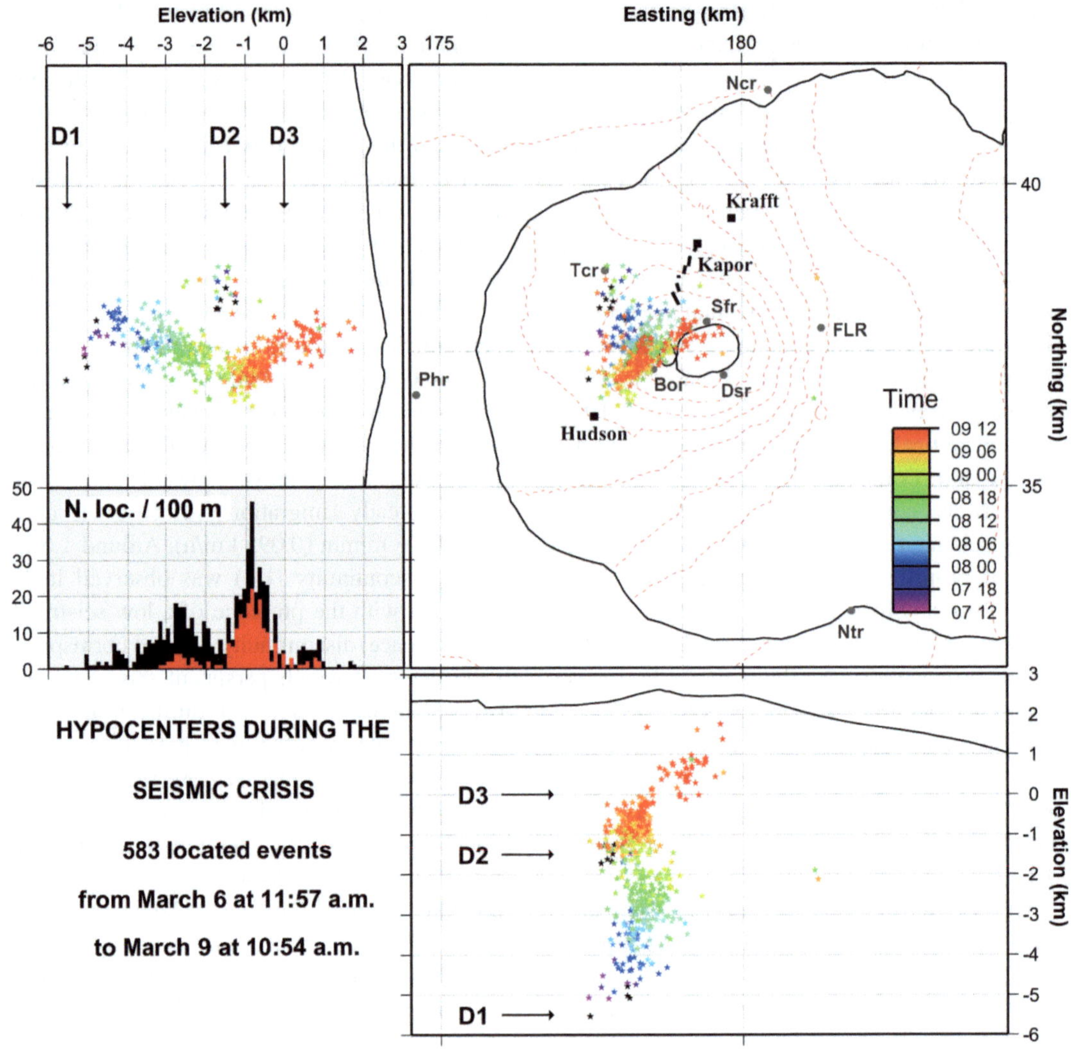

Fig. 13.8 Geographic location of the hypocenters during the seismic crisis: map view and north-south and east-west cross sections. Hypocenters are shown using different colors depending on their time of occurrence (time scale is given in day-hour). Seismic network stations are shown as *filled circles*. The fissures where the March 9 eruption broke out are shown north of the summit, and the Krafft and Kapor craters, where the activity focused after several days, are indicated as *filled squares*. The histogram inserted on the north-south cross section shows the number of hypocenters as a function of elevation, with intervals of 100 m, in *black* and superposed in *red* the same but when only considering events with a magnitude greater than 1.5

of the upward propagating conduit as activity vanishes gradually downwards from the tip. The mean migration velocities for earthquakes along that path suggest that magma migrated upwards with a velocity at least an order of magnitude lower than velocities observed for horizontal downrift migrations at Krafla (30–72 m/min) (Einarsson and Brandsdóttir 1980) and Kilauea (about 11 m/min) (Koyanagi et al. 1988; Rubin and Gillard 1998) or vertical intrusion rates below Izu Peninsula (8–16 m/s) (Hayashi and Morita 2003). The relatively low velocity migration preceding the March 9 eruption suggests a difficult propagation of the magma requiring the reopening of a closed conduit.

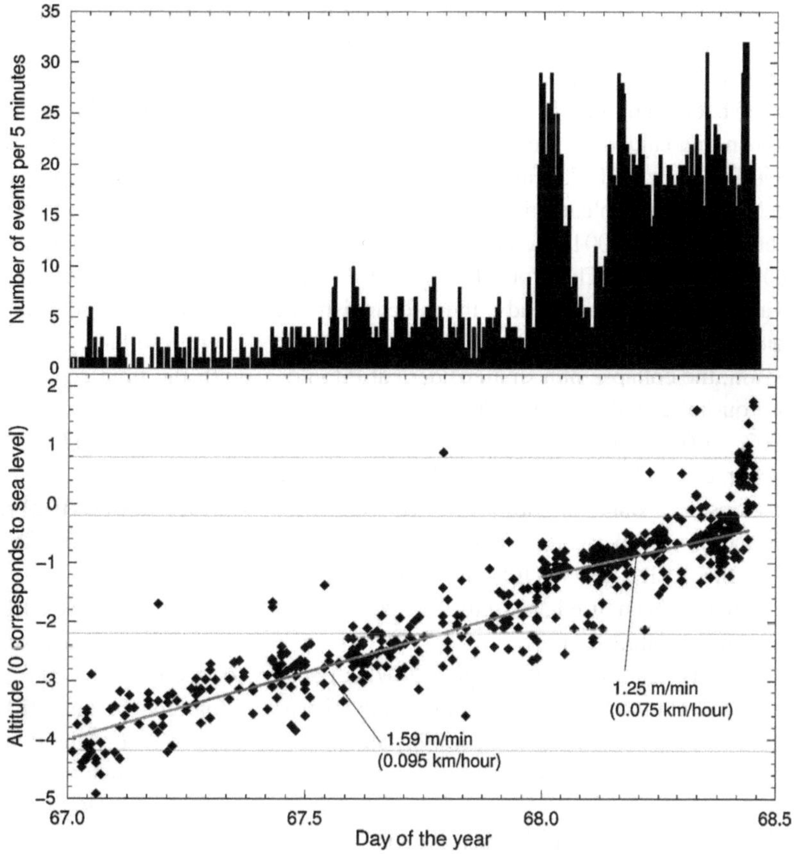

Fig. 13.9 Number of volcano-tectonic events per 5-min windows (*upper plot*) and evolution of the focal depth of the 583 located hypocenters (*lower plot*). *Continuous lines* on the lower plot are linear regressions corresponding to the mean migration rates, boundaries of the layers of the velocity model used for location are indicated by *light color horizontal lines*

One hour before the eruption, the activity moved clearly above sea level and there was no systematic migration of the hypocenters which were distributed between sea level and the summit of the central cone (2500 m). Also, the zone of seismicity above sea level was not situated in the continuous extrapolation of the one below, and corresponded to the volume commonly activated during the pre-eruptive crises of Piton de la Fournaise. Sea level appears as a major geologic discontinuity (D3) which also corresponds to a low seismicity zone. The activity above sea level was of a very short duration, lasting for only an hour, before the appearance of tremor originating from the eruption site. If we assume that magma travelled from sea level to the surface during that last hour we obtain a high vertical intrusion rate of about 2.4 km/h. An alternative explanation is that at least part of the migration above sea level may be due to stress transfer within magma already present below the summit. In this context, the seismicity below sea level reflects the upward propagation of new magma toward the surface until it reached sea level, where it increased the pressure in the shallow plumbing system leading to the eruption one hour later.

13.5 The Caldera Collapse Episode of April 2007 at the Piton de la Fournaise Volcano Studied with the GEOSCOPE Broadband RER Station

Caldera collapses are common features of many volcanoes. The calderas of explosive volcanoes are usually associated with large Plinian eruptions whereas the collapse of basaltic volcanoes is associated with large lateral eruptions or intrusions draining a magma reservoir. Only two large volcanic collapses on basaltic volcanoes have been monitored: the 1968 collapse of Fernandina caldera in the Galápagos archipelago (e.g. Filson et al. 1973), and the caldera collapse at Miyakejima volcano in Japan in 2000 (e.g. Kumagai et al. 2001; Ueda et al. 2005; Yamamoto et al. 2001). Therefore, few sets of data are available in order to study the mechanism of caldera collapse on basaltic volcanoes. For this reason, the collapse of a small caldera at Piton de la Fournaise volcano in 2007 (Michon et al. 2007, 2009, 2013; Peltier et al. 2009b; Staudacher et al. 2009; Urai et al. 2007) provided precious new data. In addition to the data from the monitoring networks of the OVPF, the data from the RER GEOSCOPE (Roult et al. 2012) station has been used to study the 2007 event (Fontaine et al. 2014).

Activity leading to the April 2007 collapse from 2006 until the caldera collapse in 2007 was quasi-continuous (Peltier et al. 2009b; Staudacher et al. 2009). Five phases can be distinguished:

(1) A 25-day eruption, in July–August 2006, at the base of the south flank of the Central Cone, with a lava flow volume estimated to be 2×10^6 m^3.
(2) A 4-month eruption, starting on August 30, 2006, in the Dolomieu crater. It progressively infilled the crater until January 1, 2007, with a volume of lava estimated to be 20×10^6 m^3.
(3) A short-lived (15 h) eruption, on February 18, 2007, in the Dolomieu crater. It emitted about 1×10^6 m^3 of lava.
(4) On March 30, 2007, a short-lived eruption (10 h) at the south-east base of the Central Cone, with an emission of 0.6×10^6 m^3 of lava.
(5) A distal eruption, starting on April 2, 2007, at low elevation (~600 m a.s.l.), 7 km east-southeast of the summit. It lasted until May 1 and emitted the largest volume of all historical eruptions (>200 × 10^6 m^3; Bachèlery et al. 2010).

Increasing deflation of the summit of the volcano was observed from April 5, suggesting that the rock column was starting to collapse. The first major collapse of the caldera roof took place in the evening of that day and was associated with a 4.8 Ms seismic event (unpublished OVPF Report) and correlated with a coeval sudden decrease in seismicity beneath the summit. Immediately after, the seismicity increased again and a new collapse increment occurred 2.5 h later. Similar cycles continued for about 48 h, with the time separating the collapse increments decreasing from about 2 h to 30 min (Michon et al. 2007, 2011).

13.5.1 Data Analysis of the Broadband Seismic RER Station

The RER GEOSCOPE station (Roult et al. 2010) was the only broadband seismic station available to record the 2007 summit caldera collapse (Fontaine et al. 2014). The STS-1 sensors are built to record the translational ground vibrations but are known to record rotational movement as well, without any possible discrimination because there is only one station (Battaglia et al. 2000; Pillet and Virieux 2007; Rodgers 1968). Figure 13.10 corresponds to the tilt recordings of the north-south component. The signature of the February 18 summit eruption is a small tilt step. A cyclonic period (called Gamède) started on February 23, with a maximum on February 25. On March 30, the large downward step indicates the starting time (16:31 GMT) for the dyke propagation, followed by the proximal eruption at 19:00. The April 2 distal eruption is not associated with any obvious step. Nevertheless, its signature corresponds to a clearly visible change of trend on April 1 at 20:26 GMT (prior to the Mw 8.1 Solomon earthquake that occurred at 20:39:59 GMT) as noticed in Roult et al. (2012) and Fontaine et al. (2014). On April 5, at 20:48 GMT, the main major event is the first of several successive collapse events producing a saw-tooth pattern.

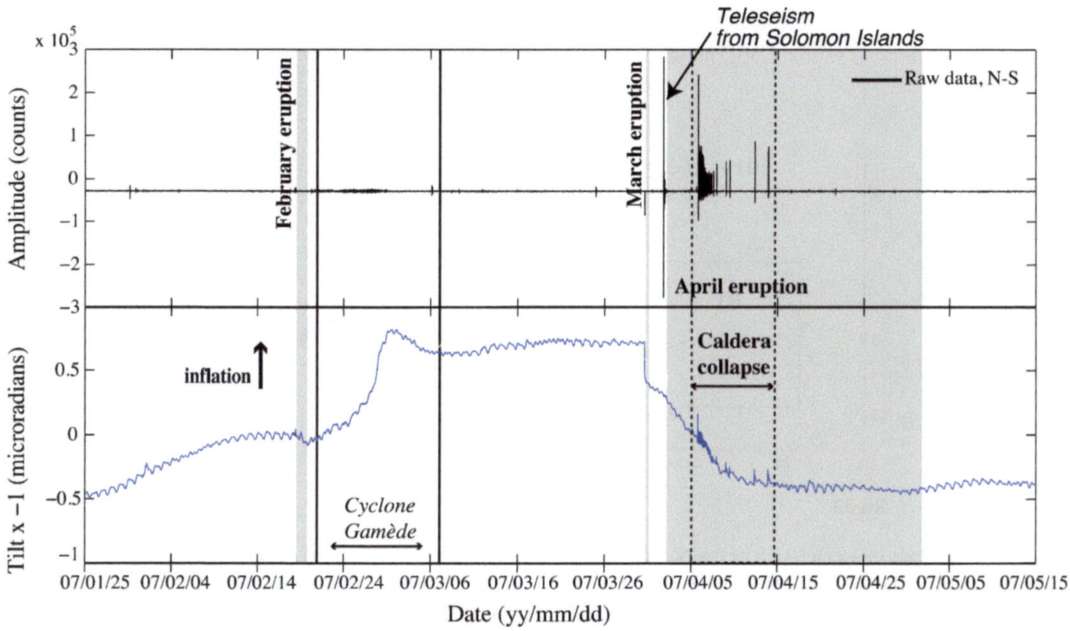

Fig. 13.10 *Top* Recording of the N-S component (raw data in counts) for the VH channel; *Bottom* N-S tilt signals retrieved from the VH channel. Two high tides are observed every day. From Fontaine et al. (2014)

13.5.2 Collapse Events Before and After the Main Collapse at 20:48 GMT on April 5

We identified several steps before the main collapse event occurring at 20:48 on April 5 2007. These 'precursor' events may be clearly identified on the 3 components. A subsequent, comprehensive analysis led to more than 50 incremental drops being identified over a period of 30 days. We observe similar features for these increments, with signal frequency content between 0.005 and 0.1 Hz. With time, the high frequency phase ($\sim$0.1 Hz) becomes greater. In order to better separate the effects from the tides and from the volcanic signal, we have computed the theoretical surface tidal acceleration for the three components (Hartmann and Wenzel 1995; Wenzel 1996). Subtracting the theoretical tidal component from the observed signal, we obtained a clearer volcanic signal (Fig. 13.11). The large step corresponds to the dike propagation onset at 16:31 preceding the March 30 eruption at 19:00 GMT (day 89). The April 2 eruption at 06:00 (day 92) is preceded by a clear trend change on April 1 at 20:26 GMT.

13.6 Tremor Location and Lava Volume Quantification

As on most volcanoes, the volcano-related seismicity is not limited to classic tectonic events but also includes various types of seismic signals such as LP events, tremors and rockfalls. The first two categories are of particular interest for understanding the behavior of volcanoes, as they are assumed to have fluids directly involved in their source processes (Chouet 1996). However, these events are usually difficult or impossible to locate using standard arrival time methods, because their onsets are emergent (LP events) or because they are stationary for a long time (tremor). To locate such events, Battaglia and Aki

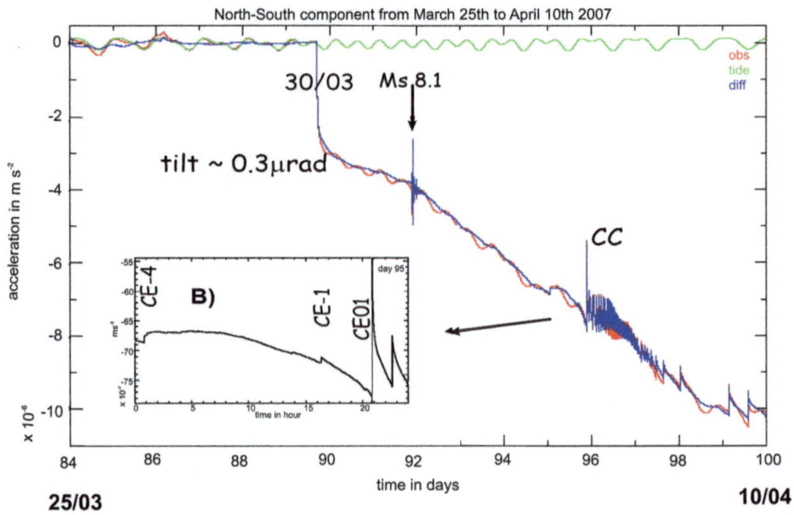

Fig. 13.11 **a** Trace in acceleration of the N-S component from March 25 (day 84) to April 10 (day 100). The theoretical Earth tide acceleration has been computed (in *green*) and removed from the raw data (in *red*) in order to highlight the volcano-tectonic effects (in *blue*): March 30 and April 2 eruptions and the collapse episode CC. **b** Enlargement of day 95 (April 5th) showing that the collapse episode of April 5 at 20:48 GMT (CE01) is preceded by various steps, early on April 5: CE-4 at 00:46 GMT, and later CE-1 at 16:18 GMT

(2003) developed a method based on seismic amplitudes corrected for station site effects using coda site amplification factors. Once corrected, the spatial distribution of amplitudes shows smooth and simple contours for many types of events (Aki and Ferrazzini 2000). Based on the simplicity of these distributions Battaglia and Aki (2003) developed inversion methods for locating their origins. To achieve this, the decrease of the amplitude as a function of the distance to the source is approximated by the decay either of surface or body waves in a homogeneous medium. The method is effective for locating rockfalls, long period events and eruption tremor sources.

The body wave amplitude location technique has been applied to several separate frequency bands (0.5–1.5, 1.5–2.5, 2.5–3.5, 3.5–4.5, 5–10 Hz and the whole frequency range) to study tremor recorded at the beginning of five eruptions which occurred between March 1998 and June 2000 (Battaglia and Aki 2003). For frequencies above 1.5 Hz, sources are generally found at shallow depth and close to the location of the eruptive vents or fissures. The best correlation between the position of the vents and tremor sources is obtained for the 5–10 Hz frequency band (Fig. 13.12), suggesting that the tremor in this band is directly generated at the eruption sites. Because of this, the method is a useful tool for locating fissures at the beginning of eruptions. As exemplified in the case of the September 1999 eruption, tremor in the 5–10 Hz band can also be used to locate the eruptive fissures minutes before their opening (Fig. 13.12). Lower frequencies seem to be related to deeper processes than the eruptions.

Battaglia et al. (2005a) located sources for tremor recorded during four eruptions between March 1998 and February 2000 with the aim of following their evolution. Tremor spatial amplitude distributions are usually simple and smooth when a single eruption site is active. Since sources for most bands above 1.5 Hz are found at shallow depth, the location technique is ineffective at Piton de la Fournaise to provide information on the deeper plumbing system. On rare occasions more complex amplitude distributions

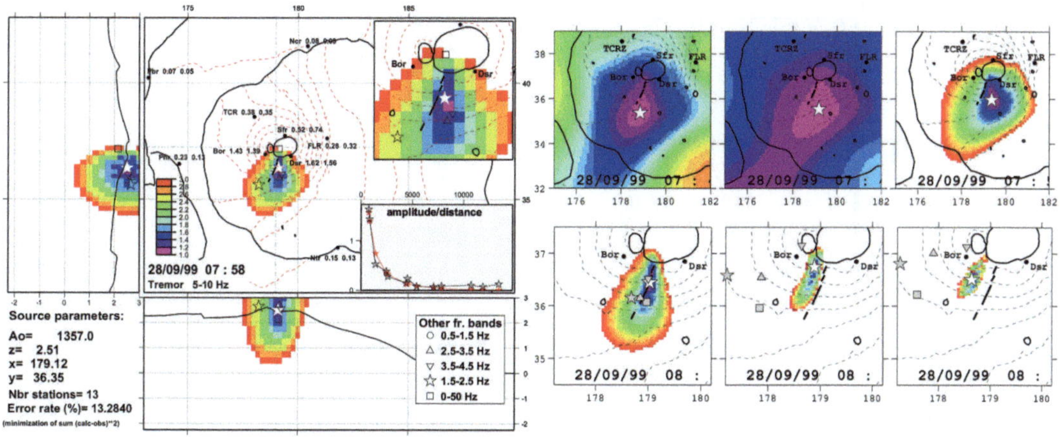

Fig. 13.12 *Left plot* location in the 5–10 Hz band obtained for the first sample of tremor recorded after the beginning of surface activity for the September 1999 eruption. Horizontal view and vertical cross-sections are shown. The best source is plotted as a *white star*, with sources for the other frequency bands plotted as other *symbols*. *Right plots* epicentral locations obtained in the 5–10 Hz band for 3 samples collected before (3 upper plots) and after (3 lower plots) the opening of the fissures

are observed which depend strongly on the frequency band. The eruption which started on March 9, 1998 was particularly interesting as it lasted for 6 months and included several eruption sites, sometimes active simultaneously. For this eruption, the 5–10 Hz frequency band is particularly efficient for mapping the eruptive activity as sources are found at the location of all the eruptive vents, including the Hudson vent and the fissures outside Enclos, where the eruptive activity was not as powerful as at the Kapor/Krafft sites (Fig. 13.13). In contrast, sources for the 1.5–2.5 Hz band are mainly generated in the area of the Kapor/Krafft vents and show no contributions from the other vents. On the other hand no migration of the tremor sources was observed in relation to the opening of fissures further downslope for the July and September 1999 eruptions.

Based on the obtained source locations, Battaglia et al. (2005b) proposed several relations for estimating the amounts of lava emitted during eruptions based on the use of tremor source amplitude. This is possible because, alongside the location of the tremor sources, the program also gives the source amplitude of the phenomenon. The authors assumed a linear or quadratic relation between the lava output and the tremor source amplitude. Both relations were calibrated using data for the March 1998 eruption and tested using the volumes emitted during the July and September 1999 and February 2000 eruptions. Battaglia et al. (2005b) found that the 5–10 Hz frequency band gives lava output much higher than for lower frequencies, suggesting either that for higher frequencies the tremor generation depends on the eruption site in addition to the lava output, or that tremor at those frequencies is not directly caused by the outflow of lava, but by gas. The best estimations were obtained using the quadratic relation for frequencies below 5 Hz, with an agreement of better than 15 % with the field observations for the July and September 1999 eruptions, and 30 % for the

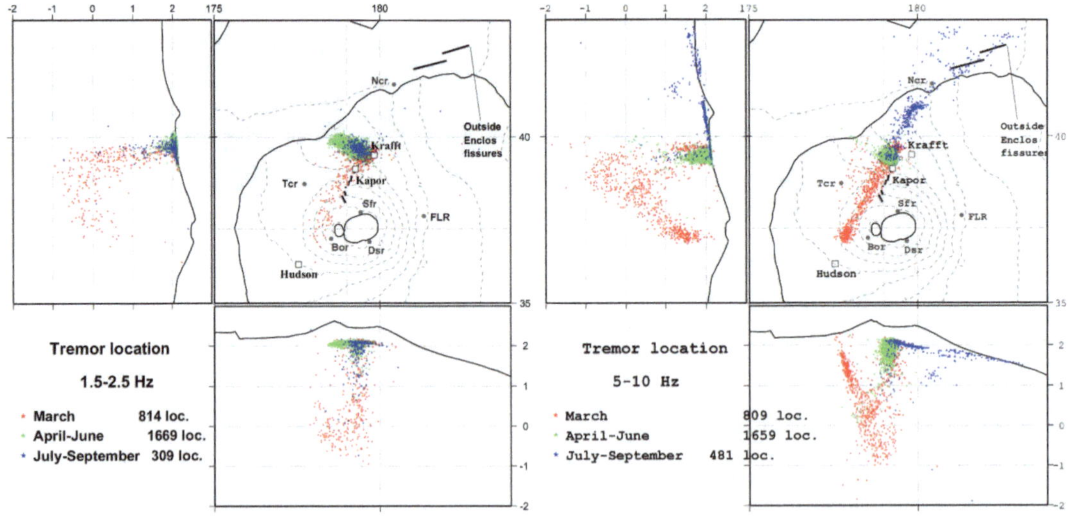

Fig. 13.13 Results from the application of the amplitude location program to measurements obtained during the 1998 eruption for tremor in the 1.5–2.5 and 5–10 Hz bands. For each plot, the locations are divided into three periods and plotted with different colors explained in the *lower left corners*. March corresponds to the period when Kapor, Krafft and Hudson vents were active, during the second period only Kapor vent was active, and July–September includes the period of activity outside the Enclos Fouqué. Note that the sources found at depth are an artifact caused by the presence of several simultaneously active vents (Kapor, Krafft and Hudson)

February 2000 eruption. The results show that the cumulative amplitude of tremor recorded during eruptions of Piton de la Fournaise is directly related to the amounts of lava erupted.

13.7 Tracking Magma Movements Before Eruptions of Piton de la Fournaise Volcano with Seismic Methods

Tracking the real-time propagation of magma through its pathway to the surface is a key issue in volcano monitoring and eruption prediction. On Piton de la Fournaise volcano, two major advances have led to a significant improvement of seismic time-lapse imaging of dyke propagation and eruption at the surface. In 2011, Taisne et al. (2011) used the broad-band seismic network of PdF (Fig. 13.14) to develop a new method for locating dyke migration using amplitude ratio of background seismic energy radiated during dyke propagation. By applying this method to the January 2010 Piton de la Fournaise volcano eruption, the authors imaged complex dike propagation dynamics which strongly differ from a model of constant velocity dike propagation (Fig. 13.14). Most strikingly and unexpectedly, they were able to show that most of the dyke transport to the surface occurs during the quiet seismic period preceding eruptions.

More recently, De Barros et al. (2013) discovered that high frequency, very shallow (100 m depth) seismicity was triggered during pre-eruptive swarms near eruptive sites up to a few hours prior to the presence of magma at the surface. Figure 13.15 shows the location of the seismic high-frequency energy radiated near-surface about 2 h before the October 2010 eruption. This eruption was finally located a few hundred meters from the predicted eruptive area where the high-frequency seismic energy was located.

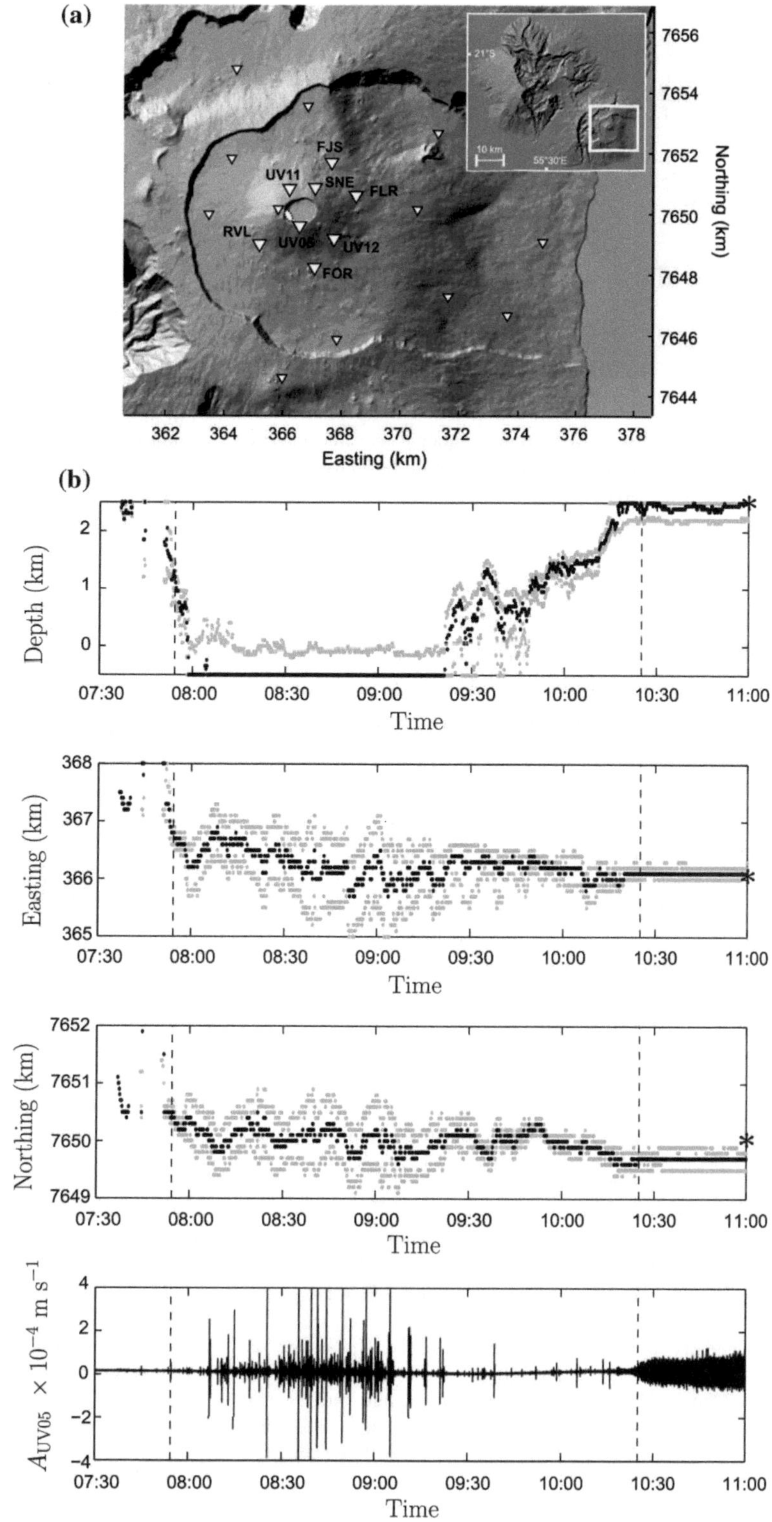

Fig. 13.14 **a** Broad-band seismic stations (OVPF/UnderVolc) used by Taisne et al. (2011) to image the dyke propagation preceding the January 2010 eruption of PdF. **b** Time-lapse 3D spatial location of the propagating dike prior to the January 2010 eruption of PdF. The *bottom figure* shows the raw seismic signal preceding the eruption (Taisne et al. 2011)

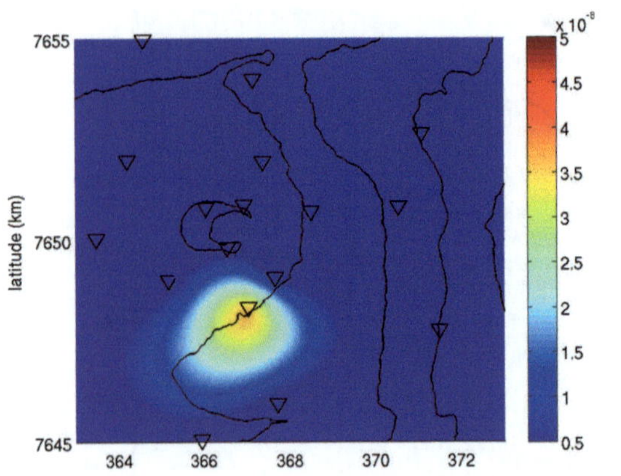

Fig. 13.15 Location of high-frequency seismic energy preceding the October 2010 of Piton de la Fournaise (De Barros et al. 2013)

In conclusion, the development of novel seismic methods, together with a dense and modern seismic network (OVPF/UnderVolc), has led to a very accurate monitoring of dyke propagation to the surface, and to improving the prediction of the location of eruptive vents. The real-time application of these methods in volcano observatories will lead to a great improvement in volcanic hazard assessment and eruption prediction.

Acknowledgments We thank the OVPF's technical staff for operating and maintaining the Piton de la Fournaise network since 1980.

References

Aki K, Ferrazzini V (2000) Seismic monitoring and modeling of an active volcano for prediction. J Geophys Res 105:16617–16640

Bachèlery P, Blum PA, Cheminée JL, Chevallier L, Gaulon R, Girardin N, Jaupart C, Lalanne FX, Le Mouel JL, Ruegg JC, Vincent PM (1982) Eruption at Le Piton de la Fournaise volcano on 3 February 1981. Nature 297(5865):395–397

Bachèlery P, Saint-Ange F, Villeneuve N, Savoye B, Normand A, Le Drezen E, Barrère A, Quod J-P, Deplus C (2010) Huge lava flows into the sea and caldera collapse, April 2007, Piton de la Fournaise volcano, IAVCEI third workshop on collapse calderas, La Réunion, pp 73–74

Battaglia J, Aki K (2003) Location of seismic events and eruptive fissures on the Piton de la Fournaise volcano using seismic amplitudes. J Geophys Res 108(B8). doi:10.1029/2002JB002193

Battaglia J, Bachèlery P (2003) Dynamic dyke propagation deduced from tilt variations preceding the March 9, 1998 eruption of the Piton de la Fournaise volcano. J Volcanol Geotherm Res 120:289–310

Battaglia J, Aki K, Montagner J-P (2000) Tilt signals derived from a GEOSCOPE VBB station on the Piton de la Fournaise volcano. Geophys Res Lett 27:605–608

Battaglia J, Aki K, Ferrazzini V (2005a) Location of tremor sources and estimation of lava output using tremor source amplitude on the Piton de la Fournaise volcano, 1. Location of tremor sources. J Volcanol Geotherm Res 147:268–290

Battaglia J, Aki K, Staudacher T (2005b) Location of tremor sources and estimation of lava output using tremor source amplitude on the Piton de la Fournaise volcano, 2. Estimation of lava output. J Volcanol Geotherm Res 147:291–308

Battaglia J, Ferrazzini V, Staudacher T, Aki K, Cheminée J-L (2005c) Pre-eruptive migration of earthquakes at the Piton de la Fournaise volcano (Reunion Island). Geophys J Int 161(2):549–558

Brandsdóttir B, Einarsson P (1979) Seismic activity associated with the September 1977 deflation of the Krafla central volcano in NE Iceland. J Volcanol Geotherm Res 6:197–212

Brenguier F, Kowalski P, Staudacher T et al (2012) First results from the UnderVolc high resolution seismic and GPS network deployed on Piton de la Fournaise Volcano. Seismol Res Lett 83(1):97–102

Chevallier L, Lalanne F-X, Bachèlery P, Vincent PM (1981) L'éruption du mois de février 1981 au Piton de la Fournaise (Ile de la Réunion, Océan Indien). Phénoménologie et remarques structurales. C R Acad Sci Paris 293(II):187–190

Chouet B (1996) Long-period volcano seismicity: its sources and use in eruption forecasting. Nat Geosci 380(6572):309–316

De Barros L, Bean CJ, Zecevic M, Brenguier F, Peltier A (2013) Eruptive fracture location forecasts from high-frequency events on Piton de la Fournaise volcano. Geophys Res Lett 40(17):4599–4603

Einarsson P, Brandsdóttir B (1980) Seismological evidence for a lateral magma intrusion during the July 1978 deflation of the Krafla volcano in NE Iceland. J Geophys Res 47:160–165

Filson J, Simkin T, Leu LK (1973) Seismicity of a caldera collapse: Galapagos Islands, 1968. J Geophys Res 78 (35):8591–8622

Fontaine FR, Roult G, Michon L, Barruol G, Di Muro A (2014) The 2007 eruptions and caldera collapse of the Piton de la Fournaise volcano (La Réunion Island) from tilt analysis at a single very broadband seismic station. Geophys Res Lett. doi:10.1002/2014GL059691

Gallart J, Driad L, Charvis P, Sapin M, Hirn A, Diaz J, de Voogd B, Sachpazi M (1999) Perturbation to the lithosphere along the hotspot track of La Reunion from an offshore-onshore seismic transect. J Geophys Res B 104(2):2895–2908

Harlow DH, Power JA, Laguerta EP, Ambubuyong G, White RA, Hoblitt RP (1996) Precursory seismicity and forecasting of the June 15, 1991, eruption of Mount Pinatubo. In: Newhall CG, Punongbayan RS (eds) Fire and mud, eruptions and lahars of Mount Pinatubo. PHIVOLCS and University of Washington Press, Philippines, pp 285–306

Hartmann T, Wenzel H-G (1995) The HW95 tidal potential catalogue. Geophys Res Lett 22:3553–3556

Hayashi Y, Morita Y (2003) An image of a magma intrusion process inferred from precise hypocentral migrations of the earthquake swarm east of the Izu Peninsula. Geophys J Int 153:159–174

Koyanagi RY, Tanigawa WR, Nakata JS (1988) Seismicity associated with the eruption. In: EW Wolfe (ed) The Puu Oo eruption of Kilauea volcano, Hawaii: episodes 1 through 20, January 3, 1983, through June 8, 1984. US Geological Survey, Professional Paper, vol 1463, pp 183–235

Kumagai H, Ohminato T, Nakano M, Ooi M, Kubo A, Inoue H, Oikawa J (2001) Very-long-period seismic signals and caldera formation at Miyake Island, Japan. Science 293:687–690. doi:10.1126/science.1062136

Lee WHK, Lahr JC (1975) HYPO71 (revised): a computer program for determining hypocenter, magnitude, and first motion pattern of local earthquakes. US Geological Survey of Open File Report, pp 75–311

Lénat J-F, Bachèlery P (1990) Structure et fonctionnement de la zone centrale du Piton de la Fournaise. In: Lénat J-F (ed) Le volcanisme de la Réunion, monographie. Centre de Recherches Volcanologiques, Clermont-Ferrand, pp 257–296

Massin F, Ferrazzini V, Bachèlery P, Nercessian A, Duputel Z, Staudacher T (2011) Structures and evolution of the plumbing system of Piton de la Fournaise volcano inferred from clustering of 2007 eruptive cycle seismicity. J Volcano Geotherm Res 202(1–2):96–106. doi:10.1016/j.jvolgeores.2011.01.008

Michon L, Staudacher T, Ferrazzini V, Bachelery P, Marti J (2007) April 2007 collapse of Piton de la Fournaise: a new example of caldera formation. Geophys Res Lett 34:L21301. doi:10.1029/2007GL031248

Michon L, Villeneuve N, Catry T, Merle O (2009) How summit calderas collapse on basaltic volcanoes: new insights from the April 2007 caldera collapse of Piton de la Fournaise volcano. J Volcano Geotherm Res 184 (1–2):138–151. doi:10.1016/j.jvolgeores.2008.11.003

Michon L, Massin F, Famin V, Ferrazzini V, Roult G (2011) Basaltic calderas: collapse dynamics, edifice deformation, and variations of magma withdrawal. 116:B03209. doi:10.1029/2010JB007636

Michon L, Di Muro A, Villeneuve N, Saint-Marc C, Fadda P, Manta F (2013) Explosive activity of the summit cone of Piton de la Fournaise volcano (La Réunion island): a historical and geological review. J Volcano Geotherm Res 264:117–133

Michon L, Ferrazzini V, Di Muro A (2016) Magma paths at Piton de la Fournaise volcano. In: Bachèlery P, Lénat J-F, Di Muro A, Michon L (eds) Active volcanoes of the southwest Indian Ocean: Piton de la Fournaise and Karthala. Active Volcanoes of the World. Springer, Berlin

Peltier A, Ferrazzini V, Staudacher T, Bachèlery P (2005) Imaging the dynamics of dyke propagation prior to the 2000–2003 flank eruptions at Piton de La Fournaise, Reunion Island. Geophys Res Lett 32(22):L22302. doi:10.1029/2005GL023720

Peltier A, Bachèlery P, Staudacher T (2009a) Magma transport and storage at Piton de La Fournaise (La Réunion) between 1972 and 2007: a review of geophysical and geochemical data. J Volcanol Geotherm Res 184(1–2):93–108. doi:10.1016/j.jvolgeores.2008.12.008

Peltier A, Staudacher T, Bachèlery P, Cayol V (2009b) Formation of the April 2007 caldera collapse at Piton de La Fournaise volcano: Insights from GPS data. J Volcano Geotherm Res 184(1–2):152–163. doi:10.1016/j.jvolgeores.2008.09.009

Peltier A, Staudacher T, Bachèlery P (2010) New behaviour of the Piton de La Fournaise volcano feeding system (La Réunion Island) deduced from GPS data: influence of the 2007 Dolomieu caldera collapse. J Volcano Geotherm Res 192:48–56

Pillet R, Virieux J (2007) The effects of seismic rotations on inertial sensors. Geophys Res Lett 171:1314–1323. doi:10.1111/j.1365-246X.2007.03617.x

Rodgers PW (1968) The response of the horizontal pendulum seismometer to Rayleigh and Love waves, tilt, and free oscillations of the Earth. Bull Seismol Soc Am 58:1385–1406

Roult G, Montagner J-P, Romanowicz B, Cara M, Rouland D, Pillet R, Karczewski J-F, Rivera L, Stutzmann E, Maggi A, GEOSCOPE-Team (2010) The GEOSCOPE program: progress and challenges during the past 30 years. Seismol Res Lett 81(3):427–452. doi:10.1785/gssrl1.81.3.247

Roult G, Peltier A, Taisne B, Staudacher T, Ferrazzini V, Di Muro A, OVPF Team (2012) A new comprehensive classification of the Piton de la Fournaise activity spanning the 1985–2010 period. Search and analysis of short-term precursors from a broad-band seismological station. J Volcano Geotherm Res 241–242:78–104. doi:10.1016/j.jvolgeores.2012.06.012

Rubin AM, Gillard D (1998) Dike induced seismicity: theoretical considerations. J Geophys Res 103:10017–10030

Staudacher T, Ferrazzini V, Peltier A, Kowalski P, Boissier P, Catherine P, Lauret F, Massin F (2009) The April 2007 eruption and the Dolomieu crater collapse, two major events at Piton de la Fournaise (La Réunion Island, Indian Ocean). J Volcano Geotherm Res 184(1–2):126–137. doi:10.1016/j.jvolgeores.2008.11.005

Taisne B, Brenguier F, Shapiro NM, Ferrazzini V (2011) Imaging the dynamics of magma propagation using radiated seismic intensity. Geophys Res Lett 38: L04304. doi:10.1029/2010GL046068

Ueda H, Fujita E, Ukawa M, Yamamoto E, Irwan I, Kimata F (2005) Magma intrusion and discharge process at the initial stage of the 2000 activity of Miyakejima, central Japan, inferred from tilt and GPS data. J Int 161:891–906. doi:10.1111/j.1365-246X.2005.02602.x

Ukawa M, Tsukahara H (1996) Earthquake swarms and dike intrusions off the east coast of Izu Peninsula, central Japan. Tectonophysics 108:323–337

Umakoshi K, Shimizu H, Matsuwo N (2001) Volcano-tectonic seismicity at Unzen Volcano, Japan, 1985–1999. J Volcanol Geotherm Res 112:117–131

Urai M, Geshi N, Staudacher T (2007) Size and volume evaluation of the caldera collapse on Piton de la Fournaise volcano during the April 2007 eruption using ASTER stereo imagery. Geophys Res Lett 34: L22318. doi:10.1029/2007GL031551

Wenzel HG (1996) The Nanogal software: Earth tide data processing package ETERNA 3.3. Bull Inf Marées Terrestres 124:9425–9439

Yamamoto E, Ukawa M, Fujita E, Okada Y, Kikuchi M (2001) Step-like tilt change occurred during the caldera-forming stage of the 2000 Miyakejima Volcanic Activity. J Geogr 110:181–190

Zecevic M, De Barros L, Bean CJ, O'Brien GS, Brenguier F (2013) Investigating the source characteristics of long-period (LP) seismic events recorded on Piton de la Fournaise volcano, La Réunion. J Volcanol Geotherm Res 258:1–11

Zobin VM, Gonzalez Amezcua M, Reyes Davila GA, Dominguez T, Cerda Chacon JC, Chavez Alvarez JM (2002) Comparative characteristics of the 1997–1998 seismic swarms preceding the November 1998 eruption of Volcan de Colima, Mexico. J Volcanol Geotherm Res 117:47–60

Static and Dynamic Seismic Imaging of Piton de la Fournaise

14

Jean Battaglia and Florent Brenguier

Abstract

Data from the Piton de la Fournaise Volcanological Observatory ("Observatoire Volcanologique du Piton de la Fournaise", OVPF) and from the Undervolc project (Brenguier et al. in Seismol Res Lett 83(1):97–102, 2012) have been used to image the interior of the volcano. Two different approaches have been carried out. One is noise-based and allows both imaging and monitoring. The other is based on P-wave first arrival times of local earthquakes to derive 3D tomographic models.

14.1 Noise-Based Seismic Imaging and Monitoring of Piton de la Fournaise Volcano

14.1.1 Introduction

Ambient seismic noise continuously travels along the surface and through the interior of the Earth. These seismic waves thus carry crucial information about the medium they propagate through. When crossing volcanoes, these waves are affected by the mechanical heterogeneities of the volcanic edifice and are also perturbed during periods of unrest by temporal changes in volcano interiors. Noise-based seismic imaging and monitoring allows the structure and temporal evolution of volcanic edifices to be viewed. In particular, it has proved successful in the detection of very small seismic velocity changes in volcanic edifices (0.1 %) that have been shown, on Piton de la Fournaise volcano (La Réunion island), to be precursors of volcanic eruptions.

Piton de la Fournaise volcano is monitored by the OVPF/IPGP volcano observatory (Battaglia et al. 2016; Chap. 13). In 2009, within the framework of the UnderVolc project (Brenguier et al. 2012), 15 new broad-band seismic stations were added to the observatory seismic network, which was composed of about 20 seismic stations (Fig. 14.1). The intense eruptive activity,

J. Battaglia (✉)
Laboratoire Magmas et Volcans, Université Blaise Pascal-CNRS-IRD, OPGC, 5 rue Kessler, 63038 Clermont-Ferrand, France
e-mail: J.Battaglia@opgc.univ-bpclermont.fr

F. Brenguier
Institut des Sciences de la Terre, University of Grenoble, Grenoble, France

P. Bachèlery et al. (eds.), *Active Volcanoes of the Southwest Indian Ocean*, Active Volcanoes of the World, DOI 10.1007/978-3-642-31395-0_14

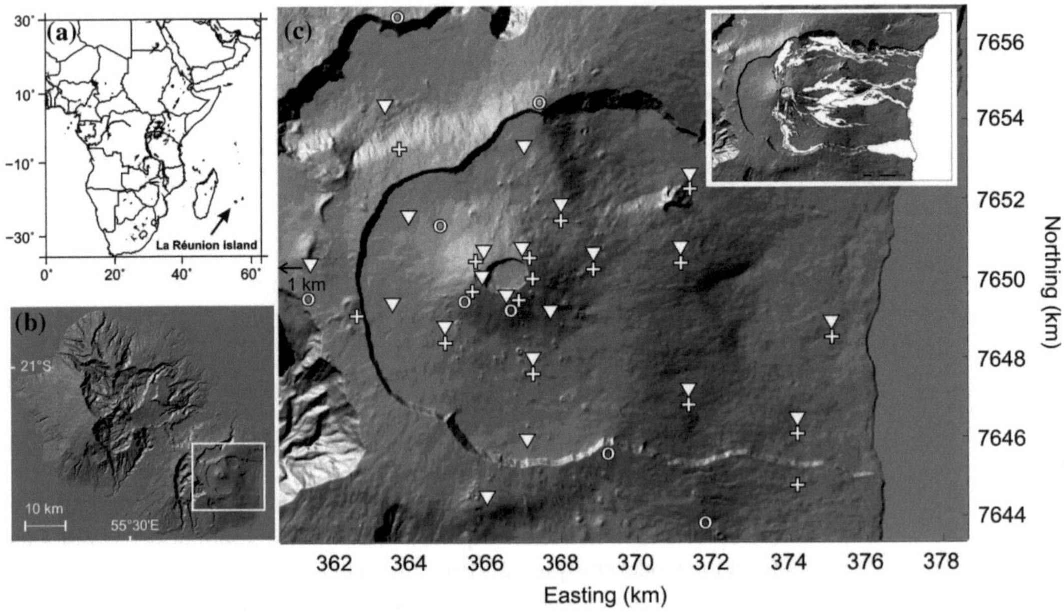

Fig. 14.1 Location of **a** La Réunion island, **b** Piton de la Fournaise volcano, and **c** the UnderVolc and observatory seismic and GPS networks. UnderVolc broad-band seismic stations are shown as *inverted triangles*, GPS stations as crosses, and observatory short-period seismic stations as *circles*. The inset map shows lava flows from 2000 to 2010 (courtesy of OVPF, T. Staudacher, Z. Servadio)

coupled with only weak tectonic activity, makes Piton de la Fournaise volcano well suited for studies focused on the processes of magma pressurization and injection and for the development of innovative imaging and monitoring methods.

More than 30 eruptions occurred between 2000 and 2010 at Piton de la Fournaise volcano (PdF). These eruptions lasted from a few hours to a few months and were associated with the emission of basaltic lava whose volume ranged from less than one to tens of millions of cubic meters (Peltier et al. 2009). The time period under consideration here (1998–2011) is bracketed by 2 major eruptions, namely the March 1998 eruption (60 million cubic meters of lava emitted), and the April 2007 eruption associated with the 300 m-high collapse of the main Dolomieu crater (240 million cubic meters of lava emitted) (Bachèlery et al. 2010).

14.1.2 Noise-Based Seismic Imaging of Piton de la Fournaise Volcano

In the field of ultrasonics, it has been shown both theoretically and experimentally that a random wavefield has correlations which, on average, take the form of the Green's function of the media (Weaver and Lobkis 2001). In seismology, recent studies also showed that the Green's function between pairs of seismographs can be extracted from cross-correlations of coda waves and ambient noise (Campillo 2006). Moreover, some authors used correlations of long ambient seismic noise sequences from regional networks in order to extract and to invert Rayleigh waves to produce high-resolution seismic images of the shallow crustal layers under these networks (Shapiro et al. 2005).

By using continuous records of ambient seismic noise from the volcano observatory short period seismic network, Brenguier et al. (2007) were able to measure Rayleigh wave velocity dispersion curves obtained from the cross-correlations of 18 months of ambient seismic noise. The noise records between 21 receivers have been correlated to reconstruct Rayleigh waves, and a sufficient signal-to-noise ratio was obtained for 210 inter-station paths. The authors used the reconstructed waveforms to measure group velocity dispersion curves at periods of between 1.5 and 4.5 s. The obtained measurements have been inverted for two-dimensional group velocity maps and finally for a 3-D S-wave velocity model of the edifice between +2 and −1 km relative to sea level. This tomography led to a high resolution image of the shallow structure of PdF volcano (Fig. 14.2). The results show a high seismic velocity body located in the central part of the volcano, which moves westward with depth (+1.3 to −1.1 km relative to sea level). It is spatially delimited by the borders of the active 8 km-wide caldera. Its geometry at 0.5 km below sea-level coincides well with the central area and the NE and SE rift zones. This structure is surrounded by a low-velocity ring which is interpreted as effusive and pyroclastic products. The high velocities are associated with intrusive magma slowly cooling within the surrounding low velocity material.

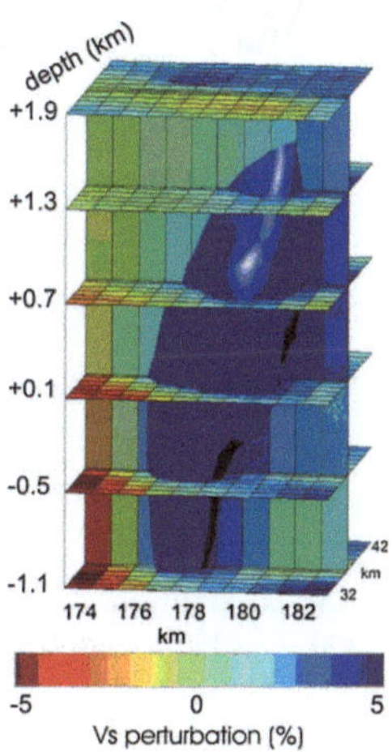

Fig. 14.2 3D tomographic image of the shallow central part of Piton de la Fournaise volcano using cross-correlations of ambient seismic noise (Brenguier et al. 2007)

14.1.3 Noise-Based Seismic Monitoring of Velocity Changes at Piton de la Fournaise Volcano

Pressurized volcanic fluids (magma, water) or gases induce deformation and thus perturbations of the elastic properties of volcanic edifices. These small perturbations can be detected as changes in seismic wave properties using repetitive seismic sources (Grêt et al. 2005; Ratdomopurbo and Poupinet 1995; Wegler et al. 2006). However, none of these approaches were suitable for providing a continuous monitoring of volcano elastic properties. In a pioneering work Sens-Schoenfelder and Wegler (2006) proposed using the repetitive waveforms of seismic noise cross-correlations to track for subsurface volcanic edifice velocity changes. In this manner, the continuous recording of ambient seismic noise allows continuous monitoring of volcano interiors.

The approach consists of measuring very small waveform time-delays in the coda of noise cross-correlations. This method is described as the so-called Moving Window Cross Spectrum (Clarke et al. 2011; Ratdomopurbo and Poupinet 1995) or Coda Wave Interferometry (Snieder et al. 2002) techniques. Coda waves (late part of seismograms) are scattered waves that travel long distances and thus accumulate time delays as a consequence of a seismic velocity change in the propagating medium. Measuring travel time perturbations in the coda thus allows for the detection of very small velocity changes that would not be detectable by a classic measure of first arrival time delays. The drawback to this approach is that it is difficult to estimate the travel paths of the scattered waves constituting the coda. However, recent promising results suggest it may be possible to produce refined 3D maps of small changes in the near future (Larose et al. 2010).

Brenguier et al. (2011) measured seismic velocity variations within the Piton de la Fournaise edifice from June 2010 to April 2011, following Brenguier et al. (2008). Continuous seismic velocity changes, as well as the daily seismicity rate, are shown on Figs. 14.3 and 14.4 for two different time periods. Figure 14.3 shows small (0.1 %) seismic velocity drops preceding eruptions between July 1999 and December 2000. On Fig. 14.4, the authors observe a velocity drop starting about 1 month before the October 14, 2010 eruption. The velocity decrease correlates in time with an increase in seismicity and has been interpreted as the deformation of the edifice induced by magma pressure buildup and injection. These velocity drops are interpreted as being associated with the deformation of the edifice induced by magma pressure build-up. Interestingly, seismic velocity increased during the eruption of October 2010, and after the short eruption of December 2010, indicating a possible mechanism of stress relaxation induced by the emptying of the magmatic reservoir. Furthermore, to improve the accuracy and thus time resolution of velocity measurements, Baig et al. (2009) developed a cross-correlation filtering method based on time-frequency transforms and phase coherence filtering.

Noise-based seismic velocity monitoring is thus a unique method that allows volcanic activity to be monitored precisely. It must, however, be noted that other phenomena perturb seismic velocities of the rock mass, such as the presence of water in the medium, temperature and barometric atmospheric pressure changes, and solid and oceanic tides. It is thus important to model these effects accurately so as to correct the observed seismic velocity changes for them in order to extract the volcanic-related signal.

In April 2007, a major eruption occurred at Piton de la Fournaise volcano (Staudacher et al. 2009) producing more than 240 million cubic meters of lava (Bachèlery et al. 2010). It was located 7 km to the south east of the summit. Clarke et al. (2013) observed an unusually large decrease in seismic velocity associated with this volcanic episode which could not be explained by the pre-eruptive edifice inflation due to magma pressure build-up, as

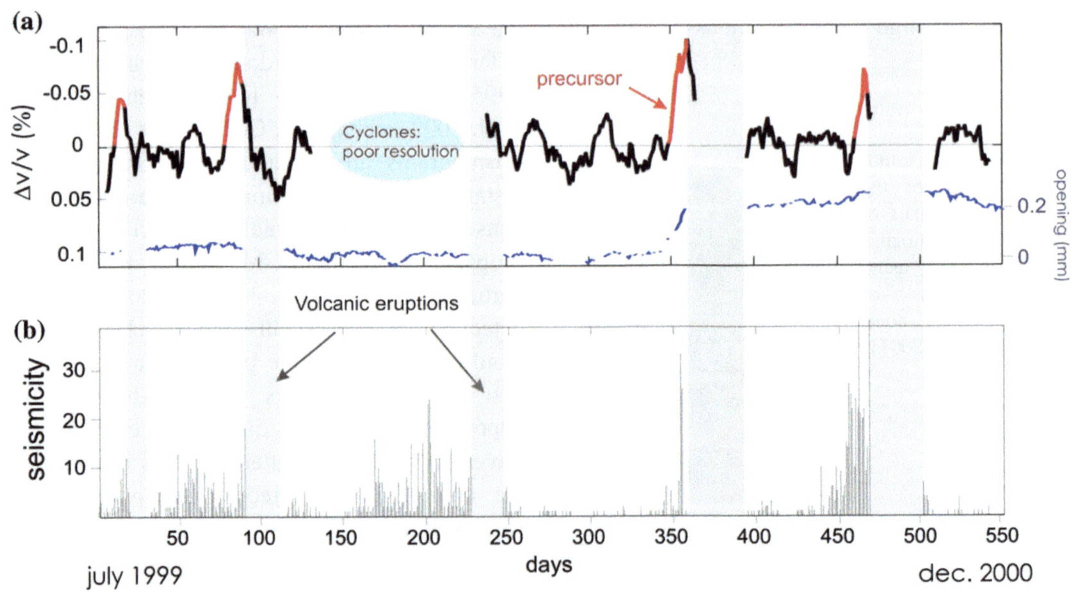

Fig. 14.3 **a** Relative velocity changes compared to extensometer (FORX). The travel time shifts are measured in the frequency range 0.1–0.9 Hz. For details, see Brenguier et al. (2008), **b** inter-eruptive seismicity (pre-eruptive swarms are excluded)

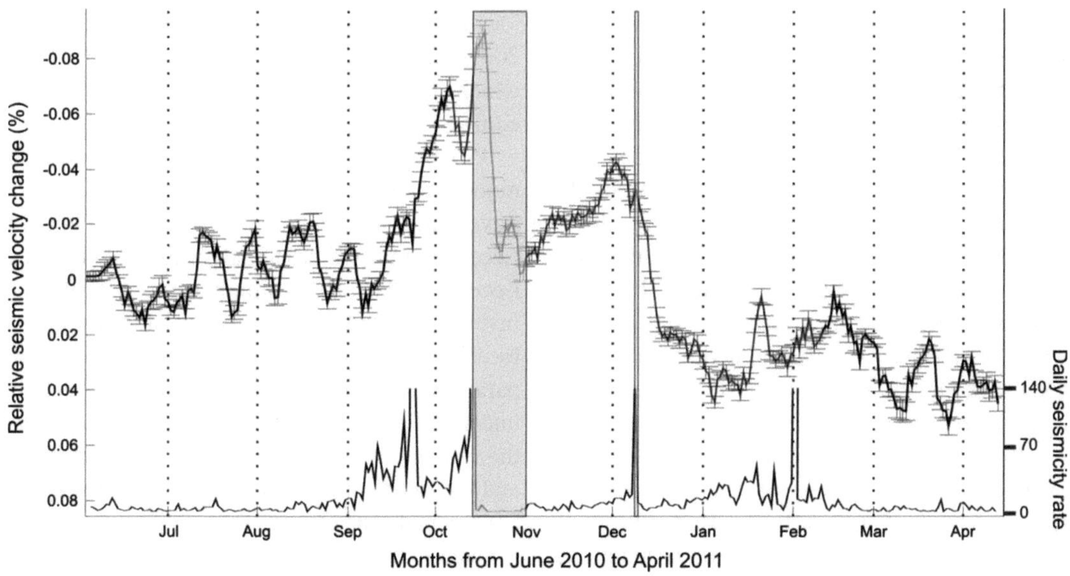

Fig. 14.4 Relative seismic velocity changes with *error bars* and daily seismicity rate between June 2010 and April 2011. Eruption periods are shown as *gray rectangles*

described by Brenguier et al. (2008). Also, Clarke et al. (2013) showed surface displacement images obtained from InSAR data inversion (Froger et al. 2016, this volume Chap. 16). These results show that between March and May 2007, a widespread volcanic edifice flank movement occurred with a maximum eastward displacement of 1.4 m. However, the timing of this movement and, in particular, its relation with the occurrence of the April 2007 eruption could not be identified due to a lack of temporal resolution of the InSAR images. Clarke et al. (2013) used high temporal resolution seismic velocity change measurements to prove that the strong drop in seismic velocities started at the time as a small eruption which preceded the main April eruption by a few days. Inferring the association between this high seismic velocity decrease and the widespread volcanic flank movement, the authors concluded that the volcanic flank movement started at the time of this small precursory eruption. Considering new simulations of volcanic edifice deformation from Got et al. (2013), it is likely that the small precursory eruption triggered a large elasto-plastic deformation of the edifice flank that released stresses and favored the horizontal migration of magma up to a significant distance along the April 2007 eruptive fissure. This study shows how noise-based monitoring has been used to explain the origin of the unusual April 2007 eruption at Piton de la Fournaise volcano and that this method can also be used to detect volcanic flank movements and possible instabilities on volcanoes.

This method is now used routinely by the Piton de la Fournaise Observatory in order to improve eruption forecasting, together with earthquake and deformation observations. Lecocq et al. (2013) developed an integrated package that includes all the processing steps required to obtain continuous seismic velocity changes from continuous seismic records.

14.2 P Wave Velocity Structure

In order to gain information about the internal structure of Piton de la Fournaise, Prôno et al. (2009) carried out a 3D tomographic inversion based on P-wave first arrival times from earthquakes recorded by OVPF. Seismic tomography is a major tool which enables information on the structure below the ground surface to be obtained using the propagation velocity of seismic waves through the medium (Aki and Lee 1976). The method is based on the fact that the velocities are representative of the physical properties of the medium which they cross. Prôno et al. (2009) used data from the short period monitoring network collected between 1996 and 1999. In addition to the common volcano-tectonic activity which is located at shallow depth, above sea level and below the central cone, these data include deeper events which preceded the March 9, 1998 eruption. This pre-eruptive swarm included a large number of events below sea level, down to 5–6 km b.s.l. (Battaglia et al. 2005). These data provide an unprecedented set of 861 events which shed light on the deep structure of the volcano, as compared to previous localised tomographic works (Brenguier et al. 2007; Lankar 1997; Nercessian et al. 1996) which only imaged the shallow part of the edifice.

Prôno et al. (2009) used a tomographic technique (Monteiller et al. 2005) based on an accurate travel-time computation using the Podvin-Lecomte finite-difference algorithm (Podvin and Lecomte 1991) and a simultaneous probabilistic inversion of both velocity models and earthquake locations. The technique enables the inversion parameters, such as the correlation length, and the uncertainties on the different a priori parameters: the velocity of the inversion cells, and the position and the initial time of the earthquakes, to be well constrained. The choice of the optimal parameters is made so that they minimize both the RMS and a cost function used to solve the inverse problem. This process allows the result with the lowest

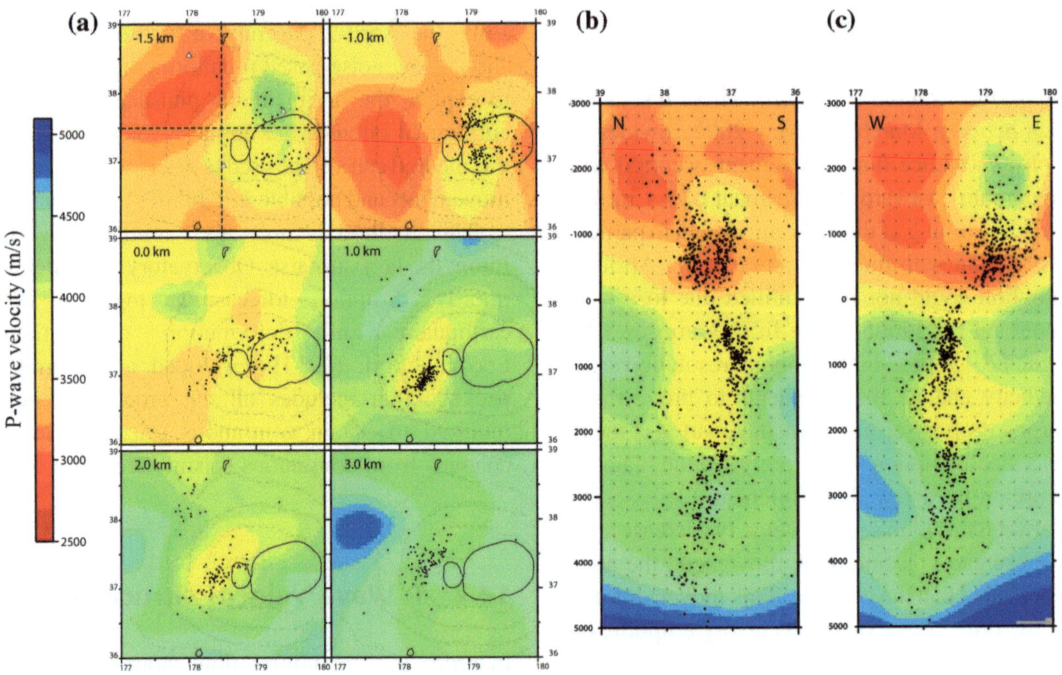

Fig. 14.5 2D sections of the P velocity model obtained for the small grid. Plot **a** shows horizontal sections extracted every kilometer: *black points* indicate the seismic hypocenters located at ±500 m from the depth of the section, *dotted lines* are the location of the vertical cross-sections. Spatial coordinates are in Gauss-Laborde kilometric scale, contours of Brory and Dolomieu craters are indicated. Plot **b** shows the north-south vertical section, and plot **c** the west-east section taken close to the principal seismic swarm, as indicated on the horizontal section at −1.5 km in plot **a**

degree of freedom, i.e. the simplest model to fit the data, to be determined. This provides high quality stable tomographic images on two grids: a coarse grid with nodes every 500 m horizontally and 250 m vertically, and a fine grid with 100 m sampling in both directions. Resolution tests (checkerboard, spike, shape test) were carried out to determine where the velocity model was well constrained. They indicated a rather good reconstruction of the medium from 2 km a.s.l to 3.5 km b.s.l., along the more or less vertical conduit defined by the seismicity below the summit caldera. Away from the conduit the resolution decreases sharply due to the poor ray coverage. The data mainly provide a reliable reconstruction of a 1D velocity model below the summit.

The obtained P-wave velocity model (Fig. 14.5) outlines the presence of several clear features. (1) Confirmation of a **high-velocity plug**, as observed by Nercessian et al. (1996), under the summit craters, above sea level,. It is interpreted as corresponding to an intrusive, solidified dyke-and-sill complex with little fluid magma storage. (2) This anomaly is surrounded by a **low-velocity ring** known to be fractured lavas and scorias and vesicular lava or scorias located on volcano flanks. (3) Below the high velocity plug, still above sea level, a **low velocity volume** is found, in which a large part of the shallow seismicity is located. This seismicity describes a V-shaped pattern with two branches having a main east-west extension, well correlated with the southern and northern borders of Dolomieu crater. (4) The base of the low velocity zone is delimited by an area of **strong vertical velocity gradient**, which is found at sea level. This confirms that a major interface in the volcanic structure is located at sea level, as already suggested by the spatio-temporal distribution of the seismicity preceding the March 1998 eruption (Battaglia et al. 2005). (5) Below sea level, about 1–2 km b.s.l., **a second zone of low velocity** is observed, with a low seismicity zone in its center. This could correspond to an area of magma storage that could feed the major eruptions of the Piton de la Fournaise volcano.

Acknowledgments We thank the OVPF's technical staff for operating and maintaining the Piton de la Fournaise network since 1980.

References

Aki K, Lee WHK (1976) Determination of three dimensional velocity anomalies under a seismic array using first P-arrival times from local earthquakes. J Geophys Res 81:4381–4399

Bachèlery P, Saint-Ange F, Villeneuve N, Savoye B, Normand A, Le Drezen E, Barrère A, Quod J-P, Deplus C (2010) Huge lava flows into the sea and caldera collapse, April 2007, Piton de la Fournaise volcano, IAVCEI third workshop on Collapse Calderas, La Réunion, pp 73–74

Baig A, Campillo M, Brenguier F (2009) Denoising seismic noise cross correlations. J Geophys Res 114 (B8):2156–2202. doi:10.1029/2008JB006085

Battaglia J, Ferrazzini V, Staudacher T, Aki K, Cheminée J-L (2005) Pre-eruptive migration of earthquakes at the Piton de la Fournaise Volcano (Reunion Island). Geophys J Int 161(2):549–558

Battaglia J, Brenguier F, Roult G (2016) Seismic monitoring at Piton de la Fournaise. In: Bachèlery P, Lénat J-F, Di Muro A, Michon L (eds) Active volcanoes of the southwest Indian Ocean: Piton de la Fournaise and Karthala, active volcanoes of the world. Springer, Berlin

Brenguier F, Shapiro NM, Campillo M, Nercessian A, Ferrazzini V (2007) 3-D surface wave tomography of the Piton de la Fournaise volcano using seismic noise correlations. Geophys Res Lett 34:L02305. doi:10.1029/2006GL028586

Brenguier F, Shapiro N, Campillo M, Ferrazzini V, Duputel Z, Coutant O, Nercessian A (2008) Towards forecasting volcanic eruptions using seismic noise. Nat Geosci 1(2):126–130

Brenguier F, Clarke D, Aoki Y, Shapiro NM, Campillo M, Ferrazzini V (2011) Monitoring volcanoes using seismic noise correlations. C R Geosci 343:633–638. doi:10.1016/j.crte.2010.12.010

Brenguier F, Kowalski P, Staudacher T et al (2012) First results from the UnderVolc high resolution seismic and GPS network deployed on Piton de la Fournaise Volcano. Seismol Res Lett 83(1): 97–102

Campillo M (2006) Phase and correlation in random seismic fields and the reconstruction of the green function. Pure Appl Geophys 163(2):475–502

Clarke D, Zaccarelli L, Shapiro N, Brenguier F (2011) Assessment of resolution and accuracy of the moving window cross spectral technique for monitoring crustal temporal variations using ambient seismic noise. Geophys J Int. doi:10.1111/j.1365-246X.2011.05074.x

Clarke D, Brenguier F, Froger J-L, Shapiro N, Peltier A, Staudacher T (2013) Timing of a large volcanic flank movement at Piton de la Fournaise volcano using noise-based seismic monitoring and ground deformation measurements. Geophys J Int 195(2):1132–1140

Froger J-L, Cayol V, Famin V (2016) The March–April 2007 eruption of Piton de la Fournaise as recorded by interferometric data. In: Bachèlery P, Lénat J-F, Di Muro A, Michon L (eds) Active volcanoes of the southwest Indian Ocean: Piton de la Fournaise and Karthala. Springer, Berlin

Got JL, Peltier A, Staudacher T, Kowalski P, Boissier P (2013) Edifice strength and magma transfer modulation at Piton de la Fournaise volcano. J Geophys Res 118:1–18. doi:10.1002/jgrb.50350

Grêt A, Snieder R, Aster R, Kyle P (2005) Monitoring rapid temporal changes in a volcano with coda wave interferometry. Geophys Res Lett 32:1–4

Lankar V (1997) Approches par tomographie sismique du Piton de la Fournaise, La Réunion. Ph.D., thesis, Inst. de Phys. du Globe, Paris

Larose E, Planes T, Rossetto V, Margerin L (2010) Locating a small change in a multiple scattering environment. Appl Phys Lett 96(20):204101

Lecocq T, Caudron C, Brenguier F (2013) MSNoise, a Python package for computing and monitoring seismic velocity changes using ambient seismic noise. Seismol Res Lett

Monteiller V, Got J-L, Virieux J, Okubo P (2005) An efficient algorithm for double-difference tomography and location in heterogeneous media, with an application to the Kilauea volcano. J Geophys Res 110: B12306. doi:10.1029/2004JB003466

Nercessian A, Hirn A, Lépine J-C, Sapin M (1996) Internal structure of Piton de la Fournaise volcano from seismic wave propagation and earthquake distribution. J Volcanol Geotherm Res 70:123–143

Peltier A, Bachèlery P, Staudacher T (2009) Magma transport and storage at Piton de La Fournaise (La Réunion) between 1972 and 2007: a review of geophysical and geochemical data. J Volcanol Geotherm Res 184(1–2):93–108. doi.org/10.1016/j.jvolgeores.2008.12.008

Podvin P, Lecomte I (1991) Finite-difference computation of traveltimes in very contrasted velocity models—a massively parallel approach and its associated tools. Geophys J Int 105:271–284

Prôno E, Battaglia J, Monteiller V, Got J-L, Ferrazzini V (2009) P-wave velocity structure of Piton de la Fournaise volcano deduced from seismic data recorded between 1996 and 1999. J Volcanol Geotherm Res 184(1–2):49–62

Ratdomopurbo A, Poupinet G (1995) Monitoring a temporal change of seismic velocity in a volcano: application to the 1992 eruption of Mt. Merapi (Indonesia). Geophys Res Lett 22(7):775–778

Sens-Schoenfelder C, Wegler U (2006) Passive image interferometry and seasonal variations of seismic velocities at Merapi Volcano. Indonesia Geophys Res Lett 33:1–5

Shapiro N, Campillo M, Stehly L, Ritzwoller M (2005) High-resolution surface-wave tomography from ambient seismic noise. Science 307(5715):1615

Snieder R, Grêt A, Douma H, Scales J (2002) Coda wave interferometry for estimating nonlinear behavior in seismic velocity. Science 295(5563):2253

Staudacher T, Ferrazzini V, Peltier A, Kowalski P, Boissier P, Catherine P, Lauret F, Massin F (2009) The April 2007 eruption and the Dolomieu crater collapse, two major events at Piton de la Fournaise (La Réunion Island, Indian Ocean). J Volcano Geotherm Res 184(1–2):126–137. doi:10.1016/j.jvolgeores.2008.11.005

Weaver RL, Lobkis OI (2001) Ultrasonics without a source: thermal fluctuation correlations at MHz frequencies. Phys Rev Lett 87(13). doi:10.1103/PhysRevLett.87.134301

Wegler U, Lühr B, Snieder R, Ratdomopurbo A (2006) Increase of shear wave velocity before the 1998 eruption of Merapi volcano (Indonesia). Geophys Res Lett 33:1–4

Ground Deformation at Piton de la Fournaise, a Review From 20 Years of GNSS Monitoring

15

Thomas Staudacher and Aline Peltier

Abstract

The record of ground deformation brings useful information to model volcano plumbing system and to preview the eruptive behavior of active volcanoes. Besides tiltmeters and extensometers, the GNSS seems to be one of the most attractive networks to detect and to track in real-time minute ground deformation. Here we report GNSS data recorded at Piton de la Fournaise and look in detail the pre-eruptive, but also the long term inter-eruptive behavior of this exceptionally active volcano. In particular baseline variations of the summit stations show a quite constant comportment before eruptions and permit reliable forecasting of eruptions. Model calculation using GNSS data are used to locate the magma reservoir and to retrace the origin and the shape of dikes that fed eruptions. Finally, for the first time the constancy of the inter-eruptive slow slipping of the east flank of Piton de la Fournaise has been attested by the GNSS network. In this chapter we present a review of 20 years of GNSS monitoring at Piton de la Fournaise and its contribution in the detection of eruption precursors and in the understanding of the volcano functioning.

15.1 Introduction

The survey of volcanoes and prediction of volcanic eruptions is still an enormous challenge, especially for the most explosive and dangerous volcanoes in regions where subduction volcanism is predominant as, for example, in the pacific volcanic belt, also called the ring of fire. Famous examples of disastrous eruptions are the 1883 Krakatoa eruption in Indonesia, where several tens of thousands of people were killed and the Minoan eruption in Santorini, Greece, where

T. Staudacher (✉) · A. Peltier
Observatoire Volcanologique du Piton de la Fournaise, Institut de Physique du Globe de Paris—Sorbonne Paris Cité, Sorbonne Paris Cité, Univ. Paris Diderot, CNRS, F-97418 La Plaine des Cafres, La Réunion, France
e-mail: thomasstaudacher1950@gmail.com

A. Peltier
e-mail: peltier@ipgp.fr

P. Bachèlery et al. (eds.), *Active Volcanoes of the Southwest Indian Ocean*,
Active Volcanoes of the World, DOI 10.1007/978-3-642-31395-0_15

3600 years ago 30 km^3 of rhyodacite magma erupted, leaving a 7 × 10 km caldera.

While hotspot type volcanism is usually less violent, explosive periods are also known. The pumice and ash flows between Saint Pierre and Saint Louis at La Réunion Island attest to these violent eruptions in the late stage of the Piton des Neiges volcano (Fig. 15.1) activity some 165 and 175 ka ago (Fretzdorff et al. 2000). Other deposits exist around the island with ages between 80 and 260 ka. In Hawaii, the Uwekahuna ash deposits, which represent 75 Mm^3 testify explosive eruptions some 2700–2100 years ago (Dzurisin et al. 1995) and in 1790 an explosive event at Kilauea killed at least 80 members of a warring party (Mastin et al. 1998), as they passed the caldera.

It is thus crucial to monitor volcanoes to predict their eruptions and mitigate the associated risk. Today numerous instruments allow the survey of volcanoes and bring critical information to scientist and authorities. Satellite and ground based measurements allow detection of increased activity of hot springs, fumaroles and gas emissions.

Seismic stations monitor volcano-tectonic earthquakes produced by stress variations in the volcanic edifice (e.g. Aki and Ferrazzini 2000). Tiltmeter, extensometer and GNSS receivers detect minute to second ground deformation variations due to pressurization of magma reservoir or due to dike injection towards the surface. These ground data are completed by radar data from satellites that can be used to map surface deformation over larger areas from space (e.g. Briole et al. 1997; Froger et al. 2004, 2016), but the data are periodic and not available in real time.

Since the creation of the volcanological observatory at La Réunion Island (**O**bservatoire **V**olcanologique du **P**iton de la **F**ournaise, OVPF) in late 1979, sixty-one eruptions occurred

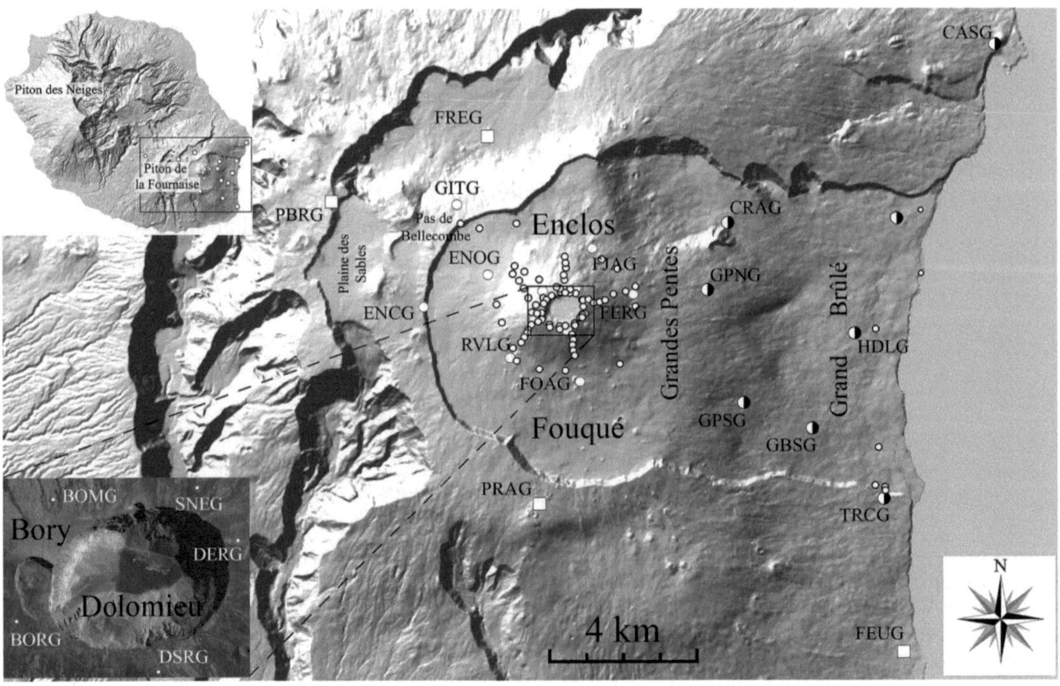

Fig. 15.1 Digital Elevation Model of the Piton de la Fournaise volcano with the location of the permanent and survey GNSS networks. *White* large dots show the permanent summit network; *white* and *black dots* represent more recent permanent stations (installed in 2009–2010) in the Grandes Pentes and Grand Brûlé areas; small *white dots* on the east coast close to TRCG indicate receivers on the up to 60 m thick 2007 lava flow; *white squares* show the location of recent permanent stations outside of the Enclos Fouqué caldera (installed in 2012–2013). Small *white dots* around the terminal cone are the sites for rapid static campaigns

at Piton de la Fournaise, which attest to the extraordinarily high eruption rate of this volcano (Staudacher et al. 2016). During a total of 1480 days of activity, these eruptions produced a volume of 660 × 10^6 m^3 of lava and covered an area of 52.8 km^2, with some of this area having multiple, overlain flows. Each eruption is usually preceded by weeks of increasing seismic activity and always by a seismic crisis lasting several minutes or hours as well as by long- and short-term ground deformation (Staudacher et al. 2003; Battaglia et al. 2005; Peltier et al. 2005; Schmid et al. 2012). Thanks to its high eruptive activity, Piton de la Fournaise is an excellent laboratory to study in detail the behavior of a volcano before, during and after eruptive events and to learn how to forecast eruptions more precisely.

15.2 The Deformation Network at Piton de La Fournaise

From the inception of the observatory in 1979 until 2004, mainly tiltmeters (Blum et al. 1981; Saleh et al. 1991; Delorme et al. 1993; Peltier et al. 2005) and extensometers (Bachèlery et al. 1998; Staudacher et al. 2003; Peltier et al. 2016, 2006) were installed on the massif of Piton de la Fournaise to monitor in continuous and in real-time the volcano deformation (see Peltier et al. 2015, 2016).

The first station monitoring deformation was a 180 m long optical tiltmeter installed in 1980 in the 4.7 km long vault at Rivière de l'Est, 8.5 km away from the summit. As temperature conditions were particularly stable in the vault and promised excellent data, a Blum type tiltmeter, a high-precision quartz made tiltmeter (Blum et al. 1981) was first installed in 1984 (Peltier et al. 2011). In Plaine des Sables and on the Piton de la Fournaise summit cone, dry tiltmeters (Blum et al. 1981; Zlotnicki et al. 1990) were set up in 1979. Fifteen geodetic points (tripods and concrete pillars) for geodetic leveling were also implemented and a high-precision continuously recording tiltmeter was installed in the Bory crater (Blum et al. 1981). From 1986 to 1988 the tiltmeter network was expanded by six more electronic tiltmeters whose data was telemetered to OPVF.

In 1980 a large fracture in the Bory crater was equipped with an extensometer that continuously measured its opening. In 1990, one-component extensometers were installed at Soufrière and at Dolomieu Sud (close to SNEG and DSRG respectively in Fig. 15.1). In 1995, 3-component extensometers were installed at Soufrière, Dolomieu Sud, Magne and Château Fort over deep cracks with 3–5 cm of opening at surface (Staudacher et al. 2003; Peltier et al. 2006). They measured variation of extension, shear and vertical components every minute. Data were transmitted to the observatory by radio. This configuration allowed monitoring of subtle pre-eruptive displacements over the cracks up to 3 months before eruption occurred. Between 1998 and 2003 clear long term opening, with rates between 0.0025 and 0.0059 mm per day, preceded each of the eight eruptions of this period (Peltier et al. 2006). This allowed forecasting of eruptions and warning of the local authorities several days to weeks before they begun. For the location of these stations see http://www.ipgp.fr/pictures_lib/1087.jpg and Peltier et al. (2016).

In 1994, the first experiments of permanent GPS measurements were undertaken at Piton de la Fournaise. However it was only in 2004 that a real continuously recording, telemetered GNSS network was installed. The network currently consists of 27 permanent stations (Fig. 15.1).

We will focus here on ground deformation monitored by the GNSS network and the application of these data to better understand the connection between the magma reservoir and the surface.

15.3 The GNSS Network

The first GNSS measurements at Piton de la Fournaise were attempted in 1994 at Bory (BORG) and "Enclos Zéro" (ENCG) as base station (Fig. 15.1). Further limited attempts were made in 2001, but the present permanent GNSS network was intensely implemented around the summit craters in 2004, followed in 2005 by stations at the base of the summit edifice, about 2 km from the summit. Presently ten stations exist on the central cone (white large dots in Fig. 15.1), five receivers were installed around the summit craters (BORG, BOMG, SNEG, DERG and DSRG) and other five receivers were installed at the base of the cone (RVLG, ENOG, FJAG, FERG and FOAG). Two stations, ENCG and GITG, were installed outside of the Enclos Fouqué caldera as reference stations. After the exceptional 2007 eruption (Staudacher et al. 2009) three GNSS receivers were installed in 2008 on the up to 60 m thick lava flow emitted during this eruption in order to follow its subsidence with time (GB1G, GB2G GB3G, small white dots on the south east coast on Fig. 15.1) together with another nearby reference station outside of the caldera at TRCG. In the framework of the "Undervolc" program (Brenguier et al. 2012; Clarke et al. 2013) five stations (CRAG, GPNG, GPSG, GBSG and HDLG, large black and white dots on Fig. 15.1) were installed between September 2009 and April 2010 in the Grandes Pentes and the Grand Brûlé areas together with a new reference station outside the caldera at CASG. They were completed in July 2011 by an additional station (GBNG) in the northern Grand Brûlé. This network has been enlarged in 2012–2013 by four receivers (FREG, PRAG, PBRG, FEUG, white squares on Fig. 15.1) outside of the caldera. Several permanent stations previously named RIVG, FORG, FJSG and BONG had to be moved for some tens of meters to several hundreds of meters. This was done in an effort to limit the visual impact in the field and to connect the GNSS receivers to the WiFi network used by OVPF for transmission. Their new names were changed from RIVG → RVLG; FORG → FOAG; FJSG → FJAG and BONG → BOMG and the new sites are 263, 23, 33, 102 m from the old ones, respectively.

The GNSS receivers used are Topcon GB1000, Trimble NetRS and NetR9; they are linked to the observatory by WiFi connection and can be configured at distance. Data acquisition is usually done every 30 s, but can be increased to 1 Hz and even to 50 Hz for the most recent NetR9 receivers. Five stations (BOMG, SNEG, DSRG, FOAG and RVLG) are operated in a RTK mode, sending their position relative to GITG every second to the observatory. GB1G and GB2G, located on the April 2007 lava flow, are equipped with Ashtech Z-Xtreme receivers. They are not linked to the observatory and data are recovered every month and a half. We use three types of antennas; Ashtech Geodetic 4 antennas (BORG, DERG and ENCG), Zephyr GNSS Geodetic II antennas (GITG, FREG, PYRG, PBRG and RERG) and Topcon PG-A1 (remaining stations). Most antennas are fixed on 2 cm diameter and 30 cm long stainless steel rods, which are cemented in massif basaltic ground or fixed on concrete pillars (BORG, DSRG and DERG). All the antennas are oriented towards the north.

Daily data files were transmitted to the observatory by Wifi and evaluated by an automated GPS calculation (B-PNAV, Briole 2007), which produces solutions for each epochs (every 1 or 30 s) but also daily mean values. The evaluation is performed using GITG as reference station (GITG: 55°41′13.90846″; −21° 13′06.00911″; 2266.663 m). In case of any problem with the GITG reference station, evaluation can be done with the ENCG station (55° 40′41.49314″; −21°14′35.93014″; 2353.356 m). Since 2012, solutions with the GAMIT software are also calculated.

The data files are stored in RINEX format in the OVPF database and acquisition is synchronized every night to data servers of Institut de Physique du Globe de Paris (IPGP). They can be downloaded via Volobsis, which is a portal to access data collected in the French volcanological and seismological observatories managed by IPGP.

In addition to the permanent network we also have 84 sites (small white dots in Fig. 15.1), equipped with stainless steel rods that are measured with rapid static surveying routinely every 6 months, or more often in case of ongoing eruptions. The stainless steel rods of 2 cm diameter are cemented in the ground or fixed on tripods or concrete pillars. The antenna is set on the stainless steel rod with an adapter, which allows a quick and precise positioning (±<0.1 cm) of the antenna for the rapid static campaigns. The sites are configured in a double circle around the summit craters and in the form of six radial transects. Further sites are along the path to Pas de Bellecombe (Fig. 15.1). The rapid static measurements are done with Ashtech Z-Xtreme receivers with a 1 Hz acquisition. For this purpose an additional session of 1 Hz acquisition at the GITG reference station is configured. Data are recorded at each site for 7 min. This allows a precision in the relative position of about 1–2 cm horizontally and 2–3 cm vertically, sufficient in regard of the tens of cm displacements associated to volcanic activity (see below).

15.4 Long-Term GNSS Variations

The automatic daily data processing gives us the position of each station (East, North and Elevation). The distances between the reference station and the permanent stations on the central cone vary between 2 and 6 km, with elevations comprised between 1950 and 2580 m, which may induce systematic errors due to local meteorological conditions. For that reason we use preferably the variation of baselines between the individual stations, which are much shorter (416–1477 m). Such a representation allows reducing systematic errors and is more precise than absolute positions.

In Fig. 15.2 we plot all presently available GNSS summit baselines. In this representation we can distinguish four different periods, a rest period before 1998, a high activity period between 1998 and 2007 characterized by 26 eruptions, a low activity period (characterized by 8 eruptions) following the Dolomieu crater collapse in April 2007 and lasting until 2010, and again a rest period since 2011.

As mentioned before, only rapid static measurements exist from 1995 to 2004; continuous measurements started only in 2004. The oldest data exist for the BORG-DERG baseline, which passes straight through the Bory-Dolomieu craters from WSW to ENE (see Fig. 15.3o). Even though only occasional data exist between 1995 and late 1997, the distance BORG-DERG did not change, indicating the absence of significant ground deformation across the summit. This is in agreement with the lack of eruptions during this period. This time is also characterized by a very small number of volcano-tectonic seismic events with an average of only 1.7 events per day from 1995 to 1997 (OVPF database; Staudacher et al. 2016). It appears that the volcanic system was at rest and, at least until November 1997, magma transfer beneath Piton de la Fournaise was low or absent (Peltier et al. 2009a). GNSS measurements recorded between March 11 and 17, 1998, immediately after the start of the March 9 eruption, indicate a small increase of the BORG-DERG baseline, probably produced by new arrival of deeper magma. Since then, this extension continues in a general trend until 2007 (Fig. 15.2) and suggests a long-term magma refilling of the reservoir beneath the summit, creating a gradual build-up of the pressure in the magma reservoir (Peltier et al. 2009a). This long-term general trend was only interrupted by rapid variations due to eruptions (see Sect. 15.6). For example, the February 13, 2000 eruption, located on the northern flank with eruptive fissures located between 2500 and 2230 m elevation induced a strain of $\sim 3 \times 10^{-4}$, which corresponds to an elongation of 44 cm of the BORG-DERG baseline, pushing the eastern Dolomieu border to the sea, the only free boundary of the volcanic massif. In November 2002, the eruption, located on the eastern flank at 1760 m elevation, induced a strain of -10^{-4}, slightly reducing the BORG-DERG baseline by 15 cm. In August 22, 2003, an eruption on the northern flank (2590–2210 m elevation) again, caused a strain of 1.8×10^{-4}, which extended the baseline across Dolomieu crater by 27 cm. From 1995 to 2007 the distance between BORG and DERG stations extended of 2.53 m.

The lack of permanent data between 1998 and 2004 prevents distinguishing which part of variation is due to pre-eruptive dilatation of the

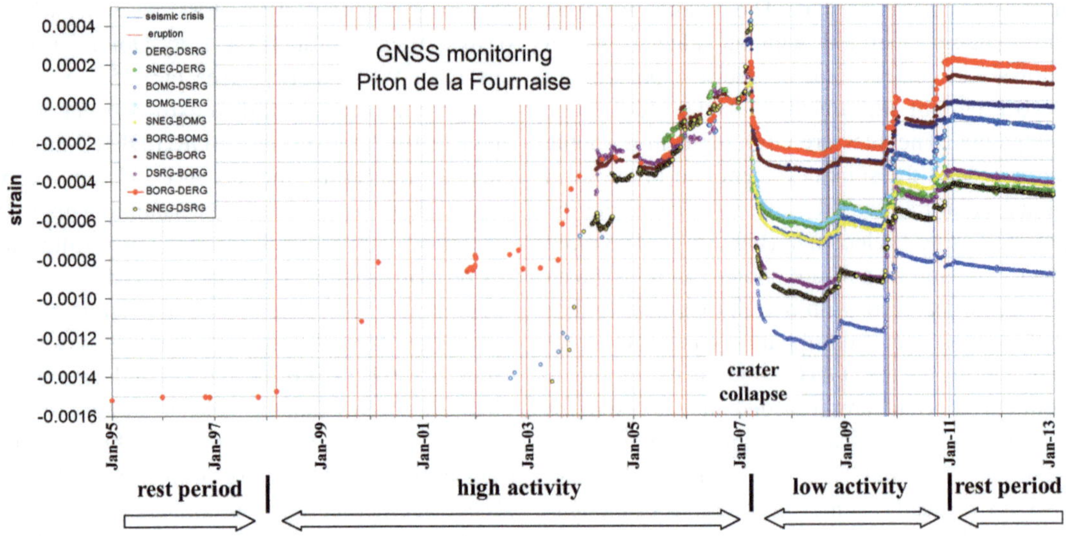

Fig. 15.2 Strain measured across the summit area from GNSS data. Data from 1995 to 2004 are from rapid static campaigns, afterward they are from daily average positions of permanent stations. Before 2007 the trend of the variation indicate a general inflation of the summit until the Dolomieu collapse in April 2007, and deflation afterwards, persisting through the time of this writing—February 2013. *Red* and *blue vertical lines* represent eruption onsets and seismic-no eruptive crises, respectively

summit area and which part is due to ground displacements during the emplacement of the eruptive dikes during this period. We have a much better view between 2004 and 2007. Since 2004, each eruption is preceded by extensions on baselines across the summit. In Fig. 15.3, we plot summit baseline changes for each eruptive period over 100 days. In each figure, we plot the same range of the ordinate, which limits unfortunately the number of curves shown due to the spread of the curves after the 2007 eruption and the Dolomieu collapse (see Fig. 15.2), but allows a better visual comparison between different eruptions. The arrows show the trend of the variation, while the numbers correspond to the average slope over all baselines available for that period. More detailed data are given in Table 15.1.

The May 2, 2004 eruption, which started below the Bory crater on the southern flank, was preceded by a slow three months average extensional strain of 0.74×10^{-6} per day (Fig. 15.3a), which speeded up in the last 10 days before eruption to 2.75×10^{-6}. Three months later, the August 12, 2004 eruption, which started in the Dolomieu crater, but propagated to the east, was preceded by 1 ½ months of extensional strain of 1.05×10^{-6} per day (Fig. 15.3b). In February 2005, six months after the last eruption, a 2 ½ months long of slow extension (Fig. 15.3c) preceded a distal eruption in Plaine des Osmondes. In the same year, in October 4, 2005, a second eruption in the Dolomieu crater was preceded by more than three months of slow extensional strain of 0.46×10^{-6} per day (Fig. 15.3d). This elongation, indicative of inflation, continued over 1 ½ months, after a short stable period leading to the November 29, 2005 summit eruption and one month later to the December 26, 2005 eruption in Plaine des Osmondes (strain of 0.63 and 0.74×10^{-6} per day, Fig. 15.3e, respectively).

They were followed in July 20 and August 30, 2006 by eruptions on the southern flank and in the Dolomieu crater after strain gradients of 0.91 and 0.81×10^{-6} per day (Fig. 15.3f), respectively, as well as in February 18 and March 30, 2007 by eruptions in the Dolomieu crater and on the south-eastern flank preceded by strain gradients of 0.79 and 0.70×10^{-6} per day respectively (Fig. 15.3g, h).

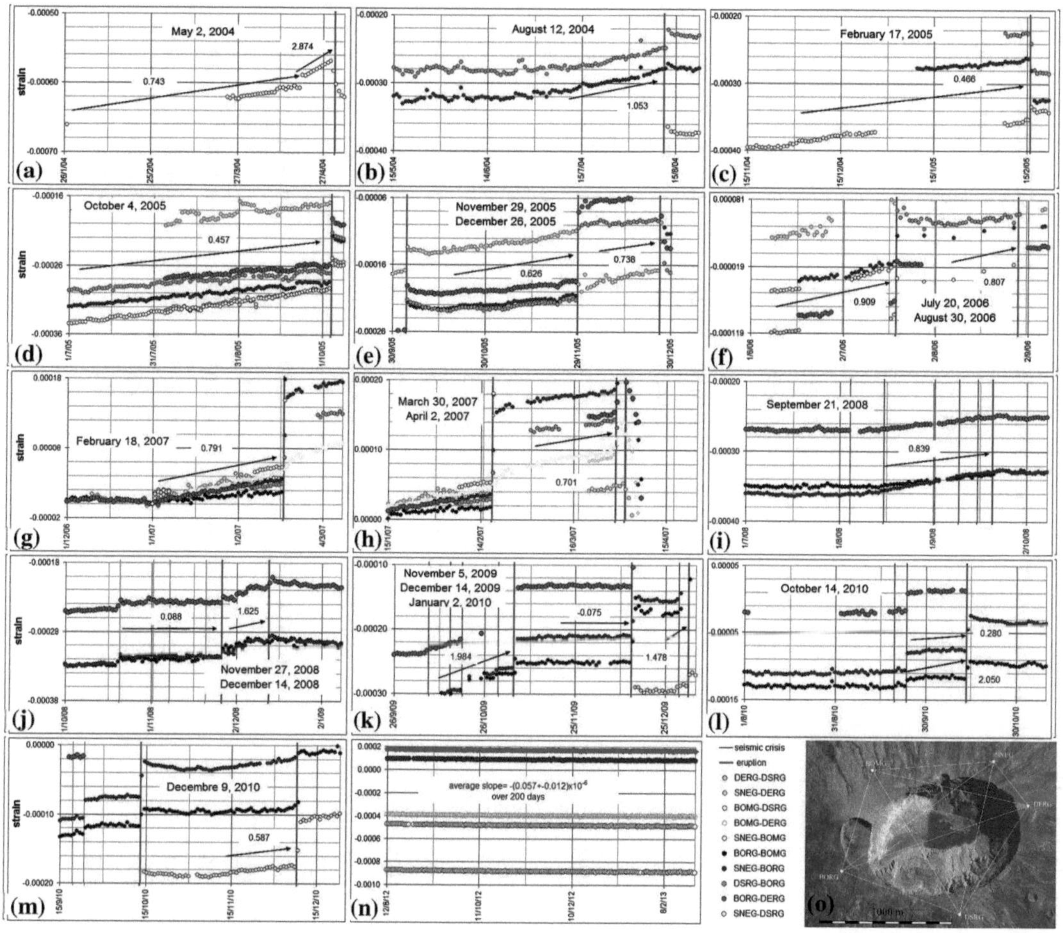

Fig. 15.3 Pre-eruptive strain measured across the summit area for 14 eruptions. For better comparison, we report on each *inset* data over 100 days and a scale range of 0.2 ‰ in ordinate. This range does not allow showing all curves after the 2007 eruption and Dolomieu crater collapse, which spread the *curves* (see Fig. 15.2). The *arrows* show the trend of the variation, while the numbers correspond to the average strain rate in microstrain of all baselines available for each period

The averaged strain rate of the baselines always ranges between 0.46 and 1.05×10^{-6} per day [except for the high strain rate of 2.9×10^{-6} per day immediately preceding the May 2, 2004 eruption (Fig. 15.3a)], with an average value of 0.73×10^{-6} per day (Table 15.1). On the BORG-DERG baseline of 1476.8 m, this corresponds to an extension of 1.08 mm/day. This behavior is quite similar to measurements with the extensometer network from 1999 to December 2003 (Staudacher et al. 2003; Peltier et al. 2006, 2016). It recorded average fissures opening rates of about 0.0035 mm per day over weeks and up to three months before eruptions begun at the northern (Magne station) and the southern base (Château Fort station) of the central cone (http://www.ipgp.fr/pictures_lib/1087.jpg). To a certain extent we may compare a GNSS baseline to a mega-extensometer, integrating the variations of all fissures and fractures between two stations on a hectometer to kilometer base.

In total, all ten eruptions that occurred after the first continuous GNSS measurements in 2004 and before the April 2007 eruption were preceded by significant extension across the summit area.

Table 15.1 Summary of GNSS data preceding 18 eruptions between May 2005 and December 2010

Date	BORG-SNEG	Period	BORG-DERG	Period	BOMG-DERG	Period	BOMG-DSRG	Period	SNEG-DSRG	Period	SNEG-DERG	Period
	1×10^{-6}/day	Days	1×10^{-6}/day	Days	1×10^{-6}/day	Days	1×10^{-6}/day	Days	1×10^{-6}/day	Days	1×10^{-6}/day	Days
Distance in 2013 (m)	1398.29		1476.84		1190.38		1252.49		1158.75		416.21	
02/05/2004	1.0111	9							1.0000	93		
12/08/2004	0.9800	45							1.4444	45		
17/02/2004	0.3973	37							0.5305	82		
04/10/2005	0.3937	95	0.3667	60					0.5579	95	0.5763	55
29/11/2005	0.5167	30	0.5333	30					0.7511	45	0.7362	45
26/12/2005	0.6273	22	0.5455	22					0.9182	22	1.1314	22
20/07/2006	0.9156	32	0.6875	32					0.9902	41	1.2462	41
30/08/2006	0.7500	20							1.1400	20		
18/02/2007	0.7617	46	0.7778	36	0.8694	36	0.7742	31	0.8634	41	1.1291	35
30/03/2007	0.6629	35	0.5556	9			0.7037	27	0.9407	27	0.9111	9
Average	0.702		0.578		0.8694		0.739		0.904		0.955	
Std. dev.	0.224		0.142				0.050		0.283		0.260	
21/09/2008	0.8000	37	0.5135	37	0.7486	37	0.9730	37	1.2730	37	0.9803	37
27/11/2008	**0.0696**	**23**	**0.0880**	**23**	**0.1029**	**35**	**0.2121**	**33**	**0.2824**	**34**	**0.0606**	**34**
14/12/2008	0.6000	16	1.1333	15	1.5125	16	2.3750	16	2.6938	16	1.7375	16
05/11/2009	1.6714	35	1.5238	21	2.8238	21						
14/12/2009	**0.1460**	**37**	**0.0541**	**37**	**0.0838**	**37**	**−0.3514**	**37**	**−0.2919**	**37**	**0.0449**	**37**
02/01/2010	1.5889	9			1.7222	9			1.2000	9		
14/10/2010	**0.2368**	**19**	**0.2105**	**19**	**0.5263**	**19**	**0.3158**	**19**	**0.2105**	**19**	**0.3174**	**19**
09/12/2010	0.6467	30	0.5000	30	0.6800	30	0.5333	30	0.6133	30	0.6163	30
Average	1.061		0.918		1.497		1.294		1.445		1.111	
Std. dev.	0.525		0.501		0.872		0.962		0.883		0.572	

(continued)

Table 15.1 (continued)

Date	DSRG-BORG	Period	DERG-DSRG	Period	BORG-BOMG	Period	SNEG-BOMG	Period
Date	DSRG-BORG 1×10^{-6}/day	Period Days	DERG-DSRG 1×10^{-6}/day	Period Days	BORG-BOMG 1×10^{-6}/day	Period Days	SNEG-BOMG 1×10^{-6}/day	Period Days
Distance in 2013 (m)	945.00		950.06		793.79		906.35	
02/05/2004	2.8736	10						
12/08/2004	0.7356	45						
17/02/2004	0.4714	7						
04/10/2005	0.2768	95	0.5700	60				
29/11/2005	0.5400	30	0.6800	45				
26/12/2005	0.3833	24	0.8227	22				
20/07/2006	0.9563	32	0.6561	41				
30/08/2006	0.5296	27						
18/02/2007	0.5957	46			0.3745	47	0.9702	47
30/03/2007	0.5481	27			0.3632	38	0.9200	35
Average	0.560		0.682		0.369		0.945	
Std. dev.	0.197		0.105		0.008		0.036	
21/09/2008	0.6243	37	0.9162	37	0.4973	37	1.0622	37
27/11/2008	**−0.0485**	**34**	**0.2753**	**34**	**−0.0800**	**35**	**0.0081**	**35**
14/12/2008	1.3438	16	2.6125	16	0.6500	16	1.5938	16
05/11/2009					1.3194	31	2.5829	35
14/12/2009	**−0.2784**	**37**	**-0.3460**	**37**	**−0.0324**	**37**	**0.2216**	**37**
02/01/2010			0.8889	9	1.1444	9	2.3222	9
14/10/2010	**0.0789**	**19**	**0.1684**	**19**	**0.1158**	**19**	**0.6368**	**19**
09/12/2010	0.4600	30	0.4667	30	0.4633	30	0.8900	30
Average	0.8094		1.221		0.815		1.690	
Std. dev.	0.4701		0.950		0.392		0.748	

Shown are the strain rates recorded for each baseline, the number of days before eruption used for their calculation and the average values for eruptions between 2004 and 2007, as well as for the post 2007 collapse eruptions. The third line shows the distance between the stations on January 1, 2007. The last line indicates the strain rate from Aug 2012 to March 2013. Data highlighted in bold mark eruptions that do not follow the general pre eruptive systematic (see text) and that are not used for the calculation of average values. More data can be requested from the authors.

Since the 2007 Dolomieu crater collapse, which in three months entailed a tremendous subsidence of the summit of 40 cm and a contraction of the crater-crossing baseline of 61 cm, the long-term trend of summit-crossing baselines indicates contraction of the whole summit area. The general inter-eruptive trend of inflation recorded before 2007 is not observed any more (Fig. 15.2). This might indicate the absence of pressure build-up, and the absence of detectable long-term magma refilling (Peltier et al. 2010). Another possibility could be that such a catastrophic event could create void space for magma to refill without causing significant ground deformation associated with the pressurization; however the observed continuous deflation over years would not support such a concept.

The pre-eruptive deformation patterns also no longer show the same consistently inflationary signal as observed before the collapse.

The September 21, 2008 eruption, located in the western part of the collapsed Dolomieu crater, was preceded by one month of extensional strain, at a rate of 0.84×10^{-6} per day (Fig. 15.3i), while the November 27, 2008 eruption, still in the Dolomieu crater, showed no significant baseline variation (Fig. 15.3j). The December 14, 2008 eruption, also located in the Dolomieu crater, on the eastern and north-eastern sides, was again preceded by a strong strain gradient of 1.6×10^{-6} per day (Fig. 15.3j). The same was true for the next eruption in November 5, 2009 on the eastern flank, close to the Dolomieu rim. It was preceded by one month of a particularly strong strain increase of 2×10^{-6} per day (Fig. 15.3k), while the December 14, 2009 eruption, on the southern flank still close to the Dolomieu crater rim, was not preceded by any inflation at all. Before the January 2, 2010 eruption, the increase started very late, only 4 days before the eruption. The following eruption, on October 14, 2010 on the southern flank, showed only a small strain variation of 0.28×10^{-6} per day. Finally the most recent eruption, on the northern flank, on December 9, 2010, was again preceded by 1 month of moderate strain variation of 0.59×10^{-6} per day (Fig. 15.3m).

In contrast to the pre-collapse period, the post-collapse period was characterized by a large number of seismic crises not followed by eruptions (Fig. 15.2 and Roult et al. 2012). Some of these crises are accompanied by ground deformation recorded by tiltmeters as well as by the GNSS stations that could be indicative of non-eruptive dike intrusions in depth. By contrast, the November 27, 2008 eruption was not preceded by any ground deformation; as it started from the same vent in the Dolomieu crater as the preceding one, it could be considered as a resumption of the September eruption, using he same path way. Several seismic crises took place before this eruption, in particular on October 21 (Fig. 15.3j). The latter produced a spread of the Dolomieu crater of 1 and 2 cm in north-south and east-west direction, respectively, which could be due to the initiation of the uprising dike. Including this ground deformation, we would get a total strain gradient of 0.36×10^{-6} per day preceding the November 27, 2008 eruption.

The October 14, 2010 eruption, on the southern flank, was preceded by an only small strain variation of 0.028×10^{-6} per day (Fig. 15.3l), however it also was preceded on August 23 by a seismic crisis, which produced respectively a north-south and east-west elongation of 6.6 and 3.8 cm, respectively, and might represent the beginning of a rising dike. The small variation recorded just before the following eruption was probably partly exhausted by this crisis. If we include its ground deformation we get a pre eruptive strain gradient of 2.05×10^{-6} per day, one of the largest rates observed. The January 2, 2010 eruption was preceded by a short but intense strain increase of 1.48×10^{-6} per day (Fig. 15.3k). In fact the only real exception to the pattern of pre-eruptive inflation in the post 2007 period is the December 14, 2009 eruption, which showed even a slight negative strain gradient of -0.075×10^{-6} per day (Fig. 15.3k) and no seismic crisis in the preceding weeks.

In Fig. 15.3n (see also Table 15.1, last line) we represent the GNSS baselines since July 15, 2012. The average strain gradient is $-0.062 \pm 0.020 \times 10^{-6}$ per day, which corresponds to an average contraction of the summit area of 0.06 mm per day (or ~2.2 cm per year). Since the 2007 Dolomieu crater collapse, except close to the eruptive periods, the summit area is still contracting. This behavior can be interpret in several ways: (1) the deep magma transfers into the shallow magma reservoir beneath the summit ended after the 2007 collapse and the pre-eruptive inflation recorded would be associated to pressurization linked to magma crystallization and/or degassing; (2) the effect of the edifice relaxation, following the 2007 collapse, hid any other deformation signals linked to deeper processes; (3) the 2007 collapse has created "void space" and magma supply from a deeper magma reservoir would not create detectable ground deformation.

15.5 East Flank Movement

Until 2009, only the deformation of the central cone had been monitored by the OVPF GNSS network. In the framework of the UNDERVOLC project (Brenguier et al. 2012), the GNSS network extended between end 2009 and early 2010 with six new stations in the eastern slopes of the volcano (at elevations comprised between 136 and 1414 m; Fig. 15.4). These new installations followed the large displacement of the eastern flank related to the April 2007 eruption of up to 1.4 m, which was detected by interferometric Synthetic Aperture Radar (InSAR) (Augier et al. 2008; Clarke et al. 2013).

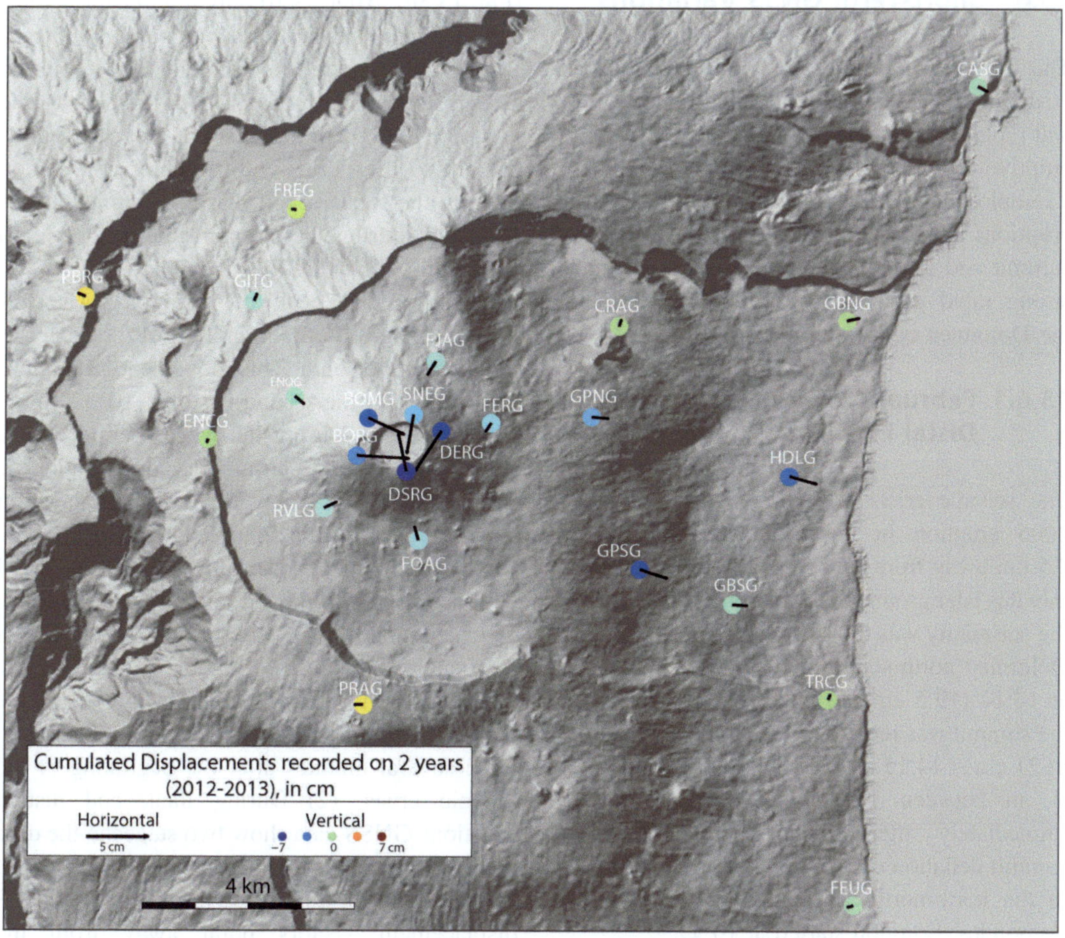

Fig. 15.4 Displacements recorded by the GNSS network on the whole Piton de la Fournaise's massif in 2012–2013

Between 2009 and 2013, these stations recorded continuous and eastward steady motion at a rate of up to 2.1 cm $year^{-1}$; this motion was accompanied by a subsidence of up to 3.5 cm $year^{-1}$ (Fig. 15.4). This motion would result from gravity processes; and even if it is not yet well constrained how this motion is accommodated in depth, it might occur along a basal fault plane as observed at Kilauea (Peltier et al. 2015). Such eastern flank instability could explain the asymmetrical deformation pattern observed during dike injections with systematic larger eastward and upward displacements to the east (Froger et al. 2004; Fukushima et al. 2005; Peltier et al. 2007, 2009a, b; Got et al. 2013).

15.6 Short-Term GNSS Variations

The long term GNSS continuous trends, previously presented, are only disrupted by short-term and rapid variations, which are directly linked to eruptive activities and caldera/pit-crater collapse. Examples of such short-term variations before eruptions are shown in Fig. 15.5 for a distant, a summit and a flank eruption, as well as for the strong rapid ground deformation associated to the Dolomieu collapse in April 2007.

15.6.1 February 17, 2005 Distal Eruption

The seismic crisis preceding the February 17, 2005 eruption in the Plaine des Osmondes, 3.5 km away from summit, started at 13:35. Ten minutes later ground deformation appeared, when the seismicity was the strongest (Fig. 15.5a). The Dolomieu north-south extension was the largest at 14:14 (19.5 cm). Simultaneously inflation of the summit was recorded, with a maximum uplift of 21 cm at 14:15 at SNEG (Fig. 15.5a) and with 9 cm between 14:20 and 14:30 at DSRG. Immediately after, seismicity decreased and summit deflation occurred, when the dike moved to the north-north-east towards the Plaine des Osmondes. Eruption started at 16:35.

15.6.2 October 4, 2005 Summit Eruption

On October 4, 2005, the seismic crisis began on 9:30. Fifteen minutes later the Dolomieu crater started to expand continuously until 10:48 with a maximum extension of 10 cm. Seismicity almost stopped at 10:00 and eruption in the Dolomieu crater begun at 10:35, while ground deformation continued until 10:45. No vertical summit inflation was recorded for this eruption (Fig. 15.5b).

15.6.3 July 20, 2006 Proximal Eruption

The seismic crisis preceding the July, 20 eruption, on the south-south-west and southern flank of Piton de la Fournaise begun on July 19 at 22:20, with a short hiatus from 22:40 to 22:50 (Fig. 15.5c). After 23:10 only few seismic events occurred and the first fissure opened at 00:06. During the first peak of seismicity, SNEG shifted 5.8 cm to the north-east, DERG 5.2 cm to the east, DSRG 5.9 cm to the south-east and BONG 3.2 cm to the north, with a center of inflation beneath the western part of the Dolomieu crater. At 22:45, during the seismic quiescence period the south shift of DSRG inverted (Fig. 15.5d), and the second peak of seismicity was followed at 23:00 by the migration of the center of inflation to the south. This movement lasted until the opening of the second fissure at 01:37 on the southern flank. During the whole period DSRG shifted 23 cm the east. This deformation is compatible with a first vertical dike propagation between 22:20 and 22:40, centered below the western Dolomieu crater, followed from 22:50 with a dike which moved laterally to the south.

In summary, (1) strong and rapid GNSS deformation (up to 1 cm min^{-1}) begin a few min/hours before the beginning of eruptions, during the propagation of the dike to the surface, often several minutes after the beginning of the seismic crisis. (2) During flank and distant eruptions GNSS data show two stages in the dike propagation to the surface (Peltier et al. 2007, 2008, 2009a, b); a first step during which GNSS displacement vectors point away from the

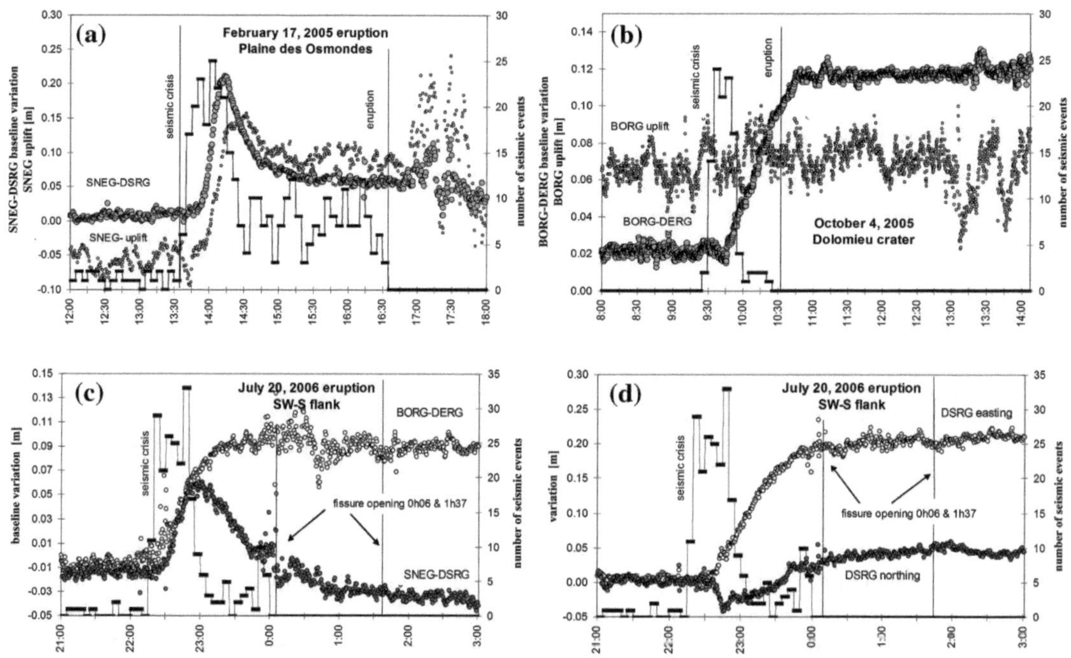

Fig. 15.5 a–c baseline and uplift variations for 3 eruptions and **d** variations of easting and northing of DSRG GNSS station for the July 20, 2006 eruption, compared to seismicity. Figure 15.5**a** shows the north-south baseline variation and variation of the z-axis before the February 17, 2005 distal eruption in Plaine des Osmondes. 15.5**b** represents the west-east baseline and uplift variation during the October 4, 2005 eruption in the Dolomieu crater. 15.5**c** shows the north-south and west-east baseline variation the July 20, 2006 summit eruption and 15.5**d** shows its easting and northing motion. *Black horizontal* bars represent the number of seismic events cumulated over 5 min

Dolomieu crater, revealing a vertical propagation of the magma centered below the Dolomieu crater; and a second step during which GNSS displacement vectors rotate and summit stations begin to subside, revealing a lateral propagation of the magma towards one of the volcano's flank. (3) During the course of eruptions, only slight summit deflation or contraction of the Dolomieu crater dimensions are recorded, linked to the drainage of the magma from the shallow reservoir to the eruptive vent.

15.6.4 Dolomieu Crater Collapse

The collapse of the Dolomieu crater floor on April 5, which occurred during the April 2007 eruption, produced strong and very rapid ground deformation of the volcanic massif (Staudacher et al. 2009). In Fig. 15.6 we report the SNEG-DERG baseline variation from April 5 to April 7. Even though all summit GNSS stations recorded similar variations, we show here only the SNEG-DERG baseline. Other GNSS antennas of the summit stations were rapidly covered during the collapse by up to 1 cm of volcanic ash which prevented correct reception of the satellite signals. From about 09:00 on April 5, the SNEG-DERG baseline (Fig. 15.6) shortened faster and faster reaching up to 13.5 cm at 20:48:30. Between 20:48:30 and 20:49:00, the baseline increased suddenly by 22 cm, simultaneously with a 4.8 M_s seismic event (at 20:48:48.40, International Seismological Centre, On-line Bulletin, http://www.isc.ac.uk). In the following 36 h, regular cycles of ground displacements of 4–7 cm were observed, simultaneously with cyclic tilt variations and increase of the amplitude in the spectrogram (Michon et al. 2007; Staudacher et al. 2009; Fontaine et al. 2014; Peltier et al. 2016). Each cycle ended by

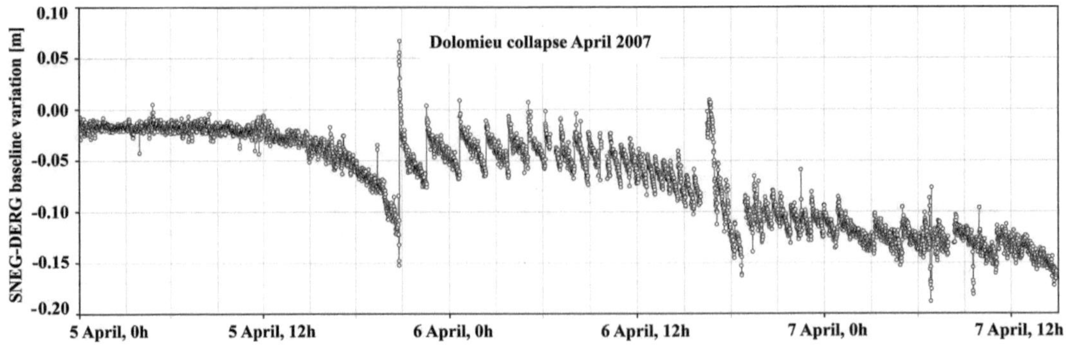

Fig. 15.6 Representation of the SNEG-DERG baseline variation during the Dolomieu crater collapse, which indicates regular baseline shortenings during periods of 30 min to 2 h, periods during which the piston located between the magma reservoir and the surface slides down. Each 30 min to 2 h sequence is followed by abrupt ricochets, due to elastic rebounds of the massif

the formation of large clouds of volcanic ash from the Dolomieu crater. The observed ground deformation is explained by successive collapse events with a piston like mechanism (Roche et al. 2001; Staudacher et al. 2009) starting with a main collapse at 20:48:48.40. The ongoing lateral eruption at low elevation created a depressurization in the magma reservoir which failed to support the rock column between magma chamber and surface. This "low pressure" produced regular downwards sliding of the "piston" and slow summit deflation, lasting between 120 min in the beginning and 25 min 24 h later, followed by sudden elastic rebounds of the massif and abrupt inflations (snap increase of the baseline). After 36 h the main collapse was achieved.

15.7 Inversion of GNSS Data—Imaging the Shallow Magma Plumbing System of Piton de La Fournaise

In the following we summarize a detailed study of the ground displacements recorded by GNSS, preceding the eruptions from 2003 to 2010 (9 summit, 10 proximal and 4 distal eruptions; Roult et al. 2012), which allow to better characterize the pressure source involved and the characteristics of the plumbing system of Piton de la Fournaise. Based on the position variations of each GNSS points, we have calculated numerical models to constrain the location, volume, depth and overpressure of magma bodies that can account for both the long-term and short-term surface deformation (Peltier 2007; Peltier et al. 2007, 2008). In our study GNSS data are embedded in a three dimensional (3D) elastic model based on a mixed boundary element method (MBEM) (Cayol and Cornet 1997). The model is combined with Sambridge's Monte Carlo inversion method (Sambridge 1999) to minimize the misfit between calculated and observed displacements.

Regarding the distribution and the timing of the ground deformation, the a priori information on the shape of the sources are the following: the short-term ground deformation that precedes the eruptions by a few minutes to hours is induced by a dike propagating first vertically from the summit reservoir and then extending laterally (Toutain et al. 1992; Peltier et al. 2005), while in the absence of any evidence of dike intrusion between eruptions (Brenguier et al. 2008; Peltier 2007), the long-term ground deformation is induced by a pressure source, most probably an over/under pressurizes magma reservoir. We modeled the short-term pressure sources as dike-shaped overpressure sources and the long-term ones as ellipsoids defined by a change in pressure (ΔP), the 3D coordinates of the center and the dimensions of the three half axes.

Modelling is done in two steps: the first, vertical propagation step is modelled from permanent GNSS data in the first stage of the magma injection. The dike is defined by six geometrical parameters (3D coordinates of the origin, length, height, and dip) and a ΔP. The second, lateral propagation step is modeled from permanent but also from rapid-static survey GNSS data spanning the whole time period of the dike emplacement (see Peltier et al. 2008 for more detail). For the second step, the elevation from which the dike deviates laterally is deduced from the first step inversion modeling. The upper part of the dike is connected to the surface at the eruptive fissure locations.

For the calculation, the edifice is assumed to be elastic, homogeneous and isotropic, with a Young's modulus of 5 GPa, and a Poisson's ratio of 0.25. The structures (topography, dikes and magma reservoir) are modeled by triangular meshes. The meshed topography has been chosen two times larger than the caldera surrounding the summit cone in order to avoid edge effects. Surface displacements are calculated at the tops of the mesh elements.

15.7.1 Long-Term Inflation Sources

The best fit models that explain most of the pre-eruptive inflations between 2003 and 2007 are similar between eruptions. The source is centered below the Dolomieu crater at a depth between ~2300 (±500) m (Peltier et al. 2008). This pressure source corresponds to a shallow magma reservoir, where magma is temporarily stored during its ascent before being erupted. Its location is consistent with the seismic low velocity zone inferred at sea level (i.e. ~2500 m depth; Nercessian et al. 1996) and topped by seismic swarms systematically recorded at 500–2500 m depth before each eruption (Nercessian et al. 1996; Massin et al. 2011).

15.7.2 Short-Term Dike Intrusions

Figure 15.7 shows an example of best fit models that fit ground displacements recorded during dike injections feeding one proximal and one distal eruption in 2004 and 2005 respectively. The systematic modeling of each eruption between 2003 and 2007 gives a range of 0.3–6 Mm^3 for the volume of the recent dike injections at Piton de la Fournaise, and a dike opening of 0.3–1.5 m. This is on the same order of magnitude as the solidified dike thicknesses observed in Dolomieu crater's outcrops (Peltier et al. 2012). The distal eruptions are fed by large dikes, whereas the summit/proximal eruptions are fed by low-volume dikes (Fig. 15.7). These results are in agreement with the inversion of INSAR data at Piton de la Fournaise (Fukushima et al. 2005; Tinard 2007), except that, because of low temporal resolution, INSAR data do not detect the first step of vertical magma migration beneath the crater, due to the closure of the subvertical part of the dike when the overpressure in the reservoir decreases (Fukushima et al. 2010).

The volume of each dike is about 5–20 % of the volume of lava extruded during the corresponding eruption; this is in agreement with a predominantly exogenous growth of the volcano (Fukushima et al. 2010; Peltier et al. 2012).

These results give us an overall picture of the Piton de la Fournaise shallow magma feeding system. It consists of a shallow magma reservoir (~2300 ± 500 m depth) and two distinct preferential paths for the dikes. The starting points of the two main dike paths are located on the western and eastern sides of the Dolomieu crater. Many fractures border the Dolomieu crater, and acts as preferential paths for ascending fluids. As proposed by Peltier et al. (2009a, b), in a state of structural stability, dikes are initiated below the western part of the Dolomieu due to a favored rupture at the top of the magma reservoir. In an unstable state, dikes are initiated below the eastern part of the Dolomieu crater, influenced by the unbuttressed eastern flank of Piton de la Fournaise, leading to distal eruptions.

Indeed recent modeling of GNSS data, considering an elasto-plactic medium, evidenced that the eastern flank accumulated stress during summit and proximal eruptions, until its plastic

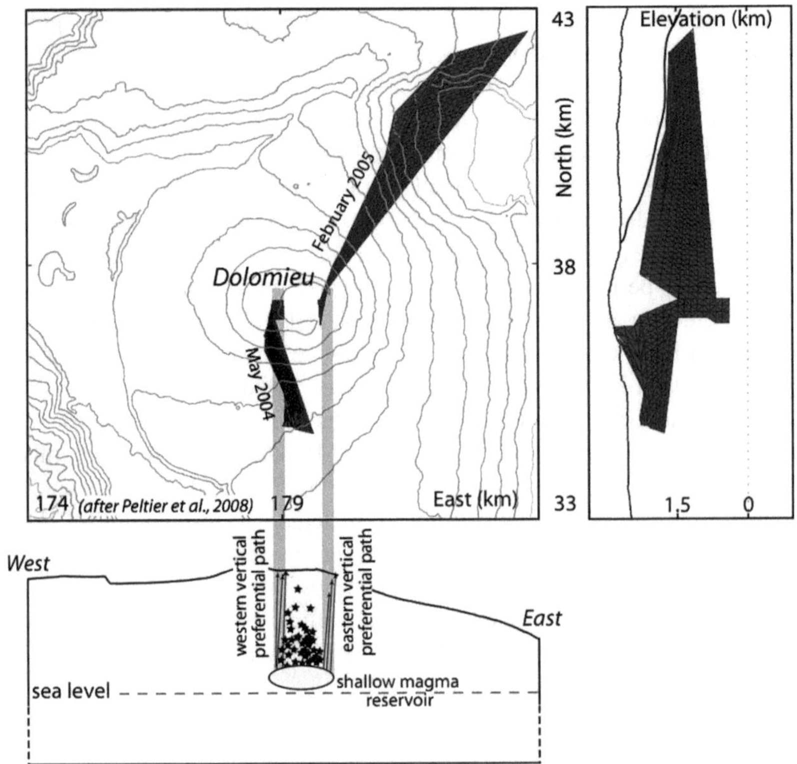

Fig. 15.7 Shape and location of the dikes modeled for the May 2, 2004 proximal eruption and the February 17, 2005 distal eruption, based on permanent and rapid static GNSS data (after Peltier et al. 2008 and Peltier et al. 2009b). *Stars* show the seismic active region above the magma reservoir. These models allow a view into the Piton de la Fournaise plumbing system, showing that the eruption close to the summit had been fed by a dike path starting below the western part of the Dolomieu crater, while the distal eruption started from the eastern part, most probably from the eastern side of the magma reservoir

threshold was reached leading to its instability and eastern flank stick-slip and distal eruption (Got et al. 2013).

15.8 Summary

Modern GNSS networks and in particular permanent differential GNSS method allow determining positions with an accuracy of <1 cm. Using baseline variations permits an even better precision on the order of several millimeter. Such data allow pursue ground deformation and via inversed model calculations getting deep views inside of volcanic massifs. Since the start of the permanent GNSS network at Piton de la Fournaise, the temporal series permit to visualize the state and evolution of the volcanic massif and to predict all (except one; Dec. 14, 2009) eruptions weeks or months before they break out by small but clearly measurable summit expansions.

- Even after the huge 2007 eruption and summit collapse, which disrupted the volcanic massif thoroughly and made the deformation patterns of the volcano more heterogeneous, almost all eruptions were preceded by clear signs of summit inflation. This allows alerting local authorities long time enough before eruptions, thus protecting inhabitants and tourists from injuries.
- Besides these long-term ground deformation, rapid ground deformation are recorded during

the dike injection itself, which allow forecasting at short-term the direction the dike will pursue. This is crucial in term of risk mitigation.

- Large rapid ground deformation of the massif was recorded by GNSS during the collapse of the Dolomieu crater in April 2007 and allowed to constrain the mechanism of the collapse.
- For the first time the more recently installed GNSS receivers on the east flank attest the slow but constant movement of the east flank of Piton de la Fournaise to the east with a rate of up to 2 cm per year.
- The good spatial coverage of the active massif and the precision of the GNSS data make model calculations possible and allowed us to clarify the geometry of the plumbing system, distinguishing the source region for eruptions at and near the summit and eruptions that occur far away from the summit. The former originate below the western part of the Dolomieu crater, the latter from the eastern part, probably because the less stable eastern flank represents a free boundary and favors the progress of dikes to the east.

Acknowledgments We are grateful to P. Boissier for his permanent commitment to bring in GNSS data from the field, the constant improvement of the automatic data processing and the maintenance of the database. P. Kowalski, F. Lauret and P. Catherine are thanked for their effort to install and make the GNSS recorder running. Asta Miklius from the HVO is warmly thanked for their extremely helpful reviews and her effort to correct the English.

References

Augier A, Froger JL, Cayol V, Fukushima Y, Tinard P, Souriot T, Mora O, Staudacher T, Durand P, Fruneau B, Villeneuve N (2008) The April 2007 eruption at Piton de la Fournaise, Réunion Island, imaged with ENVISAT-ASAR and ALOS-PALSAR data. USEReST Worhshop, Napoli

Aki K, Ferrazzini V (2000) Seismic monitoring and modelling of an active volcano for prediction. J Geophys Res 105:16617–16640

Bachèlery P, Kowalski P, Catherine P, Delmond JC, Blum PA, Croce J (1998) Precise temporal and mechanical identification of dike emplacement using deformation monitoring at piton de la Fournaise. In: Casale R, Fytikas M, Sigvaldasson G, Vougioukalakis G (eds.) Volcanic Risk—The European Laboratory Volcanoes (European Commission, EUR 18161 EN: 475–485)

Battaglia J, Ferrazzini V, Staudacher T, Aki K, Cheminee JL (2005) Pre-eruptive migration of earthquakes at the Piton de la Fournaise volcano (Reunion island). Geophys J Int 161:549–558

Blum PA, Gaulon R, Lalanne FX, Ruegg J (1981) Sur l'évidence des précurseurs de l'éruption du volcan du Piton de la Fournaise à La Réunion. C Acad SCI Paris 292D:1449–1455

Brenguier F, Kowalski P, Staudacher T, Ferrazzini V, Lauret F, Boissier P, Catherine P, Lemarchand A, Pequegnat C, Meric O, Pardo C, Peltier A, Tait S, Shapiro NM, Campillo M, Di Muro A (2012) First results from the UnderVolc high resolution seismic and GPS network deployed on Piton de la Fournaise Volcano. Seismol Res Lett 83(1). doi: 10.1785/gssrl.83.1.88

Brenguier F, Shapiro NM, Campillo M, Ferrazzini V, Duputel Z, Coutant O, Nercessian A (2008) Towards forecasting volcanic eruptions using seismic noise. Nat Geosci. doi:10.1038/ngeo104

Briole P, Massonnet D, Delacourt C (1997) Post-eruptive deformation associated with the 1986–87 and 1989 lava flows of Etna detected by radar interferometry. Geophys Res Lett 24(1):37–40

Briole P (2007) http://iterre.fr/spip/B-PNAV-calculs-de-GPS-cinematique.html

Cayol V, Cornet FH (1997) 3D mixed boundary elements for elastostatic deformation field analysis. Int J Rock Mech Min Sci 34:275–287

Clarke D, Brenguier F, Froger JL, Shapiro NM, Peltier A, Staudacher T (2013) Timing of large volcanic flank movements at piton de la Fournaise volcano using noise-based seismic monitoring and ground deformation measurements. Geophys J Int

Delorme H, Blum PA, Ruegg JC, Briole P, Toutain JP, Cheminée JL (1993) Ground deformations monitoring at piton de la Fournaise volcano (Réunion Island). Comparison of various techniques. Volcano observatories, surveillance of volcanoes and prediction of eruptions, WOVO Workshop

Dzurisin D, Lockwood JP, Casadevall TJ, Rubin M (1995) The Uwekahuna Ash Member of the Puna Basalt: product of violent phreatomagmatic eruptions at Kilauea volcano, Hawaii, between 2800 and 2100 years ago. J Volcanol Geoth Res 66:163–184

Fontaine FR, Roult G, Michon L, Barruol G, Di Muro A (2014) The 2007 eruptions and caldera collapse of the Piton de la Fournaise volcano (La Réunion Island) from tilt analysis at a single very broadband seismic station. Geophys Res Lett 41:2803–2811. doi:10.1002/2014GL059691

Fretzdorff S, Paterne M, Stoffers P, Ivanova E (2000) Explosive activity of the Reunion Island through the

past 260,000 years as recorded in deep-sea sediments. Bull Volcanol 62:266–277
Froger JL, Fukushima Y, Briole P, Staudacher T, Souriot T, Villeneuve N (2004) The deformation field of the August 2003 eruption at Piton de la Fournaise, Reunion Island mapped by ASAR interferometry. Geophys. Res. Letters 31(14). doi:10.1029/2004GL020479
Froger JL, Cayol V, Famin V (2016) The March-April 2007 eruptions of Piton de la Fournaise as recorded by interferometric data. In: Bachèlery P, Lénat JF, Di Muro A, Michon L. (eds), Active Volcanoes of the Southwest Indian Ocean: Piton de la Fournaise and Karthala. Active Volcanoes of the World. Springer-Verlag, Berlin
Fukushima Y, Cayol V, Durand P (2005) Finding realistic dike models from interferometric synthetic aperture radar data: the February 2000 eruption at Piton de La Fournaise. J Geophys Res 110. doi:10.1029/2004JB003268
Fukushima Y, Cayol V, Durand D, Massonnet D (2010) Evolution of magma conduits during the 1998–2000 eruptions of Piton de la Fournaise volcano, Réunion Island. J Geophys Res 115: doi:10.1029/2009JB007023
Got JL, Peltier A, Staudacher T, Kowalski P, Boissier P (2013) Edifice strength and magma transfer modulation at piton de la fournaise volcano. J Geophys Res. doi:10.1002/jgrb.50350
Massin F, Ferrazzini V, Bachèlery P, Nercessian A, Duputel Z, Staudacher T (2011) Structures and evolution of the plumbing system of Piton de la Fournaise volcano inferred from clustering of 2007 eruptive cycle seismicity. J Volcanol Geoth Res 202:96–106
Mastin LG, Christiansen RL, Swanson DA, Stauffer PH, Hendley JW (1998) Explosive eruptions at Kilauea volcano, Hawai'I. USGS Fact Sheet: 132-98. http://pubs.usgs.gov/fs/fs132-98/
Michon L, Staudacher T, Ferrazzini V, Bachèlery P, Marti J (2007) April 2007 collapse of Piton de la Fournaise: A new example of caldera formation. Geophys Res Lett 34:L21301. doi:10.1029/2007GL031248
Nercessian A, Hirn A, Lépine JC, Sapin M (1996) Internal structure of Piton de la Fournaise volcano from seismic wave propagation and earthquake distribution. J Volcanol Geoth Res 70:123–143
Peltier A, Ferrazzini V, Staudacher T, Bachèlery P (2005) Imaging the dynamics of dike propagation prior to the 2000–2003 flank eruptions at Piton de la Fournaise, Reunion Island. Geophys Res Lett 32. doi:10.1029/2005GL023720
Peltier A, Staudacher T, Catherine P, Ricard LP, Kowalski P, Bachèrlery P (2006) Subtle precursors of volcanic eruptions at Piton de la Fournaise detected by extensometers. Geophys Res Let. 33:l06315. doi:10.1029/2005GL02495
Peltier A, Staudacher Th, Bachèlery P, Cayol V (2009a) Formation of the April 2007 crater collapse at Piton de la Fournaise: insights from GPS data. J Volcanol Geoth Res 184:152–163. doi:10.1016/j.jvolgeores.2008.09.009
Peltier A (2007) Suivi, Modélisation et Evolution des processus d'injections magmatiques au Piton de la Fournaise, Doctorat, Université de La Réunion
Peltier A, Staudacher T, Bachèlery P (2007) Constraints on magma transfers and structures involved in the 2003 activity at Piton de la Fournaise from displacement data. J Geophys Res 112:(B03207). doi:10.1029/2006JB004379
Peltier A, Famin V, Bachèlery P, Cayol V, Fukushima Y, Staudacher T (2008) Cyclic magma storages and transfers at Piton de la Fournaise volcano (La Réunion hotspot) inferred from deformation and geochemical data. Earth Planet Sci Lett 270(3–4):180–188
Peltier A, Bachèlery P, Staudacher T (2009b) Magma transfer and storage at Piton de la Fournaise (La Réunion Island) between 1972 and 2007: a review of geophysical and geochemical data. J. Volcanol Geoth Res. 184(1–2):93–108
Peltier A, Staudacher T, Bachèlery P (2010) New behaviour of the Piton de La Fournaise volcano feeding system (La Réunion Island deduced from GPS data: influence of the 2007 Dolomieu crater collapse. J Volcanol Geoth Res 192:48–56
Peltier A, Bachèlery P, Staudacher T (2011) Early detection of large eruptions at Piton de La Fournaise volcano (La Réunion Island): contribution of a distant tiltmeter station. J Volcanol Geoth Res 199:96–104
Peltier A, Massin F, Bachèlery P, Finizola A (2012) Internal structure and building of basaltic shield volcanoes: the example of the Piton de la Fournaise terminal cone (La Réunion). Bull Volcanol 74:1881–1897
Peltier A, Poland M, Staudacher T (2015) Are Piton de la Fournaise (La Réunion) and Kīlauea (Hawai'i) really "analog volcanoes"? In AGU monograph: Hawaiian volcanoes, from source to surface, monograph number 208, ISBN 978-1-118-87204-citaccc8
Peltier A, Beauducel F, Staudacher T, Catherine P, Kowalski P (2016) Contribution of tiltmeters and extensometers to monitor Piton de La Fournaise activity. In: Bachèlery P, Lénat JF, Di Muro A, Michon L. (eds) Active volcanoes of the Southwest Indian Ocean: Piton de la Fournaise and Karthala. Active volcanoes of the world. Springer, Berlin
Roche O, van Wyk de Vries B, Druitt TH (2001) Sub-surface structures and collapse mechanisms of summit pit craters. J Volcanol Geoth Res 105:1–18
Roult G, Peltier A, Taisne B, Staudacher T, Ferrazzini V, Di Muro A and the OVPF group (2012) A new comprehensive classification of the Piton de la Fournaise eruptions spanning the 1986–2011 period. Search and analysis of eruption precursors from a broad-band seismological station, J Volcanol Geoth Res 241–242:78–104
Sambridge M (1999) Geophysical inversion with a neighbourhood algorithm—I. Searching a parameter space. Geophys J Int 138:479–494

Saleh B, Blum P, Delorme H (1991) New silica compact tiltmeter for deformations measurement. J Surv Eng 117(1), 27–35. doi: 10.1061/(ASCE)0733-9453

Schmid A, Grasso J R, Clarke G, Ferrazzini V, Bachelery P, Staudacher T (2012) Eruption forerunners from multiparameter monitoring and application for eruptions time predictability (Piton de la Fournaise). J Geophys Res [Solid Earth] 117

Staudacher Th, Bachelery P, Ricard LP, Catherine P, Kowalski P, Cheminée JL (2003) Signaux précurseurs des éruptions volcaniques: nouvelles approches de détection: variations extensométriques comme précurseurs à long terme, Rapport quadriennal 1999–2002 XXIIIe Assemblée Générale de l'Union Géodésique et Géophysique Internationale Sapporo (Japon) L'Etat de la Planète: les frontières et les défis, du 30 juin au 11 juillet 2003, pp 125–127

Staudacher T, Ferrazzini V, Peltier A, Kowalski P, Boissier P, Catherine P, Lauret F, Massin F (2009) The April 2007 eruption and the Dolomieu crater collapse, two major events at Piton de la Fournaise (La Réunion Island, Indian Ocean). J Volcanol Geoth Res 184:126–137. doi:10.1016/j.jvolgeores.2008.11.005

Staudacher T, Peltier A, Ferrazzini V, Di Muro A, Boissier P, Catherine P, Kowalski P, Lauret F, Lebreton J (2016) Fifteen years of intense eruptive activity (1998–2013) at Piton de La Fournaise volcano: a review. In: Bachèlery P, Lénat, JF, Di Muro A, Michon L (eds) Active volcanoes of the Southwest Indian Ocean: Piton de la Fournaise and Karthala. Active volcanoes of the world. Springer, Berlin

Tinard P (2007) Caractérisation et modélisation des déplacements du sol associés à l'activité volcanique du Piton de la Fournaise, île de La Réunion, à partir de données interférométriques. Août 2003–Avril 2007. PhD Thesis, Université de Clermont Ferrand, 334 p

Toutain JP, Bachelery P, Blum PA, Cheminee JL, Delorme H, Fontaine L, Kowalski P, Taochy P (1992) Real time monitoring of vertical ground deformations during eruptions at Piton de la Fournaise. Geophys Res Lett 19(6):553–556

Zlotnicki J, Ruegg JC, Bachèlery P, Blum PA (1990) Eruptive mechanism on Piton de la Fournaise volcano associated with the December 4, 1983 and January 18, 1984 eruptions from ground deformation monitoring and photogrammetric surveys. J Volcanol Geoth Res 40:197–217

The March–April 2007 Eruptions of Piton de la Fournaise as Recorded by Interferometric Data

16

Jean-Luc Froger, Valérie Cayol and Vincent Famin

Abstract

Piton de la Fournaise is one of the few volcanoes in the world to be monitored on a regular basis from space, by use of Interferometric Synthetic Aperture Radar (InSAR) data. Thirty of the 39 eruptions that occurred during the 1998–2010 period were imaged by various radar satellites. In this chapter, we present one of the best examples of surface deformation mapping achieved at Piton de la Fournaise volcano. From March 30 to May 1 2007, Piton de la Fournaise experienced a major eruptive crisis, characterized by the largest historical flank eruption (210×10^6 m^3), and by a 340 m caldera collapse. The event was captured by InSAR data from the ENVISAT and ALOS-1 satellites. Large and unusual displacements of the Central Cone and the Eastern Flank of Piton de la Fournaise occurred during the co-eruptive period and in the following months. Analysis of the InSAR displacement combined with other geophysical and field observations allows us to propose conceptual models to explain the Central Cone and the Eastern Flank displacements. The April 2007 caldera collapse induced a sudden decompression of the hydrothermal system contributing to the strong centripetal subsidence of the Central Cone. This then decreased exponentially with poro-elastic

J.-L. Froger (✉) · V. Cayol
Laboratoire Magmas et Volcans, UMR6524,
CNRS-IRD-Université Blaise Pascal, 5 rue Kessler,
63038 Clermont-Ferrand, France
e-mail: J.L.Froger@opgc.univ-bpclermont.fr

V. Cayol
Université de Lyon, Université Jean Monnet, 23 rue du Docteur Paul Michelon, 42023 Saint Etienne, France

V. Famin
Laboratoire GéoSciences Réunion, Université de La Réunion, Institut de Physique du Globe de Paris, CNRS, Sorbonne Paris Cité, 15 avenue René Cassin, 97744 Saint-Denis, France

P. Bachèlery et al. (eds.), *Active Volcanoes of the Southwest Indian Ocean*,
Active Volcanoes of the World, DOI 10.1007/978-3-642-31395-0_16

compaction and creep of the Central Cone. For the Eastern Flank, the displacement is related to the combined effect of a sill and slip on a detachment. Dense olivine-rich magma from the deeper part of the main magma storage was injected as a dyke which may have encountered a pre-existing structural discontinuity, intruded it as a sill and activated it as a detachment surface. Magma then further propagated to the April 2 eruption site.

16.1 Introduction

The development of spatial geodetic techniques has brought about considerable advances in the monitoring and interpretation of ground surface displacement in volcanic settings over the last few decades (Dzurizin 2007). Several studies have looked at static displacements occurring as an immediate response to stress variation, resulting either from the infilling or emptying of a magma source in the upper crust (Pritchard and Simons 2002; Froger et al. 2007) or from propagation of a magmatic intrusion to the surface (Sigmundsson et al. 1999; Froger et al. 2004; Fukushima et al. 2010). Time-dependant displacement has also been observed in relation to deep intrusions embedded in a visco-elastic (Newman et al. 2006; Fialko and Pearse 2012) or elasto-plastic crust (Trasatti et al. 2005), pressurisation or depressurisation of hydrothermal systems (Rinaldi et al. 2010), flank motion (Palano et al. 2009; Montgomery-Brown et al. 2011), or compaction and loading of lava flows (Stevens et al. 1997).

Piton de la Fournaise has been extensively monitored since 1980 by ground surface displacement (GPS, Electronic Distance Measurement and tiltmeters) and seismic monitoring networks operated by the Observatoire Volcanologique du Piton de la Fournaise (OVPF). It is also one of the few volcanoes in the world to be monitored on a regular basis, since 1998, by Interferometric Synthetic Aperture Radar (InSAR) data. The use of this technology is still new and it is too early to propose a review of the work done by this method at Piton de la Fournaise. Therefore, we propose to illustrate the InSAR techniques by focussing on one of the finest achievements obtained in recent years. This chapter looks at both static and time-dependant ground surface displacement recorded at Piton de la Fournaise during and after the March 30 to May 1, 2007 eruptions.

While the OVPF's GPS monitoring network, concentrated in the vicinity of the Central Cone (Fig. 16.1), gives an excellent record of the summit displacement that occurred in March–April 2007, it does not provide any constraint on the displacement over the whole edifice. Using the various acquisition geometries and sensitivities provided by the C-band ASAR and the

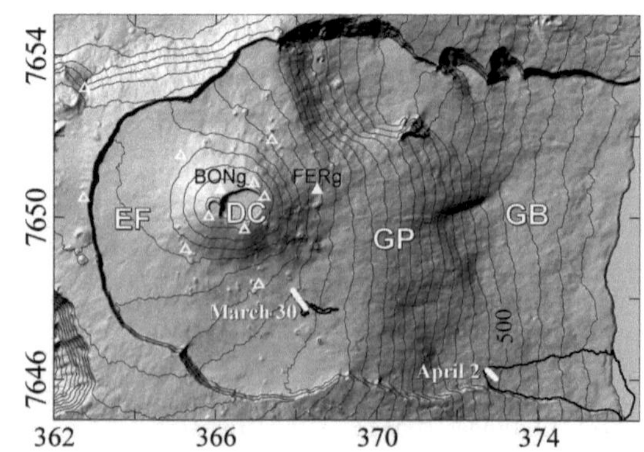

Fig. 16.1 Topography of Piton de la Fournaise with 100 m elevation contours and location of the main places discussed in text (*DC* Dolomieu Crater, *EF* Enclos Fouqué; *GB* Grand Brûlé, *GP* Grandes Pentes). *White triangles* indicate the locations of the OVPF GPS permanent stations, the *filled white triangles* indicate the location of stations discussed in text. Locations of the March 2007 and April 2007 eruptive fissures are indicated by bold white lines, while lava flow contours are in black. Coordinates are in km UTM (40 zone south)

L-band PALSAR data, we computed the EW and vertical components of the ground surface displacement, which occurred over the entire edifice during the March–April 2007 eruptions and in the following months. These results reveal extremely large and unusual time-dependant displacements of the Central Cone and the eastern flank of Piton de la Fournaise, which both continued to deform for at least a year after the end of the eruption. The characterisation of such large time-dependant displacement has essential implications not only for risk assessment at Piton de la Fournaise, but also for the understanding of deformation processes related to magma transfer in general.

16.2 Piton de la Fournaise Activity from 1998 to 2006 as Inferred from Insar Data

After five years and a half of quiescence, a new eruptive period started in March 1998. Co-eruptive displacements associated with 30 eruptions which took place between 1998 and December 2001 and between August 2003 and December 2010 were recorded by RADARSAT-1 & -2, ASAR, PALSAR and TerraSAR-X interferograms. This monitoring evidences that eruptions are associated with asymmetrical displacement with the eastern flank undergoing significant seaward displacements, much larger than the western flank, and that inter-eruptive displacement is mainly related to compaction of lava flows and dike contraction (Tinard 2007).

Co-eruptive displacements associated with eruptions from 1998 to June 2000 (Fukushima et al. 2010) and to the August 2003 eruption (Froger et al. 2004) were analysed using a 3-D boundary element models combined with a near-neighbourhood inversion method. These analysis evidence that North South trending dikes are eastward dipping and that all eruptions can be modelled by dike rooted at depths less than 1000 m below the ground, with the exception of a deep sheet intrusion in 1998. For the 1998–2000 period, intruded volume represents 17 % of the total volume of emitted magma (Fukushima et al. 2010). Repetitive intrusions along the N25–30°E, N120°E preferential intrusion directions steepen the eastern part of the Central Cone, probably explaining the unusual steep slopes (Michon et al. 2009) and increasing deviatoric stress in the Eastern Flank.

16.3 The March–April 2007 Eruptions

The March–April 2007 eruptions were preceded by 10 months of quasi-continuous activity with three summit eruptions (July–August 2006, August 2006–January 2007 and February 2007, Peltier et al. 2009a). By the end of this period the Dolomieu crater (Fig. 16.1) had become overfilled and lava flows started to spill over the crater rim at its lowest elevation. The first two months of 2007 were marked by a continuous inflation of the edifice and the Central Cone (Staudacher et al. 2009; Fontaine et al. 2014). Peltier et al. (2009b) interpreted this inflation as evidence of a recharge of the main magma storage zone generally assumed to exist beneath the Central Cone, between 0 and 1000 m above sea level (Lénat and Bachèlery 1990; Peltier 2007; Peltier et al. 2009a; Prôno et al. 2009). A deeper source of inflation was recently proposed from the RER far field very broadband seismic station (Fontaine et al. 2014). Whatever the source's depth, the March–April 2007 eruptions were heralded by a progressive increase in seismicity beneath the summit zone from the end of February to the end of March. On March 30, at 18H50 GMT an eruptive fissure opened at 1900 m above sea level (a.s.l.) at the southeastern base of the Central Cone (Staudacher et al. 2009). The fissure fed a small lava flow (<10^6 m^3, Staudacher et al. 2009, Fig. 16.1) of aphyric basalt during an inital, ~10-h-long, eruptive phase. On April 2, a new eruptive fissure opened 7 km away from the summit on the lower Eastern Flank (~600 m a.s.l., Fig. 16.1), also emitting aphyric basalt. From April 5, the OVPF seismic and GPS networks recorded a significant increase in activity below the Central Cone (Staudacher et al. 2009; Peltier et al. 2009b). The crisis climax was reached in the night between April 5 and 6 with the onset of the collapse of the Dolomieu crater. At the same time, a significant increase in eruptive activity was observed at the lower Eastern Flank eruptive fissure, together with an evolution of lava composition toward oceanite (Di Muro et al. 2014). The eruption continued until May 1 with a continuous lava emission at the Eastern Flank fissure. Considering the total emitted volume of lava (estimated at $\sim 210 \times 10^6$ m^3 by Bachèlery et al. 2010) and the magnitude of the Dolomieu collapse (340 m in depth, 100–120 $\times$ 10^6 m^3 in volume, Michon et al. 2007; Urai et al. 2007), the April 2007 crisis can be regarded as the largest eruption observed at Piton de la Fournaise during the 20th and 21th centuries.

After the end of the April eruption, no further activity occurred for about 500 days and only 8 minor eruptions have occurred since. This suggests that the system was significantly modified by the April 2007 events, as also shown by the seismic and geodetic data (Rivet et al. 2014).

16.4 Data Processing and Results

16.4.1 Computation of the Interferograms

We processed two separate interferometric datasets spanning the co-eruptive period (i.e. the period including the March 30 intrusion/eruption, the April 2 eruption and the April 6 caldera collapse) and the post-eruptive period (i.e. from May 2007 to September 2008) respectively. For the co-eruptive period, we calculated two ascending PALSAR interferograms, two ascending and one descending ASAR interferograms. Figure 16.2a, b shows the more coherent PALSAR and ASAR interferograms. For the post-eruptive period, we calculated six ascending PALSAR interferograms and 129 ASAR interferograms from ascending and descending viewing directions. All interferograms (PALSAR and ASAR) were produced with the DIAPASON© software (CNES/Altamira-Information, CNES 1996) using the two-pass method described by Massonnet and Feigl (1998). They are available in our online database (https://wwwobs.univ-bpclermont.fr/casoar).

16.4.2 Computation of the Displacement

The co-eruptive interferograms show complex displacement patterns that reflect the superimposition of different signals related to the March–April eruptions, including the March 30 intrusion/eruption, the opening of the April 2 vent and the April 5–April 9 Dolomieu collapse.

The characterisation of the individual signals is difficult to obtain from the interferograms, as each of them provides only a projection of the ground surface displacement in a specific line of sight (LOS) of the satellite. We obtained a more discriminating dataset by constructing maps of the displacement components (EW and vertical) from the independent co-eruptive interferograms using the approach proposed by Wright et al. (2004). A detailed description of the InSAR data processing performed before calculating the displacement components is presented in Froger et al. (2015).

The ASAR and PALSAR interferograms covering the one and a half-year period of quiescence after the end of the March–April 2007 eruption show that the ground surface displacement continued for at least a year, with decreasing amplitude, both at the Central Cone and in the Grandes Pentes. In order to characterise this displacement, we calculated their EW and vertical components from five different ASAR datasets (Froger et al. 2015).

We estimated the quality of the inverted co-eruptive and post-eruptive displacement components on a pixel-by-pixel basis using the total standard error as formulated by Strang (1986). We used the resulting standard error map, combined with an image of the mean

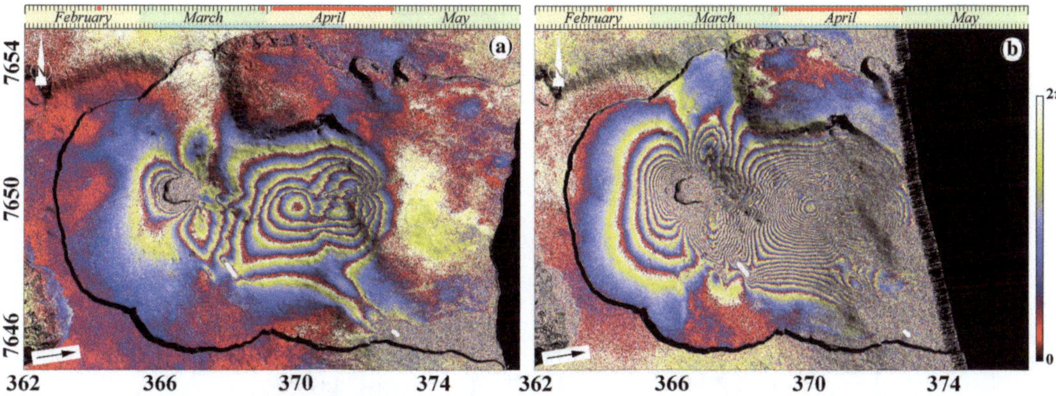

Fig. 16.2 **a** PALSAR ascending interferogram spanning 46 days from March 4 to April 19, 2007. **b** ASAR ascending interferogram spanning 35 days from March 26 to April 30, 2007. *Red horizontal bars* on the chronological scales indicate periods of volcanic activity while cyan horizontal bars indicate the time spanned by the interferograms. Phases are draped on the shaded DEM. A complete phase cycle (*red-blue-yellow*) represents a range increase between the ground surface and the satellite of 11.8 and 2.8 cm for the PALSAR and ASAR interferograms, respectively. *Black arrows* indicate LOS directions. Locations of the March 2007 and April 2007 eruptive fissures are indicated by *bold white lines*. Coordinates are in km UTM (40 zone south)

coherence of the interferograms, to create a mask where all pixels in the area of interest (Enclos Fouqué, Grandes Pentes, Fig. 16.1) with a standard error higher than 0.07 m or a mean coherence lower than 25 % are masked. Then we used the neighbouring pixel values to interpolate all the masked areas whose size did not exceed 5 × 5 pixels. The interpolation was done using a Kriging algorithm (Journel and Huijbregts 1978).

16.4.3 Co-eruptive Displacement

The co-eruptive EW and vertical components of displacement show four main signals (Fig. 16.3a, b). (**1**) The western part of the Central Cone is affected by both eastward and downward displacements indicating a centripetal subsidence (C1 in Fig. 16.3c) reaching a maximum of 50 cm at the NW rim of the Dolomieu crater. (**2**) The eastern flank of the Central Cone develops a large NS-trending inflation axis of about 4 km ending to the south at the location of the March 30 eruptive fissure (C2 in Fig. 16.3c), characteristic of a dyke intrusion (Pollard et al. 1983). Indeed, the horizontal displacement (Fig. 16.3a) is oriented eastward on the eastern side of the axis and westward on the western side. The large wavelength of displacement suggests a relatively deep source along most of its length. At its southern end, the NS axis divides into two N150°E branches and the wavelength of displacement becomes shorter, suggesting that the source of the displacements becomes shallower (C2a and C2b in Fig. 16.3c). The easternmost branch (C2b) coincides with the location of the March 30 eruptive fissure, indicating that the displacement axis is related to the March 30 dyke intrusion. Given the displacement pattern orientation, the dyke is more likely oriented NS rather than N125°E as proposed earlier (Fig. 14 in Peltier et al. 2009b). (**3**) To the east, between the foot of the Central Cone and the base of the Grandes Pentes, horizontal displacement shows a fan-shaped pattern of ~14 km^2 in the Grandes Pentes, indicating an extensive eastward displacement with a maximum of up to 1.4 m (C3 in Fig. 16.3c). The vertical displacement has a more complicated pattern, with a general subsidence of the Grandes Pentes, with a maximum of up to 33 cm, intersected in its central part by a ~N170°E trending uplift reaching 37 cm (C4 in Fig. 16.3c). These observations indicate that a large-wavelength downslope displacement of the Grandes Pentes occurred, accompanied by a shorter-wavelength uplift of its central part. (**4**) At the SE base of the

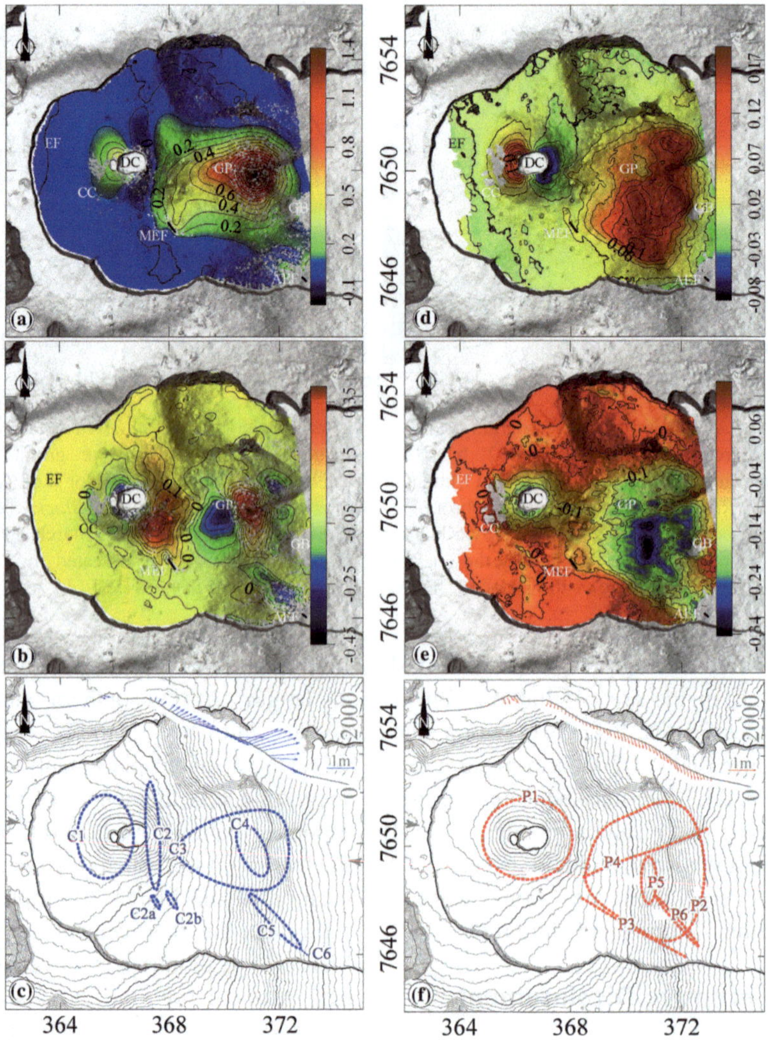

Fig. 16.3 Co- and post-eruptive displacements draped on the shaded DEM. **a** EW co-eruptive displacement, contour level interval is 0.1 m. **b** Vertical co-eruptive displacement, contour level interval is 0.05 m. **c** Sketch map of the main displacement patterns (*blue dashed lines*) referred to as C1–C6. **d** EW post-eruptive component, contour level interval is 0.02 m. **e** Vertical post-eruptive component, contour level interval is 0.05 m. **f** Sketch map of the main post-eruptive displacement patterns (*red dashed lines*) referred to as P1–P6. In **c** and **f**, *grey arrows* point to the extremities of the profile of the displacement vectors shown in the upper part of the figure. In **a**, **b**, **d** and **e**, *thick black lines* represent the location of eruptive fissures. *AEF* April 2, 2007 eruptive fissure, *MEF* March 30, 2007 eruptive fissure; *CC* Central Cone, *EF* Enclos Fouqué, *GB* Grand Brûlé, *GP* Grandes Pentes. Coordinates are in km UTM (40 zone south)

Grandes Pentes a N125–130°E trending axis (C5 in Fig. 16.3c), reaching a maximum subsidence of 28 cm (Fig. 16.3b), is aligned with the April 2 eruptive fissure (C6 in Fig. 16.3c), indicating that this displacement is related to the magma migration toward the April 2 eruption vent.

16.4.4 Post-eruptive Displacement

The post-eruptive EW and vertical displacement also shows four main signals (Fig. 16.3d, e). (**1**) The Central Cone is affected by horizontal displacement asymmetrical with respect to the crater and concentric vertical displacement, both indicating a centripetal subsidence (P1 in Fig. 16.3f). It is noteworthy that the pattern of the centripetal subsidence is limited to the Central Cone suggesting a relatively shallow source of displacement. (**2**) The Grandes Pentes display a large trapezoidal displacement pattern (P2 in Fig. 16.3f). The total area affected by the displacement is up to 20 km^2. Contrary to the co-eruptive period, the displacements are uniformly distributed eastward and downward over the entire zone, indicating an overall downslope motion. The displacement pattern is limited to the south by a sharp N120–125°E limit roughly joining the March 30 and the April 2 eruptive fissures (P3 on Fig. 16.3f). Similarly as during the co-eruptive period, the displacement does not extend to the Grand Brûlé area. (**3**) The centre of the trapezoidal displacement pattern is occupied by a NS-elongated subsidence axis (P5 in Fig. 16.3f). This subsidence axis roughly coincides with the southern half of the co-eruptive ~N170°E trending axis of uplift (C4 in Fig. 16.3c). (**4**) At the southern end of the NS subsidence axis, a ~2.5 km long, N135°E narrow subsidence axis (P6 in Fig. 16.3f), nearly coincident with C5 (Fig. 16.3c), ends close to the April 2 eruptive fissure.

16.4.5 Displacements and Intrusive Activity

Among the different displacement patterns revealed by the InSAR data, both the NS uplift axis, on the eastern flank of the Central Cone (C2 in Fig. 16.3c), and the inflation axis on the central Grandes Pentes (C4 in Fig. 16.3c) are observed only during the co-eruptive period. On the contrary, the centripetal subsidence of the Central Cone (C1 and P1 in Fig. 16.3c, f, respectively) and the large wavelength downslope displacement of the Grandes Pentes (C3 and P2 in Fig. 16.3c, f, respectively) continued long after the eruption. This observation suggests that the first two patterns are associated with the elastic response of the medium to transient co-eruptive processes, while the last two patterns could be associated with persistent processes or alternatively could reflect an inelastic response of the medium to transient processes. In the following subsections we discuss the possible causes of these displacements.

16.4.6 Displacement of the Central Cone

Displacement of the Central Cone consists of the superimposition of the centripetal subsidence observed during both co- and post-eruptive periods and the NS uplift axis (C2b in Fig. 16.3c) affecting the cone eastern flank during the co-eruptive period. The fact that the NS uplift axis ends at the location of the March 30 eruptive fissure strongly suggests that this axis was produced by the March 30 dyke.

It is likely that the centripetal subsidence observed on the western flank of the Central Cone during the co-eruptive period affected the entire Central Cone, similarly as during the post-eruptive period, but that this subsidence was partly masked on co-eruptive displacement, by the NS uplift due to the March 30 dyke. The extension of the area of centripetal subsidence does not change significantly between the co- and post-eruptive periods. This suggests that the sources responsible for the deflation in the co-eruptive period and in the post-eruptive period are possibly the same.

We provide some constraints on this source by carrying out inversion of the post-eruptive displacements. For simplicity, we assume an elastic, homogeneous and isotropic edifice, with a Young's modulus of 5 GPa, and a Poisson's ratio of 0.25 (Cayol 1996). We use a 3-D Mixed Boundary Element Method (Cayol and Cornet 1997), combined with a neighborhood inversion method (Sambridge 1999) following the approach proposed by Fukushima et al. (2005). We test a variety of source geometries including a vertical elliptical cylinder, an ellipsoid and a horizontal elliptical dislocation. We assume that first geometry is a good proxy for the brecciated column of rock that extends from the floor of Dolomieu crater to the roof of the supposed main magma storage zone (Gailler et al. 2009).

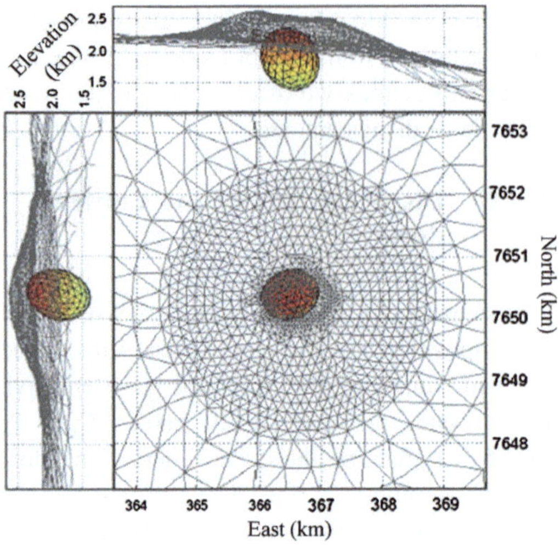

Fig. 16.4 The best-fit model (95.5 % of data explained by the model) for the Central Cone post-eruptive displacement is an ellipsoid which the main parameters (with 95 % confidence interval) are: x = 366.5 ± 0.4 km UTM, y = 7650.4 ± 0.5 km UTM, z = 1858 (+132 − 788) m a.s.l., s1 = 528 (+360 − 284) m, s2 = 474 (+426 − 128) m, s3 = 390 (+239 − 185) m. The variation of pressure is −1.7 MPa

We consider the second and third geometries as good proxies for the magma storage zone or a hydrothermal system. Ellipsoids and elliptical dislocations provide significantly better misfit values than the vertical elliptical cylinder. The most important feature highlighted by this preliminary modeling is that, whatever the geometry, a very shallow deformation source is required to explain the observed displacements (i.e. with centre depth between 1858 m a.s.l. and 2378 m a.s.l., Figure 16.4). Because we assume the co-eruptive and post-eruptive displacement sources to be very close or the same, these conclusions may hold for the source of co-eruptive deflation as well.

16.4.6.1 Post-eruptive Subsidence Decay

In Fig. 16.5, we compare the post-eruptive displacement recorded at a summit GPS station (BONg, Fig. 16.1) with the InSAR-derived displacement calculated at the pixel nearest to the station. Both times series show a similar exponentially decreasing subsidence that continues for at least a year after the end of the March–April 2007 eruptions. A vanishing deflation of the main magma storage associated with the March–April 2007 eruptions could not explain this behavior since the deflation source inferred by the numerical modeling is significantly shallower than the estimated depth of this storage zone (i.e. 460 m a.s.l., Peltier et al. 2009b). A combination of two processes can more plausibly explain the post-eruptive exponential subsidence of the Central Cone. The first process consists in the elasto-visco-plastic deformation of the Central Cone in response to the cavity created by the Dolomieu collapse. Although most creep should occur in the portion of the Central Cone between the summit and the floor of the collapsed column (at ~380 m below the summit), the strongly brecciated and cohesion-less rock column may also contribute to the deformation of the upper edifice by closing the voids created during the column collapse. The second process could be the sudden depressurization of a hydrothermal system within the Central Cone, causing a large deflation immediately after the Dolomieu collapse with an exponential decrease in the months after the eruption. Several observations support the presence of a shallow hydrothermal system beneath the Dolomieu crater and its depressurisation during the Dolomieu caldera collapse. These observations include the large positive self-potential anomaly centered on the Central Cone (Malengreau et al. 1994; Lénat et al. 2000; Lénat 2007; Barde Cabusson et al. 2012), the seismicity beneath the summit craters, attributed to hydrofracturing (Lénat et al. 2012), the series of ash plumes emissions, possibly of phreatic origin, following the April 5–9 caldera collapse (Staudacher et al. 2009; Michon et al. 2013), the SO_2 plume rising from the summit at about the same period (Gouhier and Coppola 2011), and finally the high-temperature (up to 200 °C) steam plumes emanating from the new crater in the days following the collapse (Urai et al. 2007).

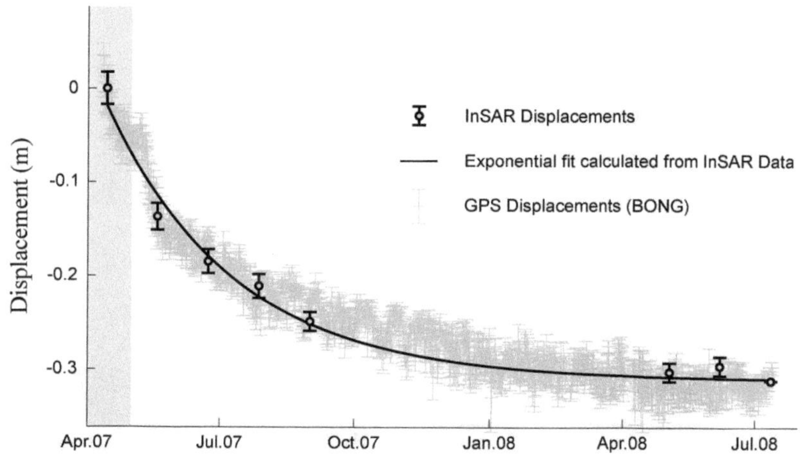

Fig. 16.5 Comparison of the measurements recorded at the summit GPS station BONg (see Fig. 16.1 for location), projected in the LOS of ASAR ascending data (in *grey*) with the ascending InSAR displacements calculated at the nearest pixel (in *black*). Exponential regression on the InSAR data (*black line*) gives a relaxation time of 61 days. The *grey box* indicates the April 2007 eruption

Such a conceptual model explains all these observations as well as the displacements affecting the Central Cone before, during, and after the 2007 eruption (Fig. 16.6). An active hydrothermal system occupies the highly porous rock column between the floor of Dolomieu crater and the roof of the magma storage zone (Lénat et al. 2012). Before the August 2006 eruption, this hydrothermal system was unsealed. The lava flow that filled up the Dolomieu crater from August to December 2006 (about 20×10^6 m^3, Peltier et al. 2009a) temporary sealed the upper boundary of the hydrothermal system. Subsequent heating of the hydrothermal fluids by the underlying magma reservoir induced a progressive pressurization of this shallow hydrothermal system and resulted in a short wavelength inflation of the Central Cone. This inflation probably contributed to the continuous inflation recorded by the GPS stations in the months preceding the March–April 2007 eruptions and previously attributed to the recharge of the main magma storage zone (Peltier et al. 2009b). When the Dolomieu collapsed, the hydrothermal system was suddenly depressurised (Fig. 16.6b), generating firstly phreatic blasts, then SO_2 and steam emissions. As a consequence of the hydrothermal system depressurisation, the Central Cone experienced a strong initial centripetal subsidence, which then decreased exponentially because of the poroelastic relaxation and/or the creep of the Central Cone.

16.4.7 Displacement of the Grandes Pentes

The InSAR-derived displacements show that Piton de la Fournaise volcano underwent a widespread seaward sliding sector of the Grandes Pentes. These results suggest that a slip surface does exist underneath the Eastern Flank of Piton de la Fournaise, as proposed long ago (e.g. Duffield et al. 1982). In addition, these results show that this slip surface has been activated during an eruption, which had not been observed yet for this volcano. A preliminary inversion (Augier 2011), confirmed by a step-by-step approach following a decision tree (Cayol et al. 2014), indicates that the slip surface is a shallow ($\sim$500 m below the ground) normal fault, dipping to the east, parallel to the local average slope. This last characteristic is also supported by the average plunge of the displacement vectors (15°, displacement profile in Fig. 16.3c). The structure responsible for the co-eruptive displacement of the Grandes Pentes is a detachment.

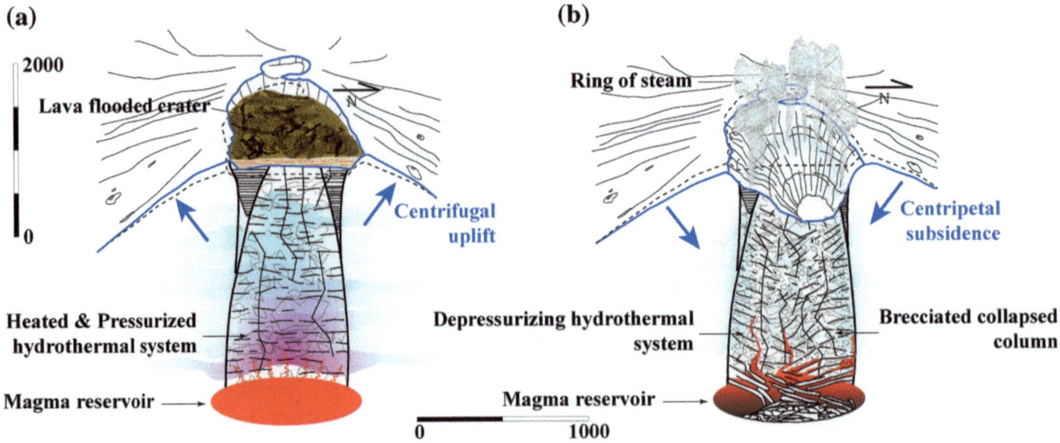

Fig. 16.6 Interpretative geological sketches illustrating: **a** the heating and pressurization of the hydrothermal system, hosted mainly in the brecciated column, by a magma reservoir in the months preceding the April 2007 eruption; **b** the depressurization of the hydrothermal system induced by the 5–9 April 2007 collapse of the Dolomieu crater

16.4.7.1 Did the March 30 Dyke Intrusion Trigger the Detachment Slip?

Since they integrate all the events, which occurred between March 30 and April 5–6, co-eruptive InSAR data do not allow an accurate chronological analysis of displacements. Observations of continuous seismic velocity changes provide more temporal constraints (Clarke et al. 2013). A strong seismic velocity reduction episode, affecting the eastern flank of Piton de la Fournaise, is observed starting on March 30. Clarke et al. (2013) explained this episode as a consequence of the widespread eastward movement of the Eastern Flank, thus suggesting the early nature of this displacement.

From a thorough analysis of the GPS displacements, Got et al. (2013), suggest that the March 30 dyke intrusion was preceded, by a first stage of magma intrusion occurring below the FERg GPS station (Fig. 16.1) in the form of a subhorizontal pressurized structure (i.e. a sill) and then by an eastward displacement at the same station, 13 and 8 min earlier, respectively. This eastward displacement recorded at the FERg station, which continued for at least 2 days with an exponentially decreasing magnitude, is the slip on the Grandes Pentes detachment also recorded on InSAR data. As the slip started to be recorded 8 min before the displacement related to the March 30 dyke intrusion, it is therefore likely that the detachment slip, similarly as indicated by the stress models of Chaput et al. (2014), reduced the minimum principal stress close to the summit, allowing the injection of the March 30 dyke.

16.4.7.2 An Intrusion Within the Central Eastern Flank

From InSAR data, the N170°E co-eruptive uplift (C4 in Fig. 16.3c), superimposed on the detachment slip, suggests an inflation source in the middle of the Grandes Pentes. It is possible that this N170°E axis represents a migration of the inflation from the FERg station toward the east, although the absence of continuous GPS stations in the Grandes Pentes does not allow this interpretation to be confirmed. From the migration of the eruptive vents, it is obvious that the propagation of magma toward the April 2 fissure interrupted the supply of magma at the March 30 fissure. Moreover, the InSAR data show that the two eruptive fissures cannot be related to a single N120°E intrusion as suggested by Peltier et al. (2009b). This probably means that another intrusion propagated from beneath the Central Cone independently of the March 30 NS dyke. While propagating toward the east, this intrusion uplifted the volcano at the FERg station, before inflating the Grande Pentes along the N170°E axis, and eventually opening up the April 2 fissure (Fig. 16.7). The N125–135°E subsiding axis, observed at the south-eastern base of the Grandes Pentes, on both the co- and post-eruptive displacements (features C5 and P6 in Fig. 16.3c, f), would then mark the final path of the magma toward the 2 April eruptive site. The subsequent emission of lava through the April 2 eruptive fissure resulted in the progressive withdrawal of magma accumulated beneath the Grandes Pentes, causing the subsidence observed on the post-eruptive displacements. A temporary magma accumulation within the Grandes Pentes between March 30 and April 2 is also supported by the anomalous depletion in Li, Cu and Ti observed in the first lavas erupted on April 2 (Vlastélic et al. 2013), which suggests that the April 2007 magma was stored in a shallow reservoir ($\sim$590 m above sea level) for some hours to days.

16.4.7.3 Sill Injection Promoting Detachment Slip

Assuming that the Eastern Flank intrusion is planar in most of its path and propagating from the magma storage zone located below the Central Cone, between 0 and 1000 m a.s.l., the interpretation of InSAR co-eruptive displacements imposes the intrusion to have a subhorizontal dip and a N170°E strike underneath the Grandes Pentes. Moreover, GPS data indicate that the March 30 uplift, at the FERg GPS station, assumed to be caused by the planar intrusion, preceded eastward displacement of the Grandes Pentes by 5 min. Based on these constraints, we suggest that during its propagation to the surface this intrusion may have encountered some pre-existing structural discontinuity, exploited it to pursue its propagation as a sill and then activated it as a detachment plane (Fig. 16.7) as indicated by the numerical models of Chaput et al. (2014).

This speculative model is supported by observations made at Piton des Neiges, the extinct La Réunion Island's volcano, where Famin and Michon (2010) reported the existence of a kilometre-scale detachment, with a low-angle dip ($\sim$15°) toward the sea. About 50–70 sills of oceanite magmas were repeatedly injected into this detachment and acted as a triggering factor to slip. It is interesting to note that the oceanite composition of magmas emitted at the April 2 vent, and more generally at the distal vents of Piton de la Fournaise, is different from that of the March 30 dyke, but similar to that of sills in the Piton des Neiges detachment (Famin and Michon 2010). Due to their high density, oceanite magmas ($d \geq 3.1$) are associated with larger overpressures and are more able to propagate along low-dipping fissures (i.e. sills) than aphyric basalts ($d \leq 2.9$). In addition, it has been proposed that intrusions composed of oceanite magmas stem from the deeper parts of the magma reservoir enriched in cumulative olivine (Peltier et al. 2008; Welsch et al. 2009). To conclude, co-eruptive surface displacements of the Eastern Flank, recorded by InSAR data during the March–April 2007 eruptions, are thus more compatible with a scenario of "sill-induced spreading" (i.e. flank collapse triggered by sill injection in a detachment) rather than "rift-induced spreading" (i.e. flank displacement pushed by a dyke).

16.4.7.4 Post-eruptive Subsidence of the Sliding Sector

One of the most significant results provided by the analysis of the post-eruptive displacements is the exponentially decrease of the downslope motion of the Grandes Pentes for at least a year after the end of the April 2007 eruption. This displacement cannot be attributed to a post-eruptive slip of the detachment, because the post-eruptive displacement vectors have a much higher dip (65–75°, displacement profile in Fig. 16.3f) than co-eruptive displacement vectors (15°, displacement profile in Fig. 16.3c). A simple computation (taking only conductive heat transfer into account) shows that it cannot either be attributed to the thermo-elastic contraction of the Grandes Pentes intrusion as unrealistic thickness of the cooling intrusion would be required.

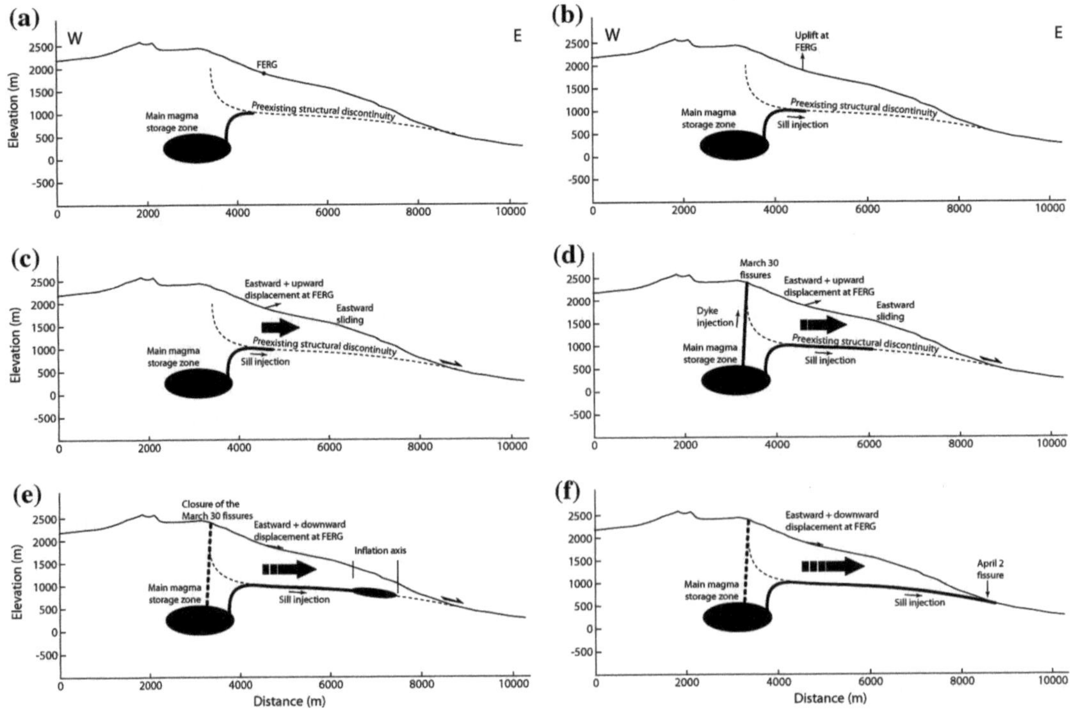

Fig. 16.7 Conceptual cross-sectional sketch illustrating the different steps of the April 2007 events in the Grandes Pentes. **a** Before March 30, 16:25, a first intrusion, initiated from the eastern wall of the main magma storage zone and propagated, as a dyke, to the surface. **b** During its propagation the intrusion encountered a pre-existing structural discontinuity, exploited it to pursue its propagation as a sill, inducing the FERg GPS station uplift at 16:25. **c** On March 30, at 16:30 the sill activated the structural discontinuity as a detachment surface. **d** On March 30, at 16:38, change in the host rock stress field, resulting from slip of the detachment, allowed a new intrusion to propagate as a dyke from the upper part of the main magma storage. **e** Between March 31 and April 2, supply of magma at the March 30 fissure is interrupted and the sill propagation is arrested, inducing magma accumulation within the Grandes Pentes and inflation at the surface. **f** The sill resumes its propagation within the eastern flank and reached the surface. The detachment slip starts to decay

We note that post-eruptive displacements in the Grandes Pentes are delimited by sharp features (P3–P4 in Fig. 16.3f) not observed on the co-eruptive displacement. One of these sharp features, the central N65–70 lineament (P4 in Fig. 16.3f), coincides with a previously recognised normal fault (Michon and Saint Ange 2008). Given the high dip of the displacement vectors (65–75°), it is possible that the other sharp feature also corresponds to a normal fault. Creep on these normal faults could have been activated in the post-eruptive period by the EW extensional stress field resulting from the co-eruptive Grandes Pentes displacement.

16.5 Main Lessons

The displacements that occurred during and after the March–April 2007 eruptions at Piton de la Fournaise were measured using both ASAR and PALSAR InSAR data. The co-eruptive displacement pattern results from the superimposition of multiple processes related to the March 30 intrusion/eruption, the April 2 eruption, and the April 5–9 collapse of the Dolomieu crater. The post-eruptive displacements show evidence of time-dependent processes acting both at the Central Cone and in the Grandes Pentes.

At the Central Cone, the observed displacements are mainly related to the decompression of a pressurized hydrothermal system during the Dolomieu collapse. The idea that the hydrothermal system has acted as a deformation source within the Piton de la Fournaise during the March–April 2007 leads us to reconsider the interpretation of the displacements recorded by the OVPF GPS network, at the Central Cone, during inter-eruptive periods. These inter-eruptive displacements are mentioned by Peltier et al. (2009a) as related to the shallow plumbing system. An alternative interpretation, however, could be that the observed pre-eruptive inflation is related to the pressurisation of the hydrothermal system in response to an increase in the heat flux originating from a magmatic source located somewhere below. After the eruption, the hydrothermal system pressure may drop gradually either because of the disruption of the hydrothermal system or simply because of the heat source withdrawal, resulting in a deflation of the Central Cone. Hence, the hydrothermal system, having a poro-elastic rheology, would induce a time dependant response to transient magma movements, similarly as reported at Campi Flegrei and Long Valley (Hurwitz et al. 2007).

The displacement measured in the Grandes Pentes is the first direct evidence of a large active seaward motion affecting the Eastern Flank of Piton de la Fournaise. This evidence is a strong argument in favor of the previously proposed interpretations of the eastern part of the Enclos Fouqué-Grand Brûlé structure resulting from successive eastward flank destabilization (Lénat et al. 1990; Labazuy 1996; Lénat et al. 2001). In April 2007, this displacement was triggered by a sill intrusion within the Grandes Pentes. While propagating to the surface the magma may have encountered a pre-existing structural discontinuity, intruded it as a sill and activated it as a detachment slip. This scenario provides an explanation for magma transfer from the main magma storage zone, beneath the Central Cone, to the April 2 eruption site. It is supported by the displacement measured early on March 30 at the easternmost OVPF GPS station (FERg, Got et al. 2013).

This interpretation of the Central Cone and the Grandes Pentes displacements provides a conceptual framework that now needs to be explored further by numerical modeling. The potential hazard posed by large flank destabilization emphasizes the need for continuous monitoring of the Grandes Pentes. At the time of the March–April 2007 eruption, OVPF geodetic and seismic monitoring network did not extend to the Grandes Pentes, and, without InSAR, the Grandes Pentes displacement would have gone unnoticed. This clearly illustrates the benefit of using satellite remote sensing data to monitor remote or dangerous areas on volcanoes.

Acknowledgements The authors would like to especially thank Fran Van Wyk de Vries for proofreading an early version of this manuscript. We thank also the JAXA and the European Space Agency for providing the ALOS-PALSAR data through the ALOS-ADEN project #3622 and the ALOS-RA4 project #1287 and the ASAR data through the AO-ENVISAT project #746. The interpretations we present in this paper owe much to the geophysical and field observations collected during the March–April 2007 eruption by the staff of the Volcanological Observatory and by our colleagues of La Réunion University. Discussions with A. Augier, L. Michon, J.-F. Lénat, P. Bachelery, A. Finizola, P. Labazuy, Ph. Mairine, A. Di Muro, A. Peltier, T. Staudacher and N. Villeneuve greatly helped us to develop these interpretations.

References

Augier A (2011) Etude de l'éruption d'avril 2007 du Piton de la Fournaise (île de la Réunion) à partir de données d'interférométrie RADAR et GPS, développement et application de procédure de modélisation. PhD thesis, Université Blaise Pascal, Clermont-Ferrand, France

Bachèlery P, Saint-Ange F, Villeneuve N, Savoye B, Normand A, Le Drezen E, Barrre A, Quod J-P, Deplus C (2010) Huge lava flow into the sea and caldera collapse, April 2007, Piton de la Fournaise. IAVCEI Third Workshop on Collapse Calderas, La Réunion, pp 73–74

Barde Cabusson S, Finizola A, Peltier A, Chaput M, Taquet N, Dumont S, Duputel Z, Guy A, Mathieu L, Saumet L, Sorbadère F, Vieille M (2012) Structural control of collapse events inferred by self-potential

maping on the Piton de la Fournaise volcano (La Réunion Island). J Volcanol Geotherm Res 209–210:9–18
Cayol V (1996) Analyse élastostatique tridimensionnelle du champ de déformations des édifices volcaniques par éléments frontières mixtes. PhD thesis, Université de Paris VII, Paris, France
Cayol V, Cornet FH (1997) 3D mixed boundary elements for elastostatic deformation fields analysis. Int J Rock Mech Min Sci Geomech Abstr 34:275–287
Cayol V, Catry T, Michon L, Chaput L, Famin V, Bodart O, Froger J-L, Romagnoli C (2014) Sheared sheet intrusions as mechanism for lateral flank displacement on basaltic volcanoes: Applications to Réunion Island volcanoes. J Geophys Res Solid Earth 119(10):7607–7635
Chaput M, Pinel V, Famin V, Michon L, Froger J-L (2014) Cointrusive shear displacement by sill intrusion in a detachment: A numerical approach. Geophys Res Lett 41:1937–1943. doi:10.1002/2013GL058813
Clarke D, Brenguier F, Froger J-L, Shapiro NM, Peltier A, Staudacher T (2013) Timing of a large volcanic flank movement at Piton de la Fournaise volcano using noise-based seismic monitoring and ground deformation measurements. GJI 195:1132–1140
CNES (1996) Philosophie et mode d'emploi de la chaîne logicielle interférométrique, DIAPASON, Toulouse, France
Di Muro A, Métrich N, Vergani D, Rosi M, Armienti P, Fougereox T, Deloule E, Arienzo I, Civetta L (2014) The shallow of Piton de la Fournaise volcano (La Réunion Island, Indian Ocean) revealed by the major 2007 caldera forming eruption. Submitted to Journal of Petrology
Duffield WA, Stieljes L, Varet J (1982) Huge landslide blocks in the growth of Piton de la Fournaise, La Réunion, and Kilauea Volcano, Hawaii. J Volcanol Geotherm Res 12:147–160
Dzurisin D (2007) Volcano deformation—geodetic monitoring techniques. Springer, Berlin 441 p
Famin V, Michon L (2010) Volcano destabilization by magma injections in a detachment. Geology 38:219–222
Fialko Y, Pearse J (2012) Sombrero uplift above the Altiplano-Puna magma body: evidence of a ballooning mid-crustal diapir. Science 338:250–252
Fontaine FR, Roult G, Michon L, Barruol G, Di Muro A (2014) The 2007 eruptions and caldera collapse of the Piton de la Fournaise volcano (La Réunion Island) from tilt analysis at a single very broadband seismic station. Geophys Res Lett 41:2803–2811. doi:10.1002/2014GL059691
Froger J-L, Fukushima Y, Briole P, Staudacher Th, Souriot Th, Villeneuve N (2004) The deformation field of the August 2003 eruption at Piton de la Fournaise, Reunion Island, mapped by ASAR interferometry. Geophys Res Lett. doi:10.1029/2004GL020479
Froger J-L, Remy D, Bonvalot S, Legrand D (2007) Two Scales of inflation at Lastarria-Cordon del Azufre volcanic complex, central Andes, revealed from ASAR-ENVISAT interferometric data. Earth Planet Sci Lett. doi:10.1016/j.epsl.2006.12.012
Froger J-L, Famin V, Cayol V, Augier A, Michon L, Lénat J-F (2015) Time-dependent displacements during and after the April 2007 eruption of Piton de la Fournaise, using interferometric data. J Volcanol Geotherm Res 296:55–68
Fukushima Y, Cayol V, Durand P (2005) Finding realistic dike models from interferometric synthetic aperture radar data: The February 2000 eruption at Piton de la Fournaise. J Geophys Res. doi:10.1029/2004JB003268
Fukushima Y, Cayol V, Durand P, Massonnet D (2010) Evolution of magma conduits during the 1998–2000 eruptions of Piton de la Fournaise volcano. J Geophys Res, Réunion Island. doi:10.1029/2009JB007023
Gailler L-S, Lénat J-F, Lambert M, Levieux G, Villeneuve N, Froger J-L (2009) Gravity structure of Piton de la Fournaise volcano and inferred mass transfer during the 2007 crisis. J Volcanol Geotherm Res 184:31–48
Got J-L, Peltier A, Staudacher T, Kowalski P, Boissier P (2013) Edifice strength and magma transfer modulation at Piton de la Fournaise volcano. J Geophys Res. doi:10.1002/jgrb.50350
Gouhier M, Coppola D (2011) Satellite-based evidence for a large hydrothermal system at Piton de la Fournaise volcano (Reunion Island). Geophys Res Lett. doi:10.1029/2010GL046183
Hurwitz S, Christiansen LB, Hsieh PA (2007) Hydrothermal fluid flow and deformation in large calderas: Inferences from numerical simulations. J Geophys Res. doi:10.1029/2006JB004689
Journel AG, Huijbregts C (1978) Mining Geostatistics. Academic Press, Waltham, 600 p
Labazuy P (1996) Recurrent landslides events on the submarine flank of Piton de la Fournaise volcano (Reunion Island). In: McGuire WJ, Jones AP, Neuberg J (eds) Volcano instability on the Earth and other planets, vol 110, pp 295–306 (J Geol Soc Lond)
Lénat J-F (2007) Retrieving self potential anomalies in a complex volcanic environment: a SP/elevation gradient approach. Near Surf Geophys 5:161–170
Lénat J-F, Bachèlery P (1990) Structure et fonctionnement de la zone centrale du Piton de La Fournaise. In: Lénat JF (ed) Le volcanisme de La Réunion. Cent. de Rech. Volcanol, Clermont-Ferrand, France, pp 257–296
Lénat J-F, Bachèlery P, Bonneville A, Galdéano A, Labazuy P, Rousset D, Vincent P (1990) Structure and morphology of the submarine flank of an active basaltic volcano: Piton de la Fournaise (Réunion Island, Indian Ocean). Oceanol. Acta, 10:211–223
Lenat J-F, Fitterman D, Jackson DB, Labazuy P (2000) Geoelectrical structure of the central zone of Piton de la Fournaise volcano (La Réunion). Bull Volcanol 62:75–89

Lénat JF, Gibert-Malengreau B, Galdéano A (2001) A new model for the evolution of the volcanic island of Reunion (Indian Ocean). J Geophys Res 106:8645–8663
Lénat JF, Bachèlery P, Peltier A (2012) The interplay between collapse structures, hydrothermal systems and magma intrusions: the case of the central area of Piton de la Fournaise volcano. Bull Volcanol 74:407–421. doi:10.1007/s00445-011-0535-3
Malengreau B, Lenat J-F, Bonneville A (1994) Cartographie et surveillance temporelle des anomalies de polarisation spontanée (PS) sur le Piton de la Fournaise. Bull Soc Géol Fr 165:221–232
Massonnet D, Feigl KL (1998) Radar interferometry and its application to changes in the Earth's surface. Rev Geophys 36:441–500
Michon L, Saint-Ange F (2008) Morphology of Piton de la Fournaise basaltic shield volcano (La Réunion Island): characterization and implication in the volcano evolution. J Geophys Res. doi:10.1029/2005JB004118
Michon L, Staudacher T, Ferrazzini V, Bachèlery P, Marti J (2007) April 2007 collapse of Piton de La Fournaise: a new example of caldera formation. Geophys Res Lett. doi:10.1029/2007GL031248
Michon L, Cayol C, Letourneur L, Peltier A, Villeneuve N, Staudacher T (2009) Edifice growth, deformation and rift zone development in basaltic setting: Insights from Piton de la Fournaise shield volcano (Réunion Island). J Volcanol Geotherm Res 184:14–30
Michon L, DiMuro A, Villeneuve N, Saint-Marc C, Fadda P, Manta F (2013) Explosive activity of the summit cone of Piton de la Fournaise volcano (La Réunion island): a historical and geological review. J Volcanol Geotherm Res 263:117–133
Montgomery-Brown EK, Sinnett DK, Larson KM, Poland MP, Segall P, Miklius A (2011) Spatiotemporal evolution of dike opening and décollement slip at Kīlauea Volcano. Hawai'i. J Geophys Res. doi:10.1029/2010jb007762
Newman AV, Dixon TH, Gourmelen N (2006) A four-dimensional viscoelastic deformation model for Long Valley Caldera, California, between 1995 and 2000. J Volcanol Geotherm Res 150:244–269
Palano M, Gresta S, Puglisi G (2009) Time-dependent deformation of the Eastern Flank of Mt. Etna: after-slip or viscoelastic relaxation? Tectonophys 473:300–311
Peltier A (2007) Suivi, Modélisation et Evolution des processus d'injections magmatiques au Piton de La Fournaise. PhD thesis, Université de La Réunion, 365 p
Peltier A, Famin V, Bachèlery P, Cayol V, Fukushima Y, Staudacher T (2008) Cyclic magma storages and transfers at Piton de La Fournaise volcano (La Reunion hotspot) inferred from deformation and geochemical data. Earth Planet Sci Lett 270:180–188
Peltier A, Bachèlery P, Staudacher T (2009a) Magma transfer and storage at Piton de La Fournaise (La Réunion Island) between 1972 and 2007: a review of geophysical and geochemical data. J Volcanol Geotherm Res 184:93–108
Peltier A, Staudacher T, Bachèlery P, Cayol V (2009b) Formation of the April 2007 caldera collapse at Piton de La Fournaise volcano: Insights from GPS data. J Volcanol Geotherm Res. doi:10.1016/j.jvolgeores.2008.09.009
Pollard DD, Delaney PT, Duffield WA, Endo ET Okamura AT (1983) Surface deformation in volcanic rift zones. Tectonophys 94:541–584
Pritchard ME, Simons M (2002) A satellite geodetic survey of large scale deformation of volcanic centres in the central Andes. Nature 418:167–170
Prôno E, Battaglia J, Monteiller V, Got J-L, Ferrazzini V (2009) P-wave velocity structure of Piton de la Fournaise volcano deduced from seismic data recorded between 1996 and 1999. J Volcanol Geotherm Res 184:49–62
Rinaldi AP, Todesco M, Bonafede M (2010) Hydrothermal instability and ground displacement at the Campi Flegrei caldera. Phys Earth Planet Inter 178:155–161. doi:10.1016/j.pepi.2009.09.005
Rivet D, Brenguier F, Clarke D, Shapiro NM, Peltier A (2014) Long-term dynamics of Piton de la Fournaise volcano from 13 years of seismic velocity change measurements and GPS observations. J Geophys Res Solid Earth 119. doi:10.1002/2014JB011307
Sambridge M (1999) Geophysical inversion with a neighbourhood algorithm—I. Searching a parameter space. Geophys J Int 138:479–494
Sigmundsson F, Durand Ph, Massonnet D (1999) Opening of an eruptive fissure and seaward displacement at Piton de la Fournaise volcano measured by RADARSAT satellite radar interferometry. Geophys Res Lett. doi:10.1029/1999GL900055
Staudacher T, Ferrazzini V, Peltier A, Kowalski P, Boissier P, Catherine P, Lauret F, Massin F (2009) The April 2007 eruption and the Dolomieu crater collapse, two major events at Piton de la Fournaise (La Réunion Island, Indian Ocean). J Volcanol Geotherm Res 184:126–137
Stevens NF, Murray JB, Wadge G (1997) The volume and shape of the 1991–1993 lava flow field at Mount Etna, Sicily. Bull Volcanol 59:1–6
Strang G (1986) Introduction to Applied Mathematics. Wellesley-Cambridge, 398 p
Tinard P (2007) Caractérisation et modélisation des déplacements du sol associés à l'activité volcanique du Piton de la Fournaise, île de La Réunion, à partir de données interférométriques. Août 2003 – Avril 2007. PhD thesis. Univerité Blaise Pascal, Clermont-Ferrand
Trasatti E, Giunchi C, Bonafede M (2005) Structural and rheological constraints on source depth and overpressure estimates at the Campi Flegrei caldera, Italy. J Volcanol Geotherm Res 144:105–118
Urai M, Geshi N, Staudacher T (2007) Size and volume evaluation of the caldera collapse on Piton de la

Fournaise volcano during the April 2007 eruption using ASTER stereo imagery. Geophys Res Lett. doi:10.1029/2007GL031551
Vlastélic I, Menard G, Gannoun M, Piro J-L, Staudacher T, Famin V (2013) Magma degassing during the April 2007 collapse of Piton de la Fournaise: the record of semi-volatile trace elements (Li, B, Cu, In, Sn, Cd, Re, Tl, Bi). J Volcanol Geotherm Res 254:94–107
Welsch B, Faure F, Bachèlery P, Famin V (2009) Microcrysts record transient convection at Piton de la Fournaise volcano (La Réunion hotspot). J Petrol 50:2287–2305
Wright TJ, Parsons BE, Lu Z (2004) Toward mapping surface deformation in three dimensions using InSAR. Geophys Res Lett. doi:10.1029/2003GL018827

Contribution of Tiltmeters and Extensometers to Monitor Piton de la Fournaise Activity

17

Aline Peltier, François Beauducel, Thomas Staudacher, Philippe Catherine and Philippe Kowalski

Abstract

Tiltmeters and extensometers were among the first instruments used to monitor volcano ground deformation and are used worldwide. Since their implementation in the early 80's at Piton de la Fournaise (La Réunion Island), tiltmeters and extensometers largely contributed to the knowledge and the forecast of eruptive activity on this volcano. These instruments allowed to evidence eruptive precursors at Piton de la Fournaise on two time scales (months/weeks and hours/mins), associated with reservoir pressurization and dike propagations, respectively, but also to follow eruptive/post eruptive relaxation and collapse events.

17.1 Introduction

In volcanic area, fluid motions in depth and/or pressure variations into the plumbing system cause stress variations inside the edifice at the origin of its deformation. At the free surface, this might induce ground uplift/subsidence, slope variations, opening/closing of pre-existing fractures… Early detection of very small changes in the ground motion of volcano edifices is thus largely used to predict their volcanic activity.

Tiltmeters were among the first instruments used to monitor volcano ground deformation (Jaggar and Finch 1929; Imamura 1930). Tiltmeters measure small changes in the tilt of volcano slopes, with a detection level that can reach 10^{-9} µrad. By contrast to other geodetic methods, like GNSS or Electronic Distance Meter, no referential nor post-processing are required to give a

A. Peltier (✉) · T. Staudacher · P. Catherine · P. Kowalski
Observatoire Volcanologique du Piton de la Fournaise, Institut de Physique du Globe de Paris, Sorbonne Paris Cité, Univ Paris Diderot, CNRS, La Plaine des Cafres, F-97418 La Réunion, France
e-mail: peltier@ipgp.fr

F. Beauducel
Institut de Physique du Globe de Paris, Sorbonne Paris Cité, Univ Paris Diderot, 1 rue Jussieu, F-75005 Paris, France

P. Bachèlery et al. (eds.), *Active Volcanoes of the Southwest Indian Ocean*,
Active Volcanoes of the World, DOI 10.1007/978-3-642-31395-0_17

physical measurement. The measurements are immediate and punctual in the space, making this instrument one of the most adapted and easiest to use for real-time monitoring; and this, even if external influences as temperature, heavy rain, (…) might perturb data interpretation (e.g. Wolfe et al. 1981). Tiltmeters are thus the most common geodetic instruments used to monitor volcano ground deformation. They have been widely installed on many volcanoes around the world, on which both rapid (minutes to hours) and slow (weeks to months) tilt variations have been detected. Rapid variations have been attributed to various sources as earth tides (e.g. Saleh 2003), earthquakes (e.g. Bonaccorso and Gambino 1997), dike propagations (e.g. Okamura et al. 1988; Toutain et al. 1992; Peltier et al. 2005; Montgomery-Brown et al. 2011), lava fountaining, explosions (e.g. Bonaccorso 2006; Bonaccorso et al. 2011), (…), whereas slow variations have been attributed to over/under pressurization of magma reservoirs (e.g. Kamo and Ishihara 1989; Beauducel and Cornet 1999; Murray and Wooller 2002), long-term magma uprising, cyclic long-term lava dome extrusion (e.g. Voight et al. 1998), geothermal fluid circulations (e.g. Battaglia et al. 2006), (…).

Even if they are less widely used than tiltmeters, extensometers are also simple and useful instruments to early detect volcano ground motion by measuring the relative displacements of the two sides of equipped pre-existing fractures (Bachèlery et al. 1998; Staudacher et al. 2003; Peltier et al. 2006).

At Piton de la Fournaise (La Réunion Island), tiltmeters and extensometers have been the first instruments of the geodetic network installed on the volcano flanks (Blum et al. 1981; Delorme 1994). This led to long time series that have been the subject of numerous studies. This chapter reviews the contribution of these two instruments in the detection of long-term and short-term eruptive precursors, used to forecast eruptions and to better constrain the dynamic of magma propagations and collapse events at Piton de la Fournaise.

17.2 Volcano Setting

With 93 eruptions observed during the last 60 years (1950–2010), Piton de la Fournaise is currently one of the most active basaltic hotspot volcanoes in the world (Roult et al. 2012; Staudacher et al. 2016, Chap. 9). Recent volcanic activity consists of fissure eruptions, located inside the summit Dolomieu crater or along one of the rift zones (N10° to the north-north east, N120° to the east, N170° to the south-south east; Fig. 17.1; Bachèlery 1981). The activity is mostly effusive or weakly explosive. Even if 97 % of the volcanic activity of the last 60 years focused inside the Enclos Fouqué caldera—which formed about 4500 years ago (Gillot and Nativel 1989; Staudacher and Allègre 1993) and inside which the current active terminal cone built—some eruptions occurred up to 10 km outside of this structure at low elevation, threatening inhabitants.

Since the establishment of a permanent volcanological observatory (Observatoire Volcanologique du Piton de la Fournaise, OVPF) in late 1979, numerous studies allowed to depict the volcano behaviour before the renew of a surface activity. Eruptions are always preceded by precursors on two time scales:

1. a long-term pattern of slight summit inflation (up to 5 cm; Peltier et al. 2008, 2009a; Schmid et al. 2012), accompanied by increasing seismicity, and lasting a few weeks/months; this phase corresponds to the pressurization of magma inside the shallow reservoir located at around 2500 m depth below the summit craters (±1000 m; Battaglia et al. 2005; Peltier et al. 2008, 2009a; Massin et al. 2011; Prôno et al. 2009).

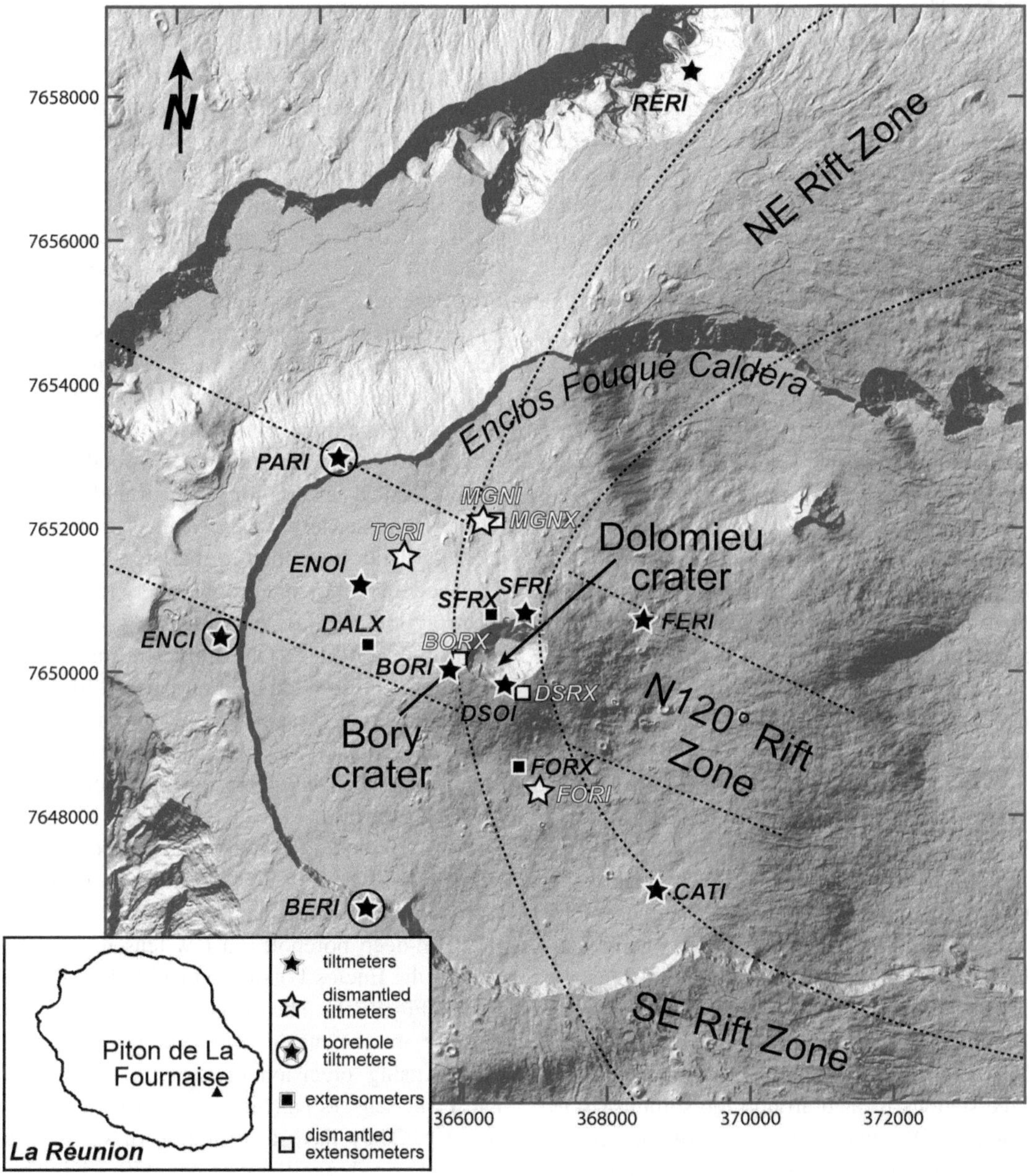

Fig. 17.1 Tiltmeter and extensometer networks of Piton de la Fournaise. Locations of surface tiltmeters (*stars*), borehole tiltmeters (*stars with circles*), and extensometers (*squares*) are shown; the dismantled stations are represented in white. Locations and names of the main structures (Dolomieu and Bory craters, rift zones, Enclos Fouqué caldera) are also reported. The *inset* shows the location of Piton de la Fournaise on La Réunion Island

2. a short-term pattern of strong ground deformation (up to few tens of cm; Peltier et al. 2008, 2009a), accompanied by intense seismicity (more than six events per min; Aki and Ferrazzini 2000; Roult et al. 2012), and lasting few min/hours; this phase corresponds to the dike propagation to the surface.

17.3 Tiltmeter and Extensometer Networks

In 2014, the OVPF's geodetic permanent network consists of twenty-four telemetered GNSS receivers, ten tiltmeter stations, and three extensometer stations [see Staudacher and Peltier (2016, Chap. 15) for a complete review concerning GNSS at Piton de la Fournaise].

The first tiltmeter station have been installed in 1980, on the outer flank of the volcano, at 8 km from the summit (RERI station, Fig. 17.1); at that time, data were not telemetered and were collected every 15 days. Until 1984, and the installation of the first two telemetered tiltmeter stations on the terminal cone, only reoccupation of a dry-tiltmeter network (twelve spirit-level tilt stations in the form of levelling crosses with 30–40 m long bases; ~5 μrad of resolution; e.g. Lénat et al. 1989a) allowed to follow the summit ground deformation. From 1984, the size of the telemetered tiltmeter network progressively increased up to reach seven instruments in 1995 and ten instruments in 2011.

Among these ten stations, seven are installed on the surface and three inside boreholes. Except the CATI station -and the old FORI and TCRI stations destroyed by lava flows- installed in natural cavities at 2–3 m depth, and the RERI station installed in a closed anthropogenic vault, the current surface stations are directly installed on the ground and are protected from weather conditions by an insulating foil and a plastic box. The seven surface tiltmeter stations consist of three stations located around the summit (BORI, SFRI, DSOI), two at the base of the terminal cone (ENOI, FERI), one at 3.5 km south-east of the terminal cone (CATI), and one at 8 km north-east of the terminal cone, outside of the Enclos Fouqué caldera (RERI; Fig. 17.1). Six of these seven surface stations (BORI, SFRI, DSOI, FERI, CATI, RERI) are equipped with BLUM-type tiltmeters (Blum 1963). These stations are currently composed of two perpendicular instruments, orientated tangentially and radially to the summit. BORI, DSOI and SFRI BLUM-type tiltmeters are doubled with pair of electrolytic SPECTRON tiltmeters (oriented NS and EW), whereas ENOI station is equipped with only a pair of electrolytic SPECTRON tiltmeters.

Sensors of the BLUM-type tiltmeters are silica horizontal pendulum sealed to the rock by a silica cone (Blum 1963; Toutain et al. 1992). With a low coefficient of thermal expansion (0.54×10^{-6} K^{-1}) and a low creep rate (-2×10^{-7} yr^{-1}), silica is well adapted to minimize thermal effects and instrumental drift. Experiments showed that the annual instrumental drift is less than 0.5 μrad (Blum 1963). BLUM-type tiltmeters measure slope tilting via the motion of a light spot on a cell, with detection limits of 0.01–0.1 μrad due to surrounding thermomechanical effects on the ground surface. These Blum-sensors were designed at IPGP but their production stopped in the late 2010s.

Since then, they are progressively replaced by electrolytic SPECTRON tiltmeters. Two types of electrolytic SPECTRON tiltmeters have been installed on Piton de la Fournaise, "SH50050 Ceramic Series" sensors at the summit (BORI, DSOI, SFRI) and "Model #554" sensors at ENOI (see www.spectronsensors.com for more details on these sensors).

In 2010 and 2011, three "new generation" tiltmeters (LILY self-levelling borehole tiltmeters from Applied Geomechanics) have been installed in 10 m-deep boreholes, a few tens meters outside of the Enclos Fouqué caldera (BERI, ENCI, PARI; Fig. 17.1). The dual-axis tiltmeters record angular movement in two orthogonal vertical planes using precision electrolytic tilt sensors. LILY tilt sensors can self-level through a range of ±10°. They have a resolution of <5 nanoradians over a dynamic range of ±330 μrad. They are watertight and mechanically stable. They are equipped with high-precision thermometer and pressure sensor (Cambell Scientific CS540) to control fluid infiltrations and/or water table variations around the instrument.

The first extensometer was installed in 1980 on a large fracture inside the Bory crater (BORX; Fig. 17.1), it was a one-component instrument monitoring the fracture opening. In 1990, two other one-component extensometers were installed close to the Dolomieu crater rims (DSRX, SFRX; Fig. 17.1). In 1995, funds have been obtained to install four other 3-component extensometer stations (two in place of SFRX and DSRX and two at the base of the terminal cone along the south and north rift zones, FORX, MGNX; Fig. 17.1). This network has been completed in 2003 by a new station, west of the terminal cone (DALX; Fig. 17.1); but in 2013 only three extensometer stations still work (DALX, FORX, SFRX). MGNX was destroyed by lava flows in August 2003. Each extensometer station is installed on fractures, of a few tens of meters long in surface, and with a surface opening of a few to a few tens of centimetres. Each 3-component station consists of three instruments measuring the extension, shear and vertical relative displacements of the two sides of the fractures. They are fixed at the surface, at 2 m depth and at about 10 m depth for FORX, SFRX and DSRX stations, respectively.

The currently working stations (DALX, FORX and SFRX) are equipped with "Telemac" instruments, REPP model. They consist of a high-resolution plastic track potentiometer with a maximum range of 25 mm (SFRX and DALX) or 50 mm (FORX) associated with a true 19 bits digitizer, giving a resolution of about 0.1 μm. All stations are protected by insulating foil and plastic box to reduce thermal effects.

Some of the old equipments were BLUM-type extensometers. They consisted of a sinusoidal current applied on a central coil, located between a couple of induction coils. The whole was fixed on one side of the fracture. The position of a ferrite rod, plunging into these coils and fixed on the opposite side of the fracture, changed with the displacement of the fracture boarder, producing small variations of electrical tension in the induction coils (Peltier et al. 2006).

During the first years of the observatory, in parallel to this permanent network, width of open fractures, located around the summit and equipped with 3 stainless nails, were also periodically measured (Lénat et al. 1989a).

Before 2012, the recording frequency of the extensometers, and the BLUM-type and SPECTRON-type tiltmeters was 1 sample/min, but since then, the recording frequency of all stations is 1 Hz. All data are transmitted in real time to the observatory. In parallel to tilt or extension measurements, each station records at the same frequency the atmospheric and/or rock temperatures, and electronic parameters (battery voltage...); temperature measurements are used to control the dependence of this parameter on the tilt daily and seasonal fluctuations whereas electronic parameters measurements allow to check in real time the status of the station.

17.4 Effects of External Parameters on the Signals

Both daily and seasonal variations are recorded on the tiltmeter and extensometer signals (Figs. 17.2 and 17.3). These are mainly linked to the sensitivity of instruments to meteorological parameters, and especially to thermal variations; daily and seasonal thermal variations act both to the electronic and to the rock dilatation.

This constitute the main limit of these instruments, that's why when it was possible tiltmeters have been installed inside natural cavities as old lava tubes (CATI, and the old FORI and TCRI stations), in closed vaults (RERI) or in boreholes (BERI, ENCI, PARI) to reduce thermal amplitude (only ~0.5 °C of seasonal thermal amplitude inside boreholes). In such environment, thermal effects on the signals are largely reduced compared to the summit stations installed directly on the ground and mainly the effect of the Earth tide is observed, which is a 0.1 μrad peak-peak signal (Fig. 17.2). Borehole installation is certainly an appropriate and effective solution to reduce surface effects and improve tilt detection sensitivity. But even for shallow borehole (10 m), we need a ground access to the site to bring the drill, compressor and equipment. That's why BERI, ENCI and PARI stations are located outside the Enclos Fouqué caldera, near paths suitable for 4WD vehicles. Nevertheless, a solution using helicopter transport has been tested at PARI site, and may allow in the future borehole installation inside the caldera.

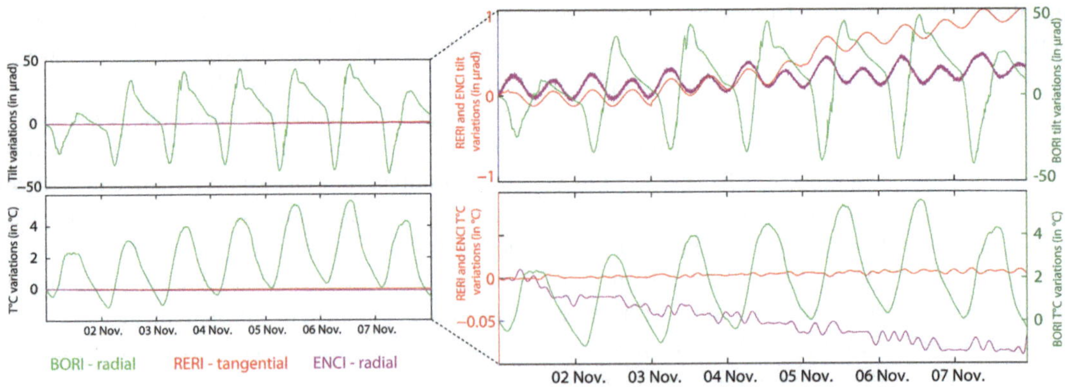

Fig. 17.2 Comparison of the tilt (*top*) and temperature (*bottom*) daily variations recorded on the BORI radial station (*in green*), the RERI tangential station (*in red*), and the ENCI radial station (*in purple*), in November 2013. To the *right*: vertical zoom with distinct scale for the BORI station and the RERI/ENCI stations

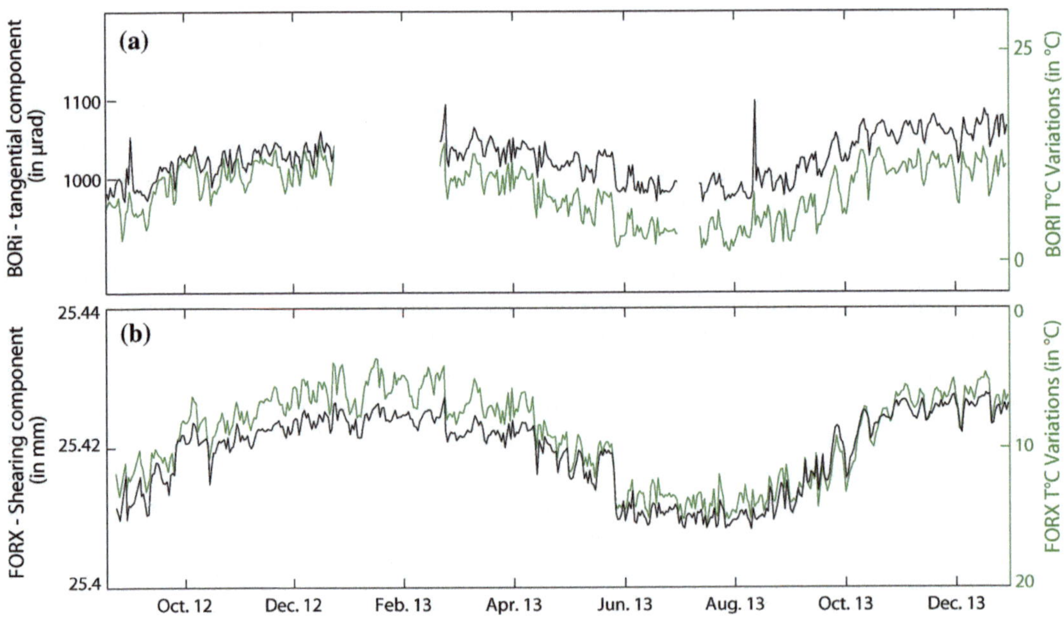

Fig. 17.3 Comparison of (**a**) the tilt (*in black*) and temperature (*in green*) seasonal variations recorded on the BORI tangential tiltmeter and comparison of (**b**) the shear motion (*in black*) and temperature (*in green*) seasonal variations recorded on the FORX extensometer in 2012–2013, a period without eruptions or intrusions

17.4.1 Seasonal Variations

During periods without any eruptive activity (as observed since end of 2010), a first order linear relationship exists between the seasonal background noise (up to 100 µrad on the summit BLUM-tiltmeter stations and up to 0.02 mm on the extensometers; Fig. 17.3) and the seasonal ground temperature variation; with a strong correlation (correlation coefficient of 0.93 on Fig. 17.3a) or anti-correlation (correlation coefficient of 0.94 on Fig. 17.3b) observed between the two datasets (Delorme 1994).

Relation between long-term tilt signals and rainfalls has also been reported on other volcanoes worldwide, but this relationship is highly nonlinear and depends on the level of water table

(Wolfe et al. 1981; Julien and Bonneton 1984). Such relationship has been evidenced on the tiltmeter signals of Piton de la Fournaise but with less influence than thermal variations.

17.4.2 Daily Variations

Tiltmeter daily variations are mainly due to thermo-mechanic effects and earth tides. Previous studies show that earth tides are visible on BLUM-tiltmeter recordings at 1/10 of μrad (Saleh 1986); this effect is thus not detectable on the surface BLUM-tiltmeters installed on Piton de la Fournaise, except on the ones installed at RERI (in a closed vault) and that are highly sensitive.

Tiltmeter variations of a sunny day, without eruptions, display the same trend as the temperature evolution, with amplitude comprised between 10 and 100 μrad for surface stations not isolated by any vault (e.g. BORI in Fig. 17.2), and less than 0.2 μrad for stations isolated by a vault (e.g. RERI in Fig. 17.2) and borehole stations (e.g. ENCI in Fig. 17.2). For the borehole tiltmeters, daily variations are strongly reduced (<0.2 μrad), sometimes less than earth tides that are visible in the time series. At 10 m depth, daily temperature variations are considerably reduced (about 0.003° only).

These daily variations are directly linked to the rock dilatation induced by the sunshine. The temperature of the dark basalt ground may rise up to 50 °C. The value of the amplitude variation on the BLUM-type tiltmeters depends on the amplitude of the temperature variation, but also on the sensitivity of the instrument, which itself depends on the period of the pendulum and on the sensitivity of acquisition system. If a delay exists between the thermal and tiltmeter waves, it is not stable with time and is most probably influenced by the environment of the station and other parameters as water infiltration... Some post-processing filtering can be applied to reduce these effects without affecting other signal (Beauducel 1992; Beauducel and Cornet 1999).

Even if these daily variations can affect the signal quality, the rapid variations linked to dike intrusions can be clearly distinguished on the signal (see Sect. 5.2; Figs. 17.4, 17.5, 17.6 and 17.7) because of the distinct time scale between the two phenomena (mins/hours for dike intrusions and a day for thermal noise).

As observed on tiltmeter signals, extensometers signals are also affected by noise due to thermo-elastic effects (up to 0.004 mm per day; Peltier et al. 2006).

17.5 Signals Associated with Volcanic and Seismic Activity

Long time series of tiltmeter and extensometer recordings at Piton de la Fournaise allow to evidence eruptive precursors on two time scales (months/weeks and hours/mins), associated with distinct processes in depth.

17.5.1 Long-Term Eruptive Precursors

At the beginning of the geodetic network, in the 80s, no significant long-term inflation of edifice has been recorded prior eruptions at Piton de la Fournaise, except in 1985–1986 (Lénat et al. 1989b). Continuous and significant long-term inflation of Piton de la Fournaise has been first evidenced after the 1998 eruption, which correspond to a renew of eruptive activity after 6 years of rest (Battaglia et al. 2005; Peltier et al. 2009a; Figs. 17.4 and 17.5).

A comparison made by Peltier et al. (2006) between the long-term trends of tiltmeter and extensometer signals for the 1997–2003 period showed that among these two methods, extensometers were the most efficient tools to detect subtle long-term precursors associated with volcano deformation. Before all post-1998 eruptions, long-term (50–130 days; Staudacher et al. 2003; Peltier et al. 2006; and even 200 days before the most recent eruptions; Schmid et al. 2012) fracture movements are recorded, with a systematic and linear opening of the fractures (Fig. 17.4) accompanied by dextral shearing on

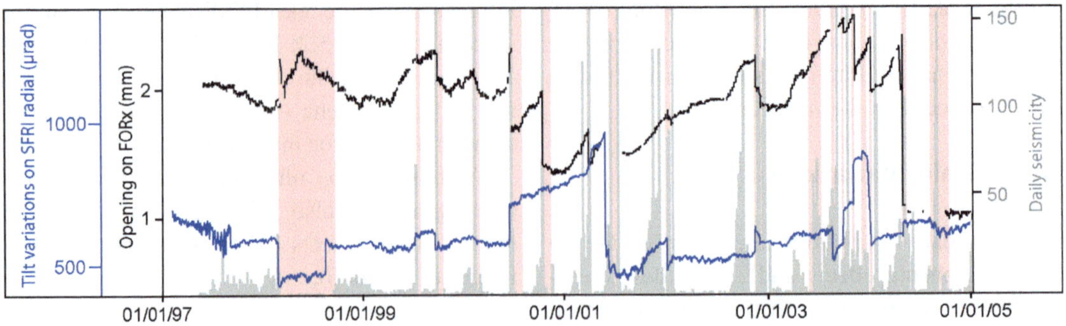

Fig. 17.4 Long-term variation recorded on the opening component of the FORX station (*in black*) and the radial component of the SFRI station (*in blue*). An increase of the signal corresponds to a fracture opening on FORX and a tilt to the north on SFRI (corresponding to a summit inflation). Daily seismicity is reported in *grey*. *Red shaded* areas represent eruptive periods

the southern station (FORX) and sinistral shearing on the northern station (MGNX), with rates comprised between 0.0025 and 0.0059 mm per day (i.e. strain variation of 1 to 5 × 10^{-4} per day). This range of strain variation is in the same order as the one deduced from GNSS data for the 2004–2010 period (Staudacher and Peltier 2016, Chap. 15). The total long-term fracture opening can reach up to 0.5 mm as observed before the August 2003 eruption (Peltier et al. 2006).

The re-opening of pre-existing fractures are attributed to the pressurization of the shallow plumbing system preceding the renew of eruptive activity. The stress changes induced by such pressurization as well as the fracture re-opening would be also at the origin of the seismic velocities decrease recorded a few days before eruptions (Brenguier et al. 2008). The shearing movement (dextral to the south and sinistral to the north) is in agreement with the gravitational instability of the eastern flank, which slides continuously down the east at a slow rate (up to 2 cm yr^{-1}; Peltier et al. 2015; Staudacher and Peltier 2016, Chap. 15). This continuous and systematic long-term inflation (Figs. 17.4 and 17.5), not observed before 1998, attested a change in the volcano behaviour after the 1998 deep refilling that accompanied the renew of the eruption activity after 6 years of rest (Battaglia et al. 2005; Peltier et al. 2009a). This long-term inflation has been interpreted as a continuous refilling of the shallow reservoir between 1998 (or 2000; when the long-term inflation signature began very clear and significant) and 2007. After 2007 and the main Dolomieu crater collapse the inter-eruptive deformation pattern was mainly dominated by relaxation of the edifice, first detected by GNSS data (Peltier et al. 2010, Staudacher and Peltier 2016, Chap. 15). This slow and slight inter-eruptive edifice relaxation was also detected by tiltmeters and extensometers (Fig. 17.5).

Note that the monitored eruptive activity preceding 1998 (i.e. 1981–1992) did not attest of such continuous refilling, and erupted magma during this period would be stored at shallow level (Lénat et al. 1989a; Peltier et al. 2009a).

Even if similar long-term trends between surface tiltmeter and extensometer data (showing summit inflation and fracture opening; Fig. 17.4) are observed before some post-1998 eruptions, long-term pre-eruptive signals are systematic and clearer on the extensometers. This behaviour can be explain both by the extensometer installations that are deeper in the ground, making them less sensitive to thermal variations, but also by the heterogeneous medium. In a high fractured and heterogeneous medium as observed on Piton de la Fournaise, the weak zones have easy fractures openings in response to edifice inflation, making the fracture monitoring a powerful technique to detect long-term eruptive precursors. For instance, the surface tiltmeter network did not record any significant slope variations before the

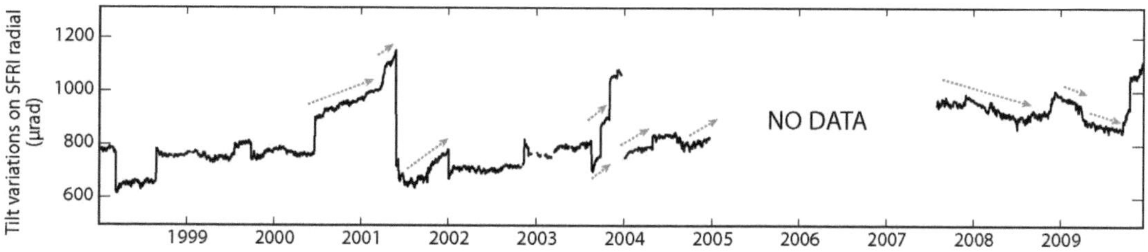

Fig. 17.5 Long-term variation recorded on the radial component of the SFRI station between 1998 and 2010. An increase of the signal corresponds to a tilt to the north (summit inflation). *Grey arrows* represent the most significant inter-eruptive trends

June 2000, June 2001, May 2003 and August 2003 eruptions, whereas the extensometer network records fracture opening up to three months before these eruptions (Peltier et al. 2006).

Extensometers demonstrate that they are useful tools to record long-term subtle forerunners. One main limit of these data is their embedding in numerical modelling to retrieve the source at the origin of the observed signals. Indeed such modelling requires a good knowledge of the orientation, the shape and in particular the depth of the equipped fracture; a deeper and longer fracture amplifies the surface movements more.

By contrast, tiltmeter data (together with GNSS data since early 2000 at Piton de la Fournaise) can be directly and easily used for this purpose. Tilt variations recorded a few weeks before the September 2003 eruption (Figs. 17.4 and 17.5) have been thus used in inversion modelling to constrain the location of the pre-eruptive source of pressure in depth (Peltier et al. 2007). This model locates an elliptical source at about 300 above sea level (i.e. 2200 m depth) below the Dolomieu crater (3 MPa of overpressure; Fig. 17.8). The depth of the modelled source preceding this eruption corresponds to the most superficial low-velocity anomaly, spreading from 0 to 1 km a.s.l., and that could be interpreted as a shallow magma storage zone.

17.5.2 Short-Term Eruptive Precursors

The largest ground deformation recorded in the summit area occurred shortly (mins/hours) before the eruption onset, during the dike propagation toward the surface (Fig. 17.4). The first published results of short-term ground deformation associated with dike propagation date back to the first eruption monitored by the observatory, on February 3, 1981. At that time only a levelling network of eight stations was set up and its re-measurement showed displacements of up to 18 ± 2 cm associated with the eruption start (Bachèlery et al. 1982). In 1983–1984, periodic re-measurements of dry tilts stations allow to better constrain short-term ground deformation. Just after the December 4, 1983 outbreak, tilt measurements showed that large deformation occurred all around the summit (up to 210 µrad) whereas nothing of significance was recorded at the base of the terminal cone, revealing that the inflation centre was located between the summit and the eruptive vent (Lénat et al. 1989a). For this eruption a permanent extensometer allow to follow the dynamic of the deformation, and showed that the summit deformed rapidly during the intrusive phase preceding of about 2 h the eruption onset. The same pattern has been detected during the 1985–1986 eruptive sequence by the first telemetered tilt stations but the few numbers of stations did not allow any trilateration to follow in real-time the inflation centre.

The first very detailed studies of the follow up in real-time of the inflation centre during a dike propagation date back to the April 18, 1990 eruption (Toutain et al. 1992), when the number of continuous and telemetered stations was enough for such calculation. Continuous tiltmeter data associated with the 1990 eruption (Toutain et al. 1992) but also associated with the post-1998 eruptions (Peltier et al. 2005; Peltier

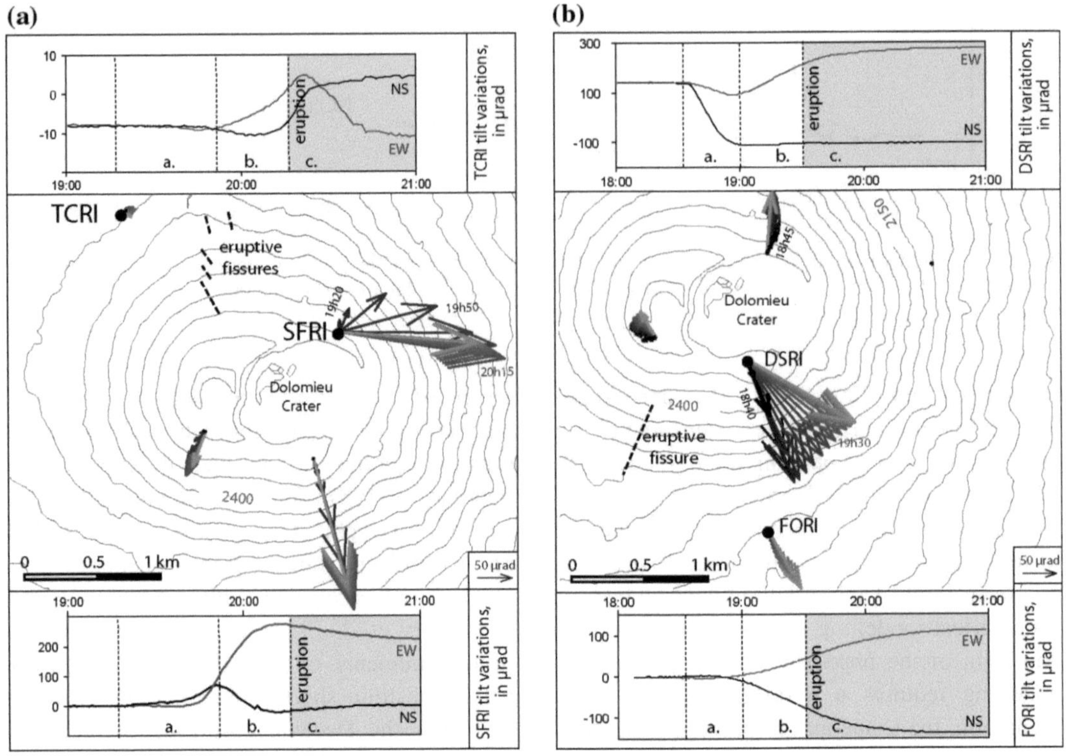

Fig. 17.6 Tiltmeter variations recorded during the dike injections preceding (**a**) the February 13, 2000 eruption and (**b**) the September 30, 2003 eruption. NS and EW components are in *black* and *grey*, respectively. Letters *a*, *b* and *c* represent the first vertical dike migration, the lateral dike migration and the eruption, respectively

et al. 2009a) show a similar intrusive scheme with a "two-stage" dike propagation (Figs. 17.6 and 17.7):

1. In the first stage, lasting 10–50 min and corresponding to the beginning of the seismic crisis (Roult et al. 2012), the intensity of the tilt increases, up to reach 1000 µrad at the summit and a few tens of µrad at the base of the cone. The orientations of tilt vectors are constant during this stage and define an inflation centre below the Dolomieu crater, revealing a vertical migration of the magma from the shallow reservoir toward the summit (Fig. 17.6). This strong summit inflation stage is accompanied by an opening of the fractures equipped by extensometers (Peltier et al. 2006; Fig. 17.7).
2. Flank eruptions are characterized by a second stage, lasting 10 min to several hours, tilt vectors rotate indicating the lateral magma propagation toward one flank of the volcano. During this stage the inflation centre migrates in the direction of the future eruptive vents (Fig. 17.6), while a general summit deflation is recorded, which usually continues during the course of the eruption. This summit deflation is generated by pressure release in the plumbing system at the apex of the summit when the dike propagates to the flank. Trends of the signals recorded by extensometer (closing/opening, dextral/sinistral shearing…) during this second stage depend on the dike location relative to the stations (Fig. 17.7; Peltier et al. 2006). But at the end of eruptions, the remaining signal on the opening component of the FORX extensometer is most often a negative signal (fracture closure; Fig. 17.4) due to the cone deflation during the course of the eruption

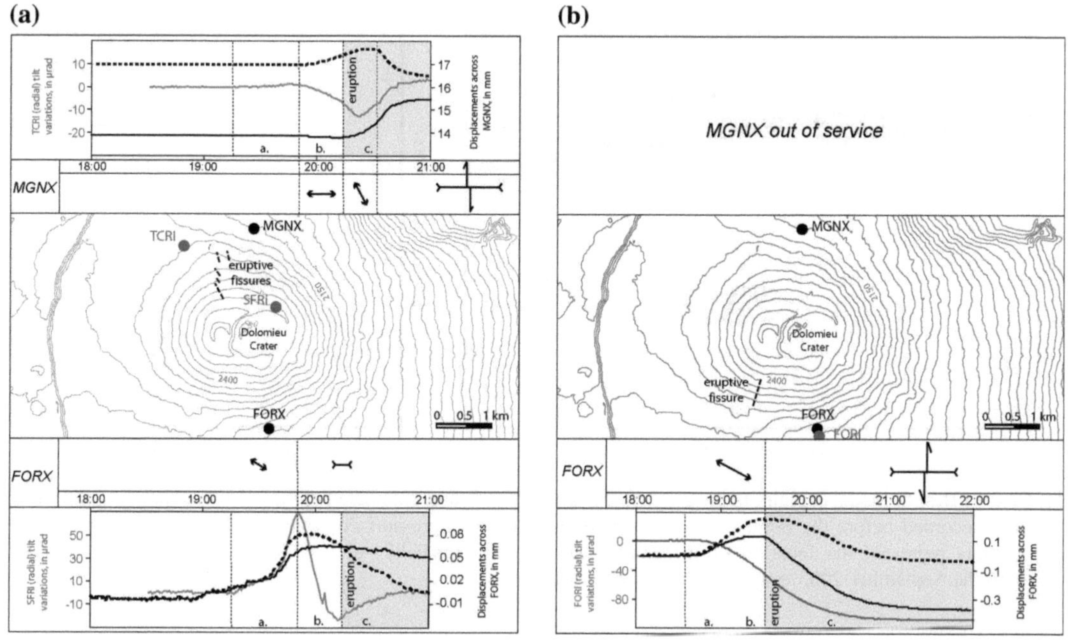

Fig. 17.7 Fracture motions recorded at FORX and MGNX during the dike injections preceding (**a**) the February 13, 2000 eruption and (**b**) the September 30, 2003 eruption. The *black dotted lines* and the *black lines* represent the opening and the shearing components, respectively; whereas the *grey lines* are comparative curves, with the signals recorded on tiltmeters. An increase of the extensometer signal corresponds to an opening (opening component) or a dextral shearing (shearing component). Letters *a*, *b* and *c* represent the first vertical dike migration, the lateral dike migration and the eruption, respectively. The scale of the motion vectors is distinct for the two fractures and is given by the curve amplitude

that tends to close pre-existing fractures. For tiltmeters, the remaining signal depends on the location and the orientation of the eruptive fissure location relative to the station.

A posteriori inversion models of the tiltmeter data associated with the 1998 eruption and the 2003 eruptions allow to constrain the shape of the associated dikes and reveal that the second step begins when dike is at ~1500–2000 m depth (Battaglia and Bachèlery 2003; Peltier et al. 2007); a depth which corresponds to a re-orientation of σ1 (from vertical to horizontal; Letourneur et al. 2008). Fukushima et al. (2010) showed that the change of dike direction from vertical to lateral is a consequence of a density balance between the host rocks and the intruded magma.

Thanks to the study of the inflation centre migration, velocities of dike migration could be estimated. A velocity decrease has been observed as soon as dike propagates away from the terminal cone; from 2.5 m sec^{-1} during the vertical migration to 0.2–0.8 m sec^{-1} during the lateral migration (Toutain et al. 1992; Peltier et al. 2005). The estimated dike velocities at Piton de la Fournaise are in the same order than the ones observed in North Iceland volcanoes (Gudmundsson 1995), but higher of one order of magnitude than observed for the Pu'u 'O'o eruption in Kilauea (Okamura et al. 1988) or the Teishi Knoll eruption (Ueki et al. 1993). The higher porosity and the higher number of fractures could explain the fast dike velocities observed at Piton de la Fournaise; pre-existing fractures being preferential and easy pathways for the magma to reach the surface (Toutain et al. 1992; Peltier et al. 2005).

Aki and Ferrazzini (2000) and Peltier et al. (2005) showed that the magma injection duration

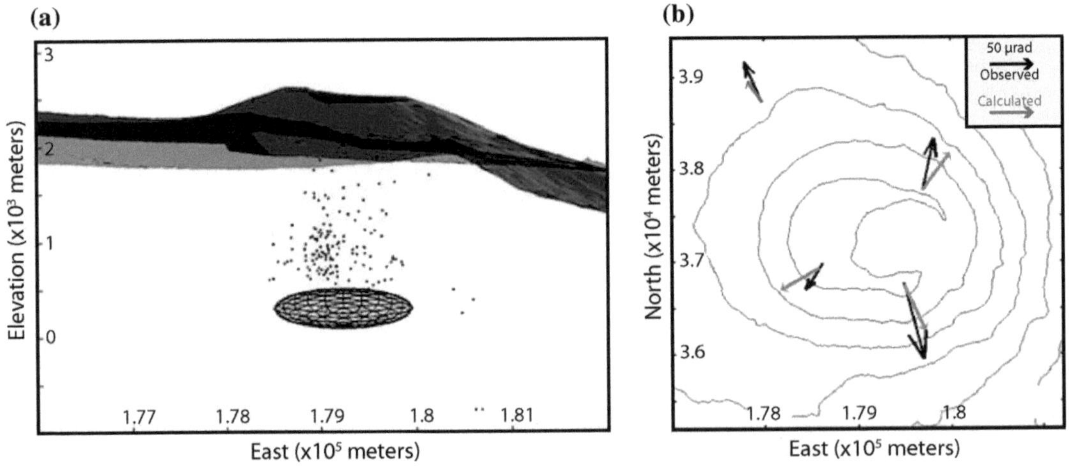

Fig. 17.8 **a** Best fit model explaining the pre-eruptive tilt variations recorded before the September 2003 eruption. Black points represent the location of the earthquakes preceding the September 2003 eruptions. **b** Observed (*in black*) and calculated (*in grey*) tilt variations induced by the over-pressurized source (3 MPa) shown in the inset (**a**). After Peltier et al. (2007)

roughly correlates with the distance of the eruptive fissures from the summit. Tiltmeters are thus very useful to easily follow in real-time the inflation centres and quickly locate which flank of the volcano is intruded by the magma and on which distance, this is crucial for risk mitigation in real-time.

17.5.3 Eruptive and Post-eruptive Summit Relaxation

The most common edifice behaviour observed since the implementation of the geodetic network at Piton de la Fournaise is the eruptive and post-eruptive summit relaxation. The first detected post-eruptive summit relaxation was observed after the first monitored eruption in February 1981. Between June 1982 and June 1983, dry-tiltmeter re-measurements showed on BORI and SFRI stations an apparent tilt of 29 and 27 (±5) µrad, respectively, in one year to the south, reflecting the summit relaxation (Lénat et al. 1989a). These movements that occurred at a steady rate stopped about 4.5 months before the next eruption, on December 4, 1983 eruption.

Since this first observation, such post-eruptive relaxation was systematically recorded by the OVPF's geodetic network (tiltmeters, extensometers, GNSS) after each eruption. Usually deflation continued during a few months, except for the 2000–2007 period, during which eruptions succeeded closely in time, and during which a significant post-eruptive relaxation was only observed after major distal eruptions (Peltier et al. 2009a; Figs. 17.4 and 17.5).

The observed post-eruptive relaxation is induced by mechanical and thermal readjustment of the summit zone and the area intruded by dike injection and is accompanied by a closure of the instrumented fractures (up to a few of 1/10 mm; Fig. 17.4).

17.5.4 Signals Associated with Pit Crater and Collapse Episodes

In April 2007, Piton de la Fournaise experienced a major collapse (Michon et al. 2007; Staudacher et al. 2009). The whole Dolomieu crater collapsed on a depth of about 340 m during successive collapse cycles spanning on three days (5-6-7 April 2007; Staudacher et al. 2009; Peltier et al. 2011; Roult et al. 2012). Continuous tiltmeters recorded specific signals associated with these cycles. At the summit, the BORI tiltmeter recorded cyclic signals lasting from 2 h at the beginning to 30 min at the end (Staudacher et al.

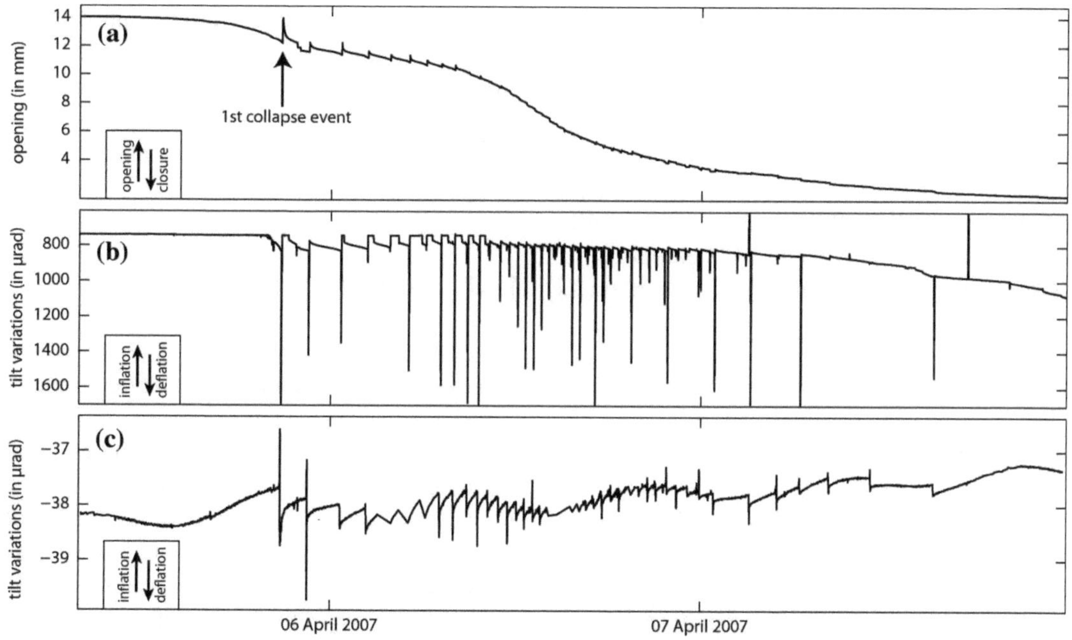

Fig. 17.9 Signals recorded during the April 2007 Dolomieu crater collapse on (**a**) the opening component of the SFRX extensometer station, (**b**) the radial component of the BORI tiltmeter station, and (**c**) the tangential component of the RERI tiltmeter station. Note that the signals of the BORI radial tiltmeter were reversed due to its reversed installation on the field

2009). According to Michon et al. (2007), Peltier et al. (2009b) and Staudacher et al. (2009), each cycle began with a progressive inward deflation of the summit and ended by sudden outward displacements (up to 150 μrad; Fig. 17.9). The rapid outward displacements were directly associated with the collapse events, which also induced up to 1.8 mm of quick openings of the SFRX fracture. Fracture openings were followed by slow closures with the same cyclicity as observed on tiltmeters (Fig. 17.9).

At the base of the cone and on the RERI distant station, the signals were reversed. At RERI, each collapse was preceded by slow tilt down to the east followed by a nearly instantaneous tilt down to the west (0.2–2 μrad; Peltier et al. 2011; Fig. 17.9). Records of the seismic station located in the same vault as the RERI station display the same cyclicity. According to Fontaine et al. (2013), each successive collapse event started with a short-period event induced by rock failure, followed by a VLP signal induced by dip-slip motion on the caldera ring fault, and ended with a ULP signal likely related to a relaxation process of the edifice. This scenario agrees with the summit tiltmeter and extensometer records. The reversed signals observed on the summit and basal tiltmeter stations support the hypothesis of a narrow and deep column of rock collapsing into the magma reservoir. The subsidence of the rock column into the magma reservoir generated inward displacements at the summit (beginning of a cycle) and an increase of pressure in the reservoir, influencing the distal RERI station with outward displacements (Peltier et al. 2011).

Since the first tiltmeter network installation in 1980, two smaller collapse events occurred in 1986 (150 m in diameter/80 m deep) and 2002 (200 min diameter/25 m deep) with pit-crater formation in the south-eastern and south-western part of the Dolomieu crater, respectively (Delorme et al. 1989; Longpré et al. 2006). In both cases, associated tiltmeter and extensometer signals recorded a few days of progressive summit deflation and a sharp summit deflation and fracture closure associated with the collapse event itself. Contrary to the 2007 Dolomieu crater collapse, only one sharp signal had been recorded revealing that these two events formed in only one main step. These two pit craters most probably collapsed in response to the rapid draining of small shallow magma storage zones; inferred at <0.4 km depth by Delorme et al. (1989) for the 1986 event using a point source model to fit the associated tiltmeter data.

17.6 Perspectives

Fluid transports from depth to surface induced ground deformation. During magma shallow intrusions or eruptions, observed surface deformation in the surrounding area (volcano central cone) is of the order of few tens to few hundred µrad of amplitude. This signal is largely greater than tilt stations background noise.

Before 2010, the tilt network configuration was mainly focused inside the Enclos Fouqué caldera, and allowed to monitor accurately shallow and short-term magma propagation, despite its low sensitivity level. Detection of deeper pressure variation requires a higher sensitivity and extended wider network. Moreover, early detection of the refilling of the shallow feeding system (<2.5 km) through deep magma ascent (−4 to −20 km below sea level) and tracing of the magma spread towards distant areas (outside the Enclos Fouqué caldera) are priority issues for understanding the behaviour of Piton de la Fournaise with major implications for risk mitigation. Furthermore, there were 3 recent eruptions ≪out the Enclos≫ in 1977, 1986 and 1998, and possible direct ascents of magma in areas outside the caldera, are potentially the most dangerous hazard for inhabitants. To address this particular objective, new generation of tiltmeter stations using shallow borehole, high-precision instruments, have been installed in 2010–2011 outside the Enclos Fouqué caldera. The network configuration should allow us in the future to better characterize and model the spatial wavelengths of the deformation field associated with the distinct eruptive stages of the volcano (pre-eruptive unrests, summit dike propagations, distal dike propagations…), but also would contribute to better forecast large distant eruptions early enough to give the inhabitants warning.

17.7 Conclusions

Since their implementation in the early 80's, tiltmeters and extensometers largely contributed to the knowledge and the forecast of eruptive activity at Piton de la Fournaise. These simple instruments, easy to use, allow to evidence eruptive precursors at Piton de la Fournaise on two time scales (months/weeks and hours/mins), associated with reservoir pressurization and dike propagations, respectively. At Piton de la Fournaise, tiltmeters appear to be more suitable to follow, study and model short-term volcano deformation linked to the dike propagation whereas extensometers, less sensitive to external factors thanks to their deeper installation, are more suitable to follow and study long-term volcano deformation.

Tiltmeters and extensometers are thus largely complementary, but also complementary to the GNSS network that develops and grows significantly since the early 2000's (Staudacher and Peltier 2016, Chap. 15).

Acknowledgements We are grateful to Alessandro Bonaccorso for the review of this chapter. This is IPGP contribution 3664.

References

Aki K, Ferrazzini V (2000) Seismic monitoring and modelling of an active volcano for prediction. J Geophys Res 105:16617–16640

Bachèlery P (1981) Le Piton de La Fournaise: Etude Volcanologique, Structurale et Pétrologique. PhD thesis, University of Clermont Ferrand, France: 1–257

Bachèlery P, Blum PA, Cheminée JL, Chevallier L, Gaulon R, Girardin N, Jaupart C, Lalanne F, Le Mouel JL, Ruegg JC, Vincent P (1982) Eruption at Le Piton de la Fournaise volcano on 3 February 1981. Nature 297(5865):395–397

Bachèlery P, Kowalski P, Catherine P, Delmond J C, Blum PA, Croce J (1998) Precise temporal and mechanical identification of dike emplacement using deformation monitoring at Piton de la Fournaise. In: Casale R et al (eds)The European laboratory volcanoes, proceedings of the 2nd workshop, Santorini, European Commission, Brussels. pp 475–485

Battaglia J, Bachèlery P (2003) Dynamic dyke propagation deduced from tilt variations preceding the March 9, 1998, eruption of the Piton de la Fournaise volcano. J Volcanol Geotherm Res 120:289–310

Battaglia J, Ferrazzini V, Staudacher Th, Aki K, Cheminée JL (2005) Pre-eruptive migration of earthquakes at Piton de la Fournaise volcano (Réunion Island). Geophys J Int 161:549–558

Battaglia M, Troise C, Obrizzo F, Pingue F, De Natale G (2006) Evidence for fluid migration as source of deformation at Campi Flegrei caldera (Italy). Geophys Res Lett 33:L01307. doi:10.1029/2005GL024904

Beauducel F (1992) Modélisation d'une crise intrusive au Piton de la Fournaise (Réunion). Atténuation des perturbations thermomécaniques sur les mesures de deformations. Rapports de DEA, Institut de physique du globe de Paris 1:47–81

Beauducel F, Cornet F (1999) Collection and three-dimensional modeling of GPS and tilt data at Merapi volcano, Java. J Geophys Res 104(B1):725–736

Blum PA (1963) Contribution à l'étude des variations de la verticale en un lieu. Ann Géophys 19:215–243

Blum PA, Gaulon R, Lalanne FX, Ruegg JC (1981) Sur l'évidence des précurseurs de l'éruption du volcan du Piton de la Fournaise à La Réunion. C Acad SCI Paris 292D:1449–1455

Bonaccorso A (2006) Explosive activity at Mt Etna summit craters and source modelling by using high-precision continuous tilt. J Volcanol Geotherm Res 158:221–234

Bonaccorso A, Gambino S (1997) Impulsive tilt variations on Mount Etna Volcano (1990–1993). Tectonophysics 270:115–125

Bonaccorso A, Cannata A, Corsaro RA, Di Grazia G, Gambino S, Greco F, Miraglia L, Pistorio A (2011) Multidisciplinary investigation on a lava fountain preceding a flank eruption: the 10 May 2008 Etna case. Geochem Geophys Geosyst 12(7):Q07009. doi:10.1029/2010GC003480

Brenguier F, Shapiro NM, Campillo M, Ferrazzini V, Duputel Z, Coutant O, Nercessian A (2008) Towards forecasting volcanic eruptions using seismic noise. Nat Geosci 1(2):126–130

Delorme H (1994) Apport des déformations à la compréhension des mécanismes éruptifs : le Piton de La Fournaise. Thèse d'état. Université de Paris VII, France: 1–613

Delorme H, Bachèlery P, Blum PA, Cheminée JL, Delarue JF, Delmond JC, Hirn A, Lépine JC, Vincent PM, Zlotnicki J (1989) March 1986 eruptive episodes at Piton de La Fournaise volcano (Réunion Island). J Volcanol Geotherm Res 36:199–208

Fontaine F, Roult G, Michon L, Barruol G, Ferrazzini V, Di Muro A, Reymond D, Peltier A, Staudacher T (2013) Understanding the dynamics of the February-April 2007 eruptions of the Piton de la Fournaise and the related caldera collapse from a single very broad-band seismic station. Abstract V41B-2784, Fall Meeting, AGU, San Francisco

Fukushima Y, Cayol V, Durand P, Massonnet D (2010) Evolution of magma conduits during the 1998–2000 eruptions of Piton de la Fournaise volcano, Réunion Island. J Geophys Res 115(B10204) doi:10.1029/2009JB007023

Gillot PY, Nativel P (1989) Eruption history of the Piton de la Fournaise volcano, Reunion Island, Indian Ocean. J Volcanol Geotherm Res 36:53–55

Gudmundsson A (1995) The geometry and growth of dykes. In: Baer G, Heiman A (eds) Physics and chemistry of dykes, pp 23–34

Imamura A (1930) Topographical changes accompanying earthquakes or volcanic eruptions. Pub Earthquake Investigation 25

Jaggar TA, Finch RH (1929) Tilt records for thirteen years at the Hawaiian Volcano Observatory. B Seismol Soc Am 19(1):38–51

Julien P, Bonneton JR (1984) Etude structurale et mesures et mesures de déformations à la Soufrière de Guadeloupe. Bull PIRPSEV 91

Kamo K, Ishihara K (1989) A preliminary experiment on automated judgement of the stages of eruptive activity using tiltmeter records at Sakurajima, Japan. IAVCEI Proc Volcanol 1:585–598

Lénat JF, Bachèlery P, Bonneville A, Tarits P, Cheminée JL, Delorme H (1989a) The December 4, 1983 to February 18, 1984 eruption of Piton de La Fournaise (La Réunion, Indian Ocean): description and interpretation. J Volcanol Geotherm Res 36:87–112

Lénat JF, Bachèlery P, Bonneville A, Hirn A (1989b) The beginning of the 1985–1987 eruptive cycle at Piton de la Fournaise (La Réunion); new insights in the magmatic and volcano-tectonic systems. J Volcanol Geotherm Res 36:209–232

Letourneur L, Peltier A, Staudacher T, Gudmundsson A (2008) The effects of rock heterogeneities on dyke paths and asymmetric ground deformation: the example of Piton de la Fournaise (Réunion Island). J Volcanol Geotherm Res 173:289–302

Longpré MA, Staudacher T, John Stix (2006) The November 2002 eruption at Piton de la Fournaise volcan, La Réunion Island: groud deformation, seismicity, and pit crater collapse. Bull Volcanol. doi:10.1007/s00445-006-0087-0

Massin F, Ferrazzini V, Bachèlery P, Nercessian A, Duputel Z, Staudacher T (2011) Structures and evolution of the plumbing system of Piton de la Fournaise volcano inferred from clustering of 2007 eruptive cycle seismicity. J Volcanol Geotherm Res 202:96–106

Michon L, Staudacher T, Ferrazzini V, Bachèlery P, Marti J (2007) April 2007 collapse of Piton de La Fournaise: a new example of caldera formation.

Geophys Res Lett 34 (L21301). doi:10.1029/2007GL031248

Montgomery-Brown EK, Sinnett DK, Larson KM, Poland MP, Segall P, Miklius A (2011) Spatiotemporal evolution of dike opening and décollement slip at Kilauea Volcano, Hawai'i. J Geophys Res 116 (B03401). doi:10.1029/2010JB007762

Murray JB, Wooller LK (2002) Persistent summit subsidence at Volcán de Colima, Mexico, 1982–1999: strong evidence against Mogi deflation. J Volcanol Geotherm Res 117(1–2):69–78

Okamura TA, Dvorak JJ, Koyanagi RY, Tanigawa WR (1988) Surface deformation during dike propagation, in Pu'u 'O'o Eruption of Kilauea Volcano, Hawaii: Episodes 1 Through 20, January 3 1983, Through June 8 1984. Edited by E. W. Wolfe. US Geol Surv Prof Pap 1463:165–181

Peltier A, Ferrazzini V, Staudacher T, Bachèlery P (2005) Imaging the dynamics of dyke propagation prior to the 2000–2003 flank eruptions at Piton de La Fournaise. Reunion Island. Geophys Res Lett 32(22):L22302. doi:10.1029/2005GL023720

Peltier A, Staudacher T, Catherine P, Ricard LP, Kowalski P, Bachèlery P (2006) Subtle precursors of volcanic eruptions at Piton de la Fournaise detected byextensometers. Geophys Res Lett 33:L06315. doi:10.1029/2005GL025495

Peltier A, Staudacher T, Bachelery P (2007) Constraints on magma transfer and structures involved in 2003 activity at Piton de la Fournaise from displacement data. J Geophys Res 112(B03207). doi:10.1029/2006JB004379

Peltier A, Famin V, Bachèlery P, Cayol V, Fukushima Y, Staudacher T (2008) Cyclic magma storages and transfers at Piton de La Fournaise volcano (La Réunion hotspot) inferred from deformation and geochemical data. Earth Planet Sci Lett 270:180–188

Peltier A, Bachèlery P, Staudacher (2009a) Magma transfer and storage at Piton de La Fournaise (La Réunion Island) between 1972 and 2007: a review of geophysical and geochemical data. J Volcanol Geotherm Res 184(1–2):93–108

Peltier A, Staudacher T, Bachèlery P, Cayol V (2009b) Formation of the April 2007 caldera collapse at Piton de La Fournaise volcano: Insights from GPS data. J Volcanol Geotherm Res 184:152–163

Peltier A, Staudacher T, Bachèlery P (2010) New behavior of the Piton de La Fournaise volcano feeding system (La Réunion Island) deduced from GPS data: Influence of the 2007 Dolomieu caldera collapse. J Volcanol Geotherm Res 192:48–56

Peltier A, Staudacher T, Bachèlery P (2011) Early detection of large eruptions at Piton de La Fournaise volcano (La Réunion Island): contribution of a distant tiltmeter station. J Volcanol Geotherm Res 199:96–104

Peltier A, Poland M, Staudacher T (2015) Are Piton de la Fournaise (La Réunion) and Kīlauea (Hawai'i) really "analog volcanoes"?. In: Carey R, Cayol V, Poland M, Weis D (eds) Hawaiian volcanoes: from source to surface. Wiley, Hoboken, NJ. doi:10.1002/9781118872079.ch23

Prôno E, Battaglia J, Monteiller V, Got JL, Ferrazzini V (2009) P-wave velocity structure of Piton de la Fournaise volcano deduced from seismic data recorded between 1996 and 1999. J Volcanol Geotherm Res 184:49–62

Roult G, Peltier A, Taisne B, Staudacher T, Ferrazzini V, Di Muro A, and the OVPF group (2012) A new comprehensive classification of the Piton de la Fournaise eruptions spanning the 1986–2011 period, Search and analysis of eruption precursors from a broad-band seismological station. J Volcanol Geotherm Res 241–242:78–104

Saleh B (1986) Développement d'une nouvelle instrumentation pour les mesures dedéformations - Applications au génie civil. PhD Thesis, University P. & M. Curie Paris VI, 214 pp

Saleh B (2003) Study of Earth tides using quartz tiltmeter. J Surv Eng 129(2):51–55

Schmid A, Grasso J R, Clarke D, Ferrazzini V, Bachèlery P, Staudacher T (2012) Eruption forerunners from multiparameter monitoring and application for eruptions time predictability (Piton de la Fournaise). J Geophys Res 117(B11203). doi:10.1029/2012JB009167

Staudacher T, Allègre CJ (1993) Ages of the second caldera of Piton de la Fournaise volcano (Réunion) determined by cosmic ray produced ^{3}He and ^{21}Ne. Earth Planet Sci Lett 119:395–404

Staudacher T, Peltier A (2016) Ground deformation at Piton de la Fournaise, a review from 20 years of GNSS monitoring. In: Bachèlery P, Lénat J-F, Di Muro A, Michon L (eds) Active volcanoes of the Southwest Indian Ocean: Piton de la Fournaise and Karthala. Active volcanoes of the World. Springer, Berlin

Staudacher T, Bachelery P, Ricard LP, Catherine P, Kowalski P, Cheminée JL (2003) Signaux précurseurs des éruptions volcaniques : nouvelles approches de détection : variations extensométriques comme précurseurs à long terme. Rapport quadriennal 1999 – 2002 XXIIIe Assemblée Générale de l'Union Géodésique et Géophysique Internationale Sapporo (Japon). pp 25–127

Staudacher T, Ferrazzini V, Peltier A, Kowalski P, Boissier P, Catherine P, Lauret F, Massin F (2009) The April 2007 eruption and the Dolomieu crater collapse, two major events at Piton de la Fournaise. J Volcanol Geotherm Res 184(1–2):126–137

Staudacher T, Peltier A, Ferrazzini V, Di Muro A, Boissier P, Catherine P, Kowalski P, Lauret F (2016) Fifteen years of intense eruptive activity (1998–2013) at Piton de La Fournaise volcano: a review. In: Bachèlery P, Lénat J-F, Di Muro A, Michon L (eds) Active Volcanoes of the Southwest Indian Ocean: Piton de la Fournaise and Karthala. Active volcanoes of the World. Springer, Berlin

Toutain JP, Bachelery P, Blum PA, Cheminee JL, Delorme H, Fontaine L, Kowalski P, Taochy P (1992) Real time monitoring of vertical ground

deformations during eruptions at Piton de la Fournaise. Geophys Res Lett 19(6):553–556
Ueki S, Hamaguchi H, Morita Y, Mishimura T, Kasahara K (1993) Magma ascent process of the 1989 submarine eruption of Ito. Int Assoc of Volcanol and Chem of the Earth's Inter. Canberra
Voight B, Hoblitt RP, Clarke AB, Lockhart AB, Miller AD, Lynch L, Mc Mahon J (1998) Remarkable cyclic ground deformation monitored in real-time on Montserrat, and its use in eruption forecasting. Geophys Res Lett 25(18):3405–3408
Wolfe JE, Berg D, Sutton GH (1981) The change in strain comes mainly from the rain: Kipapa, Oahu. Bull Seism Soc Am 71:1625–1635

A Geographical Information System for Mapping Eruption Risk at Piton de la Fournaise

18

Paule-Annick Davoine and Cécile Saint-Marc

Abstract

Piton de la Fournaise, situated on La Réunion Island, is one of the most active volcanoes in the world. It has been the subject of continuous surveillance since the start of the 1980s, in particular on account of the potential economic stakes. Spatial knowledge concerning the eruptive risk for the management of volcanic hazard is a major challenge. It requires knowledge of past events, and implies the need to relate different eruptive phenomena, such as lava flows, cones and fissures, and also earth tremors providing warning signs for eruptions. Geographical Information Systems (GIS) are widely recognised for the management of spatial data and natural hazard mapping. This chapter thus sets out to present the different stages in the design of the GIS developed in the *Observatoire Volcanologique du Piton de la Fournaise*, and its interest in the spatial analysis of volcanic hazard.

18.1 Introduction

Piton de la Fournaise, located on La Réunion Island in Indian Ocean, is one of the most active volcanoes in the world. It has been under constant surveillance since 1980, in particular on account of the potentially considerable economic stakes. Most of the eruptions occur within the *Enclos Fouqué*, an uninhabited zone, which is nevertheless widely frequented by tourists. Certain major events, however, have occurred outside the *Enclos Fouqué* zone, damaging inhabited areas, plantations and communication networks. The study of the volcanic hazard requires the efficient management of large volumes of geographical data relating to past eruptions. Indeed, the reconstruction of the history of the volcano, and the analysis of the spatial distribution of the eruptive phenomena and their evolution over time, should make it possible to identify the eruptive phenomena that are liable to recur. Phenomena relating to eruptions, such as lava flows, fissures, tephras and fractures are all hazards that need to be spatialised and analysed individually, before

P.-A. Davoine (✉) · C. Saint-Marc
Université de Grenoble-Alpes,
Laboratoire d'Informatique de Grenoble, BP 72 681,
Rue de la Passerelle,
38400 Saint-Martin-d'Hères, France
e-mail: paule-annick.davoine@imag.fr

P. Bachèlery et al. (eds.), *Active Volcanoes of the Southwest Indian Ocean*,
Active Volcanoes of the World, DOI 10.1007/978-3-642-31395-0_18

being studied conjointly so as to assess exposed zones and rank their levels of risk. They also need to be related to other phenomena constituting signs preceding eruptions, such as seismic activity, thus enabling better understanding of the way in which the volcano behaves, and the development of eruptive scenarios.

Geographical Information Systems (GIS) are widely recognised for their ability to store, process and represent spatial information relating to environmental data (Burrough 1989). They are also used for the management of volcanic hazard, and they facilitate the integration, exploration, visualisation and collation of multi-source data (Aronoff 1989; Pareschi et al. 2000). The processing of spatial data afforded by these tools then facilitates the development of maps of the exposed areas, or hazard maps.

The study of the volcanic hazard on Reunion Island required the design and development of a GIS. This tool, developed under ArcGIS software, has enabled the integration and the linking of the geographical information relevant to the knowledge of eruptive phenomena from the *Piton de la Fournaise*. However certain challenges emerged for entering data into the GIS, and later for the mapping procedure. Indeed, the information relating to past eruptions is markedly heterogeneous and scattered, on account of the diversity of the sources and data collection ways. Continuous surveillance of the activity of the *Piton de la Fournaise* has only been instated recently, dating from the establishment of the *Observatoire Volcanologique du Piton de la Fournaise* (OVPF). It is only from the 1980s that OVPF implemented systematic collection of surface data relating to eruptions: lava flows, cones, eruptive or non eruptive faults, tephras…

The aim of this chapter is to present the Geographical Information System developed in the OVPF on Reunion Island. After recalling value of GIS tools for the management of the volcanic risk, based on related works, we will describe the methodological choices made for the design of the GIS. We will then present the main cartographic productions required for the spatial analysis of the volcanic hazard.

18.2 State-of-the-Art: The Use of GIS Technology in Volcanology

18.2.1 The Interest of GIS Technology for Volcanic Hazard Assessment

It has been demonstrated that GIS technology plays a major role in the definition of volcanic hazard, in risk assessment and in the decision-making process (Pareschi et al. 2000). It is useful at every stage of volcanological observation (Esri 2009):

- Before an eruption: it enables the spatial probability of the occurrence of a volcanic event to be defined via the creation of a hazard map (for instance, the map of seismic phenomena preceding an eruption can be used by experts in tectonics to understand the location of upcoming faults). This map can be combined with the vulnerable populated areas to produce a risk map, and risk-mitigation actions can be initiated. In this way, hazard maps are a reference for policy makers to assist in land-use planning and to provide a basis for emergency plans (Pareschi et al. 2000).
- During the eruption: in association with real-time sensors, it helps to follow the evolution of volcanic indicators in an impending crisis and supports the decision-making process for emergency-response [e.g. evacuation around Mount Merapi in 2006—Esri (2009)]
- After an eruption: GIS technologies are useful to model the mechanism of eruptions, to study eruptive activity (e.g. Bellotti et al. 2010) and also to plan territorial evolutions and outcomes after eruptive activity.

GIS is a powerful tool for risk-assessment:

- It enables numerous parameters to be crossed: several raw indicators (faults, fumaroles, earthquake locations, etc.) or aggregated indicators (vent opening probability, results of simulation models, etc.) concerning a single volcanic source (e.g. Alberico et al. 2007), or resultant hazard maps concerning several volcanic sources that may impact the same area (e.g. Alberico et al. 2011).

- It enables large amounts of data to be processed: processing field data up to the last 10 ka (Lirer et al. 2010). The longer is the period of reliable data available, the more reliable is the assessment of the hazard (Pareschi et al. 2000). As a consequence the organization and storage of field and computed data in a structured information system is also a key point in the study of hazards, a function available in a GIS (e.g. Gaspar et al. 2004; Di Muro et al. 2012).
- It is suited to large-scale simulations: for example, Lirer et al. (2010) computed 100 simulations of eruptions sourced from different points to assess the volcanic hazard.
- It enables reasoning on various scales: from regions (e.g. Lirer et al. 2010) to cities (e.g. Alberico et al. 2011). Observations on different scales enable different results to be shown, mainly because they do not use the same scale and the same data (Alberico et al. 2011).
- It encourages the enhancement of the quality and homogeneity of data sets, as GIS only deal with homogeneous and validated reference data. To obtain high quality data for a GIS is often a major issue, especially in the area of risk analysis (Carrara et al. 1995; Gaspar et al. 2004).

18.2.2 Some Approaches to Quantifying Volcanic Hazard Using GIS

The classic approach to quantifying volcanic hazard consists in evaluating one hazard for each volcanic phenomenon independently (lava flows, ash fallouts, pyroclastic flows, etc.), and then to combine these different hazards in a global hazard map.

Two main kind of methods are used in the process of hazard quantification (Pareschi et al. 2000):

- using knowledge about past events via geo-spatial models, assuming that future events will share the same characteristics and dynamics as past events;
- using mathematical models to simulate the evolution of eruption processes.

These two types are often combined, either to overcome a lack of historical data for some eruptive phenomena (e.g. Alberico et al. 2007, 2011) or, less frequently, to lend weight to hazard quantification by crossing the results of these two approaches (e.g. Selva et al. 2011).

A general model for volcanic hazard quantification methods using GIS has been defined by Felpeto et al. (2007):

- Evaluation of the temporal probability for the occurrence of the hazard, for a given time period and according to past eruptions (if the historical eruptive record is too short, this step is passed over and the probability of occurrence is considered to be equal to 1);
- Definition of the source area for eruptions and computation of the probability of vent openings in that area;
- Characterization of the expected eruption (explosivity index, maximum length of lava flows, dispersion of ashes, etc.);
- Numerical simulation of each phenomenological hazard;
- Elaboration of the final hazard map.

Steps (iv) and (v) can be completely automatized and integrated directly into the GIS, as an extended tool (Felpeto et al. 2007).

Whatever the approach implemented the evaluation and management of the volcanic risk require it to be spatialised, and thus entail the use of a GIS.

18.3 Methodological Appraoch for the Design of a GIS for the *Piton de la Fournaise*

18.3.1 Volcanological Data

In the course of time, numerous observers and scientists have sought to study the different eruptions of the *Piton de la Fournaise*, either using surveillance data, or by collecting in situ data relating to surface phenomena. These studies have provided a large amount of information on early and recent volcanic activity of the volcano. Thus a first step to make the OVPF's GIS consisted in enumerating, collecting and characterising this data and converting it into coherent and homogenous layers of geographical information. Table 18.1 presents the list of data collected, its origin and sources, and its format.

A second stage consisted in checking and validating the data, and then homogenising and completing it. The difficulty at this stage resides mainly in the management of the uncertainty and incoherence of data collected, and also in the divergence of viewpoints. Because the sources are diverse, several datasets for the same

Table 18.1 List of collected data related to the eruptive phenomena of Piton de la Fournaise Volcano

Data sets		Source	Format	Data collection method
Lava flows	After 1972	OVPF/IPGP	Digital files (shp[a])	Field measurements, Photointerpretation from DEM
	1931–1984	Geosciences reunion	Digital files (mif/mid[a])	Digitalisation from paper maps
	1800–1979	OVPF/IPGP	Tracing paper, Paper maps[b]	Testimonies (Bory de Saint-Vincent 1804; Lacroix 1936), Unknown
	Before 1800 outside *Enclos*	Geosciences reunion others	Paper maps, drawings	P. Bachèlery, Bibliography, Archive material
Cones		Geosciences reunion	Digital files (mif/mid)	Photointerpretation
		OVPF/IPGP	Tracing paper, paper maps	Unknown
		OVPF/IPGP	Digital files (shp)	Unknown
Eruptive faults	After 1972 inside *Enclos*	OVPF/IPGP	Digital files (shp)	Field measurements, Photointerpretation from DEM
	Before 1972	OVPF/IPGP	Tracing paper, Paper maps	Unknown
Non-eruptive faults	After 1972 on the top part of the volcano	OVPF/IPGP	Digital files (shp)	Field measurements, Photointerpretation from DEM
Eruptions	Before 1972	OVPF/IPGP	Digital files (ascii)	Archive material, Sensors network
	After 1972	OVPF/IPGP	Text descriptions	Archive material, Global Volcanism Program
Blocks		Geosciences reunion	Digital files (raster)	Field measurements, Photointerpretation from DEM
Seisms		OVPF/IPGP	Digital files (ascii)	Seismic sensors network

[a]*Shp files* format of the GIS ArcGIS/*mif-mid files* format of the GIS MapInfo
[b]Maps on tracing paper include hand-drawn limits of cones, eruptive faults and lava flows; each eruption was drawn on a different tracing paper map

phenomenon can coexist. For example, depending on sources, a given lava flow presents different mapped itineraries, and it is hard to know which one is the correct one. The problem of incomplete data is also considerable, and needs to be dealt with prior to data collation.

Several processes need to be implemented prior to the creation of layers of geographical information:

- to ensure coherence between the different datasets and the different phenomena, geometric and topographic corrections were performed on GIS data;
- tabular data and their semantics, and also the geographical coordinates, were reviewed and homogenised before conversion into layers of geographical information;
- data in image format, such as sketches or old maps, were geo-referenced and digitised and their attributes were recorded;
- layers of geographical information relating to blocks, cones and certain lava flows, initially defined in MapInfo® format were converted to ArcGIS® format;
- by superimposing different layers of information, we were able to complete certain missing data. This was the case for instance with the geographical location of certain geo-chemical samplings, or the dates of certain eruptive events;
- data was created using photo-interpretation from orthophotos or from a Digital Elevation Model (lava flows, cones, fissures and blocks). It was then validated by comparisons of the different layers of geographical information (cones/flows/fissures) using field observations and expert judgment (OVPF and *Géoscience Réunion* laboratory researchers). Figure 18.1 describes an example of the processing chain required to integrate data in the GIS.

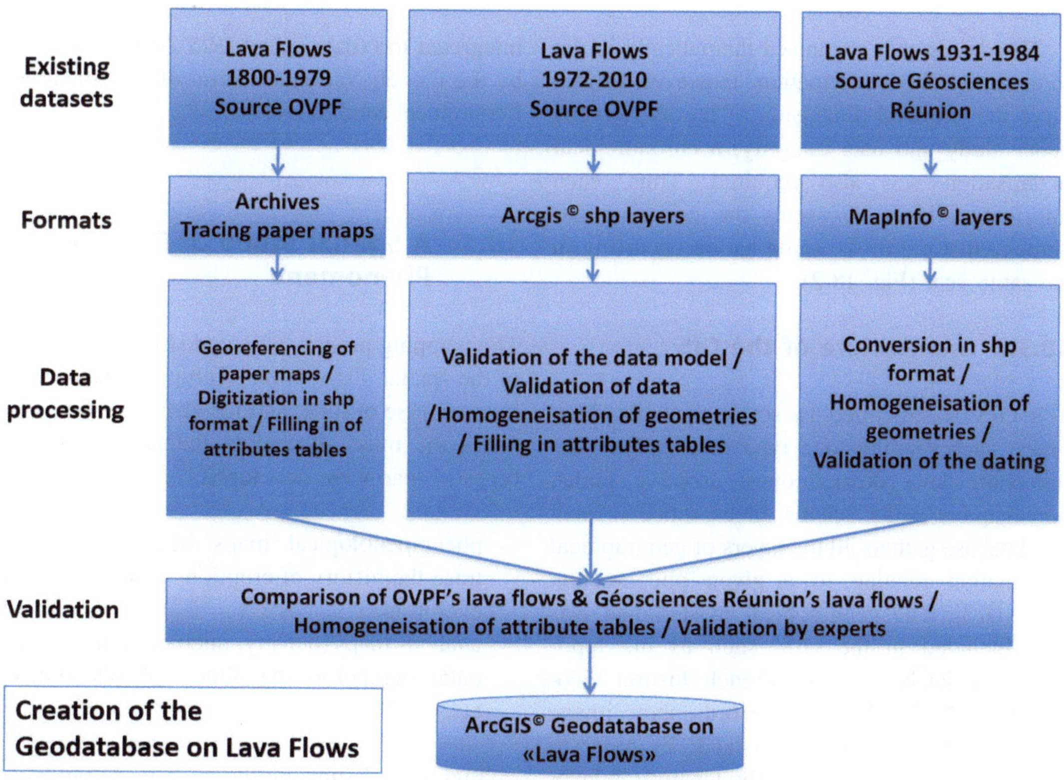

Fig. 18.1 Processing chain to integrate data on lava flows in the volcanological GIS

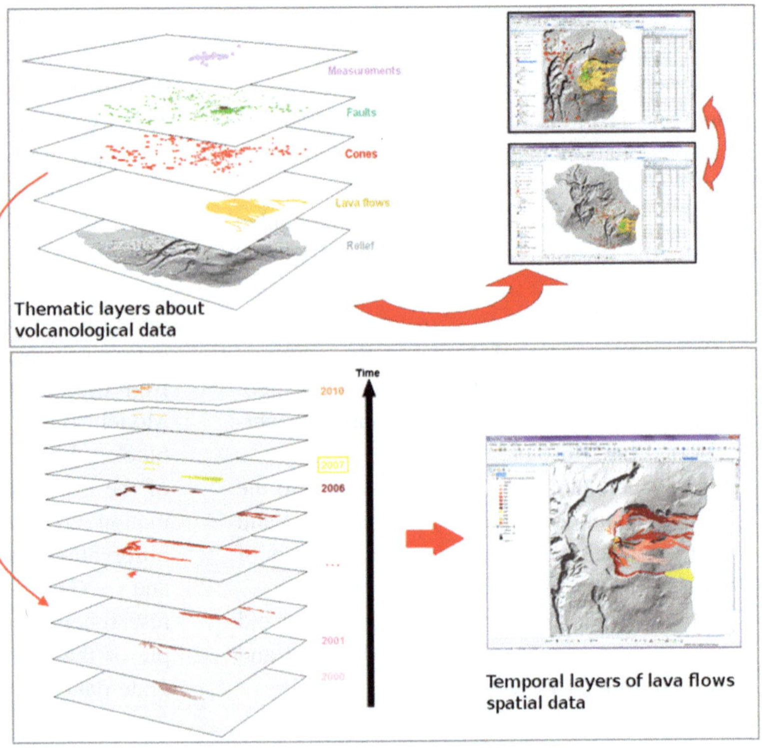

Fig. 18.2 Structuration and integration of volcanological data into the GIS

The data is thus organised thematically in the form of layers of information. However, to preserve the temporal dimension of the phenomena under study and data integrity, a chronological organisation was also required. Thus, each information layer represents the eruptive phenomena that are of the same nature occurring in the same year (Fig. 18.2).

18.3.2 Architecture of the GIS

The GIS is structured in seven thematic databases, each relating to a type of eruptive phenomenon: lava flows, cones, eruptive faults, non-eruptive faults, seisms, blocs, cinders. Each geodatabase gathers all the layers of geographical information relating to a given phenomenon (Fig. 18.3). The geographical master data was also included in the GIS, such as the topographical database of the French Institut Geographique National (IGN), the orthophoto image and the DEM of Reunion Island. In order to evaluate the vulnerability of the territory, a layer concerning territorial statistical units was integrated too (data called *IRIS 2000 ilôts,* made by the French National Institut of Statistical and Economical Studies—INSEE).

18.4 A Spatial Study of Eruptive Phenomena

The mapping process is intended for the analysis of the spatial distribution of eruptive phenomena in the course of time, and for identification of the zones the most affected in the past. Thus two types of map were developed:

- phenomenological maps which attempt to trace the history of eruptions and their spatial evolution.
- analysis maps (density, interpolation of captor data, geometric magnitude of phenomena, etc.).

Figure 18.4 sums up the main maps produced from data on eruptive phenomena.

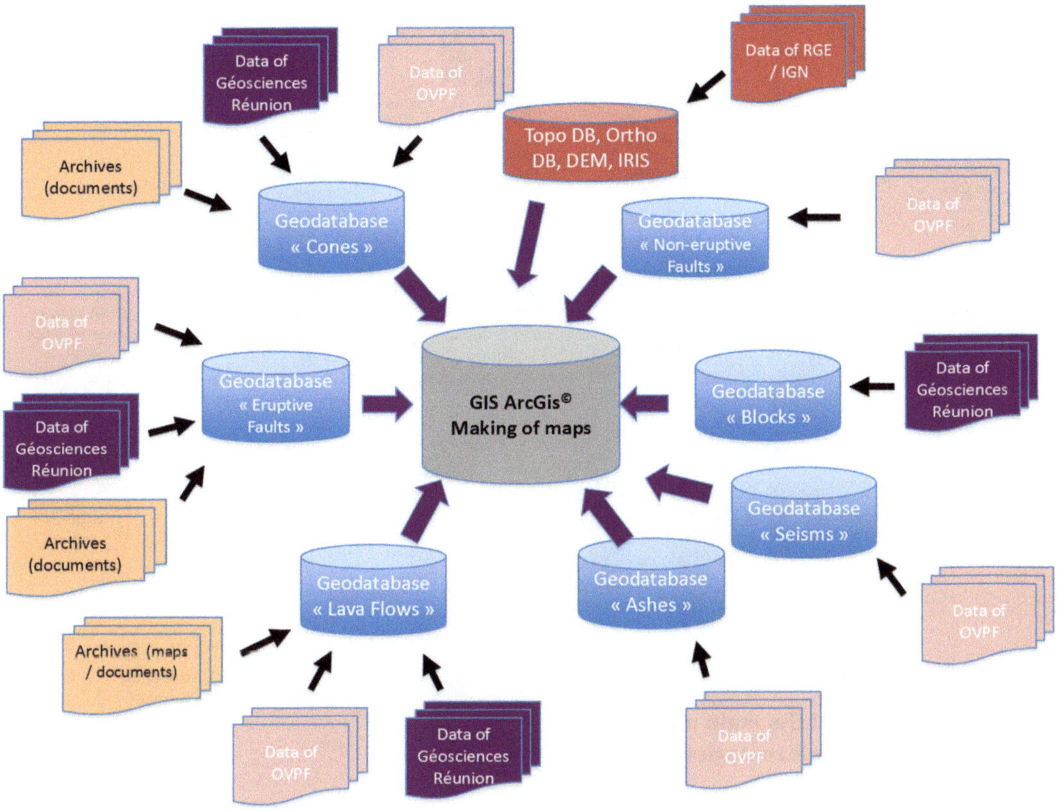

Fig. 18.3 Architecture of the OVPF's GIS

18.4.1 Phenomenological Maps

Phenomenological maps provide an overview of past events. They should answer the questions "Where?", "What?" and also "When" did the phenomena occur. The task is to map the chronology of eruptive phenomena, to identify spatial trends, and their variations over time.

Figure 18.5a shows the spatio-temporal distribution of lava flows from eruptions having occurred between 1972 and 2010. Figure 18.5b is an other type of phenomenological map: it shows the distribution of eruptive faults for each year in a synthetise way and does not show every single fault.

18.4.2 Analysis Maps

In order to evidence the numbers of phenomena observed per surface unit, density maps were developed. These maps enable a ranking of levels of exposure to different phenomena for hazardous areas. They complete the phenomenological maps. Figure 18.5c–f shows the diversity of uses enabled by density maps.

For instance, Fig. 18.5c presents the density of volcanic cones (developed using the "kernel density" tool in the Spatial Analyst extension of ArcGIS). The density map enables the identification of zones of cone concentration, and the deduction of the sectors where eruptions have

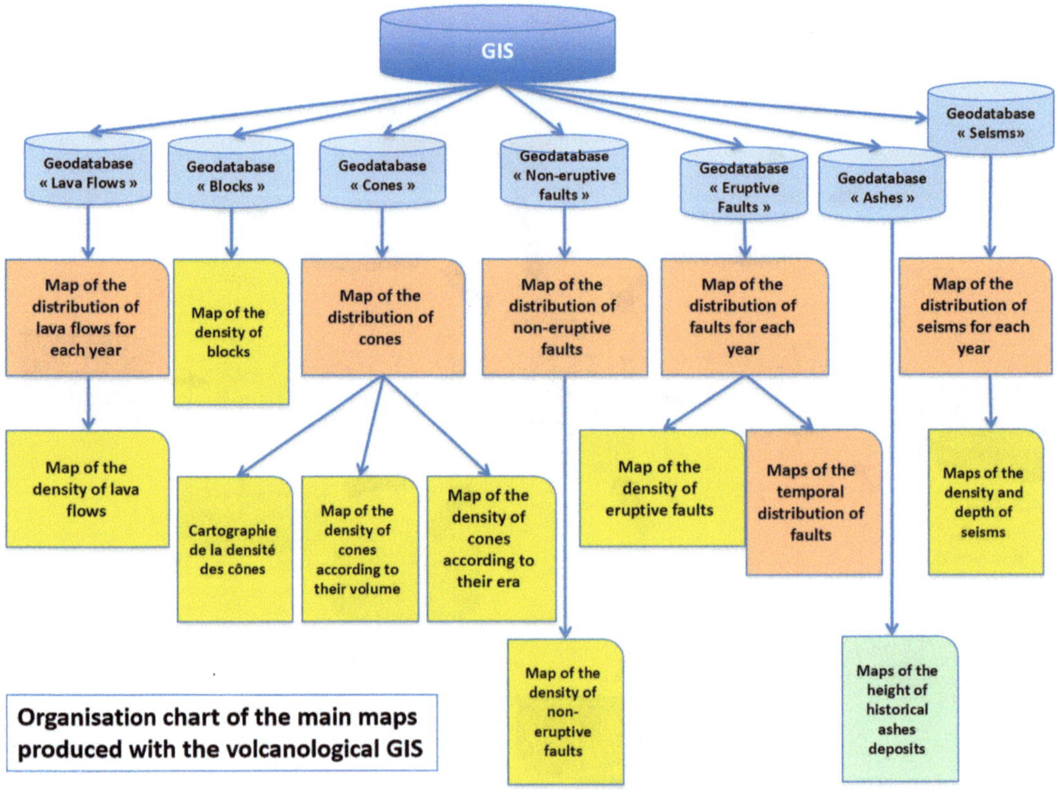

Fig. 18.4 Main maps produced for the spatial analysis of volcanic hazard

been the most numerous. It also make it possible to identify a circular "rift zone" where eruptions are concentrated, as identified by Bachèlery (1981).

Figure 18.5d shows a density map of lava flows over the last forty years. It is possible to see the areas most frequently covered by lava. It also shows how the lava is driven to the surface in three preferential directions (channels) around the central crater: to the north, the south-east and the south-west.

The combination of several density maps is also useful for the spatial apprehension of eruptive phenomena. For instance, mapping cone densities according to their age (recent and older cones, as in Bonali et al. 2011) evidences two eruptive centres: an older centre located to the north-west of the *Piton de la Fournaise*, and a present centre forming the rift zone mentioned above (Fig. 18.5e).

Density maps of different phenomena can also be crossed together, and with thematic data, to see the impact of volcanic activity on a thematic field of interest. For instance, Fig. 18.5f shows how much exposure tourist amenities are to volcanic hazard.

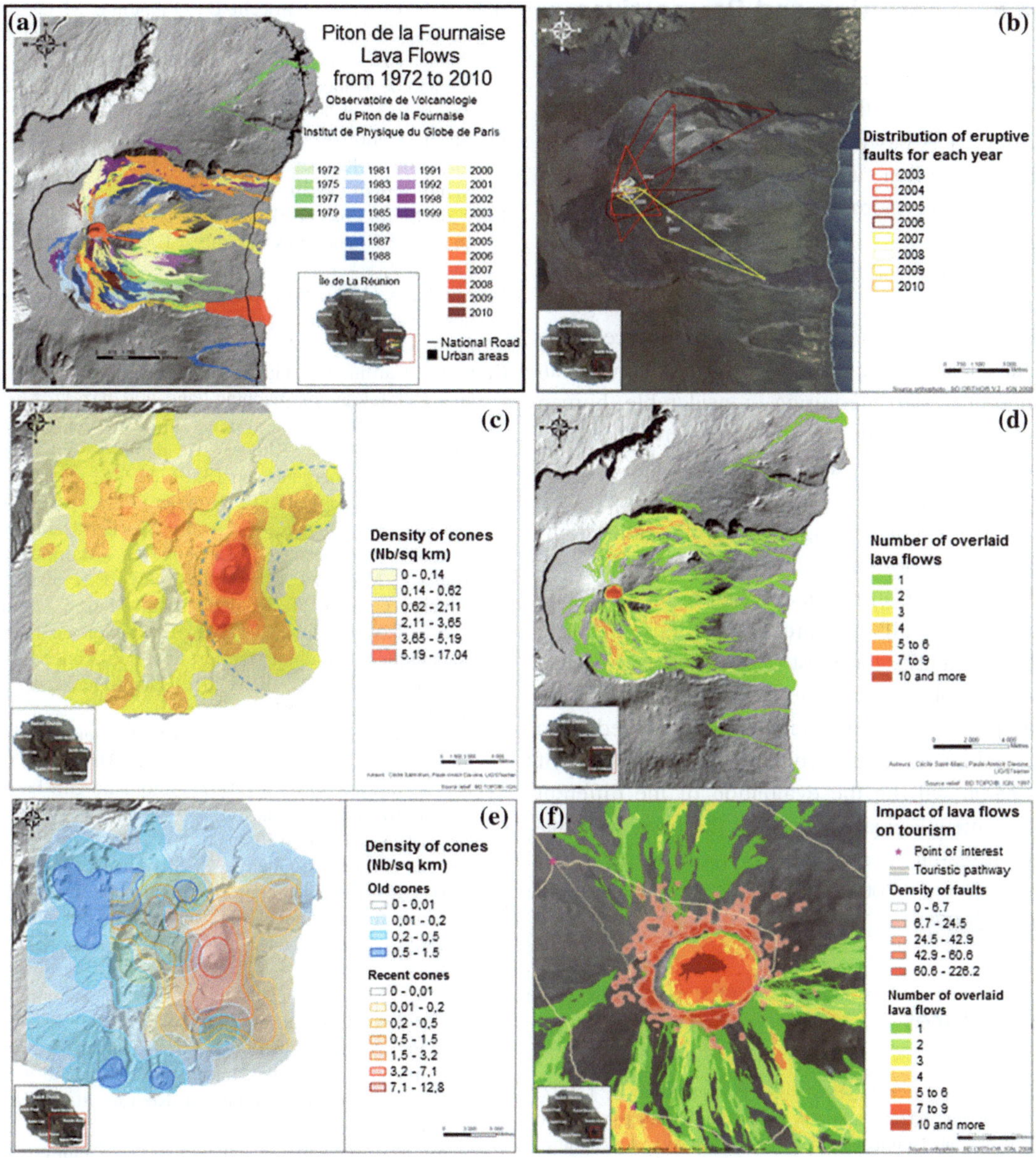

Fig. 18.5 Examples of phenomenological maps and density maps. **a** Spatial chronology of lava flows. **b** Spatio-temporal distribution of eruptive faults. **c** Density of volcanic cones. **d** Density of lava flows. **e** Overlaid density of older cones in *blue* and recent cones in *orange*, according to the classification of Bonali et al. (2011). **f** Impact of volcanism on touristic activity, crossing lava flows footprints with touristic pathways and tourist amenities

18.5 Conclusion and Perspectives

Knowledge of the history of the volcano is a prerequisite for the evaluation of the volcanic hazard. It is based not solely on the incomplete early data, often variable in quality and precision, but also on more recent and better-structured data.

In our project, the GIS environment played an important part in the integration of a range of heterogeneous data from numerous sources. The data cleansing and validation phase was an important step, requiring considerable human input. It used both the tools on offer in the Arc-GIS environment and other software of the spread-sheet type.

The resulting cartographic productions highlight the spatio-temporal distribution of volcanic phenomena. Calling on relevant aspects of seismology and different processing steps, we were able to identify the most exposed areas, to relate eruptive phenomena one to another, and thus to contribute to the enhancement of knowledge on the behaviour of this volcano. The next stage of research will be the evaluation of the different phenomenological hazards on the basis of existing research in this area (Wright et al. 1992; Pareschi et al. 2000; Alberico et al. 2007; Felpeto et al. 2007).

References

Alberico I, Lirer L, Petrosino P, Scandone R (2007) Volcanic hazard and risk assessment from pyroclastic flows at Ischia island (southern Italy). J Volcanol Geoth Res 171:118–136

Alberico I, Petrosino P, Lirer L (2011) Volcanic hazard and risk assessment in a multi-source volcanic area: the example of Napoli city (Southern Italy). Nat Hazards Earth Sci 11:1057–1070

Aronoff S (1989) Geographic information systems: a management perspective. WDL Publications, Ottawa, p 294

Bachèlery P (1981) Le Piton de la Fournaise (île de la Réunion): Etude Volcanologique, Structurale et Pétrologique. Thèse de doctorat de l'Université de Clermont-Ferrand, p 257

Bellotti F, Capra L, Sarocchi D, D'Antonio M (2010) Geostatistics and multivariate analysis as a tool to characterize volcaniclastic deposits: application to Nevado de Toluca volcano, Mexico. J Volcanol Geoth Res 191:117–128

Bonali FL, Corazzato C, Tibaldi A (2011) Identifying rift zones on volcanoes: an example from La Réunion island, Indian Ocean. Bull Volcanol 73(3):347–366

Bory de Saint-Vincent JBGM (1804) Voyage dans les quatre principales îles des mers d'Afrique. Tome II, Paris, p 231

Burrough PA (1989) Principles of geographical information systems for land resources assessment. Clarendon Press, Oxford, p 194

Carrara A, Cardinali M, Guzzetti F, Reichenbach P (1995) GIS technology in mapping landslide hazard. Geographical information systems in assessing natural hazards, vol 5. Kluwer Academic Publishers, Dordrecht, Netherlands, pp 135–175

Di Muro A, Bachelery P, Boissier P, Cyprien A, Davoine PA, Fadda P, Favalli M, Ferrazzini V, Finizola A, Leroi G, Levieux G, Mairine P, Manta F, Michon L, Morandi A, Nave R, Peltier A, Principe C, Ricci T, Roult G, Saint-Marc C, Staudacher T, Villeneuve N (2012) Evaluation de l'aléa volcanique à la réunion: rapport final de l'Année I. In: Rapport MEEDDM, Feb 2012, Saint-Denis de la Réunion, France, p 81

Esri (2009) GIS versus the volcano: USGS scientists plan life-saving disaster response, ArcNews 31(1):32–33

Felpeto A, Martí J, Ortiz R (2007) Automatic GIS-based system for volcanic hazard assessment. J Volcanol Geoth Res 166:106–116

Gaspar JL, Goulart C, Queiroz G, Silveira D, Gomes A (2004) Dynamic structure and data sets of a GIS database for geological risk analysis in the Azores volcanic islands. Nat Hazards Earth Syst Sci 4:233–242

Lacroix A (1936) Le volcan actif de l'île de la Réunion et ses Produits. Gauthier et Villard ed, Paris, p 297

Lirer L, Petrosino P, Alberico I (2010) Hazard and risk assessment in a complex multi-source volcanic area: the example of the Campania Region, Italy. Bull Volcanol 72:411–429

Pareschi MT, Cavarra L, Favalli M, Giannini F, Meriggi A (2000) GIS and volcanic risk management. In: Natural hazards, vol 21. Kluwer Academic Publishers, Pays-Bas, pp 361–379

Selva J, Orsi G, Di Vito MA, Marzocchi W, Sandri L (2011) Probability hazard map for future vent opening at the Campi Flegrei caldera, Italy. Bull Volcanol 74:497–510

Wright TL, Chu JY, Esposo J, Heliker C, Hodge J, Lockwood JP, Vogt SM (1992) Map showing lava-flow hazard zones, island of Hawaii: US Geological Survey Miscellaneous Field Studies Map MF-2193, scale 1:250,000

Perception of Risk for Volcanic Hazard in Indian Ocean: La Réunion Island Case Study

19

Rosella Nave, Tullio Ricci and Maria Giuseppina Pacilli

Abstract

In 2011 a research project on volcanic risk assessment at La Réunion Island (Project Aléa, Institute de Physique du Globe de Paris, France) was conducted in order to determine more efficient strategies to manage future volcanic crises. The project included the evaluation of volcanic scenarios through field and historical data analysis, as well as a survey on volcanic risk perception in resident population. A clear scientific information and an effective communication with public play a crucial role in risk mitigation strategies. Moreover the evaluation of the public perception during both volcanic crises and dormant periods is an important element in developing actions focused on specific social and cultural contexts. For this porpose a questionnaire was developed based on the ones used in similar researches conducted on Italian active volcanoes. Items were designed to measure variables connected with personal perception of hazard and risk, trust in mitigation actions and in information received about these aspects. In addition, specific items related to the peculiarities of La Réunion Island environment were included. A total of 2000 questionnaires were distributed taking into account factors such as the proximity to the volcano and the involvement of communities in recent volcanic emergencies. Main results coming out from this survey, if on the one hand show an adequate

R. Nave (✉)
Istituto Nazionale di Geofisica e Vulcanologia,
Osservatorio Vesuviano, Naples, Italy
e-mail: rosella.nave@ingv.it

T. Ricci
Istituto Nazionale di Geofisica e Vulcanologia,
Rome, Italy
e-mail: tullio.ricci@ingv.it

M.G. Pacilli
Dipartimento Istituzioni e Società, Università degli
Studi di Perugia, Perugia, Italy
e-mail: maria.pacilli@unipg.it

P. Bachèlery et al. (eds.), *Active Volcanoes of the Southwest Indian Ocean*,
Active Volcanoes of the World, DOI 10.1007/978-3-642-31395-0_19

residents' perception of natural hazards, on the other hand highlight their poor knowledge of the island's active volcano, a similar lacking knowledge of emergency plan for volcanic crises, but also a high confidence in scientists to provide accurate and reliable information on volcanic risk and hazards in contrast with Local Authorities. Remarkable findings of this study assess some key elements that should be considered by the institutions in charge for defining policies aimed at volcanic risk mitigation and management of future volcanic crises.

19.1 Introduction

The social dimension of risk is a complex topic that has been investigated since the early 40s, and perception of risk has become a conventional terms (Slovic 1992), defined as combination of social, economic, cultural and environmental factors (Renn et al. 1992).

For a general review of risk perception literature, in particular related to natural hazards case studies, we refer to Wachinger et al. (2010). Qualitative assessment of people's perception has been increasingly integrated in natural hazards' studies while volcanic risk perception have been widely investigated all over the world, mainly in the last twenty years. Several studies on this issue have been published (Gaillard and Dibben 2008 and references therein; Bird et al. 2009; Morin and Lavigne 2009; Njome et al. 2010; Ricci et al. 2013a, b), analyzing and assessing volcanic risk perception from different perspectives.

The main focus for several of these studies were the evaluation of people's knowledge of volcanic hazards and their perception of risk, and how such perception could be related with people's preparedness and response to emergency measures planned. Risk perception is strictly connected with risk, as a result of hazard, vulnerability and exposure (Chester et al. 2002). In particular hazard assessment also includes understanding the impact of a given hazard on the environment, as well as vulnerability assessment means also to take in account people's attitude toward risk.

This is the reason why perception of risk has been recently suggested to be included as a new factor in the Risk equation: Risk = Hazard × Vulnerability × Exposure × Perception (Greco 2012). Moreover, moving from a qualitative assessment of social dimension of risk to a more quantitative one, risk perception has been introduced in multi-criteria risk analysis framework as a weight index representing risk judgment and choices made by the population at risk (Meng et al. 2013).

The IPGP Aléa Project aim of assessing volcanic risk perception and evaluating eruptive scenarios for risk map construction not only allow the development of more efficient actions for emergency managing at La Réunion Island, but can be also considered as a new comprehensive approach toward volcanic risk mitigation.

For a detailed description of the geology and volcanic hazards of Piton de la Fournaise volcano refers to Chaps. 4 and 8 of this monograph respectively.

19.2 Methodology

19.2.1 The Survey

The study conducted at La Réunion Island is the second largest ever carried out on volcanic risk perception, in terms of questionnaires distributed, after the one conducted on Vesuvio by Barberi et al. (2008). The survey instrument is a questionnaire, a widely utilized tool in social studies adopted to provide reliable quantitative data as well as reaching a representative sample of the population and allowing comparison with similar surveys results (Bird 2009).

A 43 items questionnaire was developed starting from those already developed and tested in similar researches conducted in Italy (Barberi et al. 2008; Ricci et al. 2013a). Items were designed to measure variables such as:

Salience of the hazard: the tendency to spontaneously mention volcanic hazards as a problem relative to other concerns, and the amount of time spent thinking about the threat;
Risk perception: ratings of both the likelihood and severity of future eruptions as well as one's level of anxiety about a potential eruption and feelings of personal vulnerability to the effects of a potential eruption;
Feelings of self-efficacy: feelings of control regarding one's ability to protect him/herself and his/her family from the effects of an eruption;
Knowledge of the hazard and of hazard mitigation strategies: factual information concerning past eruption events and measures taken to protect the public from future eruptions;
Perceived preparedness of and trust in officials: confidence in the government's level of preparedness and in the ability of government officials, scientists and the media to provide accurate information about potential eruptions;
Sources of received information about volcanic hazards and the emergency plan and preferred methods of receiving such information;
Sense of community: the degree to which residents feel bonded to their community was measured using the 18 item a modified Scale of Sense of Community by Prezza et al. (1999).
Demographic questions: sex, age, highest level of education achieved, marital status, etc.

Also specific items regarding the peculiar dynamic of La Réunion Island were included. Most part of the items are 5-points Likert scale closed questions (Likert 1932), widely used in psychometric researches. There are also several open-ended questions in order to obtain descriptive data. To facilitate data entry, the questionnaire was expressly designed to be read through ICR (Intelligent Characters Recognition) and OMR (Optical Marks Recognition) methodologies.

19.2.2 Distribution Procedures of Questionnaires

A total of 2000 questionnaires were handed out in the 24 municipalities of La Réunion Island. The number of questionnaires distributed was calculated in relation to the number of residents of each municipality, assuring a minimum of 60 questionnaires distributed in the smallest towns. The distribution plan also took into account aspects such as the cities proximity to the volcano and the involvement of communities in recent volcanic emergencies, as happened at Sainte Rose in 1977 and Saint Philippe in 1986 and 2007. The distribution and collection of questionnaires took place between November 2011 and February 2012 and was designed to reach an equal percentage of men and women, and representing all classes of age. In particular, the questionnaires were distributed to junior high schools and high schools students, in each municipalities, in collaboration with School Chief Committees and selected teachers. The researchers also distributed questionnaires to students asking them to take the surveys home to be completed by their parents. Responsibles of municipalities' public offices, mostly Centres for Social Actions and Aggregation, and of local associations also collaborated in the distribution of questionnaires to the general population.

A total of 61 town offices or associations were involved. In particular 1139 questionnaires were distributed to students (667) and parents (472) in 55 junior high and high schools, 565 questionnaires were handed out in 20 municipalities' public offices and 296 through 9 local associations.

19.3 Results

19.3.1 Sample Grouping

In data elaboration, in addition to the Total sample including all the respondents, main survey results have been distinguished on the basis of the map of vulnerability to lava flow related to major probability of vent opening (Fig. 19.1; see Chap. 18). In particular results have been divided in three groups: Group A—two municipalities, Sainte Rose and Saint Philippe, the communities most exposed to volcanic and that have experienced evacuations during the crises in 1977 and 2007; Group B—six municipalities, Saint Pierre, Le Tampon, Petite Ile, Saint Joseph, Plaine des Palmistes, Saint Benoit; Group C—the remaining sixteen municipalities of the Island, Bras Panon, Cilaos, Entre Deux, Etang Salé, La Possession, Le Port, Les Avirons, Saint André, Saint Denis, Saint Leu, Saint Louis, Saint Paul, Sainte Marie, Sainte Suzanne, Salazie, and Trois Bassins.

19.3.2 Sample Characteristics

Of the 2000 questionnaires distributed, 1220 (61 %) were collected and 1204 of these (60.2 %) were suitable for the analysis. In particular, Group A—149 respondents out of 220 questionnaires handed out (67.7 %), Group B—354 respondents out of 590 questionnaires handed out (60.0 %), and Group C—701 respondents out of 1190 questionnaires handed out (58.9 %). The percentages of men and women calculated on the number of respondents to this item (N = 1125) are respectively 38.6 and 61.4 %. The age of respondents range from 10 to 98 years, with an average of 27.2 and a standard deviation of 16.4. With regard to the education level, 4.2 % have primary school license, 45.7 % have a junior high school education, 32.5 % are high school graduates, and 17.6 % have a university degree, no one of the respondents has a post university education level.

19.3.3 Hazard Salience

"Hazard salience" refers the extent to which the volcanic threat is on the minds of residents. To measure this feature one of the first questions of the survey was an open-ended item asking respondents to list the three greatest problems facing their communities.

All the answers provided by the respondents were then grouped into 20 categories. As shown

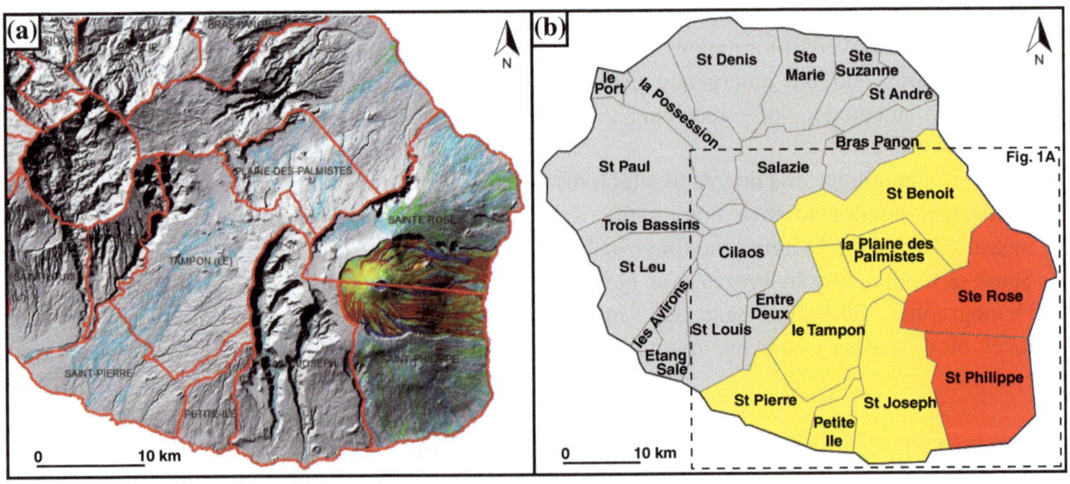

Fig. 19.1 **a** Map of vulnerability to lava flow related to the probability of vent opening on Piton de La Fournaise volcano (see Chap. 17); **b** Subdivision of the 24 Municipalities of La Réunion Island: *red* Group A; *yellow* Group B; *grey* Group C

Table 19.1 Rankings of the 5 most frequently mentioned community problems

Total sample	Group A	Group B	Group C
N = 942	N = 225	N = 287	N = 538
(1) Public services 349 (37.0 %)	(1) Social life 83 (70.9 %)	(1) Public services 110 (38.3 %)	(1) Traffic 198 (36.8 %)
(2) Social life 344 (36.5 %)	(2) Public services 47 (40.2 %)	(1) Traffic 90 (31.4 %)	(2) Public services 192 (135.7 %)
(3) Traffic 315 (33.4 %)	(3) Remoteness 42 (35.9 %)	(3) Social life 86 (30.0 %)	(3) Social life 175 (32.5 %)
(4) Remoteness 252 (26.8 %)	(4) Traffic 27 (23.1 %)	(4) Civic sense 84 (29.3 %)	(4) Remoteness 133 (24.7 %)
(5) Civic sense 221 (23.5 %)	(5) Civic sense 17 (14.5 %)	(5) Remoteness 77 (26.8 %)	(5) Civic sense 120 (22.3 %)
(15) Other natural hazards 23 (2.4 %)	(14) Volcanoes 3 (2.6 %)	(16) Other natural hazards 2 (0.7 %)	(12) Other natural hazards 21 (3.9 %)
(18) Volcanoes 3 (0.3 %)	(–) Other natural hazard 0 (0.0 %)	(–) Volcanoes 0 (0.0 %)	(–) Volcanoes 0 (0.0 %)

Rankings of "volcanoes" and "other natural hazards" are also listed for comparison
Numbers before parenthesis indicate the number of respondents while *numbers in parenthesis* indicate the percentage of residents who mentioned each problem

in Table 19.1, problems such as inefficiency of Public services, lack of Social life, Traffic, Remoteness, and Civic sense, are the issues most frequently mentioned by respondents from each municipality group followed by other categories such as Crime, Natural environment, Trash and pollution, Drugs and alcohol, Stray animals, Unemployment and so on.

Another survey item that measured the salience of the hazard asked residents to number from 1 to 7 (from most to least worrying) natural phenomena threatening La Réunion Island (Fig. 19.2).

The most indicated concerning phenomenon by 45 % (N = 117) of the Total sample respondents is Debris flow, while Volcanic eruption is mentioned only by 5.4 % (N = 14). Particularly interesting is that 9.6 % of the Total sample respondents consider Volcanic eruption as the less worrying natural phenomenon.

In results from the towns of the Group A, that have experienced volcanic emergencies due to lava flows in the last decades, the concerning for volcanic eruption is quite low (3.7 %, N = 1). This result is probably due to residents' awareness of the effects connected to effusive activity.

In the Group A, again there is a significant difference in concern for Earthquake, most probably because they link seismic activity to the volcano dynamic. The same for the high score percentage for other natural hazards, probably referred to the occurrence of fires, caused both by lava flows and, less naturally, man.

When asked to respondents if they agree in considering a volcanic event as a threat for their place of residence, only the mean value from Group A shows a remarkable difference in recognizing their own territory as exposed to a volcanic hazard. Responses are based on a 5-point Likert scale, with higher numbers indicating greater consensus to the question. In particular, the mean value for Total sample is 2.2 (N = 1159), for Group A is 3.4 (N = 148), while for Groups B and C the score is significantly lower, 2.1 (N = 341) and 2.2 (N = 667) respectively.

19.3.4 Risk Perception

Perception of Risk is a general term that sums up several aspects related to how people consider the risk connected with a particular hazard. These

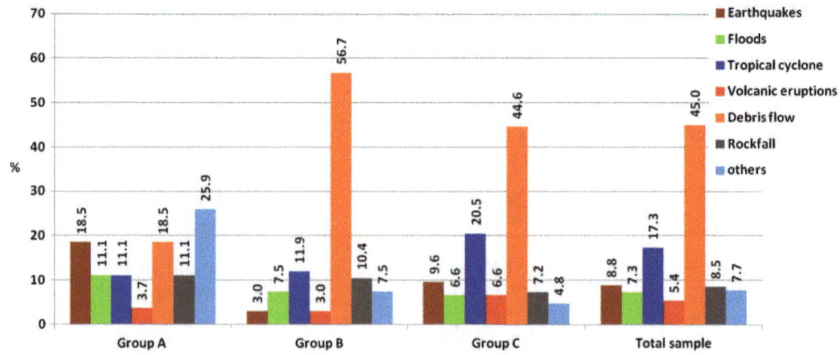

Fig. 19.2 Most mentioned natural phenomena threatening La Réunion Island

elements include the concern felt in relation to a catastrophic event, the perception of the seriousness of the consequences of such an event, and also the perceived likelihood of a disaster event. A series of items in the questionnaire have been developed to assess this complex parameter.

Residents were asked to grade their level of concern with respect to an eventual eruption and to the severity of consequences for their city, themselves and their families. The answers are on a 5-point Likert scale where responses with higher numbers indicate a greater concern about a possible event and severity of their consequences. In particular, 52.7 % (N = 621) of the Total sample respondents express "none" or "little" concern (19.0 and 33.7 % respectively), while only 24.9 % (N = 293) show "high" or "very high" levels of concern (14.3 and 10.5 % respectively). In a comparison between the three groups, only respondents from Group A show a significantly higher concern (34.9 %) than Group B and C (23.45 %).

About consequences of a volcanic eruption on themselves and their families, 16.4 % of all respondents (N = 1155) expect no serious consequences, 27.1 and 25.9 % graded "mild" or "moderate" consequences respectively, 20.6 % assume possible serious consequences while for 10 % of respondents the impact of a volcanic event could be extremely serious. Respondents from the Group A, more exposed to volcanic hazards, have a higher perception of the event's severity (34.7 %) than the other two groups (30.1 %).

Regarding the seriousness of the consequences of a possible eruption on their territory, the same trend is highlighted with little impact (24.3 %) to moderately severe (24.1 %), again with 10.6 % of all respondents (N = 1167) identifying as extremely serious also the impact of an eruption on their territory. Again, the comparison with the residents of Group A confirms an higher perception of the severity of a possible eruption impact (35.9). No meaningful differences between Group B and C were pointed out (30.8).

Associated with the perception of risk there is people's attitude to recognize the more likely volcanic hazardous phenomena and to grade problems related to their occurrence. Some questionnaire items focus on this aspect and the answers refer to a 5 points Likert scale, with higher numbers indicating a major perceived impact of a phenomenon associated with volcanic activity.

The highest mean values are the ones expressed for Gas and Ashfall (2.69/5 and 2.59/5 respectively). This is probably due to the occurrence of gas emission during effusive activity (known as volcanic fog), and above all the lava haze (laze) generated from the interaction of lava flow entering seawater. On the other hand ashfall is perceived as the most impacting effect of volcanic activity for Group C, the sector of the island more distant from Piton de la Fournaise volcano.

The comparison between results from each group confirms this outcome, while reasonably the proximity to the volcano and the recent experiences influences the higher percentage of Group A respondents scoring Lava flow as the most impacting phenomenon (Fig. 19.3).

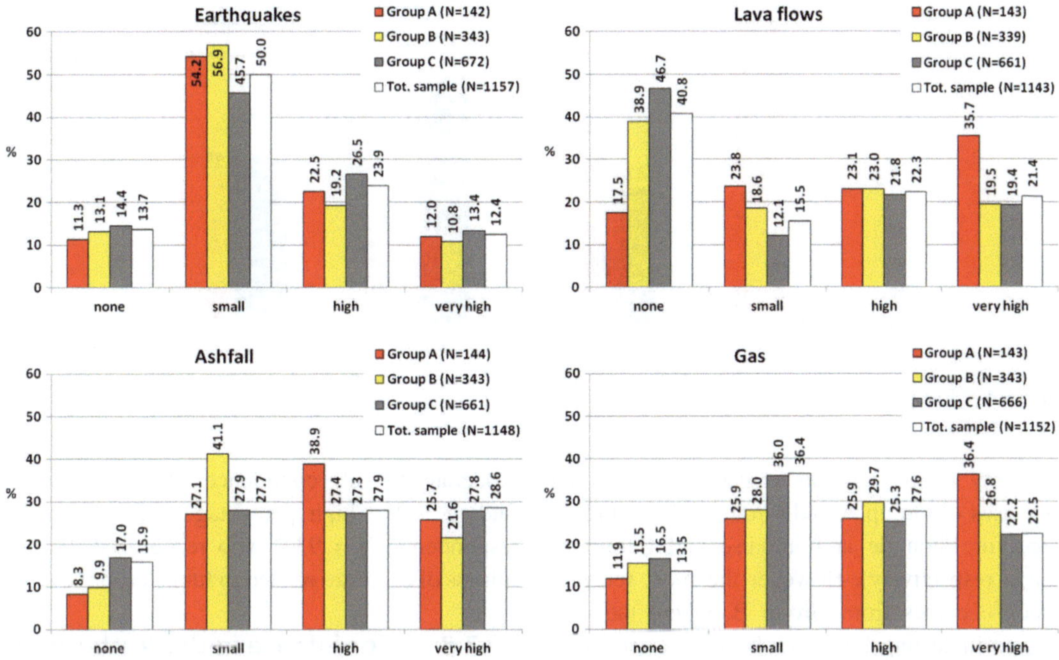

Fig. 19.3 Perceived severity of volcanic phenomena

19.3.5 Self-efficacy

Self-efficacy is recognized as an important variable in determining whether a population at risk would be able to take self-protective measures. Residents were asked to evaluate the personal ability to protect themselves against the effects of an eruption and the preparedness to cope with an emergency. Results, on a 5-point Likert scale with higher numbers representing greater sense of self-efficacy, show that 49.4 % of all respondents (N = 1169) think they are able to protect themselves and their families from volcanic threat. Group A (N = 147), the most exposed to volcanic hazards, shows a higher percentage of respondents in indicating as high to very high their ability to manage a volcanic crisis (53.7 %) with respect to 48.8 % expressed by Group B and C respondents (N = 1022).

Moreover, when asked to evaluate personal and family preparedness to face a volcanic emergency, 30.7 % of the Total sample (N = 1182) affirm to be well prepared. In particular, Group A (N = 146) shows a little higher perceived preparedness, 37.7 % with respect to 29.7 % of Group B and C respondents (N = 1036).

19.3.6 Knowledge of Risk

To assess the level of knowledge of volcanic risk respondents were asked to list the active volcanoes of the island and to indicate which volcano poses the major threat to their area. 68.4 % of the Total sample (N = 1069) indicate as active both the volcanic complex of Piton de la Fournaise and Piton des Neiges, with only 29.0 % indicating the Piton de la Fournaise as the only active volcano on the island, and 1.3 % named the Piton des Neiges. 62.8 % of all respondents (N = 1105) indicate these two volcanoes as potentially dangerous.

When asked to indicate which volcano poses the major threat to their town, 60.9 % of all respondents (N = 330) identify the Piton des Neiges as the most likely source of danger. This shows that La Réunion populations have a very low knowledge of the real volcanic risk. From results of each group it appears evident how the

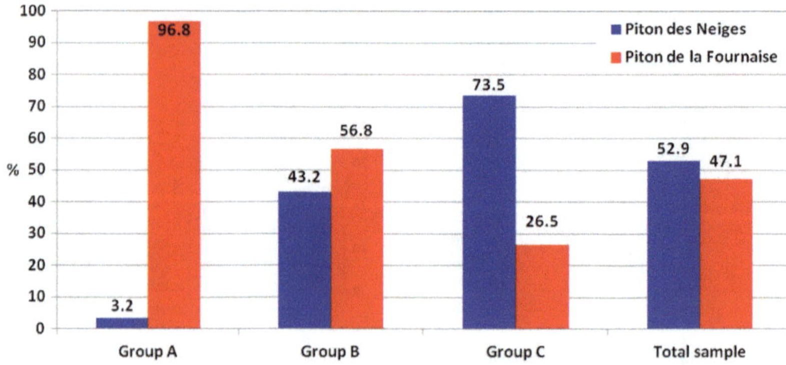

Fig. 19.4 Most threatening volcano

proximity to the volcano edifice influences the responses of Group A and Group C (with higher score for Piton de la Fournaise and Piton des Neige, respectively). However the most relevant response comes from Group B, where the percentage indicating Piton des Neiges is very high (43.2 %) considering that Group B refers to an area with probability of future eruptive vent opening (Fig. 19.4).

19.3.7 Emergency Plan: Knowledge and Confidence

Some questions were designed to assess people awareness of the existence of an Emergency Plan in case of eruption, and also to grade residents' confidence in the Plan success. Also residents' opinion on the usefulness of civil protection exercises for increasing their preparedness in managing eruptive crisis was investigated. As a result, 80.4 % of the Total sample respondents (N = 896) say to be not aware of the plan developed by the authorities in case of volcanic crisis. Results from the three groups confirm this general considerable lack of knowledge of the Plan, also and rather worrying, for the Group A residents (Fig. 19.5).

When asked to indicate what the Plan provide for, 27.7 % of all respondents (N = 472) do not know it, and this percentage is near 19.7 % for respondents of Group A (N = 76), although they experienced a partial evacuation during recent eruptive crises.

With regard to the success of an evacuation, 82.2 % of all respondents (N = 962) indicate a moderate to full confidence. Another relevant finding is the high percentage (77.2 %) of all respondents (N = 937) who recognized the usefulness of exercises related to future eruptive crisis.

19.3.8 Confidence in the Authorities

Two questions asked to express a personal opinion on preparedness of Local Authorities and Civil Protection, led by the Prefecture, in managing a volcano emergency. Total sample results indicate a general good consensus for Prefecture with a mean value of 3.3/5 (N = 1168), while a lower score is graded to Local Authorities (2.8/5; N = 1172). An important outcome to point out is the score conferred from respondents of Group A to Local Authorities' preparedness, 3.11/5, compared to the one assigned by respondents of Group B and C, 2.70/5. This result is probably due to the recent experiences of volcano emergencies they had to face and their positive evaluation of crisis management operated by Local Authorities with respect to Prefecture and Civil Protection.

Furthermore, when asked to evaluate Institutions in providing accurate and reliable information on volcanic risk, respondents express their maximum confidence in the Volcano Observatory and then in the Prefecture and Civil Protection, Local Authorities and Media (Fig. 19.6). In particular, 69.3 % of all respondents (N = 1160) indicate "A lot" or "complete confidence" in the Volcano Observatory scientists, in contrast with

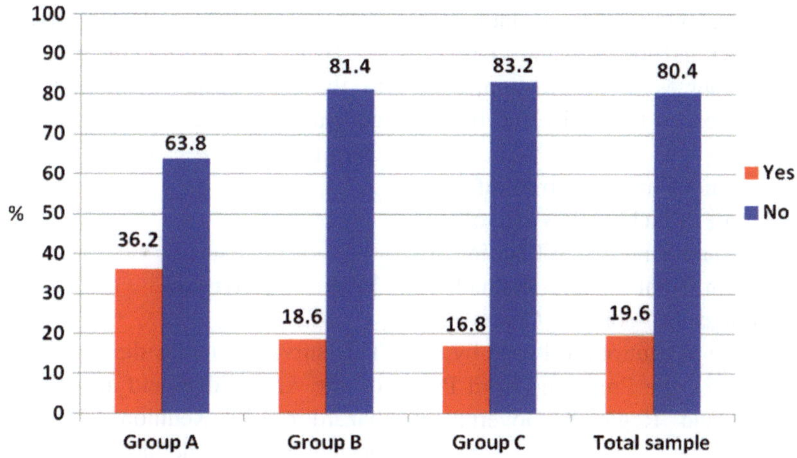

Fig. 19.5 Knowledge of the emergency plan

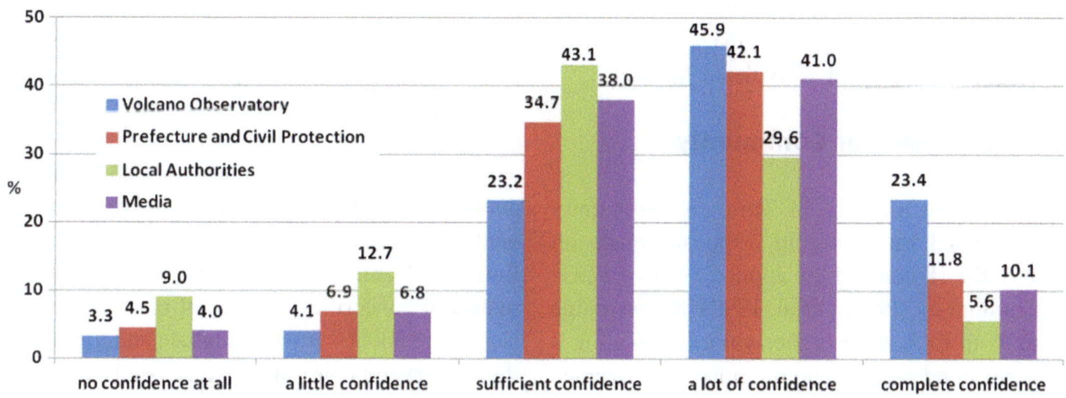

Fig. 19.6 Confidence in information sources for volcanic risk

the 53.9 % for Prefecture (N = 1161), 51.1 % Media (N = 1163), and 35.2 % for Local Authorities (N = 1152).

With regard to the item asking to indicate the favourite source of information on volcanic risk, Volcano Observatory turns out to be the most favourite source for 58.2 % of Total sample respondents (N = 1113), followed by the Local Authorities (42.4 %), Prefecture (30.5 %), Schools (28.1 %), and Regional Government (19.1 %). Interestingly, Local Authorities are indicated as the most favourite source of information on volcanic risk by Group A respondents (60.2 %; N = 137) while Volcano Observatory, the second favourite source, is preferred by only 45.3 %. This outcome put in evidence a weak impact of Volcano Observatory perceived by respondents of Saint Philippe and Sainte Rose during the recent volcanic emergencies.

19.3.9 Information on Volcanic Risk

When respondents were asked to evaluate the received amount of information on volcanic hazards, 61.7 % of the Total sample (N = 1166) consider it as sufficient or more than sufficient. However, it is to take in account the high percentage of respondents, 38.3 %, stating as poor or even lacking the information received on this aspect.

When asked to indicate sources of the received information on volcanic hazards, the role of the Television (84.2 %) and Radio (59.5 %) is recognized as predominant by respondents (N = 1166). It is a matter of concern the very low respondents percentage indicating the School (23.6 %) and the Volcano Observatory (24.2 %) as sources of information, even lower than the one related to Family and friends (28.6 %).

Regarding institutions and preferred information channels from which receiving official information on volcano Emergency Plan in case of crises, once again Total sample respondents (N = 1125) identified the Volcano Observatory (50.1 %) as favourite institution, followed by the Municipality (42.7 %), the Prefecture/Civil Protection (29.9 %) and Regional Government (17.4 %).

The favourite information channels for receiving information on volcanic activity are once more Television and Radio, with 80.2 and 65.6 % of respondents (N = 1175) respectively. It also be noted that 29.4 % indicate Internet.

19.3.10 Sense of Community

Scores of the measure of sense of community were derived by averaging responses made on each of the 18 items of the applied Scale of Sense of Community (Prezza et al. 1999). Scores range from 1 up to 4, with higher values indicating a stronger bond to one's community. The average for the entire sample (N = 890) is 2.84 with a standard deviation of 0.42.

The municipality with the highest sense of community level is Le Port (3.13), Cilaos and Salazie (3.01). The lowest result is the one of Trois Bassins (2.52). The other towns do not differ significantly from each other.

In order to carry out a psycho-social analysis on some results, two variables as the perceived confidence in the Institutions and the perceived level of their preparedness have been constructed. Then a cross-correlation between these variables and the resulted Sense of Community has been performed. The correlation shows that the increase of the sense of community, and therefore the emotional bond with their place of residence, significantly increases both confidence in the information from the Institutions (r = 0.21, $p < 0.001$) and in their preparedness in dealing with a possible volcanic crises (r = 0.20, $p < 0.001$). In addition, the preparation attributed to the Institutions is strongly associated with general confidence that respondents attribute to them (r = 0.30, $p < 0.001$).

19.4 Synthesis

This survey provides the first wide investigation of the knowledge and perception of volcanic hazard on La Réunion Island residents' connected to Piton de la Fournaise volcano. The present study has been realised during an inter-eruptive period and far from the last volcanic crisis, when people's interest and concern about the volcanic issue was not amplified by personal emotions, Media pressure, and effects of information campaign carried out only during the emergency.

The outcomes of this study provide some key parameters to take in account to improve prevention policies also aimed at a better response to future volcanic crises. In particular principal matter of concern as the role of information, the issue of residents' trust in officials and the role of the Observatory are briefly discussed and some suggestions presented below. Information received on volcanic risk has been stated as poor or non-existent for a large percentage of respondents. By coincidence, few weeks before our study the French Red Cross carried out a survey on Natural risks perception and preparedness of population at La Réunion Island (Croix-Rouge Française 2011). Even if the two different surveys' design did not allowed a meaningful quantitative comparison, a qualitative evaluation put in evidence, as common outcomes, the residents' interest to be more informed on natural hazard. That confirms the peculiar role of information in natural risks mitigation actions, and how it would be strategic the use of radio and television, and the development of informative-educational material designed to reach different social contexts. A matter of concern is the missing role of schools resulted from this study. The school system, in collaboration with scientists, has to be more and more involved in these particular actions and, above all, in promoting the knowledge of the island natural peculiar environment. Certainly the very high number of respondents identifying Piton des Neiges as an active volcano has to be a matter of concern in relation with this wrong opinion. Another cause for reflection is the almost total ignorance of the Emergency Plan, which calls for a deeper effort by institutions to increase population preparedness level for future eruptive crises, working on a more effective informative approach.

The resulting trust in authorities and scientists is an element to reflect on also according to outcomes by Wachinger et al. (2013). These results recognized that trust, or lack of trust, in authorities and experts, and personal experience of a natural hazard, impact more on people's risk perception, while others factors as culture, education, personal aspects and Media influence can act as mediators or amplifiers of the main links between experience, trust and perception. The Volcano Observatory, on the basis of the high consensus and confidence expressed by the population involved in this survey, will have to play a leading role in the development and coordination of information/education programmes to be implemented soon at La Réunion Island. Moreover, in addition to their scientific activities, volcanologists should have a more effective public role during volcanic crises.

The last remarkable element of concern is the changing attitude of volcanologists, more and more involved in integrating social aspects in volcano hazards studies, and their being recognised among the main actors in public policy (Donovan et al. 2012). This new role is driven not only by a spontaneous evolution in volcanologists' interests, but also, and above all, by a pressing request that population at risk and decision makers address to scientific community.

Acknowledgments This research has been carried out with the financial support of "Direction Régionale de l'Environnement, de l'Aménagement et du Logement de La Réunion" and the collaboration of the Observatoire Volcanologique du Piton de la Fournaise and Institute de Physique du Globe de Paris in the framework of the Aléa Project. We thank all the school heads, "Principals" for junior high school and "Proviseurs" for high schools, and teachers of science for their invaluable collaboration as well as the "Rectorat de l'académie de La Réunion" for the authorization to access schools. We also thank the responsibles and personnel of all the "Centre Communal d'Action Sociale" and local associations involved in our study. Finally, we wish to thank G. Levieux, A. Finizola, and B. Suski for their fruitful help, collaboration and logistical support before, during and after the distribution/collection stages.

References

Barberi F, Davis MS, Isaia R, Nave R, Ricci T (2008) Volcanic risk perception in the Vesuvius population. J Volcanol Geoth Res 172. doi:10.1016/j.jvolgeores.2007.12.011

Bird DK (2009) The use of questionnaires for acquiring information on public perception of natural hazards and risk mitigation—a review of current knowledge and practice. Nat Hazards Earth Syst Sci 9:1307–1325

Bird DK, Gisladottir G, Dominey-Howes D (2009) Resident perception of volcanic hazards and evacuation procedures. Nat Hazards Earth Syst Sci 9:251–266

Chester DK, Dibben CJL, Duncan AM (2002) Volcanic hazard assessment in Western Europe. J Volcanol Geoth Res 115

Croix-Rouge Francaise (2011) Enquéte sur la perception et la préparation de la population réunionnaise aux risqué naturels. Rapport, 25 Novembre 2011, p 59

Donovan A, Oppenheimer C, Bravo M (2012) Social studies of volcanology: knowledge generation and expert advice on active volcanoes. Bull Volcanol 74 (3):677–689

Gaillard JC, Dibben JL (eds) (2008) Volcanic risk perception and beyond. J Volcanol Geoth Res 172

Greco P (2012) Il fattore P. Analysis and monitoring of environmental risk 3:61–64. ISSN 2240-1520

Likert R (1932) A technique for the measurement of attitudes. Arch Psychol 140:1–55

Meng B, Liu M, Liufu HY et al (2013) Risk perception combining spatial multi-criteria analysis in land-use type of Huianan city. Saf Sci 51:361–373

Morin J, Lavigne F (2009) Institutional and people's response in the face of volcanic hazards in island environment: Case of Karthala volcano, Comoros Archipelago. Part II—deep-seated root causes of Comorian vulnerabilities. SHIMA Int J Res Island Cultures 3(1):54–71

Njome MS, Suh CE, Chuyong G, De Wit MJ (2010) Volcanic risk perception in rural communities along the slopes of mount Cameroon, West-Central Africa. J Afr Earth Sci 58:608–622

Prezza M, Costantini S, Chiarolanza V, Di Marco S (1999) La Scala italiana del senso di comunità. Psicologia della Salute 3–4:135–159

Renn O (1992) Concepts of risk: a classification. In: Krimsky S, Golding D (ed) Social theories of risk. Praeger, Westport, pp 53–79

Ricci T, Barberi F, Davis MS, Isaia R, Nave R (2013a) Volcanic risk perception in the Campi Flegrei area. J Volcanol Geoth Res 254. doi:10.1016/j.jvolgeores.2013.01.002

Ricci T, Nave R, Barberi F (2013b) Vesuvio civil protection exercise MESIMEX: survey on volcanic risk perception. Ann Geophys 56(4):S0452. doi:10.4401/ag-6458

Slovic P (1992) Perception of risk reflections on the psychometric paradigm. In: Krimsky S, Golding D (eds) Social theories of risk. Praeger, Westport, pp 117–152

Wachinger G, Renn O (2010) Risk perception and natural hazards. CapHaz-Net WP3 (report available at: http://caphaz-net.org/outcomes-results/CapHaz-Net_WP3_Risk-Perception2.pdf, retrieved 23 Sept 2014)

Wachinger G, Renn O, Begg C, Kuhlicke C (2013) The risk perception paradox—implications for governance and communication of natural hazards. Risk Anal 33 (6):1049–1065

Volcanological Map of the Plaine des Sables, Piton de la Fournaise

20

Claudia Principe, Andrea Morandi, Andrea Di Muro and Laurent Michon

The "Plaine des Sables" (PdS) is the main entry door for the visitor approaching the summit area of Piton de la Fournaise (PdF) basaltic shield volcano. The plain, located between 2300 and 2400 m asl and 5 km west of PdF summit crater, owes its name to a m-thick continuous cover of basaltic lapilli that strongly contrast with the dominantly effusive behavior of PdF. Several cones of variable volume and size and related lava flows occur scattered inside the PdS.

This map is the first detailed map of PdS and represents a first attempt to introduce synthematic units for the study of the Piton de la Fournaise volcanological district. The new cartography is offered in this first version with a relatively detailed resolution (1:5000) and on a DEM base (courtesy of N. Villeneuve, OVPF). Synthematic mapping is based on unconformity surfaces, which permit to define key geological events inside the stratigraphic succession of volcanic deposits (Salvador 1987). In the PdS case, this approach allowed grouping the whole PdS activity occurred after the collapse of the "Plaine des Sable" caldera (Plaine des Sables Synthem; <60–24 ka; Merle et al. 2010; Staudacher and Allègre 1993) by means of a first order unconformity represented by the collapse scarp itself. The eastern side of the PdS is bounded by a second scarp related to the younger (<4.8 ka) "Enclos Fouqué" caldera (Bachèlery 1981; Ort et al. 2015). Inside the PdS Synthem, the stratigraphic unit of Bellecombe ashes (#bac) groups at least three eruptive events occurred between 4880 ± 35 BP and 2855 ± 35 BP (Morandi et al. 2016; Ort et al. 2015) and represents the main depositional marker over the whole PdS area. Bellecombe ashes were emplaced during the main explosive (phreatomagmatic) event known at PdF (Bachèlery 1981; Michon et al. 2013; Morandi et al. 2016) and whose source area was partly located inside the PdS itself (Ort et al. 2015). Several cones (e.g. #c) and voluminous lava units (#ol and #oc) predate Bellecombe activity. The age of #oc lavas emitted before Enclos Fouqué collapse by central vents located east of the PdS is in the range 5.1–3.5 ka (Staudacher and Allègre 1993). Lavas range from

Electronic supplementary material The online version of this chapter (doi:10.1007/978-3-642-31395-0_20) contains supplementary material, which is available to authorized users.

C. Principe (✉) · A. Di Muro
Istituto di Geoscienze e Georisorse, CNR Pisa, Italy
e-mail: c.principe@igg.cnr.it

A. Morandi
Dipartimento di Scienze della Terra, University of Florence, Florence, Italy

A. Di Muro
Observatoire Volcanologique du Piton de la Fournaise (OVPF), IPGP, Bourg Murat, France

L. Michon
Laboratoire GéoSciences, Université de La Réunion and IPGP, Saint Denis, France

P. Bachèlery et al. (eds.), *Active Volcanoes of the Southwest Indian Ocean*, Active Volcanoes of the World, DOI 10.1007/978-3-642-31395-0_20

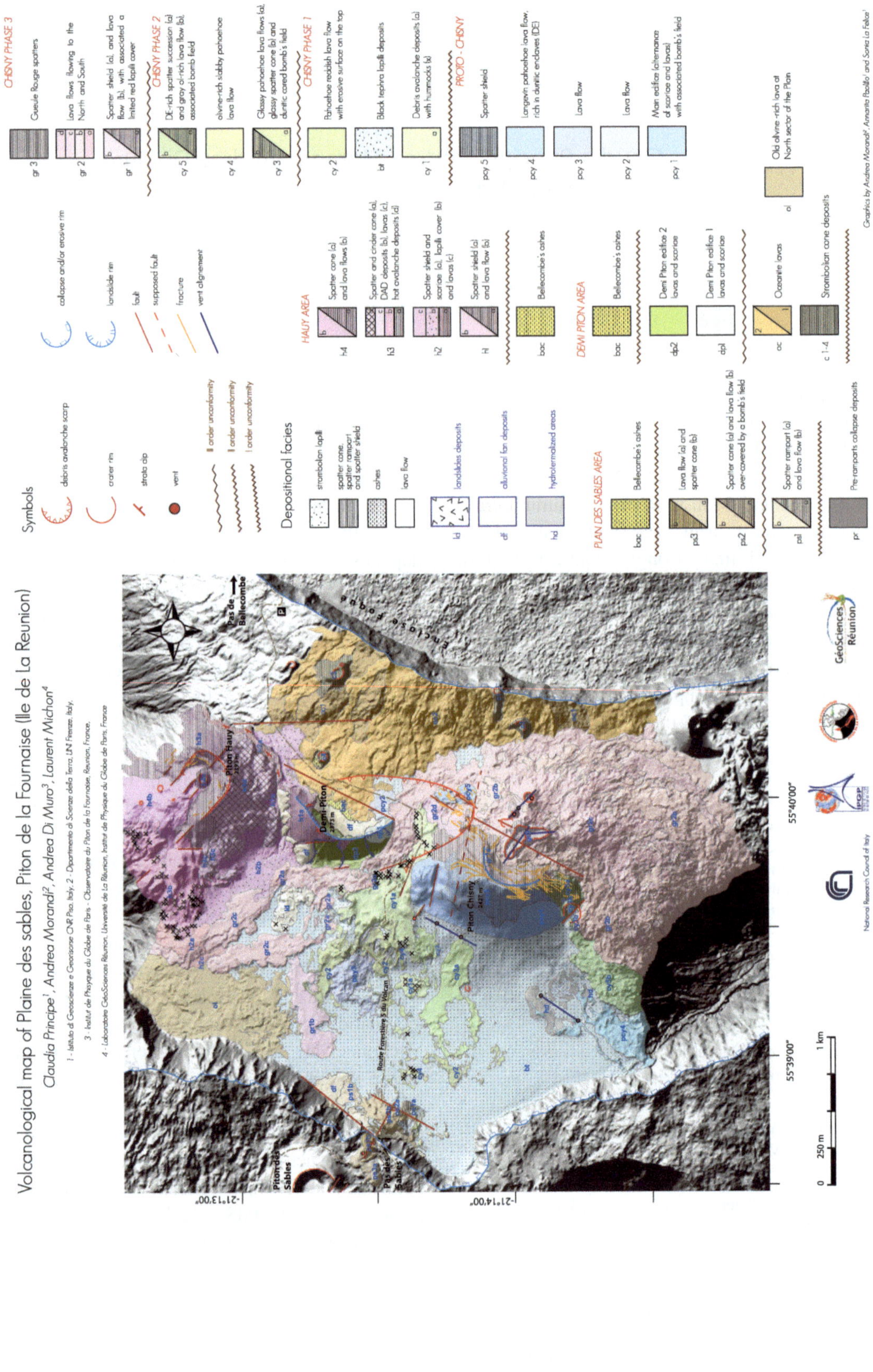

Fig. S1 Supplementary figure: Volcanological map of Plain des Sables, Piton de la Fournaise (Ile de la Réunion, France)

aphyric to olivine—rich (oceanite) basalts and span the whole range of surface textures known at PdF (Morandi et al. 2015). An outstanding feature of many lavas cropping out in the PdS is the abundance of ultramaphic (dunite-werhlite) nodules, testifying unusually deep magmatic storage levels (Bureau et al. 1998).

Two main lapilli units (Chisny Black Tephra, #bt and Haüy Yellow Tephra #h2) permit stratigraphic correlation across most of the PdS area. Beside typical monogenetic activity widespread on PdF massif, at least three large polygenetic centers can be identified inside PdS: Demi Piton, Piton Chisny, and Piton Haüy (Michon et al. 2015; Morandi et al. 2015, 2016). Their eruptive dynamics ranges from violent lava fountaining (e.g. #bt and #h2) to long lasting effusive behavior (e.g. 'Gueule Rouge' lava lake, east of Chisny Rampart; #gr3). With the notable exception of Bellecombe ashes, phreatomagmatic activity seems to play only a very minor role in the activity of the PdS. Bellecombe ashes have been grouped in a single sub-synthem, which permits to propose a relative chronology for the largest cones of the PdS, namely Demi Piton #dp (pre-Bellecombe), Haüy #h and Chisny cones #cy and #pcy (post-Bellecombe). The main age constraints for the youngest part of the volcanic sequence correspond to a radiocarbon age of 1105 ± 60 BP for a lava down the Langevin valley (Bachelery 1981; Tanguy et al. 2011) and an even younger age for the Black Tephra fall emplaced during the late activity of Chisny cone (381 ± 26 BP; Morandi et al. 2016). Mapping of Piton Chisny and Piton Haüy products highlights the presence of two Debris Avalanches deposits (DAD #cy1 and #h3b) and related scarps and hummocks, as well as the morphological evidence of a number of eruptive and non-eruptive fractures.

The unvegetated lapilli cover which gives the name to the Plaine des Sables was mainly originated from the young activity of lava fountaining of Chisny (southern part of the plain) and Haüy (northern part of the plain) eruptive centers. In the central and northern portion of the map, other minor lapilli blankets are present, but have very limited dispersion, confined close to the source areas (Morandi et al. 2016).

Our work demonstrates that the Plaine des Sables has been the site of frequent and sometimes violent eruptive activity even in recent times, shortly before the beginning of human settlement on the La Réunion island. This excentric activity of PdF was periodically able to shed ashes and lapillis all over and outside the Plain des Sables. The most voluminous eruptions emitted lavas able to propagate down the main valleys, towards the villages and the towns located at the feet of the PdF massif. Most important, this excentric activity does not correspond only to monotonous piling of lavas or mild strombolian activity from a single constant source. PdS activity indeed developed on scattered and sometimes long-living eruptive centers and was able to produce a wide span of eruptive behaviors and volcanic hazards. The Plaine des Sables must therefore be considered as one of the most active and potentially hazardous sectors of the PdF massif (Di Muro 2012 and 2015).

References

Bachèlery P (1981) Le Piton de la Fournaise (Ile de la Reunion). Etude volcanologique, structurale et petrologique du Piton de la Fournaise, Ile de La Reunion - Volcanological, structural and petrological study of Piton de la Fournaise, La Réunion island. PhD thesis, University of Clermont-Ferrand, Clermont-Ferrand, pp 255

Bureau H, Métrich N, Pineau F, Semet PM (1998) Magma-conduit interaction at Piton de la Fournaise volcano (Réunion island): a melt and fluid inclusion study. J Volcanol Geoth Res 84:39–60

Di Muro M (2012 and 2015) Evaluation de l'aléa volcanique a La Réunion. Rapport final-année I et année II. Projet BRGM/IPGP

Merle O, Mairine Ph, Michon L, Bachèlery P, Smietana M (2010) Calderas, landslides and paleo-canyons on Piton de la Fournaise volcano (La Réunion island, Indian Ocean). J Volcanol Geoth Res 189:131–142

Michon L, Di Muro A, Villeneuve N, Saint-Marc C, Fadda P, Manta F (2013) Explosive activity of the summit cone of Piton de la Fournaise volcano (La Réunion island): a historical and geological review. J Volcanol Geotherm Res 263:117–133

Michon L, Ferrazini V, Di Muro A, Villeneuve N, Famin V (2015) Rift zones and magma plumbing system of Piton de la Fournaise: how do they differ from Hawaii and Etna? J Volcanol Geotherm Res 303:112–129

Morandi A, Principe C, Di Muro A, Bachelery P (2015) Polygenetic hawaiian activity at Piton de la Fournaise volcano (Plaine des Sables; La Réunion Island). J Volcanol Geoth Res (submitted)

Morandi A, Di Muro A, Principe C, Michon L, Leroi G, Norelli F, Bachèlery P (2016) Pre-historic explosive activity at Piton de la Fournaise volcano. In: Bachèlery P, Lénat JF, Di Muro A, Michon L (eds) Active volcanoes of the Southwest Indian Ocean: Piton de la Fournaise and Karthala. Active Volcanoes of the World. Springer, Berlin

Ort M, Di Muro A, Michon L, Bachèlery P (2015) Explosive eruptions from the interaction of magmatic and hydrothermal systems during flank extension: the Bellecombe ashes of Piton de la Fournaise (La Réunion Island). Bull Volcanol (accepted)

Salvador A (1987) Unconformity-bounded stratigraphic units. Geol Soc Am Bull 98:232–237

Staudacher T, Allègre CJ (1993) Ages of the second caldera of Piton de la Fournaise volcano (Réunion) determined by cosmic ray produced ^{3}He and ^{21}Ne. Earth Planet Sci Lett. 119:395–404

Tanguy JC, Bachèlery P, Le Goff M (2011) Archeomagnetism of Piton de la Fournaise: bearing on volcanic activity at La Réunion Island and geomagnetic secular variation in Southern Indian Ocean. Earth Planet Sci Lett 303:361–368. doi:10.1016/j.epsl.2011.01.019

Part II

Karthala Volcano

The Volcanism of the Comoros Archipelago Integrated at a Regional Scale

21

Laurent Michon

Abstract

The development of the Comoros archipelago in the Mozambique channel has been diversely interpreted since the 1970s. The two end-member causes are, on the one hand, a deep mantle plume that developed a hotspot track from the Seychelles Plateau to the Grande Comore, and, on the other hand, a lithospheric deformation that reactivated transform faults and controlled the magma path. The present work first surveys the sparse geological, geophysical and geochronological data available for this archipelago, re-evaluates the age of the magmatic activity and integrates this evolution at a regional scale. Combining realistic magma production rates, the volume of each edifice and the geochronological, it is showed that the magmatic activity started first in Mayotte about 20 Ma and second, almost simultaneously, in Anjouan, Mohéli and Grande Comore about 10 Ma ago. This magmatism, coeval with magmatic periods in areas surrounding the Mozambic channel, the southern East African rift and Madagascar, is organised in three periods since Late Oligocene. Magmatic provinces are now superimposed with seismic zones and graben structures. In consequence, the Comoros archipelago is tentatively interpreted as part of the East African rift rather than related to a distinct deep mantle plume.

Keywords

Comoros archipelago · Magmatism · Hotspot · Rift · Mozambique channel

L. Michon (✉)
Laboratoire Géosciences Réunion, Université de La Réunion, Institut de Physique du Globe de Paris, Sorbonne Paris Cité, UMR 7154 CNRS, 97744 Saint Denis, France
e-mail: laurent.michon@univ-reunion.fr

P. Bachèlery et al. (eds.), *Active Volcanoes of the Southwest Indian Ocean*,
Active Volcanoes of the World, DOI 10.1007/978-3-642-31395-0_21

21.1 Introduction

The origin of intraplate magmatism and the related volcanoes is a long-lasting debate (Morgan 1972; Sleep 1990; Anderson 2000; Courtillot et al. 2003). On the one hand, non-plate boundary magmatism results from deep mantle ascent rooted in the transition zone or at the core mantle boundary, producing flood basalts and/or linear chains of extinct volcanoes (Morgan 1972; Sleep et al. 1988). On the other hand, all intraplate volcanism is explained by shallow, plate-related stresses that fracture the lithosphere and cause volcanism along these fault zones (Anderson 2000). A reconciling classification proposes that intraplate volcanoes stem from three mantle sources lying at different depths (Courtillot et al. 2003). Primary plumes (e.g. La Réunion and Hawaii) would originate at the core mantle boundary. Superswells in the lower mantle would feed secondary hotspots from the transition zone (e.g. Samoa and Tahiti). Finally, tertiary hotspots would result from asthenospheric convections caused by lithospheric deformation.

In such a scientific debate, the volcanoes of the Comoros archipelago were diversely interpreted in the past decades (Morgan 1972; Hajash and Armstrong 1972; Upton 1980; Emerick and Duncan 1982; Emerick 1985; Nougier et al. 1986; Späth et al. 1996; Class and Goldstein 1997; Class et al. 1998; Courtillot et al. 2003). The aim of this paper is first to review the different interpretations, second to re-evaluate the age of the magmatism and third to integrate it at a regional scale.

21.2 Volcanoes of the Comoros Archipelago

The Comoros archipelago, located in the northern Mozambique channel between the northern tip of Madagascar and Mozambique, is composed of 4 volcanic islands aligned along a WNW-ESE direction (Fig. 21.1). From east to west, they correspond to Mayotte, Anjouan, Mohéli and Grande Comore. Including the Geiser and Leven Banks whose volcanic origin was confirmed by the volcanic rocks dredged on the submarine flanks (Daniel et al. 1972), the overall volcanic alignment remarkably displays the different erosion stages of hotspot volcanoes (Darwin 1842, Fig. 21.1c). Geochronological data seems to corroborate the diachronous magmatic activity, starting in Mayotte at least 10.58 Ma ago, in Mohéli and Anjouan around 3.9 and 5 Ma ago, respectively, and finally in Grande Comore since 0.13 Ma (Hajash and Armstrong 1972; Emerick and Duncan 1982, 1983; Nougier et al. 1986; Debeuf 2004; Table 21.1; Fig. 21.2). It is worth noting that a syenite xenolith dated at 11.1 Ma in Anjouan (Montaggioni and Nougier 1981; Nougier et al. 1986) suggests an early phase of magmatic activity coeval with the volcanism of Mayotte (Nougier et al. 1986). For Mayotte, a maximum age of 15 Ma was estimated for the onset of the magmatic activity on the seafloor (−3300 below sea level) (Nougier et al. 1986). Magmas emitted in the Comoros archipelago belong to the typical silica-undersaturated, alkaline series (Strong 1972; Flower 1973; Nougier et al. 1986). Their isotopic composition indicates a mixing between the Indian Ocean lithosphere and deep mantle sources showing affinities with the HIMU and EM1 components (Emercik 1985; Späth et al. 1996; Class and Goldstein 1997; Class et al. 1998; Deniel 1998; Debeuf 2004). At the scale of the archipelago, lavas from the active Karthala volcano (Grande Comore), which are the less undersaturated and have a distinct isotopic signature (EM1), would result from a higher degree of mantle partial melting than for the other volcanoes (Späth et al. 1996; Class et al. 1998).

Whether the volcanic activity developed on an oceanic or continental crust remains unclear. The presence of sandstone xenoliths in lavas of Mayotte, Anjouan, Mohéli and Grande Comore was first interpreted as evidencing a continental nature of the underlying crust (Lacroix 1922; Flower and Strong 1969). However, magnetic

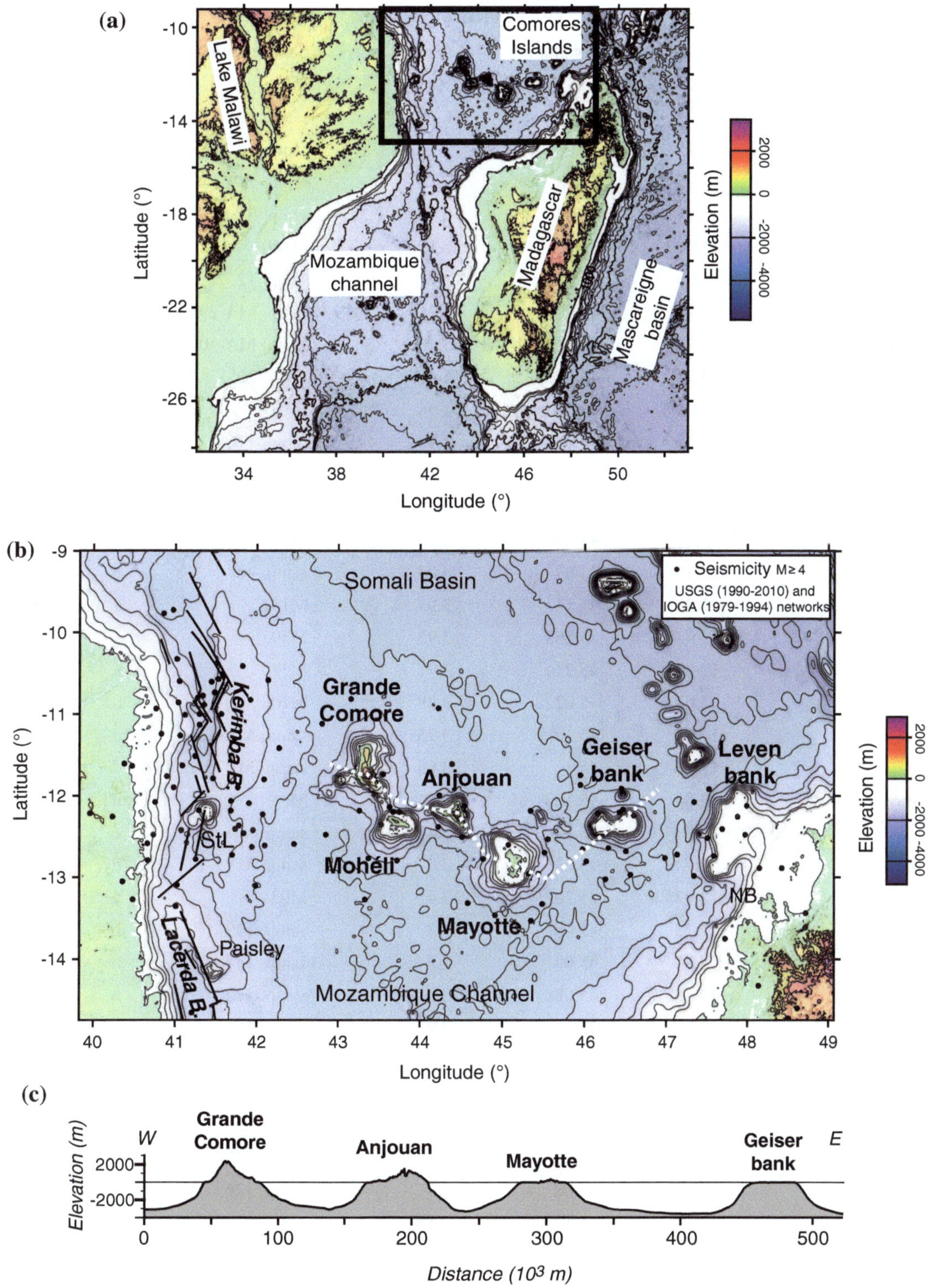

Fig. 21.1 **a** Location of the Comoros archipelago, between the northern tip of Madagascar and Africa. **b** The archipelago is composed from east to west of Mayotte, Anjouan, Mohéli and Grande Comore. Geiser and Leven banks are submarine reliefs of volcanic origin (Daniel et al. 1972). Saint Lazare (*StL*) and Paisley seamounts also correspond to submarine volcanoes related to the Kerimba and Lacerda basins (Mougenot et al. 1989). *Black lines* account for Miocene to Quaternary faults related to the Kerimba and Lacerda basins (after Mougenot et al. 1989). *NB* Nosy Be. **c** Topography along the volcanic track

Table 21.1 Geochronological dating of the volcanic rocks of the Comoros archipelago

Location	Latitude	Longitude	Age (Ma)	Sample name	Reference
Mayotte	−12.9	45.133	10.58	M41	a
Mayotte	−12.75	45.08	7.7	samp. 16	b
Mayotte	−12.75	45.08	7.7	samp. 92	b
Mayotte	−12.75	45.08	7.1	samp. 110	b
Mayotte	−12.75	45.08	6.9	samp. 31	b
Mayotte	−12.989	45.151	6.11	M140	a
Mayotte	−12.75	45.08	5.9	samp. 12	b
Mayotte	−12.71	45.1	5.28	samp. MA-29	c
Mayotte	−12.71	45.1	5.26	samp. MA-30	c
Mayotte	−12.75	45.08	5	samp. 22	b
Mayotte	−12.999	45.142	4.98	M137	a
Mayotte	−12.75	45.08	4.9	samp. 116	b
Mayotte	−12.67	45.0567	4.41	M112	a
Mayotte	−12.649	45.0385	4.36	M104	a
Mayotte	−12.673	45.05	4.2	M109	a
Mayotte	−12.75	45.08	4.03	samp. 46	b
Mayotte	−12.98	45.113	3.97	M132	a
Mayotte	−12.949	45.162	3.95	M39a	a
Mayotte	−12.881	45.214	3.91	M38	a
Mayotte	−12.647	45.022	3.79	M108	a
Mayotte	−12.94	45.1	3.75	samp. MA-81	c
Mayotte	−12.86	45.1	3.65	samp. MY13-1	d
Mayotte	−12.84	45.107	3.64	M55	a
Mayotte	−12.94	45.1	3.41	samp. MA-82	c
Mayotte	−12.74	45.23	3.4	samp. MY1-1	d
Mayotte	−12.975	45.202	3.4	M93	a
Mayotte	−12.75	45.08	3.3	samp. 149	b
Mayotte	−12.75	45.08	3.3	samp. 147	b
Mayotte	−12.75	45.08	3.3	samp. 112	b
Mayotte	−12.77	45.1	3.27	samp. MA-68	c
Mayotte	−12.72	45.08	3.25	samp. MA-25	c
Mayotte	−12.91	45.1	3.24	samp. MA-73	c
Mayotte	−12.75	45.08	3.23	samp. 03	b
Mayotte	−12.75	45.08	3.1	samp. 117	b
Mayotte	−12.75	45.08	2.7	samp. 126	b
Mayotte	−12.94	45.175	2.55	M87	a
Mayotte	−12.944	45.124	2.52	M95	a
Mayotte	−12.75	45.08	2.5	samp. 93	b
Mayotte	−12.9	45.084	2.43	M46	a
Mayotte	−12.75	45.08	2.4	samp. 138	b

(continued)

Table 21.1 (continued)

Location	Latitude	Longitude	Age (Ma)	Sample name	Reference
Mayotte	−12.737	45.144	2.39	M97	a
Mayotte	−12.777	45.23	2.32	M4	a
Mayotte	−12.77	45.18	2.31	samp. MA-37	c
Mayotte	−12.75	45.08	2.2	samp. 83	b
Mayotte	−12.75	45.08	2.15	samp. 85	b
Mayotte	−12.841	45.125	2.13	M82	a
Mayotte	−12.75	45.08	2	samp. 87	b
Mayotte	−12.67	45.09	2	M25	a
Mayotte	−12.991	45.109	1.95	M115	a
Mayotte	−12.75	45.08	1.85	samp. 06	b
Mayotte	−12.75	45.2	1.59	samp. MA-59	c
Mayotte	−12.76	45.1	1.49	samp. MA-38	c
Mayotte	−12.75	45.08	1.49	samp. 144	b
Mayotte	−12.72	45.2	1.42	samp. MY3-1	d
Mayotte	−12.809	45.189	1.41	M56	a
Mayotte	−12.736	45.19	1.3	M123	a
Mayotte	−12.804	45.193	0.8	M58	a
Mayotte	−12.762	45.099	0.78	M59	a
Mayotte	−12.783	45.234	0.15	M53	a
Anjouan	−12.15	44.3	11.1	samp. AN14	b
Anjouan	−12.15	44.3	3.9	samp. AN08	b
Anjouan	−12.24	44.32	1.52	samp. AJ-28-1	d
Anjouan	−12.15	44.3	1.5	samp. AN02	b
Anjouan	−12.17	44.41	1.18	samp. AN-11	c
Anjouan	−12.17	44.38	0.39	samp. AJ-21-1	d
Anjouan	−12.15	44.3	0.36	samp. AN01	b
Anjouan	−12.15	44.3	0.36	samp. AN04	b
Mohéli	−12.25	43.66	5	samp. MO12	b
Mohéli	−12.25	43.66	3.2	samp. MO10	b
Mohéli	−12.25	43.66	2.75	samp. RH-32	c
Mohéli	−12.33	43.84	1.89	samp. RH-22	c
Mohéli	−12.33	43.84	1.53	samp. RH−21	c
Mohéli	−12.28	43.71	1.14	samp. RH-42	c
Mohéli	−12.25	43.66	0.71	samp. RH-33	c
Mohéli	−12.29	43.77	0.62	samp. RH-36	c
Mohéli	−12.25	43.66	0.56	samp. MO04	b
Mohéli	−12.25	43.66	0.48	samp. MO05	b
Grande Comore	−11.85	43.49	0.13	samp. 35G	c
Grande Comore	−11.62	43.27	0.01	samp. GC1-1	d

a Debeuf (2004), *b* Nougier et al. (1986), *c* Emerick and Duncan (1982, 1983), *d* Hajash and Amstrong (1972)

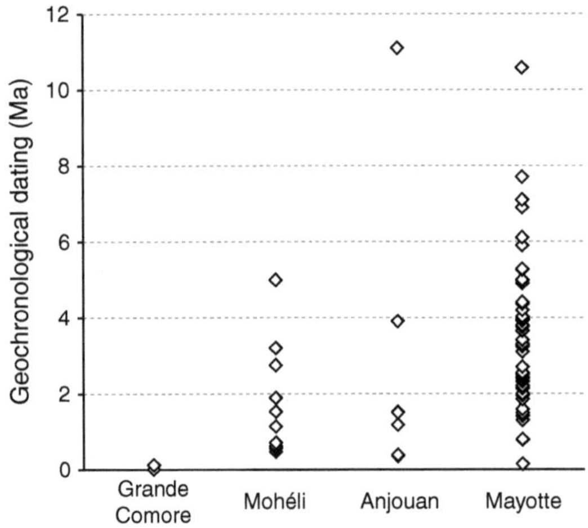

Fig. 21.2 Age of the volcanism in the Comoros archipelago. Geochronological dating from Hajash and Armstrong (1972), Emerick and Duncan (1982, 1983), Nougier et al. (1986), Debeuf (2004)

data suggest that the Comoros archipelago lie on a Late Jurassic to Early Cretaceous oceanic lithosphere resulting from the North-South opening of the Somali Basin (Rabinowitz et al. 1983; Coffin and Rabinowitz 1987). The sedimentary rocks could then correspond to xenoliths of sedimentary rocks resulting from erosion of continental units stemming from either the Davie Ridge (Leclaire et al. 1989) or the Karoo formations cropping out on the eastern or western coasts of Africa or Madagascar, respectively. In such a plausible scenario, the occurrence of the 0.175 km^2 sandstone massif cropping out in the subaerial central part of Anjouan (Debeuf 2004), about 3500 m above the sea floor, requires additional processes than upward magma transport to be uplifted to its present structural level. Whatever the nature of the crust, seismological data clearly show that the volcanic edifices stand on a seismically active zone connecting the northern extremity of Madagascar in the east to the African coast in the west (Bertil and Regnoult 1998; Rindraharisaona et al. 2013) (Fig. 21.1b). This seismicity develops in a regional E-W extensive setting (Grimison and Chen 1988; Foster and Jackson 1998; Heidbach et al. 2008).

21.3 Hotspot and Rift-Related Interpretations of the Magmatism of the Comoros Archipelago

Three main interpretations have been put forward to explain the magmatism of the Comoros archipelago. Their main difference stands on the source of the magmatism. First, the intraplate volcanism together with the island alignment was interpreted as arising from a deep mantle plume (Morgan 1972). The hotspotimpingement, about 40–50 Ma ago below the Seychelles Plateau, would have produced a NNE-SSW volcanic alignment formed by the Amirantes arc, the Farquhar isles and the Comoros archipelago (Emerick and Duncan 1982; Emercik 1985). The plume hypothesis would be supported by the lack of "any significant zone of seismicity" (pp. 417 in Emerick and Duncan 1982) and the EM1 and HIMU signatures of the volcanic rocks (Emerick 1985; Späth et al. 1996; Class et al. 1998). However, in such a model the northwest orientation of the volcanic alignment is hardly compatible with the migration of the African plate toward the NNE (Müller et al. 1993).

Thus, opposite views, involving lithospheric processes instead of deep plumes, have been invoked to explain the development of the magmatism. The Comoros archipelago would represent the surface expression of either a very slowly spreading ridge, similar to the tholeiite and alkali olivine basalt islands at the southern end of the Red Sea (Upton 1980), or the reactivation of lithospheric transform zones (Nougier et al. 1986). Whatever the lithosphere dynamics, the magmatism could be linked to the asthenosphere and be a passive response to forms of lithospheric breakup (Courtillot et al. 2003).

I showed that several lines of evidences can be used to support the hotspot or tectonic origins of the Comoros archipelago. Instead of proposing an additional interpretation, I present a re-evaluation of the age of the magmatism, I compare it to the surrounding volcanic areas and integrate it in the regional geodynamic setting.

21.4 Discussion

21.4.1 Onset of the Volcanism of the Comoros Archipelago

The age of 15 Ma proposed for the beginning of the magmatism in Mayotte has been inferred for a magma production rate of 0.5 $m^3\ s^{-1}$ (Nougier et al. 1986), which corresponds to a mean value determined for some of the most active volcanoes worldwide (between 0.26 and 0.69 $m^3\ s^{-1}$; Wadge 1980). Yet, this value is one order of magnitude larger than the magma emission rate defined for the current shield building stage of Karthala volcano (Grande Comore), between 0.04 and 0.06 $m^3\ s^{-1}$, from the activity of the last two centuries (Bachèlery 1999). Such production rates for Karthala correspond to intermediate values; between those of moderately active hotspot volcanoes like La Gomera, Terceira, Santo Antao or Sao Miguel (production rates around 1–7 × $10^{-3}\ m^3\ s^{-1}$) and those related to active to very active hotspots like Kilauea, Mauna Loa and Piton de la Fournaise (production rates between 0.3 and 1.6 $m^3\ s^{-1}$; Crisp et al. 1984; Gerlach 1990; White et al. 2006).

Assuming an average long-term production rate of the Comoros archipelago in the same range as the current magma emission rate of Karthala, 0.05 $m^3\ s^{-1}$, and considering the total volume of each edifice from the sea floor, I determined a duration of the magmatic activity of around 10, 3.5, 5.5 and 9 Ma for Mayotte, Anjouan, Mohéli and Grande Comore, respectively. Added to the oldest subaerial geochronological ages of each edifice (Emerick and Duncan 1982; Nougier et al. 1986; Debeuf 2004), these results suggest a beginning of the magmatic activity in Mayotte, Anjouan, Mohéli and Grande Comore around 20, 7, 10.5 and 9.1 Ma ago, respectively. Thus the magmatism of the Comoros archipelago probably initiated 5 Ma earlier than initially proposed. Furthermore, as mentioned above, an early activity of Anjouan likely started as early as 11.1 Ma (Nougier et al. 1986). In consequence, I propose that the magmatic activity begun first in Mayotte about 20 Ma ago and second, almost simultaneously, in Anjouan, Mohéli and Grande Comore between 9 and 11 Ma (Fig. 21.3).

21.4.2 Regional Distribution of the Volcanism

Integrated at a regional scale, the magmatism of the Comoros archipelago shows strong petrological and geochemical similarities with the undersaturated alkaline basalt and basanite series emplaced in northern Madagascar and along the Kerimba and Lacerda basins (Daniel et al. 1972; Mougenot et al. 1989; Melluso and Morra 2000). Conversely, they present distinct chemical and isotopic compositions than the Neogene magmas emitted in central Madagascar around the Alaotra graben, which are explained by smaller degrees of partial melting of an heterogeneous mantle lithosphere (Melluso et al. 2011)Fig. 21.4).

The most remarkable point is the timing of the magmatic activity in and around the Mozambique channel. Geochronological data suggest the occurrence of three periods of magmatic activity since the Upper Oligocene (Fig. 21.3). The first period, which developed in late Oligocene, is documented in the Rungwe volcanic province, the southernmost volcanic occurrence related to the East African rift (Roberts et al. 2012b), and in Central Madagascar (Piqué 1999; Bardintzeff et al. 2010). Only a few volcanic manifestations can be linked to this event. The second period of activity started around 20 Ma ago and lasted about 3–4 Ma. The magmatism resumed in the Rungwe volcanic province (Rasskazov et al. 2003) and in Central Madagascar (Piqué 1999; Bardintzeff et al. 2010). It also probably initiated in Mayotte as estimated above from the magma production rates. After a lull of 4–6 Ma, the activity drastically spread over the Mozambique channel and the surrounding areas, forming the third, largest period of activity (Fig. 21.3). The former Rungwe and Central Madagascar provinces were reactivated

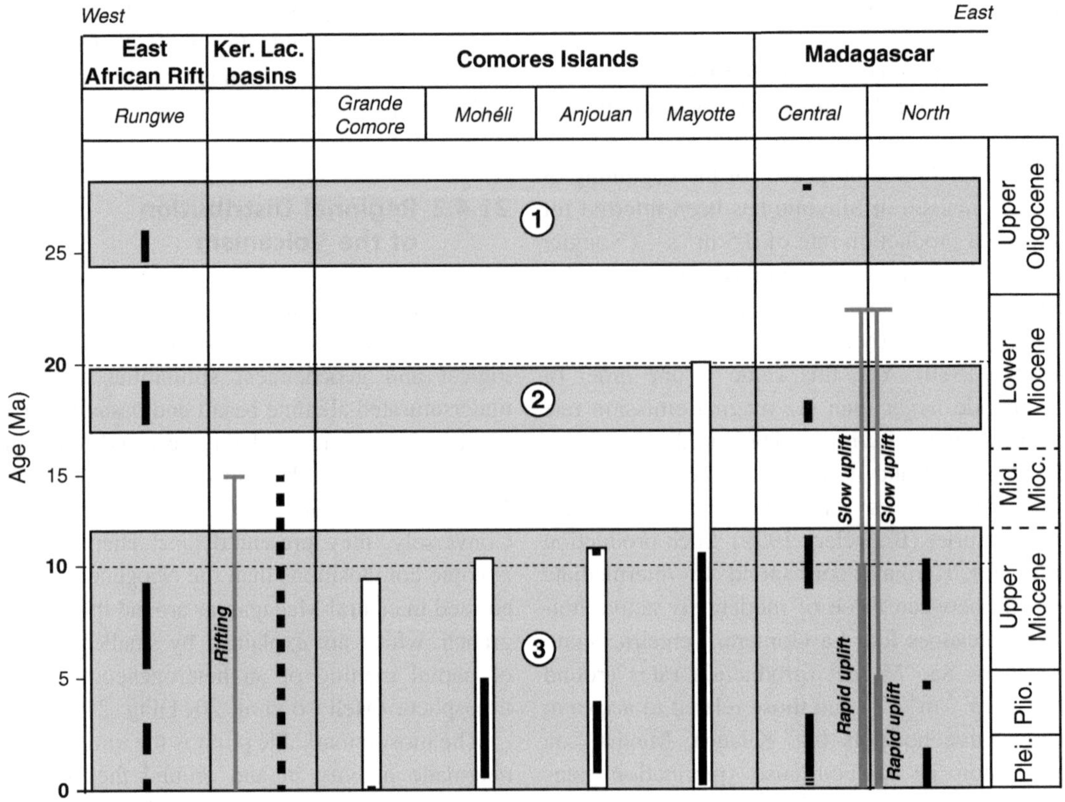

Fig. 21.3 Synthesis of the age of the volcanism (*black bars*) in the Comoros archipelago and the surrounding volcanic provinces. See Fig. 21.4 for location. Geochronological dating from Hajash and Armstrong (1972), Emerick and Duncan (1982, 1983), Montaggioni and Nougier (1981), Nougier et al. (1986), Ebinger et al. (1989, 1993), Rasamimanana et al. (1998), Piqué (1999); Bardintzeff et al. (2001, 2010), Rasskazov et al. (2003), Debeuf (2004), Cucciniello et al. (2011), Roberts et al. (2012a). *White bars* account for the newly estimated activity for the different edifices of the Comoros archipelago. 1, 2 and 3 identify the different periods of volcanism that developed at a regional scale. Rifting of the Kerimba and Lacerda basins is dated by Mougenot et al. (1986a). The uplift of Madagascar is determined from river profile analysis (Roberts et al. 2012b). See text for explanation

(Ebinger et al. 1989, 1993; Bardintzeff et al. 2010). Moreover, volcanic activity appeared almost simultaneously in the southwest and north parts of Madagascar and in the Anjouan, Mohéli and Grande Cormore (Emerick and Duncan 1982, 1983; Nougier et al. 1986; Rasamimanana et al. 1998; Debeuf 2004; Bardintzeff et al. 2010; Cucciniello et al. 2011). Additionally, oceanographic data indicate magmatic occurrences in the Kerimba and Lacerda basins since the middle Miocene (Mougenot et al. 1986a, 1989).

To sum up, the temporal distribution of volcanism of the Comoros archipelago shows striking similarities with the adjacent volcanic provinces, suggesting a narrow genetic link at regional scale.

21.4.3 The Magmatism of the Comoros Archipelago Integrated in the Regional Geodynamics

The Neogene magmatism that developed in the Kerimba and Lacerda basins, in the East African rift and in Madagascar is spatially superimposed to an active seismicity linked to an overall E-W extension (Grimison and Chen 1988; Bertil and Regnoult 1998; Foster and Jackson 1998;

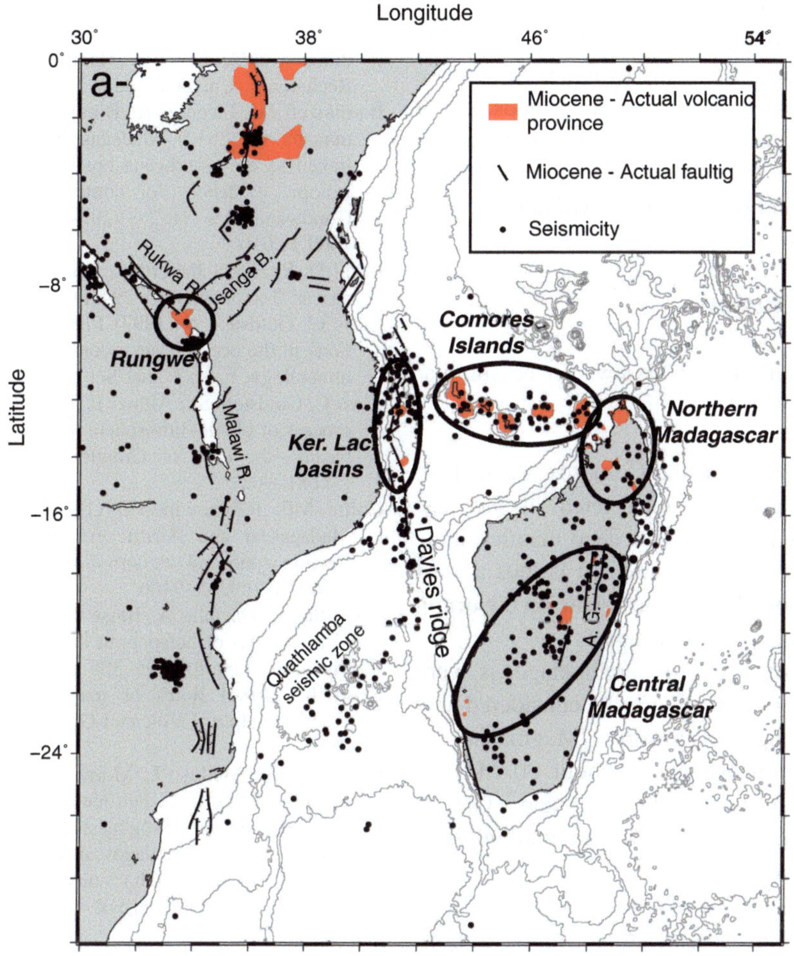

Fig. 21.4 **a** Distribution of seismic activity, volcanism and normal faults active from the Miocene to the actual. Earthquakes from the Institut et Observatoire de Géophysique d'Antananarivo (IOGA) (Bertil and Reignoult 1998). *A.G.* Alaotra graben

Delvaux and Barth 2010; Rindraharisaona et al. 2013). These seismic zones are associated with graben structures initiated in Late Oligocene or Late Miocene (Rukwa rift; Ebinger et al. 1989; Roberts et al. 2012a), Middle Miocene (Kerimba basin; Mougenot et al. 1989) and Pliocene (Madagascar; Laville et al. 1998; Kusky et al. 2010). This entire system of grabens would correspond to the southeastward continuity of the East African rift (Mougenot et al. 1986b; Piqué 1999; Stamps et al. 2008).

The Comoros archipelago is also located in an active seismic zone. However, very few data allow a precise determination of the stress regime that affected this seismic band. No obvious tectonic structures have been identified and only rare focal mechanisms suggest a present strike slip stress regime (Delvaux and Barth 2010). It is worth noting that this stress field is consistent with the Plio-Quaternary transpression described along the N150 trending faults of the Kerimba basin (Mougenot et al. 1986a, 1989). Strike slip and transpression, if any, could explain the occurrence of uplifted blocks like the sandstone one in Anjouan and extension along NE-SW faults (Debeuf 2004).

In consequence, considering the spatial and temporal distribution of the seismic activity, the magmatism and the tectonic structures, I propose that the volcanism of the Comoro Islands do not result from a distinct deep hotspot but instead primarily stems from lithospheric deformation in the general context of the East African rift.

21.5 Conclusion

The main points of this review can be summarized as follows:

1. The magmatism of the Comoros archipelago has been diversely interpreted in the past decades as the result of a deep mantle plume or lithospheric deformation. It is thus a classical example that fed the debate on the origin of intraplate ocean volcanoes.
2. The age of Comoros archipelago is very likely older than previously considered and did not migrated linearly. Magmatism in Mayotte probably started about 20 Ma ago whereas Anjouan, Mohéli and Grande Comore may develop almost simultaneously about 10 Ma ago.
3. The magmatic activity was coeval with that developed in Madagascar and in the Rungwe volcanic province. At a regional scale, three different periods of activity can be distinguished. All these provinces are spatially superimposed to a current seismic activity.
4. The Comoros archipelago is consequently interpreted as the result of a lithospheric deformation in the general context of the East African rift rather than the surface expression of a deep mantle plume.

References

Anderson DL (2000) The thermal state of the upper mantle; no role for mantle plumes. Geophys Res Lett 27(22):3623–3626

Bachèlery P (1999) Le fonctionnement des volcans boucliers. Exemples des volcans de La Réunion et de la Grande Comore. Habilitation à Diriger des Recherches, Université de la Réunion, 239 pp

Bardintzeff JM, Liégeois JP, Bonin B, Bellon H, Rasamimanana G (2010) Madagascar volcanic provinces linked to the Gondwana break-up: geochemical and isotopic evidences for contrasting mantle sources. Gondwana Res 18(2–3):295–314. doi:10.1016/j.gr.2009.11.010

Bertil D, Regnoult JM (1998) Seismotectonics of Madagascar. Tectonophysics 294(1):57–74

Class C, Goldstein S (1997) Plume-lithosphere interactions in the ocean basins: constraints from the source mineralogy. Earth Planet Sci Lett 150:245–260

Class C, Goldstein SL, Altherr R, Bachèlery P (1998) The process of plume–lithosphere interactions in the ocean basins—the case of Grande Comore. J Petrol 39 (5):881–903

Coffin MF, Rabinowitz PD (1987) Reconstruction of Madagascar and Africa: evidence from the Davie fracture zone and western Somali basin. J Geophys Res 92(B9):9385–9406

Courtillot V, Davaille A, Besse J, Stock J (2003) Three distinct types of hotspots in the Earth's mantle. Earth Planet Sci Lett 205:295–308

Crisp JA (1984) Rates of magma emplacement and volcanic output. J Volcanol Geotherm Res 20(3):177–211

Cucciniello C, Melluso L, Morra V, Storey M, Rocco I, Franciosi L, Grifa C, Petrone CM, Vincent M (2011) New Ar ages and petrogenesis of the Massif d'Ambre volcano, northern Madagascar. In: Beccaluva L, Bianchini G, Wilson M (eds) Volcanism and evolution of the African lithosphere, vol 478. Spec Pap Geol Soc Am, pp 257–281

Daniel J, Dupont J, Jouannic Ch (1972) Relations Madagascar-Archipel des Comores (Nord-Est du Canal du Mozambique). Sur la nature volcanique du Banc de Leven. C R Acad Sci Paris 274:1874–1877

Darwin C (1842) The structure and distribution of coral reefs, 2nd ed. Smith Elder and Co, London, 95 p

Debeuf D (2004) Etude de l'évolution volcano-structurale et magmatique de Mayotte (Archipel des Comores, Océan Indien): Approches structurale, pétrographique, géochimique et géochronologique. Unpublished Ph. D. thesis, Université de La Réunion, 243 pp

Delvaux D, Barth A (2010) African stress pattern from formal inversion of focal mechanism data. Tectonophysics 482(1):105–128. doi:10.1016/j.tecto.2009.05.009

Deniel C (1998) Geochemical and isotopic (Sr, Nd, Pb) evidence for plume-lithosphere interactions in the genesis of Grande Comore magmas (Indian Ocean). Chemical Geol 144:281–303

Ebinger CJ, Deino AL, Drake RE, Tesha AL (1989) Chronology of volcanism and rift basin propagation:

Rungwe volcanic province, East Africa. J Geophys Res Solid Earth 94(B11):15785–15803
Ebinger CJ, Deino AL, Tesha AL, Becker T, Ring U (1993) Tectonic controls on rift basin morphology: evolution of the Northern Malawi (Nyasa) Rift. J Geophys Res Solid Earth 98(B10):17821–17836
Emerick CM (1985) Age progressive volcanism in the Comores Archipelago and northern Madagascar. PhD thesis, Oregon State University, 195 pp
Emerick CM, Duncan RA (1982) Age progressive volcanism in the Comores Archipelago, western Indian Ocean and implications for Somali plate tectonics. Earth Planet Sci Lett 60(3):415–428
Emerick CM, Duncan RA (1983) Errata to Age progressive volcanism in the Comores Archipelago, western Indian Ocean and implications for Somali plate tectonics. Earth Planet Sci Lett 62:439
Flower MFJ (1973) Petrology of volcanic rocks from Anjouan, ComoreArchipelago. Bull Volcanol 36:238–250
Flower MFJ, Strong DF (1969) The significance of sandstone inclusions in lavas of the Comores archipelago. Earth Planet Sci Lett 7:47–50
Foster AN, Jackson JA (1998) Source parameters of large African earthquakes: implications for crustal rheology and regional kinematics. Geophys J Int 134(2):422–448
Gerlach DC (1990) Eruption rates and isotopic systematics of ocean islands: further evidence for small-scale heterogeneity in the upper mantle. Tectonophysics 172(3):273–289
Grimison NL, Chen WP (1988) Earthquakes in the Davie Ridge-Madagascar region and the southern Nubian-Somalian Plate boundary. J Geophys Res 93:10439–10450
Hajash A, Armstrong RL (1972) Paleomagnetic and radiometric evidence for the age of the Comores archipelago, west central Indian Ocean. Earth Planet Sci Lett 16(2):231–236
Heidbach O, Tingay M, Barth A, Reinecker J, Kurfeß D, Müller B (2008) The World Stress Map database release 2008. doi:10.1594/GFZ.WSM.Rel2008
Kusky TM, Toraman E, Raharimahefa T, Rasoazanamparany C (2010) Active tectonics of the Alaotra-Ankay Graben System, Madagascar: possible extension of Somalian-African diffusive plate boundary? Gondwana Res 18(2–3):274–294. doi:10.1016/j.gr.2010.02.003
Lacroix A (1922) La constitution lithologique de l'archipel des Comores. C R XIIIème congrès Int Géol Bruxelle 2:949–979
Laville E, Piqué A, Plaziat JC, Gioan P, Rakotomalala R, Ravololsonirina Y, Tidahy E (1998) Le fossé méridien d'Ankay-Alaotra, témoin d''une extension crustale récente et actuelle à Madagascar. Bull Soc Géol Fr 169(6):775–788
Leclaire L, Bassias Y, Clocchiatti M, Segoufin J (1989) La ride de Davie dans le Canal de Mozambique: approche stratigraphique et géodynamique. C R Acad Sci Paris 308:1077–1082
Melluso L, Morra V (2000) Petrogenesis of Late Cenozoic mafic alkaline rocks of the Nosy Be archipelago (northern Madagascar): relationships with the Comorean magmatism. J Volcanol Geotherm Res 96(1):129–142
Melluso L, le Roex AP, Morra V (2011) Petrogenesis and Nd-, Pb-, Sr-isotope geochemistry of the Cenozoic olivine melilitites and olivine nephelinites ("ankaratrites") in Madagascar. Lithos 127(3–4):505–521. doi:10.1016/j.lithos.2011.08.003
Montaggioni L, Nougier J (1981) Les enclaves de roches détritiques dans les volcans d'Anjouan (archipel des Comores). Origine et interprétation dans le cadre de l'évolution du canal de Mozambique. Bull Soć Géol Fr 7:595–601
Morgan WJ (1972) Deep mantle convection: plumes and plate motions. Am Ass Petroleum Geol Bull 56:203–213
Mougenot D, Virlogeux P, Vanney JR, Malod J (1986a) La marge continentale au Nord du Mozambique; résultats préliminaires de la campagne Md40/Macamo. Bull Soc Géol Fr 2(3):419–422
Mougenot D, Recq M, Virlogeux P, Lepvrier C (1986b) Seaward extension of the East-African rift. Nature 321:599–603
Mougenot D, Hernandez J, Virlogeux P (1989) Structure et volcanisme d'un rift sous-marin: le fossé des Kérimbas (marge nord-mozambique). Bull Soc géol Fr 5(2):401–410
Müller RD, Royer JY, Lawver LA (1993) Revised plate motions relative to the hotspots from combined Atlantic and Indian Ocean hotspot tracks. Geology 21(3):275–278
Nougier J, Cantagrel JM, Karche JP (1986) The Comores Archipelago in the western Indian Ocean: volcanology geochronology and geodynamic setting. J African Earth Sci 5:135–145
Piqué A (1999) The geological evolution of Madagascar: an introduction. J Afr Earth Sci 28(4):919–930
Rabinowitz PD, Coffin MF, Falvey D (1983) The separation of Madagascar and Africa. Science 220 (4592):67–69
Rasamimanana G, Bardintzeff JM, Rasendrasoa J, Bellon H, Thouin C, Gioan P, Piqué A (1998) Les épisodes magmatiques du Sud-Ouest de Madagascar (bassin de Morondava), marqueurs des phénomènes de rifting crétacé et néogène. C R Acad Sci 326 (10):685–691
Rasskazov SV, Logachev NA, Ivanov AV, Boven AA, Maslovskaya MN, Saranina EV et al (2003) A magmatic episode in the western rift of East Africa (19–17 Ma). Russ Geol Geophys 44(4):317–324
Rindraharisaona EJ, Guidarelli M, Aoudia A, Rambolamanana G (2013) Earth structure and instrumental of Madagascar: implications on the seismotectonics. Tectonophysics 594:165–181. doi:10.1016/j.tecto.2013.03.033
Roberts EM, Stevens NJ, O'Connor PM, Dirks PHGM, Gottfried MD, Clyde WC et al (2012a) Initiation of the western branch of the East African Rift coeval

with the eastern branch. Nat Geosci 5(4):289–294. doi:10.1038/ngeo1432
Roberts GG, Paul JD, White N, Winterbourne J (2012b) Temporal and spatial evolution of dynamic support from river profiles: a framework for Madagascar. Geochem Geophys Geosys 13(4). doi:10.1029/2012GC004040
Sleep NH (1990) Hotspots and mantle plumes: Some phenomenology. J Geophys Res 95(B5):6715–6736
Sleep NH, Richards MA, Hager BH (1988) Onset of mantle plumes in the presence of preexisting convection. J Geophys Res 93:7672–7689
Späth A, le Roex AP, Duncan RA (1996) The Geochemistry of Lavas from the Gomores Archipelago, Western Indian Ocean: petrogenesis and Mantle Source Region Characteristics. J Petrol 37(4):961–991
Stamps DS, Calais E, Saria E, Hartnady C, Nocquet JM, Ebinger CJ, Fernandes RM (2008) A kinematic model for the East African Rift. Geophys Res Lett 35(5): L05304. doi:10.1029/2007GL032781
Strong DF (1972) Petrology of the island of Moheli, western Indian Ocean. Geol Soc Am Bull 83(2):389–406
Upton BGJ (1980) The Comores archipelago. In: Nairn AEM, Stehli FG, Churkin M (eds) The ocean basins and margins, vol 6. Plenum Press, New-York, pp 21–24
Wadge G (1980) Output rate of magma from active central volcanoes. Nature 288:253–255
White SM, Crisp JA, Spera FJ (2006) Long-term volumetric eruption rates and magma budgets. Geochem Geophys Geosys 7(3):Q03010. doi:10.1029/2005GC001002

Structure and Eruptive History of Karthala Volcano

22

Patrick Bachèlery, Julie Morin, Nicolas Villeneuve, Hamidi Soulé, Hamidou Nassor and Ahmed Radadi Ali

Abstract

Much work has been done on Karthala volcano over the last 30 years concerning the volcanology, structural geology, petrology, geochemistry and geophysics. We focus here on the current state of knowledge about the structure of this typical shield volcano and its recent eruptive activity. Karthala is a large basaltic shield volcano with a typical "Hawaiian" shape, with two well-developed rift zones diverging from a polylobate summit caldera complex. Karthala has not been dissected by erosion, although the concave morphology of its southern and eastern flanks may have resulted from huge flank landslides. Karthala volcano is one of the most active African volcanoes. During the last two centuries, most of the eruptions have been magmatic, emitting alkali basalts. They have been characterised by eruptive fissures which opened along the rift zones, within the caldera, or at low elevation far from the rift zones and the caldera. Karthala's eruptive style is mostly effusive; however, phreatic and

P. Bachèlery (✉) · A.R. Ali
Laboratoire Magmas et Volcans, Université Blaise Pascal—CNRS—IRD, Observatoire de Physique du Globe de Clermont-Ferrand, 5 rue Kessler, 63038 Clermont-Ferrand, France
e-mail: P.Bachelery@opgc.fr

J. Morin
Laboratoire de Géographie Physique de Meudon (LGP), UMR 8591, 1 Place Aristide Briand, 92195 Meudon, France

N. Villeneuve
Observatoire Volcanologique du Piton de la Fournaise (OVPF), Institut de Physique du Globe de Paris Sorbonne Paris-Cité UMR 7154 CNRS, Université Paris Diderot, 75238 Paris, France

H. Soulé · H. Nassor · A.R. Ali
Observatoire Volcanologique du Karthala, CNDRS, Moroni, Grande Comore

P. Bachèlery et al. (eds.), *Active Volcanoes of the Southwest Indian Ocean*, Active Volcanoes of the World, DOI 10.1007/978-3-642-31395-0_22

phreatomagmatic explosions also occur. Karthala erupted regularly (with an average frequency of one eruption every 6–8 years) over the past 100 years until the phreatic eruption of 1991. Since then, its activity has increased, with four eruptions from 2005 to 2007, two of them with significant impacts on the local population. The two phreatomagmatic to violent-strombolian eruptions in 2005 were more explosive and long-lasting than the preceding eruptions, projecting ashes and volcanic debris onto the eastern part of the island, affecting as many as 245,000 people. In the recent past, the 1977 eruption was the most destructive of Karthala's historic lava flows. But in 1858, a lava flow travelled 13 km from the upper part of the north rift zone to the western coast, going close to the capital Moroni. Wherever the location of the eruptive vent, lava flows can reach inhabited areas and the sea in a few hours.

22.1 Introduction

Karthala volcano is certainly not among the best-known shield volcanoes on Earth; much less works have been published on it compared to Hawaii, La Réunion, the Azores or the Galapagos. This is probably the result of its moderate volcanic activity, but also of the relative isolation of the Comoro Islands.

Karthala is a basaltic shield volcano that forms the southern two-thirds of Grande Comore Island (Ngazidja Island) in the Indian Ocean (11.75°S and 43.38°E—Fig. 22.1). After Piton de la Fournaise in La Réunion Island, Karthala is the second most active volcano in the Indian Ocean. The similarities and differences between these two volcanoes must be regarded as revealing significant differences in their deep roots (magmatic sources, plumbing system) and geodynamic context.

Karthala is one of the largest active volcanoes in the world and forms the highest point (summit elevation 2361 m) of the Comoros archipelago. Since the first written reports mentioning eruptive activity at Karthala around the beginning of the twentieth century, only a few studies have dealt with the structural characteristics, eruptive behaviour or petrological and geochemical features of Karthala volcano. Among the pioneering works are those of Voeltzkow (1906), Lacroix (1916, 1922, 1938), Pavlovsky and de Saint-Ours (1953), de Saint-Ours (1960), Flower and Strong (1969), Esson et al. (1970), Strong (1972), Strong and Jacquot (1970), Newitt (1974), Upton (1982), and Krafft (1982, 1983). Geochronological data concerning the four islands of the archipelago have been published by Hajash and Armstrong (1972), Emerick and Duncan (1982, 1983), Nougier et al. (1986), and Debeuf (2004).

In 1988, the *Centre National de Documentation et de Recherche Scientifique des Comores* (CNDRS) established a volcano observatory on Karthala (OVK), with the scientific and technical support of the *University of La Réunion* and the *Institut de Physique du Globe de Paris*. This was the starting point for a new series of geological and geophysical studies of the volcanoes of Grande Comore. The volcanological and stratigraphic features were described in the *volcano-tectonic map of Grande Comore Island* (Bachèlery and Coudray 1993). Geophysical studies, mostly focused on the summit zone of Karthala volcano, were presented in papers by Lénat et al. (1998), Bossu (2000), Rowland and Garbeil (2000), Savin et al. (2001, 2005), and Prévot (2010). The geochemistry of the lavas has been studied by Kaneoka et al. (1986), Reisberg et al. (1993), Späth et al. (1996), Class and Goldstein (1997), Class et al. (1998), Bourdon et al. (1998), Claude-Ivanaj et al. (1998), Deniel (1998), Class et al. (2005, 2009). Petrological and geochemical

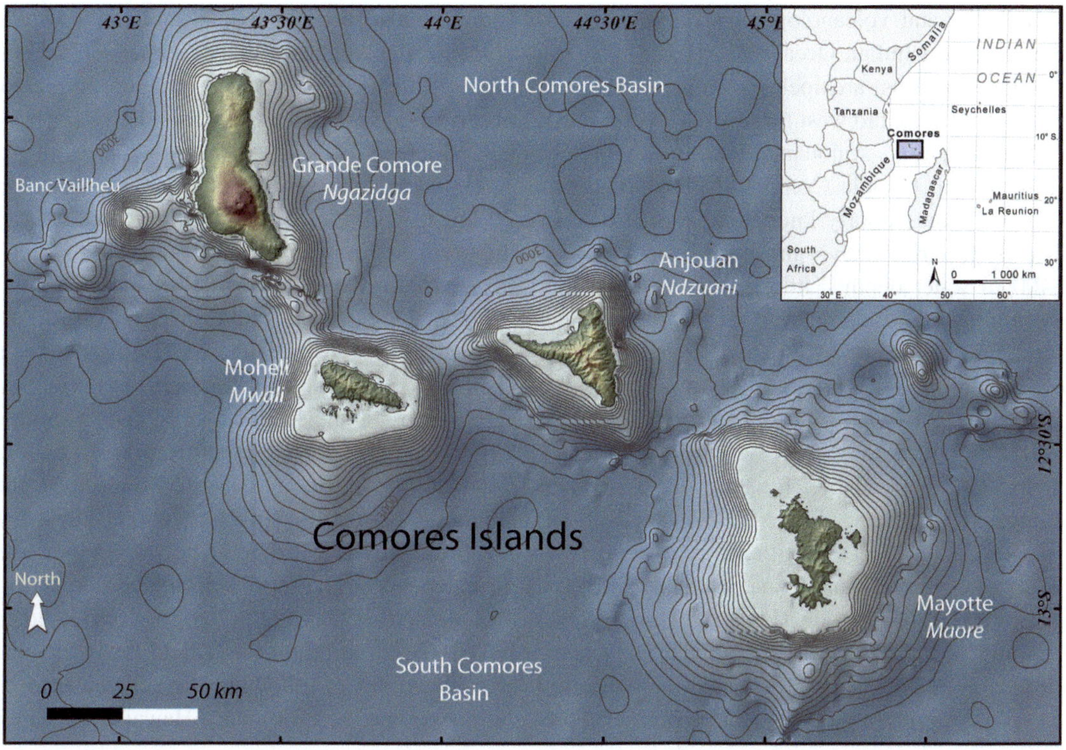

Fig. 22.1 The Comoros archipelago. (Bathymetry from *GEBCO* (General Bathymetric Chart of the Oceans)—http://www.gebco.net/, topography from *SRTM* (Shuttle Radar Topography Mission)—http://www2.jpl.nasa.gov/srtm/). Isobaths every 200 m

aspects are discussed by Bachèlery and Hémond (2016, Chap. 23).

The first eruption recorded by the seismic and deformation networks of OVK was in 1991 (Bachelery et al. 1995). Four other eruptions then occurred between 2005 and 2007. The April and November 2005 eruptions were highly explosive phreatomagmatic to violent-strombolian eruptions (Smietana 2007). They dispersed ash and scoriaceous pumice over the island, and posed major threats to the population and local economy. This helped to raise awareness of the risks associated with the eruptions of Karthala volcano. Several studies on this topic ensued (Morin et al. 2009; Morin and Lavigne 2009; Morin 2012; see also Morin et al. 2016, Chap. 25). Reports on the recent eruptive activity of Karthala volcano are available on the *Global Volcanism Program* website (*Smithsonian Institution*), at: http://www.volcano.si.edu/volcano.cfm?vn=233010.

Here are presented the key features of the structure and the main phases of eruptive activity of Karthala volcano.

22.2 The Tectonic Context of Grande Comore

The tectonic context in which the Comorian volcanism developed, and the various hypotheses concerning its origin, are discussed by Michon (2016, Chap. 21).

Grande Comore is the westernmost and youngest island of the Comoros archipelago (Fig. 22.1). The islands (10–13°S; 43–46°E) are located in the Mozambique Channel, approximately midway between two continental blocks: the northern tip of Madagascar and the East African continent. It is one of the most seismically and volcanologically active areas in the southwest Indian Ocean.

The Comoros archipelago consists of four islands; from east to west these are Mayotte (also called Maore), Anjouan (Nzwani), Moheli (Mwali) and Grande Comore (Ngazidja), together with some poorly known submarine banks. The islands rise above an abyssal plain 3000–3400 m water depth. Little is known about the morphology and the nature of the submarine formations of Grande Comore. No geochemical or geochronological data is available for the submarine part where the oldest rocks for each volcanic edifice may exist. No detailed bathymetric data is available. Existing data show an elongated edifice, like the island itself (Fig. 22.1), but with a significant extension to the west, where a reef-bank (Banc Vailheu) outcrops occasionally at low tide (Figs. 22.1 and 22.2). Nothing is known about this submarine extension, nor the bulge east of La Grille volcano. Steep slopes, mainly observed in the east of the island, and to a lesser extent south of Karthala volcano, may indicate the existence of ancient flank landslides, a type of structure typically observed on basaltic shield-volcanoes (Bachèlery and Villeneuve 2013).

Grande Comore is 65 km in length and about 20 km in width, elongated along a N–S axis to the north of Karthala volcano and along a NW–SE axis to the south of Karthala's summit (Fig. 22.3). This elongated shape is due to its construction by two major shield volcanoes, La Grille in the north and Karthala in the south. Geological studies (Bachèlery and Coudray 1993) have led to the identification of an old volcanic massif, the massif of M'Badjini, in the southernmost part of Grande Comore, but it has not been proven whether or not M'Badjini was an independent centre of volcanic activity.

La Grille volcano (1087 m) is considered to have passed through the shield stage to the post-shield stage, following the scheme of evolution proposed for the Hawaiian Islands

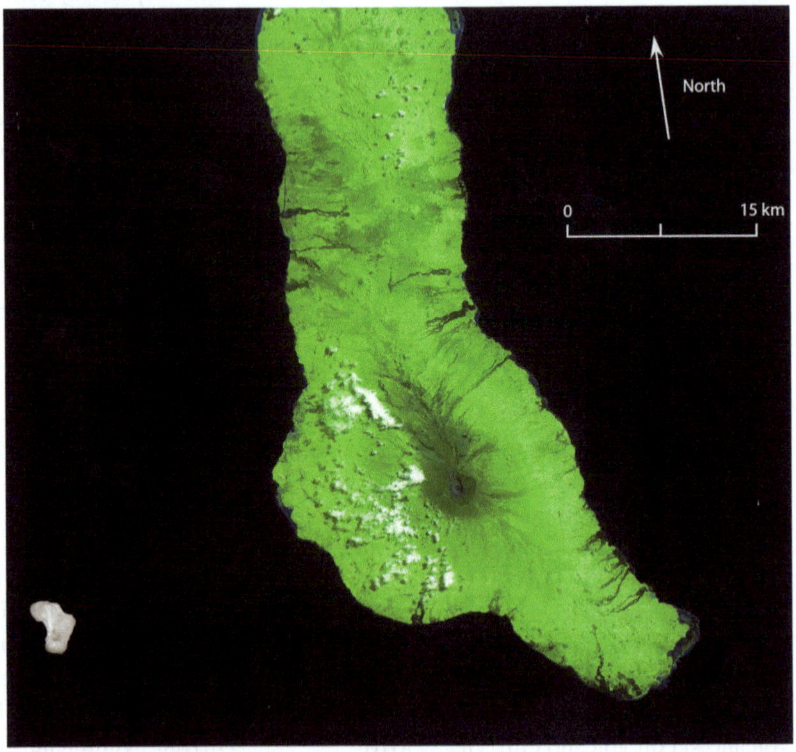

Fig. 22.2 Grande Comore island from space. SPOT 1 XS image acquired 21 March 1992. Young lava flows (in *black*), rift zones and summit caldera of Karthala volcano are clearly visible. La Grille volcano occupies the northern end of the island. Southwest of Grande Comore, the "Banc Vailheu" is emergent at very low tide (*grey area*)

Fig. 22.3 The summit of Karthala volcano from La Grille volcano (*Photo* J. Morin). The picture shows the north rift zone of Karthala volcano. Scoriaceous cones in the foreground belong to the La Grille volcano. They are characteristic of the morphology of the summit area of La Grille, whose last eruption dates back only a few hundred years

(Peterson and Moore 1987). The La Grille morphology is little affected by erosion. Numerous cinder cones, more or less aligned along fissure-like axes, cover the top of the volcano (Fig. 22.3). Based on morphological considerations, the degree of alteration of the volcanics, the vegetation and soil cover, Bachèlery and Coudray (1993) defined four volcanic units of different relative ages from 'ancient' to 'recent'. No clear evidence for the existence of a summit caldera can be highlighted. Bachèlery and Coudray (1993) proposed that the flat summit of La Grille volcano might represent the remnant of a large summit caldera, now completely buried beneath a cap formed by the younger lavas and pyroclastic deposits.

La Grille lavas are mainly silica-undersaturated basanites, with rare alkali basalts and nephelinites (see Bachèlery and Hémond 2016, Chap. 23). Numerous Strombolian cones occupy the top and flanks of the volcano. Rift zones are less pronounced than on Karthala, but alignments of cinder cones suggest past rift zones with a north–south orientation to the south of La Grille summit, and with a NNW–SSE orientation on the western flank.

Along the north coast, several tuff-rings result from violent hydromagmatic eruptions. The change from shield-building stage to post-shield stage of basaltic shield-volcanoes, as proposed for the Hawaiian Islands evolution, is frequently marked by the formation of scattered cinder cones, in response to an increase in viscosity and gas content of the more alkalic and evolved magmas, and deeper magma storage. This activity is commonly restricted to the summit part of the volcano, and the caldera ceases to be visible when cinder cones and large quantities of pyroclastic material cover the top of the shield volcano (Bachèlery and Villeneuve 2013, and references therein). La Grille volcano seems to be at a more advanced evolutionary state relative to Karthala volcano, with less frequent eruptions, more alkalic magmas and a completely filled caldera. However, petrological and geochemical features of La Grille magmas reflect their primitive character, and different partial melting processes from those inferred for Karthala volcano (see Bachèlery and Hémond 2016, Chap. 23).

La Grille volcano has erupted several times in the last few thousand years, and a further eruption cannot be ruled out, considering the relatively recent age of some of the last eruptions of this volcano (conventional radiocarbon ages of 1300 ± 65 years BP and 740 ± 130 years BP—from ^{14}C dating of charcoal collected below scoriaceous tephra layers—in Bachèlery and Coudray (1993)—2σ calibrated ages (Calib 7.0) are respectively AD 635–886 and AD 1029–1424). These recent ^{14}C ages and the remarkably fresh morphology of some lava flows testify to very recent volcanic events. The discovery of fragments of pottery in lapilli fallouts to the south of La Grille shows that the first inhabitants of the island witnessed eruptions of this volcano.

22.3 Stratigraphy and Morphology of Karthala Volcano

Karthala has erupted frequently in historical times, most recently in 2007. Unfortunately, there is a serious lack of geochronological data that can be used to put absolute time constraints on the Karthala units. Only two K–Ar ages have been published. One is for the recent formations of the western slope of the volcano (0.01 ± 0.01 Ma—Hajash and Armstrong 1972), the other for a more ancient lava flow south of the island (0.13 ± 0.02 Ma—Emerick and Duncan 1982); the latter probably belongs to the M'Badjini massif.

Bachèlery and Coudray (1993) established a relative chronology by using criteria such as the alteration, soil and vegetation development, and petrographic considerations. They recognized three units from the present pyroclastic fall and lava flows: "Karthala actuel", the youngest formations, with unaltered lava flow surfaces and no vegetation, corresponding to eruptions mostly younger than 1850; "Karthala récent", which have unaltered or little altered lava flow surfaces with identifiable morphological characters that can be used to estimate their relative age—due to the tropical climate they are largely covered by vegetation except at high altitude; and "Karthala ancien", the oldest and ubiquitously-altered formations, with no remnant of lava flow surface textures, having developed instead a soil cover.

Most of the flanks of the volcano are covered with "Karthala récent" units, and very old units are probably not visible due to the absence of deep valley erosion. "Karthala ancien" is only exposed on the eastern flanks of Karthala volcano, from the coast up to about 1900 m altitude.

The same type of altered lava flows characterizes the southernmost region in the M'Badjini massif (Fig. 22.4). M'Badjini massif is distinguished by its morphology, superficial alteration, but also thanks to a relatively dense network of fractures and intrusions (dykes) with orientations that differ from the Karthala SE rift zone

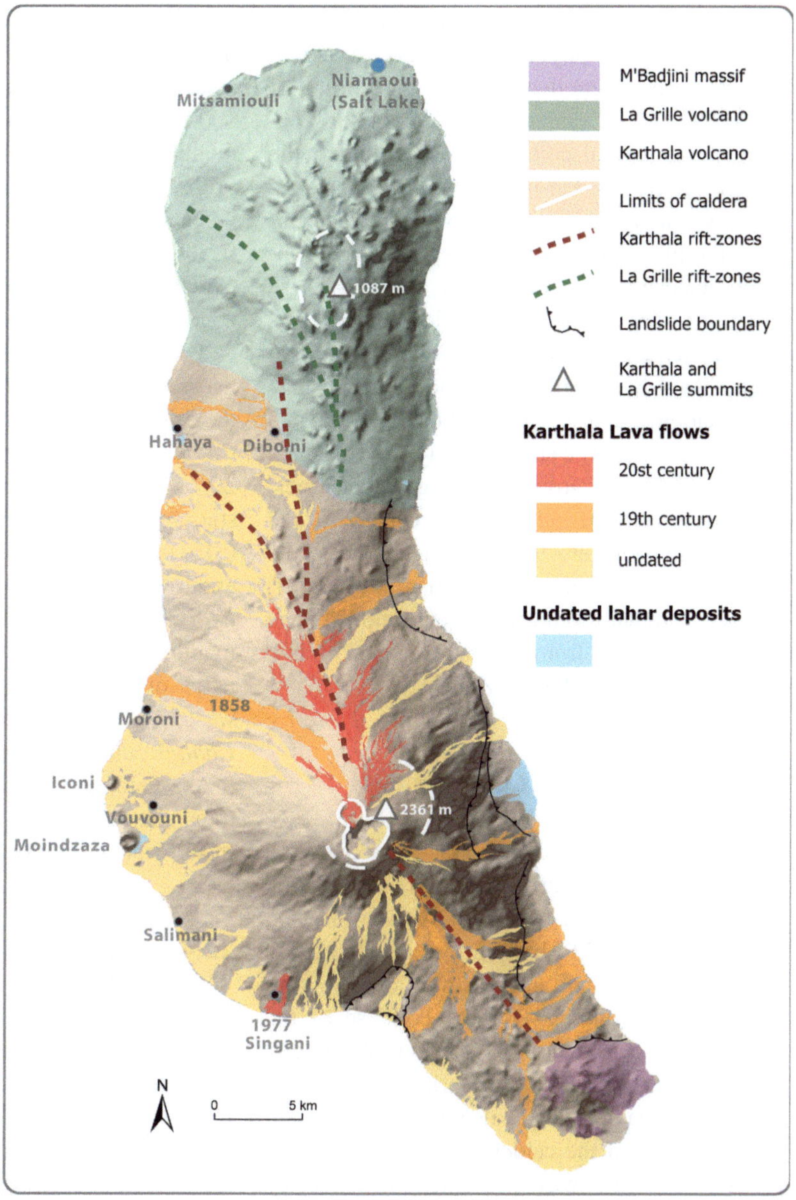

Fig. 22.4 Simplified volcanological map of Karthala volcano (Morin 2012, from Bachèlery and Coudray 1993)

(Bachèlery and Coudray 1993). Structural data tends to distinguish M'Badjini from Karthala, but the lavas of the M'Badjini massif are similar to those of the "Karthala ancien". They differ from the formations of "Karthala actuel" and "Karthala récent" by the presence of plagioclase phenocrysts, which have not been observed in the more recent lavas, and by their surface alteration.

22.4 Main Structural Features of Karthala Volcano

Unquestionably, the most prominent structural feature is the two rift zones extending on opposite sides of the summit. Rift zones are axes of major intrusive and effusive activities, and consist of alignments of cinder and spatter cones along eruptive fissures. At Karthala, they form a ridge-like topography. The second structural feature is the caldera that occupies the summit of the volcano. The caldera results from the coalescence of several collapse units including the present pit craters. On the south and east flanks of the volcano, areas with steep slopes and concave morphology are interpreted as large subaerial landslide amphitheatres. All of these structures, the basaltic composition of the lavas and the frequency of the eruptions, indicate that Karthala volcano is still in a shield-building stage of development. The lack of both erosion and soil over a very large surface area of the volcano is typical of rapid growth. During recent times, pit craters that form and are filled repeatedly, and magma injection into the rift zone, indicate the probable existence of a shallow magma chamber. Lavas are emitted from central vents and rift zone fissures, but also from eccentric vents at low altitude away from the rift zone.

22.4.1 The Caldera Complex

The Karthala caldera complex is approximately elliptical (3.5 × 2.8 km—Fig. 22.5). It was first described by Lacroix (1938), and Pavlovsky and de Saint-Ours (1953), but the first detailed study was that of Strong and Jacquot (1970) who proposed a chronology for the collapse of the different units. Newitt (1974), Bachèlery and Coudray (1993), Mouginis-Mark and Rowland (2001), and Nassor (2001) also depicted the main characteristics of the caldera, but the most comprehensive work was published recently by Poppe (2012).

Karthala caldera appears to have formed in a similar manner to typical Hawaiian calderas, such as *Mokuaweoweo* caldera on Mauna Loa or Kilauea caldera. The caldera results from the coalescence of several sub-circular collapse units with diameters varying from 1300 to 1800 m. The caldera occupies the summit zone of Karthala volcano, with the main pit-crater, called *Choungou-Chahale*, open in the middle (Fig. 22.5), at the junction of the two well-developed rift zones. Its elongated shape corresponds exactly to the rift zone directions. The caldera rim is over 100-m-high in some places, but the caldera wall is totally buried in the north, at the *Porte d'Itsandra*. From an orientation analysis of the main volcano-tectonic lineaments of the caldera (N130°–N140° and N340°–N10°), Bachèlery and Coudray (1993) and Poppe (2012) argue for a rift zone control on the collapse structures.

Eruptions within the caldera take place from fissures or vents opened on the northern part of the caldera floor (1965, 1972, 2007) or within the *Choungou-Chahale* crater (April and November 2005, May 2006). Lénat et al. (1998) provide a study of Karthala caldera using the self-potential (SP) method. Positive SP anomalies reaching about 1000 mV have been interpreted as upward hydrothermal circulation driven by heat sources mainly located below the northern and western parts of the caldera. Thus, the northern and western parts of the caldera, and the *Choungou-Chahale,* currently appear as the most active parts of the caldera complex. The most recent structures, the strongest hydrothermal activity and the most recent summit eruptions are located here.

22.4.2 Choungou-Chagnoumeni Pit-Crater

In the northern lobe of the caldera, there is a small pit-crater (Figs. 22.5 and 22.6), the *Choungou-Chagnoumeni* ("new crater" in Swahili—about 200 m in diameter). Its formation was related to the eruption of 1918. From a depth of about 150 m in 1936, it was gradually infilled to a depth of 50–60 m in 1952, about 30 m in 1965, and only 10 m after the eruption of 1972. The crater was the site of the January 2007 eruption, and was completely infilled by the lava emitted during this eruption, which slightly overflowed across its western edge. Active fumaroles exist along its southern edge, where the crater wall is highest (approximately 10 m), and along the 2007 eruptive fissure crossing the crater.

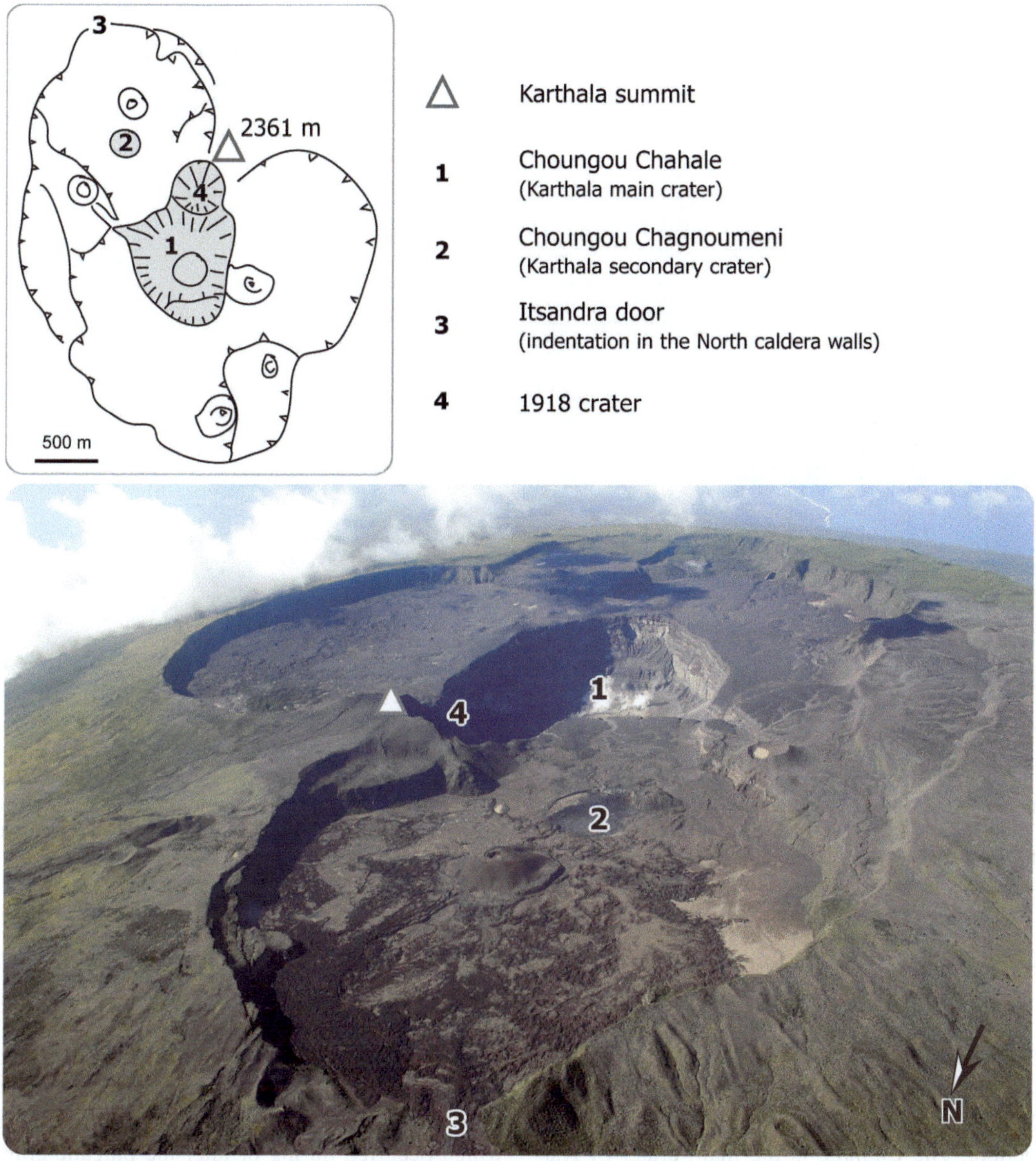

Fig. 22.5 Aerial oblique view of Karthala summit caldera (*photo* J. Morin). The lobes and terraces shaping the caldera complex are visible. Chougou Chahale lies in the middle of the caldera with Choungou Chagnoumeni in the north. The caldera floor is widely covered by the tephra fallout from the 2005 eruptions. The black lava flow of 2007 is visible in the Chougou Chagnoumeni. *Upper left* Structural outline of the Karthala caldera

Another pit-crater is clearly identifiable on the aerial photograph of 1961, in the northern lobe of the caldera, to the northwest of *Choungou-Chagnoumeni*. It had a diameter of approximately 300 m, and was less than 10 m deep at the time. It was completely obliterated by lava flows emitted during the 1972 eruption.

Fig. 22.6 The northern lobe of Karthala caldera in January 2007 (*Photo* P. Bachèlery). Choungou-Chagnoumeni pit-crater is clearly visible in the middle of the picture, to the north of the Chougou-Chahale crater. Choungou-Chagnoumeni is filled by lava flows emitted during the January 13 2007 eruption. A *line of spatter-cone* marks the unusual orientation of this eruptive fissure across Choungou-Chagnoumeni. The scoria cone NE of Choungou-Chagnoumeni was formed in September 1972. A scoria cone of unknown age is cut by the S–SW pit-crater rim. The caldera floor and surrounding areas are largely blanketed by ash-falls from the two explosive eruptions of 2005

22.4.3 Choungou-Chahale Central Crater

Choungou-Chahale ("ancient crater" in Swahili —overall dimensions: 1.4–0.8 km, and about 250 m depth) pit-crater occupies the centre of the caldera. The *Chahale* is the main crater of Karthala volcano and has been the most active in historical times. It results from the coalescence of several units, a combination of collapses and crater openings during violent explosive eruptions. In the north wall, just below the summit of the volcano, the crater opened by the phreatic eruptions of 1918 is still easily recognizable (Fig. 22.5). South of *Chahale,* a platform at mid-depth (Fig. 22.7) is the only remnant of the crater that existed prior to the phreatic eruption of 1918 that enlarged and lowered the crater.

The *Chahale* crater has experienced frequent morphological changes during the last 150 years. The earliest descriptions date back to 1871, with the description of a crater which was less deep and narrower than today (von der Decken, in Newitt 1974). The first well-described changes are those which occurred during or after the phreatic eruption of 25 and 26 August 1918. This followed a short fissure eruption that occurred from August 11 to 13 on the northern rift zone of the volcano, and another, bigger one, which took place from February to April 1904 on the southeast rift zone. After the 1918 eruption, the diameter and the shape of *Choungou-Chahale* changed with the opening of a new crater (approximately 450 m in diameter) north of *Chahale,* possibly following an explosion, and the collapse of the *Choungou-Chahale* floor by approximately 150 m.

Fig. 22.7 The Choungou-Chahale crater on May 24, 2006 (*Photo* J. Morin). The crater is partly filled by a frozen lava lake produced during the 2005 eruptions. Beyond the crater to the south, the platform corresponds to the floor of the crater which preceded the 1918 phreatic eruption. *Light-colored* thick lava flows in the crater walls mark ancient pit-craters filled by lavas characteristic of a slow cooling favouring plagioclase crystallisation as microlites in the groundmass. Dykes and fractures cut the crater walls (not visible at the photo scale). Photo taken from the summit of Karthala, looking south

A water lake has formed periodically in the *Choungou-Chahale*. This seems to happen when the depth exceeds 300 m. The presence of greenish lakes has been noted in the past, as in the years 1947–1948 (Poisson 1949; Battistini 1967) and in April 1952 (Pavlovsky and de Saint-Ours 1953; de Saint-Ours 1960). The new crater which formed in the floor of *Choungou-Chahale* during the phreatic eruption of 1991 was also occupied by a bluish-green water lake. Frequent and significant variations in the lake level (of a metric order) could be observed between 1991 and 1997 thanks to the periodic appearance or disappearance of an island in its centre. They did not seem to correlate with the rate of rainfall recorded on Grande Comore, but more likely with changes in the rate of seismicity recorded under Karthala. Variations in the lake level have been interpreted as a rise of the water table in a geothermal system (Savin et al. 2001). The lake disappeared during the eruption of April 2005, returned a few weeks later, and disappeared once more during the November 2005 eruption.

22.4.4 Signs for the Probable Existence of an Older Caldera

About four kilometres northwest of the present caldera, small relict areas of high relief, surrounded by recent lava flows and a more or less flat topography, indicate the probable presence of an ancient collapse boundary. From their alteration and mineralogy, Bachèlery and Coudray (1993) propose that these rocks belong to the "Karthala ancien" unit, which is widely represented on the eastern flank of the volcano.

In addition, the morphology of the summit area of the volcano shows a clear break in slope between the altitudes of 2200 and 2300 m, from the north of the present caldera round towards the western and southern flanks of the volcano. This structure has been interpreted as the trace of an ancient caldera, probably corresponding to the "Karthala ancien", now almost completely buried under the recent lava flows.

22.4.5 Rift Zones and Links with La Grille Volcano

Like Hawaiian volcanoes such as Mauna Loa, the shape of the Karthala shield is controlled by the presence of two prominent opposing rift zones, one trending north and the other southeast. Rift zones are known to be preferential channelways that bring magma to the surface. At the surface, they are marked by alignments of spatter ramparts, spatter cones and cinder cones. Rare open cracks can also be seen, mainly in the upper part of the rift zone, near the summit caldera border.

During the recent period, the functioning of the two rift zones has differed slightly. Between 1848 and 1880, eruptions took place along both the north and southeast rift zones, and the quantities of magma emitted by the two structures were equivalent. From 1904 to 1972, volcanic activity was localized along the north rift zone and its extension within the caldera. Consequently, the northern flank of Karthala looks morphologically much younger and devoid of vegetation compared to the southern flank. "*La Soufrière*" fumarole, located 2 km north of the caldera at 2030 m a.s.l., along the north rift, has been active for over 50 years, which helps reinforce this impression of youth. Activity at "*La Soufrière*" results in the continuous emission of sulphur dioxide at a temperature of between 93° and 96°. During the past 170 years, the north rift zone has been slightly more productive than the southeast rift zone (49 and 43 % respectively of the total volume of magma produced at Karthala volcano). The magma emitted by the two rift zones is very similar, with a minor tendency towards more olivine- and pyroxene-rich magmas along the southeast rift zone compared to the north rift zone.

The southeast rift zone extends more than 20 km on land. Its submarine extension is not precisely known. Using the GEBCO (General Bathymetric Chart of the Oceans) bathymetry database (Fig. 22.1), it can be estimated at between 5 and 10 km. The few available bathymetric data between Grande Comore and Moheli show the existence of a NNW–SSE ridge extending the southeast rift zone of Karthala, but there is no evidence that this ridge is really a continuation of the rift zone. The rift zone is rectilinear, with an orientation of N145°. This eruptive axis was active on numerous occasions in the past, sometimes with cracks extending over several kilometres long. A fumarolic area, assumed to be responsible for the death of 17 people in 1903, was also reported 1600 m southeast of the caldera (Krafft 1983). Near the M'Badjini massif, recent eruptive fissures are shifted southwest relative to the axis of the rift zone and are locally oriented N160°.

The north rift zone extends about 25 km from the northern end of the caldera (*Porte d'Itsandra*), with an average orientation of N350°. In its middle portion, about 15 km from the summit, it divides into two branches, one extending into the western flank of the La Grille massif, while the other bends considerably to the northwest. The north rift zone has been active frequently during historical times with some very long eruptive fissures opening up at times, for example a 6-km-long crack which opened in 1904. There is no obvious extension of these two branches of the north rift zone below sea level.

The current activity of Karthala is not only limited to the summit caldera and the rift zones. Eruptive fissures open occasionally on the flanks, radially from the summit of the volcano. The 1977 eruption (Krafft 1982) is the most recent example of this type of event. Only the eastern flank of Karthala appears relatively spared by the eruptions, with eruptive fissures being very rare in this sector of the volcano. Some of the eruptions on the west flank, like the one in 1977 (Fig. 22.4) correspond to "eccentric" eruptions as defined by Rittman (1963). We note a clear tendency for the open eruptive fissures on the west flank of Karthala to cluster along an east–west axis. The submarine ridge that emerges at the *banc Vailheu* reef appears to extend this axis (Figs. 22.1 and 22.2).

22.4.6 Flank Landslides

The eastern flank of Karthala is characterized by unusually steep slopes, sometimes greater than 30° (Fig. 22.8). Many curvilinear escarpments, draped by recent lava flows, can be identified. These structures, extending over 30 km along the east coast, are extended to the south by an oblique wall dipping into the ocean (the Foumbouni fault). These structures probably represent respectively the headwall and the lateral boundary of a slump or a landslide affecting the east flank of the volcano. The submarine morphology immediately offshore also shows the presence of anomalously steep slopes (Fig. 22.1). Rocks exposed within this steep-sloped area, or in the Foumbouni fault belong to the "Karthala ancien" and "M'Badjini" units. Recent lava flows that cover the scarps are not affected by any gravity tectonics. Thus, the landslide structures on the eastern flank of Karthala appear to be old structures, possibly now largely inactive.

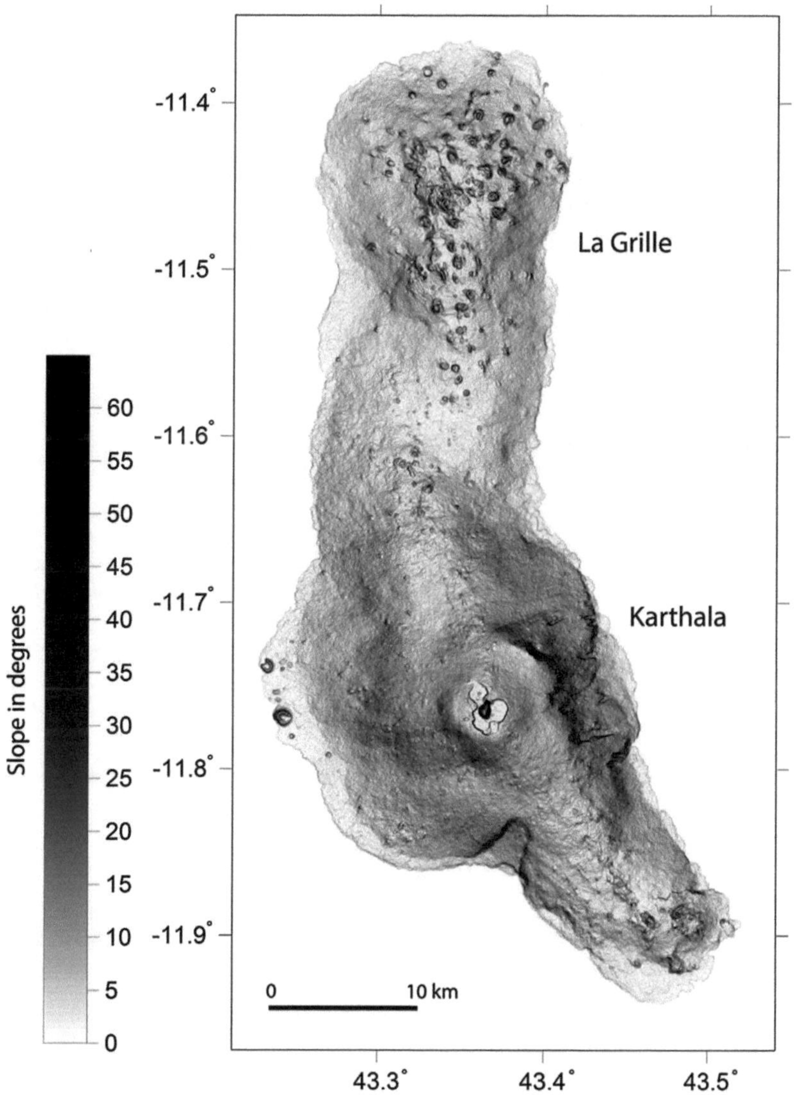

Fig. 22.8 SRTM-derived slope map for Grande Comore island (Digital Terrain Model at 1° — approximately 30 m). Steep slopes are clearly visible on the eastern flank of Karthala, and an amphitheatre-like scar is also present in the south-western flank of the volcano. This reflects the existence of flank landslides in the recent past of the volcano. The summit of La Grille is scattered with scoria cones, while the slopes of Karthala show much smoother morphologies. Slope values in degrees

A small amphitheatre-like scarp also exists south of the Karthala summit. Although it is almost entirely buried by sub-historical flows, its morphology and the age of the formations exposed in the cliffs allow this landslide to be linked to the "Karthala recent".

22.5 Recent Eruptive Activity

The eruptive activity at Karthala volcano has many similarities to that of volcanoes in Hawaii, La Réunion or the Galapagos islands. Unfortunately, the historical chronicles that document this past activity are rare. The first written accounts of eruptions and maps of lava flows (Fig. 22.9) were published in the early 20th century (Lacroix 1920, 1938). Following this, only occasional descriptions of certain eruptions were published (Upton et al. 1974; Krafft 1982; Bachèlery et al. 1995). The first geological map of Grande Comore was compiled from 1985 to 1993 and published by Bachèlery and Coudray (1993). It contains a complete map of the lava flows, eruptive cones and fissures for the whole island (Karthala and La Grille), and a compilation of data on the known Karthala eruptions. This database is supplemented

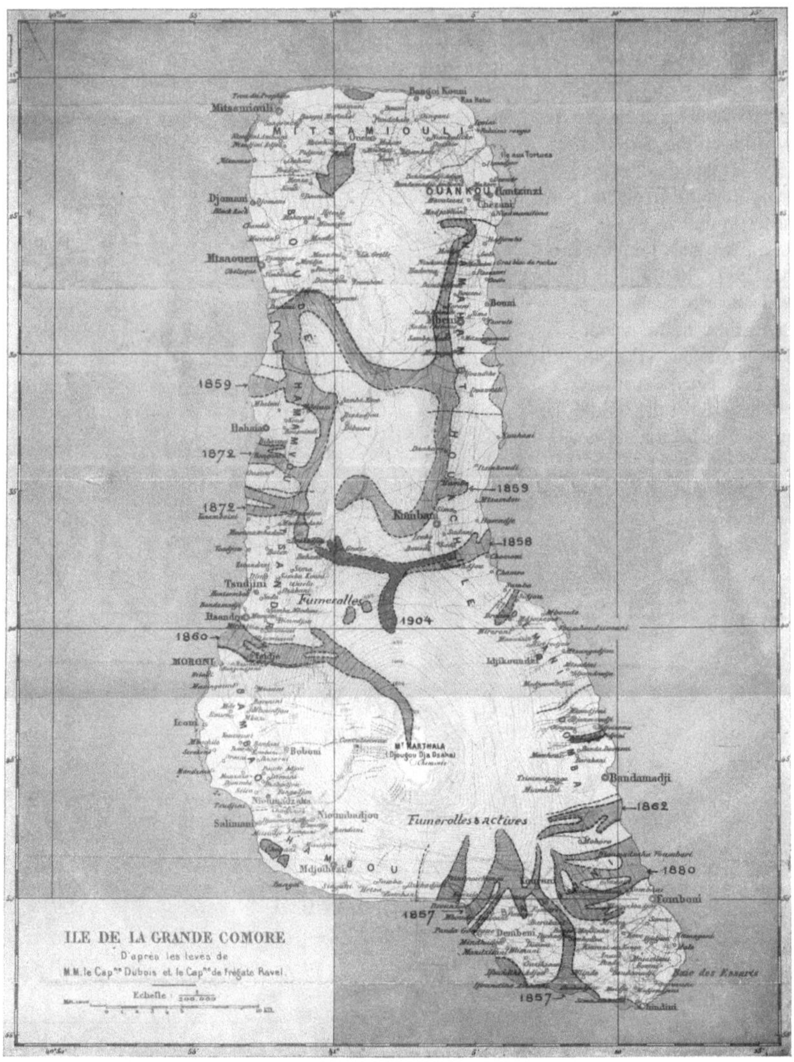

Fig. 22.9 Map of the lava flows in Grande Comore published by Lacroix (1938). The lava flow of 1860 (in fact 1858—von der Decken 1871) leaves the caldera to flow down the northern flank and enters the sea at a point which is now part of the city of Moroni, the capital of the Comoros. This map is not completely accurate. The lava flow from 1904, for instance, is incorrectly positioned. The 1918 lava flow is not shown

by a few dates of eruptive events from ^{14}C on carbonised wood. Morin et al. (2016, Chap. 25) discuss eruptive activity, volcanic hazards and risk in greater depth.

Karthala has erupted regularly in the last century, the only period for which its activity is fairly well known. Fourteen eruptions are listed from 1904 to today (Table 22.1), averaging one every six to eight years, but with a relative variability over time (fourteen years of inactivity before and after the phreatic eruption in 1991). The average magma emission rate is evaluated at 1.5 million m^3 $year^{-1}$ for historic times (since 1800–1850—Bachèlery and Coudray 1993). This makes Karthala volcano one of the most active African volcanoes, and the second most active volcano in the Indian Ocean, after Piton de la Fournaise on La Réunion Island. During the last two centuries, most of the eruptions were magmatic, emitting alkali basalts. They were characterised by eruptive fissures opened along the rift zones (as in 1857, 1858, 1859, 1860, 1862, 1872, 1882, 1904 and 1918), within the caldera (1948, 1952, 1965, 1972), or at low elevation far from the rift zones and the summit caldera, as in 1977 (Krafft 1982; Bachelery and Coudray 1993).

Karthala's eruptive style is mostly effusive, although phreatic and phreatomagmatic explosions have occurred frequently as in 1918, 1948, 1952, 1991 and 2005. Pyroclastic deposits at the summit of the volcano, in and around the caldera, demonstrate the importance of these explosive events in the very recent past of Karthala (Nassor 2001).

On April 5, 1977, a vent opened at 360 m altitude, just 1 km north of the village of N'Djoyesi, about twenty kilometres south of Moroni. An 'a'ā lava flow, 400–1500-m-wide, traveled nearly 3 km before entering the ocean. It partially affected the villages of Singani and Hetsa, destroying 293 houses and 566 meters of road. More than 4000 people were evacuated to Moroni (Krafft 1982).

This eruption is a typical example of an eruption outside of the rift zones, on the lower western slopes of the volcano. This type of eruption is characterized by the brief existence of a short eruptive fissure on which a scoriaceous cone of some tens to hundreds of meters-high built up quickly. As in the case of the 1977 eruption, magmas characterizing such low-altitude eruptions are generally very rich in olivine and pyroxene phenocrysts, with a very high output rate of emission. The opening of such eruptive fissures, at low altitude and feeding fluid and fast moving lava flows, means that this type of eruption can be considered to be a high potential threat for the local population.

Fourteen years passed before the next eruption, one of the longest periods of quiescence for the volcano during the 20th century. The July 11, 1991 eruption was the first seismo-volcanic event recorded by the new monitoring network at Karthala volcano. After an intense seismic crisis (4000 earthquakes per day from July 4 to 10), the eruption started in the *Choungou-Chahale*. The deposits are mainly present on the floor and on the western walls of the caldera. They comprise blocks, lapilli and ash, showing an absence of juvenile magma. A strong sulphur smell was noticeable in Moroni leading to the displacement of 1000 inhabitants. The eruption lasted less than one day and was purely phreatic (Bachèlery et al. 1995). A new crater formed, approximately 280 m in diameter and 43 m deep, in the bottom of the *Choungou-Chahale*. The crater was partly filled by a water lake that changed colour from red-orange just after the eruption to green about two months later, and then dark blue (Fig. 22.10).

By mid-2003, a strong increase in seismic activity was recorded by the Karthala Volcano Observatory (OVK—Soulé 2003). This activity continued at a variable rate during 2004 and early 2005. In March 2005, 40 earthquakes on average were recorded every day. An eruption began on April 16 around 11:00 (local time), after a seismic crisis of three hours. In the afternoon of April 16, and the next day, grey plumes and imposing ash-falls (accompanied by a strong sulphur smell) were observed, especially in the east of the island. Local authorities evacuated 10,000 people. On April 17, at the beginning of the afternoon, an aerial survey showed a column of ash and gas rising out from the *Choungou-*

Table 22.1 Chronology and main characteristics of eruptions at Karthala volcano since 1808 (after Bachèlery and Coudray 1993; Morin et al. 2009)

Year	Date	Location	Eruption type	Covered area (10^6 m^2)	Estimated volume (10^6 m^3)	Felt seismicity	Damage
2007	13 Jan	Choungou Chagnoumeni	Magmatic—summit	0.03	0.1	Yes	Yes, building collapse after earthquake
2006	28 May–1 June	Choungou Chahale	Magmatic—summit	0.2	18	Yes	no
2005	25 Nov–8 Dec	Choungou Chahale	Phreatomagmatic then magmatic—summit (collapse?)	>1/2 of the island covered by ashes	>50	Yes	yes
2005	16–20 Apr	Choungou Chahale	Phreatomagmatic then magmatic—summit (collapse?)	2/3 of the island covered by ashes	>50	Yes	yes
1991	11 July	Choungou Chahale	Phreatic—summit (collapse?)	–	No magma	Yes	yes
1977	5–10 Apr	South West flank	Magmatic—eccentric	1.8	10.8	Yes	Yes, Singani and Hetsa villages ruined
1972	8 Sept–5 Oct	Caldera	Magmatic—summit	2.5	12	–	–
1965	12 July	Caldera and Choungou Chagnoumeni	Magmatic—summit	0.05	0.15	–	–
1959	*1 June*	*Choungou Chahale*	*Magmatic—summit*	–	–	–	–
1952	10–14 Feb	Choungou Chahale	Magmatic *and phreatomagmatic* —summit	–	–	–	–
1948	13–16 June	Choungou Chahale	Magmatic *and phreatomagmatic* —summit	16	6	yes	yes
1948	22 Apr–4 May	Choungou Chagnoumeni	Magmatic—summit	–	–	–	–
1941	–	*Caldera*	*Magmatic—summit*	–	–	–	–
1926-29	*Several times*	*Caldera*	*Magmatic—summit*	–	–	–	–
1918	25–26 Aug	Choungou Chahale	Phreatic—summit (collapse)	25	No magma	Yes	Yes
1918	11–13 Aug	North rift zone	Magmatic—lateral	2.7	10	Yes	–
1910	–	*North rift zone*	*Magmatic—lateral*	–	–	–	–
1904				11	44	Yes	Yes, 1 killed

(continued)

Table 22.1 (continued)

Year	Date	Location	Eruption type	Covered area (10^6 m^2)	Estimated volume (10^6 m^3)	Felt seismicity	Damage
	25 Feb to Apr	North rift zone	Magmatic *and phreatomagmatic* —lateral				
1903	–	South East rift zone	Gas emission	–	–	–	Yes, 17 killed
1883	*March 1883–1884*	*South East rift zone*	*Magmatic—lateral*	–	–	–	*Yes, 1 village*
1880	–	South East rift zone	Magmatic—lateral	2.4	10	Yes	Yes
1876	–	South East rift zone	Magmatic—lateral	4	17	–	Yes, 1 village ruined north of Foumbouni
1872	–	North rift zone	Magmatic—lateral	1.6	7.2	–	–
1865	–	–	*Magmatic*	–	–	*Yes*	–
1862	–	South East rift zone	Magmatic—lateral	–	–	–	–
1860	Dec	South East rift zone	Magmatic—lateral	5.5	30	–	Yes
1859	–	North rift zone	Magmatic—lateral	3.9	20	–	–
1858	–	Caldera and North rift zone	Magmatic *and phreatomagmatic* —lateral	12.5	63	Yes	–
1857	–	South East rift zone and caldera	Magmatic—lateral and summit	10	56	–	Yes
1855	June or July	South East rift zone	Magmatic *and phreatomagmatic* —lateral	–	–	–	Yes, 30 homes ruined
1850	–	*West–South West flank*	*Magmatic—eccentric*	–	–	–	–
1848	–	South East rift zone	Magmatic—lateral	3	16	–	–
1833	–	*Caldera?*	*Magmatic*	–	–	–	–
1830	–	–	*Magmatic*	–	–	–	–
1828	May	–	*Magmatic*	–	–	–	–
1821	–	–	*Magmatic*	–	–	–	–
1814	–	–	*Magmatic*	–	–	–	–
1808	–	–	*Magmatic*	–	–	*Yes*	–

Eruptions emphasized in italic are uncertain

Fig. 22.10 Evolution of the in crater after the July 1991 phreatic eruption (*Photos* P. Bachèlery). **a** Photo was taken on 11 July 1991, a few hours after the phreatic explosion. A white vapour cloud covers the crater floor, accompanied by a loud noise of boiling water. **b** In late July 1991, after dissipation of the cloud, a *red-orange* coloured lake appeared. Several areas are still boiling. **c** The lake in June 1993. The level has dropped considerably and an island appears in the centre of the crater. Water is *greenish*. Its banks are *yellowish* after precipitation of compounds (Fe hydroxides?) that coloured the lake *red-orange* just after the eruption. **d** In 2003, the water level in the lake is higher. The water has turned a *dark blue* hue

Chahale, but also emissions of lava into the *Choungou-Chahale*. While the ash-fall continued, causing panic among the villagers in the east of the island (lightning and rumbles accompanying heavily laden ash plumes contributed greatly to inspiring fear), the first mud flows were also reported in the coastal villages in the west of the island. In the afternoon of April 17, ash-falls affected much of the north and west of the island, including the capital, Moroni, and Hahaya airport. The explosive activity fluctuated then declined in the night and early on April 18. A new aerial survey provided observations on the active lava lake at the bottom of *Choungou-Chahale*, but this activity stopped during the day of April 19. The water returned to the bottom of *Choungou-Chahale* around May 8 to form a new green lake. This eruption lasted only three days, but was marked by the accompanying falling ash and mudflows.

After a period of low seismicity from May to August 2005, a new crisis was recorded on August 25 and 26, with 190 earthquakes. After this crisis, seismicity continued until the end of October and early November. After a seismic crisis at about 1:20 on November 24, a new eruption began the same day at 17:20 (local time). As during the April eruption, a large, dark ash plume spread out over Karthala. From MODIS images (NASA), the top of the plume was estimated to be at an altitude of 11.6 km, with a lateral E-W extension of up to 280 km. From the start, this eruption was visible from Moroni, with rumbles and lightning in the plume, while major ash-falls affected a large part of the island, first to the east and southeast and, from November 25 at 5:00 am (local time), over the southwest region and Moroni, affecting 244,520 people (Bull. Global Volcanism Network 30:11-2005). Ash fallouts were very intense, largely obscuring sunlight and forcing residents to stay home or wear dust masks. The paroxysmal stage stopped during the day of November 25 and the next day an active lava lake was observed in the *Choungou-Chahale*. On November 30, 80 % of its surface was frozen, and by December 5 only two small cones were emitting projections. The eruption stopped on December 8.

Following these eruptions, the top of the volcano was covered with a large amount of blocks, bombs and ash. For months and years after the eruptions, fallout deposits were remobilized by rain, creating lahars. Lahars were more frequent on the western flank of Karthala, where high precipitation and strong rain showers are frequent.

In May 2006, OVK recorded an increase in seismic activity, with twenty events a day, associated with unusual seismic signals [LP (Long Period)-like signals but with a frequency reaching 4–5 Hz instead of the usual LP 1 Hz frequency on Karthala]. From May 20 to May 27, scientists working on top of Karthala felt numerous earthquakes, heard noises related to the strongest felt earthquakes, and observed an increase in fumarolic activity around *Choungou-Chahale*'s floor. From the evening of May 26, continuous vibrations were felt. An impending eruption was strongly suspected but OVK was not able to confirm this because of technical problems in its monitoring network (Morin et al. 2009). On May 28, approximately two hours of seismic crisis, with both SP (short Period) and LP signals, were monitored. Around 21:07, a tremor was recorded, confirming that a magmatic eruption had begun. From the coast, a red cloud was visible above the volcano. From May 28 at 23:00 to May 29 at 09:30, an SO_2 plume about 3 km high extended up to 60 km north-westward. In the morning of May 29, an aerial survey of the volcano showed that the eruption was contained within *Choungou-Chahale*, causing no threat to the population. A lava pond formed in the *Choungou-Chahale*, fed by a huge lava fountain 30–50 m high (Morin 2012; Fig. 22.11). On June 1, the pond's surface was dark, and the fountain, still active on May 31, had disappeared. The *Choungou-Chahale* floor 'grew' a hundred meters in elevation, indicating an estimated 18×10^6 m^3 lava production during this eruption.

During late 2006 to early 2007, the OVK network again detected earthquakes. On January 12, 2007, Grande Comore inhabitants felt $M > 4$ earthquakes. A seismic crisis began at 00:27 (local time) on 13 January, followed by an eruption at 02:50. A red glow was visible over Karthala. The tremor decreased after only a few hours only and disappeared around 15:00. While the eruptive activity was already over, a seismic crisis continued with events of $M > 3.5$ recorded and felt until January 15. The frequency and intensity of seismicity decreased until January 18, and this activity continued at a low level for a few weeks. It was only possible to locate the exact place of the eruption on January 17. The January 2007 eruption occurred in the *Choungou-Chagnoumeni*. The pit-crater was completely filled by 100,000 m^3 of lava and pyroclastites.

Like most of the eruptions that have occurred during the last two centuries, the May 2006 and January 2007 eruptions were purely magmatic.

The 2007 eruption ended a three-year cycle of activity, with two particularly violent eruptive episodes. This series of four eruptions raised awareness of the dangers posed by Karthala volcano to the population of Grande Comore (see Morin et al. 2016, Chap. 25).

22.6 Summary

Karthala volcano is one of the worlds' largest active alkaline basalt shield volcanoes, and one of the least known. This basaltic shield volcano shares many similarities with the archetypal volcanoes of Hawaii or La Réunion. Structures like the summit caldera and the two prominent rift zones contribute to shaping the general morphology of the volcano and of Grande

Fig. 22.11 Eruptions at Karthala volcano in 2005 and 2006. **a** Explosive activity on April 17, 2005 observed from the west flank of the volcano at approximately 2000 m in altitude (*Photo* D. Hoffschir). Blocks falling from the growing plume can be observed. **b** Lava lake observed from the north in the Choungou Chahale during a survey on April 18, 2005 (*Photo* H. Soulé). **c** Lava lake in the Choungou Chahale on November 26, 2005 (*Photo* C. Roche). **d** Active lava lake observed in the Choungou Chahale on May 29, 2006 (*Photo* J. Morin)

Comore island. Eruptive activity is dominated by frequent eruptions emitting basaltic lava flows mostly from fissure vents. Phreatic or phreatomagmatic explosions were common at the summit crater *Choungou Chahale* during the short historical period (since 1800–1850).

Despite its relatively low production rate compared with Hawaiian volcanoes, or its neighbour, Piton de la Fournaise on La Réunion Island, Karthala's structures are very similar to those of Mauna Loa in Hawaii. Numerous eruptive fissures are concentrated along preferential zones of fissuring, called rift zones, tens of kilometers long by only a few kilometers wide. A large summit caldera complex lies at the summit of the volcano, at the intersection of the two rift zones. Coalescent vertically-subsided blocks and graben-like structures form the caldera, the main crater lying in the centre. During the last few decades the eruptive activity has diversified, with an eruption along a radial crack open at low altitude (1977), a phreatic explosion (1991) and phreatomagmatic to violent-strombolian eruptions and lava lakes (2005–2007) at the summit crater. This cycle of eruptions is a reminder that Karthala volcano is potentially dangerous to the local population.

Aknowledgments The authors wish to thank Sebastien Zaragosi for providing the frame of the Fig. 22.1. We are grateful to Fran Van Wyk de Vries for proofreading this manuscript. This paper benefited from the constructive comments of L. Michon and JF Lénat.

References

Bachèlery P, Coudray J (1993) Carte volcano-tectonique (1/50000e) de la Grande Comore et notice explicative. Edited by the French Embassy in Moroni, Comores, and The University of La Réunion, St. Denis de La Réunion

Bachèlery P, Hémond C (2016) Geochemical and Petrological aspects of Karthala. In: Bachèlery, P., Lénat J-F, Di Muro A, Michon L (eds) Active Volcanoes of the Southwest Indian Ocean: and Karthala. Active Volcanoes of the world. Springer, Berlin (Chapter 23)

Bachèlery P, Villeneuve N (2013) Hot Spots and large Igneous Provinces. In: Shroder JF (Editor-in-chief), Owen LA (Volume Editor). In treatise on geomorphology, vol 5, Tectonic geomorphology, Academic Press, San Diego, pp 193–233

Bachèlery P, Ben Ali D, Desgrolard F, Toutain JP, Coudray J, Cheminée JL, Delmond JC, Klein JL (1995) L'éruption phréatique du volcan Karthala (Grande Comore) en Juillet 1991 : première crise sismo-volcanique suivie depuis l'implantation d'un réseau de surveillance. Comptes Rendus Académie Sciences Paris, t 320, Issue 8, Série II a, 691–698

Battistini R (1967) Le volcan actif de la Grande Comore. Revue de Géographie de Madagascar, No 10(11):41–77

Bossu R (2000) A simple approach to constrain the position and the geometry of seismogenic structures: Application to the Karthala volcano (Grande Comores Island). Mozambique J Seismolog 4:41–48

Bourdon B, Joron JL, Claude-Ivanaj C, Allègre CJ (1998) U-Th-Pa-Ra systematic for the Grande Comore volcanic: melting processes in an upwelling plume. Earth Planet Sci Lett 164:119–133

Class C, Goldstein SL (1997) Plume-lithosphere interactions in the ocean basins: constraints from the source mineralogy. Earth Planet Sci Lett 150:245–260

Class C, Goldstein SL, Altherr R, Bachelery P (1998) The process of plume—lithosphere interactions in the Ocean Basins—the case of Grande Comore. J Petrol 39(5):881–903

Class C, Goldstein SL, Stute M, Kurz MD, Schlosser P (2005) Grand Comore Island: a well-constrained "low $^3He/^4He$". Earth Planet Sci Lett 233:391–409

Class C, Goldstein SL, Shirey SB (2009) Osmium in Grande Comore lavas: a new extreme among a spectrum of EM-type mantle endmembers. Earth Planet Sci Lett 284:219–227

Claude-Ivanaj C, Bourdon B, Allègre CJ (1998) Ra-Th-Sr isotope systematic in Grande Comore Island: a case study of plume-lithosphere interaction. Earth Planet Sci Lett 164:99–117

de Saint-Ours J (1960) Etudes géologiques dans l'extrême Nord de Madagascar et de l'Archipel des Comores. Thèse de Doctorat, Rapport Service géologique de Tananarive no 221, Tananarive, Madagascar, p 205

Debeuf D (2004) Etude de l'évolution volcano-structurale et magmatique de Mayotte (Archipel des Comores, Océan Indien). PhD thesis, University of La Réunion, p 277

Deniel C (1998) Geochemical and isotopic (Sr, Nd, Pb) evidence for plume-lithosphere interactions in the genesis of Grande Comore magmas (Indian Ocean). Geochem Geol 144:281–303

Emerick CM, Duncan RA (1982) Age progressive volcanism in the Comores Archipelago, western Indian Ocean and implications for Somali plate tectonics. Earth Planet Sci Lett 60:415–428

Emerick CM, Duncan RA (1983) Errata. Earth Planet Sci Lett 62:439

Esson J, Flower MFJ, Strong DF, Upton BGJ, Wadsworth WJ (1970) Geology of the Comores Archipelago, Western Indian Ocean. Geol Mag 107-6:549–557

Flower MFJ, Strong DF (1969) The significance of sandstone inclusions in lavas from the Comores Archipelago. Earth Planet Sci Lett 7:47–50

Hajash A, Armstrong RL (1972) Paleomagnetic and radiometric evidence for the age of the Comores Islands, West Central Indian Ocean. Earth Planet Sci Lett 16:231–236

Kaneoka I, Takaoka N, Upton BGJ (1986) Noble gas systematics in basalts and a nodule from Réunion and Grande Comoro islands, Indian Ocean. Chem Geol 59:35–42

Krafft M (1982) L'éruption volcanique du Karthala : Avril 1977 (Grande Comore, océan Indien). Comptes Rendus Académie Sciences Paris 294:753–758

Krafft M (1983) Guide des volcans de la Grande Comore [unpublished], p 94

Lacroix A (1916) La constitution des roches volcaniques de l'archipel des Comores. Comptes Rendus Acad Sci Paris, t 163(9):213–219

Lacroix A (1920) Une éruption du volcan Karthala à la Grande Comore en août 1918. Comptes Rendus Acad Sci Paris, t 171:5–10

Lacroix A (1922) La constitution lithologique de l'archipel des Comores. C.R. XIIIème congrès Int. Géol. Bruxelles 2:949–979

Lacroix A (1938) Le volcan actif de l'île de La Réunion (supplément) et celui de la Grande Comore. Gauthier Villars Ed., Paris p 57

Lénat JF, Robineau B, Durand S, Bachèlery P (1998) A self-potential survey of the summit zone of Karthala volcano (Grande Comore). Comptes Rendus Acad Sci Paris, Earth Planet Sci 327:781–788

Michon L (2016) The volcanism of the Comores archipelago integrated at a regional scale. In: Bachèlery P, Lénat J-F, Di Muro A, Michon L (eds) Active volcanoes of the southwest Indian Ocean: and Karthala. Active volcanoes of the World. Springer, Berlin (Chapter 21)

Morin J (2012) Gestion institutionnelle et réponses des populations face aux crises volcaniques: études de cas à La Réunion et en Grande Comore. PhD thesis, University of La Réunion, p 358 + annexes

Morin J, Lavigne F (2009) Institutional and social responses to hazards related to Karthala Volcano,

Comoros; part I: the deep-seated root causes of Comorian vulnerabilities. Shima Int J Res Island Cult 3:54–71

Morin J, Lavigne F, Bachèlery P, Finizola A, Villeneuve N (2009) Institutional and social responses to hazards related to Karthala Volcano, Comoros; part I: analysis of the May 2006 eruptive crisis. Shima Int J Res Island Cult 3:33–53

Morin J, Bachèlery P, Soulé H, Nassor H (2016) Volcanic and crisis management on Grande Comore island. In: Bachèlery P, Lénat J-F, Di Muro A, Michon L (eds) Active Volcanoes of the Southwest Indian Ocean: Piton de la Fournaise and Karthala. Active volcanoes of the World. Springer, Berlin (Chapter 25)

Mouginis-Mark PJ, Rowland SK (2001) The geomorphology of planetary calderas. Geomorphology 37:201–223

Nassor H (2001) Contribution à l'étude du risque volcanique sur les grands volcans boucliers basaltiques: le Karthala et le Piton de la Fournaise. PhD thesis, University of La Réunion

Newitt M (1974) The Karthala Caldera, Grande Comore, an historical note. Bull Volcanol 38:493–494

Nougier J, Cantagrel JM, Karche JP (1986) The Comores archipelago in the western Indian Ocean: volcanology, geochronology and setting. J Afr Earth Sci 5–2:135–145

Pavlovsky R, de Saint-Ours J (1953) Etude géologique de l'archipel des Comores. Haut Commissariat de Madagascar et Dépendances, Direction des Mines et de la Géologie, Service Géologique Tananarive, vol. Travaux du bureau géologique numéro 51, p 55

Peterson DW, Moore JG (1987) Geologic history and evolution of geologic bib168 concepts, Island of Hawaii. US Geol Soc Prof Pap 1350:149–189

Poisson C (1949) Mission scientifique au volcan Karthala. Le naturaliste malgache, Tananarive

Poppe S (2012) Caldera collapse on basaltic shield volcanoes: analogue models compared to the Karthala caldera complex, Grande Comore. Master in Geology, Ghent University, Belgium

Prévot T (2010) Etude multidisciplinaire du système hydrothermal du volcan Karthala (Comores). MSc apprenticeship report, University Blaise Pascal, France, pp 50

Reisberg L, Zindler A, Marcantonio F, White W, Wyman D, Weaver B (1993) Os isotope systematics in ocean island basalts. Earth Planet Sci Lett 120:149–167

Rittman A (1963) Les volcans et leur activité, Masson Ed., Paris, p 461

Rowland SK, Garbeil H (2000) The slopes of oceanic basalt volcanoes. Remote Sens Act Volcanism AGU Monogr 116:223–247

Savin C, Ritz M, Join JL, Bachèlery P (2001) Hydrothermal system mapped by CSAMT method on Karthala volcano, Grande Comore Island (Indian Ocean). J Appl Geophys 48:143–152

Savin C, Grasso JB, Bachèlery P (2005) Seismic signature of a phreatic explosion: hydrofracturing damage at Karthala volcano, Grande Comore Island, Indian Ocean. Bull Volc 67:717–731

Smietana M (2007) Etude pétrologique et volcanologique des dépôts des quatre dernières éruptions du Karthala. MSc apprenticeship report, University of La Réunion, France, pp 51

Soulé H (2003) Proposition d'un plan d'urgence pour la gestion des crises volcaniques en Grande Comore. Ya Mkobe 10:59–73

Späth A, Le Roex AP, Duncan RA (1996) The geochemistry of lavas from the Comores Archipelago, western Indian Ocean: petrogenesis and mantle source region characteristics. J Petrol 37(4):961–994

Strong DF (1972) The petrology of the lavas of Grande Comore. J Petrol 13:181–217

Strong DF, Jacquot C (1970) The Karthala Caldera, Grande Comore. Bull Volcanol 34:663–680

Upton BGJ (1982) Oceanic Islands. In: Nairn P, Stehli FG (eds) Ocean basins and their margins, Indian Ocean, vol 6(13). Plenum Press, New York, pp 585–648

Upton BGJ, Wadsworth WJ, Latrille E (1974) The 1972 eruption of Karthala Volcano, Grande Comore. Bull Volcanol 38:136–148

Voeltzkow A (1906) Die Comoren, Zeitschrift der gesellschaft für Erdkunde zu Berlin, pp 606–630

von der Decken CC (1871) Reisen in Ost-Afrika. 2, book 5, Leipzig, pp 241–243

Geochemical and Petrological Aspects of Karthala Volcano

23

Patrick Bachèlery and Christophe Hémond

Abstract

The eruptive activity at Karthala consists of hypersthene-normative and mildly-alkalic basalts and some hawaiites, with MgO contents varying from approximately 4 % to more than 20 % for oceanites and ankaramites. Most of La Grille volcano's products are basanites and olivine nephelinites, often with a significant proportion of olivine. Although these rocks are silica under-saturated, modally they rarely contain crystals of feldspathoïds. Olivine and augite are the dominant phenocrysts in the Karthala and La Grille lavas (more than 40 % vol. in oceanites and ankaramites). Plagioclase is also present in some phenocryst-rich samples, mainly from the oldest units of the "Karthala ancien" and "M'Badjini" massifs. Plagioclase also occurs in some recent aphyric lavas of the summit caldera. The consistent variations in the Karthala lava compositions observed for both major and trace element contents are in agreement with crystal fractionation or accumulation. Conversely, La Grille lavas do not follow clear fractionation trends. Large variations in incompatible trace elements for a given MgO %, negative correlation between incompatible trace elements and SiO_2, and K depletion are symptomatic of magma generation driven by variable degrees of partial melting of a source containing residual amphibole (and/or phlogopite). Mantle-normalized trace elements patterns for Karthala lavas show a pattern typical of Oceanic Island Basalts generated from a mantle source containing garnet and amphibole. The

P. Bachèlery (✉)
Laboratoire Magmas et Volcans, Université Blaise Pascal—CNRS—IRD, Observatoire de Physique du Globe de Clermont-Ferrand, 5, rue Kessler, 63038 Clermont-Ferrand, France
e-mail: P.Bachelery@opgc.fr

C. Hémond
UMR CNRS 6538 Domaines Océaniques, Université de Brest, UBO Institut Universitaire Européen de la Mer, Place Copernic, 29280 Plouzané, France

P. Bachèlery et al. (eds.), *Active Volcanoes of the Southwest Indian Ocean*, Active Volcanoes of the World, DOI 10.1007/978-3-642-31395-0_23

Comoros plume represents a "low $^3He/^4He$" hotspot dominated by recycled ^4He-rich material. Grande Comore Island is located on the ~140 Ma-old oceanic crust of the Somali Basin. All Grande Comore melts represent mixtures between plume-derived and peridotitic lithosphere-derived melts. The compositional variability of alkali basalts from Karthala is consistent with the mixing of plume—and lithosphere-derived melts, with a higher contribution of an EM1 mantle source in the "Karthala actuel" and "Karthala recent" lavas. The "Karthala ancien" and "M'Badjini" units require a source containing a predominantly HIMU component, some enriched EM1, and smaller amounts of a depleted DM component. La Grille lavas more likely reflect a source containing a significant involvement of a DM component, in addition to a HIMU component, and a lower degree of melting.

23.1 Introduction

Karthala volcano is the active volcano on Grande Comore, the youngest of the Comoro Islands. The Comoros archipelago is situated at the northern entrance of the Mozambique Channel, between 10°21′ and 13°03′ south and 43°15′ and 45°20′ east. It comprises four volcanic islands (Grande Comore [Ngazidga], Anjouan [Ndzuani], Moheli [Mwali], and Mayotte [Maore]) distributed along a 300 km-long WNW trend (see Bachelery et al. 2016, Chap. 22, Fig. 22.1). This chapter proposes a synthesis of the main geochemical and petrological studies of Karthala volcano, as well as some aspects of La Grille volcano, the second volcanic edifice of Grande Comore Island. More general considerations, concerning the whole archipelago, will also be addressed.

The volcanic units referred to here are those proposed by Bachèlery and Coudray (1993). Very few age determinations are available for the Grande Comore volcanoes, and the lava units cannot be distinguished by petrography alone. A relative chronology has been established on the basis of morphological criteria and on the alteration of the lava flows. "Karthala actuel" unit groups together the historical lava flows (since 1850). "Karthala récent" corresponds to unaltered to little altered lava flows of unknown age, while "Karthala ancien" and "M'Badjini" units are the oldest units, strongly altered, mainly outcropping on the eastern and southern flanks of Karthala volcano. M'Badjini may once have been a separate volcano from Karthala volcano (see Bachèlery and Coudray 1993 and Bachèlery et al. 2016, Chap. 22, for more details).

Lavas from the two volcanoes which comprise Grande Comore have distinct petrological and geochemical characteristics, primarily illustrated by the significant difference in alkali/silica ratios between the two alkaline series. The first works on the petrography of the lavas of the Comoros archipelago were by Voeltzkow (1906), Bauer (1911) and Lacroix (1916, 1922, 1938). More than two decades later, petrological aspects were also broached by de Saint Ours (1960), Thomson and Flower (1971), Strong (1972), and Upton et al. (1974). More recently, Bachèlery and Coudray (1993), Desgrolard (1996), Coltorti et al. (1999), and Radadi Ali (2012) have detailed the petrology of the lavas and xenoliths. The origin of the magmas, their source and their evolution were also studied from a geochemical perspective by Nougier et al. (1986), Class (1994), Späth et al. (1996), Class and Goldstein (1997), Claude-Ivanaj et al. (1998), Bourdon et al. (1998), Class et al. (1998, 2005, 2009), Deniel (1998), Debeuf (2004), and Pelleter et al. (2014).

The alkali basalts of Karthala usually contain a significant proportion of olivine and clinopyroxene phenocrysts (more than 10 %). In this

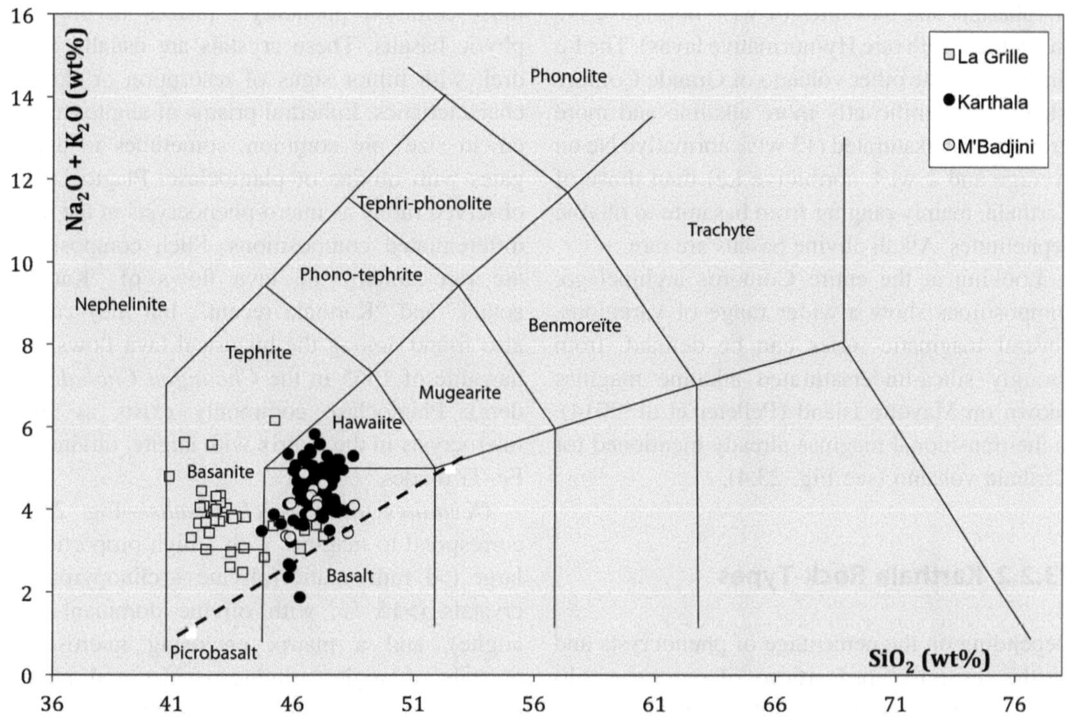

Fig. 23.1 Total alkalies-silica diagram showing fields of alkaline rock suites from Grande Comore Island. *Fine black lines* delineate the IUGS volcanic rock-type classification redrawn from Le Maitre (1989). *Dashed curve* separates the fields of alkali and subalkali series of Macdonald and Katsura (1964) for Hawaiian basalts. Note that most primitive basic compositions of Grande Comore lavas are either alkaline (Ne-normative) or subalkaline (Hy-normative). Data from Bachèlery and Coudray (1993), Späth et al. (1996), Desgrolard (1996), Class et al. (1998), Deniel (1998)

chapter, we used the rock names recommended by the IUGS (Le Maitre 1989) based on the total alkali/silica diagram for lavas, and on the modal composition of granular rocks. However, as for the Piton de la Fournaise volcano (Boivin and Bachèlery 2009), we choose to use the term oceanite, instead of picritic basalt or picrite, to refer to basalts enriched in MgO by mechanical accumulation of olivine crystals (as opposed to primitive liquids rich in MgO). The term oceanite (Lacroix 1912, 1923) defines a melanocratic variety of picritic basalt made up of numerous large crystals (>20 %) of olivine and a lesser proportion of clinopyroxene, in a matrix enclosing microscopic crystals of augite, olivine, oxides and plagioclase.

Olivine-rich basalts (oceanites) and ol+cpx-rich basalts are characterised by more than 20 % vol. of olivine and/or clinopyroxene phenocrysts. In the Karthala series, most of the historical lavas are alkali basalts with a mean MgO content of about 4–8 %. Olivine-rich basalts (oceanites) and ol+cpx-rich basalts are characterised by higher MgO contents (>8–22 %). Some samples plot in the trachy-basalt field (hawaiite) of the TAS diagram (Fig. 23.1).

23.2 Petrography

23.2.1 General Aspects

In all the petro-geochemical studies, the moderately alkaline to transitional character of Karthala's magmas, first mentioned by Strong (1972), is highlighted. The Karthala lavas are mainly compose of alkali basalts. They are saturated to slightly undersaturated basalts, picrobasalts and hawaiites (4 wt% normative Ne on average, with rare Hy-normative lavas). The La Grille lavas, the other volcano of Grande Comore Island, are significantly more alkaline and more strongly undersaturated (13 wt% normative Ne on average and 2 wt% normative Lc) than those of Karthala, mainly ranging from basanite to olivine nephelinites. Alkali olivine basalts are rare.

Looking at the entire Comoros archipelago, compositions show a wider range of variations. Several magmatic series can be defined, from strongly silica-undersaturated alkaline magmas known on Mayotte Island (Pelleter et al. 2014), to the transitional magmas already mentioned for Karthala volcano (see Fig. 23.4).

23.2.2 Karthala Rock-Types

Depending on the percentage of phenocrysts and on the respective proportions of pyroxene, olivine and plagioclase, several facies have been identified (Bachèlery and Coudray 1993). Olivine, clinopyroxene and plagioclase are generally euhedral to subhedral (polyhedral or skeletal crystals), but rounded and fractured olivine xenocrysts, sometimes with kink bands, can be present in the most porphyric lavas, showing signs of non-equilibrium with the host melt.

Karthala lavas are aphyric basalts, olivine basalts (± clinopyroxene), olivine-rich basalts (oceanites), pyroxene-rich basalts (ankaramites) and plagioclase-phyric basalts. All these mineral parageneses reflect the complexity of the differentiation process involved.

Alkali aphyric basalts **and** ***alkali porphyric basalts*** are the most common rock-types at Karthala volcano. The aphyric basalts (less than 5 % mafic phenocrysts—Fig. 23.2a) contain occasional phenocrysts (mm in size) and/or micro-phenocrysts of clinopyroxene and olivine. The olivine basalts, olivine-augite basalts, and augite-olivine basalts are more porphyric (5–15 % mafic phenocrysts—Fig. 23.2b). They contain olivine and/or clinopyroxene phenocrysts, accompanied rarely by small amounts of plagioclase. Clinopyroxene and olivine are the most common phenocryst phases in the porphyric basalts. These crystals are usually euhedral, with minor signs of resorption or skeletal characteristics. Euhedral prisms of augite (mm to cm in size) are common, sometimes in aggregates with olivine or plagioclase. Plagioclase is observed rarely as micro-phenocrysts in the more differentiated compositions. Such compositions are rare among the lava flows of "Karthala actuel" and "Karthala recent", but they can be also found among the historical lava flows (e.g. hawaiite of 1965 in the *Choungou Chahalé* caldera). Plagioclase commonly exists as small microcrysts in the matrix with augite, olivine and Fe–Ti oxides.

Oceanites (***olivine-rich basalts***—Fig. 23.2c) correspond to magmas with a high proportion of large (>1 mm) mafic (olivine + clinopyroxene) crystals (>15 %, with olivine dominant over augite), and a matrix enclosing microscopic crystals of augite, olivine, oxides and plagioclase. Variations in the crystal content of oceanites are due to different degrees of crystal settling or incorporation of olivine and/or augite antecrysts or xenocrysts in a rising basaltic magma. In the TAS diagram, oceanites plot in the "picrobasalt" field (SiO_2 <47 %, MgO >18 %, total alkali <2 %), but they do not correspond to true Mg-rich liquids. While olivine is predominant in the olivine-rich basalts, clinopyroxene commonly occurs with the olivine. Only a few samples, mainly from the M'Badjini massif, contain olivine phenocrysts alone. Some samples contain rare small plagioclase phenocrysts.

Ankaramites (***pyroxene-rich basalts***—Fig. 23.2d) are less frequent than oceanites. These melanocratic porphyric rocks (>15 % of mafic megacrysts) have a higher proportion of clinopyroxene than olivine. They contain pyroxene crystals of more than 1 cm in size.

Oceanites and ankaramites can be found among the lava flows derived from eruptive fissures and scoria cones at low altitude along the rift zones, and more commonly from adventive cones scattered on the western slopes of the volcano (e.g. lava flow from the Singani eruption, 1977). They are uncommon at the top of the

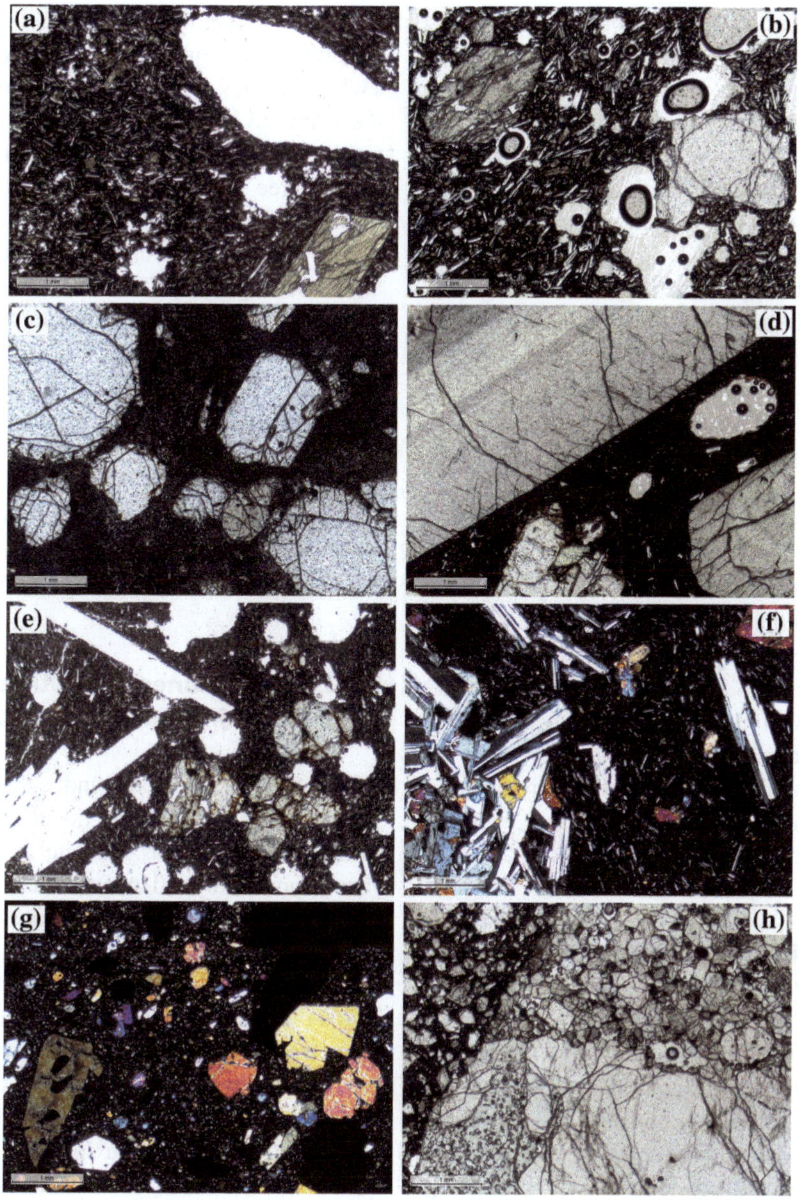

Fig. 23.2 Sequence of photomicrographs in plane-polarized light (a, b, c, d, e, and h) and cross-polarized light (f and g) showing typical rock samples from the Grande Comore magmatic series. **a** Aphyric basalt from the 1972 eruption of Karthala volcano. Few clinopyroxene phenocrysts are present. **b** Porphyric basalt from the 1977 eruption of Karthala volcano. Olivine and clinopyroxene appear as sub-euhedral phenocrysts. **c** Oceanite from the NW edge of the summit caldera of Karthala. **d** Ankaramite with large clinopyroxene phenocrysts and smaller olivine crystals; "M'Badjini" massif. Note the zonation in the clinopyroxene crystal. **e** Plagioclase-phyric basalt from the "Karthala ancien" unit—Edge of the old caldera of Karthala, alt. 1871 m. Olivine crystals and rare clinopyroxene crystals occur with the large plagioclase laths. **f** Plagioclase-phyric basalt from the "Karthala ancien" unit, ravine N'Gnachentje, east flank of Karthala. The lava encloses a gabbroic xenolith. **g** Basanite from the recent lava flow of Chioundi cone, NE of La Grille volcano. **h** Basanite enclosing a peridotite xenolith. Olivine and clinopyroxene crystals are disseminated in the groundmass of the basanite. In the peridotite, there are large crystals of olivine (and orthopyroxene) adjacent to a mosaic of smaller crystals of olivine and clinopyroxene. Lava flow by Choua-Chandroude Island, La Grille volcano

volcano, but not completely absent. They are also present among the "Karthala ancien" and "M'Badjini" units. The distinction between oceanites and ankaramites is in the relative proportions of olivine and clinopyroxene they contain, although there is a continuum in composition between these two groups of melanocratic lavas. They have incorporated varying amounts of olivine and clinopyroxene crystals during their storage or during their transfer to the surface. Their composition reflects this mixing between an "aphyric basalt" end-member and two "olivine" and "augite" end-members.

Plagioclase-phyric basalts (Fig. 23.2e, f) are only found in the "M'Badjini" and "Karthala ancien" units, at the southern end and eastern flank of Karthala volcano, respectively. Plagioclase is the main mineral phase, frequently associated with clinopyroxene and olivine crystals. Plagioclase crystals are millimeters to centimeters in size. They can occur as squat prisms isolated in the matrix, with many inclusions (mafic minerals and glass), particularly along their edges, but they more commonly form radiating groups in a doleritic texture. Gabbro xenoliths may be present in these plagioclase-rich lavas (Fig. 23.2f).

23.2.3 Main Distinctions with La Grille Lavas

Olivine is a ubiquitous mineral in La Grille lavas. Often present in large quantities, the crystals can be euhedral or skeletal, but most are fragmented crystals with signs of resorption resulting from disequilibrium with their host lava (Fig. 23.2g). Olivine is often the only mineral phase present as megacrysts in these lavas; clinopyroxene is rare, and plagioclase never present as phenocrysts. Plagioclase is also often absent from the groundmass, where nepheline may crystallize. These lavas, unlike those of Karthala, often contain xenolithic enclaves of ultramafic rocks (dunites, wehrlites, and lherzolites). The desegregation of these enclaves releases many xenocrysts into the lava (olivine, clinopyroxene, orthopyroxene, and spinel).

Mantle-derived spinel lherzolite and wehrlite xenoliths occur in some La Grille lava flows. Two main sites exist: one near the north coast of the island at Gula Ivoini, the other on the east coast, by N'Droude Island. Coltorti et al. (1999) define three groups on the basis of modal compositions and textures: one group of lherzolites and two groups of wehrlites. Spinel is the only aluminous phase. Textures vary from protogranular to porphyroclastic. In the protogranular lherzolite, olivine and orthopyroxene occur as large crystals (up to 2 mm). Smaller crystals of clinopyroxene and spinel are interstitial or are included within the orthopyroxene. Xenoliths with a porphyroclastic texture are characterized by large crystals of olivine and orthopyroxene, with small neoblasts of olivine and clinopyroxene (Coltorti et al. 1999—Fig. 23.2h).

23.2.4 Sandstones and Other Continental Rock Inclusions

Xenoliths of sandstone and other continental rocks have been regularly documented from the Grande Comore lavas. Granite and quartz were first reported by Vienne (1900), and Lacroix (1922) gave detailed descriptions of metamorphic and plutonic rocks found at Anjouan and Mayotte. More recently, Flower and Strong (1969) described sandstone inclusions in the three westernmost islands of the archipelago, and Montaggioni and Nougier (1981) from those of Anjouan. In fact, sandstone xenoliths, as well as crystals of quartz, appear in lavas from all four islands of the archipelago (Debeuf 2004). The most frequent type of sedimentary inclusion in the Comorian lavas is quartzarenite.

The presence of these enclaves was used as an argument in the debate about the nature of the crust underlying the archipelago, suggesting that the Comoros could be underlain by continental crust. According to Nougier et al. (1986), the Comoros are located in the transitional zone between oceanic and continental crust. However, based on the works of Rabinowitz (1983), Coffin et al. (1986), Coffin and Rabinowitz (1987), a consensus exists as to the existence of a Mesozoic oceanic crust (165–130 Ma) in the western Somali Basin. Although sandstone enclaves are common in the Comorian lavas, helium isotope ratios measured on olivine phenocrysts from Grande Comore do not reflect any significant crustal contamination (Class et al. 2005).

23.3 Mineral Compositions

As described above, olivine and clinopyroxene are the dominant phenocryst phases in Karthala lavas, and olivine in La Grille lavas. Plagioclase is also present in some phenocryst-rich samples, mainly from the oldest units of the Karthala and "M'Badjini" massifs. Mineral compositions are from the work of Class (1994), Desgrolard (1996), and Radadi Ali (2012).

23.3.1 Olivine

Olivine is the main phase in the Karthala lavas. It is always present in the microlites, and is rarely absent as a phenocryst phase. The composition of the olivine crystals ranges from 47 to 88 Fo mol % (Desgrolard 1996). The microlites (<Fo80) are statistically less Mg-rich than the phenocrysts (mostly between Fo83 and Fo87). This abundance of forsterite in olivine phenocrysts is consistent with the low differentiation of the lava.

Olivine phenocrysts are generally equant to tabular and subhedral to euhedral. Most are < 5 mm in the longest dimension, but rare centimeter-sized olivine megacrysts are present in some oceanites. Olivine crystals frequently contain spinel inclusions. They are optically unzoned, except for slight iron enrichment over a few tens of microns near the crystal rims. Kink-banded olivine crystals occur more frequently in the oceanites than in aphyric and olivine basalts.

Electron microprobe analyses show that the cores contain slightly more MgO than the rims. A detailed study of large crystals in the basalts indicates that most are in equilibrium with the host melt (Desgrolard 1996). The composition of the olivine crystal cores ranges from Fo87 to Fo83 (NiO wt% = 0.15–0.35), with no obvious difference between the olivines from olivine-rich basalts and those from aphyric basalts, or those of the rare xenolithic enclaves of ultramafic rocks (dunite, wehrlite). This indicates that these crystals were in equilibrium with primitive or little-evolved liquids.

Olivine-liquid thermometers (Leeman and Scheidegger 1977) applied to some Karthala lavas (Desgrolard 1996) give equilibrium temperatures of between 1190 and 1220 °C and between 1125 and 1172 °C respectively for olivine-rich basalts and aphyric basalts. Similar results were obtained by Radadi Ali (2012).

23.3.2 Clinopyroxene

Clinopyroxene is a very common mineral phase in the magmas of the more recent Karthala units. Its frequency is lower in the lavas of "M'Badjini" and "Karthala ancien", and clinopyroxene is rare in those of La Grille. In Karthala lavas, clinopyroxene is always present in the groundmass, and as euhedral phenocrysts in the ankaramites, olivine-rich basalts and aphyric basalts. It frequently appears in olivine basalts. Analyses of clinopyroxene phenocrysts plot in the Mg-rich augite and augite fields of the Enstatite-Ferrosilite-Wollastonite (En-Fs-Wo) diagram, with diopsidic compositions for some phenocryst cores. Compositions are grouped in a small domain around an average composition of En46-Fs9-Wo45. Most of the clinopyroxene crystals display concentric and/or hourglass zoning. The zoning generally corresponds to a decrease in Mg and Cr content and a simultaneous increase in Fe and Ti content from the core to the rim.

The evolution of the Al/Ti and Al^{VI}/Al^{IV} ratios in clinopyroxene are commonly used to determine the pressures of crystallization, with the ratios increasing with increasing

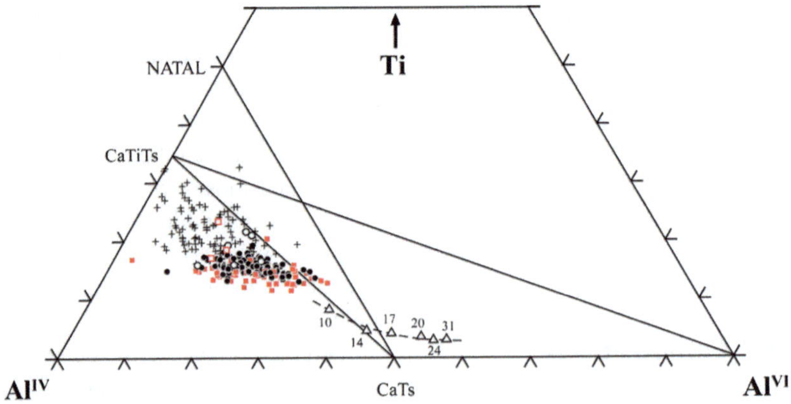

Fig. 23.3 Evolution of the clinopyroxene type I (*square*) and II (*round*) crystals in the Al^{IV}-Al^{VI}-Ti diagram. *Solid symbols* and *open symbols* denote crystal cores and rims, respectively. The microlites are also shown with the symbol *plus*. The *triangles* indicate experimental alkaline clinopyroxene compositions for different pressures (in GPa) (in Radadi Ali 2012)

crystallisation pressure. Desgrolard (1996) and Radadi Ali (2012) use these proxies to show that most of the Karthala lava flows derived from a polybaric fractionation process (Fig. 23.3). Radadi Ali (2012) recognizes two types of zonation from a study of clinopyroxenes from the 1977 eruption: (a) normal zonation reflecting a process of fractional crystallization (type I), and (b) oscillatory zonation reflecting more complex changes in the chemical composition of the host melt (type II). Based on measured concentrations of Ti, Al^{IV} and Al^{VI}, and experimental data, Desgrolard (1996) determines a crystallization pressure of 8–10 kb for the earliest pyroxenes to shallow level conditions for the majority of the crystals.

From a study of the clinopyroxenes, Späth et al. (1996) suggest that the fractional crystallization processes involved in the differentiation of Comorian lavas were not confined to shallow, sub-volcanic levels, but may have started at depths corresponding to pressures as high as 15 kbar. Such evidence includes the occurrence of complex clinopyroxene crystals with corroded, green cores that are in compositional disequilibrium with their tan-coloured mantles and rims, and have Al^{VI}/Al^{IV} ratios (~0.29–0.55) suggestive of a relatively high-pressure origin. Such crystals have never been described in the Karthala lavas.

23.3.3 Plagioclases

While plagioclase microlites appear ubiquitously in the matrix of the Karthala lavas, few lavas contain plagioclase phenocrysts. In the most recent lavas, they appear only as micro-phenocrysts in the most evolved compositions (hawaiites). Plagioclase crystals of a few mm in size are more common in the lava flows from the "Karthala ancien" and "M'Badjini" units.

Plagioclase crystals show no obvious signs of compositional zoning. They occasionally occur in small aggregates, sometimes with augite.

Plagioclase compositions range from An88.8 Ab10.6 Or0.6 to An55.4 Ab41.6 Or2.9 for phenocrysts and micro-phenocrysts. The rims and microlites contain slightly more sodium and potassium (An68.7 Ab29.8 Or1.4 to An49.6 Ab47.5 Or2.9 and An52.3 Ab42.1 Or5.6, respectively). Plagioclase from the "M'Badjini" and "Karthala ancien" lavas are, on the whole, less calcic than those of the "Karthala actuel" and "Karthala recent" units.

23.3.4 Oxides

Euhedral chromian spinel commonly occurs as inclusions in olivine, although it is also present as micro-phenocrysts in some of the most mafic samples. These chromian spinels are regarded as having crystallized at elevated temperatures. They may represent a liquidus phase. Their Cr_2O_3 composition varies from 49 wt% to 19 wt %. They are characterized by high MgO (between 8 and 12 wt%) and Al_2O_3 (between 11 and 16 wt%) contents.

Titanomagnetites are a typical phase of the groundmass with MgO, MnO and Al_2O_3 contents up to 3.7, 4.4 and 0.8 wt%, respectively (Radadi Ali 2012). Their composition is rather homogenous, corresponding to Mt8-33 on the magnetite-ulvöspinel line of the TiO_2-FeO-Fe_2O_3 diagram. Ilmenite is only observed in some gabbros.

23.4 Geochemistry

23.4.1 Comorian Magmatic Suites

Taking all compositions in the Comoros archipelago into consideration, three trends can be defined on a diagram of total alkali versus silica (Fig. 23.4). Karthala lavas are clearly moderately alkalic, while La Grille lavas are more strongly alkalic and undersaturated. The most silica-undersaturated lavas from the Comoros archipelago correspond to the highly silica-undersaturated trend formed by the lavas of the southern part of Mayotte Island, where basanites, nephelinites, melilitites and phonolites have been recognized, while a moderately silica-undersaturated trend with alkali basalts, basanites, intermediate lavas and phonolites characterise the northern part of the island (Pelleter et al. 2014). Anjouan and Moheli defined intermediate trends from basanites to phonolites, lying between the two trends for Mayotte. Mafic rocks from Mayotte show the widest range of compositions from basalts/basanites and olivine nephelinites to olivine melilitites.

The lavas emitted by the two Grande Comore volcanoes result from a low degree of differentiation, in contrast to lavas on the other islands of the archipelago, where trachytic and phonolitic compositions are present. Only basanites and basic nephelinites are present at La Grille, and Karthala emits mainly alkali basalts and rare hawaiites.

23.4.2 Major and Trace Element Variations

La Grille lavas, including the aphyric lavas, are more basic in average bulk composition than Karthala lavas (Strong 1972). Lavas from Karthala show less variation in SiO_2 content (45–48.5 wt%) than those from La Grille (from approximately 41–48 wt%). As olivine is a major phase in both Karthala and La Grille lavas, chemical variations can be well illustrated by means of MgO variation diagrams.

Karthala lavas have compositions that are controlled by the crystallisation and accumulation of olivine (+ clinopyroxene), resulting in almost constant major element and compatible and moderately incompatible trace element abundances for a given MgO content. They define linear trends in oxide and trace element versus MgO diagrams (Fig. 23.5), in good agreement with such a fractionation/accumulation evolution process. On most diagrams, a change in slope is observed at 7–8 wt% MgO, possibly indicating a transition between the cumulative (Ol $\pm$ Cpx) and non-cumulative lavas.

Conversely, La Grille lavas do not follow clear fractionation trends. Larger variations in incompatible trace elements at a given MgO % and a negative correlation between incompatible trace elements and SiO_2 content characterize La Grille lavas. These different behaviours are ascribed to a difference in the composition of the primary magmas of the two neighbouring volcanoes (Strong 1972; Deniel 1998). The variations observed for La Grille lavas suggest that processes other than mineral fractionation/accumulation are predominant in determining the major element variations. Higher MgO content for La Grille lavas is in agreement with the mineralogy of the lavas (high proportion of olivine). Geochemical features of La Grille lavas

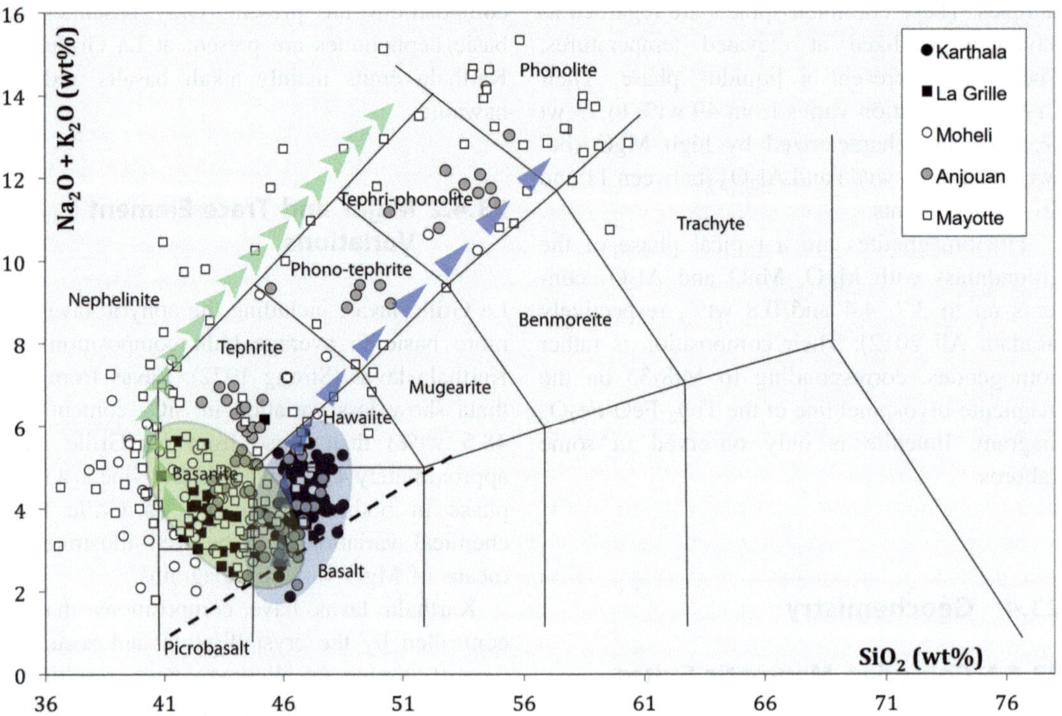

Fig. 23.4 Total alkalies-silica diagram showing fields of alkaline rock suites from the Comoros Archipelago. *Fine black lines* delineate the IUGS volcanic rock-type classification redrawn from Le Maitre (1989). *Dashed curve* separates the fields of alkali and subalkali series of Macdonald and Katsura (1964) for Hawaiian basalts. Blue and green ovals respectively delineate the compositional fields for Karthala and La Grille volcanoes. *Blue arrows* and *green arrows* respectively show the moderately silica-undersaturated and the highly silica-undersaturated trends for Mayotte lavas. Compositions from Anjouan and Moheli lie between the two Mayotte trends. Data from Strong (1972), Nougier et al. (1986), Bachèlery and Coudray (1993), Späth et al. (1996), Desgrolard (1996), Class et al. (1998), Deniel (1998), Debeuf (2004), Pelleter et al. (2014)

(Mg# of 69–71, Ni content >300 ppm, and Ni/MgO ratios of 25–27) reflect their primitive character (Späth et al. 1996).

As MgO contents are not much higher than 8 % in the non-cumulative Karthala lavas, it seems likely that the Karthala alkali basalts do not represent primitive magmas. The coherent variations in major and compatible trace elements indicate that the evolution of the Karthala suite can be ascribed to low-pressure crystal fractionation/accumulation of olivine and clinopyroxene phases. Lava flows from Karthala volcano show decreasing CaO and transition elements like Ni, Cr, Co, as well as Sc, contents with decreasing MgO content, while SiO_2, Al_2O_3, Fe_2O_3, Na_2O, K_2O, TiO_2, P_2O_5 and incompatible trace elements increase. For example, the concomitant decrease of the CaO/Al_2O_3 ratio with the decrease in MgO contents is a strong argument for the major removal of Ca-rich pyroxene. This is also well illustrated by the evolution of the composition of Karthala lavas in the Na_2O versus MgO diagram (Fig. 23.6). The good correlation of Sc/Yb with CaO/Al_2O_3 (upper right corner of Fig. 23.6) confirms that clinopyroxene fractionation plays a major role in the chemical evolution of these lavas.

Most porphyric lavas have been interpreted as being derived from basalts through olivine and/or clinopyroxene accumulation of antecrysts or xenocrysts, rather than accumulated phenocrysts. These lavas are relatively uncommon in the Comorian volcanic products. Less than 20 % of

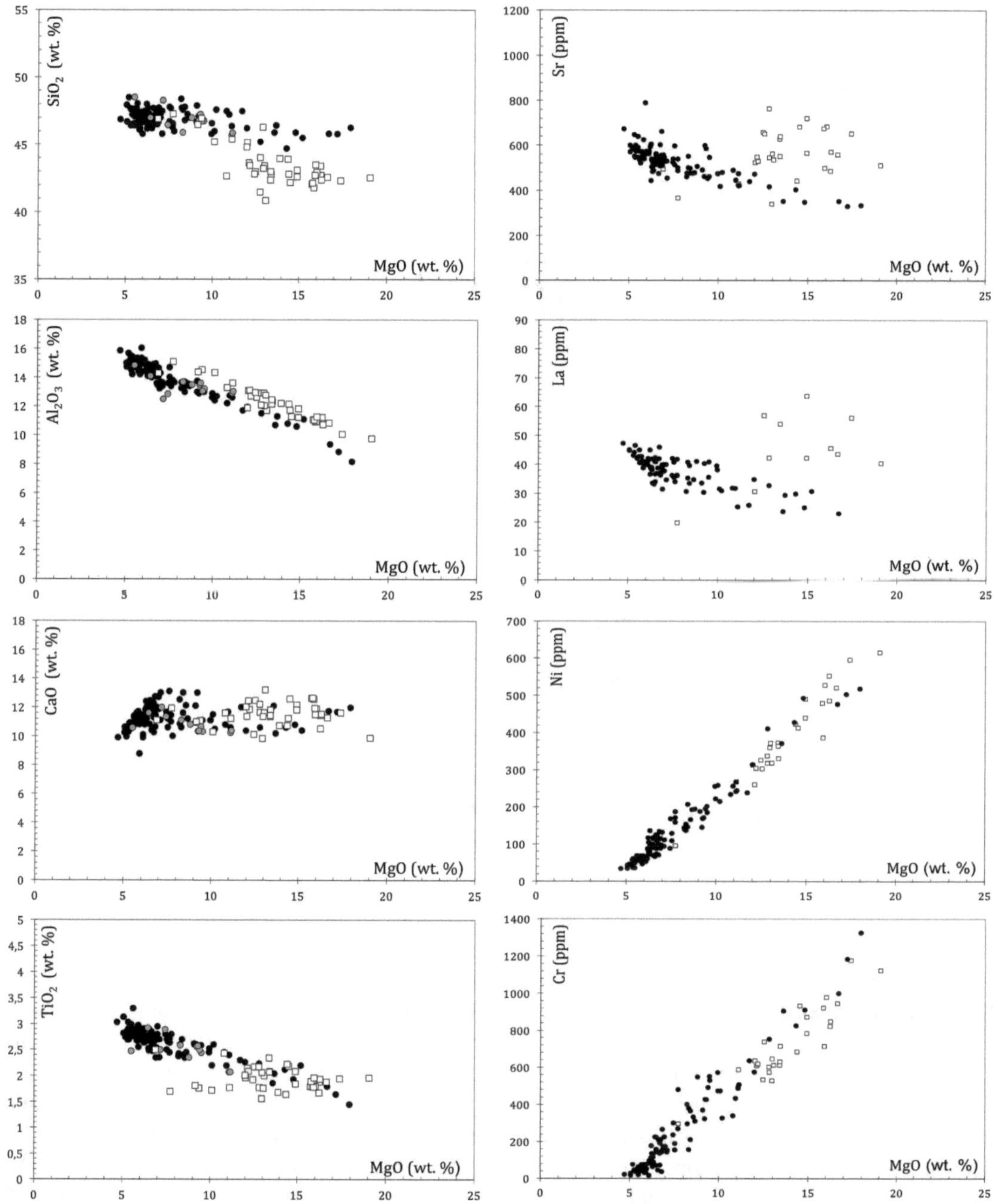

Fig. 23.5 Major (wt%) and trace (ppm) element versus MgO wt% diagrams for Grande Comore lavas. *Filled circles* Karthala lavas, *open circles* "M'Badjini" lavas, *open squares* La Grille lavas. Data from Bachèlery and Coudray (1993), Späth et al. (1996), Desgrolard (1996), Class et al. (1998), Deniel (1998)

Karthala lavas are highly porphyric (>20 vol. % of megacrystals). They are mainly found among the products issued from the vents at low altitude, and porphyric lavas have erupted only twice during the very short historical record, once in 1872 and the other in 1977.

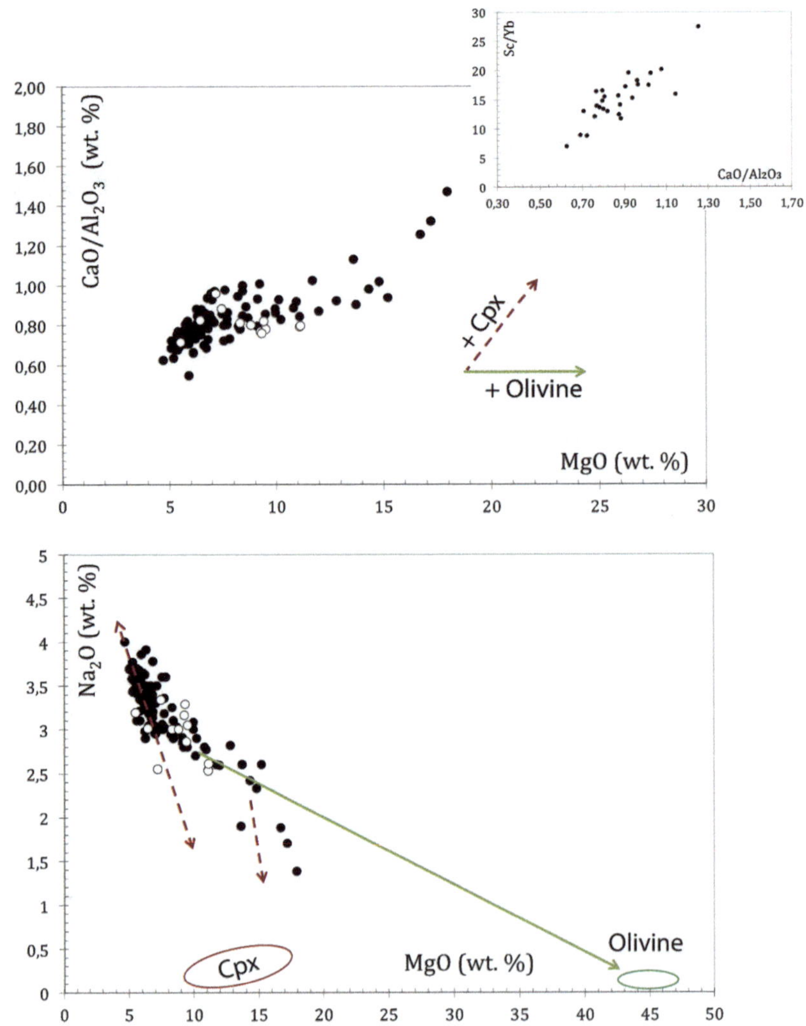

Fig. 23.6 Clinopyroxene fractionation, and olivine and clinopyroxene accumulation illustrated in CaO/AL_2O_3 versus MgO (*top*) and Na_2O versus MgO (*bottom*) diagrams for Karthala (*closed circles*) and M'Badjini (*open circles*) lavas. Lavas compositions from Bachèlery and Coudray (1993), Class et al. (1998), Deniel (1998), Desgrolard (1996), Späth et al. (1996). Clinopyroxene and olivine compositions from Desgrolard (1996). Plot of the Sc/Yb and CaO/Al_2O_3 ratios for Karthala lavas (*upper right corner*). Data from Class et al. (1998), Deniel (1998)

23.4.3 Mantle Source

It is not possible to evoke the mantle source of Karthala lavas without discussing that of its neighbour, La Grille volcano. Based on trace element and radiogenic isotope compositions (Figs. 23.7 and 23.8), Class and Goldstein (1997), Class et al. (1998), Claude-Ivanaj et al. (1998), and Deniel (1998) show that Grande Comore lavas reflect the interaction of a mantle plume with oceanic lithosphere. La Grille lavas exhibit the most radiogenic $^{143}Nd/^{144}Nd$ (0.51283–0.51291) and the least radiogenic $^{87}Sr/^{86}Sr$ (0.70315–0.70336), $^{206}Pb/^{204}Pb$ (18.98–19.31) and $^{208}Pb/^{204}Pb$ (38.80–39.15). "Karthala ancien" and "M'Badjini" lavas have the most radiogenic $^{206}Pb/^{204}Pb$ (19.47–19.86) and intermediate $^{87}Sr/^{86}Sr$ and $^{143}Nd/^{144}Nd$. "Karthala actuel" and "Karthala recent" units exhibit the most radiogenic $^{87}Sr/^{86}Sr$ (0.70354–0.70396), the least radiogenic $^{143}Nd/^{144}Nd$ (0.51265–0.51276) and intermediate $^{206}Pb/^{204}Pb$ (Class et al. 1998; Deniel 1998).

Deniel (1988) proposes that the EM1 and HIMU components are contained within the mantle plume and that the DMM component is located in the oceanic lithosphere. Späth et al. (1996), Class and Goldstein (1997), Class et al. (1998) favour a homogeneous plume with a dominant EM1 signature, and a DMM-HIMU

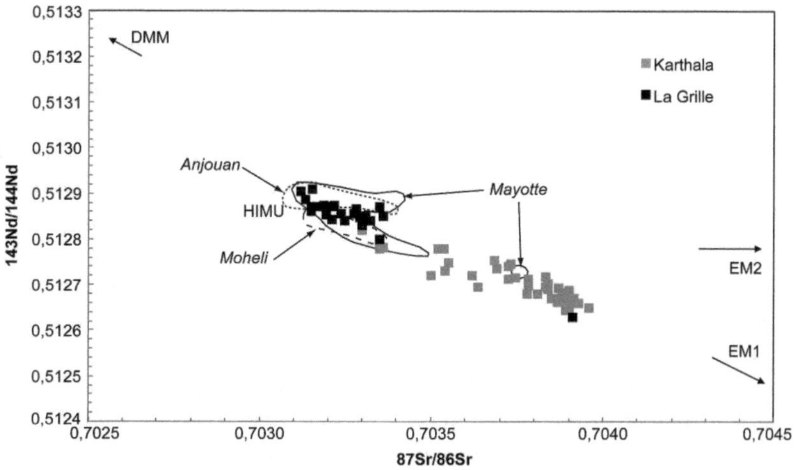

Fig. 23.7 Sr and Nd isotope variations in lavas from Karthala (*black squares*) and La Grille (*grey squares*) volcanoes in Grande Comore. Composition fields for lavas from Mayotte (*solid lines*), Anjouan (*dotted line*) and Mohéli (*dashed line*) are drawn. Data from Georoc database (http://georoc.mpch-mainz.gwdg.de/georoc/) and Pelleter et al. (2014). The sample from La Grille volcano with a high $^{87}Sr/^{86}Sr$ and low $^{143}Nd/^{144}Nd$ ratio corresponds to a lava flow emitted at the extension of the north rift zone of Karthala volcano. Karthala lava compositions close to the La Grille field belong to "M'Badjini" and "Karthala ancien" units

oceanic lithosphere. Other geochemical parameters and tracers were used in parallel to depict source composition and melting mechanism of the present activity of the Comoros hotspot. Again using trace elements, radiogenic isotopes and U series, Claude-Ivanaj et al. (1998) suggest that the heterogeneity is due to the lithospheric mantle (DMM-HIMU) whereas the plume is homogenous, as seen in Karthala lavas for $^{87}Sr/^{86}Sr = 0.704$, (Th/U) Kapa ratio = 2.8 and (^{226}Ra)/Ba = 5000.

Alkali basalts from Karthala are seen as a variable mixture between plume- and lithosphere-derived magmas. In Sr-Nd–Pb isotope space, the plume end-member lies on the so-called LoNd (low $^{143}Nd/^{144}Nd$) mantle array, considered as a mixing of EM1 and HIMU mantle components (Class et al. 1998; Deniel 1998). Deniel (1998) compares the lavas of the "Karthala ancien" and "M'Badjini" with those of the present Karthala volcano, showing a higher contribution of EM1 mantle source in the most recent products of Karthala than in the past. "Karthala ancien" and "M'Badjini" lavas result from a source containing more of the HIMU component and less EM1, with some DM component. Class et al. (1998) mention two samples of the "M'Badjini" massif whose isotopic compositions are very different to Karthala and other M'Badjini samples, with low-Pb and intermediate Sr- and Nd-isotope ratios that resemble those of La Grille.

La Grille lavas show a more limited range of variations in Sr, Nd and Pb isotopes, at the low-Sr, high-Nd and low-Pb end of the Grande Comore compositional spectrum (Späth et al. 1996; Class and Goldstein 1997; Deniel 1998). They reflect the significant involvement of a DM component, in addition to a HIMU component (with a less clear EM1 contribution). La Grille lavas represent an end-member of the island's compositional array (Fig. 23.7).

If the Dupal isotope anomaly derives from thermal erosion of the Gondwanan lower continental crust within the upper mantle, as found recently for the St. Paul—Amsterdam plateau (Janin et al. 2012), this raises the question of its influence on the Comoros plume impinging on the thinned old crust of the Mozambique canal. The EM1 flavour of the magma source may not have a deep origin, and might instead support the involvement of the upper mantle in the genesis of the magma beneath Grande Comore.

The mineralogy of the mantle source has also been discussed. The existence of residual amphibole (and/or phlogopite) in the lithospheric component of the source of the Comorian magmas has been proposed by Späth et al. (1996), Class and Goldstein (1997), Claude-Ivanaj et al. (1998), Class et al. (1998), and Deniel (1998). The main arguments lie in trace element systematics, such as the relative depletion in K, Rb and MREE. This is particularly pronounced in La Grille lavas, also characterized by the decrease in K/Nb ratio with decreasing degrees of partial melting in the La Grille suite, whereas the K/Nb ratio is relatively constant in the Karthala suite (Class et al. 1998). Smaller or no K depletion in Karthala lavas would thus imply a lower contribution from the amphibole-bearing source (Deniel 1998).

This identification of amphibole in the source region implies that metasomatism took place prior to melting. The existence of hydrous phases lowers the solidus of the lithospheric mantle sufficiently for it to melt when heated up by conduction from under-plated plume material. An episode of metasomatism can be seen in La Grille mantle xenoliths. Coltorti et al. (1999) study mantle-derived spinel lherzolite and wehrlite xenoliths from La Grille volcano. They concluded that there was metasomatised oceanic mantle beneath Grande Comore, due to the migration of a Na-rich carbonatitic melt. The interstitial glass contained in the lherzolite and wehrlite xenoliths has very variable compositions, from very undersaturated to highly oversaturated in silica, and with very high Na contents (up to 14 wt%). Based on the alkali element enrichment (Na) and trace element anomalies (Nb, Zr, and Ti), these authors propose that the metasomatism was due to an "ephemeral" carbonatitic melt. However, no hydrated phases have been found in these mantle xenoliths.

The coupling of isotopes and incompatible trace element ratios in La Grille lavas supports the idea that La Grille magmas formed from variable and low degrees of partial melting of a metasomatised, amphibole-bearing lithospheric mantle source, whereas Karthala lavas represent variable mixtures of plume- and lithosphere-derived melts formed by higher degrees of partial melting. However, Class et al. (2009) discuss the origin of the La Grille melts based on Os isotopes (see also Reisberg et al. 1993). They propose instead that all Grande Comore melts, including the La Grille magmas, represent mixtures between plume-derived and peridotitic lithosphere-derived melts, in order to explain the higher $^{187}Os/^{188}Os$ ratios they observe compared to direct lithospheric mantle melts. The EM1 contribution may also come from that source.

Class et al. (2005, 2009) show that good correlations exist between Sr–Nd isotopes and He and Os isotope ratios. This agrees with the assumption of a mixing of lithosphere-dominated and plume-dominated melts to form the Grande Comore isotope array. The lithosphere-dominated end-member has $^{3}He/^{4}He$ = $6.9 \pm 0.15\ R_A$ ($^{3}He/^{4}He \approx 103{,}400 \pm 2{,}200$), while the plume-dominated end-member is characterized by low $^{3}He/^{4}He = 5.2 \pm 0.13\ R_A$ ($^{3}He/^{4}He \approx 137{,}000 \pm 3{,}400$). Thus, the plume end-member of the Comorian volcanism represents a notably «low $^{3}He/^{4}He$» hotspot, dominated by recycled ^{4}He-rich material (Craig and Rison 1982; Kaneoka et al. 1986; Class et al. 2005). Class et al. (2005) also suggest, based on helium isotope ratios, that the metasomatism of the lithosphere beneath Grande Comore reflects the influence of low degree melts during ocean crust formation.

Mantle-normalized trace elements patterns for Karthala lavas show a typical Ocean Island Basalt pattern (Fig. 23.8). Späth et al. (1996) and Deniel (1998) infer the presence of garnet in the mantle source of Karthala lavas, based on the low HREE concentrations and strongly fractionated REE patterns (Tb/Yb_N = 1.7–2.2 for Karthala lavas, and Tb/Yb_N = 1.5–2.3 for La Grille lavas —Class et al. 1998). From trace element contents and U series, Claude-Ivanaj et al. (1998) conclude that partial melting occurred at the base of old garnet and amphibole-bearing metasomatised

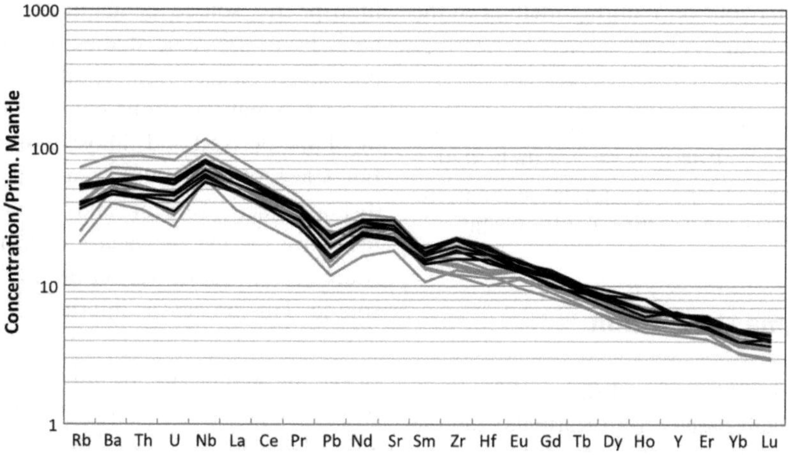

Fig. 23.8 Conventional primitive-mantle normalized trace element diagram of some representative Karthala (*black lines*) and La Grille (*grey lines*) samples. Data from Georoc database (http://georoc.mpch-mainz.gwdg.de/georoc/)

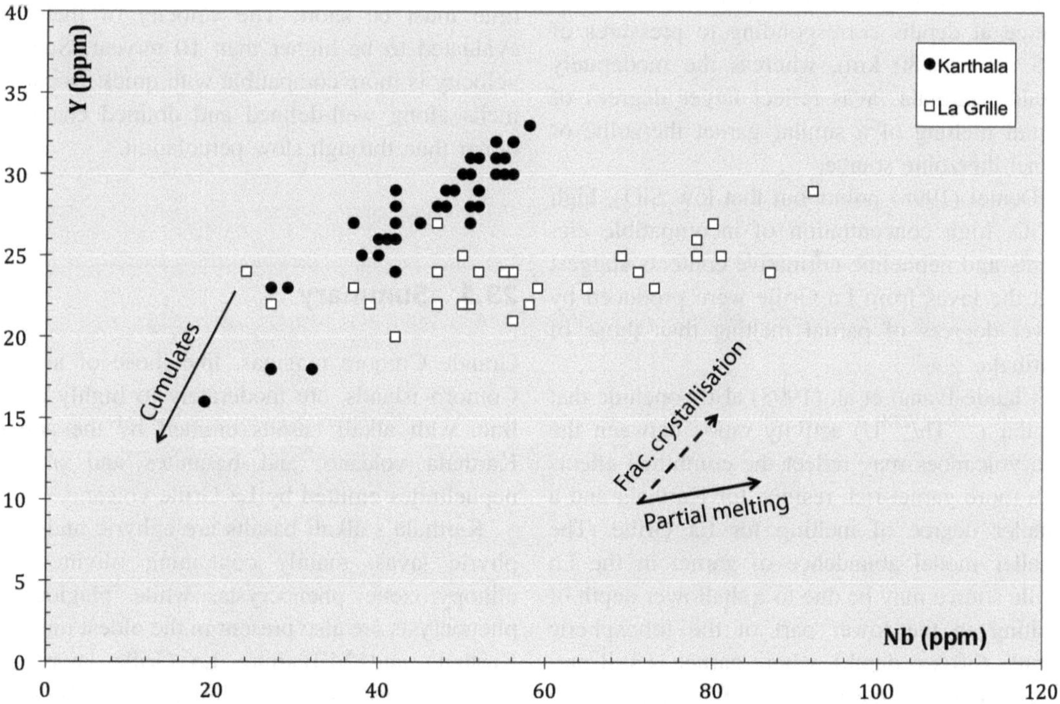

Fig. 23.9 Y versus Nb diagram for Grande Comore lavas. Data from Class et al. (1998), Späth et al. (1996) (redrawn from Class et al. 1998)

lithosphere at depths greater than 80 km. Class et al. (1998) discuss the presence of residual garnet in the source using the Y versus Nb diagram (Fig. 23.9). While Karthala compositions form a trend highlighting the incompatible behaviour of Y and Nb during fractional crystallisation of olivine + clinopyroxene, those from La Grille are strongly contrasting with the more compatible behaviour of Y during melting in the presence of garnet ($D_Y^{Ol/melt} < 0.01$; $D_Y^{Cpx/melt} \approx 0.5-0.7$; $D_Y^{Grt/melt} \approx 5-10$; $D_Y^{Amph/melt} \approx 0.5-0.7$—White 1997) in the solid residue. However, for Bourdon et al. (1998), as the Karthala source is located at greater depth, the garnet mode in the residue should be greater than for La Grille.

23.4.4 Magma Generation

For Späth et al. (1996), the chemical differences between the alkali basalts from Karthala, and the basanites and nephelinites from La Grille may be interpreted as the result of decreasing degrees of partial melting. The more silica-undersaturated basanitic and nephelinitic La Grille lavas may have been produced by relatively small degrees of partial melting of a garnet lherzolite mantle source at depths corresponding to pressures of >25 kbar (~80 km), whereas the moderately alkalic Karthala lavas reflect larger degrees of partial melting of a similar garnet lherzolite or spinel lherzolite source.

Deniel (1998) points out that low SiO_2, high Na_2O, high concentration of incompatible elements and nepheline normative contents suggest that the lavas from La Grille were produced by lower degrees of partial melting than those of Karthala.

Claude-Ivanaj et al. (1998) also conclude that similar ($^{230}Th/^{238}U$) activity ratios between the two volcanoes may reflect the combined effects of a more garnet-rich residue for Karthala and a smaller degree of melting for La Grille. The smaller modal abundance of garnet in the La Grille source may be due to a shallower depth of melting in the lower part of the lithospheric mantle (80 km depth), where garnet is still present. Therefore, the characteristics of La Grille basanites are consistent with such lavas being generated by small-degree melts of a relatively deep, metasomatised lithosphere.

Class and Goldstein (1997) and Class et al. (1998) also show that basanites from La Grille reflect variable degrees of partial melting of a lithospheric source containing residual amphibole. From the same studies, it has been shown that alkali basalts from Karthala volcano show compositional variations consistent with mixing of plume-derived and lithosphere-derived melts. For Class et al. (1998), Karthala alkali basalts display trace element and isotope variations revealing mixing between a deep mantle plume source and some La Grille-like lithospheric source. This process is inferred to have occurred at the base of the lithosphere where plume melts lower the solidus of the lithosphere and act as metasomatic agents, increasing the temperature and triggering melting. The lithospheric melts then contaminate the primary plume melts. Karthala lavas are a mixture of plume melts and high degree lithospheric melts.

The Ra–Th-U disequilibria (Claude-Ivanaj et al. 1998) indicate that the magma transfer time must be short. The velocity of melts is evaluated to be higher than 10 m/year. Such a velocity is more compatible with quick ascent of melts along well-defined and drained channels rather than through slow percolation.

23.5 Summary

Grande Comore magmas, like those of all the Comoro islands, are moderately to highly alkaline, with alkali basalts emitted by the active Karthala volcano, and basanites and olivine nephelinites emitted by La Grille volcano.

Karthala's alkali basalts are aphyric and porphyric lavas, mainly containing olivine and clinopyroxene phenocrysts, while plagioclase phenocrysts are also present in the oldest units of Karthala and M'Badjini. La Grille lavas are undersaturated, with olivine as the only phenocryst phase. High Mg numbers and the occasional presence of mantle-derived xenoliths in La Grille lavas are indications that these lavas are more primitive than Karthala lavas, and may have passed from the mantle to the surface without being stored in a magma reservoir.

Trace element and isotope compositions show a lithosphere-dominated geochemical signature. Magma generation is driven by variable degrees of partial melting of a source containing residual amphibole (and/or phlogopite) and garnet, with the Karthala source probably located deeper than the La Grille source.

Interactions between melts deriving from a metasomatised lithosphere and a shallow mantle plume have been discussed. The alkali Karthala basalts indicate a more pronounced involvement of the mantle plume component. The HIMU component may be present within the DM MORB mantle as known from the global MORB data set. The origin of the EM1 remains controversial: either introduced into the Comorian magmas through delamination or thermal erosion of the lower continental crust during the break-up of Gondwana, that is consistent with the geographic position of the Comoros archipelago near to the continental margins of Africa and Madagascar, or derived from the deeper mantle, source of the thermal plume.

Acknowledgments The authors are grateful to Fran Van Wyk de Vries for proofreading this manuscript. This paper benefited from the constructive comments of A. Di Muro.

References

Bachèlery P, Morin J, Villeneuve N, Soulé H, Nassor H, Radadi Ali H (2016) Structure and eruptive history of Karthala volcano. In: Bachèlery P, Lénat J-F, Di Muro A, Michon L (eds) Active volcanoes of the southwest Indian Ocean: Piton de la Fournaise and Karthala. Active volcanoes of the world. Springer, Berlin

Bachèlery P, Coudray J (1993) Carte volcano-tectonique (1/50000e) de la Grande Comore et notice explicative. Edited by the French Embassy in Moroni, Comores, and The University of La Réunion, St. Denis de La Réunion

Bauer M (1911) Gesteinsproben des Wittu Insel, von Zanzibar, Archipel von den Comoren, Madagascar, Ceylon, etc. In: Voeltzkow A., Reise in Ost Afrika in den jahren 1903-1905. Bd J, Absteill 1911, 17–51

Boivin P, Bachèlery P (2009) Petrology of 1977 to 1998 eruptions at Piton de la Fournaise, La Réunion Island. J Volcanolo Geoth Res 184:109–125. doi:10.1016/j.jvolgeores.2009.01.012

Bourdon B, Joron JL, Claude-Ivanaj C, Allègre CJ (1998) U-Th-Pa-Ra systematics for the Grande Comore volcanic: melting processes in an upwelling plume. Earth Planet Sci Lett 164:119–133

Class C (1994) A geochemical study of Grande Comore Island: Indian Ocean hotspot source compositions and plume-lithosphere interactions. PhD thesis, University of Karlsruhe, 217 p

Class C, Goldstein SL (1997) Plume-lithosphere interactions in the ocean basins: constraints from the source mineralogy. Earth Planet Sci Lett 150:245–260

Class C, Goldstein SL, Altherr R, Bachelery P (1998) The process of plume—lithosphere interactions in the ocean basins—the case of Grande Comore. J Petrol 39(5):881–903

Class C, Goldstein SL, Stute M, Kurz MD, Schlosser P (2005) Grand Comore Island: a well-constrained "low $^3He/^4He$" mantle plume. Earth Planet Sci Lett 233:391–409

Class C, Goldstein SL, Shirey SB (2009) Osmium isotopes in Grande Comore lavas: a new extreme among a spectrum of EM-type mantle endmembers. Earth Planet Sci Lett 284:219–227

Claude-Ivanaj C, Bourdon B, Allègre CJ (1998) Ra-Th-Sr isotope systematic in Grande Comore Island: a case study of plume-lithosphere interaction. Earth Planet Sci Lett 164:99–117

Coffin MF, Rabinowitz PD, Houtz RE (1986) Crustal structure in the western Somali Basin. Geophys J R Astr Soc 86:331–369

Coffin MF, Rabinowitz PD (1987) Reconstruction of Madagascar and Afrika : evidence from the Davie Fracture Zone and western Somali Basin. J Geophys Res 92(B9):9385–9406

Coltorti M, Bonadiman C, Hinton RW, Siena F, Upton BGJ (1999) Carbonatite metasomatism of the oceanic upper mantle: evidence from clinopyroxenes and glasses in ultramafic xenoliths of Grande Comore, Indian Ocean. J Petrol 40:133–165

Craig H, Rison W (1982) Helium 3: Indian Ocean hotspots and the East African Rift, EOS. Trans Am Geophys Union 63(45):1144

Debeuf D (2004) Etude de l'évolution volcano-structurale et magmatique de Mayotte (Archipel des Comores, Océan Indien). PhD thesis, University of La Réunion, 277 p

Deniel C (1998) Geochemical and isotopic (Sr, Nd, Pb) evidence for plume-lithosphere interactions in the genesis of Grande Comore magmas (Indian Ocean). Geochem Geol 144:281–303

de Saint-Ours J (1960) Etudes géologiques dans l'extrême Nord de Madagascar et de l'Archipel des Comores. Thèse de Doctorat, Rapport Service Géologique de Tananarive n° 221, Tananarive, Madagascar, 205 p

Desgrolard F (1996) Pétrologie des laves d'un volcan intraplaque océanique : le Karthala, Ile de la Grande Comore – (R.F.I. des Comores). PhD Thesis, University Joseph Fourier - Grenoble 1, 176 p.+annexes

Flower MFJ, Strong DF (1969) The significance of sandstone inclusions in lavas from the Comores Archipelago. Earth Planet Sci Lett 7:47–50

Janin M, Hémond C, Maia M, Nonnotte P, Ponzevera E, Johnson KTM (2012) The Amsterdam–St. Paul Plateau: a complex hot spot/DUPAL-flavored MORB interaction. Geochem Geophys Geosyst 13:Q09016. doi:10.1029/2012GC004165

Kaneoka I, Takaoka N, Upton BGJ (1986) Noble gas systematics in basalts and a dunite nodule from Réunion and Grande Comoro Islands, Indian Ocean. Chem Geol 59:35–42

Lacroix A (1912) Sur la constitution minéralogique des volcans de l'île de la Réunion. Comptes Rendus de l'Académie des Sciences Paris, CLV, 538

Lacroix A (1916) La constitution des roches volcaniques de l'archipel des Comores. Comptes Rendus Académie Sciences Paris, t 163, n° 9, 213–219

Lacroix A (1922) La constitution lithologique de l'archipel des Comores. C.R. XIIIème congrès Int. Géol. Bruxelles 2, 949–979

Lacroix A (1923) Océanites. Minéralogie de Madagascar, vol. III, Paris, pp 49–50

Lacroix A (1938) Le volcan actif de l'île de La Réunion (supplément) et celui de la Grande Comore, Gauthier Villars Ed., Paris, 57 p

Leeman WP, Scheidegger KF (1977) Olivine/liquid distribution coefficients and test for crystal-liquid equilibrium. Earth Planet Res Lett 35(2):247–257

Le Maitre RW (1989) Igneous rocks. Recommendations of the International Union of Geological Sciences Subcommission on the Systematics of Igneous Rocks. IUGS and Blackwell Scientific Publications, Oxford

Macdonald GA, Katsura T (1964) Chemical composition of Hawaiian lavas. J Petrol 5:82–133

Montaggionni L, Nougier J (1981) Les enclaves de roches détritiques dans les volcans d'Anjouan (archipel des Comores). Origine et interprétation dans le cadre de l'évolution du canal de Mozambique. Bulletin de la Société Géologique de France 7:595–601

Nougier J, Cantagrel JM, Karche JP (1986) The Comores archipelago in the western Indian Ocean: volcanology, geochronology and geodynamic setting. J Afr Earth Sc 5–2:135–145

Pelleter AA, Caroff M, Cordier C, Bachèlery P, Nehlig P, Debeuf D, Arnaud N (2014) Melilite-bearing lavas in Mayotte (France): an insight into the mantle source below the Comores. Lithos. doi:10.1016/j.lithos.2014.09.012

Radadi Ali A (2012) Le rôle de la fugacité d'oxygène dans l'évolution des laves du Karthala (Grande Comore). PhD thesis, University Blaise Pascal, Clermont-Ferrand

Späth A, Le Roex AP, Duncan RA (1996) The geochemistry of lavas from the Comores Archipelago, western Indian Ocean: petrogenesis and mantle source region characteristics. J Petrol 37(4):961–994

Strong DF (1972) The petrology of the lavas of Grande Comore. J Petrol 13:181–217

Rabinowitz P (1983) The separation of Madagascar and Africa. Science 220:67–69

Reisberg L, Zindler A, Marcantonio F, White W, Wyman D, Weaver B (1993) Os isotope systematics in ocean island basalts. Earth Planet Sci Lett 120:149–167

Thompson RN, Flower MFJ (1971) One-atmosphere melting and crystallization relations of lavas from Anjouan, Comores archipelago, Western Indian ocean. Earth Planet Sci Lett 12:97–107

Upton BGJ, Wadsworth WJ, Latrille E (1974) The 1972 eruption of Karthala Volcano, Grande Comore. Bull Volcanologique 38:136–148

Vienne E (1900) Colonies et Pays de Protectorates Mayotte et Comores. Exposition Universelle de 1900, Paris, p 200

Voeltzkow A (1906) Die Comoren, Zeitschrift der gesellschaft für Erdkunde zu Berlin, 606–630

White WM (1997) Geochemistry, on-line textbook. Johns Hopkins University Press, Baltimore

24 Groundwater Prospection in Grande Comore Island—Joint Contribution of Geophysical Methods, Hydrogeological Time-Series Analysis and Groundwater Modelling

Anli Bourhane, Jean-Christophe Comte, Jean-Lambert Join and Kassim Ibrahim

Abstract

In Grande Comore island, groundwater is the only safe water resource for drinking but only one third of the population have access to this resource. All existing wells are drilled in the volcanic aquifers of the coastal zone. Among them, about one third provide groundwater of acceptable salinity (less than 1 g/L), one third provide water of salinity comprised between 1 and 3 g/L, and the remaining third is generally disused due to salinities higher than 3 g/L. To date, inland groundwater resources have been largely overlooked. The development of groundwater in Grande Comore requires an improved understanding of the complex young volcanic aquifers, quantitatively and qualitatively. This work applies an integrated hydrogeological methodology aiming at improving the conceptual understanding of Grande Comore volcanic aquifers in both coastal and inland areas, and proposing a robust approach for prospecting and managing the groundwater resources. This methodology comprised (1) a review of the current hydrogeological knowledge regarding the structure, properties and conceptualisation of the volcanic aquifers, (2) the spatial characterisation of both aquifer structures and seawater intrusion in coastal areas, through

A. Bourhane (✉) · J.-L. Join
Laboratoire Géosciences Réunion, Université de La Réunion, Institut de Physique du Globe de Paris, CNRS UMR, 7154 La Réunion, France
e-mail: anli.bourhane@univ-reunion.fr

J.-C. Comte
Groundwater Research Group — SPACE, Queen's University Belfast, Northern Ireland, UK

J.-C. Comte
School of Geosciences, University of Aberdeen, Scotland, UK

K. Ibrahim
Faculty of Sciences and Techniques, University of the Comoros, Moroni, Union of the Comoros

P. Bachèlery et al. (eds.), *Active Volcanoes of the Southwest Indian Ocean*, Active Volcanoes of the World, DOI 10.1007/978-3-642-31395-0_24

the implementation of geophysical surveys comprising electrical resistivity tomography (ERT) and time-domain electromagnetic soundings (TDEM), (3) the characterisation of coastal groundwater dynamics through the acquisition of high temporal resolution heads and salinity measurements in three wells from contrasted hydrogeological settings, (4) the quantification of the impact of volcanic heterogeneity on coastal groundwater salinity through numerical groundwater modelling and (5) the exploration of the poorly known inland aquifers through the interpretation of ERT and TDEM investigations together with a discussion of the hydrogeological analogy with the better known volcanic island of La Réunion. Results reveal a strong potential for both further developing coastal aquifers and initiating the prospection of inland aquifers, and provide methods for improving the management of existing groundwater infrastructures.

24.1 Introduction

In Grande Comore, the most populated island of the Comoros archipelago (around 382 070 inhabitants in 2010), water access is a major issue affecting the whole territory. More than half of the population is concentrated in coastal areas, at less than 5 km from the shoreline. Currently, coastal wells supply 20 localities representing about 35 % of the island's population (Mohamed and Othman 2006). About 65 % of the population do not have a permanent access to groundwater and still use complementary rainwater collected from houses roofs in domestic tanks. Rainwater however is only sufficient during the wet season. During the dry season (generally from May to September), the water use is rationed, particularly in families living in the north-east and the south-east of the island where the dry period is longer.

More than 30 % of groundwater hydraulic facilities are contaminated by relatively high concentrations of saltwater (total dissolved solids TDS > 3 g/L) and are consequently disused by local populations. Only 15 wells among 54 existing wells provide a water of suitable quality for drinking. Due to those exceptional hydrogeological conditions, the local drinkable water salinity guideline (often expressed in TDS) is usually taken at 3 g/L in Grande Comore Island, instead of 1 g/L as recommended by the World Health Organisation (WHO) standards (WHO 2003). This local guideline value is based essentially on tastes considerations by the population and corresponds approximately to 5 mS/cm in water electrical conductivity. Wells with higher salinity are commonly still used for the irrigation, livestock drinking or washing.

In this work, an integrated hydrogeological methodology is applied in the aim at advancing the conceptual understanding of Grande Comore volcanic aquifers in both coastal and inland areas, and proposing a robust approach for prospecting and managing the groundwater resources.

24.2 Methodology

A review of the available literature provided the current status of knowledge about the Grande Comore hydrogeological settings. Then, multi-techniques ground geophysical investigations carried out in 3 coastal zones enabled imaging the complexity of the geometry of the seawater interface in Grande Comore. These methods included Time Domain Electro-Magnetics (TDEM) soundings and Electrical Resistivity

Tomography (ERT) profiles. Secondly, a hydrogeological time series analysis is performed using dataloggers recordings from three coastal wells for understand the impact of the tidal fluctuations on the groundwater. Finally, a 3D numerical model using the code SEAWAT (Guo and Langevin 2002) was applied to investigate the theoretical effects of local aquifer heterogeneities associated to typical various volcanic structures on the patterns of seawater intrusion in the basal aquifer.

24.2.1 Geophysical Investigations

Electrical and electromagnetic methods are particularly adapted for studying the geological heterogeneities and salinity contrasts in both volcanic and coastal aquifers (e.g., Albouy et al. 2001; Revil et al. 2012). On Grande Comore Island, geophysical investigations were performed in 3 coastal and 2 altitude zones and included ERT and TDEM soundings. Those techniques are sensitive to the geological properties (such as the clay content), and hydrogeological variables (such as the water saturation and salinity) and have the ability to provide information with a much higher spatial resolution than hydrological measurements carried out at defined location (Huisman et al. 2004).

ERT. This technique provides a 2D or 3D continuous characterization of subsurface electrical resistivity (Griffiths et al. 1990; Dahlin 1993). 10 ERT profiles were carried out using an ABEM Lund System using up to 64 electrodes at 20 m-spacing distributed on a maximum of 8 cables and a maximum AB spacing of 1260 m, which resulted to a maximum depth of investigation of about 315 m. 2D measurements were taken with the Wenner-Schlumberger array, which has a relatively good signal/noise ratio (Dahlin and Zhou 2004) and is sufficiently sensitive to the geometrical features of seawater interface in coastal groundwater (Comte 2008; Comte et al. 2010). Measured resistivities were inverted using the 2D code RES2DINV v3.55 (Loke 2006) using the robust L1-norm method with incorporation of the topography.

TDEM. This technique is commonly used in hydrogeological and environmental investigations (Danielsen et al. 2003), including in volcanic islands contexts (Descloitres 1998; Boubekraoui et al. 1998). 15 TDEM soundings with 50 m-side square loops were used in a coincident array (transmitter loop = receiver loop). The investigation depths varied between 25 and 75 m depending on the geoelectrical properties of each site. Sounding were inverted using the 1D inversion code TEM Res. v8 (AEMR 2012).

24.2.2 Time Series Analysis

On Grande Comore Island, three study sites were monitored using tidal gauges and automatic data loggers installed in the wells at different periods both monitoring groundwater hydraulic head, electrical conductivity (later converted to salinity) and temperature with a constant time step of 30 min. Measured hydraulic heads were corrected from the variations of the atmospheric pressure using a complementary barometric logger.

24.2.3 Groundwater Modelling

The numerical code SEAWAT-2000 (Guo and Langevin 2002) integrated to the interface Visual MODFLOW (Schlumberger Water Services) was applied to simulate in 3D the groundwater flow and seawater intrusion for 4 theoretical volcanic aquifers models. SEAWAT uses the finite-difference approach to solve for variable-density groundwater flow resulting from freshwater/seawater interactions in porous media. The 3D rectangular models had 92,000 rectangular cells with a total length of 5 km perpendicular to the coastline (500 m in the sea and 4.5 km inland), a total width of 2 km parallel to the coastline and a max depth of 300 m below the mean sea level. Models were run in transient mode with stationary boundary conditions until reaching pseudo-steady state conditions, i.e. until spatiotemporal stabilisation of groundwater heads and salt concentrations.

24.3 General Settings of Grande Comore

24.3.1 *Climate and Hydrology*

The Comoros archipelago has a humid tropical climate, including two seasons: a hot and wet season from October to April and a cool and dry season from May to September. The annual rainfall varies from less than 1.5 m in the north-east up to more than 4 m in the south-west of the island (western flank of the Karthala volcano). The latter is attributed to the combination of dominant western winds (Kashikazi from the NW and Kusi from the SW) with the orogenic effect of the Mount Karthala during the austral summer. Despite a high rainfall, the runoff is very low (estimated annually at 5 % of rainfall; Marini 1990) and there are no permanent rivers. This low runoff is explained by a high infiltration rate supported by the high permeability of the subsoil resulting from the dense and wide fracturing of the lava flows, and the presence of lava tubes, caves or scoria layers regularly intercalated between lava flow units.

24.3.2 Geology

The Comoros volcanic archipelago is located midway between the west coast Madagascar and the African east coast, at the northern extremity of the Mozambique Channel (Michon 2016, Chap. 21). It covers a total area of 2033 km^2 including four mean islands, in line with an approximate SE–NW younging volcanic trend. The three western islands belong to the Union of the Comoros of which the westernmost island of Grande Comore is geologically the most recent [most of the rocks have less than 1 Ma], the largest (1024 km^2) and the highest (2361 m). The volcanism is still active at the Karthala volcano (Bachèlery et al. 2016, Chap. 22). The island is composed of three main shield volcanoes of dominant basaltic type (Bachèlery and Coudray 1993): (1) the massif of the M'Badjini (743 m), which covers a limited area at the south-east of the island, is the oldest (Miocene) and is characterised by both its high density of fracturing and its alteration, the latter reflected by erased 'aa' and 'pahoehoe' lava flows surface features and deep ferralitic weathering; (2) the massif of La Grille (1087 m), which covers the northern half part of the island, is of intermediate age (Middle Pleistocene), and is less weathered with a morphology characterised by gentle slopes and scattered by several cinder cones; (3) the massif of the Karthala, which covers a surface of about two thirds of the island in its centre, is an active shield volcano (Quaternary) that recorded several activity phases during the last century and is morphologically characterised by the presence of two rifts zones on both sides of interlocked calderas series.

24.3.3 Hydrogeology

24.3.3.1 Seawater Intrusion in Coastal Aquifers and Challenges of Prospecting and Managing the GroundwaterResources in Young Volcanic Islands

Groundwater salinisation caused by seawater intrusion is a major cause of degradation of water quality in the world (Meybeck and Helmer 1989; Vengosh 2003). This phenomenon is particularly problematic in volcanic islands where freshwater resources are found in limited quantities. The insular freshwater lens is in contact with seawater and this theoretical interface is controlled by a hydrodynamic balance according the Ghyben-Herzberg principle (Ghyben 1888; Herzberg 1901). In a global context of climate change and sea level rise, insular groundwater resources become increasingly affected by seawater intrusion (Bates et al. 2008).

In young volcanic islands, because of the dominant fracture porosity, basaltic lava flows constitute aquifer units of exceptionally high permeability. This results in very low groundwater hydraulic gradients, typically 1 ‰, and in a very thin fresh water lens lying on a saltwater edge throughout a wide and shallow mixing zone, which deepens inland with a gentle slope of typically 2–5 % (Join 1991). Geologically, coastal volcanic aquifer units are typically composed of an accumulation of elongated and narrow basaltic flows with low dipping angles seaward. Inland, a staged and channelled aquifer system overlays a deep and thick saturated zone called the basal aquifer. The basal aquifer is generally well known only in the coastal fringe, up to a distance of approximately 2 km from the shoreline (Coudray et al. 1990). Beyond this distance, borehole information is scarce due to the high topography requiring deep and expensive boreholes. Recent lava flows units commonly have a high porosity and an anisotropic hydraulic behaviour due to preferential water flow in scoria beds or lava tubes. Thin tuff palaeo-soils and clastic deposits in old valley (alluviums, lahars) are the only potentially occurring impermeable layers. The high degree of geological heterogeneity is responsible for complex marine contamination patterns.

Quantifying the groundwater resources through the implementation of pumping tests on the wells is very challenging in recent volcanic environments because: (1) due to the very high aquifer permeabilities (and, in Grande Comore, because of also both the large diameter of traditional wells and the technical characteristics of the pumping equipment), it is difficult to generate significant water table drawdowns; and (2) the high risk of seawater invasion caused by pumping. The application of alternative and multidisciplinary approaches is required to overcome the limitations and uncertainties of traditional investigation methods implemented individually and/or on limited sample points.

This study proposes to assess the impact of natural settings and anthropogenic influence on volcanic groundwater quantity and quality exemplified by the case of Grande Comore Island. An integrated investigation approach is applied, including (1) a review of both the existing hydrogeological knowledge of Grande Comore and the generic conceptual model of young volcanic island aquifers, (2) the hydrostructural interpretation of ground geophysical investigations (resistivity imaging and electromagnetic soundings), (3) the analysis of time-series of groundwater heads and salinity with regards to oceanic, geologic and abstraction conditions, and (4) the application of a theoretical numerical model to confirm the impact of typical heterogeneities and recharge conditions on seawater intrusion.

24.3.3.2 Current Status of Hydrogeological Knowledge and Gaps

The renewable annual groundwater volume of Grande Comore is estimated between 0.5 and 2×10^9 m^3 (Marini 1990; Ragot 2007; Mohamed 2012). This large renewable volume would result from a high rate of effective infiltration, estimated according to different authors between 57 and 63 % of the annual rainfall (Boinali 1982; Marini 1990; Ibrahim 2009). The combined annual abstraction of the currently 11 operating wells (out of the 54 existing) can be estimated at about 7×10^6 m^3 (Ragot 2007), which constitutes only 0.4–1.4 % of the renewable annual groundwater resource.

In the coastal zone, the hydrodynamic properties of the basal aquifer have been provided by pumping tests carried on a number of the wells performed y the United Nation Development Program (UNDP) by Marini (1990). Obtained parameters display exceptionally high values of transmissivities, hydraulic conductivities, diffusivities and well productivities (Table 24.1). Most of the results, however, would not be representative of the most recent and highly productive units, where it is difficult to obtain significant drawdowns required for a reliable interpretation of the pumping tests. Aquifer storativities appear relatively low reflecting unconfined to semi-confined conditions. Those properties remain comparable with those of the recent volcanic aquifers of La Réunion (Massari 1990; Join 1991; Join et al. 1988; Folio 2001). Aquifers hydrodispersive properties that control solute transport, such as salt water, are not known.

The lowest productivities are found in the southern part of the island (massif of M'Badjini),

Table 24.1 Well and aquifer hydrodynamic properties of the basal aquifer in the costal zone of Grande Comore (after Marini 1990; Ibrahim 2009) and comparison with typical properties in La Réunion volcanic aquifers (from Massari 1990; Join 1991; Join et al. 1988; Folio 2001)

Parameter	Range Grande Comore	Range La Reunion
Transmissivity (m^2/s)	1.0×10^{-2}–3.0×10^{-1}	1.0×10^{-3}–2.8×10^{-1}
Hydraulic conductivity[a] (m/s)	1.6×10^{-3}–2.7×10^{-1}	1.0×10^{-7}–2.0×10^{-2}
Secure production yield[b] (m^3/h)	10–165	80–200
Specific yield (m^2/s)	2.8×10^{-3}–1.8×10^{-2}	1.0×10^{-3}–6.0×10^{-1}
Storativity (−)	2.2×10^{-5}–1.0×10^{-2}	1.0×10^{-4}–1.4×10^{-3}
Diffusivity (m^2/s)	19–896	5–200

[a]Estimated from the mean water column in each well
[b]Well yield per meter of drawdown

where appreciable drawdowns can be observed during pumping tests, providing between 3 and 20 m^3/h. Those wells are poorly influenced by tidal fluctuations and provide fresh to slightly brackish groundwater. In contrast, the recent volcanic units of La Grille and Karthala generally display much higher productivities but their exploitation is limited by the pump capacities and the vulnerability to marine salinisation. The high aquifer diffusivities result in both a low attenuation and a short time-lag of the tidal signal, which enhance the seawater encroachment, further exacerbated by a large tidal range (up to 3 m).

Water table elevations, close to the sea level everywhere in the coastal zone, are not known accurately because of (i) the absence of accurate measurement of well elevations and (ii) the influence of tidal fluctuations. They have been estimated indirectly inland by geophysical surveys (VES by UNDP 1987; CSAMT by Savin et al. 2001; ERT by Join et al. 2013).

In terms of water quality, currently only 11 wells among 54 existing wells provide groundwater with a suitable salinity for drinking (i.e., TDS < 1 g/L). Water physico-chemical analyses by SCP (2009) confirmed both the widespread saline contamination of groundwater at public fountains and the bacteriological contamination of rainwater domestic tanks, while the Moroni distribution network provide water of acceptable quality. Groundwater was measured at a generally slightly basic pH of ~8 and low turbidity.

To date, perched aquifers in the altitude areas have not yet been developed for water supply, and those aquifers' hydraulic properties remain largely unknown.

24.3.3.3 Hydrogeological conceptual model—Analogy with La Réunion Island

The first deep geophysical surveys performed on the flanks of the Karthala volcano allowed discussing for the first time the analogy of the hydrogeological conditions between the Karthala massif and the better known La Réunion Island (Savin et al. 2001). According to Join et al. (2005), the hydrogeological context of both La Réunion and Grande Comore island can be decomposed into two hydrogeological domains based on two types of methodological approaches of investigation: the littoral domain and the inland domain.

The littoral domain. The geographical zone corresponding to the littoral domain is defined arbitrarily as the area where the basal groundwater can be prospected by conventional drilling, i.e. where the unsaturated zone is less than 300 m thick. In that domain, hydraulic gradients are very low, groundwater heads are close to the mean sea level and both heads and salinity are strongly influenced by tidal fluctuations. The tidal forcing is higher in Grande Comore than in La Réunion, where tidal ranges reach 3 and 1.2 m, respectively. In Grande Comore, a correlation was observed between submarine groundwater discharge and the presence of old valleys filled by recent and more permeable lava flows, particularly on the southern and northern coasts of the island (UNDP 1987), which indicates preferential flow channels. At the Prophet's hole beach (2.5 km North of Mitsamiouli), a significant submarine discharge is in line with a rift zone oriented in SSE-NNO direction (UNDP 1987). These observations confirm that fresh groundwater flow is strongly influenced by 3D permeability contrasts in the volcanic formations.

The inland domain. Beyond 300 m of unsaturated zone, were drilling investments are exceptional, the existence of a staged aquifer system is evidenced in La Réunion by hundreds of springs emerging in the deep erosion canyons. Join and Coudray (1993) distinguish three types of aquifers: (1) superficial perched aquifers, (2) intermediate perched aquifers and (3) the deep saturated level i.e., the basal aquifer accessible in the littoral domain. In Grande Comore however, where the climate is drier, deep emergence levels are not accessible due to the lack of significant erosion. Only the perched aquifers associated to the superficial levels (Type 1) can be observed in the altitude zones. Such perched systems are known in the older volcanic massifs of M'Badjini and La Grille, where weathered lava flows and pyroclastic deposits can form relatively impermeable layers capable to support groundwater systems. Around the summit zone of the Karthala volcano, deep geophysical investigation (Savin et al. 2001) have revealed a bell-shaped piezometric surface of the basal groundwater, which can be explained by an increase of the hydraulic gradient towards the centre of the island due to a higher recharge, a lower flow section, and possibly also, as observed in La Réunion Island, lower permeabilities associated to both a hydrothermally weathered core of the volcanic massif (zeolitisation) and dykes (Bret et al. 2003).

24.3.4 Water Uses

In the late sixties, the first development of groundwaterresources was initiated by the French Geological Survey (BRGM) with borehole drilling in coastal areas (UNDP 1987). In 1965–1967, a 35 m-deep well (TP5) providing fresh groundwater (salinity ~200 mg/L) was drilled in the Karthala basalts near Vouvouni, about 7 km south of Moroni and 2 km from the coast. In 1967, a shallow well was also drilled in the sand dunes near Mitsamiouli to supply that town, but progressive saline and bacteriological contaminations led to its abandon in 1970. In the early 1970s, two deep boreholes were drilled near Hahaya to supply the Moroni International Airport and a public fountain for the population of the area, providing water with TDS of about 1 g/L. In 1977, a limited water conveyance network was completed to supply the Moroni agglomeration area (the most important urban area of the Comoros archipelago) from the well TP5 at a secure pump rate of 165 m^3/h.

During the eighties, a major groundwater prospection funded by the United Nations Development Programme supported the drilling of 44 new wells in the coastal areas (UNDP 1987). Secure yields were estimated on 20 of them. Values varied from 2 m^3/h in the island's oldest part to 165 m^3/h in the island's most watered part. The mean secure yield was estimated at 50 m^3/h. In rural areas, community water conveyance networks were progressively developed to supply some villages from the UNDP wells. Away from the coast in the mountain zones, six little springs, with discharges ranging from 0.5 to 2.5 m^3/d in dry season, and from 2.5 to 6 m^3/d in wet season, were also identified and collected (UNDP 1987): two are located in the weathered massif of M'Badjini, and four discharge from the massif of La Grille.

In 2000, an additional well (SHA) was drilled by the French Geological Survey (UNDP 2000) for complementing the supply of the village of Hahaya and the nearby airport. It reaches a depth of 87 m and abstracts a highly mineralised groundwater [1.9 mS/cm] at a mean rate of 50 m^3/h.

In 2009, in the Oishili region, a community self-funded initiative allowed the rehabilitation and equipment of an existing UNDP coastal well (ONU 40) including the installation of a water conveyance network supplying six villages (Ibrahim 2009). The well is strongly affected by marine influence [3 mS/cm] and the secure production is limited to 10 m^3/h.

Apart from the aforementioned sectors, most of the villages have today no permanent source of water. The population still use rainwater or buy water (from Vouvouni wells) sold throughout the island by tank trucks.

The widespread salinity problem in Grande Comore has limited the development of groundwater resources in the last decades. Frequently, local consumers (organised in local council communities) abandon the groundwater facilities because of both the mediocre quality and the cost of the water. In Oishili community self-funded network, the water price was fixed at 3 €/m^3 (vs. 1.40 €/m^3 in La Réunion Island or 2 €/m^3 in the nearby Mayotte Island) which seems relatively excessive for rural populations. Across the island, the price of water varies depending on the cost of infrastructures, the resource availability and the supply mode (e.g., 0.40 €/m^3 in Mitsoudje versus 2.5 €/m^3 in Ouroveni). In this delicate context, the management of water supply can lead to frequent tensions among different villages due to (i) such disparities in cost and quality, (ii) local governance dysfunctions and (iii) technical limits (Mohamed 2012). Thus, the groundwater access in Grande Comore remains not only a public health issue but also a real socio-economic challenge.

Currently, two programmes funded respectively by the African Development Bank (ADB) and the French Development Agency (AFD) both aim at prospecting new groundwater resources in Grande Comore. The later is considering drilling a number of exploration boreholes followed by large diameter wells for supplying rural communities in both coastal and mountain areas (Join et al. 2013).

24.4 Results

24.4.1 Aquifers Structure from Geophysical Investigations

Multi-techniques ground geophysical investigations carried out in 2012 allowed confirming the different types of groundwater occurrence in Grande Comore. The littoral domain is characterized by variable resistivities resulting from different degrees of weathering and seawater intrusion (Fig. 24.1).

The older units of the Mbadjini massif (Fig. 24.2a) appears relatively homogeneous with relatively low resistivities (50–200 Ω.m) resulting from deep weathering of the basalts. Along the coast, the seawater interface is dipping with a high angle inland as a result of relatively low aquifer permeabilities, which is consistent with a relatively high hydraulic gradient imaged. Some more recent lava flows (>200 Ω.m) can locally overlay the weathered basalt and may constitute local perched aquifers.

The intermediate age to recent units of La Grille and Karthala volcanoes of the littoral domain appear more heterogeneous with highly contrasted resistivities. On the western flank of the recent Karthala volcano (southwestern and wettest part of the island; Figs. 24.2b and c), at shallow depths, the unsaturated basalts and the fresh groundwater are characterised by high resitivities (>1000 Ω.m and 100–1000 Ω.m, respectively). At greater depths, the presence of saline/brackish groundwater results in very low resistivities (<100 Ω.m for brackish water and <10 Ω.m for salt water). On transects normal to the coast (Fig. 24.2b), the interface between fresh and saline groundwater displays a very low dipping angle towards the island resulting from low hydraulic gradients and high permeabilities. This angle is expected even lower in drier areas and likely higher in the slightly older and more weathered units of La Grille volcano. In such

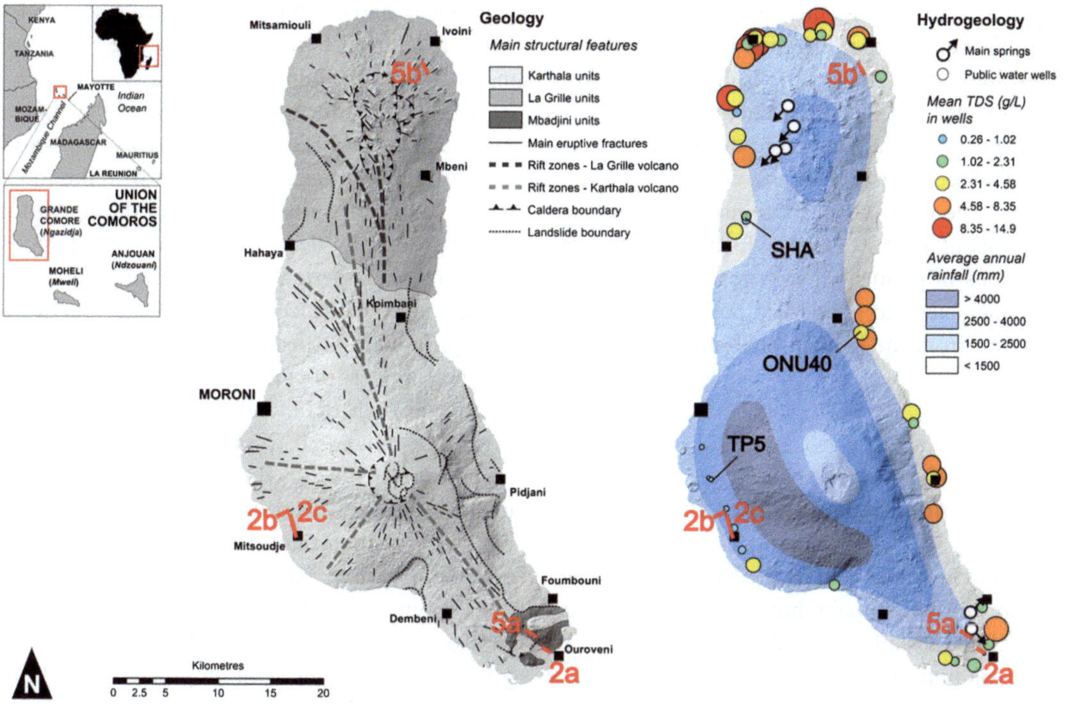

Fig. 24.1 Physical setting of Grande Comore Island (Ngazidja), in the Mozambique Channel—Main geological, hydrogeological and hydroclimatic features. The geophysical surveys presented in Figs. 24.2a–c and 24.5a, b as well as studied wells SHA, TP5 and ONU40 are located

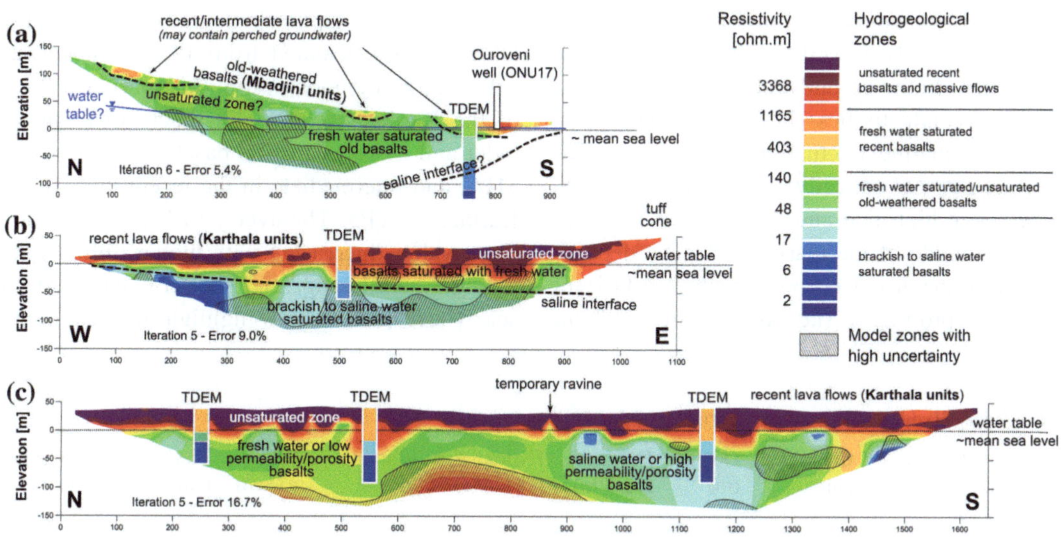

Fig. 24.2 Typology of the basal volcanic aquifers in Grande Comore interpreted from combining ERT and TDEM results—**a** old and weathered basalts of the Mbadjini volcano; **b** recent basalts of the Karthala volcano (section normal to the coast); **c** same as **b** but section parallel to the coast. The profiles are located on Fig. 24.1 (2a–c)

recent volcanic areas, this low dipping angle of the seawater interface explains the apparent wide transition zone between saline and fresh groundwater and the occurrence of high degrees of salinities at relatively large distances from the sea. Transects parallel to the coast (Fig. 24.2c) reveal large resistivity variations in the saturated zone (about 10–500 Ω.m), which suggests large variations of permeability, porosity and/or salinity associated to possible preferential flows resulting from the depositional structure of lava flows.

24.4.2 Hydrodynamics and Groundwater Time-Series

The temporal behaviour of the basal groundwater of the coastal domain have been investigated through the installation of data loggers recording the evolution of groundwater heads, electrical conductivity and temperature on three selected wells: SHA, TP5 and ONU40. The SHA well has been monitored between Oct. 2012 and Aug. 2013, TP5 from Aug. 2013, and ONU40 from 2009. The hydrogeological characteristics of those well are summarised in Table 24.2.

In those coastal wells, data logger records show that both groundwater heads (H) and electrical conductivities (EC) are strongly controlled by marine fluctuations (Fig. 24.3). The groundwater EC (directly correlated with TDS) increases at high tide and decreases at low tide. The magnitude of the marine influence depends on both (i) the distance from the coast and (ii) the hydraulic properties (i.e. diffusivities) of the geological units, which are typically correlated with the degree of weathering causing a decrease in aquifer permeability.

At a larger time scale, it appears that EC in SHA and ONU40 are influenced by the variations of the tidal range in relation with the moon phases. In both wells, it increases during spring tides periods and decreases during neap periods. In spring tides periods, the tidal amplitudes are higher and enhance the seawater invasion through a wide dispersion of the mixing zone.

The impact of the pumping also appears clearly on both SHA and TP5 wells; a low and rapid drawdown (≈0.15 m) is observed when the pump starts. This drawdown is not clearly visible in ONU40, likely because of the very high permeability of the young lava flow. In SHA well, the pump appears clearly operating only at high tides when EC is also slightly higher.

Tidal efficiencies (i.e. the ratio between groundwater vs. sea level amplitudes) of 69, 44 and 34 % were measured in ONU40, TP5 and SHA, respectively (Table 24.2). As predicted by the model of Jacob (1950), tidal efficiencies decrease with both the distance from the coast and the aquifer diffusivity (the latter being correlated with the aquifer's degree of weathering). Diffusivities calculated from Jacob's model are very high in the most recent (historic) lava flows of the Karthala at ONU40, much lower in the older and more weathered basalts of La Grille at SHA, and intermediate in the recent units of the Karthala at TP5. The average EC measured also show values increasing with higher diffusivities, younger bedrock age and shorter distance to the sea. These observations highlight the role of the

Table 24.2 Hydrogeological characteristics of the three monitored wells

Well	Locality	Geologic unit	Depth (m)	Dist. to sea (km)	Hydraulic grad. (‰)	Rainfall (m/y)	TDS (mS/cm)	Tidal effic. (%)	Diffusivity (m^2/s)
SHA	Hahaya	La Grille	87	2.5	1.4	<1.5	1.9–2	34	200
TP5	Vouvouni	Karthala recent	28	2	0.3	>4	0.3	44	265
ONU40	Oishili	Karthala historic	76	1.5	n.a	~2	3.5–4	69	345

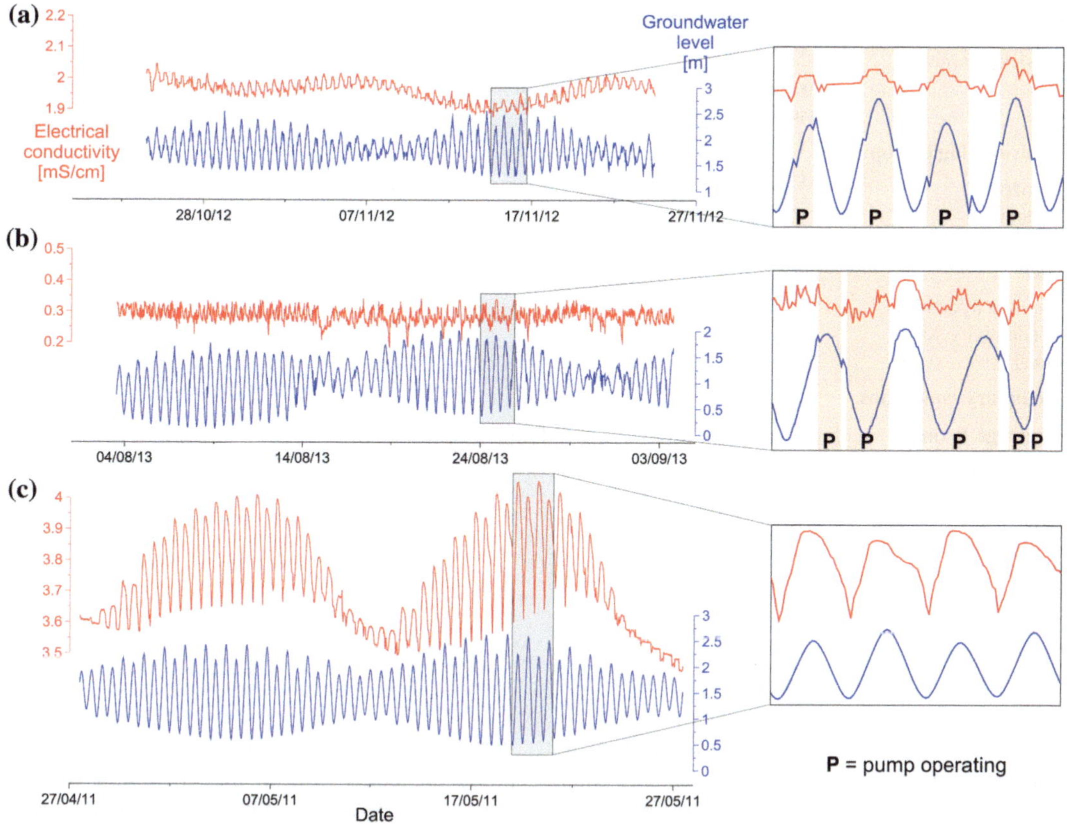

Fig. 24.3 Extracts of dataloggers records (water table and electrical conductivity) in the coastal wells **a** well SHA at Hahaya, **b** well TP5 at Vouvouni and **c** well ONU 40 at Oishili

rocks age (i.e. weathering) in both the propagation of the tidal wave and the saline intrusion.

24.4.3 Theoretical Simulation of the 3D Impact of Volcanic Structures on Groundwater Flow and Saline Intrusion

A 3D numerical model was applied to investigate the theoretical effects of local aquifer heterogeneities associated to typical various volcanic structures on the patterns of seawater intrusion in the basal groundwater. The model domain was defined as a rectangular block accounting for the coastal domain on the flank of a volcanic island. An initial theoretical homogeneous aquifer model was considered, from which three typical aquifer heterogeneities were tested: (1) a paleo-valley 100 m-deep, 700 m-wide at surface filled with massive lava flow; (2) a palaeovalley with the same dimensions but filled with highly fractured and permeable lava flow; and (3) a natural surface trench (ravine) supporting rainfall infiltration during heavy rain and flood events. The model hydrogeologic parameters and boundary conditions applied (Table 24.3) are representative of the youngest volcanic units of the Karthala volcano in Grande Comore.

The homogeneous model (model a) results in a smooth and regular transition zone between fresh and saline water displaying a low dipping angle towards the island (Fig. 24.4a). The thickness of freshwater lying on transition levels reaches about 30 m below sea level at 1.2 km from the shoreline, and about 75 m at 5 km. The local presence of a palaeovalley infilled with a massive, low permeability isotropic lava flow

Table 24.3 Hydrogeologic parameters applied to the numerical model

Model parameters	Model a—homogeneous volcanic formations	Model b—massive lava flow infilling model a	Model c—fractured lava flow infilling model a	Model d—infiltration trench on model a
Hydrogeologic properties				
Horizontal hydraulic conductivity Kh (m/s)	10^{-3}	10^{-5}	10^{-2}	10^{-3}
Vertical hydraulic conductivity Kv (m/s)	10^{-4}	10^{-5}	10^{-2}	10^{-4}
Boundary conditions				
Recharge R (mm/y)	From 60 on the coast to 475 at 5 km from the coast			
Entering upgradient flux F (m^3/lin. m/y)	3736	3736	3736	1869
Trench infiltration I (m^3/lin. m/y)	0	0	0	415
Sea level (0 m) boundary conductance (m^2/s)	6	6	6	6

(Kh/100, model b) results in a deepening of the transition zone beneath the palaeovalley, where freshwater thickness appears almost double ($\sim$50 m) at 1.2 km from the shoreline (Fig. 24.4b). In contrast, the local presence of a palaeovalley infilled with a high permeability isotropic lava flow (Kh*10, model c) results in the thinning of the freshwater levels, which disappear on a distance from the coast up to 2 km, particularly on the borders of the palaeovalley (Fig. 24.4c). The width of the transition zone along the coast appears also significantly increased. The presence of an infiltration trench (ravine) perpendicular to the coast (model d), which locally supports an increase in freshwater aquifer recharge during flooding events (here half of the lateral flux reflecting the recharge in high elevation zones is redistributed along the trench), also results in the thickening of the freshwater levels, comparable to the effect of a massive lava flow (Fig. 24.4d). Given the large diversity of volcanic units expected in young volcanic environment such as in Grande Comore, those theoretical simulations confirms the significant lateral variation of salinities observed in water wells along the coast in both La Grille and Karthala recent volcanoes (see Fig. 24.1). Those simulations supports the hypothesis that in addition to the regional impact on seawater intrusions of both the age of the volcanic units and the recharge, local structures can also potentially impact the seawater intrusion at the local/field scale.

24.4.4 Structure of the Perched Aquifers from Geophysical Investigations

The explosive activity of the Grande Comore volcanism due to phreatomagmatic events is responsible of thick and weathered pyroclastic deposits expected to bear perched aquifers. The altitude domain (Figures 24.5a and b) is characterized by the frequent occurrence of relatively shallow (typically 0–50 m deep) low resistivity layers (10–150 Ω.m) attributed to weathered pyroclastics and ashes emitted from the large

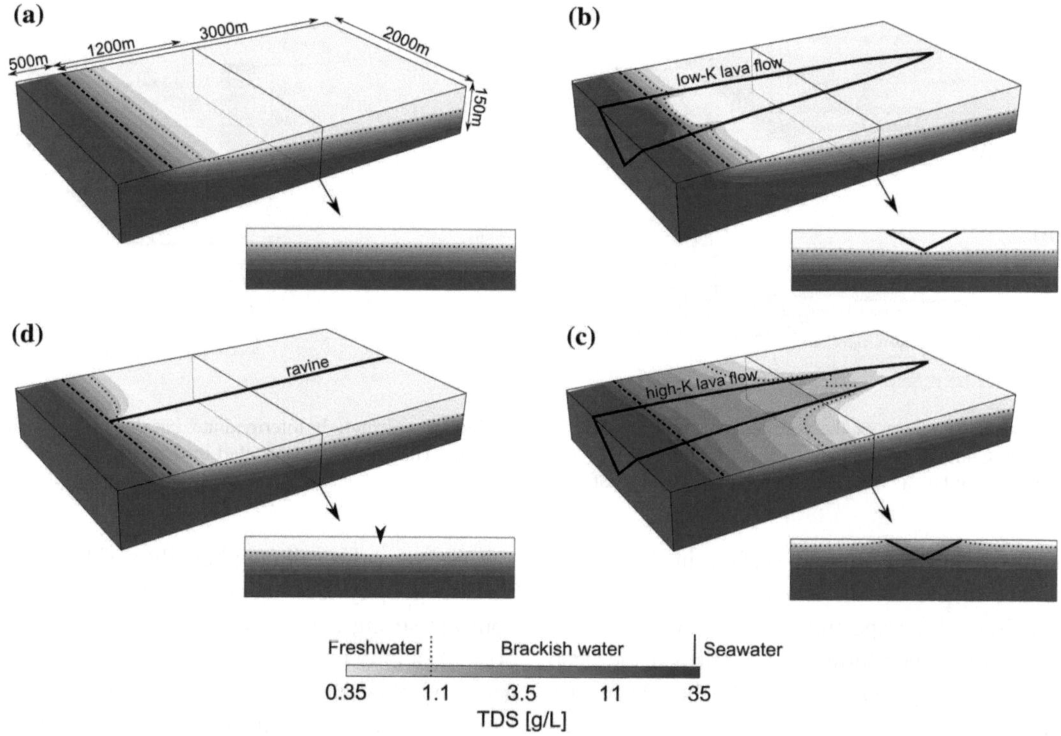

Fig. 24.4 Impact of local heterogeneities on seawater intrusion in the coastal zone: **a** homogeneous model; **b** palaeovalley infilled with lower permeability (K) lava flow; **c** palaeovalley infilled with higher permeability lava flow; **d** local infiltration in ravine. The top face is at mean sea level elevation. *Dashed line* is the *coastline* and *dotted line* is the contour 1 g/L corresponding to the WHO (2003) max salinity standard for drinking water. Full simulated domain is 5000 × 2000 × 300 m; vertical exaggeration x3

numbers of tuff cones present in high elevation zones. Those low permeability layers overlay variably weathered lava flows, depending on the age of the massif, i.e. in order of increasing age: the Karthala, La Grille and M'Badjini. They are frequently overlain by more or less continuous high resistivity recent lava flows (>200 Ω.m), which constitute potential aquifers suitable for local water supplies.

Because of the scarcity of exploration boreholes, the hydrodynamic properties of these perched aquifer systems are unknown. During the UNDP project, an exploration well was performed in Inane, at 500 m of elevation, in the massif of M'Badjini. This 50 m-deep well intercepted a perched and low mineralised groundwater (TDS = 100 mg/L) between 34 and 42 m of depth. This well is currently collapsed (Fig. 24.5a).

24.4.5 Intermediate Staged Aquifers and Basal Aquifer of the Altitude Domain

The presence of a bell-shaped piezometric surface of the basal groundwater in the centre of the Karthala volcano was confirmed by both CSAMT investigations (Savin et al. 2001) and deep vertical electrical soundings (UNDP 1987). In La Réunion Island, the current challenges consist of evaluating those groundwater resources and the possibility of exploitation of a groundwater body estimated at about a thousand

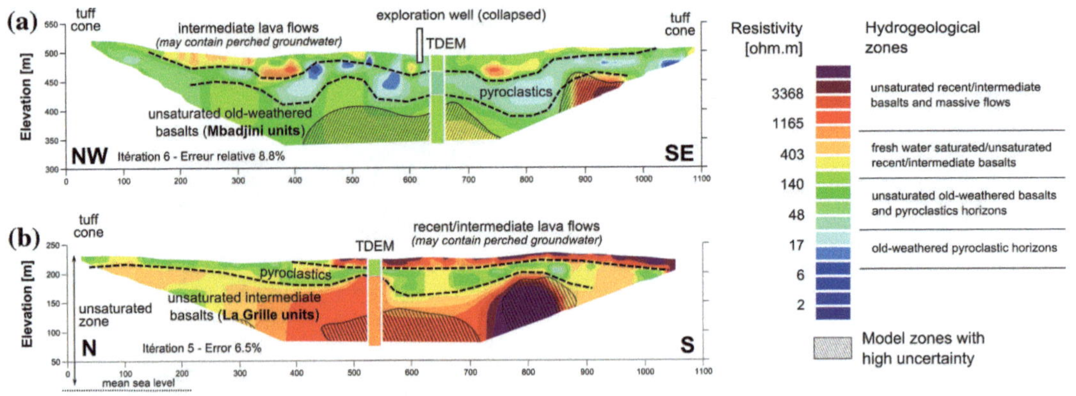

Fig. 24.5 Typology of the perched volcanic aquifers in Grande Comore interpreted from combining ERT and TDEM results—**a** old and weathered basalts of the M'Badjini volcano; **b** intermediate basalts of La Grille volcano. The profiles are located on Fig. 24.1 (4.5a, b)

metres below the ground surface in the mountainous interior of the island (Join et al. 1996). Geophysical prospecting by CSAMT coupled with springs monitoring has confirmed the deep groundwater potential in both the Piton des Neiges and Piton de la Fournaise volcanoes. In Grande Comore Island, the presence of an inland groundwater resource in the massif of La Grille is supported by four permanent mountainous springs (Maoueni, Helendje, Ivembeni and N'Tsorale). However, this deep potential is not accessible by traditional borehole investigations. On the basis of the Canaries experience (Ecker 1976; Custodio and Cabrera 2012), thousands of metres long horizontal galleries starting from the external flank of the volcano could also be applied in Grande Comore (Join et al. 2013).

24.5 Discussion

The Grande Comore case study illustrates the diversity of natural controls on both groundwater occurrence and vulnerability to marine intrusion in volcanic islands. In this work, the main aquifer types in Grande Comore were captured by geophysical investigations. Results well support a hydrogeological conceptual model similar to La Réunion Island (Join et al. 2005) with specificities mainly related to the lack of erosion valleys in Grande Comore.

Results confirm the applicability of ERT and TDEM geophysical techniques to investigate both the strong aquifer heterogeneity and variations of salinity in the coastal zones, as well as the geometry of perched reservoirs in the inland zone. However, the sole application of those methods carries a number of uncertainties and limitations, e.g. (i) for ERT, the poor resolution of the method in low resistivity material such as below the saline interface and (ii) for TDEM, both the 1D nature of the results that can only account for stratified media and the poor resolution when the unsaturated zone is thick and resistant. The combination of ERT and TDEM however helps in constraining the results interpretation, particularly in areas where a single technique has a poor resolution (Albouy et al. 2001).

Hydrogeological structures and heterogeneities have been observed at different scales. At regional scale (island scale), the patterns of seawater invasion is controlled by both the recharge from rainfall and the degree of weathering. Aquifers located in the more weathered oldest units (such as the massif of M'Badjini), are generally less vulnerable to seawater intrusion compared to the most recent units (such as the Massif of Karthala). At local scale, geophysical results revealed a large diversity of geological heterogeneities and salinity variability. Groundwater modelling confirms that in the coastal zone the presence of channels of recent high-permeability lava flows enhances the saline invasion, while the presence of permanent or temporary streams supporting freshwater recharge locally push back the saline interface. Coastal tuff and hyaloclastic cones, due to their low permeabilities, are liable to constitute impermeable hydraulic barriers that can limit the marine impact. Inland, perched aquifers are associated with impervious layers of pyroclastics that are robustly imaged as layered conductor anomalies by both ERT and TDEM.

Complementary, groundwater time-series acquisition constitutes an efficient tool for monitoring and interpreting the groundwater behaviour at high temporal resolution (Johnson et al. 2012; Lewis and Allen 2009), and provides useful data for resource management. In the three monitored wells, it revealed that groundwater salinity is controlled by tidal efficiency, recharge, abstraction and the aquifer degree of weathering. In all cases, high tides lead to systematic peaks of salinity and the highest mean salinity values are recorded during the spring tides periods.

As a result of those hydrogeological conditions, the marine contamination in Grande Comore is generally higher in both the youngest formations and the driest regions. This is particularly the case for localities situated in the east of Grande Comore (less than 1.5 m/y of rainfall), which are the most affected by the drinking water problems.

24.6 Conclusions and Recommendations for Groundwater Prospection and Management

Results of this study allow proposing in Grande Comore both a prospection methodology compatible with the young volcanic environment and a better groundwater management approach.

The prospection and management of basal aquifers can be carried out following a downscaling approach:

1. Regionally: mapping of lower elevation areas far from the coast to identify sectors of potential low salinity (this can be supported by regional (island-scale) salinity extrapolations and/or groundwater modelling)
2. Locally: (i) in recent volcanic areas, implementation of ground geophysical investigations (ERT/TDEM) to identify favourable structures for freshwater (higher resistivity channels in the saturated zone, dipping of the saltwater wedge); and (ii) in old weathered volcanic areas, implementation of ground geophysical investigations (ERT/TDEM) to identify the minimum depth of the saline interface
3. Optimisation of well design: (i) low groundwater penetration, high diameter, horizontal drains in young massifs of La Grille and Karthala; (ii) optimum groundwater penetration above saline water, high diameter, well fields, moderate yields, etc., in the old weathered massif of M'badjini
4. Optimisation of pumping with regard to the tidal forcing by an alternating and dynamic automated pumping when groundwater salinity is naturally lower (i.e. during low tides)

The prospection and management of inland aquifers requires a more specific approach that focuses on different geological targets:

1. Locally: prospection of perched superficial aquifers through implementation of ground geophysical investigations (ERT/TDEM) in cones area to identify (i) in the young massifs of La Grille and Karthala, recent thick and wide resistive lava flows overlying low resistivity pyroclastic levels; (ii) in old weathered massifs (M'badjini), recent thick and wide resistive lava flows overlying either low resistivity weathered basalts or pyroclastic levels
2. Optimisation of well design to penetrate the perched aquifer full thickness, including horizontal drains if required
3. Regionally: implementation of island-scale deep geophysical surveys (ex CSAMT/SP/deep ERT/VES) to evaluate the shape of the basal aquifer water table and identify sectors of shorter horizontal distance for drilling horizontal galleries.

The development of drinking water supply in Grande Comore is a primordial for public health and socio-economic stability. Apart from Moroni urban area, rural localities are supplied by community conveyance networks. Currently, water access disparities (in terms of water quantity, quality and price) can be observed throughout the island (Mohamed 2012). The mobilisation of available groundwater resources should lead to both standardising of the water access in all the rural regions and initiating a sustainable socio-economic development in local communities.

Acknowledgments The ground geophysical investigation programme was funded by the French Development Agency (AFD) through the Grande Comore Groundwater Development Programme. We acknowledge the Comoros Department of Energy, Mining and Water (DGEME) for technical assistance on fieldwork through the involvement of 10 temporary workers; Hamid Soule, manager of the Karthala Volcano Observatory and PhD student in the Réunion Laboratory of Geosciences for great logistic support; Mahabadi Boinali and Said Ahmed Othman, hydrogeologists at DGEME for helpful and constructive discussions; the NERC-UPGro programme for financial contribution to the instrumentation of TP5 well; and finally the Air Austral Company for offering a weight excess for freight of geophysical equipment. Finally, we acknowledge the review of H. Celle-Jeanton which contributed in improving the final manuscript.

References

Albouy Y, Andrieux P, Rakotondrasoa G et al (2001) Mapping coastal aquifers by joint inversion of DC and TEM soundings-three case histories. Ground Water 39:87–97

Bachelery P, Coudray J (1993) Carte géologique de la Grande Comore au 1/50 000 ème, avec notice explicative, 39 p

Bates B, Kundzewicz ZW, Wu S, Palutikof J (2008) Climate change and water. IPCC Tech Pap VI x + 200 pp

Boinali M (1982) Géologie et esquisse hydrogéologique de l'archipel volcanique des Comores. 45p

Bret L, Join J-L, Legal X et al (2003) Argiles et zéolites dans l'altération d'un volcan bouclier en milieu tropical (Le Piton des Neiges, La Réunion). C R Geosci 335:1031–1038

Coudray J, Mairine P, Nicolini E, Clerc JM (1990) Approche hydrogéologique. Volcanisme Lî La Réun. Cent. Rech. Volcanol, Clermont-Ferrand

Custodio E, Cabrera MC (2012) The Canary Islands. Water Agric. Environ. Spain Can We Sq. Circ. Chapter 22 p 281

Comte JC (2008) Apport de la tomographie électrique à la modélisation des écoulements densitaires dans les aquifères côtiers - Application à trois contextes climatiques contrastés (Canada, Nouvelle-Calédonie, Sénégal). Université d'Avignon

Comte JC, Banton O, Join JL, Cabioch G (2010) Evaluation of effective groundwater recharge of freshwater lens in small islands by the combined modeling of geoelectrical data and water heads. Water Resour Res 46 doi:10.1029/2009WR008058

Ecker A (1976) Groundwater behaviour in Tenerife, volcanic Island (Canary Islands, Spain). J Hydrol 28:73–86. doi:10.1016/0022-1694(76)90053-6

Dahlin T (1993) On the automation of 2D resistivity surveying for engineering and environmental applications, Ph.D. thesis, Dept. Eng. Geol., Lund Univ., Lund, Sweden. Ph.D Thesis, Lund University

Dahlin T, Zhou B (2004) A numerical comparison of 2D resistivity imaging with 10 electrode arrays. Geophys Prospect 52:379–398. doi: 10.1111/j.1365-2478.2004.00423.x

Danielsen JE, Auken E, Jørgensen F (2003) The application of the transient electromagnetic method in hydrogeophysical surveys. J Appl Geophys 53:181–198

Folio JL (2001) Distribution de la perméabilité dans le massif du Piton de la Fournaise : apport à la connaissance du fonctionnement hydrogéologique d'un volcan-bouclier. La Réunion

Griffiths DH, Turnbull J, Olayinka A. (1990) Two-dimensional resistivity mapping with a computer-controlled array. First Break. doi: 10.3997/1365-2397.1990008

Ghyben WB (1888) Nota in verband metde voorgenomen putboring nabij Amsterdam-Tijdsch. Van Kouinglijk Inst Van Ing 8–22

Guo W, Langevin CD (2002) A computer program for simulation of three-dimensional variable-density ground-water flow. Tech. Water-Resour. Investig. Book 6 Chapter A7 77 P

Herzberg A (1901) Die Wasserversorgung einiger Nordseebäder. J Gasbeleucht Wasserversorg 815–819

Huisman JA, Bouten W, Ferre TP (2004) Bridging the Gap Between Geophysical Measurements and Hydrological Modelling. AGU Fall Meet Abstr 1:1

Ibrahim K (2009) Etude hydrogéologique de l'aquifère côtier dans la région de Oichili en Grande Comore—Mémoire de Master 2 Université d'Avignon, Université de la Réunion, 62 p. Hydraulique Sans Frontières, Comoros

Jacob CE (1950) Flow of groundwater, Engineering Hydraulics H. Rouse. Wiley, New York, pp 321–386

Johnson TC, Slater LD, Ntarlagiannis D et al (2012) Monitoring groundwater-surface water interaction using time-series and time-frequency analysis of transient three-dimensional electrical resistivity changes. Water Resour Res 48:W07506. doi:10.1029/2012WR011893

Join JL (1991) Caractérisation hydrogéologique du milieu volcanique insulaire, le Piton des Neiges: Ile de La Réunion. Université Montpellier II

Join JL, Comte JC, Bourhane A (2013) Tests méthodologiques en géophysique en vue de l'implantation de forages d'eau sur l'île de La Grande Comore—Projet Pilote de gestion du service public de l'eau en milieu rural sur l'île de La Grande Comore. 64p

Join JL, Coudray J (1993) Caractérisation géostructurale des nappes d'altitude en milieu insulaire, Ile de la Réunion. Géodinamica Acta 243–254

Join JL, Courteaud M, Robineau B et al (1996) The problem of prospecting and development of inland groundwater resources in young tropical volcanic islands. IAHS-AISH Publication International Association of Hydrological Sciences, Kingston, pp 405–411

Join JL, Folio JL, Robineau B (2005) Aquifers and groundwater within active shield volcanoes. Evolution of conceptual models in the Piton de la Fournaise volcano. J Volcanol Geoth Res 147:187–201

Join JL, Pomme JB, Coudray J, Daesslé M (1988) Caractérisation des aquifères basaltiques en domaine littoral. Impact d'un récif corallien. Hydrogéologie 107–115

Lewis K, Allen JI (2009) Validation of a hydrodynamic-ecosystem model simulation with time-series data collected in the western English Channel. J Mar Syst 77:296–311

Loke MH (2006) RES2DINV ver. 3.55, Rapid 2-D resistivity & IP inversion using the least-squares method, 139 pp., Geotomo Software, Penang, Malaysia, http://www.geoelectrical.com/download.html

Marini D (1990) Résultats et interprétations d'une campagne de pompages d'essais sur des puits dans les aquifères de base, Grande Comore. 122 p

Massari M (1990) Étude du pouvoir épurateur de divers matériaux de l'Ile de la Réunion vis à vis d'effluents domestiques et industriels. Université de Provence

Meybeck M, Helmer R (1989) The quality of rivers: from pristine stage to global pollution. Palaeogeogr Palaeoclimatol Palaeoecol 75:283–309. doi:10.1016/0031-0182(89)90191-0

Mohamed I (2012) L'eau en Grande Comore : étude de cas d'un petit espace insulaire abondamment arrosé mais en situation de pénurie d'eau. Université de La Réunion, Thèse de Doctorat

Mohamed SH, Othman SA (2006) Etude causale sur la disponibilité de la ressource en eau et la sécurité de l'approvisionnement en eau potable aux Comores (Version finale). Rapport PNUD 16(10/2006):42p

Ragot JM (2007) Termes de références pour l'étude des ressources en eaux souterraines de l'Union des Comores. 37p

Revil A, Karaoulis M, Johnson T, Kemna A (2012) Review: Some low-frequency electrical methods for subsurface characterization and monitoring in hydrogeology. Hydrogeol J 20:617–658. doi:10.1007/s10040-011-0819-x

Savin C, Ritz M, Join JL, Bachelery P (2001) Hydrothermal system mapped by CSAMT on Karthala volcano, Grande Comore Island, Indian Ocean. J Appl Geophys 48:143–152. doi:10.1016/S0926-9851(01)00078-7

SCP (2009) Etat des lieux des conditions d'accès à l'eau potable sur l'île de Grande Comore. 49p

UNDP (1987) Recherche et mise en valeur des eaux—Rapport technique—Perspectives de mise en valeur des eaux souterraines pour l'alimentation en eau des agglomérations de l'île de Ngazidja. New York

Vengosh A (2003) 9.09—Salinization and saline environments. In: Editors-in-chief: Holland HD, Turekian KK (eds) Treatise geochemistry. Pergamon, Oxford, pp 1–35

WHO (2003) Total dissolved solids in drinking-water. Background document for preparation of WHO Guidelines for drinking-water quality. (WHO/SDE/WSH/03.04/16). World Health Organization, Geneva

25 Volcanic Risk and Crisis Management on Grande Comore Island

Julie Morin, Patrick Bachèlery, Hamidi Soulé and Hamidou Nassor

Abstract

While the volcanology of Grande Comore has been widely described, issues of crisis management raised by a vulnerability census have only been treated superficially, and disaster-related risk-reduction work is almost non-existent. Four recent eruptions, in April 2005, November 2005, May 2006 and January 2007, led to a renewed scientific interest in the volcano's eruptive history and mechanisms of eruption, as well as in volcanic risk assessment and crisis management in the Comoros Islands. The crisis management system built up over the last few decades has been much improved over the last few years. However, this system is still weak due to structural vulnerabilities in the country. In this chapter we describe the risk and crisis management protocols, explain how they have been implemented over the years, and suggest ways to improve the handling of the next volcanic crisis.

25.1 Introduction

The Comoros are amongst the top six countries worldwide to be most exposed to natural disasters (ISDR 2009). This classification takes into account the number of inhabitants, occurrence of disasters, availability of infrastructures to handle these disasters, and the level of preparedness of the State. Cyclones were responsible for the highest number of deaths (559) from natural disasters on Grande Comore Island from 1900 to 2014, followed by epidemics (86), volcanic hazards (19) and floods (6) (CRED 2014). Volcanic eruptions and cyclones are the most recurrent hazards, generating disasters which

J. Morin (✉)
Laboratoire de Géographie Physique de Meudon (LGP), UMR 8591, 1 place Aristide Briand, 92195 Meudon, France
e-mail: julieapi@yahoo.fr

P. Bachèlery
Laboratoire Magmas et Volcans, Université Blaise Pascal—CNRS—IRD, Observatoire de Physique du Globe de Clermont-Ferrand, 5 rue Kessler, 63038 Clermont-Ferrand, France

H. Soulé · H. Nassor
Observatoire Volcanologique du Karthala, CNDRS, Moroni, Grande Comore

P. Bachèlery et al. (eds.), *Active Volcanoes of the Southwest Indian Ocean*, Active Volcanoes of the World, DOI 10.1007/978-3-642-31395-0_25

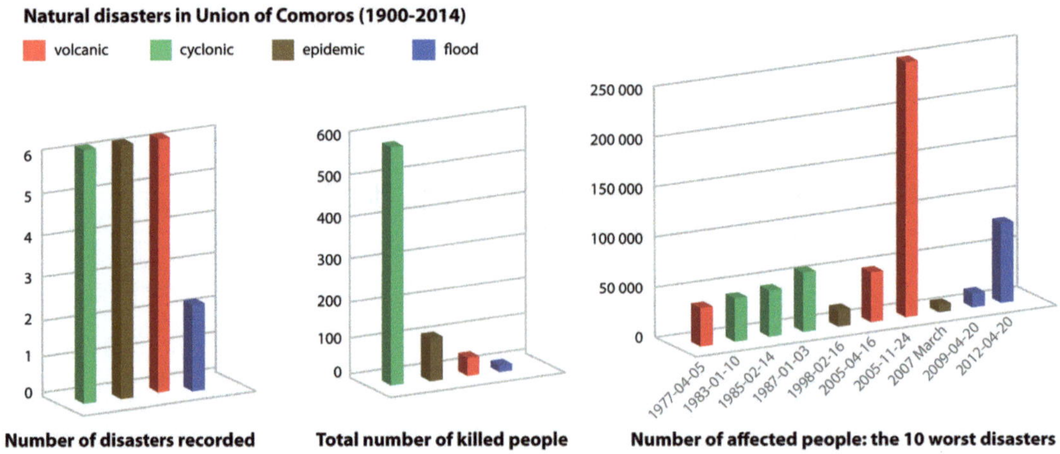

Fig. 25.1 Main natural disasters from 1900 to 2014 in the Comoros (data from EM-DAT 2014—an event is considered as a disaster when at least 10 people are killed, or 100 people affected, or when there is a call for foreign assistance, or a declaration of a "state of emergency")

affected a huge number of the local population (respectively 309,200 and 115,352—Fig. 25.1). The November 2005 Karthala eruption tops the ranking of events, affecting the highest number of people in the Comoros (245,000). According to the EM-DAT database (CRED 2014), and noted by Lenhardt and Oppenheimer (2014) in their synthesis on volcanism in Africa, this eruption also ranks number six in the world in terms of people affected, up to 2014. Furthermore, the floods, which have affected about 70,000 people on Grande Comore since 2009, are partly due to the soil sealing effects by the deposits from the 2005 eruption. The majority of the 320,000 inhabitants are exposed to volcanic hazards (PIROI 2005a), and generally have no means to protect themselves, thus leaving them highly vulnerable (Morin et al. 2009; Morin 2012). Thus volcanic risk is a major concern on this island. Only hazards generated by Karthala volcano are considered in this paper, as there is only a very low probability that the older M'Badjini and La Grille volcanoes (see Bachèlery et al. 2016, Chap. 22) would erupt.

The eruptive frequency of Karthala has been between 5.3 and 8.0 years since the 19th century (Morin et al. 2009; Bachèlery et al. 2016, Chap. 22). This frequency increased in the 21st century, with four eruptions in 21 months (April 2005, November 2005, May 2006 and January 2007). The Comorian population and authorities had to face unknown and/or forgotten hazards, and they realized what a major threat volcanism is to the island. Therefore, greater preparedness and capacity to face volcanic crisis might be expected a few years on. However, this chapter presents a literature review that highlights strong weaknesses in the volcanic risk and crisis management from 2005 to 2014. Some improvements are proposed here to enhance the Comorian risk and crisis management capacities.

25.2 Grande Comore Island: The Entirety of the Territory Exposed to Volcanic Hazards

Basaltic shield volcanoes are considered as being "quiet" volcanoes associated with a relatively low level of risk, as the occurrence of violent explosive eruptions is unusual, despite the high eruptive frequency. Karthala volcano covers two thirds of the surface area of Grande Comore Island, and both effusive and explosive eruptions may occur anywhere on the edifice (see Bachèlery et al. 2016, Chap. 22), within the caldera (summit eruptions), along the rift zones (lateral eruptions), or on the flanks at low elevation (eccentric eruptions).

Eruption locations seem to correspond to eruptive cycles. Most historic eruptions took place along the rift zones while more recent eruptions have occurred mainly inside the caldera —exclusively inside the caldera in the 21st century. The highest hazard areas are thus close to the caldera, while remote areas and areas protected by relief are on the whole spared from hazards (Fig. 25.2a, b). In terms of risk, the rift zones' ability to divert magma far away from the summit is a major concern. For instance, in 1859, a fissure opened on the north rift zone 25 km from the summit.

25.2.1 Hazards and Risks Associated with Magmatic Eruptions

Most of the eruptions during the last two centuries were magmatic. They almost all generated 'a'ā lava flows (Krafft 1983) which have covered nearly 25 % of the island since the 19th century (Nassor 2001). Most of the inhabited areas are prone to lava flow hazards. Thick ash layers from lava fountains may also affect inhabited zones near eruptive cones [e.g. 10 cm of deposits in proximal villages in 1977 (Krafft 1982)] while bombs and ash falls threaten larger areas further away from the eruptive fissure. The geological map of Karthala (Bachèlery and Coudray 1993) shows a large number of eruptions both along the rift zones and at low elevations on the flanks. Even when they occur within the caldera, lava flows can pour out through the "*Porte d'Istandra*" (see Bachèlery et al. 2016, Chap. 22), and eventually reach the coast, as in 1860 (Lacroix 1920). In 1972, lava flows starting from the northern lobe of the Karthala caldera flooded the northwest flank above Moroni. The threatened capital city was only spared because of low eruptive volume that limited the length of the lava flow (Bachèlery and Coudray 1993). Lava flows average 5.4 km in length, 4.65 m in thickness, and 19.7×10^6 m^3 in volume. A lava flow would need to be 14 km long to reach the capital, Moroni, from the summit. Currently, only 10 % of the lava flows exceed 9 km in length. On the other hand, 50 % of their starting points are located below 1170 m. From this altitude, and considering their 5.4 km average length, lava can easily reach the coast, flowing across urbanized areas (Morin 2006). The most significant historic case is the 1977 lava flow (see Bachèlery et al. 2016, Chap. 22). In addition of this "everywhere-starting-point-problem", numerous lava tube systems facilitate the flow of lava downstream, potentially threatening even villages located far away from the lava vent (Pavlovsky and Saint-Ours 1953). The Hahaya lava flow thus ran a distance of 27 km before reaching the current international airport, under which 32 tubes have been detected within a distance of less than 3 km (Krafft 1983). Lava can flow fast: one person may have been killed in 1904 by a lava flow (Bachèlery and Coudray 1993). In terms of risk, the worst-case scenario would be a lava flow reaching Moroni, where almost all the main resources are concentrated (Fig. 25.2c). In 1858, a lava flow travelled 13–14 km from the uphill North rift zone to the western coast, passing close to the capital. Moroni, meaning "city into the fire", was probably named after that event (Guébourg 1995). Based on the current urban development, 2076 buildings and 28 % of Moroni's surface area would be covered by an event of this nature if it happened today (Morin 2012; Fig. 25.3). The national Plan for preparedness and response to emergencies (*PNPRU*, see below) gives a 50 % probability that such an eruption will occur in the next few decades. It would affect 75,000 people and would "*certainly lead to socio-economic chaos*" and a strong weakening of livelihoods even for people not directly threatened by the eruption, while a lava flow in a rural area might affect around 10,000 people (Union des Comores 2007).

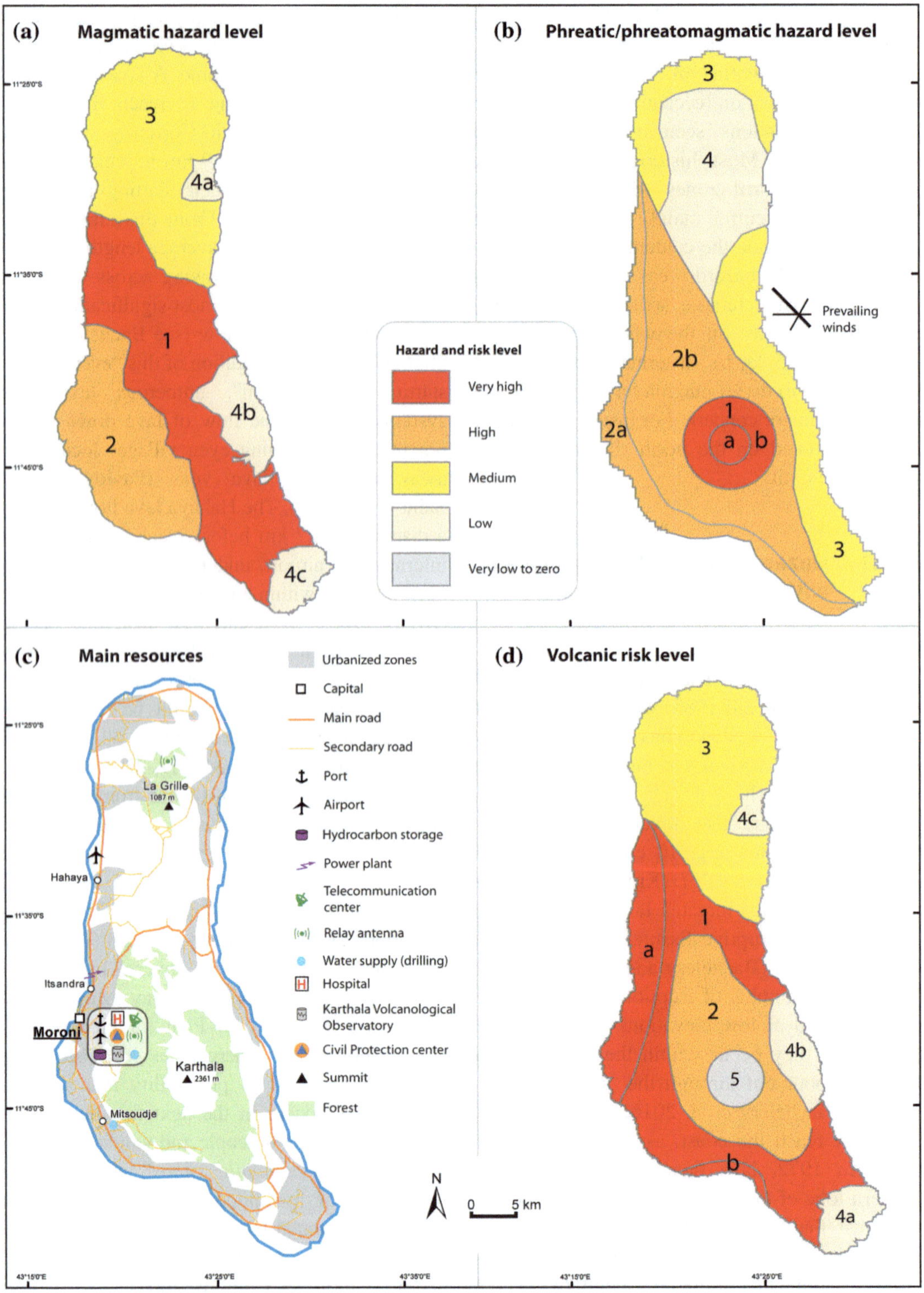

Fig. 25.2 Hazard, resource and risk map—**a** magmatic hazard map for Grande Comore based on four criteria: slope angle, eruptive frequency, eruptive cone density, lava flow simulations with TRIFLUX software (Nassor 2001); **b** ash and block fall hazard map on Grande Comore, after Nassor (2001); main resources (Morin 2012); volcanic risk map, after Nassor (2001)

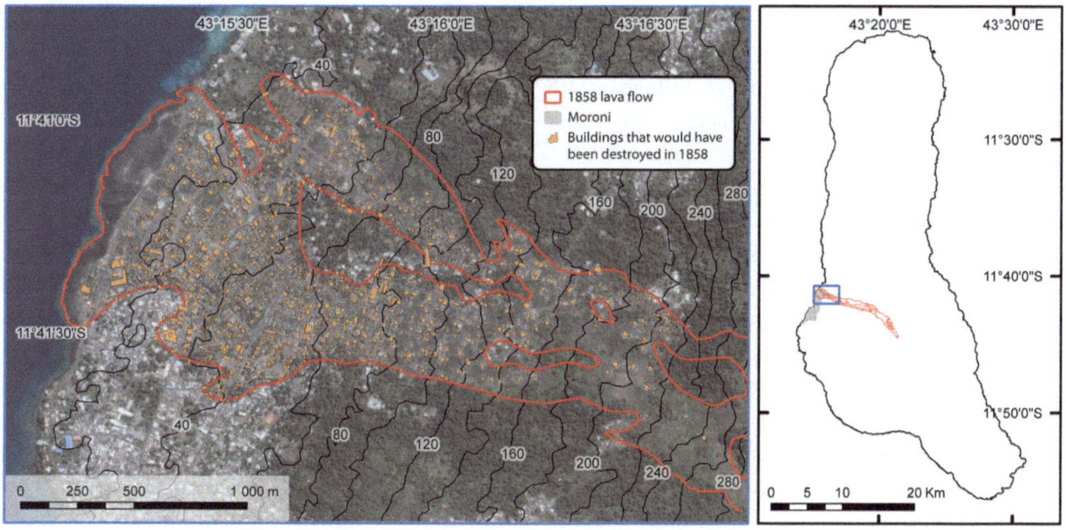

Fig. 25.3 Parts of the city of Moroni that would have been impacted by the 1858 lava flow if this flow had happened in the early 21st century

25.2.2 Hazards and Risks Associated with Explosive Eruptions

Karthala's eruptive style is mostly effusive, although phreatic and phreatomagmatic eruptions also occur (see Bachèlery et al. 2016, Chap. 22). They are mainly located within the caldera, and sometimes along coastal areas where they create hyaloclastite cones, such as Iconi and Moindzaza (Bachèlery and Coudray 1993). They could occur on the flanks if a magma intrusion met groundwater, but this has not been historically described for Grande Comore (Nassor 2001). Paroxysmal eruptions have also occurred, as in 4695, 3960 and 2390 B.P. (Nassor 2001).

Phreatic and phreatomagmatic eruptions represent the greatest volcanic hazards for a shield volcano. Consecutive aerial fallouts (Fig. 25.4) can affect large areas, or even the whole island. The Comorian population discovered this hazard when explosions in April and November 2005 generated massive ash falls. The 1991 explosion [1.5×10^6 m^3 of deposits covering 12×10^6 m^2: (Bachèlery et al. 1995)] affected only the summit zone, which is the area of highest fallout hazard (Fig. 25.2b). In November 2005, ash falls have

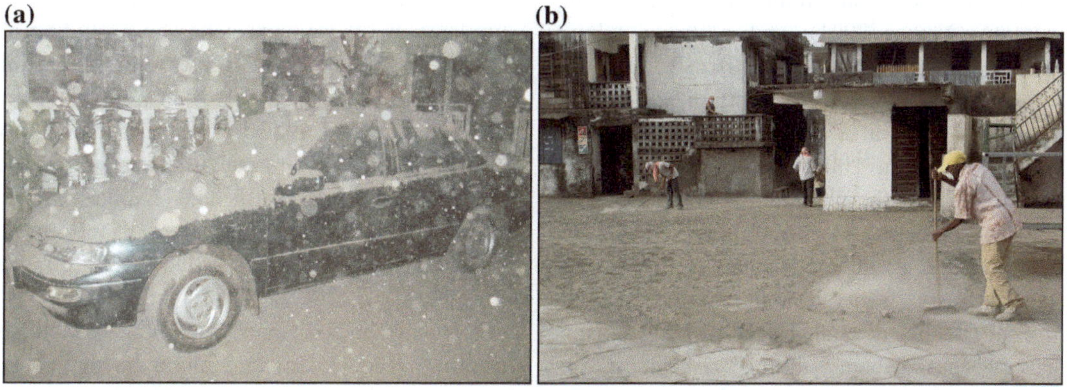

Fig. 25.4 Ash fall and ash clearing in November 2005, in Moroni (Soule 2005)

covered three quarters of the island (Smithsonian National Museum of Natural History 2005) and affected 245,000 inhabitants in 76 villages, with 175,000 people left without access to water supplies (OCHA-GVA 2005).

Fallout deposits can subsequently be mobilised by rain, creating lahars. As a consequence of the 2005 eruptions, lahars became the main volcanic hazard on the island, with high frequency and repeated damage, especially along the island's southwest coast, where the total precipitation is higher and the rain showers are stronger than elsewhere on the island (Morin et al. 2009). This new hazard still affects hundreds of homes and thousands of people, and now plays an important role in the daily life and culture of Grande Comore's inhabitants (see Morin and Gaillard 2012; Morin 2012), but it has not yet been assessed. Soule (2011) mapped the main active lahar channels in 2010, but this work is incomplete, and does not indicate the potential hazardous areas in the event of a new explosive eruption and/or lahars.

25.2.3 Other Hazards and Risks

With either effusive or explosive activity, or independent of an eruption, gas, seismicity and landslides pose additional threats.

Gas was reported to have killed 17 people on the south-east rift zone in 1903 (Bachèlery and Coudray 1993). With no such tragic consequences, gas was smelt on the western coast in 1991, one and a half hours after the 11 July eruption (Nassor 2001), and a strong sulphur smell was noted during the 2005 ash falls.

Low magnitude volcanic earthquakes (usually measuring <2.5 on a duration magnitude scale, but up to 4.3 in 2005) have sometimes been strongly felt by the population. The 1991 eruption was preceded by 30,000 earthquakes, and thousands of people evacuated, more or less spontaneously, from the south of the island to Moroni (Bachèlery et al. 1995). In 2007, seismic activity caused one building to collapse and damaged others in Moroni's old town. Frightened residents slept outside for several nights, fearing the collapse of their homes, as high frequency and magnitude seismicity was felt. Previous damage to buildings by seismic activity were attributed by Krafft (1982) to a one-hour-long seismic swarm during the 1977 eruption, which cracked water tanks in the south-west of the island, corresponding to an intensity of 5 on the MSK scale. Grande Comore, also subject to regional tectonic activity, does not have any seismic zoning. Lastly, flank landslides at different scales can generate large debris flows and debris avalanches. When they reach the sea they can potentially cause a tsunami. This hazard cannot be ignored, although the occurrence probability is low (Hartnady 2005a, b). Thus while it is scientifically known, it is not assessed, and is not taken into consideration in the risk management procedure because of its low frequency and the impossibility of managing its massive consequences.

25.2.4 Risk Assessment and Mapping

Krafft (1982, 1983) was the first to address the issue of volcanic risk on Grande Comore by showing the impact of eruptions and the reactions of the population. In 1993, Bachelery and Coudray proposed the first *risk map*, which is actually a hazard map. Bachèlery (1999) then presented a hazard-centred synthesis of the risks. Nassor (2001), and Nassor and Soule (PNUD 2002) completed this risk analysis, taking into account the main issues but avoiding these issues' vulnerabilities. Morin and Lavigne (2009) supplemented these approaches by providing an analysis of the vulnerability factors that limit the feasibility of volcanic risk management. They subsequently used a local example to show how socio-economic factors can interfere with good risk management (Morin and Gaillard 2012; Morin 2012). Up to now, there has been no complete risk assessment carried out on Grande Comore.

The current risk map (Fig. 25.2d) includes resources (Fig. 25.2c) that might be threatened by vents and fissures opening, lava flows and tephra. The summit of Karthala, La Grille and

M'Badjini volcanoes, and a small enclave on the eastern flank from Idjikoundzi to Bandamadji, are 'no risk' to 'low risk' areas. The highest risk area includes the western coast, where the main resources are concentrated (especially area '1a' which includes Moroni city and Hahaya airport) and two west-east transverse zones to the north and south of Karthala that contain main national roads.

This risk mapping should be updated using better data on hazards and by introducing a vulnerability analysis. However, knowledge of Karthala's eruptive history and behaviour is limited due to the absence of old archives in this traditionally oral society, and to the small number of eruptions recorded since the start of monitoring in 1988. Information on hazards is even more sparse (e.g. no landslide or lahar hazard maps) and the existing maps (Fig. 25.2a, b) should be greatly improved. The magmatic hazard map drawn up by Nassor (2001) includes both lava flows and magmatic deposits around eruptive cones. Pélé's hair hazards (never mentioned for Grande Comore) and lava tubes (not included in the lava flow simulations) are not considered in this map. Note that the resolution of the DEM (90 m SRTM) used by Nassor (2001) to simulate lava flows is not sufficient to obtain accurate modelling results. Work currently carried out on Grande Comore by Belgium researchers on lava flow and lahar modelling (Mossoux et al. 2014a, b) could help to refine these hazard and risk maps. In addition, the "Pyroclastic fallout" hazard assessment does not include periodic variations in wind direction. The prevailing wind direction indicated by Nassor (2001)—south-easterly winds—is based on the isopachs for ancient deposits and the direction taken by some volcanic plumes. Meanwhile, Guébourg (1995) provides wind frequency histograms that emphasize the primacy of *Kusi* over the other winds at low altitude. *Kusi* is a south-westerly wind that makes up 70 % of the winds from May to August. At higher altitude plumes may reach the inversion layer where wind directions are reversed (Tulet and Villeneuve 2011), potentially reaching Madagascar and hindering air traffic.

More importantly, this map takes into account the presence of resources, without considering the main vulnerability factors, such as isolation, that contribute greatly to increasing the risk (as there are only a few, potholed, poorly connected roads, which can easily become rutted or saturated). For example, one can question the real degree of risk in 'zone 4b', which is landlocked, near the craters, and which was affected by significant ash fall in 2005. The risk map has thus been based solely on the hazards, without taking the vulnerabilities and capacities into consideration. Vulnerability studies are highly desirable for the depiction of damage scenarios, improving risk assessment, and understanding the potential limitations for managing crises.

25.2.5 Risk Mitigation Measures

Land use is characterized by anarchic and fast urban development and the expansion of Moroni in areas prone to volcanic hazards. While Grande Comore has only a few settlements (Fig. 25.2c), meaning that absolute material losses would be limited, the ratio of loss to national wealth would be likely to be very high, even if only one location was affected by the volcano. The only hazard protection in place is the preventive covering of water tanks against ash falls. Further, there is no regulatory framework governing the right to public information on Grande Comore, although outreach is supposed to encourage the adoption of good behaviour in the face of risk and crisis. Since 2006, outreach campaigns have been led by the local volcano observatory (Observatoire Volcanologique du Karthala—OVK), particularly asking the population to respect the OVK monitoring network, which has been vandalized numerous times since 2000. The establishment of risk reduction programs by international partners resulted in the development of ad hoc outreach campaigns about volcanic risk prevention within the communities. Finally, there is no formal insurance system. People pull through disasters thanks to solidarity (donations from other villagers, from overseas relatives, and humanitarian aid). Therefore, the lack of human, technical and financial resources at all levels does not enable concrete action to be taken to face a crisis (see Morin and Lavigne 2009).

25.3 Crisis Management at Karthala Over the Last Three Decades of Eruptions

25.3.1 Theoretical Volcanic Crisis Management Capacities at the Moment of the 2006 Eruption

25.3.1.1 Volcanic Crisis Monitoring and Forecasting

The OVK was set up in 1988 at the CNDRS (Centre National de Documentation et de Recherche Scientifique in Moroni). The observatory is scientifically supervised by the OVPF (Observatoire Volcanologique du Piton de la Fournaise, itself part of the IPGP—Institut de Physique du Globe de Paris) and LGSR (Laboratoire Géosciences Réunion). The OVK was essentially funded by the French international program for scientific cooperation, with the Comorian authorities furnishing a part-time salary for the OVK head, premises in the CNDRS building, and human resources to help physically set up the OVK network on the volcano—including the presidential helicopters during the first few years. However, the insufficient resources allocated did not allow for the development and maintenance of a consistent network. The OVK was unable initially to make much progress in its understanding of the volcano as only a few eruptions were recorded. Nevertheless, the first eruption forecast and followed by the OVK, in 1991, provided the observatory with a full opportunity to play its role in providing information and advice to the authorities and the island's population. The eruptions from 2005 to 2007 subsequently promoted the OVK to a major role in Comorian society. A monthly public newsletter synthesizing activity is now published by the OVK to keep both the population and the financial partners informed.

25.3.1.2 Crisis Planning

Since 2004, Comorian authorities have adopted the PNPRU [National Emergency Preparation and Response Plan (Union des Comores 2004)] which aims to "ensure preparedness and response to emergencies that are coordinated and effective, using all available means within state and non state actors, in compliance with international humanitarian principles". The plan describes the specific role of each actor involved in the crisis management process at a national scale for any type of event (a crisis linked to geophysical, meteorological, biological, chemical, and transport conditions). Villages do not have official local management plans. Although there is no mention in the official documents, crisis management is in the joint hands of institutions and village leaders.

Village leaders (notables) and religious leaders dictate how their people should behave in all circumstances. Legal authorities recognize the legitimacy of these traditional actors, on whom they rely to carry out their plans. Pending any measures taken by the authorities when a crisis occurs, management is improvised using common sense and based on pre-existing social codes (priority given to family and neighbours etc.). This tends to prove that management planning is too "top-down". Moreover, almost all of the existing measures are based on the French system (as France was the former colonial power) while it is not always adapted to the local context (Morin 2012).

It should be noted that insurance and post-crisis recovery plans do not exist for Grande Comore. The return to normal life is managed mainly on a household and commune scale, sometimes with the aid of international programs when major disasters provoke food and sanitation emergencies.

The post-crisis phase is also characterized by the absence of official feedback on crisis occurrence and management—although the Operational

Centre for Rescue and Civil Protection (COSEP) team has initiated some informal brainstorming sessions after major crises since its creation in January 2007.

25.3.1.3 Launching Alerts

Different alert levels have been defined by the Karthala Plan, a 37-page PNPRU appendix specifically dedicated to the management of volcanic crises. Specific operational measures and emergency instructions for the public are associated with each alert level, as on most volcanoes around the world (Table 25.1). In the event of signs of unrest, the OVK passes information to the authorities and offers to put out the appropriate volcanic alert level. The alert is then theoretically disseminated as described in Fig. 25.5.

It is generally difficult to implement these theoretical management procedures because of a severe lack of resources due to the deep structural crisis encountered by the country for decades, a lack of volcanic crisis experience, and a low degree of prevention and preparedness within the government and population.

25.3.2 Conception and Growth of the Comorian Crisis Management System (1977–2005)

Eruptions over the last few decades have gradually influenced the crisis management procedures, improving the current crisis response capabilities (forecasting, warning, crisis management, and evacuations). When the 1977 eruption occurred, Comorian authorities did not possess any crisis or disaster management structures. The citizens themselves were asked to provide the means for evacuating the 4000 threatened people. Thanks to a national effort of solidarity, everybody was evacuated on time and taken to Moroni (SEAN 02:03 and 02:04 in Global Volcanism Program 2005). This eruption resulted in awareness about the importance of creating a risk management system. Krafft (1982) asked for a detailed volcanological study, risk mapping, evacuation planning, and for the inception of a volcanological observatory with at least three seismic monitoring stations to allow people to be alerted in the event of an impending eruption. The OVK was set up a few years later, incorporating both seismic and ground deformation monitoring (4 stations).

The 1991 eruption was important in terms of Civil Protection because it was the first to be monitored by the OVK by means of the deformation and seismic networks established in 1988. Chouaybou (2010) highlighted the disarray of displaced populations, disappointed with the evacuation and housing conditions, who returned home without any certainty that danger was over. In 1992, a report on regional security in the *Commission de l'Ocean Indien* (COI 1992) stressed the need to set up risk and crisis planning, provide human and material resources for Civil Protection, and strengthen the Karthala monitoring network.

In mid-2003, the OVK recorded a strong increase in seismic activity, provoking a false alert reported by the media. There were no consequences for the population. It provided, however, an opportunity to raise the level of awareness about the serious need to implement civil protection measures (Soule 2003), which had already been underlined by Nassor (2001). Meanwhile the ISDR and OCHA, as part of the International Strategy for Disaster Reduction and the impending Hyogo Conference, established national emergency preparedness platforms in East Africa and the Indian Ocean. From December 2003—during the workshops dedicated to the emergency platform creation, the Comorians conceived the idea for a national Plan for preparedness and emergency and response to emergencies (PNPRU). This plan, adopted in November 2004 (Union des Comores 2004), includes an appendix, the Karthala Plan, specifically dedicated to the management of volcanic crises. The national authorities also established a National Directorate of Civil Protection (DNPC), responsible for the assessment of needs and crisis management in the event of a serious incident.

Table 25.1 Volcanic alert levels and related actions defined in the PNPRU (National Emergency Plan)

Alert levels	Associated phenomena	Actions to be taken by crisis stakeholders	Actions to be taken by population
GREEN	✓ Nothing to report ✓ Volcano quiet, seismicity at 'normal' level	✓ Develop planning tools and training ✓ Reinforce monitoring network ✓ Educate the public	✓ Keep informed ✓ Memorize operational information (evacuation roads, meeting point, shelter location…) ✓ Remove vulnerable goods from high hazard areas
YELLOW	✓ Increase in seismicity, fumarole activity and/or water level in Chahale lake (if present). ✓ Ground deformation	✓ Reinforce monitoring network ✓ More communication between authorities and scientists	
ORANGE	✓ Seismic activity acceleration ✓ Large and rapid deformation ✓ Higher magnitude volcano-tectonic event	✓ Convene the volcanic crisis management committee ✓ Public announcement through the media ✓ Karthala summit access forbidden ✓ Notice to VAAC (air navigation safety) ✓ Civil Protection corps on alert ✓ Mobilization of evacuation logistics ✓ Implementation of specific measures in the most vulnerable regions	✓ Carefully follow the official information ✓ Follow authorities' instructions ✓ Evacuation training to safe places ✓ Stay on alert and be prepared to evacuate taking the minimum of belongings
RED **1st type**	✓ Magmatic or phreatic or phreatomagmatic activity at the summit ✓ High seismic and volcano-tectonic activity (M>4 -5)	✓ Selective evacuation of areas based on the eruptive activity ✓ Implementation of preventive measures against gas, ash and block fallouts, and obscurity in inhabited areas ✓ Reinforce communication to the population ✓ Notice to VAAC (air navigation safety)	✓ Follow authorities' instructions ✓ Keep informed about developments ✓ People in 'at risk' zones evacuate to safe place or to meeting points to be transported to shelters ✓ Urban population has to keep calm, stay at home or workplace, and keep informed
2nd type	✓ Fissure opening on the flanks with magma emission ✓ Eruptive activity in coastal areas	✓ Selective evacuation of larger areas based on the eruptive activity ✓ Establish a security perimeter and regulate traffic ✓ Notice to VAAC (air navigation safety)	

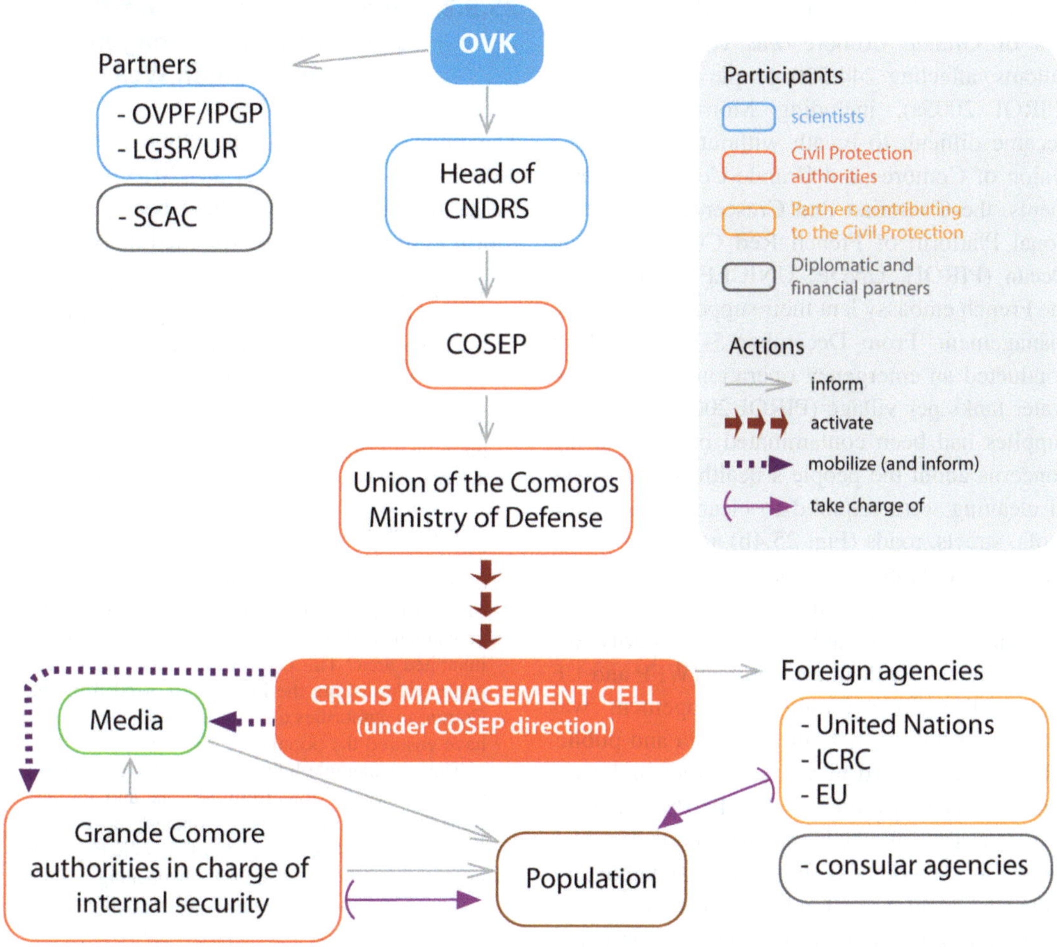

Fig. 25.5 Alert launching process in the event of an eruption [data from PNPRU, Union des Comores (2007)]. *SCAC* is the French cooperation office, *UR* is the University of La Reunion, *ICRC* the International Committee of the Red Cross, *EU* refers to the European Union. Other acronyms are explained earlier in the text

The DNPC is coordinated by the Prefect of the area concerned, and is under the supervision of the Ministry of Defence of the Union des Comores. The creation of a future Operational Centre for Rescue and Civil Protection (COSEP) was also discussed.

While planning efforts were under way, seismicity was still being recorded. In March 2005, an average of forty earthquakes was recorded daily. Between March 8 and April 4, the OVK published seven warning bulletins about the possibility of an impending eruption. The PNPRU had not yet been tested and no operational resources were deployed to face the April 2005 eruption. For the first time in living memory, the Comorian population faced ash falls and lahars. In October the seismicity was once again high. The OVK published bulletins recommending people not to climb Karthala. When the volcano erupted on November 24, an operational command post was set up at the National Center of Emergencies Management (*CNOU*) at the Moroni "*gendarmerie*", directed chiefly by COSEP, in accordance with procedures described in the PNPRU. A few months later the CNOU was definitively replaced by COSEP, which was conceived after the 2003 seismic crisis, but had not yet been formalized when the

November 2005 eruption occurred. Three quarters of Grande Comore was covered by ash fallouts, affecting 244,520 people in 76 villages (PIROI 2005a), including Moroni, where it became difficult to breath without a mask. The Union of Comores and Grande Comore governments, the Comorian Red Crescent, the International Platform of French Red Cross in Indian Ocean (PIROI), UNDP, UNICEF, ECHO, and the French embassy lent their support to the crisis management. From December 5–9, the PIROI conducted an emergency operation to clean two water tanks per village (PIROI 2005b), as water supplies had been contaminated by ash, raising concerns about the people's health. Eight weeks of cleaning were required to clear the ash from roofs, streets, roads (Fig. 25.4b) and from all the water tanks. After this eruption, a study revealed that 100,000 people living on the slopes of Karthala were vulnerable to food insecurity due to the effects of volcanic eruptions (WFP and UE 2006). The 2005 eruption management was strongly criticized by both the media and public, who felt abandoned by the authorities and powerless to face an eruption. The PNPRU had not been strictly followed and was considered as being too theoretical by some stakeholders. However, it brought about, for the first time, an improved coordination of the stakeholders, which was an important step in terms of crisis management on Grande Comore (Morin 2012). While the DNPC did not show any response capacity in November 2005, the creation of the COSEP was ratified on May 6, 2006 (Decree 06-051 /PR). The centre located at the old Moroni airport was placed under the control of the Chief of Staff of the National Development Army of Comoros (AND).

The volcano erupted again in May 2006. This eruption was purely magmatic and confined to the summit. Yet it was the source of major concern for the authorities and the population. The way in which this eruption was managed highlighted many weaknesses in the volcanic management system and enabled improvements to be proposed.

25.3.3 Feedback Analysis on the 2006 Eruption Management: Summary of Main Weaknesses

The physical aspects of the 2006 eruption are described by Bachèlery et al. (2016, Chap. 22). Crisis management aspects have been detailed by Morin et al. (2009) as partly cited below:

> When Karthala erupted on the evening of 28 May 2006, two of the three summit seismic stations were inoperative due to technical problems and lack of repair, so that – without being able to triangulate seismic activity or to corroborate data streams - it was almost impossible for the OVK to build up a picture of precursors or of the volcanic activity, to issue advice or warnings before the eruption, or to deliver any detailed information after it was clear that an eruption was in progress. The uncertainty about the exact location and type of the eruption was one of the focal points for the May 2006 crisis. What would have happened if the volcanism had been a vent opening near an inhabited area? The OVK would have been incapable of furnishing the information needed by the Comorian authorities to make decisions that would have ensured the population's safety.
>
> The considerable light cast upon Karthala (from the lava) was visible from Moroni and its surroundings. Within a few minutes, dozens of frightened people converged at the OVK to gather information about what was happening, while mosques and stadiums were invaded by thousands of people praying together. Most of them had taken the eruptive plume of smoke, coloured red from the lava lake below it, to be a huge lava fountain hundreds of meters high. This misinterpretation fuelled fear in the eruption's first hours. All night long, dozens of members of the government, national and international institutions, media and the local people came to the OVK seeking information. The situation was complicated by the volcano erupting during the transition between the old and newly elected national governments of the Union of Comoros. Ministers in the previous government had not yet completely relinquished their duties while the incoming ministers had not yet been completely established in their offices. That doubled the number of government members who came to the OVK to inquire about the situation. If PNPRU procedures had been followed, the OVK would have had a single mediator with the Comorian authorities through the Emergency Operations headquarters. In consequence, the continual flood of official and unofficial people to the OVK became difficult for the OVK team to manage.

In the early morning of 29 May, a South African plane used by AMISEC (the African Union Mission in the Comoros, sent to ensure that the presidential elections went smoothly) allowed one of us to fly over Karthala. While no extra caldera activity was observed, it was evident that a lava lake had formed in Choungou Chahale. Even now that the eruptive activity and location were better known and reported, anxiety persisted because the OVK could not state how the volcanic activity would evolve and how long it would last. Everybody remembered the 1977 Singani eruption, fearing a repeat of that event. People were also afraid of ash fallouts, as had happened during the last two eruptions, in 2005. Several felt continual tremors, contributing to their concern. The stream of people coming to the OVK stayed constant for the next few days while the OVK spent time denying false rumours that appeared in the media. During this time, other AMISEC flights helped the OVK to observe the volcanic activity's evolution. The eruption finally ended on 1 June without further consequence.

This summary introduces the limitations of Comoros' crisis management at all levels including the OVK's limited capacities, the disorganisation of the Comorian authorities, transmission of false and alarmist information by the media, deference of crisis management to foreign groups, and poor risk perception contributing to crisis development.

In contrast to the previous eruptions, the May 2006 eruption should have remained an eruptive crisis in the scientific sense alone because the hazard was confined within the caldera and did not present a threat to people or infrastructure at any time. Nevertheless, an emergency management crisis appeared as soon as the volcano erupted, resulting in panicked reactions and major difficulties in applying the PNPRU.

25.4 Is Grande Comore Ready to Face Future Volcanic Eruptions?

25.4.1 A Too Weak and Slow Evolution of the Crisis Management Capabilities?

The 2006 eruption was an "ordinary" volcanic crisis with no damage, but delicate to manage. Yet it revealed limits and weaknesses of the management that constituted crisis-prone factors. Table 25.2 summarizes the difficulties revealed by feedback on the May 2006 eruption, and then indicates if these difficulties were encountered during other recent eruptions (1977, April 2005, November 2005, and 2007). Almost all factors are common to the four eruptions that occurred during the 2000s. The 1977 eruption is different because the OVK, Civil Protection instances, and the media reaction were not yet in existence. The evolution of these factors from 2006 to December 2011, and again from 2012 to 2014, is analysed and summarized with mathematical signs: "+" when an improvement was noticed, "−" for a degradation, "=" when we did not notice any evolution [for the complete details about these failure factors, see Morin et al. (2009) and Morin (2012)].

25.4.1.1 A Weak Evolution from 2006 to 2014

During the 2006–2011 period, one failure factor deteriorated, 43 % of the factors remained the same and 46 % improved (Table 25.2). This improvement is due to the stakeholders' raised awareness and their experience following the 2005, 2006 and 2007 eruptions. Many risk reduction programs were implemented as a result of these eruptions. However, 60 % of the factors still have strong to very strong 'crisis-making' potential, 32 % have moderate potential, and only one factor has low potential. This shows that the evolution of crisis management capacities since 2006 had been moderate. Most failures identified in 2006 remained strong enough to provoke a crisis outbreak. From 2012 to 2014, about a half of the failure factors improved, but the crisis management capacities are still weak (Table 25.2).

The OVK has taken great strides during the past five years, with the monitoring network expanding widely in 2007 (new inclinometer, seismic and CO_2 stations—Fig. 25.6), a new room at the OVK office in 2008, provision of a vehicle by UNDP, and the installation of 25 solar panels to provide autonomous energy for the OVK (up to 2007 the frequent power interruptions prevented continuous monitoring of the volcano). The United Nations also financed two workers (*UNVs—United Nations Volunteers*) at

Table 25.2 Synthesis of main failure factors detected during the May 2006 volcanic crisis management and their evolution (updated from Morin 2012)

Comparison to other crises and evolution from 2006 to 2011

Symbol	Meaning	Symbol	Meaning	Symbol	Meaning
(hatched)	Irrelevant for this period	X	Common factor	?	No data
+	Improvement	-	Deterioration	=	Unchanged situation

Crisis-prone potential (CPP) for each factor in December 2011

Colour	Meaning	Colour	Meaning
Black	Maximum	Red	High
Yellow	Moderate	Green	Low

Actors' category	Failure factors for each actors' category during 2006 crisis management	Valid during other eruptions: 1977	April 2005	Nov. 2005	2007	Evolution: May 2006- Dec. 2011	Jan. 2012 - Dec. 2014	CPP
Volc. Observatory	Insufficient human resources		X	X	X	+	+	Moderate
	Insufficient training		X	X	X	=	=	High
	Lack of material and financial resources		X	X	X	+	=	High
	Monitoring network periodically ineffective		X	X	X	+, then -	=	Maximum
	No data available for the eastern flank		X	X	X	+	=	High
	Scientific analysis capacity dependent on the IPGP and OVPF external assistance		X	X	X	=	=	High
	No preventive information (outreach) to population		X	X	X	=	+	Moderate
Comorian authorities	Lack of a basic knowledge about volcanic risk	X	X	X	X	+	=	Moderate
	Lack of basic knowledge about crisis management organization	?	X	X	X	+	+	Moderate
	Political system in favour of a lack of coordination	NO	X	X	X	=	=	High
	No professional Civil Protection corps	X	X	X	X	+	+	Moderate
	No budget dedicated to Civil Protection	X	X	X	X	+	+	High
	Inertia related to the existence of multiple procedures of Civil Protection		X	X	X	=	+	High
	National Emergency Plan (PNPRU) in the main unknown and non-mastered		X	X	X	?	+	High
	No updated data in the PNPRU		X	X	X	+	=	High
	Lack of emergency stocks described in the PNPRU		X	X	X	=	+	Maximum
	Crisis management dependent on external aid	NO	X	X	NO	=	=	High
	Emergency management based on international aid mainly by UNO and EU	X	X	X	X	=	=	High
	CRCo (Comorian Red Crescent) with very limited resources, dependent on PIROI and IFRC (Regional and International institutes of Red Cross and Crescent)		X	X	X	=	+	High
	Aerial surveillance with foreign armed forces only	NO	X	X	X	+		Low
Media	Media hype		X	X	X	+		Moderate
	Lack of a basic knowledge about volcanic risk		X	X	X	+	+	High
	Lack of basic knowledge about crisis management organization		X	X	X	+	+	Moderate
	Sensationalism / Limited Ethics		?	?	?	?		High
	Whole areas of the island not covered by the media		X	X	X	=	=	Maximum
Population	Spontaneous massive evacuation of population	X	X	X	X	=		Maximum
	Lack of a basic knowledge about volcanic risk and ways to face it	X	X	X	X	+	=	Moderate
	Negative perception of crisis management	X	X	X	X	=		Moderate

the OVK (an electronics engineer and a geophysicist), but in return the SCAC withdrew its temporary scientist position. The current staff is insufficient to deal with a crisis over time. After the SCAC's withdrawal, no other partners gave a sufficiently substantial assistance to allow the monitoring network to be maintained. It became difficult, in particular, to organize the technical work led by the OVPF and IPGP scientists within the framework of the original OVK-IPGP partnership. UNDP supported two programs: "reinforcement of the management capacities in the face of natural and climatic hazards" and "integration of risk management within development policies", providing financial supporting for the monthly network maintenance tasks and the two newly-created VNU salaries. These programs have mainly been implemented by the Civil Protection bodies, which sometimes provide support to the OVK. Unfortunately, the monitoring network was once again vandalized, and elements stolen, in 2011 and 2012. On November 24, 2014, only three seismic stations in the network were operational, and seismic data could not be processed due to technical difficulties, preventing earthquake locations (the data of only one seismic station were properly teletransmitted to the OVK). A few days later, two OVPF/IPGP engineers repaired part of the network, enabling teletransmission of data from four stations. However, the launching of false alarms due to electronic problems forced the OVK to unplug the phone alarm which is supposed to alert the OVK staff in the event of unusual activity.

Despite all the efforts made the last few years, the OVK is regularly 'deaf' to advice, and partners are increasingly reluctant to finance the rehabilitation of the network since their investments are continuously destroyed by vandalism. In 2012, with the support of UNDP, the OVK and COSEP scheduled an outreach campaign in a dozen villages to solve this problem (it was sensitive enough to be placed on the agenda of the Council of Ministers in late 2011 and mentioned by the President during its *Aid el Kebir* greetings in 2012). Outreach on one hand and protective and repressive actions against theft on the other hand are required to protect the

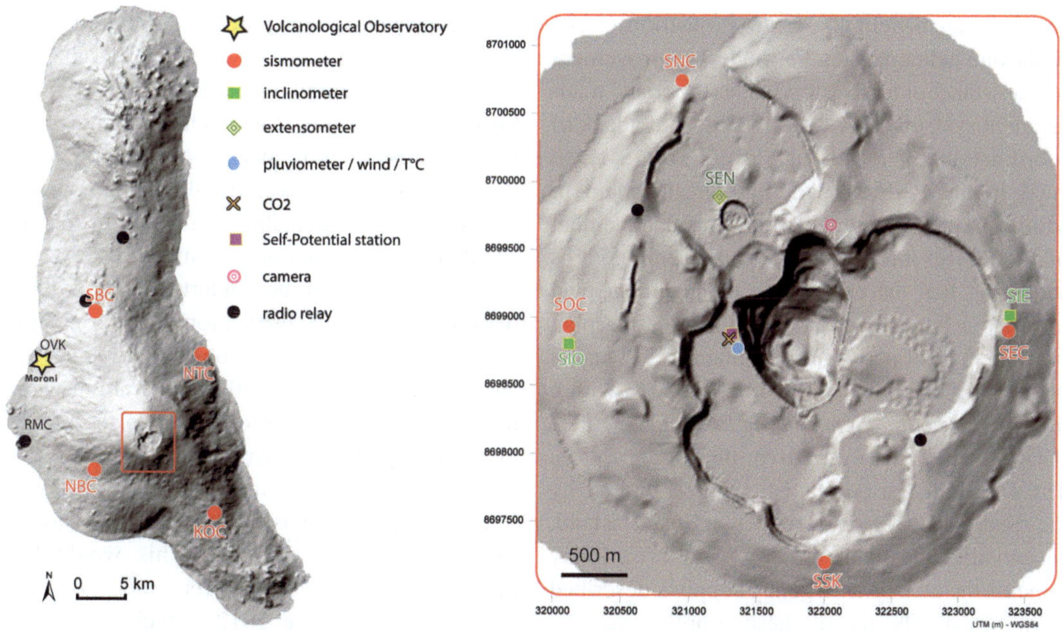

Fig. 25.6 Karthala monitoring network in 2011 (from GPS data by the OVK)—constant improvements and damage to the network are observed, as the network map changes every few weeks to months

monitoring network. Despite the obvious interest from the communities, and the commitment expressed for protecting monitoring equipment at the top of the volcano, this raised awareness had mixed results since further damage was recorded a few months later. To make the OVK viable in the long term, the Comorian authorities need to invest in the observatory, including creating permanent jobs for qualified persons (including a seismologist or volcanologist). The sustainable recruitment of the two UNV positions created and funded by UNDP is strongly encouraged by the new CNDRS Director. In October 2014, the government approved the funding of monitoring equipment to replace that stolen or vandalized in 2012. This decision has not yet been converted into action, but it is a first and strong sign of the government's commitment, despite its limited resources. Although the Comorian government has expressed its willingness to be more involved in volcanic risk reduction since 2013, concrete measures are still needed. Up to now, human, technical and financial resources have been insufficient, so it has not been possible to acquire or process volcanological data.

Progress has been made by the Comorian authorities, including the deployment of resources that can benefit the Civil Protection (especially one helicopter) and a modest improvement in the stakeholders' knowledge of crisis management systems. However, there is still a lack of coordination on a national level. This is even cited in the PNPRU, where it is attributed to administrative disorganization, mismanagement of human resources, and lack of training; it is surprising that the emergency management plan includes a list of reasons that prevent this plan from being properly implemented. The fusion between COSEP and DNPC in 2012/2013 within the new *Direction Générale de la Sécurité Civile* (DGSC) may help improve the organization. The DGSC has a permanent budget granted by the State and regional offices in each island. Meanwhile, the main critical point even today remains the lack of emergency supplies described in this plan. The list of these supplies was not updated when the PNPRU was updated in November 2007. On the ground, the Red Crescent, supported by the Red Cross, formed a volunteers' network. The volunteers own an emergency kit (a satellite phone to contact OVK and COSEP, a red flag, a whistle, a megaphone, a first aid kit, and in the most exposed villages plastic sheeting to cover the tanks) which is supposed to be used as a primary reaction tool in the event of a crisis. Risk and crisis management policy is still almost entirely dependent on foreign aid, which has implications for resource sustainability. Despite progress in many areas, a lot of weaknesses remain, resulting in a high potential for crisis creation. It is necessary to train the people in charge of the management (teach them the basis of crisis management and how to use the PNPRU), and to have access to sufficient funding (to run the civil protection corps under good conditions and to provide the necessary supplies for emergency management).The media have also made progress, especially through crisis communication training for 60 journalists. However, questions asked during the conference «*Maîtrise et Valorisation du Karthala*», which took place in 2008 in Moroni, showed that certain journalists had a very poor understanding of volcanic risk. Media coverage also remains technically difficult or non-existent in some areas, particularly in the east of the island.

Information about volcanic risks was spread to the public through outreach campaigns. Surveys, however, show that people's behaviour in the event of a crisis would still not be appropriate, with a strong inclination towards massive spontaneous evacuation. Complementary outreach work should be undertaken to avoid a biased perception. Sensitization of the population to these issues should be considered, along with alternative proposals. In 2006, several interviews revealed the importance of media in the process of informing the population of Grande Comore during a crisis. The possibility of mass awareness information being passed via the television or radio should be considered. This would help prevent false volcanic alerts, such as those launched twice since 2012, which were afterwards attributed to domestic fires.

25.4.1.2 Current Volcanic Crisis Management Capabilities on Grande Comore

The 2005 eruptions showed that Grande Comore can be wholly and frequently exposed to volcanic hazards. Efforts to raise the capacity to face crises to a higher level should have been deployed more widely. The May 2006 eruption attested that these efforts had been insufficient. The feedback on this eruption provided valuable information about the weaknesses that could increase the propensity for a crisis in the event of an eruption. This analysis could be applied to any other type of natural or technological hazard. The recommendations for reducing volcanic risk unfortunately remain mostly the same in 2014 as in 2006. Due to a lack of resources, the actual risk management is very different from the theoretical risk management described in the PNPRU. Crisis responses are most often funded by very limited means and considered only on a short-term scale, through direct emergency responses after an event. There is a predominant lack of preventive measures, despite awareness of threats. These responses reveal strong pre-existing vulnerabilities that go beyond the present economic and politic climates and the issue of volcanic eruptions (Morin and Lavigne 2009). It also revealed a major gap in training of crisis managers and the island's population. In this context, any eruption of Karthala should be considered as an operational crisis because national capacities are exceeded even in the absence of "problematic" events. Thus Morin et al. (2009) and Morin (2012) reach similar conclusions to those expressed by Chester (1993) regarding crisis management capabilities in the 1970s: they are still really poor. This analysis is unfortunately common to the other African volcanic areas (see Lenhardt and Oppenheimer 2014).

25.5 Summary and Future Perspectives

25.5.1 Recommendations for More Efficient Risk and Crisis Management Capabilities

To improve dealing with a volcanic crisis (or any other hazard), the Comorian need to think about:

- A clearer institutional framework, including clarification of the responsibilities applied at the different government levels (Union/Islands)
- A better monitoring network at the OVK, with resources to maintain it
- The adoption by the OVK of crisis management protocols proposed by IAVCEI (Geist and Garcia 2000; Newhall et al. 1999)
- Trained and competent people at all levels within the crisis management system
- Sustained and appropriate awareness campaigns for both authorities and population, with the help of media
- Mapping of hazards, resources, vulnerability, and risk
- Regulatory rules: for example the maintenance of water supply tanks should be mandatory
- Optimized use of NGO's capabilities
- Better Civil Protection institutions
- Crisis communication training for journalists and risk managers

The recommendations by Lenhardt and Oppenheimer (2014) should also been followed: (1) increase educational levels, (2) build a clear institutional framework and use NGO capacities, (3) maintain and develop the monitoring network.

As all stakeholders express a strong willingness to improve the risk management system, all Grande Comore needs now are concrete resources to obtain successful risk and crisis management capabilities.

25.5.2 Future Prospects

Fortunately, many projects and dynamics have got underway during the last few months or are about to start up:

- A project on climate change and disaster management funded by the GEF (Global Environment Facility) is scheduled to start in 2015. It will provide a support component for the observatory, with the possibility of reinforcing the monitoring network.
- The current Comorian-New-Zealand project for prospecting for geothermal resources in the Comoros offers, from 2014, funding opportunities for volcano monitoring; discussions are underway.
- The recent resumption of IPGP technical support missions through La Reunion Regional funding will regularly upgrade the monitoring network. A project late in 2014 allowed peripheral stations near the villages to be activated, whereas before they were usually located on the slopes. A new strategy which is still being discussed plans to keep most of the monitoring stations on the lower zones, near the villages, where the stations are relatively safe, and to abandon the summit stations, which are often subject to theft or vandalism.
- The development and validation of a national strategy for risk and disaster management in 2014, as part of a UNDP/DGSC program. A consultation will be launched in 2015 for the elaboration of a Disaster Risk Management national law project.
- The membership of the Comoros within the United Nations International Strategy for Disaster Reduction may bring funding opportunities in the field of natural hazards in general, and volcanic hazards in particular. The Comoros are involved in both the Arab and African platforms for risk management and disaster reduction. An international center for the management of natural disasters involving five countries in the region will soon open under the auspices of the World Bank and UN-Habitat.
- Finally, the civil society seems to have taken up the issue. An entire issue of a Comorian magazine was dedicated to Karthala in October 2014 (Al-watwan 2014), explaining the crisis management organization and dealing with legends and eruptive history. The changing way of life may also help face a crisis, for example using mobile phones, which are becoming more common in the Comoros, to spread messages of an alert.

Thanks to these recent changes, it is now possible to envisage a significant improvement in the volcanic risk management over the next decade.

Acknowledgments The authors wish to thank Nicolas Villeneuve for his review, kindly suggestions and comments. We are grateful to Fran Van Wyk de Vries for proofreading this manuscript.

References

Al-watwan (2014) Volcanisme - lettre des fonds du magma. Octobre 2014:19

Bachèlery P (1999) Le fonctionnement des volcans-boucliers. Exemple des volcans de la Réunion et de la Grande Comore. Mémoire d'Habilitation à Diriger des Recherches. H.D.R., LSTUR, Université de La Réunion, Saint-Denis de La Réunion, France

Bachèlery P, Ben Ali D, Desgroland F, Toutain JP, Coudray J, Cheminee JL, Delmond JC, Klein JL (1995) L'éruption phréatique du Karthala (Grande Comore) en juillet 1991. Comptes Rendus De L'Academie Des Sciences 320(II–a):691–698

Bachèlery P, Coudray J (1993) Carte volcano-tectonique (1/50000e) de la Grande Comore et notice explicative, 1 carte : en coul.; 140 × 90 cm, pliée 30 × 20 cm. Saint-Denis de La Réunion, France: République Fédérale Islamique des Comores, Centre national de

documentation et de recherches scientifiques/ République Française, Université de La Réunion, Département Sciences de la Terre
Bachèlery P, Morin J, Villeneuve N, Soulé H, Nassor H, Radadi Ali H (2016) Structure and eruptive history of Karthala volcano (Chapter 22). In: Bachèlery P, Lénat J-F, Di Muro A, Michon L (eds) Active volcanoes of the Southwest Indian Ocean: Piton de la Fournaise and Karthala. Active Volcanoes of the World. Springer, Berlin
Chester D (1993) Volcanoes and society. Edward Arnold, London
Chouaybou K (2010) Les risques volcaniques en Grande Comore, Géographie, Université Paris 8, Paris, France
COI (1992) Rapport d'Expertise Sécurité Régionale - Programme PARSEC, 27: La République Fédérale Islamique des Comores
CRED (2014) EM-DAT: The OFDA/CRED International Disaster Database (www.em-dat.net). Université Catholique de Louvain, Brussels, Belgique
Geist D, Garcia MO (2000) Role of science and independent research during volcanic eruptions. Bull Volc 62:59–61
Global Volcanism Program (2005) Karthala. Monthly Reports, Smithsonian National Museum of Natural History, USA
Guébourg J-L (1995) Espace et pouvoirs en Grande Comore. L'Harmattan, Paris
Hartnady C (2005a) De la possibilité d'apparition de tsunamis sur la côte est-africaine et les îles de l'Océan indien. La prévention des catastrophes en Afrique, SIPC Informations. (5 juillet 2005):27–30
Hartnady C (2005b) Continental slope landslide- and oceanic island volcano-related tsunami potential in the Western Indian Ocean. Paper read at Terra Nova—East African Rift, Mbeya, Tanzania
ISDR (2009) Global assessment report: risk and poverty in a changing climate—mortality risk index, 2, UN
Krafft M (1982) L'éruption volcanique du Karthala: avril 1977 (Grande Comore, océan Indien). Comptes Rendus De L'Academie Des Sciences de Paris 294:753–758
Krafft M (1983) Guide des volcans de la Grande Comore, 94. Nancy
Lacroix A (1920) Une éruption du volcan Karthala, à la Grande Comore, en août 1918. Comptes Rendus des Séances de l'Academie des Sciences 171:5–10
Lenhardt N, Oppenheimer C (2014) Volcanism in Africa: geological persepctives, hazards, and societal implications. In: Ismail-Zadeh A, Urrutia Fucugauchi J, Kijko A, Takeuchi K, Zaliapin I (eds) Extreme natural hazards, disaster risks and societal implications. Cambridge University Press, Cambridge
Morin J (2006) Evaluation du risque volcanique en Grande Comore. Caractérisation des aléas, étude de perception, analyse du système de gestion. Mémoire de Master Pro GCRN, Université Paul Valéry Montpellier III, Montpellier, France
Morin J (2012) Gestion institutionnelle et réponses des populations face aux crises volcaniques : études de cas à La Réunion et en Grande Comore., Ecole Doctorale Sciences Technologies Santé - UMR CNRS IPGP 7154., Université de La Réunion, Saint-Denis, La Réunion
Morin J, Gaillard JC (2012) Lahar hazard and livelihood strategies on the foot slopes of Mt Karthala volcano, Comoros. In: Wisner B, Gaillard JC, Kelman I (eds) Handbook of hazards and disaster risk reduction. Routledge, London, pp 705–706
Morin J, Lavigne F (2009) Institutional and social responses to hazards related to Karthala volcano, Comoros—part II: the deep-seated root causes of Comorian vulnerabilities. SHIMA Int J Res Island Cultures 3(1):54–71
Morin J, Lavigne F, Bachèlery P, Finizola A, Villeneuve N (2009) Institutional and social responses to hazards related to Karthala volcano, Comoros—part I: analysis of the May 2006 eruptive crisis. SHIMA Int J Res Island Cultures 3(1):33–53
Mossoux S, Feltz A, Poppe S, Canters F, Kervyn M (2014a) A flexible open-source toolkit for lava flow simulations. In: EGU General Assembly 2014, Geophysical Research Abstracts, Vienna
Mossoux S, Feltz A, Poppe S, Canters F, Kervyn M (2014b) Calibration of Q-LavHA—a quantum GIS plugin for lava flow simulation (Abstract). In: Cities on volcanoes 8, Yogyakarta, Indonesia
Nassor H (2001) Contribution à l'étude du risque volcanique sur les grands volcans boucliers basaltiques: le Karthala et le Piton de la Fournaise. Dr., LSTUR, Université de La Réunion, Saint-Denis de La Réunion, France
Newhall C, Aramaki S, Barberi F, Blong R, Calvache M, Cheminee J-L, Punongbayan RS, Siebe C, Simkin T, Sparks S, Tjetjep W (1999) Professional conduct of scientists during volcanic crises. Bull Volc 60:323–334
OCHA-GVA (2005) Comoros: Karthala Volcano OCHA Situation Report No. 2. UN Office for the Coordination of Humanitarian Affairs, cited 21 Apr 2005. Available from http://reliefweb.int/report/comoros/comoros-karthala-volcanic-eruption-ocha-situation-report-no-2
Pavlovsky R, Saint-Ours J (1953) Etude géologique de l'archipel des Comores., 55. Tananarive, Madagascar: Haut Commissariat de Madagascar et Dépendances, Direction des Mines et de la Géologie, Service Géologique Tananarive
PIROI (2005a) Point de situation, actualisé le 12 décembre 2005 -Eruption du Karthala - Croix-Rouge Francaise, Croissant-Rouge Comorien. PIROI, Moroni, Union des Comores
PIROI (2005b) Sécurisation des points d'eau suite à l'éruption du volcan Karthala en Grande Comore - Rapport final d'opération. PIROI, Moroni, Union des Comores
PNUD (2002) Rapport sur les risques naturels et la protection civile en Grande Comore, 38: PNUD
Smithsonian National Museum of Natural History (2005) Karthala (Comoros Islands) Eruption on 24 November

2005; big evacuation and one fatality. Bull Glob Volcanism Netw 30(11):2–4

Soule H (2003) Proposition d'un plan d'urgence pour la gestion des crises volcaniques en Grande Comore. Ya Mkobe 10:59–73

Soule H (2011) Cartographie préliminaire des chenaux d'écoulement des lahars actifs en 2010 (reconnaissance de terrain et extrapolation de la tête des chenaux à partir de la carte topographique de Grande Comore au 50 000e) (Unpublished)

Tulet P, Villeneuve N (2011) Large scale modelling of the transport, chemical transformation and mass budget of the sulfur emitted during the April 2007 eruption of Piton de la Fournaise. Atmos Chem Phys 11:4533–4546

Union des Comores (2004) Plan National de Préparation et de Réponse à l'urgence, 68 (+ annexes : Plan Karthala 37p. + Plan cyclone 15p.). Moroni, Union des Comores

Union des Comores (2007) Plan National de Préparation et de Réponse à l'urgence, 86. Moroni, Union des Comores

WFP, UE (2006) Comoros: comprehensive food security and vulnerability analysis (CFSVA), vol 62. United Nations World Food Programme, Rome, Italy

Index

P. Bachèlery et al. (eds.), *Active Volcanoes of the Southwest Indian Ocean*,
Active Volcanoes of the World, DOI 10.1007/978-3-642-31395-0

MIX
Papier aus verantwortungsvollen Quellen
Paper from responsible sources
FSC® C105338

If you have any concerns about our products,
you can contact us on
ProductSafety@springernature.com

In case Publisher is established outside the EU,
the EU authorized representative is:
Springer Nature Customer Service Center GmbH
Europaplatz 3, 69115 Heidelberg, Germany

Printed by Libri Plureos GmbH
in Hamburg, Germany